T0318393

Practical Chemical Thermodynamics for Geoscientists

Practical Chemical Thermodynamics for Geoscientists

Bruce Fegley

With contributions by Rose Osborne

AMSTERDAM • BOSTON • HEIDELBERG • LONDON
NEW YORK • OXFORD • PARIS • SAN DIEGO
SAN FRANCISCO • SINGAPORE • SYDNEY • TOKYO

Academic Press is an Imprint of Elsevier

Academic Press is an imprint of Elsevier
225 Wyman Street, Waltham, MA 02451, USA
The Boulevard, Langford Lane, Kidlington, Oxford, OX5 1GB, UK

Copyright © 2013 Elsevier Inc. All rights reserved.

No part of this publication may be reproduced or transmitted in any form or by any means, electronic or mechanical, including photocopying, recording, or any information storage and retrieval system, without permission in writing from the Publisher. Details on how to seek permission, further information about the Publisher's permissions policies, and our arrangements with organizations such as the Copyright Clearance Center and the Copyright Licensing Agency, can be found at our website: www.elsevier.com/permissions.

This book and the individual contributions contained in it are protected under copyright by the Publisher (other than as may be noted herein).

Notices

Knowledge and best practice in this field are constantly changing. As new research and experience broaden our understanding, changes in research methods, professional practices, or medical treatment may become necessary.

Practitioners and researchers must always rely on their own experience and knowledge in evaluating and using any information, methods, compounds, or experiments described herein. In using such information or methods they should be mindful of their own safety and the safety of others, including parties for whom they have a professional responsibility.

To the fullest extent of the law, neither the Publisher nor the authors, contributors, or editors, assume any liability for any injury and/or damage to persons or property as a matter of products liability, negligence or otherwise, or from any use or operation of any methods, products, instructions, or ideas contained in the material herein.

Library of Congress Cataloging-in-Publication Data

Fegley, Bruce.

Practical chemical thermodynamics for geoscientists / Bruce Fegley, Jr. ; with contributions by Rose Osborne.
p. cm.
Includes bibliographical references and index.
ISBN 978-0-12-810270-1 (paperback)
1. Thermochemistry. 2. Chemical equilibrium. 3. Thermodynamics. I. Osborne, Rose. II. Title.
QD511.F44 2013
541'.36–dc23

2011052540

British Library Cataloguing-in-Publication Data

A catalogue record for this book is available from the British Library.

For information on all Academic Press publications visit our website at http://store.elsevier.com.

Printed in the United States of America
12 13 14 15 16 17 9 8 7 6 5 4 3 2 1

Working together to grow
libraries in developing countries

www.elsevier.com | www.bookaid.org | www.sabre.org

ELSEVIER BOOK AID International Sabre Foundation

Contents

In gratitude to the professors who taught me thermodynamics:
H. K. Bowen, Robert L. Coble, and Thomas B. Reed
(3.00 Thermodynamics of Materials)
Carl Garland and John S. Lewis (5.60 Physical Chemistry)
Clark C. Stephenson (5.71 Chemical Thermodynamics)
Patrick M. Hurley (12.40 Chemistry of the Earth)

Preface

In this work, when it shall be found that much is omitted, let it not be forgotten that much likewise is performed...

—Samuel Johnson (1755)

If during the course of this book we help disclose to the student some of the beauty and simplicity of the thermodynamic method, if we convince a few practical chemists of the extreme practicality of the results of thermodynamic calculations, if we contribute in some measure toward making chemistry an exact science, our task is rewarded.

—Gilbert N. Lewis and Merle Randall (1923)

I decided to write this book because of my research and teaching experience, which involve the application of chemical thermodynamics to problems in the earth and planetary sciences. During the past 18 years, I have taught Thermodynamics and Phase Equilibria, a one-semester course for advanced undergraduates and incoming graduate students in the Department of Earth and Planetary Sciences at Washington University. Chemical thermodynamics is also extensively used in my one-semester courses Planetary Geochemistry and Earth System Science for advanced undergraduates and incoming graduate students.

I wrote this book for two groups of readers. The first group is the advanced undergraduates or graduate students in the earth and planetary sciences who are taking a course in chemical thermodynamics. The second group is the researchers in earth and planetary sciences who want a review of one or more topics in chemical thermodynamics or who need some thermodynamic data to help them solve problems related to their work. I hope that students and researchers with interests in the earth and planetary sciences but who are formally in related disciplines such as astronomy, chemistry, and physics, will also find this book a useful resource.

The organization and style of the book reflect my biases and opinions about the good and bad points of the many thermodynamic monographs and texts that I have used at one time or another over the past 30 years. Most chapters here begin with an overview of the subjects covered in each section of the chapter. All chapters except the first one have worked examples in the text, a number of original figures and/or tables, and about 20 problems. The references to articles and books cited in all chapters are collected together into one list. Some of the citations are to classic works by the "pioneers" of thermodynamics but the more recent literature is covered as well. The sidebars with brief biographies of scientists who developed classical thermodynamics and/or applied it in the earth and planetary sciences are intended to show a little of the human side of the development of thermodynamics.

I personally find worked examples extremely useful for learning and to illustrate the methods involved in solving the "real-world" problems that one encounters. Consequently, the book contains a large number of examples. A few examples are long because they deal with the complicated type of real-world problems that are seldom treated in most texts.

This book emphasizes the fundamental principles and their application to practical problems in calculations that students can do by hand or by using readily available computer programs such as spreadsheets and curve-fitting programs. However, the use of specialized programs for doing thermodynamic calculations is not covered. I believe it is a mistake to teach the use of computer programs in an introductory course. Students must first learn the fundamental principles and use them to solve problems themselves. Only after this firm foundation is established can students learn how to use specialized computer programs for thermodynamic calculations with critical judgment. Instead of training students to use computer programs (and many times to blindly trust the results because they do not understand how they were obtained), we should be training them to write the next generation of programs.

Many of the examples and problems involve non-SI units. Anyone reading the older scientific literature (i.e., pre-1970) inevitably encounters units such as calories, atmospheres, psi, mmHg, Amagats, and other officially disapproved units. It is important for students to have experience converting between these units and SI units so that they understand the older literature.

I dislike the qualitative figures schematically showing the alphabet variables A, B, and C in many thermodynamic texts. Thus, this book has many quantitative figures with numerical axes, and many of the figures here refer to concrete examples and to real compounds. For example, Figures 4-6 and 6-6 show the heat capacity, enthalpy, and entropy for corundum and molten alumina. Unfortunately, it is impossible and sometimes impractical to use quantitative figures throughout and to totally avoid the use of the dreaded A, B, and C.

The problems in this book have been tested in my courses and fall into three categories: (1) examples from everyday life, (2) research done by some of the scientists who made important contributions in thermodynamics, and (3) problems of current interest in the earth and planetary sciences. A solutions manual gives answers to all of the problems in the book.

It is very useful to have reliable thermodynamic data at hand for solving problems and examples in the book and for scientific research. For this purpose, the book contains tabular data throughout the text and two short appendices with thermodynamic data for some elements, inorganic compounds, minerals, and some organic compounds at 298.15 K and as a function of temperature. However, only a relatively small number of compounds can be included in the data tables without making the book unreasonably long. I selected compounds of broad interest in the geosciences. For example, Tables 3-3 and 3-4 give heat capacity equations for many common minerals in the Earth's crust and for many important gases in planetary atmospheres. The thermodynamic data in the book focus on: (1) important gases for atmospheric chemistry in the terrestrial and other planetary atmospheres; (2) carbon, nitrogen, sulfur, and phosphorus compounds involved in the biogeochemical cycles of these elements; and (3) common terrestrial minerals and minerals found in lunar samples and meteorites.

A number of colleagues have assisted me in one way or another during preparation of this book and I am glad to acknowledge their help. My collaborator Rose Osborne made important contributions in several areas. She had the thankless task of drawing (and redrawing) many of the figures from my sketches and descriptions; she prepared a number of the biographies and was responsible for finding portraits to accompany them; she was responsible for many of the computations involved in preparation of the thermodynamic data appendices, and she provided solutions for a number of the problems. Laura Schaefer, who has worked with me during the past eight years, provided invaluable assistance on many aspects of the book. Many of the students in my classes have asked questions, found errors, and made comments that have helped me during the preparation of this book. In this regard, in particular I want to acknowledge the help of Brian Shiro. The Washington University library system provided invaluable assistance in finding and obtaining numerous articles and books, and

I especially thank Nada Vaughn (former head of interlibrary loans), Rob McFarland (chemistry librarian), and Ben Woods (former chemistry library assistant) for their efforts. Frank Cynar, my original editor at Academic Press, and subsequent editors at Elsevier have shown the patience of Job in waiting for delivery of the book manuscript. Heather Tighe (project manager) and Katy Morrissey (editorial project manager) at Elsevier helped Rose and me to transform the manuscript into a book. And as always, Katharina Lodders, my wife and colleague, has provided valuable advice and support during the past years while I was "finishing" the manuscript.

Although I have tried to eliminate all errors, this is impossible to do in any book, especially in a thermodynamics book with a large number of equations and numerical values. I also aimed to include a wide range of examples and problems in the earth and planetary sciences, but I have undoubtedly slighted one discipline or another to some extent. Much effort went into the preparation of thermodynamic data tables throughout the book and the appendices, but still, some values may be undoubtedly incorrect because of sign errors, mistakes in transcription, or use of the "wrong" reference. I hope that readers will bear in mind Dr. Samuel Johnson's often repeated quote and kindly advise me of the errors, miscalculations, mistakes, and omissions that they may find.

CHAPTER

1 Definition, Development, and Applications of Thermodynamics

A theory is the more impressive the greater the simplicity of its premises is, the more different kinds of things it relates, and the more extended is its area of applicability. Therefore the deep impression which classical thermodynamics made upon me. It is the only physical theory of universal content concerning which I am convinced that, within the framework of the applicability of its basic concepts, it will never be overthrown. ...

—Albert Einstein (1949)

I. HISTORICAL DEVELOPMENT OF CHEMICAL THERMODYNAMICS

Thermodynamics is the scientific study of the relationships between heat and other forms of energy. From 1790 to about 1850, experimental and theoretical work by Sadi Carnot, Emile Clapyron, Rudolf Clausius, Count Rumford, James Joule, Hermann Helmholtz, G. H. Hess, Lord Kelvin (William Thomson), and Robert Mayer led to a quantitative understanding of the transformation of heat into work and the maximum amount of work that can be obtained from an engine. These relationships form the *first and second laws of thermodynamics*.

Scientists began to use the word *thermodynamics* in scientific articles and books in the mid-19th century. For example, while he was developing the second law of thermodynamics (1849–1854), William Thomson (later Lord Kelvin) wrote several papers published in the *Transactions of the Royal Society of Edinburgh* about "fundamental principles of general thermo-dynamics" and about a "perfect thermo-dynamic" engine. The great English geologist Sir Roderick Impey Murchison (1792–1871) referred to "the principles of thermo-dynamics" in the 1867 edition of *Siluria*, which summarized his work on the Silurian system in Great Britain and other countries. By the early 1870s the hyphen was dropped and James Clerk Maxwell wrote about the "First Law of Thermodynamics" in his book *The Theory of Heat*.

From about 1850 to 1890, experimental studies by Marcellin Bertholet, Henri Le Chatelier, Robert Kirchhoff, Walther Nernst, Julius Thomsen, J. H. van't Hoff, and other scientists developed many applications of thermodynamics to chemistry. During the same period, theoretical work originated by J. W. Gibbs and later developed by J. D. van der Waals and his colleagues, along with experimental studies by H. W. B. Roozeboom, led to an understanding of phase equilibria in single-component and multicomponent systems.

From about 1890 to 1920, experimental studies by Fritz Haber, G. N. Lewis, Walther Nernst, Wilhelm Ostwald, and others led to an understanding of the driving force of chemical reactions and

Copyright © 2013 Elsevier Inc. All rights reserved.

finally to the prediction of chemical equilibria. The two landmark books in this era were Haber's 1905 book, *Thermodynamics of Technical Gas Reactions,* and *Thermodynamics* by G. N. Lewis and M. Randall (1923). Around 1908 Haber synthesized ammonia from nitrogen and hydrogen. This breakthrough led to a Nobel Prize in Chemistry for Haber (1918). Carl Bosch, who developed Haber's work into industrial NH_3 synthesis at the German chemical company BASF, later shared the Nobel Prize in Chemistry in 1931.

During the first half of the 20th century, W. F. Giauque demonstrated the validity of the third law of thermodynamics (Nernst's heat theorem) and developed magnetic cooling (adiabatic demagnetization). He was awarded the 1949 Nobel Prize in Chemistry for this work. During the same period, Sir Ralph Fowler, E. A. Guggenheim, Joseph Mayer, Maria Goeppert Mayer, and Richard C. Tolman developed statistical mechanics, which Gibbs originated around 1900. Statistical mechanics is used to calculate the thermodynamic properties of materials from the properties of their constituent atoms and molecules.

Three figures stand out in the era since World War II. The Norwegian chemist Lars Onsager (1903–1976) won the 1968 Nobel Prize in Chemistry for his development of the reciprocal relations that describe the mutual transport of mass and heat during *irreversible processes*, such as a sugar cube dissolving in a cup of coffee. Irreversible processes are ubiquitous in nature and are important not only for everyday events such as sugar dissolving in coffee or tea, but also for the operation of thermocouples (simultaneous conduction of electricity and heat), chemical weathering of minerals, metamorphic reactions, and gas-grain reactions in the solar nebula (some combination of mass, electrical charge, and heat transport in these three cases). The Russian chemist Ilya Prigogine (1917–2003), who did most of his work in Belgium, won the 1977 Nobel Prize in Chemistry for his research in nonequilibrium thermodynamics. Prigogine developed a nonlinear theory to describe systems (such as living organisms) that are far from equilibrium. His work shows how order can develop from chaos. Harold Urey (1893–1981), who won the 1934 Nobel Prize in Chemistry for his discovery of deuterium, pioneered the application of chemical thermodynamics to cosmochemistry and developed methods to compute equilibrium isotopic fractionations. An important application of the latter work is the use of isotopes, such as oxygen isotopes, as geothermometers.

II. PIONEERING APPLICATIONS IN THE GEOSCIENCES

Going back to the 19th century, geoscientists became aware of the field of thermodynamics at an early stage of its development. As we discuss in Chapter 4, the German mineralogist and physicist Franz Neumann (1798–1895) developed a "rule" for calculating the heat capacities of chemical compounds that is known as the Neumann-Kopp rule (actually a good approximation under the right circumstances). During the mid-19th century, the French chemist and mineralogist Henri Sainte-Claire Deville (1818–1881) conducted pioneering experiments on the thermal dissociation of steam and other compounds.

At the end of the 19th century, thermodynamics began playing an increasingly important role in the geosciences. The German salt industry hired the Dutch physical chemist J. H. van't Hoff to study the origin of the Stassfurt salt deposits in Saxony. These deposits date from the Zechstein period, about 250 million years ago, and were enormously valuable at the time for the production of halite, magnesium salts, potassium salts, and bromine. Van't Hoff's study of the phase equilibria during evaporation of seawater was a major advance in understanding the origin of these and other evaporate deposits. At about the same time, the American geologist and physicist Carl Barus (1856–1935) made measurements of the specific heat, volume change on melting, and change of the melting point with pressure for diabase. Barus also studied the pressure-volume-temperature (*PVT*) properties of gases and the use of different alloys for high-temperature thermocouples.

Two other major advances in the application of chemical thermodynamics to geosciences took place at about the same time. One was the foundation in 1905 of the Geophysical Laboratory of the Carnegie Institution in Washington, D.C. The personnel in the Geophysical Laboratory, in particular Norman L. Bowen (1887–1956), pioneered the experimental study of igneous phase equilibria. The other advance was V. M. Goldschmidt's (1888–1947) study of metamorphic phase equilibria, notably the reaction

$$\mathrm{CaCO_3\ (calcite) + SiO_2\ (quartz) = CaSiO_3\ (wollastonite) + CO_2\ (gas)} \tag{1-1}$$

which takes place when a silica-bearing limestone is altered to wollastonite. Goldschmidt used Nernst's heat theorem (1906) to determine the *P-T* boundary between calcite + quartz and CO_2 + wollastonite from the meager experimental data available at the time (1912). As part of this work Goldschmidt developed the *mineralogical phase rule*, which states that the number of minerals in a rock is less than or equal to the number of components.

III. THERMODYNAMICS VERSUS KINETICS

Today, chemical thermodynamics is an integral part of modern geosciences, with applications ranging from the geochemistry of the Earth's interior to marine chemistry, biogeochemistry, air chemistry, and planetary science. Some examples of what thermodynamics can tell us include:

- The composition of a volcanic gas, such as those erupted at Kilauea on Hawaii.
- The crystallization sequence of minerals during cooling of an igneous magma or during evaporation of seawater.
- The condensation sequence of minerals in the solar nebula and expected in refractory inclusions in meteorites, such as the Ca,Al-rich inclusions in the Allende CV3 chondrite.

However, thermodynamics, although extremely useful, has its limitations. Two examples of what thermodynamics cannot and does not tell us are the rate of a chemical reaction and the mechanism by which a reaction proceeds. These two topics fall into the domain of chemical kinetics. The difference between kinetics and thermodynamics is easily illustrated by considering formation of corundum (Al_2O_3) grains in the solar nebula.

Corundum is found in some of the Ca,Al-rich inclusions in Allende and other meteorites. Chemical and mineralogical evidence indicates that the corundum formed via gas-solid reactions in the solar nebula, the cloud of gas and dust from which the solar system originated. Chemical thermodynamic models of nebular chemistry predict that corundum formed via the reaction

$$\mathrm{2\ AlO\ (g) + H_2O\ (g) = Al_2O_3\ (corundum) + H_2\ (g)} \tag{1-2}$$

as the nebular gas cooled. Reaction (1-2) is a *net thermochemical reaction* that shows the overall change of reactants into products, and it predicts the temperature at which corundum forms (commonly called the *condensation temperature*) as a function of the total pressure. On the other hand, Eq. (1-2) does not express the actual course of events or tell us how fast things happen as atoms and molecules react to form corundum.

However, we can attempt to answer these questions using information from laboratory studies of metal oxidation and reactions in flames. This research gives information about the *elementary reactions* that actually take place between atoms and molecules. The overall result of a sequence of elementary reactions is simply the net thermochemical reaction. In the case of Eq. (1-2), a plausible set of elementary reactions is

$$\mathrm{AlO\ (g) + AlO\ (g) + M \rightarrow Al_2O_2\ (g) + M} \tag{1-3}$$

$$\mathrm{H_2O\ (g) + M \rightarrow H\ (g) + OH\ (g) + M} \tag{1-4}$$

$$\mathrm{OH\ (g) + Al_2O_2\ (g) \rightarrow Al_2O_3\ (g) + H\ (g)} \tag{1-5}$$

$$\mathrm{H\ (g) + H\ (g) + M \rightarrow H_2\ (g) + M} \tag{1-6}$$

$$\mathrm{Al_2O_3\ (g) = Al_2O_3\ (corundum)} \tag{1-7}$$

Reactions (1-3) to (1-7) add up to Eq. (1-2), which is the *net thermochemical reaction*. The *M* in several reactions is a third body, which is any other gas. In the absence of kinetic data we cannot say anything about the rate of Eq. (1-2). Although it is possible to estimate the reaction rate, we will not do so here. Instead we will embark on our study of chemical thermodynamics and learn how to apply it to a wide variety of problems in the earth and planetary sciences.

CHAPTER

Important Concepts and Mathematical Methods

2

Mathematics is a language.

—Remarks attributed to Willard F. Gibbs at a Yale faculty meeting about the relative merits of mathematics and languages (Lewis and Randall, 1923)

In this chapter we lay the foundation for our study of thermodynamics. Several terms used throughout the book are introduced. Then, the nature of temperature and pressure and their measurement are explored. Pressure and temperature are used in almost every thermodynamics problem we solve, so it is important to have a practical understanding of these properties and how they are measured. Finally, we describe some basic concepts in calculus.

I. DEFINITIONS

Before we can begin discussing the laws and applications of thermodynamics, we have to understand the language that we will be using. This section presents and defines some unfamiliar words, and some familiar words are given specific meanings to avoid ambiguity. Other definitions will be given throughout the book when we need to use them and are not given here.

System. A system encompasses everything being studied and can be defined in such a way as to make it easier to solve the problem at hand. The rest of the universe outside the system is the *surroundings.* Some examples of a system are the air + fuel mixture inside the cylinder of a car engine, a balloon filled with helium gas, the coffee in a coffee cup, an electric or magnetic field in an otherwise empty container, or a mineral assemblage in a rock. The boundary between a system and its surroundings can be real, like the cup holding the coffee, or it can be imaginary. The example of a mineral assemblage in a rock illustrates a system with an imaginary boundary because we chose only the assemblage as the system, even though it is an integral part of the rock. A system also need not contain matter, as in the example of an electric or a magnetic field; however, in this book we will generally consider systems that involve matter in some form (solid, liquid, or gas). A system can be closed or open; a *closed system* (e.g., a sealed bottle) does not exchange matter with its surroundings, but an *open system* (e.g., an active volcano erupting magma and volcanic gases) does. Both closed and open systems can exchange energy with their surroundings; an *isolated system* does not exchange anything with the surroundings.

Homogeneous system. Only one phase with uniform properties is present in a homogeneous system. For instance, the water in a glass and the air inside a balloon are two examples of

Copyright © 2013 Elsevier Inc. All rights reserved.

homogeneous systems. Even though air is composed of a number of gases, it has a uniform composition and properties and acts as a homogeneous system.

Heterogeneous system. Two or more phases are present in a heterogeneous system. For example, a soft-drink bottle containing soda and air is a heterogeneous system because two phases (gas and liquid) are present. Likewise, a glass of a soft drink containing air, the soda, and an ice cube is also a heterogeneous system because three phases (gas, liquid, solid) are present. Other examples of heterogeneous systems are crystals + melt in a cooling magma, rock + fluid in a hydrothermal system, or dust + gas in the solar nebula. We discuss heterogeneous systems and phase equilibria in more details in Chapters 7 and 10–12.

State. The properties of a system determine its state. For example, the temperature (T), pressure (P), and volume (V) determine the state of a gas. An equation of state mathematically relates the properties of a system. The ideal gas law ($PV = nRT$) is an equation of state that relates T, P, V, and n, the number of moles, using R, the ideal gas constant. For reference, a mole of a substance is Avogadro's number ($N_A = 6.02214199 \times 10^{23}$ or 6.022×10^{23} to three decimal places) of particles of that substance. For example, a mole of helium gas is 6.022×10^{23} atoms of helium, and a mole of O_2 gas is 6.022×10^{23} oxygen molecules.

Extensive variable. An extensive variable depends on the size of a system or on the quantity of material in a system. For example, the volume of a gas is an extensive variable because it depends on the size of the container holding the gas. Likewise, the mass of a rock is an extensive variable. The kinetic energy ($= \frac{1}{2}mv^2$) of a system is also an extensive variable because it depends on the mass (m) as well as the velocity (v). In general, the energy of a system is an extensive variable.

Intensive variable. An intensive variable is characteristic of a system and does not depend on the size of the system. The temperature of a gas is an intensive variable because it does not depend on the size of the container holding the gas. Refractive index, viscosity, and surface tension are also intensive variables. Specifying the amount of material present (e.g., dividing by the mass) transforms extensive variables into intensive variables.

State variable. A state variable depends only on the difference between the initial and final states of the system. It does not depend on the path between the initial and final states. The schematic diagrams in Figure 2-1 illustrate this concept. These diagrams show the initial (P_A, V_A) and final (P_B, V_B) states of a system and four different paths between the two states. A *path* is the route from the initial to the final state of a system—for example, how the volume and pressure of the air + fuel mixture inside the cylinder of a car engine change as the piston moves. Pressure and volume are state variables. Thus, the volume change ($\Delta V = V_B - V_A$) and the pressure change ($\Delta P = P_B - P_A$) between the initial state A and the final state B are the same in all four cases shown in Figure 2-1, even though the paths between the initial and final states are different.

All the thermodynamic functions that we use throughout this book are state variables. This is very important, because it means that changes in these functions do not depend on the way a mechanical process or chemical reaction occurs. However, as described in Chapter 3, heat and work are not state functions. The heat flow and the work done in a mechanical process, such as compression or expansion of the air + fuel mixture in the cylinder of a car engine, or during a chemical reaction such as iron rusting, depend on the path taken.

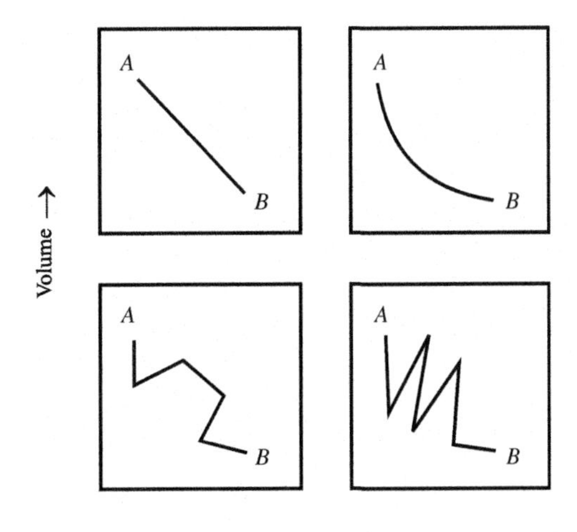

FIGURE 2-1

Volume plotted as a function of pressure. The change in volume $\Delta V = V_B - V_A$ and the change in pressure $\Delta P = P_B - P_A$ are the same in each case and are path independent.

II. PRESSURE AND TEMPERATURE

Temperature and *pressure* are two variables of great importance in thermodynamics and are both *intensive variables* (i.e., their values do not depend on the size of the system in question). For example, if you have a drinking glass and a bucket sitting on top of your desk, the temperature and pressure in each is the same, even though the volumes and masses are different. Let's take a moment to talk about temperature and pressure.

A. What is temperature?

When we think of temperature in our everyday lives, it relates to the relative hotness or coldness of an object. However, what makes an object hot? The temperature of a body is related to the average kinetic energy of the atoms or molecules that make up the body by the equation

$$T = \frac{2}{3} K \frac{1}{k} \tag{2-1}$$

where K is the average kinetic energy and k is Boltzmann's constant ($k = 1.381 \times 10^{-23}$ J K^{-1}). *Kinetic energy* is the energy of motion and is related to the velocity of an object by the equation

$$K = \frac{1}{2} mv^2 \tag{2-2}$$

where m is mass (kg) and v is velocity (m s^{-1}). Equation (2-1) shows that by putting energy into a system, we can raise the temperature. Conversely, as energy dissipates, temperature falls. We can put energy into a system by heating it (e.g., heating a pan of water on top of a stove) or by other methods. Joule's experiments, discussed in Chapter 3, showed that mechanical agitation of water would increase the water's temperature. Water molecules gain kinetic energy as they move around, so water will be slightly warmer after it is stirred. Any process that increases the average velocity of particles in a system increases the temperature of the system.

What happens if two systems of different temperatures contact one another? Imagine putting a hot spoon that just came out of the dishwasher into a drawer on top of a cold spoon. The spoons are now touching each other. Since the molecules in the two spoons can collide with each other, those from the higher-temperature spoon will transfer some of their kinetic energy to the molecules in the lower-temperature spoon during collisions. In this way, the warmer spoon will cool off while the cooler spoon warms up. This process will continue until both spoons have the same temperature. Each spoon is a system, and when the temperatures of both spoons stop changing, the two systems will be at *thermal equilibrium*.

Imagine that a system A, spoon or otherwise, is in contact with a system B, and system B is in contact with system C. If B is in thermal equilibrium with both A and C, then A must be in thermal equilibrium with C as well. This is the *zeroth law of thermodynamics*, and it is the basis of temperature measurement. We discuss another important concept, the difference between heat and temperature, in Chapter 3 as part of our discussion of the first law.

B. Temperature measurement

Generally, temperature is measured indirectly by measuring its effects. Most materials expand when heated, and each material has its own characteristic thermal expansion coefficient (α). Most familiar temperature-measuring devices are based on this principle. Table 2-1 lists average thermal expansion coefficients for some materials used in thermometers and for some common geological materials. Volumetric and linear thermal expansion coefficients express the fractional change in volume or length, respectively, with temperature.

Figure 2-2 shows a simplified diagram of a liquid-in-glass thermometer, which is an evacuated tube filled partly with alcohol or mercury and sealed at both ends. The liquid in the tube expands on heating and fills more of the tube. The tube has marks that relate liquid volume (proportional to liquid column length) with temperature via the thermometer equation,

$$t(^\circ\text{C}) = \frac{100(l - l_0)}{(l_{100} - l_0)} \tag{2-3}$$

The calibration of a mercury-in-glass thermometer illustrates the use of Eq. (2-3). The thermometer is immersed in a mixture of ice + water at the melting point of ice (0°C), and the distance from the bottom of the bulb to the top of the mercury column is measured. This is the distance l_0. A similar measurement is made when the thermometer is immersed in boiling water at 1 atmosphere pressure. This is the distance l_{100}. Then the temperature (t°C) corresponding to distance l can be calculated and the temperature scale can be etched on the thermometer. Note that when $l = l_0$, Eq. (2-3) gives $t = 0$°C, and that when $l = l_{100}$, Eq. (2-3) gives $t = 100$°C.

Another familiar device is made of two strips of metal with different coefficients of thermal expansion. Figure 2-3 is a schematic diagram of this device, which is used in household thermostats. Two thin strips of metal are fused together in a coil. The metal with the larger thermal

Table 2-1 Some Typical Thermal Expansion Coefficients (α) for Different Materials[a]

Volumetric Coefficients		
Substance	**α (K^{-1}) × 10^5**	***t* (°C)**
He (gas)	366.1	0
methanol (CH_3OH)	149	20
Hg (liquid)	18.11	20
Fe (metal)	4.61	800
Fe (liquid)	14.4	1535 (m.p.)
fayalite (Fe_2SiO_4)	3.19	20–900
forsterite (Mg_2SiO_4)	4.36	23–1600
spinel (γ-Mg_2SiO_4)	1.89	24–750
pyrope ($Mg_3Al_2Si_3O_{12}$)	3.15	25–700
grossular ($Ca_3Al_2Si_3O_{12}$)	2.69	25–675
graphite	3.15	800
diamond	0.88	800
α-quartz (SiO_2)	2.43	25–500
stishovite (SiO_2)	1.86	27–420
coesite (SiO_2)	0.69	20–1000
high cristobalite (SiO_2)	0.6	400–1200
water ice	15.9	0
water[b]	–6.814	0
diopside ($CaMgSi_2O_6$)	3.33	24–1000
clinoenstatite ($Mg_2Si_2O_6$)	3.33	20–700
orthoenstatite ($Mg_2Si_2O_6$)	4.23	20–1084
perovskite ($MgSiO_3$)	2.2	25–108
kyanite (Al_2SiO_5)	2.6	25–800
andalusite (Al_2SiO_5)	2.48	25–1000
sillimanite (Al_2SiO_5)	1.46	25–1000

Linear Coefficients		
Substance	**α (K^{-1}) × 10^5**	***t* (°C)**
aluminum	2.31	25
copper	1.65	25
iron	1.18	25
granites and rhyolites	(0.8±0.3)	20–100
basalt, gabbro, diabase	(0.54±0.1)	20–100
Pyrex glass	0.32	19–350
silica glass	0.054	0–1000

[a] *The values listed are multiplied by 10^5, that is, the thermal expansion coefficient for He gas is 3.661×10^{-3} and that for coesite is 6.9×10^{-6}. The data are from Fei (1995), Skinner (1966), Smyth et al. (2000), Yang and Prewitt (2000), and Lide (2000).*
[b] *Water has a negative thermal expansion coefficient below 4°C, that is, it contracts when heated.*

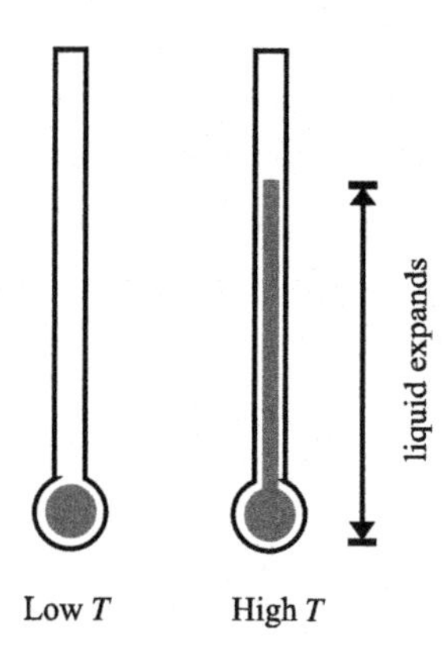

FIGURE 2-2

Liquid expanding inside a mercury-in-glass thermometer.

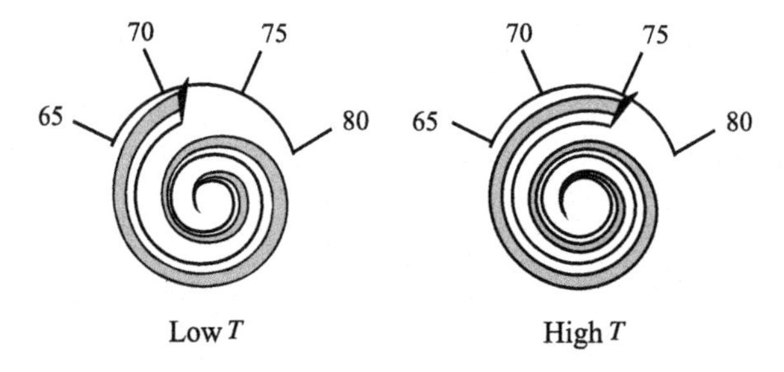

FIGURE 2-3

The bimetallic metal coil winding inside a thermostat.

expansion coefficient is on the outside of the coil. As this system warms up, the outside layer expands more than the inside layer and forces the coil to tighten. A pointer attached to the end of the coil moves in the direction of the curve (to the right in our picture) as the coil gets hotter. Again, there are marks that correlate a given amount of curvature in the coil with a given temperature.

Thermal expansion is not the only temperature-dependent property that is measured. The frequency of the light given off by a hot body varies with temperature and is measured with an optical pyrometer. Temperature is calculated by comparison to the frequency of the radiation that would be given off by an ideal black body. A *black body* is a theoretically perfect absorber and emitter of radiation. The radiation emitted through a pinhole in a hot cavity (e.g., a metal sphere) gives a good approximation to an ideal black body.

As you might expect, temperature affects the electrical properties of materials as well. The temperature-dependent electrical resistance of platinum metal wire in a platinum resistance thermometer or of a semiconducting metal oxide bead in a thermistor is used to measure temperature. *Thermocouples* determine temperature by measuring the temperature-dependent electromotive force (EMF) in volts developed between two different metals. A different EMF is generated at each

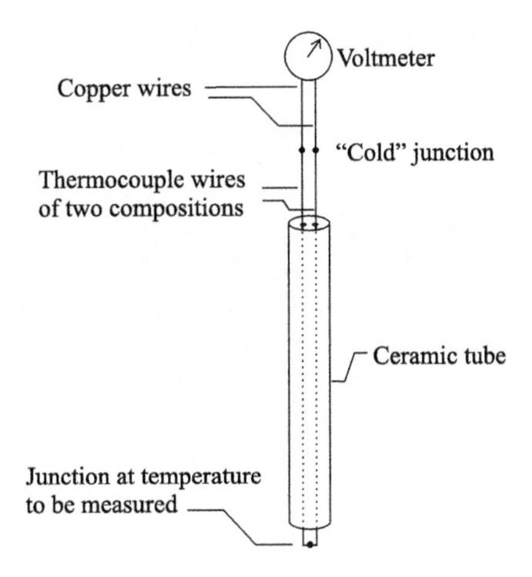

FIGURE 2-4

A sketch of a thermocouple.

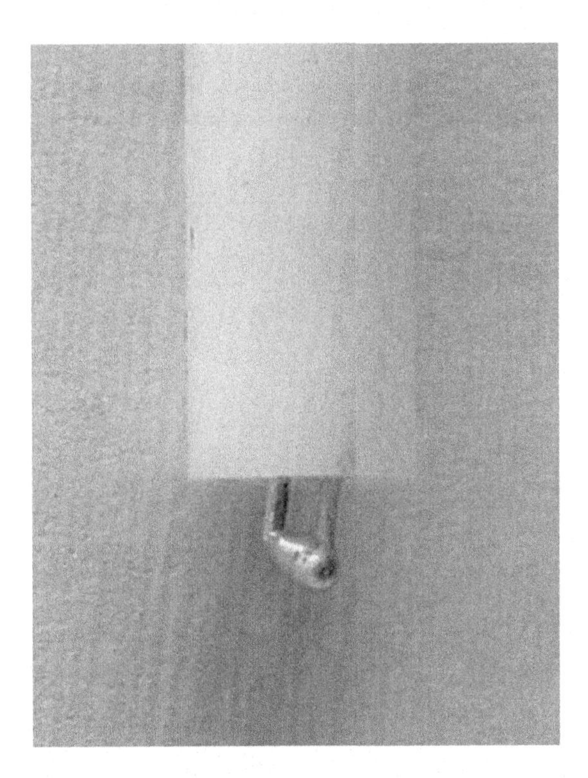

FIGURE 2-5

Photograph of the welded bead at the end of a thermocouple.

temperature by each different pair of metals. The use of thermocouples is widespread in industry and research laboratories.

Figure 2-4 shows a schematic diagram of a thermocouple. The thermocouple is made of two wires that have different compositions and are welded together at one end. Figure 2-5 shows such a weld. The wires are inside an electrically insulating tube (e.g., plastic or ceramic) so that they do not touch each other at any point other than the welded spot. This part is exposed to the system of which the temperature is being measured. The other ends of the wires connect to copper leads at the "cold" junction, which is a reference point. The cold junction is inside an ice bath at 0°C or is left at room temperature, in which case a correction must be made. The copper leads connect to an instrument such as a potentiometer or a digital voltmeter that measures EMF. Each EMF corresponds to a particular temperature. Commercial thermocouple meters typically store emf-T data for different types of thermocouples in memory. The emf-T data for the type of thermocouple connected to the meter are selected by flipping a switch. These meters also automatically correct for the nonzero temperature at the cold junction and convert the EMF readings into temperatures. Table 2-2 lists common thermocouple types and their temperature ranges. Figure 2-6 shows a plot of T versus EMF for K- and S-type thermocouples.

As you can see, thermocouples are useful because they can measure a wide range of temperatures. They also have a small mass, which means that they reach thermal equilibrium relatively quickly and are used to follow a change in temperature.

The low mass of a temperature-measuring device is important for another reason: If you were to put a large, cold thermometer into a small, hot sample, you would not get a very accurate reading. This is because the thermometer would effectively lower the temperature of the sample! You can see that this

Table 2-2 Some Common Types of Thermocouples

Type	Composition of Wires	*T*-Range (°C)	Guidelines for Use
K	chromel (Ni-Cr alloy)/ alumel (Ni-Al alloy)	−270 to 1372	Oxidizing and inert gases.
T	Cu/constantan (45% Ni-55% Cu alloy)	−270 to 400	Mild oxidizing, reducing vacuum, or inert. Good where moisture is present.
J	Fe/constantan (45% Ni-55% Cu alloy)	−210 to 1200	Reducing, vacuum, or inert.
B	Pt-30% Rh/Pt-6% Rh	0 to 1820	Oxidizing or inert. Do not insert in metal tubes.
R	Pt/Pt-13% Rh	−50 to 1768	Oxidizing or inert. Do not insert in metal tubes.
S	Pt/Pt-10% Rh	−50 to 1768	Oxidizing or inert. Do not insert in metal tubes.
C	W/W-26% Re	0 to 2320	Vacuum, inert, or hydrogen.
G	W-5% Re/W-26% Re	0 to 2320	Vacuum, inert, or hydrogen.
D	W-3% Re/W-25% Re	0 to 2320	Vacuum, inert, or hydrogen.

Data from The Temperature Handbook, *Omega Engineering, Inc., Stamford, CT, 1995, and the NIST ITS-90 thermocouple Web pages (http://srdata.nist.gov/its90/main/).*

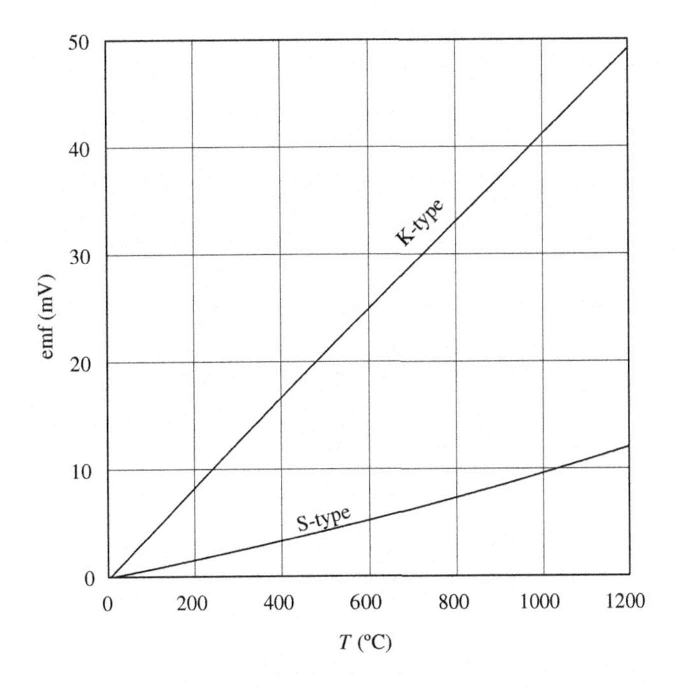

FIGURE 2-6

Electromotive force plotted versus temperature for K- and S-type thermocouples.

might be a problem during the measurement of minute temperature changes or of the temperatures of very low-mass systems.

C. Development of the international temperature scale

Another difficulty in temperature measurement has been deciding on a scale to use. Scientists want to be able to compare their results, so it is necessary to have one standard. What should that standard be based on? Historically, there has been little agreement, so a number of different temperature scales (and thermometers) have been used (see Partington, 1949). In late 16th- and early 17th- century Italy, Galileo used a *thermoscope*, a constant-volume gas thermometer. This device consisted of an air-filled bulb with a vertical tube below it dipping into a vessel of water or wine. As the bulb's temperature varied, the fluid column either rose or fell; however, the readings also depended on atmospheric pressure because the bottom vessel was open to air. Sealed alcohol in glass thermometers with etched scales were used at the Florentine Academy in Italy and by Ferdinand II, Duke of Tuscany, in the 1640s. The latter is credited with their invention. The fixed points for these thermometers were those of a mixture of ice + salt and blood heat. In the 1660s, Robert Hooke (1635–1703) and Robert Boyle (1627–1691) in England used thermometers with the melting point of ice, 0°C (Hooke), or the freezing point of oil of anise, about 17°C (Boyle), as the lower fixed points.

DANIEL GABRIEL FAHRENHEIT (1689–1736)

Fahrenheit began to take an interest in making scientific instruments in 1701, after the premature deaths of his parents, when he had to take up a trade. Beginning in 1707, he traveled extensively to meet and observe other instrument makers. He set up his own business in Amsterdam in 1717. In about 1708 he began making the thermometers that have made his name famous. He believed them to be graduated on Ole Roemer's scale; however, there was a misunderstanding about the upper fixed point, so Fahrenheit's scale was in fact unique. (Roemer had not published a description of his scale, but Fahrenheit watched him graduating thermometers during a visit.)

Fahrenheit made all sorts of equipment, including a thermometer that could be used to measure atmospheric pressure. He knew that the boiling points of liquids varied with atmospheric pressure, so he made a device that allowed a person to read the atmospheric pressure directly from a measurement of the boiling point of water.

Fahrenheit became a member of the Royal Society in 1724, though he had no formal scientific training and published very little. This was common at the time. Instrument makers and scientists worked closely together because the design and quality of the instruments were critical for allowing the researchers to carry out their work.

WILLIAM JOHN MACQUORN RANKINE (1820–1872)

William John Macquorn Rankine was a Scottish engineer. During his youth, poor health meant that he had to be educated at home by his father and by private tutors. He later attended college but did not complete a degree. In spite of this fact, he became professor of civil engineering and mechanics at the University of Glasgow in 1855. He was also one of the founders and first president of the Institution of Engineers in Scotland and a Fellow of the Royal Society.

Like many of his time who contributed to the field of thermodynamics, Rankine worked for a railroad. He developed a technique for laying out circular curves of track and investigated such topics as the best shape for wheels and how to prevent axles from breaking. He also penned a popular series of engineering textbooks and worked on plans for the water supply of the city of Glasgow.

Rankine developed his own absolute temperature scale that was based on the Fahrenheit scale because he felt reluctant to give up traditional British units of measurement in favor of the Celsius scale and the metric system. He even wrote a song about the issue of the encroaching metric system, titled "The Three-Foot Rule." Its lyrics went, in part, "A party of astronomers went measuring of the Earth, And forty million mètres they took to be its girth; Five hundred million inches, though, go through from pole to pole; So let's stick to inches, feet, and yards and the good old three-foot rule."

Modern temperature scales date to the early 18th century. In 1702, Ole Christian Roemer (1644–1710), the Danish astronomer, adopted a temperature scale based on two fixed points: the temperature of ice or snow and the boiling point of water. In 1714, Daniel Gabriel Fahrenheit, whose life is briefly described in the sidebar, constructed the first useful sealed mercury thermometer and took 0°F as the temperature of a mixture of solid NH_4Cl, ice, and water (−18°C). He arbitrarily divided the interval between this temperature and that of the melting point of ice into 32 degrees. Fahrenheit measured the boiling point of water to be 212°F. The Fahrenheit scale is still used in daily life in the United States. The Rankine temperature scale, named after the Scottish engineer William J. M. Rankine (see his biography in the sidebar), is the absolute Fahrenheit scale. The Rankine scale is related to the Fahrenheit scale by the equation $T\,(°\mathrm{R}) = T\,(°\mathrm{F}) + 459.67°$, and it is still used today by engineers.

The Centigrade scale, which is based on the freezing (0°C) and boiling (100°C) points of air-saturated water, dates to the 1740s, when centigrade scales (= 100 degrees) were proposed by Anders Celsius (see sidebar) and Carl Linnaeus (1707–1778). The temperatures of the freezing and boiling points are arbitrarily chosen; in fact, Celsius proposed 100° for the freezing point and 0° for the boiling point of water. Linnaeus proposed the reverse, and his Centigrade scale was the one used. In 1954, Linneaus's Centigrade scale was replaced by a new scale named after Celsius, which is not a centigrade scale but is based on the triple point of water.

ANDERS CELSIUS (1701–1744)

Anders Celsius was a Swedish astronomer who became a professor of mathematics and astronomy at the University of Uppsala, where his father and grandfather had also taught astronomy. After being appointed professor in 1730, Celsius went abroad to round out his education. He joined a French group in planning and executing a trip to Lapland to measure meridian lines. Their findings confirmed Isaac Newton's hypothesis that Earth is somewhat flattened at the poles. This was a daring undertaking for several reasons, not least of which was at that time in Sweden and elsewhere, the Copernican view of the solar system was sacrilegious.

Celsius had an ability to win friends and garner support. After his travels, he was able to obtain permission and funding for an observatory at the university. His other achievements in the field of astronomy included measuring the magnitudes of the stars in Aries and writing a discourse on observations by himself and others of the aurora borealis.

Celsius's name is best known for the 100° temperature scale. Celsius was not the first to propose such a scale, but his observation that two "constant degrees," or fixed points, were enough to define a temperature scale led to the general acceptance of the 100° scheme.

Before describing this revision, we introduce the *Kelvin,* or *absolute,* temperature scale used in thermodynamics. William Thomson (Lord Kelvin; see sidebar) proposed this scale in 1848 and discussed it in several later papers. The Kelvin scale is related to the Celsius (and Centigrade) scales by the equation $T\,(^{\circ}\mathrm{C}) = T\,(\mathrm{K}) - 273.15$, where the temperature of the freezing point of air-saturated water is 273.15 K. The Kelvin scale is based on the second law of thermodynamics and, like the Centigrade scale, was originally defined to have a 100° interval between the freezing and boiling points of water.

WILLIAM THOMSON, LORD KELVIN (1824–1907)

Lord Kelvin was born William Thomson in Belfast, Ireland, though he was of Scottish decent. He entered the University of Glasgow at the age of 10 and the University of Cambridge at the age of 17. He became professor of natural philosophy at the University of Glasgow in 1846 and taught there until his retirement in 1895. At that time, he enrolled himself as a research student and continued his association with the university until his death.

Kelvin's long friendship with James Joule began in 1847 at an historic meeting of the British Association. Kelvin and Joule discovered the principle of Joule-Kelvin cooling of gases, which is important for refrigeration. In the early 1850s Kelvin developed the thermodynamic (Kelvin) temperature scale and formulated the second law of thermodynamics contemporaneously with Rudolf Clausius. Kelvin was the leading British physicist throughout his lifetime. Queen Victoria knighted him in 1866 for his work on the first transatlantic telegraph cable, and she made him a peer in 1892. Kelvin is buried in Westminster Abbey, next to Sir Isaac Newton.

In 1854, Kelvin pointed out that this was a clumsy definition and instead proposed to define the temperature of a single fixed point and the size of the degree. His suggestion was not adopted at the time. The Nobel Prize–winning chemist William F. Giauque (1895–1982) revived Kelvin's proposal in 1939. By then, it was clear that it was experimentally very difficult to reproduce the freezing point of air-saturated water within a few hundredths of a degree. Different laboratories reported values ranging from 273.13 K to 273.17 K for the freezing point (0°C). These difficulties led to problems at low temperatures, where an uncertainty of 0.01 degree gave larger and larger errors as lower temperatures were achieved.

The Tenth General Conference on Weights and Measures adopted the Kelvin-Giauque proposal in 1954. The Kelvin and Centigrade (renamed Celsius) temperature scales are based on a single fixed point, the *triple point of water*. This is the equilibrium temperature for ice + water + water vapor. It was set at 0.01°C (273.16 K). The size of the degree was set as one Kelvin (1 K), which is 1/273.16 of the thermodynamic (Kelvin) temperature of the triple point of water. One degree Celsius has the same size as one Kelvin.

Although these definitions unambiguously define the Kelvin and Celsius temperature scales, it is not possible for every laboratory to maintain a triple point cell as a temperature standard. An international temperature scale that allows users all over the world to calibrate their thermometric devices to various "fixed points" is needed.

The Seventh General Conference of Weights and Measures adopted the first international temperature scale in 1927. This specified the temperatures of "fixed points," such as the freezing point of gold, and the instruments and methods for making temperature measurements with as close an approximation to thermodynamic temperatures as could be done at that time. The freezing point and the melting point are the same at equilibrium, but the temperature scale is specified in terms of the one that can be determined more accurately. The international temperature scale was revised in 1948, 1968, 1976, and most recently when the Eighteenth General Conference of Weights and Measures adopted the International Temperature Scale of 1990 (ITS-90). Table 2-3 lists temperatures of fixed points on ITS-90. These temperatures are as close as possible to the actual thermodynamic values. As mentioned earlier, the Kelvin scale is based on the second law of thermodynamics. It does not depend on the physical properties (e.g., thermal expansion coefficient, electrical properties) of any given material. We return to these issues in Chapter 6 when we discuss the second law of thermodynamics, entropy, and heat engines. Table 2-4 lists conversion factors between common temperature scales. Further information about the history of thermometry, the development of temperature scales, and modern methods of temperature measurement are given in volume 1 of Partington (1949), by Quinn (1983), and in Zemansky (1957).

D. What is pressure?

Pressure is defined as force per unit area, and *force* is given by mass times acceleration. Thus, we can calculate the pressure (P) exerted by this book or another object lying on a table by multiplying the mass (m) of the object by the average acceleration of gravity ($g = 9.81$ m s^{-2}) and dividing by the surface area (A) of the bottom of the object.

Example 2-1. The dictionary in Figure 2-7 has the dimensions 26 by 22 by 13 centimeters and mass 4 kg. The average acceleration due to gravity (g) is 9.81 m s^{-2}. The pressure exerted is

$$P = \frac{F}{A} = \frac{mg}{A} \tag{2-4}$$

Table 2-3 Defining Fixed Points of the ITS-90

Fixed Point	Temperature (K)
Vapor pressure equation of He[a]	3 to 5
Triple point[b] of equilibrium H_2	13.8033
Triple point of Ne	24.5561
Triple point of O_2	54.3584
Triple point of Ar	83.8058
Triple point of Hg	234.3156
Triple point of H_2O	273.16
Melting point of Ga	302.9146
Freezing point of In	429.7485
Freezing point of Sn	505.078
Freezing point of Zn	692.677
Freezing point of Al	933.473
Freezing point of Ag	1234.93
Freezing point of Au	1337.33
Freezing point of Cu	1357.77

[a]*The light isotope ^{3}He is used from 0.65 K to 3.2 K, and the heavier, more abundant isotope ^{4}He is used from 1.25 K to 2.18 K and 2.18 K to 5.0 K.*
[b]*The triple point is the temperature and pressure at which the gas, liquid, and solid phases all exist in equilibrium.*

Table 2-4 Temperature Conversion Factors

$T_K = t_{°C} + 273.15$
$t_{°C} = (t_{°F} - 32)/1.8$
$T_K = (t_{°F} + 459.67)/1.8$
$T_K = T_{°R}/1.8$
$T_{°R} = t_{°F} + 459.67$

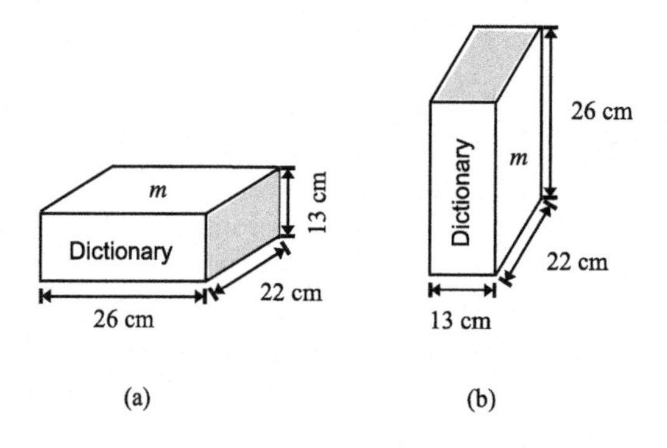

FIGURE 2-7

A dictionary standing on its side (a) and standing upright (b).

When the book is lying on its side (a), the area of the bottom of the book is $0.0572\ m^2$. When it is standing on end (b), the area is only $0.0286\ m^2$, or half of what it was before. Therefore,

$$P_1 = \frac{4\ \text{kg} \times 9.81\ \text{m/s}^2}{0.0572\ \text{m}^2} = 686\ \text{Pa} \tag{2-5}$$

$$P_2 = \frac{4\ \text{kg} \times 9.81\ \text{m/s}^2}{0.0286\ \text{m}^2} = 1372\ \text{Pa} = 2P_1 \tag{2-6}$$

However, in the two cases the force (about 39.2 N) is the same. The SI unit of pressure is the pascal (Pa). A *pascal* is one newton per meter squared ($N\ m^{-2}$). A *newton* is the SI unit of force and is equal to one kilogram meter per second squared ($kg\ m\ s^{-2}$).

In thermodynamics, we consider the pressure exerted by a gas more often than that exerted by a solid mass. Gas pressure is a statistical property because it depends on averaging values such as velocity over a very large number of bodies. Gases exert pressure in a way that is similar to solid bodies in that the pressure can still be described as force per unit area. A collection of gas particles in a balloon is moving around in all directions. The force of a great many collisions of gas particles against the sides of the balloon is what keeps the balloon inflated. The pressure inside the balloon is equal to the average amount of force per area exerted against the sides of the balloon, which will depend on the average velocity of the particles, their mass, and on how many collisions occur. The average number of collisions in turn depends on the gas density or number of particles (N) per volume (V). The kinetic theory of gases shows us that

$$P = \frac{mN\langle v^2\rangle}{3V} \tag{2-7}$$

for an ideal gas, where m is the average mass of the particles, N is the number of particles present, v is their average velocity, and V is the volume of their container. We discuss the relationship between P and V of a gas when we describe Boyle's law later in this chapter.

Example 2-2. At 300 K, N_2 molecules have an average velocity of about $517\ m\ s^{-1}$. What pressure would 0.04 moles of N_2 gas at 300 K exert inside a 1 liter cylinder? The atomic weight of N is 14.00674 $g\ mol^{-1}$, so the molecular weight of N_2 is 28.01348 $g\ mol^{-1}$.

$$P = \frac{mN\langle v^2\rangle}{3V} = \frac{(28.01348\ \text{g/mol})/N_A(0.04\ \text{mol} \times N_A)(517\ \text{m/s})^2}{3(1\ \text{L})}$$

$$= 99836\ \frac{\text{g m}^2}{\text{L s}^2} \times \frac{1\ \text{kg}}{1000\ \text{g}} \times \frac{\text{J s}^2}{\text{kg m}^2} \times \frac{9.8696 \times 10^{-3}\ \text{L atm}}{\text{J}} \times \frac{1.01325\ \text{bars}}{\text{atm}} \tag{2-7}$$

$$= 1\ \text{bar}$$

E. Pressure measurement

One of the first devices used to measure pressure was the *manometer*, a U-shaped tube partially filled with a fluid such as water, oil, or mercury, as pictured in Figure 2-8. One end of the tube connects to the system under observation. The other end is either open to the air, in which case the pressure measured is

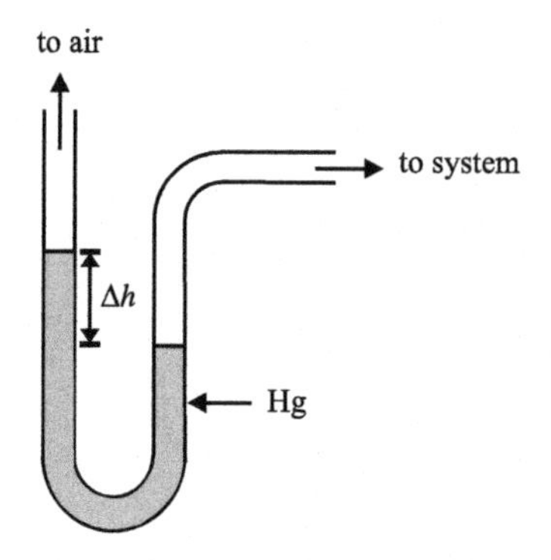

FIGURE 2-8

Pressure measurement with a manometer.

relative to the ambient pressure, or is a sealed vacuum, in which case an absolute pressure measurement is achieved (as in a barometer). A change in the pressure of the system changes the height of the mercury in the tube. The changes in height are measured visually or more precisely using a small telescope on a calibrated stand (a cathetometer). The change in height (h) is related to a change in pressure by

$$\Delta P = P_{system} - P_{ref} = \rho g \Delta h \tag{2-8}$$

where ρ is the density of the fluid in the column, g is the acceleration due to gravity, and P_{ref} is the atmospheric pressure (in the case of an open-ended manometer) or zero (for an absolute manometer that has a sealed vacuum at one end).

Example 2-3. In the late 19th century, the French physicist Émile Amagat (1841–1915) used very long, open-ended mercury manometers to measure pressures of compressed gases in his experiments. What is the gas pressure corresponding to a mercury column height of 200 meters ($\Delta h = 200$ m), assuming constant density (ρ) for Hg and constant acceleration due to gravity throughout the column and a reference pressure of one bar? At room temperature and ambient pressure ρ (Hg) = 13.534 g mL^{-1} = 13.534 kg L^{-1}. Substituting into Eq. (2-8) and converting to consistent units we find

$$\begin{aligned} P_{system} &= \rho g \Delta h + P_{ref} = (13.534 \text{ kg L}^{-1})(9.81 \text{ m s}^{-2})(200 \text{ m}) + 1 \text{ bar} \\ &= (26,554 \text{ kg m}^2 \text{ L}^{-1} \text{ s}^{-2})(9.8696 \times 10^{-3} \text{ L atm J}^{-1})(1.0133 \text{ bar atm}^{-1}) + 1 \text{ bar} \\ &= 267 \text{ bars} \end{aligned}$$

Figure 2-9 shows a *Bourdon gauge*. This device is used on compressed gas tanks in laboratories. The Bourdon gauge contains a curved tube, which uncoils as the pressure in it increases. The end of the tube connects to a pointer that indicates the pressure on a scale.

There are several different units for pressure. We will usually talk about bars or atmospheres. One *bar* is equal to 10^5 Pa or 10^2 kPa. One *atmosphere* is equal to 1.01325 bars. As mentioned earlier, the pascal is the SI unit of pressure and is used in most countries in the world. For example, TV weather reports in Canada give atmospheric pressure in kilopascal (kPa). Other common units for measuring pressure are millimeters mercury (mmHg), used in the older scientific literature, or inches of mercury, used for TV weather reports in the United States. You are probably already familiar with the unit pounds per square inch (psi), used for measuring the air pressure in tires. As shown in Figure 2-9, the gauges on

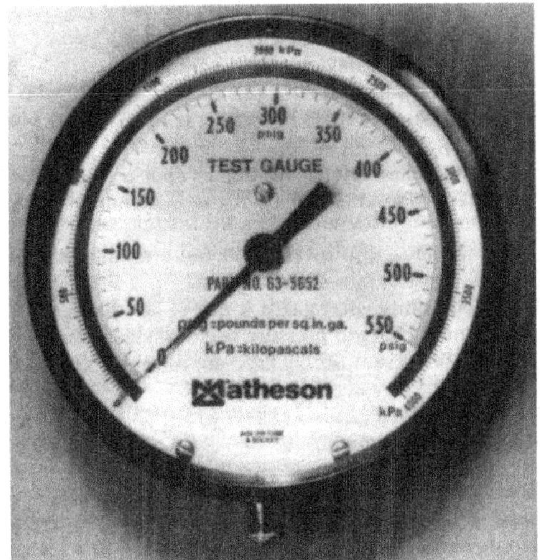

FIGURE 2-9

Photographs showing a Bourdon pressure gauge.

compressed gas tanks give pressure per square inch gauge (psig), which is the pressure (in psi) above ambient pressure (about 14.70 psi). Table 2-5 lists conversion factors for pressure measurement units. Throughout the rest of this book, *pressure* and P refer to pressure in bar unless otherwise stated. Absolute and relative presssure measurements, including in high vacuum systems, are given in Le Neindre and Vodar (1975) and by Tilford (1992).

III. BOYLE'S LAW

Robert Boyle published the results of his experiments using the newly invented mercury barometer and vacuum pump in the early 1660s. He found that at constant temperature the volume of a fixed mass of gas is inversely proportional to the pressure

$$PV = C \tag{2-9}$$

where P is pressure, V is volume, and C is a constant. An equivalent equation is

$$P_1V_1 = P_2V_2 \tag{2-10}$$

where P_1 and V_1 are the initial pressure and volume and P_2 and V_2 are the final pressure and volume of the gas, or

$$P = C/V \tag{2-11}$$

The PV product has units of energy, which are the same units used for work. Pressure is force per unit area (F/A), so $PV = (F/A)(V) = (F)(V/A) =$ force times length $=$ work (or energy). A table listing conversion factors for different units used for energy and work is in Chapter 3.

Table 2-5 Pressure Conversion Factors

	Pa	bar	dyn cm^{-2}	kg cm^{-2}
1 Pa =	1	10^{-5}	10	1.01972×10^{-5}
1 bar =	10^5	1	10^6	1.01972
1 dyn cm^{-2} =	0.1	10^{-6}	1	1.01972×10^{-6}
1 kg cm^{-2} =	9.80665×10^4	0.980665	9.80665×10^5	1
1 atm =	1.01325×10^5	1.01325	1.01325×10^6	1.03323
1 mmHg = 1 torr =	133.322	1.3332×10^{-3}	1,333.2	1.3595×10^{-3}
1 psi =	6,894.757	0.068947	68,947.57	7.0307×10^{-2}
	atm	**mmHg**	**psi**	
1 Pa =	9.8692×10^{-6}	7.5006×10^{-3}	1.45038×10^{-4}	
1 bar =	0.98692	750.062	14.5038	
1 dyn cm^{-2} =	9.8692×10^{-7}	7.5006×10^{-4}	1.45038×10^{-5}	
1 kg cm^{-2} =	0.9678	735.565	14.2233	
1 atm =	1	760.0	14.6959	
1 mmHg = 1 torr =	1.316×10^{-3}	1	0.0193367	
1 psi =	0.068046	51.7151	1	

The value of g used is 9.80665 m s^{-2}.
Here psi denotes pounds per square inch absolute.

Table 2-6 gives the molar *PV* product for air at several pressures at constant temperature. This is simply the product of the molar volume times the pressure. The molar volume $V_m = V/n$, where *n* is the number of moles. Under the conditions listed in Table 2-6, the *PV* product for air is constant within 0.03% of the one bar value. The deviations from Boyle's law occur because air is not an ideal gas. In fact, no gas behaves ideally under all conditions, but several gases (e.g., air, H_2, He, Ne, Ar, O_2, N_2) approach ideality at normal temperatures as pressure is reduced to lower and lower values. Table 2-7 lists extrapolated values of molar *PV* products. The extrapolated values are almost the same for many gases. The mean molar *PV* product, 22.4140 atm liter mol^{-1}, gives the molar volume of an ideal gas at one atmosphere pressure and 273.15 K. The molar volume of an ideal gas at one bar pressure and 273.15 K is slightly larger because one bar is a slightly lower pressure than one atmosphere and is 22.7110 liter-bar mol^{-1}.

The variation of pressure and molar volume at 273.15 K for an ideal gas and for dry air at 300 K, a nearly ideal gas, are shown in Figure 2-10. This type of curve is an *isotherm*. An isotherm shows the

Table 2-6 *PV* Product for Air at 350 K

P (bar)	*V* (liters mol^{-1})	*PV*
1	29.1008	29.1008
5	5.8207	29.1035
10	2.9109	29.1090

Source: Thermodynamic Properties of Air *by Sychev et al. (1987).*

Table 2-7 PV Product (atm L mol^{-1}) for Selected Gases at One Atmosphere and Extrapolated to Zero Pressure, 273.15 K (0°C)

Gas	μ (g mol^{-1})	ρ (g L^{-1})	$(PV)_1$	$(1+\lambda)^a$	$(PV)_0$
H_2	2.01588	0.089873	22.4303	0.99929	22.414
He	4.002602	0.17846	22.429	0.99954	22.418
Ne	20.1797	0.9002	22.417	0.99941	22.404
Ar	39.948	1.78394	22.393	1.00090	22.413
O_2	31.9988	1.42897	22.3929	1.00091	22.413
N_2	28.0134	1.2505	22.402	1.0006	22.415
Air	28.970	1.2928	22.405	1.0006	22.418
NO	30.0061	1.3402	22.389	1.0011	22.414
N_2O	44.0128	1.9786	22.244	1.0078	22.418
NH_3	17.0305	0.77138	22.0780	1.0152	22.414
CO	28.0101	1.2504	22.401	1.0005	22.412
CO_2	44.0095	1.9770	22.261	1.0069	22.414
CH_4	16.0425	0.7174	22.362	1.0024	22.416
C_2H_2	26.0373	1.1747	22.165	1.010	22.387
C_2H_4	28.0532	1.26099	22.2470	1.0075	22.414
C_3H_8	44.0956	2.005	21.993	1.0204	22.441
CH_3Cl	50.488	2.3065	21.889	1.0234	22.402
H_2S	34.081	1.5379	22.161	1.0104	22.391
SO_2	64.064	2.9267	21.890	1.0238	22.410
PH_3	33.99758	1.5305	22.213	1.0097	22.429
HCl	36.461	1.6394	22.240	1.0074	22.405
HBr	80.912	3.6375	22.244	1.0093	22.451
				Mean $\pm$ 1σ =	22.414 $\pm$ 0.014

[a] $(1+\lambda) = (PV)_0/(PV)_1$ ratio.
Sources: Coplen (2001), Din (1962), Partington (1949), Pickering (1928).

correlated variation of the volume and pressure of a constant amount of gas at constant temperature. We will come back to isothermal curves later in this chapter when we discuss the *PVT* surface for an ideal gas.

Example 2-4. A piston in a glass cylinder contains one liter of air at one bar pressure (see Figure 2-11). A pressure gauge is attached to the cylinder to measure the air pressure, and the volume of the cylinder can be read from markings on the side. The cylinder and piston are in a water bath kept at constant temperature of 25°C (298.15 K). The piston is lowered until the pressure of the air in the cylinder is 2 bars. What is the final volume of the air? Using Eq. (2-9) we find

$$PV = C$$

$$(1\ \text{bar})(1\ \text{liter}) = 1\ \text{bar} \times 1\ \text{liter} = C$$

$$1\ \text{bar} \times 1\ \text{liter} = (2\ \text{bars})(V)$$

$$V = (1\ \text{bar} \times 1\ \text{liter})/(2\ \text{bars}) = 0.5\ \text{liter}$$

Thus, the volume is halved when the pressure is doubled.

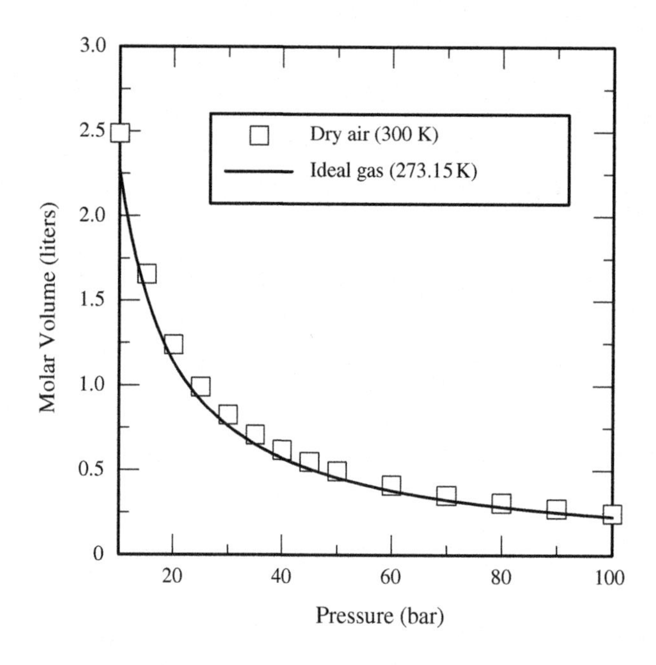

FIGURE 2-10

Isotherms for air (300 K) and an ideal gas (273.15 K).

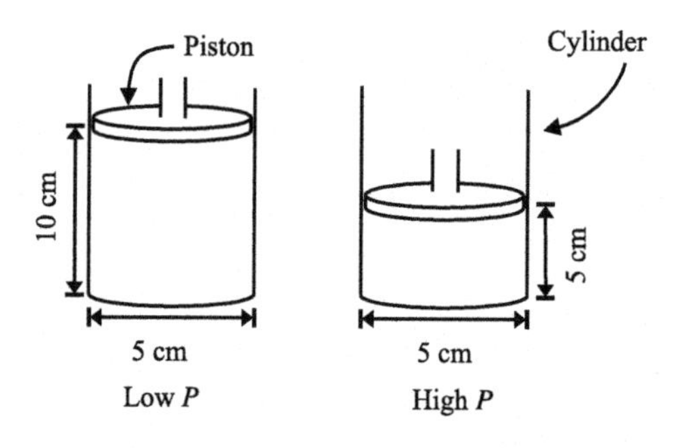

FIGURE 2-11

A cartoon of a piston compressing gas inside a cylinder.

IV. CHARLES'S OR GAY-LUSSAC'S LAW

The knowledge that air expands when heated dates back to the ancient Greeks and Romans. Hero of Alexandria (AD 50) used heated air to open the doors of a temple by kindling a fire on an altar containing a concealed air reservoir. The heated air then drove water into buckets, which pulled down cords to open the doors (Partington, 1949).

Table 2-8 α_0 Values for Helium

P (mmHg)	α_0 ($\times 10^4$ °C^{-1})
1116.5	36.58
1102.9	36.58
760.1	36.59
520.5	36.60
504.8	36.59
0 (linear fit)	36.61

In 1787, Jacques-Alexandre-César Charles (1746–1823) discovered that the volume of a gas is proportional to temperature. Later, in 1802–1808, Joseph Gay-Lussac (1778–1850) measured the variation of volumes of permanent gases (N_2, O_2, H_2) with temperature at constant pressure. His results are given by the equation

$$V = V_0(1 + \alpha_0 t) \tag{2-12}$$

where V_0 is gas volume at 0°C (273.15 K), t is temperature in degrees Celsius, α_0 is the coefficient of thermal expansion, and V is gas volume at another temperature t. Gay-Lussac found α_0 varied from 37.40×10^{-4} °C^{-1} to 37.57×10^{-4} °C^{-1} for various gases with an average value

$$\alpha_0 = \frac{(V - V_0)}{V_0 t} = \frac{1}{267} = 37.45 \times 10^{-4}\,°\mathrm{C}^{-1} \tag{2-13}$$

Subsequent work by the French chemist Henri Victor Regnault (1810–1878) in 1847, and later by many other workers, improved on these results. For example, Regnault found

$$\alpha_0 = \frac{1}{272.5} = 36.70 \times 10^{-4}\,°\mathrm{C}^{-1}$$

for air, which is close to the modern value of

$$\alpha_0 = \frac{1}{273.15} = 36.61 \times 10^{-4}\,°\mathrm{C}^{-1}$$

This value is based on determinations of the thermal expansion of gases as a function of pressure and temperature. An example of measurements for He, compiled by Partington (1949), is in Table 2-8 and shows how the value of α_0 is computed by linear extrapolation of measurements at several low pressures to the limiting value of zero pressure.

Example 2-5. You are blowing up balloons for a New Year's party and fill a balloon with one liter of air indoors at 25°C (t_1). If you hang the balloon outside the front door, where it is only 10°C (t_2), what will its volume (V_2) be? Rearranging Eq. (2-12) and solving for V_2,

$$V_2 = \frac{V_1(1 + \alpha_0 t_2)}{(1 + \alpha_0 t_1)} = 0.95\ \mathrm{L}$$

In other words, the balloons deflate slightly when hung outside.

V. DALTON'S LAW

In the preceding example, we considered the behavior of a balloon filled with air, which is a mixture of gases. The total pressure exerted by a mixture of ideal gases in a container is equal to the sum of the pressures each gas would exert alone in the container. That is, if a quantity of one gas in a one-liter container produces 0.5 bar pressure and a quantity of a second gas in another one-liter container produces 1.5 bars pressure, the two quantities of gas mixed together in a one-liter container would produce 2.0 bars of pressure. Dalton discovered this rule in 1810.

Example 2-6. Dry air is composed of N_2 (78.084 volume %), O_2 (20.946 volume %), Ar (0.934 volume %), CO_2 (0.036 volume %), and smaller amounts of many other gases (see Table 2-11). Calculate the partial pressures and mole fractions of N_2, O_2, Ar, and CO_2 in a sample of dry air collected at an altitude of 5.5 kilometers and with a total pressure of 0.505 bar. The total pressure of the sample of dry air is

$$P_T = P_{N_2} + P_{O_2} + P_{Ar} + P_{CO_2} \tag{2-14}$$

Equation (2-15) is Dalton's law. The abundances of these four constituents are in volume percent, which is the same as mole percent. Thus, we know the relative molar abundances of the four gases. The partial pressures (in bars) of the four gases are

$$\begin{aligned}
P_{N_2} &= \frac{78.084}{100} P_T = 0.78084 \times P_T = (0.78084)(0.505) = 0.394 \\
P_{O_2} &= \frac{20.946}{100} P_T = 0.20946 \times P_T = (0.20946)(0.505) = 0.106 \\
P_{Ar} &= \frac{0.934}{100} P_T = 9.34 \times 10^{-3} \times P_T = (9.34 \times 10^{-3})(0.505) = 4.72 \times 10^{-3} \\
P_{CO_2} &= \frac{0.036}{100} P_T = 3.60 \times 10^{-4} \times P_T = (3.60 \times 10^{-4})(0.505) = 1.82 \times 10^{-4}
\end{aligned} \tag{2-15}$$

The sum of the partial pressures of the four gases adds up to 0.505 bar, the total pressure. The mole fraction of any gas is equal to the number of moles of that gas divided by the total number of moles of all gases in the system. Therefore, the mole fraction of each of the four gases is simply the volume percent divided by 100, or 0.78084 for N_2, 0.20946 for O_2, and so on.

It is inconvenient to express the abundances of minor and trace gases in air or in other gas samples as percentages because of the very small numbers involved. The abundances of these gases are expressed as parts per million by volume (ppmv), parts per billion by volume (ppbv), and parts per trillion by volume (pptv). Taking the abundance of CO_2 in dry air as an example, 0.036% CO_2 = 360 ppmv CO_2 = 360,000 ppbv CO_2 = 360,000,000 pptv CO_2. The choice between ppmv, ppbv, or pptv is made such that the resulting abundance is an easily expressed value, that is, the CO_2 abundance would be given as 360 ppmv.

VI. AN IDEAL GAS THERMOMETER

Charles's or Gay-Lussac's law is the basis for ideal gas thermometry. If Eq. (2-12) remained true at all temperatures, then the volume V of an ideal gas would become zero at $t = -1/\alpha_0$. This temperature is called *absolute zero* and is –273.15°C because $\alpha_0 = 36.61 \times 10^{-4}\,°C^{-1}$. We can rewrite Eq. (2-12) by substituting in $T_0 = 1/\alpha_0 = 273.15$ K, which yields

$$V = V_0\left(1 + \frac{t}{T_0}\right) \tag{2-16}$$

Defining the absolute temperature (T) on the Kelvin scale, and substituting $t = T - T_0$ yields

$$V = V_0\left[1 + \frac{(T - T_0)}{T_0}\right] = \frac{V_0 T}{T_0} \tag{2-17}$$

which can be used to measure temperature using a nearly ideal gas such as He, H_2, or N_2 at low pressure. Between 1907 and 1912, the geologists Arthur L. Day (1869–1960) and Robert B. Sosman (1881–1967) used a N_2 gas thermometer up to 1600°C in their work at the Geophysical Laboratory of the Carnegie Institution in Washington, D.C. They measured the temperatures of several fixed reference points, and their results were unchallenged for over 40 years. Their work established the Geophysical Laboratory Temperature Scale. This laid the foundation for the pioneering studies of igneous phase equilibria by Norman L. Bowen (1887–1956) and his colleagues over the next 50 years. Throughout the rest of this book, unless stated otherwise, temperature and T refer to absolute temperature in Kelvins, not to degrees Celsius or Fahrenheit.

VII. IDEAL GAS EQUATION OF STATE (THE IDEAL GAS LAW)

We can now combine Boyle's law (the variation of volume with pressure at constant temperature) with Charles's or Gay-Lussac's law (the variation of volume with temperature at constant pressure) to get the equation of state for a hypothetical ideal gas:

$$\frac{P_1 V_1}{T_1} = \frac{P_2 V_2}{T_2} = \text{a constant} = R, \text{the ideal gas constant} \tag{2-18}$$

where the subscripts 1 and 2 indicate the initial state and the final state, respectively. Evaluating Eq. (2-18) by using the extrapolated, zero pressure value for PV (22.71098 liter-bar mol^{-1}) at 273.15 K from Table 2-9 gives a value of the ideal gas constant R:

$$R = \frac{PV_m}{T} = \frac{22.71098}{273.15} = 0.083145 \text{ liter-bar mol}^{-1}\text{ K}^{-1} \tag{2-19}$$

Equation (2-18) can be rewritten in the familiar form of the ideal gas law:

$$PV = nRT \tag{2-20}$$

using $V = nV_m$, or as $PV_m = RT$ for one mole of gas. We will use Eq. (2-20) throughout the book.

Table 2-9 Molar Volume of an Ideal Gas at 273.15 K

V (liters mol^{-1})	P
22.41400	1 atm
22.71098	1 bar

Table 2-10 Values of the Gas Constant R

Value of R	Units
0.0820578	(liter atm)/(degree mole)
82.0578	(cm^3 atm)/(degree mole)
0.083145	(liter bar)/(degree mole)
83.145	(cm^3 bar)/(degree mole)
8.31447	joules/(degree mole)
8.31447×10^7	ergs/(degree mole)
1.98721	calories/(degree mole)

Depending on the units used for P and V, the ideal gas constant has different units. However, all these values for the ideal gas constant are equivalent if the proper conversion factors are used. Table 2-10 gives values of R in different units.

VIII. THE *PVT* SURFACE FOR AN IDEAL GAS

Figure 2-12 shows the *PVT* surface for an ideal gas. Temperature and molar volume are at the base of the surface and pressure increases upward. The surface is computed with Eq. (2-20) and shows how pressure, temperature, and volume vary for one mole of an ideal gas. Figure 2-13 shows two-dimensional projections of this surface onto the *PV*, *PT*, and *VT* planes. The short dashed lines on the *PVT* surface in Figure 2-12 and in the plots in Figure 2-13 are *isotherms*. Isotherms show the variation of pressure and volume at constant temperature, that is, Boyle's law. The long dashed lines on the *PVT* surface in Figure 2-12 and in the plots in Figure 2-13 are *isochores*. They show the variation of pressure and temperature at constant volume. The solid lines on the *PVT* surface in Figure 2-12 and the plots in

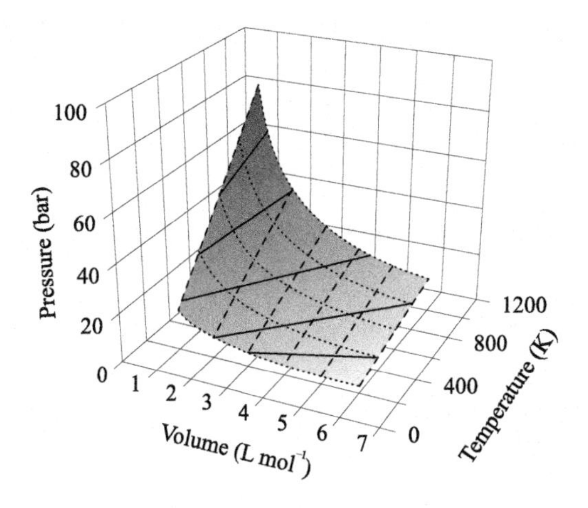

FIGURE 2-12

A three-dimensional graph of the *PVT* surface of an ideal gas.

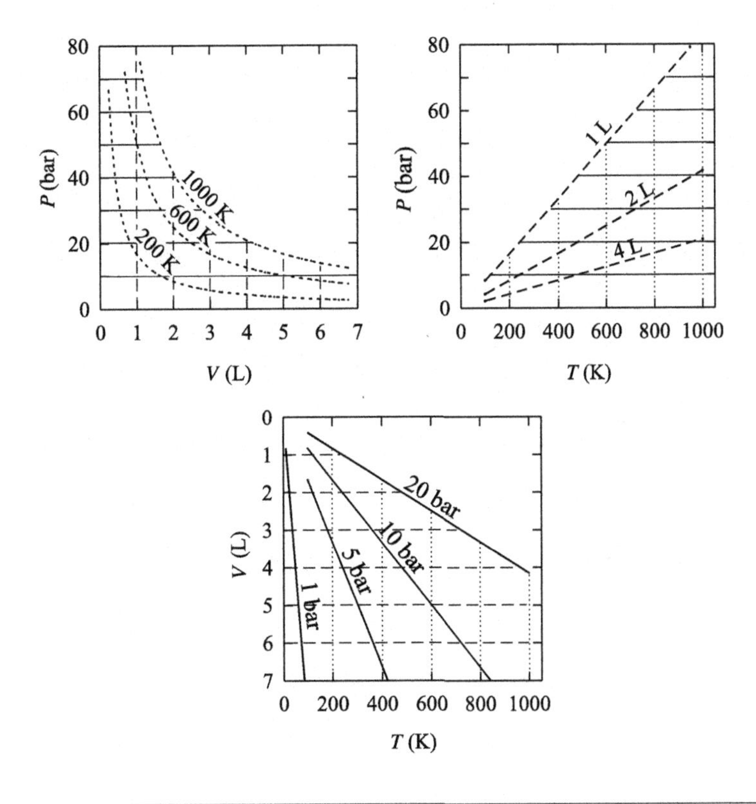

FIGURE 2-13

Two-dimensional slices through the ideal gas *PVT* surface in Figure 2-12.

Figure 2-13 are *isobars* and they show the variation of volume and temperature at constant pressure, that is, Charles's law.

IX. COMPLETE DIFFERENTIALS AND THE *PVT* SURFACE OF AN IDEAL GAS

The *PVT* surface shows that the volume of an ideal gas is a function of both temperature and pressure (see Figures 2-12 and 2-14), which is mathematically written as $V = V(T, P)$. Thus, the volume changes as we move from one temperature and pressure point to another (unless we move along an isochore, a constant volume line). A differential (i.e., an immeasurably small) change in volume (dV) when we move from temperature T and pressure P to a new temperature $T + dT$ and a new pressure $P + dP$ is given by

$$dV = \left(\frac{\partial V}{\partial T}\right)_P dT + \left(\frac{\partial V}{\partial P}\right)_T dP \tag{2-21}$$

The subscripts P and T mean that pressure and temperature are constant during differentiation. Equation (2-21) is the *complete differential* for V. It shows that the overall (or complete) differential

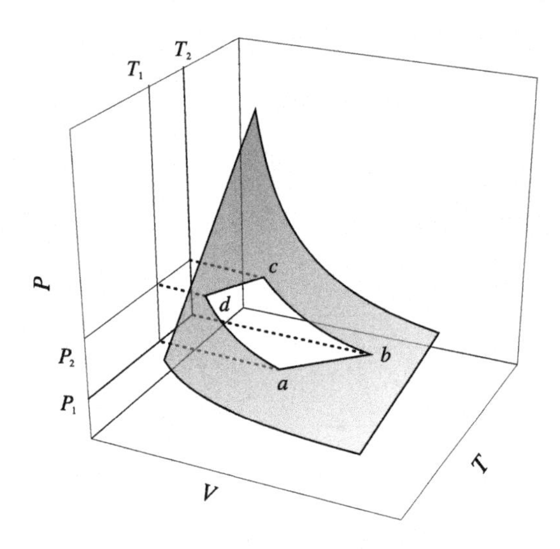

FIGURE 2-14

Region *abcd* is an infinitesimally small region on the *PVT* surface of an ideal gas.

change in V is equal to the differential change of V with T at constant P multiplied by the differential change in T, plus the differential change of V with P at constant T multiplied by the differential change in P. We used the ideal gas law to explain Eq. (2-21), but this equation is general and applies to all materials for which volume is a function only of P and T.

Figure 2-14 helps us understand Eq. (2-21). The region *abcd* is an infinitesimally (i.e., an immeasurably) small region on the *PVT* surface, so the changes in volume, temperature, and pressure as we move from point a to point c are given by dV, dT, and dP, respectively. The differential change in volume (dV) is the difference between the volume at points c and a:

$$dV = V_c - V_a \tag{2-22}$$

From Figure 2-14 we can see that both the temperature and the pressure of the gas have changed in moving from point a to point c. We can rewrite Eq. (2-22) in terms of isobaric (constant pressure) and isothermal (constant temperature) changes that, when added together, give us the same result as Eq. (2-22). We can do this because volume is a state function and depends only on the initial and final states of the system (in this case, the ideal gas of which the *PVT* surface we are considering). The values of state functions are path independent, so we can write

$$dV = V_c - V_a = (V_b - V_a) + (V_c - V_b) \tag{2-23}$$

Figure 2-14 shows that points a and b are both at the initial pressure P_1. Hence, the volume change between these two points is an isobaric change. In other words, the change in volume between points b and a is the first part of the differential in Eq. (2-21):

$$(V_b - V_a) = dV \text{ (at constant } P) = \left(\frac{\partial V}{\partial T}\right)_P dT \tag{2-24}$$

From Figure 2-14 we can also see that points b and c are both at the final temperature T_2. Hence, the volume change between these two points is an isothermal change. Thus, the change in volume between points b and c is the second part of the differential in Eq. (2-21):

$$(V_c - V_b) = dV \text{ (at constant } T) = \left(\frac{\partial V}{\partial P}\right)_T dP \tag{2-25}$$

When we substitute Eqs. (2-24) and (2-25) into Eq. (2-23), we get Eq. (2-21):

$$dV = V_c - V_a = (V_b - V_a) + (V_c - V_b) = \left(\frac{\partial V}{\partial T}\right)_P dT + \left(\frac{\partial V}{\partial P}\right)_T dP \tag{2-21}$$

Once we have mathematical expressions for the derivatives in Eq. (2-21), we can evaluate dV.

Example 2-7. Use the ideal gas law to evaluate Eq. (2-21) and derive an expression for dV_m, the complete differential of V_m, the molar volume of an ideal gas. The ideal gas law is

$$PV_m = RT \tag{2-26}$$

Next, we rearrange this to solve for V_m and find the two derivatives by first differentiating V with respect to P at constant T and then differentiating V with respect to T at constant P:

$$V_m = \frac{RT}{P} \tag{2-27}$$

$$\left(\frac{\partial V_m}{\partial T}\right)_P = \frac{\partial}{\partial T}\left(\frac{RT}{P}\right)_P = \frac{R}{P} \tag{2-28}$$

$$\left(\frac{\partial V_m}{\partial P}\right)_T = \frac{\partial}{\partial P}\left(\frac{RT}{P}\right)_T = -\frac{RT}{P^2} \tag{2-29}$$

We get the desired result by substituting Eqs. (2-28) and (2-29) back into Eq. (2-21):

$$dV_m = \left(\frac{R}{P}\right)dT - \left(\frac{RT}{P^2}\right)dP \tag{2-30}$$

Equation (2-30) can be solved numerically if we desire to do so. Finally, we could use the same basic process to derive expressions for dV for a real gas using a real gas equation of state, such as those discussed later in Chapter 8.

If we are moving along one of the isochoric (constant volume) lines on the *PVT* surface, then by definition volume is constant ($dV = 0$) and we are measuring the variation of pressure and temperature at constant volume for an ideal gas. Equation (2-21) then becomes

$$0 = \left(\frac{\partial V}{\partial T}\right)_P dT + \left(\frac{\partial V}{\partial P}\right)_T dP \tag{2-31}$$

Rearranging yields the equation

$$\left(\frac{\partial P}{\partial T}\right)_V = -\frac{(\partial V/\partial T)_P}{(\partial V/\partial P)_T} \tag{2-32}$$

Equation (2-32) expresses the change in pressure with temperature at constant volume. The numerator in Eq. (2-32) is equal to $V\alpha$, where α is the *isobaric thermal expansion coefficient*

$$\alpha = \frac{1}{V}\left(\frac{\partial V}{\partial T}\right)_P = \left(\frac{\partial \ln V}{\partial T}\right)_P \tag{2-33}$$

with units of inverse degrees (K^{-1}). Equation (2-33) shows that α gives the fractional volume change per degree temperature change at constant pressure (i.e., along an isobar). The denominator in Eq. (2-32) is equal to $-\beta V$, where β is the *isothermal compressibility coefficient*

$$\beta = -\frac{1}{V}\left(\frac{\partial V}{\partial P}\right)_T = -\left(\frac{\partial \ln V}{\partial P}\right)_T \tag{2-34}$$

with units of inverse pressure (bar^{-1}). Beta gives the fractional decrease in volume per bar pressure increase at constant temperature (i.e., along an isotherm). Rewriting Eq. (2-32) gives

$$\left(\frac{\partial P}{\partial T}\right)_V = -\frac{V\alpha}{V\beta} = \frac{\alpha}{\beta} \tag{2-35}$$

the *thermal pressure coefficient* for a material, which has units of bar K^{-1}.

Although α and β were defined in our discussion of the *PVT* surface for an ideal gas, they are properties characteristic of all solids, liquids, and gases. Likewise, Eq. (2-35) is a general equation that applies to all substances.

Example 2-8. Someone plans to use an Hg-filled Pyrex glass thermometer with a maximum temperature of 100°C to measure the boiling point of water at the Dead Sea, which is about 400 m below sea level. Is this a good idea? We can answer this question by evaluating Eq. (2-35). First we must find atmospheric pressure at the Dead Sea. The variation of atmospheric pressure in Earth's troposphere (the lower 10–12 km of the atmosphere) is approximately

$$P\ (\text{bar}) = 1.014 - 1.119 \times 10^{-4} \times \text{altitude (m)} \tag{2-36}$$

Substituting in the Dead Sea altitude of −400 m, we get

$$P\ (\text{bar}) = 1.014 - 1.119 \times 10^{-4} \times (-400\ \text{m}) = 1.06\ \text{bars} \tag{2-37}$$

The boiling point of water at 1.06 bar pressure is about 101°C. The thermometer approximates a constant volume system because the Pyrex glass has a α value much less than that of mercury

Table 2-11 Composition of Dry Air

Gas	Volume % in Air	Molecular wt. (g/mole)
N_2	78.084	28.014
O_2	20.946	31.998
Ar	0.946	39.948
CO_2	0.036	44.009
Average molecular wt. of dry air		28.970

Table 2-12 Some Typical Isothermal Compressibilities (β) for Different Materials[a]

Substance	β (bar^{-1})×10^7
methanol (CH_3OH)	1214
Hg (liquid)	40.1
α-Fe (metal)	5.9
fayalite (Fe_2SiO_4)	9.1
forsterite (Mg_2SiO_4)	8.1
spinel (γ-Mg_2SiO_4)	4.7
pyrope ($Mg_3Al_2Si_3O_{12}$)	5.7
grossular ($Ca_3Al_2Si_3O_{12}$)	6.0
graphite	29.6
diamond	2.2
α-quartz (SiO_2)	27.0
stishovite (SiO_2)	3.2
coesite (SiO_2)	10.4
water ice (−16°C)	112
water (0°C)	509.8
diopside ($CaMgSi_2O_6$)	8.8
clinoenstatite ($Mg_2Si_2O_6$)	9.6
orthoenstatite ($Mg_2Si_2O_6$)	8.7
perovskite ($MgSiO_3$)	3.9
kyanite (Al_2SiO_5)	5.2
andalusite (Al_2SiO_5)	6.6
sillimanite (Al_2SiO_5)	5.8

[a] *The values listed are generally the initial compressibility and are multiplied by 10^7, that is, β for methanol is 1.214×10^{-4} and that for graphite is 2.96×10^{-6}. With the exception of ice and water, the data are generally at room temperature (20–25°C) and are taken from Birch (1966), Knittle (1995), Smyth et al. (2000), Yang and Prewitt (2000), and Lide (2000).*

(see α values in Table 2-1). The average thermal expansion coefficient of Hg is 1.81×10^{-4} K^{-1}, and Table 2-12 gives the isothermal compressibility of Hg as 4.01×10^{-6} bar^{-1}, which yields

$$\left(\frac{\partial P}{\partial T}\right)_V = \frac{\alpha}{\beta} = \frac{1.81 \times 10^{-4}\ \text{K}^{-1}}{4.01 \times 10^{-6}\ \text{bar}^{-1}} = 45.1\ \text{bar K}^{-1} \tag{2-38}$$

This plan is not a good idea because the 1 degree increase will burst the thermometer.

X. THERMAL EXPANSION COEFFICIENT

As discussed earlier in Section II-B, every material (solid, liquid, gas) has a characteristic α value. As shown in Table 2-1, different forms of the same element (e.g., diamond and graphite) or the same compound (e.g., quartz, coesite, and stishovite) also have different α values.

Thermal expansion coefficients are not constant but vary with temperature and pressure. The pressure dependence of α is discussed in Chapter 8. At present, we are concerned with the temperature dependence. Books and scientific papers use different polynomial equations to compute α as a function of temperature. A common equation is

$$\alpha(T) = a_0 + a_1 T + a_2 T^{-2} \tag{2-39}$$

The three constants ($a_0 \sim 10^{-6} - 10^{-5}$, $a_1 \sim 10^{-9} - 10^{-8}$, and $a_2 \leq 0$) in Eq. (2-39) are determined from measurements of volume as a function of temperature. However, in many cases an average or temperature-independent thermal expansion coefficient is used, for example,

$$V = V_{ref}\left[1 + \alpha_{av}(T - T_{ref})\right] = V_{ref}\exp\left[\alpha_o(T - T_{ref})\right] \tag{2-40}$$

where V is the volume at any temperature T, V_{ref} is the volume at a reference temperature, exp stands for an exponential, and α_{av} and α_o denote average or temperature-independent thermal expansion coefficients, respectively. The following example illustrates data evaluation.

Example 2-9. Fiquet et al. (1999) measured the molar volume of lime (CaO) from 300 K to 3000 K and calculated α over the same temperature range. A subset of their data is in the top two rows of the following table. Use these data to find an equation for α as a function of temperature.

T(K)	298	778	1292	1711	2073	2473	2673	3073
V_m (cm^3)	16.783	17.136	17.528	17.869	18.183	18.557	18.784	19.160
ln V_m	2.8204	2.8412	2.8638	2.8831	2.9005	2.9208	2.9330	2.9528
α ($\times 10^5$)	4.08	4.32	4.58	4.79	4.98	5.08	5.18	5.48
Fiquet	4.19	4.41	4.64	4.84	5.00	5.08	5.18	5.46

Equation (2-35) shows that we want to compute the natural (base e) logarithm of V_m, find an equation for this, and differentiate it to get an equation for α. Our calculations are in the third and fourth rows of the table and in the two equations that follow.

$$\ln V_m = 2.8086 + 3.9332 \times 10^{-5} T + 2.5134 \times 10^{-9} T^2 \tag{2-41}$$

$$\alpha = (\partial \ln V/\partial T)_P = 3.9332 \times 10^{-5} + 5.0268 \times 10^{-9} T \tag{2-42}$$

The last row gives the results of Fiquet et al. (1999) for α from fitting 53 data points. We used V_m here, but relative volumes could also be used because α is computed from the derivative.

XI. COMPRESSIBILITY COEFFICIENT

Table 2-12 lists isothermal compressibility coefficients for many of the materials included in Table 2-1. These data show that different forms of the same element (e.g., diamond and graphite) or the same compound (e.g., Al_2SiO_5 as kyanite, andalusite or sillimanite) have different β values. Compressibilities also vary with temperature and pressure. We briefly consider the latter variations here; we discuss the temperature and pressure dependence in more detail in Chapter 8. The empirical equation

$$\ln(V/V_0) = -b_1 P + b_2 P^2. \tag{2-43}$$

is often used to represent the decrease of volume V with pressure relative to the initial volume V_0 at one bar pressure. The two constants ($b_1 \sim 10^{-7} - 10^{-5}$, and $b_2 \sim 10^{-12} - 10^{-11}$) are derived from a least-squares fit to the data. Differentiating Eq. (2-43) then gives an equation for the compressibility:

$$\beta = -(\partial \ln V/\partial P)_T = b_1 + 2b_2 P \tag{2-44}$$

The *isothermal bulk modulus* $K = \beta^{-1}$ is also used for thermodynamic calculations at high pressures. Typically, bulk moduli are given in units of Gigapascal (GPa).

Example 2-10. Bridgman (1911) measured the effect of pressure on the volume of mercury and calculated its isothermal compressibility from his data. A subset of his results for Hg at 22°C is in the first three rows of the following table. Use these data to derive an equation for β as a function of pressure, and the isothermal bulk modulus at zero pressure (K_0).

P (kg/cm^2)	0	2000	4000	6000	8000	10,000	12,000
P (bar)	0	1961.3	3922.7	5884.0	7845.3	9806.6	11,768
V/V_0	1	0.99232	0.98517	0.97857	0.97250	0.96703	0.96213
ln(V/V_0)	0	−0.00771	−0.01494	−0.02166	−0.02789	−0.03353	−0.03861
β (×10^7)	40.8	38.1	35.5	32.8	30.1	27.5	24.8
lit. (×10^7)	40.4	37.7	35.0	36.4	29.6	26.9	24.2

Equation (2-36) shows that we want to compute the natural logarithm of V/V_0, find an equation for this, and differentiate it to get an equation for β. Bridgman's pressures (kg cm^{-2}) were converted to bars by multiplying them by 0.980665. Our calculations are in the fourth and fifth rows of the table and in the two equations that follow.

$$\ln V/V_0 = -4.08275 \times 10^{-6} P + 6.79665 \times 10^{-11} P^2 \tag{2-45}$$

$$\beta = -(\partial \ln V/\partial P)_T = 4.08275 \times 10^{-6} - 13.5933 \times 10^{-11} P \tag{2-46}$$

The last row (literature × 10^7) gives results from Birch (1966) for β, which agree within 2.5%. Lit. refers to the values from the literature (Birch 1966).

The isothermal bulk modulus at zero pressure is simply

$$K_0 = 1/\beta_0 = (4.08275 \times 10^{-6})^{-1} = 24.49 \text{ GPa} \tag{2-47}$$

versus 24.75 GPa from Birch (1966). We could also find an equation for $V/V_0 = V(T, P)$, differentiate this, divide each term by the relative volume (V/V_0), and then fit an equation to β.

XII. SOME FURTHER NOTES ABOUT DIFFERENTIALS

Equation (2-21), the complete differential for V, has a simple geometric interpretation. When a property depends on three or more independent variables, a simple geometric interpretation is no longer possible, but the complete differential has the same general form. Imagine an air-filled balloon that contains n moles of air and has a tiny leak in it. The volume of the air in the balloon (V) depends on temperature and pressure, of course, but now the number of moles of gas present is also changing because the gas inside the balloon is gradually escaping. Thus, the volume depends on T, P, and the number of moles (n), and the complete differential for V is

$$dV = \left(\frac{\partial V}{\partial T}\right)_{P,n} dT + \left(\frac{\partial V}{\partial P}\right)_{T,n} dP + \left(\frac{\partial V}{\partial n}\right)_{P,T} dn \tag{2-47}$$

In Eq. (2-47), as in Eq. (2-21), the subscripts on $(\partial V/\partial T)$, $(\partial V/\partial P)$, and $(\partial V/\partial n)$ refer to the variables held constant during differentiation. Thus, volume is differentiated with respect to temperature while holding pressure and number of moles constant in the derivative $(\partial V/\partial T)_{P,n}$. This derivative gives the volume change with temperature at constant pressure and composition.

A. Perfect (or exact) differentials

We need to consider the conditions for a differential of a function to be a *perfect differential*. This is an important mathematical concept for thermodynamic functions because a function that is a perfect differential is a state function and is path independent.

In general, we can define a variable U that is a function of x, y, and z, which mathematicians write as $U(x, y, z)$. The complete differential of U is then

$$dU = \left(\frac{\partial U}{\partial x}\right)_{y,z} dx + \left(\frac{\partial U}{\partial y}\right)_{x,z} dy + \left(\frac{\partial U}{\partial z}\right)_{x,y} dz \tag{2-48}$$

For dU to be a perfect differential, the following relations must be true:

$$\frac{\partial^2 U}{\partial x \partial y} = \frac{\partial^2 U}{\partial y \partial x} \tag{2-49}$$

$$\frac{\partial^2 U}{\partial y \partial z} = \frac{\partial^2 U}{\partial z \partial y} \tag{2-50}$$

$$\frac{\partial^2 U}{\partial x \partial z} = \frac{\partial^2 U}{\partial z \partial x} \tag{2-51}$$

Conversely, if we have a differential

$$dU = Xdx + Ydy + Zdz \tag{2-52}$$

dU will be a perfect differential if there is a function $U(x, y, z)$ such that

$$X = \left(\frac{\partial U}{\partial x}\right)_{y,z} \quad Y = \left(\frac{\partial U}{\partial y}\right)_{x,z} \quad Z = \left(\frac{\partial U}{\partial z}\right)_{x,y} \tag{2-53}$$

$$\frac{\partial X}{\partial y} = \frac{\partial Y}{\partial x} \quad \frac{\partial Y}{\partial z} = \frac{\partial Z}{\partial y} \quad \frac{\partial Z}{\partial x} = \frac{\partial X}{\partial z} \tag{2-54}$$

Example 2-11. These three exercises illustrate the preceding concepts.

(a) Use Eqs. (2-49) to (2-51) to determine if Eq. (2-55) is a perfect differential:

$$U = U(x, y, z) = x^2y + y^2 + xz \tag{2-55}$$

$$\left(\frac{\partial U}{\partial x}\right)_{y,z} = 2xy + z \quad \left(\frac{\partial^2 U}{\partial x \partial y}\right)_z = 2x \quad \left(\frac{\partial^2 U}{\partial x \partial z}\right)_y = 1 \tag{2-56}$$

$$\left(\frac{\partial U}{\partial y}\right)_{x,z} = x^2 + 2y \quad \left(\frac{\partial^2 U}{\partial y \partial x}\right)_z = 2x \quad \left(\frac{\partial^2 U}{\partial y \partial z}\right)_x = 0 \tag{2-57}$$

$$\left(\frac{\partial U}{\partial z}\right)_{x,y} = x \quad \left(\frac{\partial^2 U}{\partial z \partial x}\right)_y = 1 \quad \left(\frac{\partial^2 U}{\partial z \partial y}\right)_x = 0 \tag{2-58}$$

Equation (2-55) is a perfect differential because Eqs. (2-49) to (2-51) are satisfied.

(b) Test if the differential dU is a perfect differential.

$$dU = zdx + z^2dy + (x + 2\ yz)dz \tag{2-59}$$

In this case we want to evaluate Eqs. (2-53) and (2-54):

$$\left(\frac{\partial X}{\partial y}\right) = \frac{\partial z}{\partial y} = 0 = \left(\frac{\partial Y}{\partial x}\right) = \frac{\partial (z^2)}{\partial x} = 0 \tag{2-60}$$

$$\left(\frac{\partial Y}{\partial z}\right) = \frac{\partial (z^2)}{\partial z} = 2z = \left(\frac{\partial Z}{\partial y}\right) = \frac{\partial (x + 2yz)}{\partial y} = 2z \tag{2-61}$$

$$\left(\frac{\partial Z}{\partial x}\right) = \frac{\partial (x + 2yz)}{\partial x} = 1 = \left(\frac{\partial X}{\partial z}\right) = \frac{\partial (z)}{\partial z} = 1 \tag{2-62}$$

Equation (2-59) is also a perfect differential because Eqs. (2-53) and (2-54) are true.

(c) Determine if Eq. (2-63) is a perfect differential.

$$dU = x^2dx + xydy \tag{2-63}$$

In this case we find that

$$\left(\frac{\partial X}{\partial y}\right) = \frac{\partial (x^2)}{\partial y} = 0 \neq \left(\frac{\partial Y}{\partial x}\right) = \frac{\partial (xy)}{\partial x} = y \tag{2-64}$$

so Eq. (2-63) is not a perfect differential because Eq. (2-54) is not satisfied.

B. Chain rule

The chain rule is a useful tool for working with differentials. If y is a function of x and x is a function of a third variable t, the chain rule states that the derivative of y with respect to t is

$$\left(\frac{dy}{dt}\right) = \left(\frac{dy}{dx}\right)\left(\frac{dx}{dt}\right) \tag{2-65}$$

Example 2-12. Find (dy/dt) given $y = x^2$ and $x = 3t + 1$. Combining the two functions gives

$$y = x^2 = (3t+1)^2 = 9t^2 + 6t + 1 \tag{2-66}$$

We now evaluate Eq. (2-65),

$$\left(\frac{dy}{dx}\right) = \frac{d(x^2)}{dx} = 2x \qquad \left(\frac{dx}{dt}\right) = 3 \qquad \left(\frac{dy}{dt}\right) = 18t + 6 = \left(\frac{dy}{dx}\right)\left(\frac{dx}{dt}\right) \tag{2-67}$$

and find the desired expression for (dy/dt).

C. Cyclic rule

Another rule that we will use later in the book is the *cyclic rule*. If we have a closed system described by three variables, defining the values of two variables determines the third. For example, as discussed earlier, the ideal gas equation allows us to determine the volume of a gas once we know its temperature and pressure. If the state of the system is changed without changing the volume (e.g., heating an ideal gas inside a metal tank with a very small thermal expansion coefficient so the volume of the tank and hence of the gas remains constant), we can use Eq. (2-21) to derive

$$\left(\frac{\partial V}{\partial P}\right)_T dP = -\left(\frac{\partial V}{\partial T}\right)_P dT \tag{2-68}$$

Rewriting by dividing through by dT yields

$$\left(\frac{\partial V}{\partial P}\right)_T \left(\frac{\partial P}{\partial T}\right)_V = -\left(\frac{\partial V}{\partial T}\right)_P \tag{2-69}$$

which can be rearranged to give

$$\left(\frac{\partial V}{\partial P}\right)_T \left(\frac{\partial P}{\partial T}\right)_V \left(\frac{\partial T}{\partial V}\right)_P = -1 \tag{2-70}$$

Equation (2-70) is the cyclic rule. This example uses P, V, and T, but the rule is valid for a system described by any three state variables. We will use the cyclic rule later when we discuss Maxwell's relations in Chapter 8.

PROBLEMS

1. State whether density and mass are intensive or extensive variables.
2. Calculate the temperature of helium atoms that are energetic enough to have escape velocity (11.186 km s^{-1}) from Earth.
3. What is the temperature of a perfect vacuum? Why is the temperature of interstellar space 3.2 K and not absolute zero?

4. The relationship between temperature and emf for a K-type thermocouple is given by

$$T\ (^\circ\mathrm{C}) = 6.3386 + 23.3985 \times \mathrm{emf}\ (\mathrm{mV}) + 0.01796 \times \mathrm{emf}^2\ (\mathrm{mV}^2)$$

when the temperature of the cold junction is 0°C. If the thermocouple meter reads 25 mV (i.e., 0.025 volts) what is the temperature being measured? You can check your answer by looking at Figure 2-6.
5. The temperature at Earth's core-mantle boundary is 3930 K. What is this in degree Fahrenheit?
6. What temperature has the same value on both the Celsius and Fahrenheit temperature scales?
7. The vapor pressure of a pure substance depends only on its temperature. Vapor pressure thermometers are used to measure low temperatures in chemistry laboratories. Calculate the temperature (°C) of a liquid NH_3 vapor pressure thermometer that reads 256.50 mmHg and has a vapor pressure relationship (*T* in Kelvins) of

$$\log_{10} P_{\mathrm{NH_3}}(\mathrm{atm}) = 5.201 - \frac{1248}{T}$$

8. The average surface pressure at sea level on Earth is 1 atm. Calculate the average atmospheric column density (kg m^{-2}) above the ocean, assuming $g = 9.81$ m s^{-2}. (The *atmospheric column density* is the mass of an atmospheric column with an area of 1 m^2 that extends from Earth's surface to the top of the atmosphere.)
9. Diamond anvil cells press two gem-quality diamonds against one another with a screw mechanism and generate pressures into the megabar range. Figure 2-15 is a schematic drawing of a diamond anvil cell. Assuming that the high-pressure diamond faces have diameters of 100 μm, what pressure (in bars) is generated by applying 30 pounds of force? (A 110 lb. person could easily apply this force by pressing down on an object.)
10. The air pressure in the tires of a sport utility vehicle is measured to be 28 psi on a hot day in August ($T = 30$°C). What will the air pressure be in the tires on a cold day in January ($T = 0$°C)? Give your results in bar, and assume no leakage and ideality.
11. The H-H bond energy in H_2 is 432,071 joules per mole of H_2. Calculate the H-H bond energy in terms of (a) calories and (b) ergs. *Hint:* Look at Table 2-10, which gives values of *R* in various units.
12. In September 1804, Gay-Lussac ascended to 7 km in a balloon to measure temperature, pressure, and humidity and to take air samples. The atmospheric temperature is −30.4°C and the pressure is

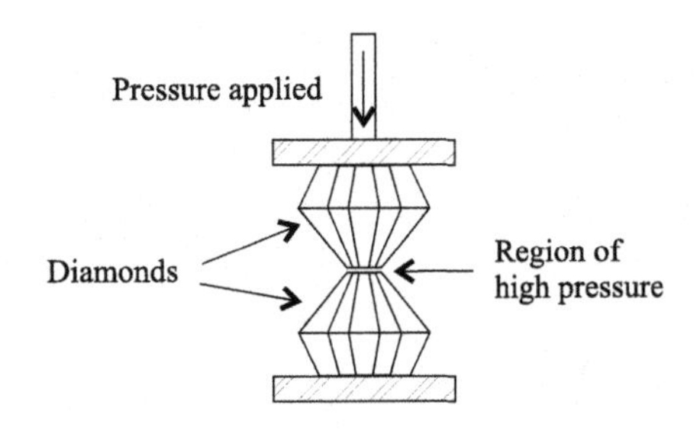

FIGURE 2-15

A schematic drawing showing the use of a diamond anvil press.

0.231 bar at 7 km altitude. Assuming dry air with an average molecular weight of 28.97 g mol^{-1} and ideality, calculate the density of air at 7 km altitude (see Table 2-11).

13. The thermal expansivity (α) and isothermal compressibility (β) of liquid water are $\alpha = 257.21 \times 10^{-6}$ K^{-1} and $\beta = 45.247 \times 10^{-6}$ bar^{-1}, respectively, at 25°C. What pressure develops in a tube completely filled with water by heating the tube one degree to 26°C?

14. An iron rod that is 1 m long at 25°C is heated to 100°C. What is its final length? Hint: for a gas $V = V_0(1 + \alpha_0 t)$ and for a solid $L = L_0(1 + \alpha_0 t)$.

15. Write the complete differential of G where $G = G(H, S)$. Also, apply the cyclic rule.

16. Use the chain rule to find $(\partial y/\partial c)$ where $y = 8x + 7x^2$, $x = 9 + 3t$, and $t = 0.5\ c$.

17. Skinner (1966) gives the following data for thermal expansion of diamond from 20°C to the indicated temperature in percent. Use these data to compute an equation for the thermal expansion coefficient of diamond.

Temp (°C)	100	200	400	600	800	1000	1200
% increase	0.031	0.091	0.268	0.489	0.747	1.038	1.236

18. Birch (1966) gives $(V_0 - V)/V_0$ versus pressure (kg cm^{-2}) for the isothermal compression of fluorite (CaF_2) at room temperature. Find an equation for the isothermal compressibility β (in bar^{-1}) and compute the zero pressure bulk modulus K_0 (in GPa).

P (kg cm^{-2})	5000	10,000	15,000	20,000	25,000	30,000
$(V_0 - V)/V_0$	0.00585	0.01147	0.01695	0.02213	0.02735	0.03234

19. Spectroscopic observations of the atmospheres of other planets give gas abundances as column densities (molecules cm^{-2}). Column densities are reported as kilometer-Amagats or centimeter-Amagats. One Amagat is the number of molecules per cm^3 in an ideal gas at STP, standard temperature (273.15 K) and pressure (1 atm). The H_2 abundance in Neptune's atmosphere is 400 km Amagats. Convert this into molecules cm^{-2}.

20. On first approximation, the sun emits light as a black body, and its surface temperature can be computed from the solar flux at the sun's surface using the Stefan–Boltzmann equation

$$F = \sigma T^4$$

(a) Compute the solar flux at the sun's surface using the solar flux at the top of Earth's atmosphere (1367.6 W m^{-2}), the solar radius (695,950 km), and the average distance of Earth from the sun (149.598×10^6 km).

(b) Use your result from part (a) to calculate the surface temperature of the sun. The Stefan–Boltzmann constant is 5.67051×10^{-8} W m^{-2} K^{-4}.

CHAPTER

The First Law of Thermodynamics

3

Die Energie der Welt ist konstant. *(The energy of the universe is constant.)*
—Rudolf Clausius (1865)

Es ist unmöglich, eine Maschine zu bauen, die fortwährend Wärme oder äußere Arbeit aus Nichts schafft. *(It is impossible to construct a machine which continuously produces heat or external work out of nothing.)*
—Walther Nernst (1924)

In this chapter we discuss the first law of thermodynamics, which is the principle of the conservation of energy applied to heat and work. The first law is integral to our daily life because the operation of automobile and truck engines, equipment in factories, and air conditioning and heating systems is based on it.

This chapter is divided into five sections. In Section I, we briefly review the history of some of the key experiments and ideas that led to the development of the first law. In Section II, we present the first law of thermodynamics and discuss some of its implications. In doing so, we introduce several new variables: internal energy (E), which is a state function; work (w) and heat (q), which are path-dependent functions; enthalpy (H), which is a state function; and the constant-pressure and constant-volume heat capacities (C_P and C_V). We also discuss pressure-volume work and the concept of reversibility. In Section III, we apply the first law to reversible processes involving ideal gases. Section IV describes adiabatic processes in planetary atmospheres. Section V is a discussion of other types of work.

I. HISTORICAL OVERVIEW OF IDEAS ABOUT HEAT AND WORK

A. The caloric theory of heat and its demise

Today it might seem obvious to us that heat is different from temperature. However, from ancient times until the late 18th century most scientists and philosophers made no distinction between the two phenomena. Sir Francis Bacon (1561–1626) was a notable exception because he concluded that heat is a form of motion. His ideas were later accepted by Robert Boyle and Sir Isaac Newton (1642–1727) as well as a few others, but the prevailing view was that heat was a weightless fluid known as *caloric*. The word *caloric* is derived from *calor*, which is the Latin word for heat. The amount of caloric in an object determined how hot or cold the object was. For example, a hotter object was believed to contain more caloric than a cooler object, and two objects with the same temperature were thought to have the same amount of caloric as one another. In 1801, John Dalton (1766–1844), the English chemist (and the discoverer of the law of partial pressures), described caloric as follows: "The most probable opinion concerning the nature of caloric is that of its being an elastic fluid of great subtility, the particles of

Copyright © 2013 Elsevier Inc. All rights reserved.

which repel one another, but are attracted by all other bodies." As we will see, the caloric theory of heat persisted until the experiments of James Joule in the 1840s.

Even though Dalton's opinions were the majority view, ideas about heat and temperature had started to change in the late 18th century. At this time, the Scottish chemist Joseph Black (1728–1799) made the important distinction between temperature and heat. As discussed in Chapter 2, temperature is an *intensive property*; that is, the temperature of an object does not depend on its size or mass. In contrast, the quantity of heat in an object is an *extensive property*, which therefore does depend on the object's size or mass. Black recognized this and showed that different materials had different *specific heats*; for example, a different quantity of heat is needed to raise the temperature of one gram of diamonds by 1°C than is needed to raise the temperature of one gram of water by 1°C. Furthermore, Black showed that two objects were in thermal equilibrium when they had the same temperature, regardless of their specific heat or mass. In other words, if you have a gram of diamonds and a kilogram of water (preferably vice versa!) and both have a temperature of 25°C, the water and diamonds are thermally equilibrated, even though they have different specific heats and different masses. Black also distinguished between *sensible heat*, which causes a rise in temperature, and *latent heat*, which causes a phase change such as melting at constant temperature. He experimentally determined that the amount of heat needed to melt ice was 79.4 times greater than the amount of heat needed to raise the temperature of an equal mass of water by 1°C. For reference, the modern value of this ratio, which is the latent heat of fusion of ice, is 79.73 calories per gram (cal g^{-1}), equivalent to 333.58 joule per gram (J g^{-1}). Despite all his insights about heat and temperature, Black accepted the flawed caloric theory of heat.

Some of the most cogent attacks on the caloric theory were made by Benjamin Thompson (Count Rumford of the Holy Roman Empire). Rumford is noted for his eventful life (see sidebar and the collected works of Rumford edited by Sanborn Brown (1968–1970)) and for his keen observations of the world around him. In 1798, while supervising the boring of a cannon at the arsenal in Munich, Rumford noticed that the process generated enough heat to boil water. He also noticed that the heat produced was proportional to the mechanical work done. This proportionality, called the *mechanical equivalent of heat,* is the amount of work required to produce a given amount of heat.

However, before proceeding further with Rumford's story, we should describe the units that are used to measure work and heat. The unit of *work* in the SI system is the *joule* (J), which is one newton meter (1 N m = 1 kg m^2 s^{-2}). *Calories* (cal) are the traditional unit for measuring heat in thermochemistry. A *calorie* is the amount of heat that, when absorbed by one gram of liquid water at one atmosphere pressure, increases the temperature of the water by 1°C. The temperature range used for this measurement is commonly from 14.5°C to 15.5°C, and the amount of heat absorbed is then called the 15° *calorie*. The modern experimental value for the mechanical equivalent of heat is 0.2389 cal J^{-1}. In 1948, by international agreement, this value was set equal to 0.2390 cal J^{-1} and the quantity of heat involved is called the *thermochemical calorie,* which is defined as exactly 4.184 J. This conversion factor (1 calorie = 4.184 joules) is the one we will use throughout this book.

Based on his cannon boring experiments, Rumford estimated a mechanical equivalent of heat that is equivalent to 0.183 calories per joule (cal J^{-1}, although he did not use these units). As mentioned, Rumford claimed that the heat produced was due to the work done. However, his experiments were not enough to dispel the caloric theory. The advocates of the caloric theory explained Rumford's results by claiming that some of the caloric originally in the metal was squeezed out of it, thus producing heat, when small metal turnings were shaved off during the cannon boring. To counter this claim, Rumford then used a blunt rod, which produced fewer metal turnings, to bore a cannon. He found that the same

COUNT RUMFORD (BENJAMIN THOMPSON) (1753–1814)

Count Rumford's life would make a good adventure movie. Rumford began life as Benjamin Thompson of Massachusetts. In his youth he was a merchant, a student at Harvard, a merchant again, and an itinerant schoolteacher. He married a wealthy widow in 1772 and inherited her fortune at her death in 1792. Thompson moved to London in 1776 to escape the American Revolution because he had been spying for the British government. He was elected a Fellow of the Royal Society in 1779, was knighted by King George III in 1784, and in the same year entered the service of the Elector of Bavaria, who became the patron of his scientific endeavors. In his capacity as Bavarian minister of war, Rumford had the duty of overseeing the Munich arsenal. One of his greatest scientific contributions came because of watching the boring of a cannon there.

Rumford then undertook experiments that led to the demise of the old caloric theory. In 1791 he became Count Rumford of the Holy Roman Empire. Shortly thereafter, he endowed the Rumford Fund of the American Academy of Arts and Sciences in Boston. Rumford helped found and endow the Royal Institution in Great Britain (1800) but ended his association with it two years later. He married Antoine Lavoisier's widow in 1805 and lived in France until his death. Throughout his life Rumford was an avid inventor who developed the double boiler, a kitchen range, a drip coffeepot, and improvements for chimneys and fireplaces. The Rumford Fund of the American Academy of Arts and Sciences helped support the work of P. W. Bridgman, the Nobel Prize–winning physicist. The Rumford Medal is one of the highest awards of the Royal Society; its recipients working in thermodynamics include Victor Regnault, Sir James Dewar, Kammerlingh Onnes, Fritz Haber, and Ilya Prigogine.

amount of work, whether done with a blunt rod or the regular boring tool, produced the same amount of heat. However, advocates of the caloric theory still refused to accept Rumford's results, and they claimed that the heat produced was due to reaction of the metal surfaces with the surrounding air during boring of the cannon.

Further support for the idea that mechanical work could be converted into heat came in 1799 from experiments done by Sir Humphry Davy (1778–1829). He rubbed together two blocks of ice that were initially at 29°F (−1.7°C) until they melted. He ended up with water at 35°F (+1.7°C). This was surprising, because according to the caloric theory of heat, liquid water has a greater capacity for caloric than ice. Therefore, when ice is melted, the water produced should absorb caloric, and the temperature should drop. This is contrary to what Davy observed. Davy repeated his experiment in a partial vacuum inside a cooled bell jar and found that the ice again melted by friction. He decided that the heat that melted the ice and raised the temperature of the water must have been generated by friction. Davy's experiments should have been the nail in the coffin of the caloric theory, but it somehow survived until Joule's impressive series of experiments about 40 years later.

B. Robert Mayer's work

The next major advance in our understanding of heat and work was made in 1842, when Robert Mayer (see sidebar) published his paper on the mechanical equivalent of heat. A few years earlier, while serving as ship's doctor on a voyage to the Dutch East Indies (modern Indonesia), Mayer developed the idea that a fixed total yield of energy is obtained by eating food and that this energy is divided between work done by a person and body heat produced. Starting from this insight, Mayer then thought about the relationship among heat, kinetic energy, and potential energy. In his 1842 paper, Mayer concluded that the amount of heat needed to raise the temperature of a mass of water from 0°C to 1°C was equal to the amount of energy released when an object of equal mass fell from a height of 365 meters. In the following example, we show how to convert this estimate for the mechanical equivalent of heat into units of calories per joule.

JULIUS ROBERT VON MAYER (1814–1878)

As the ship's doctor aboard the Dutch ship *Java*, Julius Robert von Mayer noticed that his patients' venous blood was unusually red in tropical climates. It was already known that the human body maintained its temperature by the slow combustion of food and that the blood leaving the lungs carried oxygen to the parts of the body where the combustion took place. Mayer's observation that blood returning to the lungs was still red indicated that less oxygen had been used than would have been used in a cooler climate. From this, he deduced that less work needed to be done to warm the body when the outside air was at a higher temperature.

Mayer hypothesized that a fixed amount of energy was produced by the body from the combustion of a given amount of food and that this energy was then divided between body heat produced and the physical work performed by the body. He quantified his results by calculating a mechanical equivalent of heat of 0.281 cal/J, but the scientific community of the time paid little attention to his findings. This led Mayer to suffer a nervous breakdown. However, later in life the importance of his work was finally recognized. He received the Copley Medal of the Royal Society in 1871. Mayer died of tuberculosis in 1878.

Example 3-1. We need some fundamental concepts from physics to do this conversion. The incremental amount of work (δw) done by application of a force (F) over a distance (dr) in the direction of the force is given by

$$\delta w = F dr \tag{3-1}$$

From Newton's second law, force F is mass (m) times acceleration (a):

$$F = ma \tag{3-2}$$

which, upon substitution, becomes

$$F = mg \tag{3-3}$$

if the only acceleration term is that due to the gravity (g) of Earth's gravitational field. Mayer expressed his result in terms of a mass falling 365 meters and an equal mass of water being heated. The actual mass itself is unimportant because it will cancel out of our equations, but for illustration we assume that a 1 kg mass falls and that 1 kg of water is heated. The potential energy (E_P) released by a 1 kg mass falling a distance dr is equal to the work that has to be done to raise the same mass the same distance and is given by

$$E_P = (m)(g)(dr) = (1\text{ kg})(9.81\text{ m s}^{-2})(365\text{ m}) = 3580.65\text{ kg m}^2\text{ s}^{-2} = 3580.65\text{ J} \tag{3-4}$$

where we used the definition of a joule given earlier in this chapter to simplify units. Earlier we also defined the 15°C calorie as the amount of heat needed to raise the temperature of one gram of water by 1°C, from 14.5°C to 15.5°C. The definition of the 15°C calorie is thus the specific heat of water in this temperature interval. However, as shown in Chapter 4, the specific heat of water varies with temperature, and it actually takes 1008.7 cal to raise the temperature of 1 kg of water from 0°C to 1°C. Thus, Mayer's result can be expressed as

$$1008.7\text{ cal} = 3580.65\text{ J} \tag{3-5}$$

which gives 0.282 cal J^{-1} for his estimate of the mechanical equivalent of heat. Note that we would have obtained the same result by assuming a 1 g mass and 1 g of water, or indeed any two equal masses for the falling object and the water.

Finally, before moving on, we should consider how to handle a similar calculation if a force F is not applied in the same direction as dr. This situation is pictured in Figure 3-1. In this case, the force along dr is $F(\cos\theta)$ and the incremental amount of work done is

$$\delta w = F(\cos\theta)dr \tag{3-6}$$

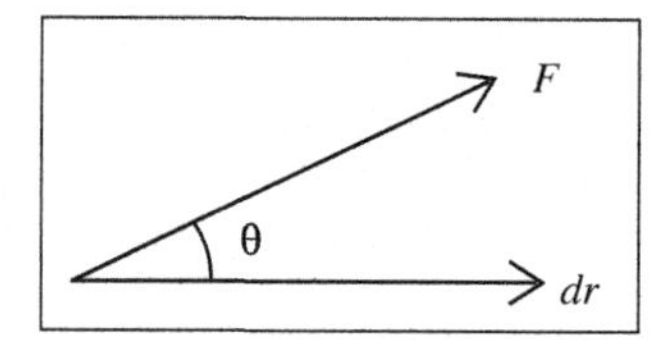

FIGURE 3-1

A schematic diagram illustrating that when force is applied in a direction other than that of *dr*, $dw = F(\cos\theta)dr$, where θ is the angle between the force and the incremental length *dr*.

where θ is the angle between the force and the distance over which it is applied. In general, for a Cartesian coordinate system (i.e., a coordinate system with three axes perpendicular to each other), the incremental amount of work done is given by

$$\delta w = F_x dx + F_y dy + F_z dz \quad (3\text{-}7)$$

where F_x represents the component of total force in the x direction, F_y is the component in the y direction, and F_z the component in the z direction.

C. Joule's experiments on the mechanical equivalent of heat

James Joule was one of the most remarkable amateur scientists of the 19th century (see sidebar). Starting about 1840, Joule began a series of ingenious experiments in which different types of work were converted into heat and the proportionality between the two was measured. In other words, Joule measured the mechanical equivalent of heat given by the ratio of work done to the heat produced. These experiments included the electrical heating of a wire coil immersed in water, the heating produced by compressing air into a cylinder that was immersed in water, and the heating produced by

Table 3-1 Joule's Determinations of the Mechanical Equivalent of Heat

Type of Work	Work (ft-lb.)[a]
Turning paddle wheels in water (1849)	772.7
Turning paddle wheels in mercury (1849)	774.1
Rubbing together two iron blocks immersed in mercury (1849)	775
Turning paddle wheels in water (1845, 1847)	781.5
Turning paddle wheels in sperm whale oil (1845, 1847)	782.1
Turning paddle wheels in mercury (1845, 1847)	787.6
Compressing gas into a cylinder immersed in water (1843)	798
Electrical current through a wire coil immersed in water (1843)	838
Mean value (±1 sigma) for the amount of work done	**789±22**
The modern experimental value is equal to	777.72
The defined thermochemical calorie gives	777.65

[a]*From Joule (1850), "On the Mechanical Equivalent of Heat,"* Phil. Trans. Roy. Soc. London ***140****, 61, and references therein. The amount of work done is the number of foot pounds needed to raise the temperature of one pound of water by 1°F—in other words, foot pounds per BTU.*

JAMES PRESCOTT JOULE (1818–1889)

Joule was possibly the greatest amateur scientist of the 19th century. He was the son of a wealthy Manchester, England, brewer and conducted his experiments in the family brewery. John Dalton, who developed the atomic theory and Dalton's law, educated Joule and his brother. Joule published his first scientific paper about electric motors at the age of 19. His research on electricity led to Joule's law ($q = I^2Rt$), which gives the amount of heat generated by a current I flowing through a wire with resistance R for time t.

His electrical research led Joule to study the transformation of work into heat. He found that a given amount of work generates the same amount of heat, regardless of how the work is done. Joule measured the mechanical equivalent of heat to be 0.241 cal/J, which is quite close to the modern accepted value of 0.239 cal/J. His experimental work established the first law of thermodynamics. Joule's free expansion experiment and his later work on gas expansion with William Thomson (Kelvin), known as the Joule–Thomson (Kelvin) effect, laid the foundation for much of modern refrigeration technology. Joule's scientific papers on these and other topics are remarkable for their insights into the basic laws of nature. They are also entertaining reading.

Perhaps the most remarkable thing about Joule was his talent for making very precise measurements with rudimentary equipment. His experimental accuracy is legendary. His colleague, Kelvin, said of him, "His boldness in making such large conclusions from such very small observational effects is almost as noteworthy and admirable as his skill in extorting accuracy from them."

paddle wheels stirring water, sperm whale oil, or mercury. A summary of Joule's results is given in Table 3-1; you can see that some of the experiments Joule did were more accurate than others. All these data, however, led Joule to conclude that the amount of heat produced in his experiments depended only on the amount of work done—not on how the work was done or on the type of work.

Joule's most famous and probably most accurate experiments were his paddle wheel experiments done in 1849. Figure 3-2 is a schematic diagram of these experiments in which falling weights were used to turn paddle wheels inside a tank of water. The tank contained stationary metal vanes to prevent rotation of water in the tank and to increase the frictional heating. The paddle wheels were positioned between the vanes; for example, from bottom to top the arrangement would be vane, paddle wheel,

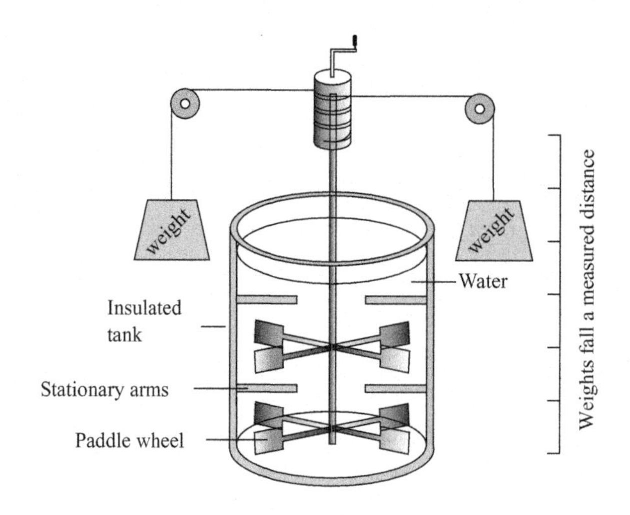

FIGURE 3-2

A schematic diagram of Joule's paddle wheel experiment. When the handle was released, the weights fell, unwinding the cord and turning the paddles. This process was repeated several times and the resulting temperature increase of the water was measured.

vane, paddle wheel, and so on. This entire apparatus was inside an insulated wood container (not shown) to prevent heat exchange with the surroundings. A system such as Joule's water tank, which is thermally insulated from the surroundings, is called an *adiabatic system*. The word *adiabatic* is derived from the Greek word *adiabatos*, which means impassable. Thus, an adiabatic wall is impassable to heat, an adiabatic system is perfectly thermally insulated from its surroundings, and an adiabatic process has no heat flow between the system and its surroundings.

The work done in Joule's experiments was calculated by multiplying the mass of the weights by the local acceleration of gravity and by the total distance they fell. The heat produced was calculated by measuring the temperature increase of the known mass of water contained in the insulated tank. Based on his paddle wheel experiments with water, Joule concluded that "the quantity of heat capable of increasing the temperature of a pound of water (weighed *in vacuo*, and taken at between 55° and 60°) by 1°Fahr. requires for its evolution the expenditure of a mechanical force represented by the fall of 772 lb. through the space of one foot." As shown in Table 3-1, Joule computed similar values for the amount of work needed to heat one pound of water from his other experiments.

Example 3-2. Using the energy conversion factors given in Table 3-2, we can convert Joule's results into more convenient units. Joule measured the work done in foot pounds (ft-lb) and gives the mechanical equivalent of heat in terms of *British thermal units* (BTU), which is the amount of heat needed to increase the temperature of one pound of water by 1°F. He found that one BTU was equivalent to 772 ft-lb. From Table 3-2, we see that 1 BTU = 252.0 cal and that 772 ft-lb = (1.356)(772) = 1046.8 J. Equating the two expressions we find that

$$1 \text{ BTU} = 252.0 \text{ cal} = 1046.8 \text{ J} = 772 \text{ ft-lb} \tag{3-8}$$

In other words, Joule's result for the mechanical equivalent of heat is equal to 252.0 cal/1046.8 J = 0.241 cal J^{-1}, which is within 1% of the modern value of 0.239 cal J^{-1}.

Table 3-2 Energy Conversion Factors[a]

	J	cal	erg	eV	liter-bar
1 J =	1	0.239006	10^{7}	1.0364269×10^{-5}	10^{-2}
1 cal =	4.184	1	4.184×10^{7}	4.336410×10^{-5}	4.184×10^{-2}
1 erg =	10^{-7}	2.390057×10^{-8}	1	$1.0364269 \times 10^{-12}$	10^{-9}
1 eV =	96,485.3415	23,060.5501	9.648534×10^{11}	1	964.853415
1 L bar =	100.000	23.90057	10^{9}	1.0364269×10^{-3}	1
1 L atm =	101.325	24.2173	1.01325×10^{9}	1.0501595×10^{-3}	1.01325
1 BTU =	1,054.35	251.9957	1.05435×10^{10}	1.0927567×10^{-2}	10.5435
1 KWH[a] =	3.600×10^{6}	860,420.65	3.600×10^{13}	37.311367	3.600×10^{4}
1 ft-lb =	1.355818	0.324048	1.355818×10^{7}	1.4052062×10^{-5}	1.355818×10^{-2}
1 cm^{-1} =	11.96266	2.85914	1.196266×10^{8}	1.2398422×10^{-4}	0.1196266
	liter-atm	**BTU**	**KWH**	**ft-lb**	**cm^{-1}**
1 J =	9.8692327×10^{-3}	9.484517×10^{-4}	2.77778×10^{-7}	0.737562	8.359345×10^{-2}
1 cal =	4.1292795×10^{-2}	3.968322×10^{-3}	1.162222×10^{-6}	3.085960	0.349756
1 erg =	9.869232×10^{-10}	9.48452×10^{-11}	2.77778×10^{-14}	7.375621×10^{-8}	8.359345×10^{-9}
1 eV =	952.23628	91.511682	2.680148×10^{-2}	71,163.9331	8,065.5424
1 L bar =	0.986923	9.484517×10^{-2}	2.77778×10^{-5}	73.75621	8.359345
1 L atm =	1	9.610186×10^{-2}	2.814586×10^{-5}	74.73348	8.470106
1 BTU =	10.405625	1	2.928750×10^{-4}	777.6486	88.136752
1 KWH =	35,529.2376	3,414.426	1	2.655224×10^{6}	3.009364×10^{5}
1 ft-lb =	1.338088×10^{-2}	1.285928×10^{-3}	3.766161×10^{-7}	1	0.113338
1 cm^{-1} =	0.1180623	1.134595×10^{-2}	3.322961×10^{-6}	8.823205	1

[a] *eV = electron volts per mole; KWH = kilowatt hour; cm^{-1} = cm^{-1} per mole*

D. Helmholtz and the conservation of energy

While Joule was working in England, Hermann von Helmholtz (see sidebar), then a physician in the Prussian army, was extending the principle of conservation of energy to include heat and work. In the late 17th century, Isaac Newton and the German mathematician Gottfried Wilhelm Leibnitz (1646–1716) had stated that the sum of kinetic energy and potential energy are conserved. Their work established the principle of the conservation of energy in Newtonian mechanics. However, the extension of this principle to heat and work was not generally appreciated until the work of Mayer, Joule, and Helmholtz. Leibnitz's statement, "The force [energy] of a moving body is proportional to the square of its velocity, or to the height to which it would rise against gravity," was quoted by Joule in his 1850 paper, *On the Mechanical Equivalent of Heat.* Joule quoted this statement to show he was extending this principle to heat and work.

HERMANN VON HELMHOLTZ (1821–1894)

At a time when science was intimately linked to philosophy, Helmholtz was initially educated as an army doctor. His medical training, along with a firm belief that everything in the world could be understood through scientific experimentation, prompted some of his later discoveries in the field of thermodynamics.

At that time it was already recognized that the heat produced by a living body was generated by the combustion of food. However, some people thought there was an additional vital force in living things not present in inanimate objects that could not be explained by scientific postulates. Helmholtz argued against these Vitalists by reasoning that if such a force existed and did not have a source governed by the laws of physics, it could be harnessed to produce a perpetual-motion machine. Perpetual-motion machines had already been accepted as impossibilities, so Helmholtz determined that all the heat generated by a living body must come from ordinary chemical reactions and mechanical forces. Furthermore, he stated that this must be the case for all heat and that no heat could be destroyed.

Helmholtz also performed experiments to measure the speed of nerve impulses and to study the nature of vision. He sought to defeat philosopher Immanuel Kant's argument that space, time, and causality do not really exist but are mental constructions. He hoped that by showing that perceptions by the organs of the body are created through chemical reactions and mechanical forces, he could prove that they do, in fact, relate to the outside world.

Helmholtz, like Mayer, was interested in metabolism and set up a small laboratory in his army barracks for his physiological experiments. As a result of these studies he became interested in the conservation of energy. In 1845, Helmholtz published a report, *The Theory of Animal Heat,* and in 1847 he delivered his now-famous paper *Über die Erhaltung der Kraft* (*On the Conservation of Forces* [Energy]) to a meeting of the Physical Society in Berlin. Helmholtz based his arguments on the idea that "it is not possible by any combination whatever of natural bodies to derive an unlimited amount of mechanical force [i.e., energy]."

More detailed historical descriptions of the development of the first law are given by Caneva (1993), Cardwell (1989, 1990), Ihde (1984), and Laidler (1993). The historian Stephen Brush gives English translations of Robert Mayer's 1842 paper ("Bemerkungen über die Kräfte der unbelebten Natur") and of Hermann Helmholtz's 1847 paper ("Über die Erhaltung der Kraft") in his book on development of the kinetic theory of gases (Brush 1965). The concept of energy, the conservation of energy, and the first law of thermodynamics are discussed by Bridgman (1941).

In his paper, Helmholtz showed that the impossibility of a perpetual-motion machine and the constant mechanical equivalent of heat determined by Joule were both consequences of the broader principle of the conservation of energy. His reasoning can be understood using this hypothetical (and impossible) example: Suppose that the mechanical equivalent of heat was not a constant and that more heat could be generated by spinning paddle wheels in water than was needed to run an engine driving the paddle wheels. If this were so, we could constantly generate heat by spinning the paddle wheels in water, use some of this heat to run an engine to drive the paddle wheels, and use the excess heat to run an electrical generator or another piece of machinery. Helmholtz's great insight was that we cannot do something like this because of the *principle of conservation of energy.* This principle states that energy can be converted from one form into another, but it cannot be created or destroyed, and whenever one quantity of energy is produced, an exactly equivalent amount of another type (or types) of energy is consumed.

II. THE FIRST LAW OF THERMODYNAMICS

A. The definition of energy and different forms of energy

Although we all expend and use energy every day, the concept of energy is hard to define in a rigorous and quantitative manner. This problem is not unique to us or to our time. The British physicist Thomas Young (1773–1829), whose many accomplishments included deciphering the Rosetta Stone, was apparently the first person to use the term *energy* in a scientific context (1807) to refer to the product mv^2 (i.e., twice the kinetic energy). However, the term *force* was generally used to denote energy until the 1850s, when Rankine referred to kinetic energy as actual or sensible energy and used the term *potential energy* in its present sense. Lord Kelvin (William Thomson) introduced the use of the term *kinetic energy* in its present sense a few years later.

So what is the definition of energy? In *The Feynman Lectures on Physics,* the Nobel Prize–winning physicist Richard Feynman (1918–1988) states that "It is important to realize in physics today, that we have no knowledge of what energy is." Energy is a fundamental concept, yet it is an abstract concept. Definitions of energy that are commonly found in dictionaries and physics books state that energy is the ability to do work, the capacity to do work, or the potential to exert a force. Energy has also been called the ability of a system to cause changes in itself or in its surroundings, or the unchangeable power of doing work. The unit of energy in the SI system is the joule, which is also the unit of work and heat. As mentioned earlier, one joule is defined as one newton meter, which has the dimensions of force times length.

There are many different forms of energy: atomic, chemical, electrical, heat, kinetic, magnetic, mechanical, potential, radiant, and solar. We cannot directly measure any of these forms of energy; rather, we must measure physical parameters from which the energy involved in a process can be calculated. We can, however, transform the different forms of energy into one another; for example, we can burn coal, gas, or oil to heat a steam turbine to drive an electrical generator to make electricity to run the electrical appliances in our home. Energy is conserved during all these transformations, and the apparent losses that occur are simply energy conversion into forms other than the form of energy desired in that particular transformation. For example, about 90% of the energy output of an incandescent light bulb is heat, and only 10% is light.

B. Nuclear reactions and the conservation of energy

Before proceeding with the statement of the first law, we should mention the interconvertibility of energy and mass, first proposed in Einstein's theory of special relativity and summarized in his famous equation

$$E = mc^2 \tag{3-9}$$

where E is energy, m is mass, and c is the speed of light in a vacuum. This interconvertibility is of great importance in nuclear reactions, but it is not important in chemical reactions where the mass changes involved (if any) are far too small to be measurable. For example, we can consider the predicted change in mass (Δm) of a chemical system associated with a change in energy (ΔE) of 100 kJ, which is comparable to the typical strength of a chemical bond:

$$\Delta m = \frac{\Delta E}{c^2} \cong \frac{10^5\ \text{J}}{(3.0 \times 10^8\ \text{m s}^{-1})^2} \cong 10^{-12}\ \text{kg} \tag{3-10}$$

This is such a small predicted change that it does not need to be considered in chemistry for practical purposes. However, the mass changes involved in nuclear reactions and in the masses of subatomic particles that are accelerated to near the speed of light in particle accelerators are of course much greater. Nevertheless, the principle of conservation of energy is still obeyed, and there are no known exceptions to it.

Radioactive beta decay is perhaps the best example showing that the conservation of energy is obeyed in all known processes. In the late 1920s, measurements apparently showed that neither energy nor momentum was conserved in beta decay. Some physicists speculated that the conservation laws were not obeyed in beta decay. However, in 1930 the Austrian physicist Wolfgang Pauli (1900–1958) proposed that an as yet unknown particle was also emitted during beta decay. The properties of this particle were such that when they were taken into account, energy and momentum were conserved in beta decay. Pauli's suggestion was further developed in 1934 by the Italian physicist Enrico Fermi (1901–1954), who named the particles *neutrinos*. It took until 1956, when neutrinos were first observed, for verification of Pauli's and Fermi's suggestions and for proof that conservation of energy and momentum were obeyed in beta decay.

C. Mathematical statement of the first law

We can use the principle of conservation of energy to define a new thermodynamic function E, the *internal energy* of a system. Internal energy can be in many forms, such as chemical bond energy, electrical energy developed by a chemical reaction, or the increased kinetic energy of particles in a gas, liquid, or solid. The internal energy does not, however, include the gravitational potential energy of a system or the kinetic energy of the center of mass of the system as it moves through space. Thus, the internal energy is not the total energy of a system. The relationship among the internal energy (E), heat (q), and work (w) of a system was studied by Mayer, Joule, and Helmholtz. Their results can be summarized as

$$dE = \delta q - \delta w \tag{3-11}$$

which is a mathematical expression of the *first law of thermodynamics*. Equation (3-11) is interpreted as follows: An infinitesimal change in the internal energy (dE) of a closed system is the sum of two terms: the gain (or loss) of an infinitesimal amount of heat (δq) by the system, and the performance of an infinitesimal amount of work (δw) by (or on) the system. The internal energy of the system can increase (dE is positive) or decrease (dE is negative), depending on the magnitude and sign of the δq and δw terms. If the system gains heat from the surroundings, δq is positive; if the system loses heat to the surroundings, δq is negative; and if there is no gain or loss of heat between the system and the surroundings, δq is zero. The latter case is an adiabatic system. If the system does work on the surroundings, δw is positive; if the surroundings do work on the system, δw is negative; and if no work is done, δw is zero. As we will see later in this chapter, this convention for the sign of δw is consistent with how we think about work being done during the compression or expansion of gases.

The internal energy E is a state function and is path independent. The differential dE is a perfect differential and its integral value (denoted ΔE) depends only on the internal energy difference between the initial and final states of a system and not on how we go from the initial state to the final state. The fact that E is path independent also means that in general, $\Delta E = 0$ around a cycle, which is mathematically expressed as

$$\oint dE = 0 \tag{3-12}$$

In contrast, both δq and δw are path dependent. We use δq and δw instead of dq and dw to stress that δq and δw are not perfect differentials. Other texts use other symbols to make the same point. By convention, the integrals of δq and δw are written q and w, respectively. We will illustrate the path independence of ΔE and the path dependence of q and w later in this chapter, but for now we want to relate ΔE to the heat and work terms in Joule's experiments.

D. Use of the first law to analyze Joule's paddle wheel experiments

We can analyze Joule's paddle wheel experiments using the first law of thermodynamics once we specify the system, its surroundings, the boundary between the two, and the change in state of the system. The insulated water tank (and all the machinery in it) are the system. There was no gain or loss of water or other material from the tank, so the system is a closed system. Furthermore, Joule observed very little gain or loss of heat between the tank and its surroundings, so he also had an adiabatic system ($\delta q = 0$). The falling weights, pulleys, and the other machinery outside the insulated tank are in the surroundings. The boundary between the system and the surroundings is taken as the outside wall of the insulated tank (see Figure 3-2).

Joule observed that the temperature of the water in the tank increased after the paddle wheels were spun by the falling weights. He did his experiments at ambient pressure, which was a constant pressure of one atmosphere. Thus, the integral (or overall) change in state of the system was a temperature increase at constant pressure of the water, metal tank, and internal machinery. For simplicity, in the following equation we label all this as water:

$$\text{water (lower } T, P = 1 \text{ atm)} = \text{water (higher } T, P = 1 \text{ atm)} \tag{3-13}$$

and we call the initial state A and the final state B. The higher temperature of the system (i.e., of the water + tank + internal machinery) at the end is due to an increase in the internal energy of the system. The internal energy change of the system between its initial state and its final state is

$$\int_A^B dE = E_B - E_A = \Delta E = q - w \tag{3-14}$$

Both dq and q are zero for an adiabatic system such as Joule's insulated water tank. Thus, the change in internal energy and the amount of work done are equal in Joule's experiments:

$$\Delta E = q - w = 0 - w \tag{3-15}$$

$$\Delta E = -w \tag{3-16}$$

The change in internal energy (ΔE) is positive because, as described, we define the work w done by the surroundings on the system as having a negative value. In other words, the system gains energy at the expense of the surroundings.

E. Pressure-volume work and indicator diagrams

Historically, the first and second laws of thermodynamics were applied by engineers to the development of better steam engines. Many of the applications of the first law that we encounter in the geological sciences concern the behavior of compressible fluids and gases. Furthermore, the operation of car and truck engines, air conditioners, refrigerators, and other types of engines and compressors involves the compression and expansion of fluids and gases. Thus, we now consider how to define work done by compression and expansion of gases and fluids. This type of work is known as *pressure-volume work*, *PV work*, or *hydrostatic work*.

Figure 3-3 shows a gas confined in a cylinder by a frictionless piston. A weight with mass m sitting on the piston provides an external applied force (F_{ap}) on the piston. The weight of the overlying atmosphere (if any) and of the piston itself are additional external forces. However, for simplicity we assume that these forces are very small by comparison with that provided by the weight itself. The piston has a cross-sectional area A, and thus the applied pressure (P_{ap}) due to the weight of mass m sitting on the piston is

$$P_{ap} = mg/A = F_{ap}/A \tag{3-17}$$

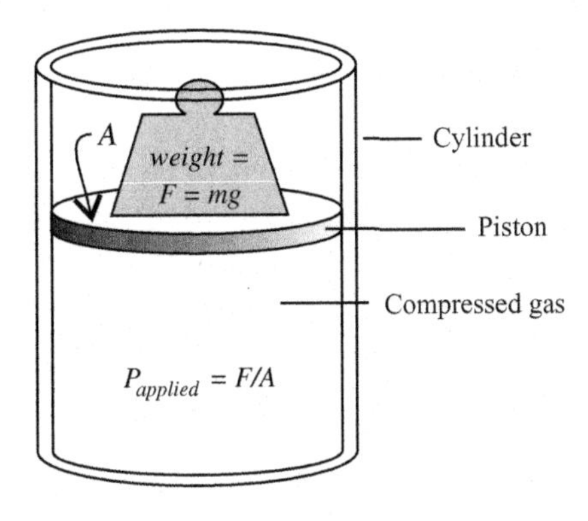

FIGURE 3-3

Illustration showing that the gas in the cylinder is compressed by pressure applied by the weight on the piston.

When we compress the gas by moving the piston downward an infinitesimal distance dr, the incremental work done (δw) is given by

$$\delta w = F_{ap} \times dr \tag{3-1}$$

$$\delta w = (F_{ap}/A) \times Adr = P_{ap} \times Adr = P_{ap}dV \tag{3-18}$$

We have now equated δw to $P_{ap}dV$. In other words, we derived a relationship between the incremental work done (δw) and the compression ($-dV$) or expansion ($+dV$) of fluids.

Example 3-3. A fluid is contained in a cylinder by a piston of 4 cm (0.04 m) radius with a weight sitting on it (e.g., see Figure 3-3). Upon heating, the fluid expands by one liter against a constant applied pressure of one bar. Calculate the following: (a) the work done in liter-bar units and in joules, (b) the applied force in newtons, and (c) the mass of the weight sitting on the piston. Assume the average acceleration of gravity (g) is 9.81 m s^{-2}.

(a) First, integrate Eq. (3-18) to find an expression for w, the work done:

$$\int \delta w = w = \int_{V_1}^{V_2} P_{ap}dV = P_{ap}(V_2 - V_1) = P_{ap}\Delta V \tag{3-19}$$

Substituting the numerical values for applied pressure and the volume change into Eq. (3-19) gives the work done as one liter-bar, which equals 99.99 joules (see Table 3-2).

(b) The applied force in newtons F_{ap} is easily calculated by rearranging Eq. (3-17) and converting the applied pressure from bar into pascals:

$$F_{ap} = A \times P_{ap} = \pi \times (0.04)^2 \times (10^5) = 502.7 \text{ N} \tag{3-20}$$

(c) Finally, the mass of the weight sitting on the piston is calculated by rearranging Eq. (3-3):

$$m = \frac{F_{ap}}{g} = \frac{502.7}{9.81} = 51.2 \text{ kg} \tag{3-21}$$

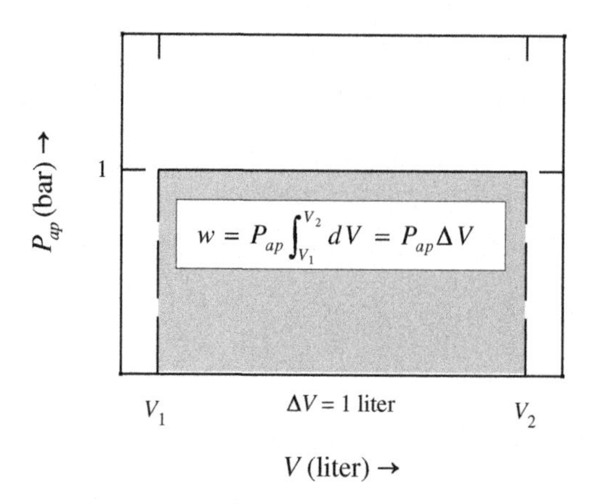

FIGURE 3-4

An indicator diagram showing that the work done is the shaded area under the *PV* line. This diagram is for the problem worked in Example 3-3, which is a fluid expanding by one liter under a constant applied pressure of one bar.

because $1 \text{ N} = 1 \text{ m kg s}^{-2}$. (You can go back to Section II-D of Chapter 2 or to the description of SI units at the beginning of the book to remind yourself of the definition of a newton.)

Equation (3-19) states that the amount of work done during expansion or compression of a fluid at a constant applied pressure is the integral of $P_{ap}dV$, which is the area under a curve or line segment on a graph of pressure versus volume. This point is illustrated in Figure 3-4, which shows the work done during the change of state considered in Example 3-3. This type of plot is called an *indicator diagram* and is now widely used to measure the performance of engines. However, indicator diagrams were originally a closely guarded trade secret. They were developed by John Southern, an assistant to James Watt (1736–1819), and were used by the company of Watt & Boulton to measure the performance of their steam engines.

We interpret Figure 3-4 as follows: The amount of work done during the expansion of the fluid from point 1 with volume V_1 to point 2 with a larger volume V_2 is the shaded area under the horizontal line. This change in state was done isobarically, that is, at constant pressure, which was one bar. Thus, in this case, we have a horizontal line, but in general the applied pressure can vary and does not have to remain constant.

F. A graphical illustration that work and heat are path dependent

When we introduced the first law, we stated that internal energy is a state function and is path independent, but that work and heat are not state functions and are path dependent. We will now use the indicator diagram in Figure 3-5 to demonstrate these important points.

Figure 3-5 shows an air-standard Otto cycle, which is an idealized model of the operation of spark-ignition, four-stroke gasoline engines. It is named after the German inventor Nicolaus August Otto (1832–1891), who invented one of the first practical internal combustion engines. The four points *A, B, C, D* shown on Figure 3-5 represent four different states of the system as it goes through the Otto

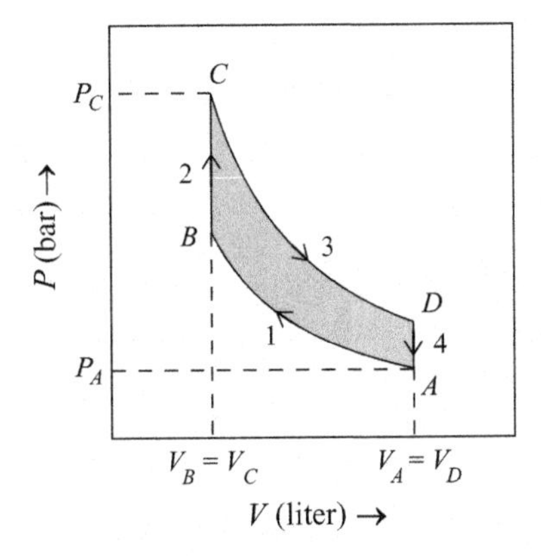

FIGURE 3-5

An indicator diagram for the air-standard Otto cycle that is described in Section II-F. The shaded area inside the cycle shows the work done.

cycle. The curves *AB, BC, CD,* and *DA* represent the four processes that occur during the cycle and approximate the actual operation of a gasoline engine.

The first step of the cycle is an adiabatic compression from point *A* to point *B* as the piston moves inward in the cylinder. During this step the volume decreases from V_A to V_B and the pressure increases from P_A to P_B. The work done during the first step (w_1) is

$$\int \delta w_1 = w_1 = -\int_{V_A}^{V_B} PdV \tag{3-22}$$

and is a negative quantity because dV is negative. As shown in Figure 3-5, both the pressure and volume vary during this step. In the next sections we will show how to evaluate the PdV integral in each step of the Otto cycle and other cycles for the special case of reversible processes involving an ideal gas. However, we do not need this information now to demonstrate that the internal energy is path independent and that work and heat are path dependent. All we need to know is that the work done during the first step of the cycle is the area under curve AB.

The second step in the cycle is a constant-volume (isochoric) increase in pressure (and temperature) from point *B* to point *C*. During this step the volume stays constant, so $V_B = V_C$, while the pressure increases from P_B to P_C. This step is an approximation of the spark ignition and subsequent explosion of the air-fuel mixture. No work is done during the second step, which is an isochoric process ($dV = 0$), so

$$\int \delta w_2 = w_2 = \int_{V_B}^{V_C} PdV = 0 \tag{3-23}$$

The third step in the cycle is an adiabatic expansion from point *C* to point *D* as the piston moves outward in the cylinder. During this step the volume increases from V_C to V_D, and the pressure decreases from P_C to P_D. The work done during this step (w_3) is

$$\int \delta w_3 = w_3 = \int_{V_C}^{V_D} PdV \tag{3-25}$$

which is equal to the area under curve *CD*. The amount of work w_3 in this step is a positive quantity because dV is positive.

The fourth and last step in the cycle is an isochoric decrease in pressure (and temperature) from point *D* back to point *A*. During this step the volume stays constant, so $V_D = V_A$, while the pressure decreases from P_D to P_A. This step is an approximation of the opening of the exhaust valve in the engine. No work is done during this step, which is an isochoric process ($dV = 0$), so

$$\int \delta w_4 = w_4 = \int_{V_D}^{V_A} PdV = 0 \tag{3-25}$$

The total amount of work done by the system (i.e., the engine) around the cycle is

$$w_{cycle} = w_1 + w_2 + w_3 + w_4 = w_3 - w_1 > 0 \tag{3-26}$$

which is equal to the shaded area inside the cycle. In general, the integral of δw around a cycle is not zero, which is expressed by the equation

$$\oint \delta w \neq 0 \tag{3-27}$$

which holds for any cycle. However, the net work done by the system around a cycle may be either a positive or a negative quantity. In this example, the net work done by the system (the engine) is positive.

The change in the internal energy of the system around the cycle is simply the sum of the internal energy changes for each of the four steps of the cycle:

$$\Delta E_{cycle} = (E_B - E_A) + (E_C - E_B) + (E_D - E_C) + (E_A - E_D) = 0 \tag{3-28}$$

This change has to be zero because of the principle of conservation of energy (i.e., the first law). Otherwise, it would be possible to construct a perpetual-motion machine by starting with the system at state *A* with internal energy E_A and going through a cycle that would restore the system to the same state with excess internal energy that could be used to do more work than was needed to run the Otto cycle.

The net heat transfer for the system around the cycle can be calculated from the sum of the heat transfers for each step of the cycle. This net transfer is

$$q_{cycle} = q_1 + q_2 + q_3 + q_4 = q_2 - q_4 > 0 \tag{3-29}$$

because steps one and three are adiabatic steps ($q_1 = q_3 = 0$). The heat transfers in the two isochoric steps (q_2 and q_4) can be calculated from the first law:

$$q_2 = \Delta E_2 - w_2 = \Delta E_2 = (E_C - E_B) \tag{3-30}$$

$$q_4 = \Delta E_4 - w_4 = \Delta E_4 = (E_A - E_D) \tag{3-31}$$

We can also calculate the net heat transfer by the system around the cycle by using the first law and the changes computed above for the internal energy and work around the cycle:

$$\Delta E_{cycle} = 0 = q_{cycle} - w_{cycle} \tag{3-32}$$

$$q_{cycle} = w_{cycle} \tag{3-33}$$

In general, the integral of δq around a cycle is not zero, which is expressed by the equation

$$\oint \delta q \neq 0 \tag{3-34}$$

which holds for any cycle. However, the net heat transfer by the system around a cycle may be either a positive or a negative quantity.

G. The definition of a reversible process

In our discussion of pressure-volume work in Section II-E, we talked about the pressure applied (P_{ap}) by a weight sitting on an ideal frictionless piston. However, we did not explicitly specify the pressure of the gas inside the cylinder. When additional weight is put onto the piston, there is a sudden increase in the pressure of the gas inside the cylinder as it adjusts to the applied pressure, and the piston moves inward. Conversely, when some weight is removed from the piston, there is a sudden decrease in the pressure of the gas inside the cylinder as it adjusts to the applied pressure, and the piston moves outward. As the gas adjusts its pressure to equal the applied pressure, the intermediate states (of pressure, volume, and temperature) that it passes through are out of equilibrium and are not described by the equation of state of the gas. The response to a large change in the applied pressure is a disequilibrium process. It is also an irreversible process. However, we can imagine that smaller and smaller changes in the weight on the piston will lead to smaller and smaller changes in the gas pressure inside the cylinder.

If the pressure (P) of the gas inside the cylinder is always very close to the applied pressure P_{ap}, we have a special case that is called a *reversible process*. In this case all the intermediate states of the system are equilibrium states. As a result, we can calculate the variation of pressure during the reversible expansion or compression of a gas (or another fluid) confined in a cylinder by a piston during each step of a reversible cycle like the Otto cycle, or indeed for any reversible change in state, by using the equation of state for the gas or fluid. The discussion in this chapter and Chapter 4 focuses on reversible changes of state for ideal gases. Reversible changes in state of real gases, liquids, melts, and solids that obey other equations of state can be treated in an analogous fashion. A reversible process is a key concept of general applicability in thermodynamics, and we need to carefully define it because we will use reversible processes throughout our discussion of thermodynamics.

A *reversible process* is one in which all the changes in the system and the surroundings can be restored by a reversal of the direction of the process without producing any other changes in the system or surroundings. To satisfy this definition, a reversible process must be infinitely slow (i.e., quasistatic), must be carried out with infinitesimal steps, and must not have dissipative effects (e.g., friction, viscosity, inelasticity, electrical resistance, or magnetic hysteresis). All reversible processes are quasistatic (infinitely slow), but not all quasistatic processes are reversible.

You have probably guessed (correctly) that a reversible process is a hypothetical concept like an ideal gas. No gas is exactly ideal, but at ambient temperature and low pressure the behavior of air, N_2,

O_2, H_2, He, Ar, and other gases is very close to the predicted behavior of an ideal gas. Likewise, it is possible to carry out processes under conditions that approximate the hypothetical reversible process. However, you should realize that all natural processes, including life, are irreversible to a greater or a lesser extent.

Over the years, scientists have thought of different examples to illustrate a reversible process, such as the reversible isothermal expansion of a gas confined in a cylinder by an ideal frictionless piston. All these examples have in common the concepts of being infinitely slow, the absence of any dissipative effects, and the use of infinitesimal changes due to the addition or subtraction of a tiny fraction of the total weight that is generating an applied pressure on a piston: a large number of small weights that comprise a large weight sitting on a piston, or a bag of sand grains sitting on a piston, or a container of mercury sitting on a piston. In each case we can reversibly expand the gas by slowly and separately removing the small weights, sand grains, or drops of mercury, one at a time, and placing them at the same height at which they were removed from the piston. During each stage of the expansion the applied pressure and the gas pressure inside the cylinder are essentially identical within a very small difference. Also, the pressure change in each step is much smaller than the initial and final pressure in each step. These conditions are met throughout the expansion as long as the final weight remaining on the piston is very much larger than that of a single weight, a single sand grain, or a single drop of mercury that is the incremental change in weight in each step.

In these examples of a reversible process, we can restore the system and surroundings to their initial state without making any other changes in them by slowly and separately replacing the weights, sand grains, or drops of mercury on the piston as it returns to the same height where each weight, sand grain, or drop of mercury was originally removed from it. Although none of these examples can be carried out in practice, they nevertheless illustrate the concept of a reversible process and the conditions that it must satisfy.

H. Constant-pressure processes and enthalpy, a new state function

We are led to the concept of enthalpy by considering a process that takes place at constant pressure P, such as hot exhaust gases coming out of the tailpipe of a car. In this case, the work done by the hot exhaust gas (the system) on the surroundings as the gas undergoes a change of state from its initial state (P, V_1) to its final state (P, V_2) is

$$w = \int \delta w = \int_1^2 PdV = P\int_1^2 dV = P(V_2 - V_1) = P\Delta V \tag{3-19}$$

The internal energy change ΔE for this change in state is found by integrating Eq. (3-11):

$$\Delta E = \int_1^2 dE = (E_2 - E_1) = q_P - P(V_2 - V_1) \tag{3-35}$$

where q_P is the heat transferred at constant pressure P. Rearranging Eq. (3-35) gives

$$(E_2 + PV_2) - (E_1 + PV_1) = q_P \tag{3-36}$$

Constant-pressure processes are fairly common (e.g., pressure variations due to weather are generally 1% or less of one bar). Thus, it is convenient to define a new function for considering constant-pressure processes. This function was named *enthalpy* by the Dutch physicist Kammerlingh Onnes (1853–1926). Enthalpy is represented by the symbol H and is defined as

$$H = E + PV \tag{3-37}$$

Combining Eq. (3-36) with Eq. (3-37), we see that the enthalpy change equals the heat flow for a constant-pressure process,

$$H_2 - H_1 = \Delta H = q_P \tag{3-38}$$

Because E, P, and V are all state functions, enthalpy is also a state function. Willard Gibbs (1839–1903), the American mathematician who developed chemical thermodynamics, called enthalpy "the heat function at constant pressure," and the term *heat function* is used in some older texts. Enthalpy is also commonly known as the *heat content*. We will use *enthalpy* much more frequently than *internal energy* and will discuss experimental measurements of enthalpies (heat contents) and enthalpies (heats) of reactions in the next two chapters.

I. Heat capacity and specific heat

At the beginning of this chapter we mentioned Joseph Black and his realization that different materials have different specific heats. Now we return to this topic and give a mathematical definition of specific heat and heat capacity. In general, the *heat capacity* (C) of a substance is the ratio of the heat added to the substance divided by the resultant temperature change:

$$C = \frac{q}{\Delta T} \tag{3-39}$$

or if we consider infinitesimal temperature changes,

$$C = \frac{\delta q}{dT} \tag{3-40}$$

Heat capacity (C) has units of joule per degree (J K^{-1}) and is an extensive variable. The molar heat capacity (C_m) with units of J mol^{-1} K^{-1} is an intensive variable. *Specific heat* (c) is defined as heat capacity per unit mass, has units of J g^{-1} K^{-1} or kJ kg^{-1} K^{-1}, and is an intensive variable.

Not surprisingly, different materials and different phases of the same material have different heat capacities. This can be seen from the data in Table 3-3 by comparing heat capacity values at 298.15 K for common silicates or for the silica polymorphs. The equations in Table 3-3 are modified Maier-Kelley equations, which typically have errors less than 1%. In Chapter 4, we describe how various types of calorimetry are used to measure enthalpies and/or heat capacities as a function of temperature. We also discuss equations used to represent heat capacity and enthalpy data.

A second important point is that heat capacity values of all materials are different for heating at constant volume or at constant pressure. For heating at constant volume ($dV = 0$), the internal energy change of a material is given by

$$dE = \delta q_V - \delta w = \delta q_V - PdV = \delta q_V \tag{3-41}$$

Table 3-3 Heat Capacity Data for Selected Minerals

$C_P = a + b \times 10^{-3}\,T + c \times 10^{5}\,T^{-2} + d \times 10^{-6}\,T^2$ (J mol^{-1} K^{-1})							C_P at 298.15 K
Name	**Ideal Formula**	***a***	***b***	***c***	***d***	***T* Range (K)**	
acmite	$NaFeSi_2O_6$	199.398	61.972	−42.664	0	298–1300	169.88
albite	$NaAlSi_3O_8$	225.965	137.906	−51.499	−45.879	298–1400	205.07
analbite	$NaAlSi_3O_8$	227.112	139.710	−53.0170	−48.549	298–1400	204.81
almandine	$Fe_3Al_2Si_3O_{12}$	349.996	278.071	−70.257	−118.996	298–1200	343.29
andalusite	Al_2SiO_5	170.021	31.738	−50.213	−2.723	298–2000	122.60
kyanite	Al_2SiO_5	168.266	38.152	−51.07	−4.940	298–2000	121.58
sillimanite	Al_2SiO_5	164.342	36.674	−46.952	−2.735	298–2000	123.72
anhydrite	$CaSO_4$	112.040	45.002	−22.002	4.970	298–1000	101.23
anorthite	$CaAl_2Si_2O_8$	240.809	120.470	−55.274	−36.082	298–1000	211.34
		340.184	−34.396	−184.727	32.355	1000–1830	
anthophyllite	$Mg_7Si_8O_{22}(OH)_2$	790.329	349.379	−196.270	−108.941	298–1200	664.02
calcite	$CaCO_3$	99.717	26.923	−21.578	0	298–1200	83.47
chrysotile	$Mg_3Si_2O_5(OH)_4$	224.309	411.109	−47.347	−220.134	298–900	274.05
clinoenstatite	$MgSiO_3$	82.098	60.409	−14.645	−19.762	298–1600	82.12
cordierite	$Mg_2Al_3(AlSi_5O_{18})$	546.430	196.287	−140.578	−39.723	298–1700	443.28
diopside	$CaMgSi_2O_6$	180.264	125.007	−40.873	−53.713	298–1000	166.78
dolomite	$CaMg(CO_3)_2$	163.131	113.445	−33.534	−19.363	298–1100	157.51
fayalite	Fe_2SiO_4	176.029	−8.822	−38.896	24.717	298–1490	131.84
ferrosilite	$FeSiO_3$	124.287	14.573	−33.775	−0.022	298–800	90.63
forsterite	Mg_2SiO_4	143.967	39.362	−32.525	−5.672	298–1800	118.61
glaucophane	$Na_2Mg_3Al_2Si_8O_{22}(OH)_2$	604.359	662.577	−118.639	−258.336	298–1200	645.48
grossular	$Ca_3Al_2Si_3O_{12}$	434.692	71.341	−111.701	−0.955	298–1200	330.22
halite	$NaCl$	45.153	17.966	0	0	298–1074	50.51

hedenbergite	$CaFeSi_2O_6$	193.714	92.654	−38.909	−25.516	298–1300	175.30
jadeite	$NaAlSi_2O_6$	184.548	88.561	−43.427	−24.520	298–1300	159.93
kaliophillite	$KAlSiO_4$	95.151	137.075	−11.762	−34.742	298–810	119.70
(high)	$KAlSiO_4$	177.650	0	0	0	810–1800	
kaolinite	$Al_2Si_2O_5(OH)_4$	253.174	243.864	−62.670	−135.240	298–800	243.37
leucite (hex)	$KAlSi_2O_6$	148.421	134.299	−21.650	0	298–955	163.8
(cubic)	$KAlSi_2O_6$	216.965	6.559	−41.478	5.986	955–1900	
microcline	$KAlSi_3O_8$	186.734	224.548	−38.142	−97.261	298–1000	202.13
		372.256	−73.966	−257.871	37.705	1000–1400	
monticellite	$CaMgSiO_4$	135.699	69.798	−27.735	−25.980	298–1100	123.00
muscovite	$KAl_2(AlSi_3O_{10})(OH)_2$	293.544	404.513	−64.139	−180.070	298–1000	325.99
nepheline (I)	$NaAlSiO_4$	27.813	295.151	−0.018	0.240	298–467	115.81
(II)	$NaAlSiO_4$	112.090	67.110	0	0	467–1180	
(III)	$NaAlSiO_4$	172.000	5.520	0	0	1180–1500	
carnegieite	$NaAlSiO_4$	116.130	85.950	−20.000	0	298–966	119.26
(cubic)	$NaAlSiO_4$	152.320	28.830	0	0	966–1799	
paragonite	$NaAl_3Si_3O_{10}(OH)_2$	299.282	359.578	−64.685	−137.508	298–800	321.50
phlogopite	$KMg_3(AlSi_3O_{10})(OH)_2$	393.489	232.196	−88.455	−96.313	298–1000	342.42
pyrope	$Mg_3Al_2Si_3O_{12}$	392.730	158.528	−98.2506	−50.1958	298–1500	325.76
sanidine	$KAlSi_3O_8$	193.981	213.559	−39.950	−91.822	298–1000	204.55
		355.389	−46.375	−230.833	25.792	1000–1473	
siderite	$FeCO_3$	48.658	112.093	0	0	298–458	82.44
silica polymorphs							
coesite	SiO_2	30.780	77.382	−4.515	−37.260	298–1000	45.46
		102.459	−34.291	−97.486	12.030	1000–1800	
cristobalite (α)	SiO_2	17.908	88.115	0	0	298–523	44.94
(β)		72.753	1.300	−41.320	0	523–1800	
quartz (α)	SiO_2	44.603	37.754	−10.018	0	298–844	44.60

(Continued)

Table 3-3 Heat Capacity Data for Selected Minerals *(continued)*

	$C_P = a + b \times 10^{-3} T + c \times 10^5 T^{-2} + d \times 10^{-6} T^2$ (J mol^{-1} K^{-1})						C_P at
Name	**Ideal Formula**	***a***	***b***	***c***	***d***	***T* Range (K)**	**298.15 K**
(β)		58.928	10.031	0	0	844–1800	
stishovite	SiO_2	54.001	31.262	−16.968	−13.986	298–1000	42.99
		87.419	−20.571	−60.766	8.809	1000–1800	
tridymite (α)	SiO_2	13.682	103.763	0	0	298–390	44.62
(β)		57.07	11.046	0	0	1000–1800	
sodalite	$Na_4(AlSiO_4)_3Cl$	921.519	258.438	−169.463	49.411	298–1000	812.33
spodumene (α)	$LiAlSi_2O_6$	145.161	170.286	−27.83	−64.065	298–1200	158.93
(β)		180.153	101.378	−40.165	−27.284	298–1700	162.77
staurolite	$Fe_4Al_{18}Si_8O_{46}(OH)_2$	1268.977	1339.179	−308.254	−499.308	298–900	1277.1
talc	$Mg_3Si_4O_{10}(OH)_2$	770.955	−994.917	−207.481	909.560	298–800	321.77
tephroite	Mn_2SiO_4	139.731	61.527	−24.390	−21.352	298–1524	128.74
tremolite	$Ca_2Mg_5Si_8O_{22}(OH)_2$	980.336	−216.427	−255.410	303.213	298–1000	655.44
wollastonite	$CaSiO_3$	99.053	39.677	−20.998	−12.054	298–1400	86.19
pseudowollastonite							
	$CaSiO_3$	101.580	30.010	−19.630	−7.710	298–1821	87.92
zoisite	$Ca_2Al_3Si_3O_{12}(OH)$	467.179	12.564	−114.174	88.349	298–1200	350.34

and the heat capacity is given by

$$C_V = \left(\frac{\delta q}{dT}\right)_V = \left(\frac{\partial E}{\partial T}\right)_V \quad (3\text{-}42)$$

The subscript V on the partial derivatives indicates that volume remains constant ($dV=0$) during differentiation. In contrast, work is done during heating at constant pressure. In this case the first law and Eq. (3-37) show that the heat flow is

$$\delta q_P = dE + \delta w = dE + PdV = dH \quad (3\text{-}43)$$

Thus, the constant-pressure heat capacity C_P is given by

$$C_P = \left(\frac{\delta q}{dT}\right)_P = \left(\frac{\partial H}{\partial T}\right)_P \quad (3\text{-}44)$$

where the subscript P on the partial derivatives indicates that pressure is held constant during differentiation. Molar C_P and C_V values are denoted by $C_{P,m}$ and $C_{V,m}$, respectively.

J. Relationship between C_V and C_P

We noted earlier that no work is done during heating at constant volume because $PdV=0$. All the heat added to the system is used to raise the temperature of the substance being heated. Thus, we expect that in general C_P will be larger than C_V because, under constant pressure, the heat added to a substance also has to do work to expand the material at constant pressure. One exception is water at 4°C, for which C_P and C_V are equal.

We can calculate $C_P - C_V$ starting from Eq. (3-44), the definition of C_P, and combining this with Eq. (3-37), the definition of enthalpy:

$$C_P = \left(\frac{\partial H}{\partial T}\right)_P = \left(\frac{\partial [E+PV]}{\partial T}\right)_P = \left(\frac{\partial E}{\partial T}\right)_P + P\left(\frac{\partial V}{\partial T}\right)_P \quad (3\text{-}45)$$

We want to substitute for the term $(\partial E/\partial T)_P$ in Eq. (3-45). We can easily do this by expressing internal energy E as a function of temperature and volume because dE is a perfect differential:

$$dE = \left(\frac{\partial E}{\partial V}\right)_T dV + \left(\frac{\partial E}{\partial T}\right)_V dT \quad (3\text{-}46)$$

Then, using the right side of Eq. (3-46) to substitute for dE in $(\partial E/\partial T)_P$, we get

$$\begin{aligned} \left(\frac{\partial E}{\partial T}\right)_P &= \frac{\partial}{\partial T}\left[\left(\frac{\partial E}{\partial V}\right)_T dV + \left(\frac{\partial E}{\partial T}\right)_V dT\right]_P \\ &= \left(\frac{\partial E}{\partial V}\right)_T \left(\frac{\partial V}{\partial T}\right)_P + \left(\frac{\partial E}{\partial T}\right)_V \left(\frac{\partial T}{\partial T}\right)_P \quad (3\text{-}47) \\ &= \left(\frac{\partial E}{\partial V}\right)_T \left(\frac{\partial V}{\partial T}\right)_P + \left(\frac{\partial E}{\partial T}\right)_V \end{aligned}$$

We can substitute this expression for $(\partial E/\partial T)_P$ back into Eq. (3-45) to get

$$C_P = \left[\left(\frac{\partial E}{\partial T}\right)_V + \left(\frac{\partial E}{\partial V}\right)_T \left(\frac{\partial V}{\partial T}\right)_P\right] + P\left(\frac{\partial V}{\partial T}\right)_P \tag{3-48}$$

This can be simplified using Eq. (3-42), which is the definition of C_V, to substitute for $(\partial E/\partial T)_V$:

$$C_P - C_V = \left(\frac{\partial V}{\partial T}\right)_P \left[P + \left(\frac{\partial E}{\partial V}\right)_T\right] \tag{3-49}$$

Equation (3-49) gives the difference between C_P and C_V.

K. The meaning of the difference between C_P and C_V

We have already stated that we expect C_P to be larger than C_V, so the difference from Eq. (3-49) should be positive. But exactly what do the terms on the right side of Eq. (3-49) mean?

The first term on the right side is $(\partial V/\partial T)_P$. This expresses the change in volume with a change in temperature at constant pressure. Most materials expand on heating, so $(\partial V/\partial T)_P$ is usually a positive number. (Two notable exceptions are liquid water, which contracts on heating in the 1–4°C range, and zirconium tungstate ZrW_2O_8, which contracts on heating over a very large temperature range of 0.3–1050 K.)

The second term on the right side of Eq. (3-49) is total pressure, which is a constant (either at about one bar or some other controlled value). The third term in Eq. (3-49) is $(\partial E/\partial V)_T$, which is the change in the internal energy with a change in volume at constant temperature. This derivative is the *internal pressure* and it represents the internal cohesive (or attractive) forces in a gas, liquid, or solid. In 1843, Joule evaluated this term for gases in another classic experiment.

L. Joule's free expansion experiment

Figure 3-6 schematically illustrates Joule's experiment. Container A holds dry air at 22 atmospheres that is allowed to expand into the evacuated container B when stopcock D is opened. The whole

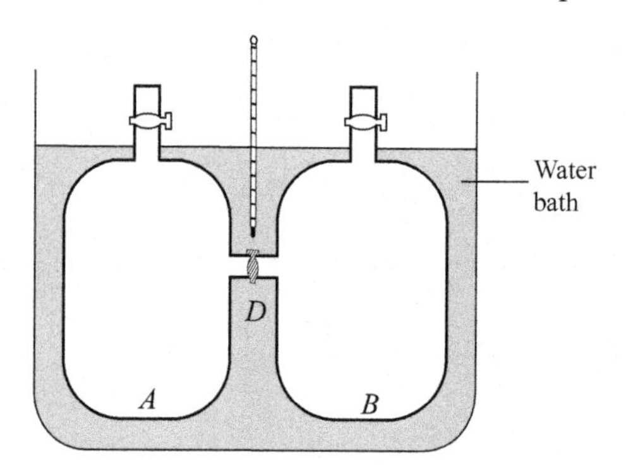

FIGURE 3-6

A schematic diagram of Joule's free expansion experiment. Air under pressure in cylinder A expands into evacuated cylinder B when valve D is opened. The apparatus is submerged in a water bath. The water is stirred and its temperature measured before and after the expansion.

apparatus is submerged in a water bath. The bath is stirred and its temperature is measured before and after the gas expansion. Joule found that “no change of temperature occurs when air is allowed to expand in such a manner as not to develop mechanical power” (i.e., when no work is done). From the first law,

$$dE = \delta q - \delta w \tag{3-11}$$

Joule observed no change in the temperature of the water bath, so the gas did not gain or lose heat ($\delta q = 0$). The gas freely expanded into a vacuum, so the applied pressure $P_{ap} = 0$, and no work was done ($\delta w = P_{ap} dV = 0$). Thus, from Eq. (3-11), the change in internal energy (dE) was also zero. As a result, Eq. (3-46) can be set equal to zero:

$$dE = \left(\frac{\partial E}{\partial V}\right)_T dV + \left(\frac{\partial E}{\partial T}\right)_V dT = 0 \tag{3-50}$$

Equation (3-50) can be evaluated easily because the isothermal conditions mean $dT = 0$, but the expansion means that dV was positive. Hence, the only way for the first term to be zero is that

$$\left(\frac{\partial E}{\partial V}\right)_T = 0 \tag{3-51}$$

Joule and Kelvin subsequently showed that Eq. (3-51) is only valid for ideal gases. Their work is discussed in Chapter 8. For now, the key point is that an ideal gas obeys Eq. (3-51).

III. SOME APPLICATIONS OF THE FIRST LAW TO IDEAL GASES

This section describes some applications of the first law to a number of thermodynamic processes in which one or another variable is held constant. We then have to decide how to calculate the changes (if any) in internal energy, work, heat, enthalpy, temperature, pressure, or volume that take place in the different types of processes. We start with the difference in constant-pressure and constant-volume heat capacities because this information is needed to do calculations for isothermal, isochoric, and adiabatic processes.

A. $C_P - C_V$ for ideal gases

Earlier we showed that the difference between C_P and C_V for any substance is given by

$$C_P - C_V = \left(\frac{\partial V}{\partial T}\right)_P \left[P + \left(\frac{\partial E}{\partial V}\right)_T\right] \tag{3-49}$$

From Joule's free expansion experiment, we know that $(\partial E/\partial V)_T = 0$, so we can simplify Eq. (3-49) to

$$C_P - C_V = P\left(\frac{\partial V}{\partial T}\right)_P \tag{3-52}$$

for ideal gases. Equation (3-52) can be rewritten by using the ideal gas law ($PV = nRT$) to solve for V and evaluating the term $(\partial V/\partial T)_P$:

Table 3-4 Heat Capacity Data for Selected Gases

$C_P = a + b \times 10^{-3} T + c \times 10^{5} T^{-2} + d \times 10^{-6} T^2$ (J mol^{-1} K^{-1}) [a]					Properties at 298.15 K and 1 bar				
Gas	a	b	c	d	C_P/R [b]	C_V/R	$\frac{C_P}{C_V} = \gamma$	$\frac{C_P - C_V}{R}$	Ref. for γ
Air	26.917	5.899	0.270	0.000	3.506	2.501	1.402	1.005	1
H_2	27.280	3.264	0.502	0.000	3.468	2.468	1.405	1.000	1
He	20.786	0.000	0.000	0.00	2.500	1.500	1.667	1.000	2
CO	24.217	11.405	1.568	−2.706	3.505	2.505	1.399	1.000	1
CO_2	46.789	7.714	−11.701	0.000	4.466	3.452	1.294	1.014	1
CH_4	23.640	47.860	−1.920	0.000	4.286	3.282	1.306	1.004	3
C_2H_2	43.764	31.072	−7.505	−6.178	5.296	4.313	1.228	0.983	3
C_2H_4	9.247	128.922	−0.773	−44.194	5.226	4.201	1.244	1.026	3
C_2H_6	2.842	177.864	1.599	−58.315	6.416	5.385	1.191	1.032	3, 27°C
N_2	28.577	3.766	−0.502	0.000	3.503	2.500	1.401	1.003	1
N_2O	47.921	7.503	−11.274	0.000	4.646	3.574	1.300	1.072	4, 15°C
NO	24.936	11.664	−1.516	−2.907	3.592	2.565	1.400	1.027	4, 15°C
NO_2	46.133	6.912	−11.024	0.000	4.472	–	–	–	–
NH_3	23.750	40.474	0.478	−8.161	4.421	3.346	1.321	1.075	3, 27°C
O_2	31.043	3.650	−2.915	0.000	3.533	2.531	1.396	1.002	1
O_3	31.611	44.034	−3.240	−19.362	4.74	3.67	1.29	1.07	4[c]
H_2O	25.800	18.025	2.356	−2.554	4.241	3.189	1.33	1.052	4, 100°C [c]

HF	27.236	2.548	0.98	0.428	3.504	–	–	–	–
Ne	20.786	0.000	0.000	0.000	2.500	1.500	1.667	1	2
SO_2	49.83	4.734	−11.183	0.000	4.792	3.729	1.285	1.063	4, 15°C
H_2S	24.605	26.856	1.914	−5.820	4.119	3.074	1.340	1.045	4, 15°C
OCS	44.021	17.61	−6.514	−4.437	4.997	–	–	–	–
CS_2	55.608	4.126	−10.716	0.000	5.498	4.452	1.235	1.046	4, 15°C
Cl_2	36.61	1.079	−2.72	0.000	4.083	3.013	1.355	1.070	4, 15°C
HCl	24.835	7.896	1.829	−1.243	3.504	2.503	1.40	1.001	4, 15°C
Ar	20.786	0.000	0.000	0.000	2.500	1.500	1.667	1	1
HBr	27.454	4.633	0.098	0.000	3.505	2.504	1.4	1.001	4, 20°C
Kr	20.786	0.000	0.000	0.000	2.500	1.500	1.667	1	2
Xe	20.786	0.000	0.000	0.000	2.500	1.500	1.667	1	2
HCN	35.687	17.869	−4.332	−3.195	4.484	3.423	1.31	1.061	4, 65°C [c]

References:

1. *Hilsenrath, J. et al (1955). Tables of thermal properties of gases; comprising tables of thermodynamic and transport properties of air, argon, carbon dioxide, carbon monoxide, hydrogen, nitrogen, oxygen, and steam.* National Bureau of Standards circular *564, 488 pp.*
2. *Ideal gas assumed.*
3. *Din, F (Ed.) (1961).* Thermodynamic Functions of Gases, *3 volumes, Butterworths Scientific Publications, London.*
4. *Partington, J. R., and Shilling, W. G. (1924).* The Specific Heats of Gases. *D. Van Nostrand Company, New York.*

[a] *Temperature range for fits is 298–2000 K, with the following exceptions: He and Ne, 298–20,000 K; CH_4, 298–1200 K; C_2H_4, 298–1500 K; C_2H_6, 298–1000 K; O_3, 298–1000 K; Ar, 298–8300 K; Kr, 298–7100 K; and Xe, 298–5800 K. The different polynomial fits have maximum errors of 0.3–4%.*
[b] *The tabulated properties at 298.15 K are for the real gas.*
[c] *Heat capacities for steam and HCN are given at 100°C and 65°C, respectively. The γ value for O_3 was measured at an unspecified temperature.*

$$\left(\frac{\partial V}{\partial T}\right)_P = \left(\frac{\partial[nRT/P]}{\partial T}\right)_P = \frac{nR}{P} \tag{3-53}$$

Substituting Eq. (3-53) into Eq. (3-52) then gives

$$C_P - C_V = P\left(\frac{\partial V}{\partial T}\right)_P = P\left(\frac{nR}{P}\right) = nR = 8.314n \text{ J mol}^{-1}\text{ K}^{-1} \tag{3-54}$$

If we consider one mole of gas and molar heat capacities, we have $C_{P,m} - C_{V,m} = R$. Molar C_P and C_V values for gases are tabulated in Table 3-4. Methods for measuring the heat capacity of low-pressure (nearly ideal) gases are described in Chapter 4.

B. Reversible isothermal expansion and compression

A gas (or other fluid) absorbs heat from the surroundings to balance work done during isothermal expansion and rejects heat to the surroundings during isothermal compression. In both cases $dE = 0$ because $dT = 0$ for an isothermal process.

Figure 3-7 is the indicator diagram for the reversible isothermal expansion at 298.15 K of an ideal gas ($V_1 = 0.05$ liter, $P_1 = 8$ bar to $V_2 = 0.40$ liter, $P_2 = 1$ bar). All points on the PV curve in Figure 3-7 are equilibrium states of the gas and are given by the ideal gas law.

Example 3-4. We want to calculate the work done, ΔE, ΔH, and the heat absorbed by the system for the reversible isothermal expansion in Figure 3-7. The work done by the system is

$$\int \delta w = w = \int_{V_1}^{V_2} P dV = \int_{V_1}^{V_2} nRT\frac{dV}{V} = nRT\ln\left(\frac{V_2}{V_1}\right) \tag{3-55}$$

where the ideal gas law was used to evaluate the integral. Using the ideal gas law, we find the number of moles of gas is $n = 1.614 \times 10^{-2}$ mol. Substituting numerical values into Eq. (3-55),

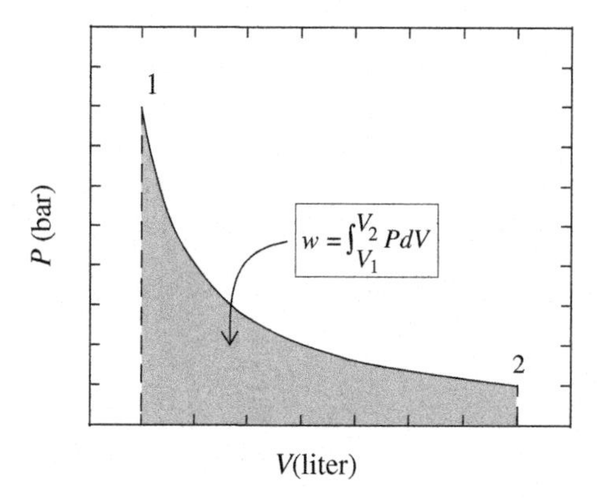

FIGURE 3-7

An indicator diagram for the reversible isothermal expansion of an ideal gas from its initial state (P_1, V_1) to its final state (P_2, V_2). The work done by the system (i.e., the gas) during the expansion is equal to the area under the PV curve and is a positive quantity.

$$w = (1.614 \times 10^{-2})(8.314)(298.15)\ln\left(\frac{0.40}{0.05}\right) = 83.20 \text{ J} \tag{3-56}$$

The internal energy change is calculated by rearranging Eq. (3-42) to give

$$\Delta E = \int dE = \int C_V dT = 0 \tag{3-57}$$

where C_V is the heat capacity at constant volume. Equation (3-57) shows that $\Delta E = 0$ because $dT = 0$ for a reversible isothermal expansion. Rearranging Eq. (3-44) gives ΔH as

$$\Delta H = \int dH = \int C_P dT \tag{3-58}$$

The $\Delta H = 0$ because the expansion is isothermal ($dT = 0$). The heat q absorbed by the system must balance the work done by the system to satisfy the first law (Eq. 3-11). Thus, $q = 83.20$ J and is different than ΔH because the isothermal expansion is not done at a constant pressure.

C. Reversible constant-volume (isochoric) processes

No PV work is done during constant-volume processes (because $dV = 0$). Equation (3-57) gives ΔE, which the first law (Eq. 3-11) shows must equal the heat flow into or out of the system.

D. Reversible adiabatic processes

A gas (or other fluid) cools on expansion if it cannot absorb heat from the surroundings to balance the work it does during the expansion. Conversely, a gas (or other fluid) heats up during compression if it cannot reject heat into the surroundings. These are two examples of adiabatic processes. A refrigerator keeps the food inside it cold and prevents heat from entering. An oven keeps the food inside it hot and prevents heat from escaping. These are two examples of adiabatic systems. As mentioned earlier, the word *adiabatic* is used in thermodynamics to describe systems that are perfectly thermally insulated from their surroundings. The word *adiabatic* is also used to describe changes (or processes) in which no heat enters or leaves the system. In an 1870 paper Rankine was the first scientist to use the term *adiabatic change*.

The key feature of an adiabatic process is that there is no heat flow ($\delta q = 0$), so, if only PV work is considered, the first law (Eq. 3-11) becomes

$$dE = -\delta w = -PdV \tag{3-59}$$

Now if we use Eq. (3-57) to substitute for dE in Eq. (3-59),

$$C_V dT = -PdV \tag{3-60}$$

and apply the ideal gas law to substitute for P, we obtain

$$C_V dT = -\frac{nRT}{V} dV \tag{3-61}$$

Once this is rearranged to

$$\frac{C_V}{T}dT = -\frac{nR}{V}dV \tag{3-62}$$

we can integrate from the initial state (V_1, T_1) to the final state (V_2, T_2), which gives

$$\int_{T_1}^{T_2} \frac{C_V}{T}dT = C_V \ln\left(\frac{T_2}{T_1}\right) = \int_{V_1}^{V_2} \left(-\frac{nR}{V}\right)dV = -nR\ln\left(\frac{V_2}{V_1}\right) \tag{3-63}$$

if we assume C_V is independent of temperature, which is an approximation. Equation (3-63) can also be written as

$$C_V \ln\left(\frac{T_2}{T_1}\right) = nR\ln\left(\frac{V_1}{V_2}\right) \tag{3-64}$$

Another equivalent formulation is

$$\left(\frac{T_2}{T_1}\right)^{C_V} = \left(\frac{V_1}{V_2}\right)^{nR} \tag{3-65}$$

Finally, a third equivalent expression is

$$\left(\frac{T_2}{T_1}\right) = \left(\frac{V_1}{V_2}\right)^{nR/C_V} \tag{3-66}$$

Equations (3-64) through (3-66) give the relationship between temperature and volume for n moles of an ideal gas undergoing a reversible adiabatic process (expansion or compression). If we had one mole of gas, Eqs. (3-64) to (3-66) would be written with $C_{V,m}$, the molar constant-volume heat capacity. As we will discuss later in Section IV, convection in Earth's lower atmosphere, called the *troposphere*, is an adiabatic process. The lower atmospheres of Venus, Jupiter, Saturn, Uranus, Neptune, and of Saturn's satellite Titan are also convective. So, we have important and practical reasons for studying the behavior of gases during an adiabatic process.

Earlier in this chapter, we showed that $C_P - C_V = nR$ for n moles of an ideal gas. Thus,

$$\frac{C_P - C_V}{C_V} = \frac{C_P}{C_V} - 1 = \frac{nR}{C_V} \tag{3-67}$$

Defining the *adiabatic exponent* (γ) as

$$\gamma = \frac{C_P}{C_V} = \frac{C_{P,m}}{C_{V,m}} \tag{3-68}$$

we can combine Eqs. (3-67) and (3-68) to get

$$\frac{nR}{C_V} = \gamma - 1 \tag{3-69}$$

So, Eq. (3-66) can be rewritten as

$$\left(\frac{T_2}{T_1}\right) = \left(\frac{V_1}{V_2}\right)^{\gamma-1} \tag{3-70}$$

From the ideal gas law,

$$\frac{T_2}{T_1} = \frac{P_2V_2}{P_1V_1} = \left(\frac{V_1}{V_2}\right)^{\gamma-1} \tag{3-71}$$

and Eq. (3-71) becomes

$$\left(\frac{P_2}{P_1}\right) = \frac{V_1}{V_2}\left(\frac{V_1}{V_2}\right)^{\gamma-1} = \left(\frac{V_1}{V_2}\right)^{\gamma} \tag{3-72}$$

Equivalently,

$$P_1V_1^{\gamma} = P_2V_2^{\gamma} \tag{3-73}$$

$$PV^{\gamma} = C \tag{3-74}$$

where C is a constant, for the pressure and volume change of an ideal gas during a reversible adiabatic process. Equation (3-74) is *Poisson's equation*. It is similar to Eq. (2-9), Boyle's law, except for the exponent γ. Both equations represent reversible, polytropic (i.e., constant heat capacity) expansions (PV^k = a constant), where $k=0$ for reversible isobaric expansion, $k=1$ for reversible isothermal expansion, and $k=\gamma=C_P/C_V$ for reversible adiabatic expansion. Now let's consider some applications of the various equations.

Figure 3-8(a) shows an isotherm and an adiabat. The C_P/C_V ratio (γ) is always greater than unity, so an adiabat is always steeper than an isotherm. We will now compare work done and changes in the internal energy, enthalpy, and heat for the isotherm and adiabat in Figure 3-8(a).

Example 3-5. Calculate the final pressure P_2 and temperature T_2, work done, ΔE, and ΔH after the reversible adiabatic expansion of air from $V_1=1$ liter, $P_1=15$ bars, $T_1=298.15$ K to $V_2=1.5$ liters. We assume air is ideal with $\gamma=1.402$, $C_V=2.501R$, and $C_P=3.506R$ (see Table 3-4). These values vary by ~0.15% over the ΔT used. We find P_2 from Eq. (3-73):

$$P_2 = \frac{(15\text{ bars})(1\text{ L})^{1.402}}{(1.5\text{ L})^{1.402}} = 8.50\text{ bars} \tag{3-75}$$

We next use the ideal gas law to find the moles of gas (n) and the final temperature T_2:

$$T_2 = \frac{P_2V_2}{nR} = \frac{(8.5\text{ bars})(1.5\text{ L})}{(0.605\text{ mol})(0.083145\text{ L bar/mol K})} = 253.46\text{ K} \tag{3-76}$$

Heat flow $q=0$ for an adiabatic process, so $w=-\Delta E=-nC_{V,m}\Delta T$ from Eqs. (3-16) and (3-57):

$$w = -0.605 \times 2.501 \times R \times (253.46 - 298.15) = 562.2\text{ J} \tag{3-77}$$

The $\Delta E=-w=-562.2$ J mol^{-1} from Eq. (3-16). The ΔH from Eq. (3-58) is

$$\Delta H = 0.605 \times 3.506 \times R \times (253.46 - 298.15) = -788.1\text{ J} \tag{3-78}$$

Because H is a path-independent state function, ΔH for any reversible adiabat is the sum of the changes along an isotherm (0) and isobar ($C_P\Delta T$) between the same initial and final points.

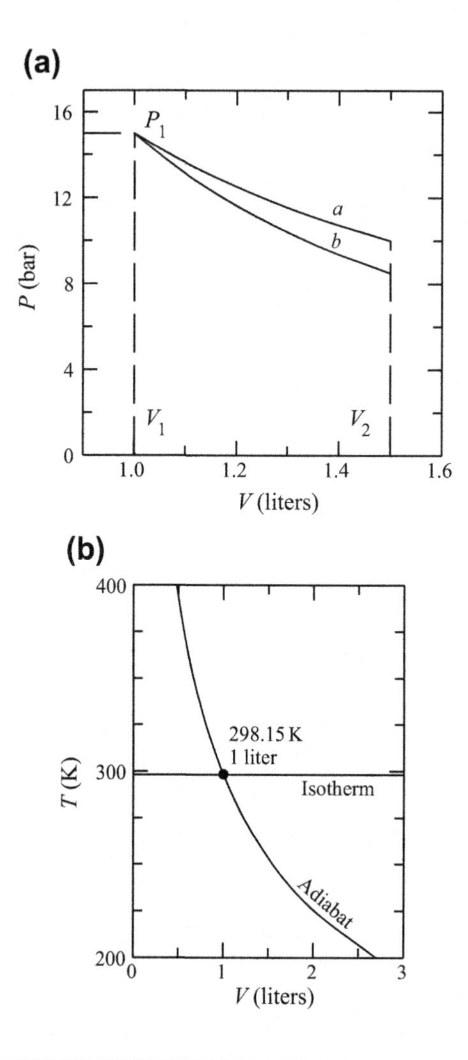

FIGURE 3-8

An indicator diagram comparing the amount of work done for two different *PV* paths. (a) Path *a* is a reversible isothermal expansion. (b) Path *b* is a reversible adiabatic expansion.

Our results for reversible isothermal expansion of air are $w = 608.1$ J, $\Delta E = 0$, $\Delta H = 0$, and $q = 608.1$ J. (See Example 3-4 for the methods.) We find that more work is done in this case.

In general, more work is always done along a reversible isotherm than a reversible adiabat between the same points. This can be seen from Figure 3-8(a) because the area under curve a, the reversible isotherm, is greater than the area under curve b, the reversible adiabat.

Figure 3-8(b) shows the reversible isotherm and adiabat from Figure 3-8(a) plotted as a function of temperature and volume. The isotherm is a horizontal line that is intersected by the adiabat at 298.15 K and 1 liter, the initial values in Example 3-5. Figure 3-8(b) shows that adiabats are always steeper than isotherms.

E. The amount of work done in reversible and irreversible processes

Now we consider irreversible adiabatic expansion of air from the same initial conditions.

Example 3-6. Calculate the final pressure P_2 and temperature T_2, work done, ΔE, and ΔH for irreversible adiabatic expansion of air from $V_1 = 1$ L, $P_1 = 15$ bars, $T_1 = 298.15$ K to $V_2 = 1.5$ L, $P_2 = 8.5$ bars. We assume air is ideal with $\gamma = 1.402$, $C_V = 2.501R$, and $C_P = 3.506R$.

We chose P_2 to be the same as in the reversible adiabatic expansion. The work done is given by Eq. (3-19) because the applied pressure was instantaneously reduced to a constant value of 8.5 bars before the gas could expand:

$$w = P_{ap}\Delta V = (8.50 \text{ bar})(0.50 \text{ L})(99.991 \text{ J/L bar}) = 425.0 \text{ J} \tag{3-79}$$

This is less than the 562.2 J work for the reversible adiabatic expansion. Equation (3-16) gives $\Delta E = -w = -425.0$ J. We use Eq. (3-57) to compute T_2 via

$$T_2 = T_1 + \frac{\Delta E}{nC_{V,m}} = 298.15 - \frac{425.0 \, J}{(0.605)(2.501R)} = 264.4 \text{ K} \tag{3-80}$$

Now that we know ΔT we use Eq. (3-58) to compute ΔH:

$$\Delta H = C_P\Delta T = nC_{P,m}\Delta T = (0.605)(3.506R)(-33.75 \text{ K}) = -595.18 \text{ J} \tag{3-81}$$

Less work is always done in an irreversible expansion than in a reversible expansion. Also, the gas is not cooled as much by the irreversible expansion. An irreversible process always gives less work than a reversible process because more energy is dissipated as heat.

IV. ADIABATIC PROCESSES AND THE THERMAL STRUCTURE OF PLANETARY ATMOSPHERES

The troposphere (i.e., the lower atmospheres closest to the surface or below the radiative-convective boundary for the Jovian planets) of Venus, Earth, Jupiter, Saturn, Uranus, Neptune, and Saturn's largest satellite, Titan, are convective. As a consequence, the tropospheres of these bodies have thermal profiles very close to adiabatic profiles.

We illustrate this concept by considering a warm, rising gas parcel in a planetary atmosphere. The parcel has the same pressure as its surroundings, but there is no heat exchange ($\delta q = 0$). As the gas rises, it does work against the planet's gravitational field and its internal energy declines:

$$dE = -\delta w = -PdV \tag{3-59}$$

We can use Eq. (3-59) to derive the variation of T and V or T and P or P and V during adiabatic expansion of the gas parcel if we assume that this occurs reversibly. However, for comparison with measurements of the thermal structure of planetary atmospheres, we also want to derive the variation of temperature with altitude (z). We use Eq. (3-42) to substitute for dE in Eq. (3-59):

$$C_V dT = -PdV \tag{3-82}$$

and then use the ideal gas law to evaluate dV

$$dV = d(nRT/P) = -(nRT/P^2)dP + (nR/P)dT \tag{3-83}$$

We now use Eq. (3-83) to substitute for dV in Eq. (3-82) and obtain

$$C_V dT = -P[-(nRT/P^2)dP + (nR/P)dT] \quad (3\text{-}84)$$

which simplifies to

$$C_V dT = (nRT/P)dP - nRdT \quad (3\text{-}85)$$

Equation (3-85) can be rearranged to give

$$(C_V + R)dT = C_P dT = VdP \quad (3\text{-}86)$$

The equation of hydrostatic equilibrium gives the pressure decrease (dP) with increasing altitude (dz) in a planetary atmosphere:

$$dP = -\rho g dz \quad (3\text{-}87)$$

where ρ is the mass density of the gas parcel. Substituting for dP using Eq. (3-87) we get

$$C_{P,m} dT = -V_m \rho g dz \quad (3\text{-}88)$$

per mole of gas. Rearranging to solve for the variation of temperature with altitude yields

$$\frac{dT}{dz} = -\frac{\rho g V_m}{C_{P,m}} = -\frac{\mu g}{C_{P,m}} = -\frac{g}{c_p} \quad (3\text{-}89)$$

where μ is the molecular weight, c_p is the specific heat of the gas, and g is the gravitational acceleration. Equation (3-89) is the *dry adiabatic lapse rate* (K km^{-1}). On Earth g increases only 0.5% from the equator to poles. The variation of g with altitude z in kilometers is given by

$$g_z = g_0\left[1 - \left(3.14 \times 10^{-4}\right)z\right] \quad (3\text{-}90)$$

(for $z << R_E = 6371$ km), and where g_0 is the value at sea level and 45° latitude (9.806 m s^{-2}).

Example 3-7. Venus's atmosphere is mainly CO_2 (96.5 volume %) and N_2 (3.5 volume %) with trace amounts of other gases. Venusian "air" has $\mu = 43.44$ g mol^{-1} and behaves as an ideal gas. The observed variation of temperature and pressure with altitude in the lowest 10 kilometers of the atmosphere is listed in Table 3-5. The average acceleration due to gravity is 8.87 m s^{-2} and is taken as

Table 3-5 Thermal Structure of Venus's Atmosphere

Altitude (z, km)[a]	T (K)	P (bar)
0	735.3	92.1
2	720.2	81.09
4	704.6	71.20
6	688.8	62.35
8	673.6	54.44
10	658.2	47.39

[a] *0 km taken as 6052.0 km radius.*

constant with latitude and altitude in this example. Calculate the dry adiabatic lapse rate (K km^{-1}) in Venus's near-surface atmosphere.

We need to evaluate Eq. (3-89), and we have all the necessary information except the mean molar heat capacity ($C_{P,m}$) of Venusian atmospheric gas. This is given by the weighted average of the molar heat capacities of CO_2 and N_2 (from Table 3-4):

$$C_{P,m} = C_{P,m}(CO_2)X_{CO2} + C_{P,m}(N_2)X_{N2} = 46.152 + 7.576 \times 10^{-3}\ T - 11.309 \times 10^5\ T^{-2} \tag{3-91}$$

At the average surface temperature of 735.3 K, $C_{P,m}$ for Venusian gas is 49.631 J mol^{-1} K^{-1} and the dry adiabatic lapse rate is

$$\frac{dT}{dz} = -\frac{\mu g}{C_P} = -\frac{(43.44\ \text{g mol}^{-1})(8.87\ \text{m s}^{-2})}{(49.631\ \text{J mol}^{-1}\ K^{-1})} = -7.76\ \text{K km}^{-1} \tag{3-92}$$

The observed temperature gradient from 0 km to 10 km (−7.73 K km^{-1}) is nearly identical to this.

V. OTHER TYPES OF WORK

An important point is that the first law (Eq. 3-11) is perfectly general and can be modified to include other types of work terms. Until now we have focused on *PV* work because this is one of the most common (if not the most common) types of work that we encounter in natural processes involving gases, liquids, melts, and solids. However, a significant number of processes of interest also involve other types of work. Table 3-6 lists some of the common types of work that are often included in the δw term in Eq. (3-11). The infinitesimal amount of work done in each case is the product of an intensive variable (e.g., pressure, force, voltage, magnetic field strength) and the differential of an extensive variable (e.g., volume, length, electrical charge, or magnetization of a solid).

For example, cloud condensation in Earth's atmosphere, any of the Jovian planets (Jupiter, Saturn, Uranus, Neptune), or Saturn's satellite Titan involves chemical work due to the phase change from vapor to liquid or vapor to solid and must be included in Eq. (3-11). The latent heat of vaporization released by condensation of water vapor to water droplet clouds or the latent heat of sublimation released by condensation of water vapor to water ice clouds warms the surrounding air and alters the

Table 3-6 Types of Work

Type	Intensive Variable	Extensive Variable	Differential Term for First Law
pressure – volume	pressure, P	volume, V	PdV
mechanical	force, F	distance, r	$F(\cos\theta)dr$
gravitational	weight, mg	height, h	$mgdh$
electrochemical	voltage, ε	charge, Z	εdZ
magnetic	magnetic field strength, $\mathbf{H}$	magnetization, $\mathbf{M}$	$\mu_o \mathbf{H} d\mathbf{M}$
chemical (phase change)	latent heat of sublimation, L_S	moles, n	$L_S dn$
	latent heat of vaporization, L_V	moles, n	$L_V dn$

thermal profile of Earth's troposphere from the dry adiabatic lapse rate. Likewise, the latent heat effects due to condensation of ammonia ice clouds in the atmospheres of the Jovian planets or due to condensation of liquid CH_4 droplet clouds in Titan's atmosphere alters the thermal profiles of the atmospheres of these objects. These effects are incorporated into Eq. (3-11) by including new work terms that give the chemical work done due to formation of dn moles of icy cloud particles from condensable vapor ($L_s dn$) or due to formation of dn moles of liquid cloud droplets from condensable vapor ($L_v dn$). The L_s and L_v terms are the latent heat (enthalpy) of sublimation and the latent heat (enthalpy) of vaporization. In general, adiabatic expansion and cooling of rising atmospheric gas containing a condensable vapor (e.g., water vapor, NH_3, or CH_4) leads to a warmer thermal profile known as a *wet adiabat* or a *wet lapse rate*.

A second example is the electrochemical work done by passage of an electrical charge dZ through a galvanic cell (e.g., a battery) with a voltage ϵ ($\delta w = \epsilon dZ$). We discuss electrochemical work in Chapter 10, where we describe chemical equilibria. A third example is the magnetic work done when a magnetic solid, such as a piece of iron metal, is placed into a magnetic field. We describe magnetic work in Chapter 4, where we discuss magnetic heat capacity, and in Chapter 9, where we discuss the third law of thermodynamics and methods of producing low temperatures. A fourth example is the work done in lifting a mass m to a height h in the gravitational field of Earth or another planet with gravitational acceleration g ($\delta w = mgdh$). This has important consequences for the decrease of atmospheric pressure with altitude in planetary atmospheres and for the isostatic compensation of loads (e.g., ice sheets and mountains) on planetary crusts. Finally, centrifugal work ($\delta w = -\omega^2 rdr$, where ω is the angular velocity in radians per second and r is the distance from the rotational axis) can also be considered. Centrifugal work is important for isotopic and molecular separation processes.

PROBLEMS

1. Show that Eq. (3-55) given the work for a reversible isothermal expansion (or compression) can also be written as $w = nRT \times \ln(P_1/P_2)$.
2. A rollercoaster car is pulled 10 m along a track that angles upward from level ground at 45°. If the car and its two passengers have a total mass of 180 kg, how much work is done to raise the car?
3. A candy bar contains 280 kcal (one dietary calorie is equal to one chemical kilocalorie) of energy. (a) Convert this value to joules. (b) If a 140 lb. person ate the candy bar and converted all of its energy into work, how high could she climb? There are about 4.45 newtons per pound.
4. Annual electricity consumption in the United States was about 1.1 trillion kilowatt hours in 2001 (*Wall Street Journal*, 8/16/01). How many joules is this?
5. How many joules does a 60-watt light bulb consume in one year if it is on continuously?
6. The experimental conditions in one of Joule's paddle wheel experiments, in which falling weights turned paddles inside an insulated water-filled tank, were as follows: total mass of weights = 26.22 kg, height of fall = 1.524 m, number of times weights fell = 21, increase in temperature of water = $\Delta T = 0.313$ K, and heat capacity of water and tank = 6305 cal K^{-1}. Use these data to calculate the mechanical equivalent of heat (joules per calorie).
7. In 1895, while dedicating a statue of Joule in the Manchester, England, city hall, Lord Kelvin told the story of meeting Joule and his new bride in Martigny, Switzerland, where Joule tried to measure the temperature increase at the bottom of a waterfall. Assuming that the

constant-volume specific heat of liquid water is 4.178 J g^{-1} K^{-1} and that the process is adiabatic, calculate the increase in temperature caused by water falling 100 meters. Assume $g = 9.81$ m s^{-2}.

8. Figure 3-9 illustrates a cycle on a *PV* diagram. An ideal gas (the system) undergoes an isobaric expansion from point *A* to point *B*. Then the pressure drops from point *B* to point *C* while a constant volume is maintained. The gas is then compressed isobarically back to its original volume between point *C* and point *D*. Finally, the pressure is increased from point *D* to point *A*, while volume remains constant. Find the changes in internal energy, heat, and work for each step and for the cycle for the ideal gas. All steps are performed reversibly. The temperature at point *A* is 298.15 K. The constant-pressure and constant-volume heat capacities of the gas are $5/2R$ and $3/2R$, respectively.

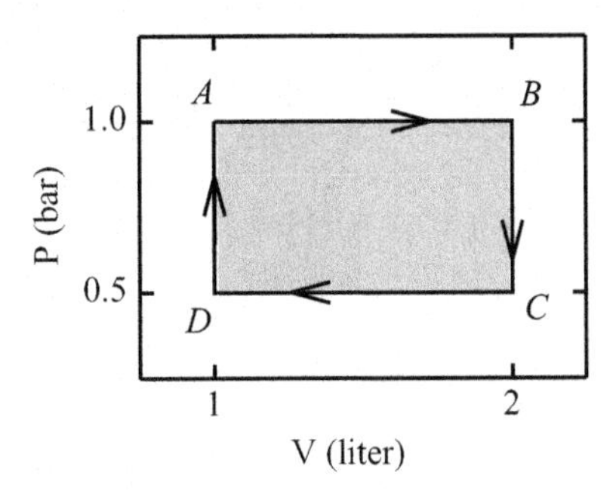

FIGURE 3-9

The indicator diagram for Problem 8.

9. Calculate C_P for air at 1500 K.

10. Calculate the final temperature, ΔE, ΔH, q, and w for the reversible adiabatic expansion of one mole of air originally at 300 K and one bar that is expanded to a pressure of 0.4 bar. You can assume that the heat capacities remain constant with temperature.

11. The enthalpy change for the reaction

$$H_2O \text{ (liquid)} = H_2O \text{ (gas)}$$

at 100°C and 1 atmosphere (i.e., the normal boiling point of water) is $\Delta H = 40{,}656.3$ J mol^{-1}. The specific volumes (i.e., per gram) of liquid water and steam are 1 cm^3 and 1,677 cm^3, respectively, under these conditions. Calculate the work done, ΔE, and ΔH, assuming ideality and neglecting the volume of the liquid (i.e., $dV \cong$ volume of the gas) per mole of water vaporized.

12. The enthalpy change for the reaction

$$H_2O \text{ (ice)} = H_2O \text{ (liquid)}$$

at 0°C and 1 atmosphere (i.e., the normal freezing point of water) is $\Delta H = 6009.5$ J mol^{-1}. The specific volumes of ice and liquid water are 1.091 cm^3 g^{-1} and 1.000160 cm^3 g^{-1}, respectively, under these conditions. Calculate w and ΔE for ice melting.

13. In one of Count Rumford's experiments, the work equivalent to that done by one horse in 2.5 hours was sufficient to boil 26.58 pounds of water originally at 32°F. Use Watt's estimate that the power of one horse is equivalent to 33,000 foot pounds per minute to calculate the mechanical equivalent of heat from Rumford's experiment. Express your answer as joules per calorie. The specific heat of water at room temperature is 4.1801 J g^{-1} K^{-1}.

14. Show that $(\partial H/\partial V)_T = 0$ for an ideal gas.

15. As stated in Section III-D, an adiabat is steeper than an isotherm. Prove this for yourself by deriving (dP/dV) for an isotherm $(PV = C)$ and an adiabat $(PV^{\gamma} = C)$ where C is a constant and $\gamma = C_P/C_V$. Show your work.

16. How much energy is required to lift 1 km^3 of magma ($\rho = 2.77$ g cm^{-3}) from 60 km depth to Earth's surface? You can neglect the small variation in g over this depth. For reference, the 1952 eruption of Kilauea volcano in Hawaii released ~1.8×10^{17} J of energy (MacDonald 1972).

17. Neptune's atmosphere is convective with an adiabatic temperature gradient below the 80 millibars ($T = 52$ K) level. Calculate the value of the dry adiabatic gradient $(\partial T/\partial z)$ at 400 K, assuming ideality, an average $g = 11.0$ m s^{-2}, and constant $C_{P,m} = 29.22$, 20.79, and 40.32 J mol^{-1} K^{-1} for H_2, He, and CH_4, which are 80%, 19%, and 1% of its atmosphere, respectively.

18. Calculate ΔE, ΔH, q, and w for the reversible adiabatic expansion of Venusian atmospheric gas from 735 K, 92.1 bars (0 km altitude) to 47.4 bars (10 km altitude). Also calculate T at 10 km. You can assume ideality and use a constant $C_{P,m} = 49.0$ J mol^{-1} K^{-1}.

19. Meteorologists define the potential temperature (θ) as the temperature that a dry air parcel would have if it were adiabatically compressed (or expanded) to 1000 millibars pressure

$$\theta = T\left(\frac{1000}{p}\right)^{R/C_P} \quad (3\text{-}93)$$

where p is the pressure in millibars (mb) and $R/C_P = 0.285$ for dry air. Calculate the potential temperature of a dry air parcel at 2 km altitude (2°C, 795 mb) in Earth's atmosphere.

20. The adiabatic exponent γ can be measured from the reversible adiabatic expansion (or compression) of a gas. Calculate γ for CO_2 from the change in state as follows:

$$CO_2 \text{ (2 bars, 300 K)} \rightarrow CO_2 \text{ (0.75 bar, 240 K)}$$

CHAPTER

4 Thermal Properties of Pure Substances and Some Applications

Les atomes de tous les corps simples ont exactement la même capacité pour la chaleur. *(The atoms of all simple bodies have exactly the same capacity for heat.)*

—Alexis Petit and Pierre Dulong (1819)

The constant-pressure and constant-volume heat capacities (C_P and C_V), and enthalpy (H) were introduced and defined in Chapter 3. In this chapter, we discuss these functions in more detail because heat capacity and enthalpy, along with entropy, introduced in Chapter 6, are fundamental thermal properties that are used to characterize the behavior of materials during heating, cooling, and phase changes. Heat capacity and entropy also provide information about the structure and behavior of materials at the atomic or molecular level.

This chapter is divided into three sections. In Section I, we review some basic concepts and definitions, discuss different processes that contribute to the heat capacity of a material, present the basic equations that relate to heat capacity and enthalpy, and discuss empirical polynomial equations used to represent enthalpy and heat capacity. Section II gives an overview of *calorimetry,* which is the measurement of heat effects due to change of state, chemical reaction, or formation of a solution. We review different types of calorimeters used to measure enthalpies and heat capacities. In Section III, we discuss thermal properties of solids and liquids, including organics, water, molten metals, oxides, and silicates. We review common types of heat capacity anomalies exhibited by solids, discuss calculation of heat capacity and enthalpy for these anomalies, discuss calculation of heat capacity and enthalpy for phase transitions such as solid-state phase transitions, melting, and vaporization, and discuss methods for estimating heat capacity and enthalpy of solids when no experimental data are available.

I. SOME BASIC CONCEPTS ABOUT HEAT CAPACITY AND ENTHALPY

A. Mean and true heat capacity

In Chapter 3, we defined the heat capacity (C) of a substance as the ratio of heat added to the substance divided by the resultant temperature change. Equation (3-39) is actually the definition of the mean heat capacity ($\overline{C}$):

$$\overline{C} = \frac{q}{\Delta T} \tag{3-39}$$

Copyright © 2013 Elsevier Inc. All rights reserved.

The mean heat capacity approaches the true heat capacity (C), given by Eq. (3-40):

$$C = \frac{\delta q}{dT} \tag{3-40}$$

as the heat added $q \rightarrow 0$ and the temperature interval $\Delta T \rightarrow 0$ (i.e., as both q and ΔT become infinitesimally small). The correction of the mean heat capacity, which is the quantity that is usually measured, to the true heat capacity (either C_V or C_P), which is the desired quantity, is done during the data analysis. The corrections are described in McCullough and Scott (1968).

B. Heat capacity contributions

The heat capacity of a material is generally due to the uptake of kinetic energy by its atoms and molecules. In most cases, the heat capacity (C_P) of a material is the sum of two contributions. One contribution is the constant-volume heat capacity C_V. The other contribution is that due to thermal expansion, which is simply ($C_P - C_V$).

Gases

In Chapter 3 we showed that the constant-volume heat capacity C_V is defined as

$$C_V = \left(\frac{\delta q}{dT}\right)_V = \left(\frac{\partial E}{\partial T}\right)_V \tag{3-42}$$

Thus, the translational, rotational, vibrational, and electronic energy levels of a gas all contribute to its heat capacity, which can be written as

$$C_P = (C_P - C_V) + C_{trans} + C_{rot} + C_{vib} + C_{el} \tag{4-1}$$

We now analyze these different contributions to the total heat capacity. From Chapter 3 we know that $(C_P - C_V) = R$ for one mole of an ideal gas, and the data tabulated in Table 3-4 show this relation is valid within a few percent for common gases. Translational heat capacity C_{trans} is due to the three-dimensional motion of a gas through space and is $1.50R$ per mole (~ 12.5 J mol^{-1}). The rotational, vibrational, and electronic heat capacity contributions are temperature dependent because the spacing between the rotational, vibrational, and electronic energy levels of a gas are different. These contributions to the heat capacity of an ideal gas are calculated from its energy levels and molecular geometry using statistical mechanics and quantum mechanics. We are not concerned with these computations here but note that for most monatomic and diatomic gases (with exceptions noted below) the spacing between the rotational, vibrational, and electronic energy levels corresponds to temperatures of ~ 10 K (ΔE_{rot}), ~ 1000 K (ΔE_{vib}), and about 10,000 K (ΔE_{el}), respectively, at which each contribution becomes important.

Using this generalization and the principle of equipartition of energy, we can estimate the heat capacities of monatomic and diatomic gases at room temperature and compare the results to the data in Table 3-4. Using Eq. (4-1), we estimate $C_{P,m}/R = 2.50$ (or $C_{P,m} \sim 20.79$ J mol^{-1}) for monatomic gases because we expect that only C_{trans} and ($C_P - C_V$) will contribute to the heat capacity at room temperature. In fact, this is the case for the noble gases (see Table 3-4) and many other monatomic gases (e.g., Br, Hg, I, Na, and K, to name just a few). But other monatomic gases have larger C_P values at room temperature because of low-lying electronic energy levels (e.g., Co, F, Fe, Ge, Nb), with Ge (g) having $C_{P,m}/R \sim 3.70$ at 298 K.

Diatomic gases can rotate about two axes perpendicular to their atomic bond. Thus, C_P for a diatomic gas also includes a rotational heat capacity contribution $C_{rot} = 2(\frac{1}{2}R) = R$. We estimate $C_{P,m}/R = 3.50$ ($C_{P,m} \sim 29.10$ J mol^{-1} K^{-1}) at room temperature for a diatomic gas with translational and rotational contributions to its heat capacity. Hydrogen has a slightly lower $C_{P,m}$ at room temperature and is an exception. Hydrogen is also notable because it behaves as two different gases, known as *ortho-*H_2 and *para-*H_2, at cryogenic temperatures. The two forms of hydrogen have parallel nuclear spins (ortho-H_2) or antiparallel nuclear spins (para-H_2). The rotational energy level separation in H_2 is comparable to the thermal energy at cryogenic temperatures due to the low moment of inertia of the H_2 molecule. However, only odd rotational energy levels ($J = 1, 3, 5, \ldots$) are filled in ortho-H_2, whereas only even rotational energy levels ($J = 2, 4, 6, \ldots$) are filled in para-H_2. As a consequence, ortho-H_2 and para-H_2 have different physical properties, including vapor pressure, boiling point, triple point, thermal conductivity, heat capacity (C_p), entropy (S), and spectra.

The typical spacing between vibrational energy levels indicates that the vibrational heat capacity contribution should not be important until ~1000 K for most diatomic gases. A number of diatomic molecules (e.g., CO, N_2, HF, HCl, HBr, HI), including several listed in Table 3.4, follow our expectations. However, other diatomics, notably Cl_2, Br_2, and I_2, have larger heat capacities because of low-lying vibrational energy levels ($\Delta E_{vib} \sim 810$, 465, and 310 K, respectively). The electronic heat capacity is important at room temperature for NO with one unpaired electron and O_2 with two unpaired electrons because of the very low-lying electronic energy levels ($\Delta E_{el} \sim 10$ and 175 K, respectively) in these two molecules.

The heat capacities of triatomic and polyatomic gases generally have significant C_{vib} terms, and we cannot estimate their heat capacities with any reliability. However, at ambient pressure, the measured heat capacities of many inorganic and organic gases are only about 1–2% different than those calculated for the ideal gas. Thus, the computed values suffice for many applications.

Liquids

The heat capacity (C_P) of liquids (e.g., water, other solvents, liquid metals, molten oxides, and silicates) is basically the sum of the constant-volume heat capacity C_V and the thermal expansion term ($C_P - C_V$). The heat capacity of liquid metals and of molten oxides and silicates is apparently constant with temperature. However, more accurate data on water, liquefied gases, and several elements that melt at relatively low temperatures show that the heat capacity of liquids usually displays more complex behavior. The C_P of Hg and several other liquid elements (Au, Bi, Ga, In, Li, Pb, Rb) is highest at the melting point and decreases toward the boiling point. In contrast, the C_P of liquefied gases such as Ar, H_2, N_2, Ne, and Xe is lowest at the melting point and increases toward the boiling point. The C_P curves of water, liquid Cs, K, Na, and O_2 are concave shaped, whereas that of liquid S is convex shaped. In general, the variations in C_P of a liquid are small relative to the heat capacity, so the assumption of a constant C_P is adequate as a first approximation.

One final point deals with the heat capacity of volatile liquids, such as alcohol or gasoline, that evaporate easily because of their high vapor pressure. Measurements of the heat capacity of volatile liquids do not give C_P or C_V but instead yield C_σ, which is the heat capacity along the saturation curve. The reason for this is that the pressure and volume of a volatile liquid vary with temperature. The difference between C_P and C_σ is generally small, and $(C_P - C_\sigma)/C_\sigma$ is ~1% at the triple point and

~25% near the boiling point of volatile liquids. In other words, $C_P - C_\sigma$ is only significant when the vapor pressure of a liquid is large. We can use n-heptane (C_7H_{16}), which is a constituent of gasoline, as an example to illustrate this point. At 0°C, the vapor pressure of n-heptane is 0.0153 bar and its $C_{P,m} = C_{\sigma,m} = 216.0$ J mol^{-1} K^{-1}. At 60°C the vapor pressure is 0.2765 bar, $C_{P,m} = 238.9$ and $C_{\sigma,m} = 238.8$ J mol^{-1} K^{-1}, which is a 0.04% difference. At 100°C the vapor pressure is 1.060 bars, $C_{P,m} = 257.1$ and $C_{\sigma,m} = 256.8$ J mol^{-1} K^{-1}, which is a 0.12% difference. The distinction between C_P and C_σ for liquid metals, molten oxides, and silicates is negligible until very high temperatures.

Solids

The heat capacity (C_P) of solids is the sum of several different contributions that vary with temperature and with the type of material (metals such as iron, semiconductors such as pyrite, and insulators such as forsterite). In general, the heat capacity of a solid can be written as

$$C_P = C_{lat} + (C_P - C_V) + C_E + C_M + C_\lambda + C_{Sch} \tag{4-2}$$

although not every component contributes to the heat capacity of every solid. The *lattice heat capacity* (C_{lat}) is generally the largest term. It arises from vibrations of the atoms (or molecules) that make up a solid. To first approximation, $C_{lat} \sim 3R$ per gram atom (i.e., per mole of atoms) for solids at room temperature, with notable exceptions being B, Be, diamond, and graphite, which have much smaller C_{lat} values. The thermal expansion term ($C_P - C_V$) is the next most important contribution to the heat capacity for most nonmetals and is ~5% of C_P at room temperature. The *electronic heat capacity* (C_E) or electronic component of solids generally does not become significant until high temperatures, but metals have significant electronic heat capacities at low temperatures. Other contributions to the heat capacity of solids can be made by magnetic transitions (C_M), ordering and disordering effects in crystal lattices (C_λ), and transitions between electronic energy levels with $\Delta E \sim kT$ in compounds with *d* and *f* electrons such as those formed by transition metals, lanthanides, and actinides. These transitions are known as *Schottky transitions,* and the Schottky heat capacity contribution is C_{Sch}. Schottky transitions generally occur at temperatures close to absolute zero and are difficult to identify. For example, Krupka et al. (1985) found that bronzite ($Mg_{0.85}Fe_{0.15}SiO_3$) has a heat capacity anomaly at 10–15 K. Other measurements ruled out a magnetic transition at these temperatures, and the anomaly can be fit with transitions between electronic energy levels of the *d* electrons in iron. Thus, it is believed to be a Schottky anomaly. We will discuss some of these heat capacity contributions in more detail in Section III.

C. The relationship between heat capacity and enthalpy

Most heat capacities are measured under constant pressure and are thus C_P values. In some cases, C_V values of gases are measured. We will review these measurements when we discuss the thermal properties of gases. We recall that C_P is defined by Eq. (3-44):

$$C_P = \left(\frac{\delta q}{dT}\right)_P = \left(\frac{\partial H}{\partial T}\right)_P \tag{3-44}$$

Rearranging and integrating gives this expression for the corresponding enthalpy difference:

$$\int_{T_1}^{T_2} dH = (H_{T_2} - H_{T_1}) = \int_{T_1}^{T_2} C_P dT \tag{4-3}$$

between any two temperatures. As discussed shortly, enthalpies, rather than heat capacities, of solids are often measured at high temperatures. It is then convenient to rewrite Eq. (4-3) as

$$\frac{d}{dT}(H_{T_2} - H_{T_1}) = C_P \tag{4-4}$$

Equations (4-3) and (4-4) are general equations for all materials at high or low temperatures. The temperature T_1 is a standard reference temperature, which is 0 K for C_P data below 298.15 K and is 298.15 K for C_P data above 298.15 K. (Throughout the rest of this book we will simply write 298 K; however, this should be interpreted as 298.15 K, and all our computations are done with respect to 298.15 K.) Sometimes T_1 is taken as 273.15 K for C_P data obtained using ice calorimeters, which are described in Section II. The $(H_{T_2} - H_{T_1})$ function is generally written as $(H_T - H_{T_{ref}})$ and is called *enthalpy*, or the *enthalpy function* or the *heat content*. The latter term is mainly found in older literature and should be avoided because of the implied equality between ΔH and q, which only holds at constant pressure, as shown by Eq. (3-38).

In principle, the C_P integral in Eq. (4-3) is easily evaluated. In practice, the evaluation becomes complex if the C_P data are at low temperatures, one or more phase changes occur, and/or heat capacity contributions due to one or more of the factors we've discussed are important within the temperature range considered. We illustrate the use of Eq. (4-3) next.

Example 4-1. Table 3-4 gives this Maier-Kelley fit for $C_{P,m}$ of dry air from 298 K to 2000 K:

$$C_{P,m} = 26.917 + 5.899 \times 10^{-3}T + 0.27 \times 10^5 T^{-2} \text{ J mol}^{-1} \text{ K}^{-1} \tag{4-5}$$

This equation is plotted in Figure 4-1, which also shows a subset of the data used to derive the fit. The maximum error of this fit for reproducing the molar C_P data for air is 0.7% (see Table 3-4). The enthalpy of air over the 298–2000 K range is obtained by the integration of Eq. (4-5):

$$H_T - H_{298} = 26.917T + \frac{1}{2}5.899 \times 10^{-3}T^2 - 0.27 \times 10^5 T^{-1} + d \text{ J mol}^{-1} \tag{4-6}$$

The constant d in Eq. (4-6) is evaluated from the condition that $H_T - H_{298}$ is 0 at 298 K, giving

$$H_T - H_{298} = 26.917T + 2.9495 \times 10^{-3}T^2 - 0.27 \times 10^5 T^{-1} - 8{,}197 \text{ J mol}^{-1} \tag{4-7}$$

rounded to the nearest joule, for the enthalpy function of dry air, valid from 298 K to 2000 K. Equation (4-7) is plotted in Figure 4-2 along with a subset of the enthalpy data of Hilsenrath et al. (1955). The fit agrees very well with the tabulated enthalpy data; for example, Eq. (4-7) gives $(H_{2000} - H_{298}) =$ 57,422 J mol^{-1} versus 57,416 J mol^{-1} from the tabulated enthalpy data. This is an insignificant error of 0.01%, which is much less than the error of the C_P fit itself.

D. Empirical heat capacity equations

As mentioned in Chapter 3, Maier-Kelley equations (or the modified four-term Maier-Kelley equations) are only approximate. They typically have errors less than 1%, but sometimes the errors may be a few percent (as for CH_4 in Table 3-4). In general and as illustrated by Example 4-1, the error in the enthalpy calculated from the integrated form of the Maier-Kelley equation is significantly smaller than that for the C_P fit itself. As a consequence, very accurate enthalpy measurements (e.g., by drop calorimetry, which we describe in a moment) are needed to obtain accurate C_P data at high temperatures. Whether or not C_P data are measured directly, more complex polynomials are needed to represent heat capacity data

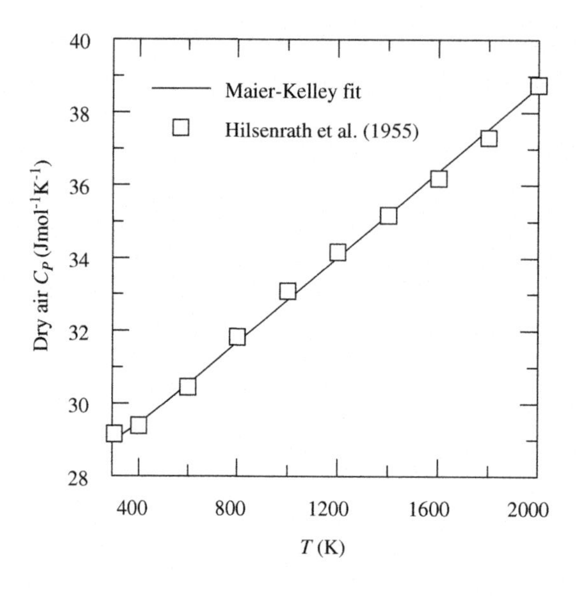

FIGURE 4-1

A comparison of a Maier-Kelley fit for the molar C_P of dry air with a subset of the data from Hilsenrath et al. (1955) used to derive the fit. See Example 4-1.

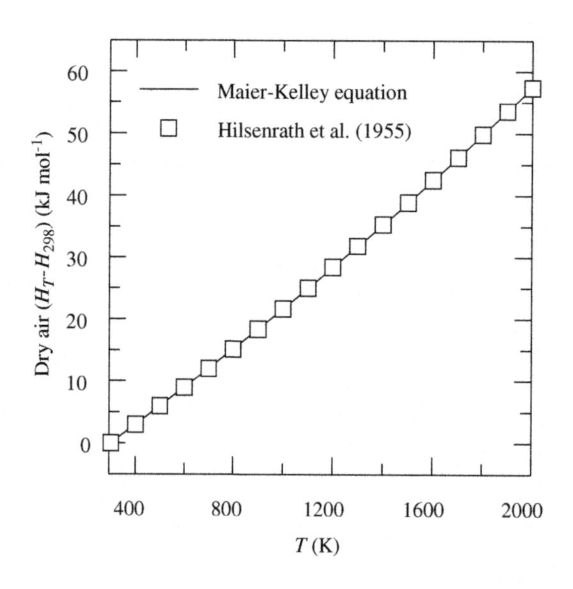

FIGURE 4-2

A Maier-Kelley equation for the molar enthalpy ($H_T - H_{298}$) of dry air is compared to a subset of the enthalpy data from Hilsenrath et al. (1955). The enthalpy equation is the integral of the C_P equation and is not derived from the enthalpy data. See Example 4-1.

with greater accuracy, especially over large temperature ranges. Thermodynamic data compilations use polynomials with six or more coefficients to represent C_P data. However, the Maier-Kelley equations have two advantages that the more accurate fits do not have—namely, they are easy to use in examples and problems, and they can be extrapolated to somewhat higher temperatures beyond the fit range. Before discussing thermal properties of solids in more detail, we give an overview of calorimetric measurements of heat capacity and enthalpy.

II. AN OVERVIEW OF CALORIMETRY

A. Basic definitions and concepts

The word *calorimeter* was coined in the 1780s by Antoine Lavoisier (1743–1794), the French chemist who founded modern chemistry. Lavoisier derived the term *calorimeter* from two words: *calor*, the Latin word for heat, and *metron*, the Greek word for measure. A calorimeter is an instrument used to measure the heat involved in a known *change of state* of a material. The change of state may be a change in temperature, pressure, stable phase, chemical composition due to some reaction, or any other change that involves heat transfer. Here we discuss calorimetric measurement of heat effects due to temperature and/or phase changes. In Chapter 5, we discuss calorimetric techniques used to measure heats (enthalpies) of chemical reactions.

Calorimetric measurements are either done directly or indirectly. The direct measurements are done by comparison with a known amount of *electrical energy* in joules, given by

$$\textit{Electrical energy}\ (\mathrm{J}) = \textit{volts}\ (\mathrm{V}) \times \textit{current}\ (\mathrm{A}) \times \textit{time}\ (\mathrm{s}) = \textit{watts}\ (\mathrm{W}) \times \textit{time}\ (\mathrm{s}) \tag{4-8}$$

$$\textit{Electrical energy}\ (\mathrm{J}) = [\textit{current}\ (\mathrm{A})]^2 \times R\ (\Omega) \times \textit{time}\ (\mathrm{s}) \tag{4-9}$$

The volt (V), the ampere (A), the ohm (Ω), and the watt (W) are the SI units for electric potential (EMF), electric current, resistance, and power, respectively. The ampere is a basic SI unit. Volts, ohms, and watts are derived SI units, related to each other as follows: $\mathrm{V} = \mathrm{W} \times \mathrm{A}^{-1}$, $\Omega = \mathrm{V} \times \mathrm{A}^{-1}$, and $\mathrm{W} = \mathrm{J} \times \mathrm{s}^{-1}$. The following example illustrates the use of Eqs. (4-8) and (4-9).

Example 4-2. A calorimeter is heated by passing a current of 0.075 amperes through its heater ($\Omega = 100$ ohms). How much energy is delivered in 30 seconds time, how much electrical power is this equivalent to, and what is the voltage drop across the heater?

(a) The electrical energy delivered is given by Eq. (4-9):

$$\textit{Energy} = (0.075\ \mathrm{A})^2(100\ \Omega)(30\ \mathrm{s}) = 16.875\ \mathrm{J} \tag{4-10}$$

(b) The electrical power is given by electrical energy per unit time and is

$$\textit{Power} = (16.875\ \mathrm{J})/(30\ \mathrm{s}) = 0.5625\ \mathrm{W} \tag{4-11}$$

The voltage drop across the heater is calculated by rearranging Eq. (4-8) or from one of the relationships between the derived SI units. The latter approach gives

$$\textit{Voltage} = (0.5625\ \mathrm{W})/(0.075\ \mathrm{A}) = (100\ \Omega)(0.075\ \mathrm{A}) = 7.50\ \mathrm{V} \tag{4-12}$$

Indirect calorimetric measurements are done by comparison with a material with well-known thermal properties, such as ice, liquid water, or corundum (Al_2O_3). Both direct and indirect measurements require corrections, called *heat-leak corrections*, for unwanted heat exchanges between the calorimetric system and the surroundings. The heat leak limits the accuracy of most calorimetric measurements and accounts for most of the complexity of calorimeter design.

B. Basic types of calorimeters

The subject of calorimeter design and use is important in thermodynamics, but it is too detailed and large to cover here in any detail. Some calorimeters are designed to measure heat capacity, and some are designed to measure enthalpy. Calorimeters can be classified into other categories as well, but it is not useful for us to do this here. In the next sections, we give brief descriptions of different types of calorimeters commonly used to measure the heat capacity or enthalpy of solids, liquids, and gases. In the reference list for this book we give several references that provide comprehensive and detailed descriptions of calorimeter design and use.

C. Ice calorimeters

The *ice calorimeter* was first developed in the mid-18th century and it is still used today. The ice calorimeter is an *isothermal calorimeter* because it measures the enthalpy ($H_T - H_{273}$) of a heated sample by the amount of ice melted to ice water at a constant temperature of 0°C, when the heated sample is dropped into the calorimeter. In other words, the ice calorimeter operates on the principle that latent heat causes a phase change at constant temperature. There are two key conditions for the proper operation of an ice calorimeter. First, all the heat given off by the heated sample must be transferred to the ice and used to melt it. This condition implies that very pure ice must be used in the calorimeter. Second, there cannot be another heat source or a heat sink in the system. Another heat source would melt more ice, whereas a heat sink would freeze some of the ice water. In both cases the enthalpy measurement would be incorrect by the amount of heat from the extraneous heat source (or sink).

Joseph Black used an ice calorimeter in his research on specific heats. His ice calorimeter was basically a block of ice into which he dropped hot objects. Lavoisier and his friend, the mathematician Pierre Simon Laplace (1749–1827), developed an improved version of Black's ice calorimeter and used it to measure animal heat, specific heats, and heats of chemical reactions.

Their ice calorimeter, illustrated in Figure 4-3, consists of three containers, one inside the other. The innermost container is the sample container, or receiver. It is made of iron wire mesh. Lavoisier and Laplace placed heated objects or small animals in the receiver. The heat from the samples or small animals is transferred to the surrounding container. This is an ice-filled vessel, which is the actual calorimeter. In turn, the calorimeter is completely surrounded by the ice jacket, which is an even larger container filled with ice. Both of the ice-filled containers have metal screen–covered outlets at the bottom to retain pieces of ice while letting out the ice water formed by melting. The ice water that's formed is weighed. The enthalpy of the heated object or the body heat of the animal is calculated from the known latent heat of fusion of ice and the mass of ice melted to water. The principles involved in the operation of this type of ice calorimeter and some of the calculations involved are illustrated in the following hypothetical example.

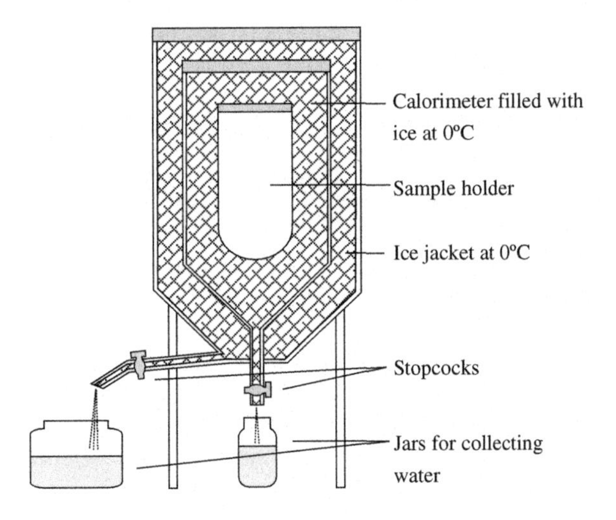

FIGURE 4-3

A schematic diagram of the Lavoisier-Laplace ice calorimeter.

Example 4-3. Suppose we want to measure the molar enthalpy ($H_{873} - H_{273}$) of copper metal. We take a piece of high-purity copper metal that weighs exactly 0.200 kilograms and heat it to exactly 600°C in a furnace. (We give the furnace temperature in degrees Celsius because the temperature controllers and sensors on most furnaces read in degrees Celsius, not Kelvins.) The sample is heated in an inert atmosphere such as high-purity argon or nitrogen to prevent any oxidation of the copper metal. Alternatively, the sample can be heated in air if it is hermetically sealed in a container that does not react with copper and has very well-known thermal properties. In the latter case the empty container is dropped into the calorimeter to determine its enthalpy prior to dropping the sample plus container into the calorimeter. We also have to make sure the entire piece of metal is heated to exactly 600°C. In other words, the sample has to be inside the isothermal hot zone in the furnace, without any temperature gradients. After we are sure that the entire piece of Cu is thermally equilibrated at 600°C, we rapidly drop it into the ice calorimeter. This is done by dropping the sample after pulling aside metal shutters at either end of a tube connecting the furnace and calorimeter. The shutters are radiation shields that prevent ice in the calorimeter being melted by heat from the furnace.

The hot copper causes some of the ice to melt to water at 0°C. The liquid water flows out the tube at the bottom of the calorimeter and is weighed. In this experiment 148.644 grams of ice melts to water. The enthalpy of the 0.200 kg of copper metal is calculated as follows:

$$(H_{873} - H_{273}) = \Delta_{fus}H_{ice} \times \frac{m_{ice}}{18.015} = 6009.5 \times \frac{148.644}{18.015} = 49.585 \text{ kJ} \quad (4\text{-}13)$$

where $\Delta_{fus}H$ is the heat of fusion of ice (6009.5 J mol^{-1}), and 18.015 is the molecular weight of water. Equation (4-13) gives the enthalpy (in kilojoules) of 0.200 kg of copper relative to 273 K. However, the standard temperature for thermodynamic data tabulations is 298 K, and molar values are listed. In principle, we could correct for the enthalpy difference between 298 K and 273 K by dropping the same mass of copper at 298 K into the ice calorimeter and use the $(H_{298} - H_{273})$ value to calculate $(H_{873} - H_{298})$ using the atomic weight of Cu (63.546 g mol^{-1}):

$$(H_{873} - H_{298}) = (H_{873} - H_{273}) - (H_{298} - H_{273}) = 49.585 - 1.905 \text{ kJ} \quad (4\text{-}14)$$

$$(H_{873} - H_{298})_m = \frac{47.68}{(200/63.546)} = 15.149 \text{ kJ mol}^{-1} \quad (4\text{-}15)$$

where the subscript m indicates a molar value. In practice, the correction to 298 K is done mathematically using either the fit to the high-temperature enthalpy data or low-temperature heat capacity data that extend beyond 298 K.

We can determine $(H_T - H_{273})$ at different temperatures by making measurements up to the melting point of copper metal. The data are then fit to using a Maier-Kelley equation or a more complex polynomial fit. If we now modify things slightly so that the copper is completely contained inside a nonreactive container with a higher melting point, we can also determine the $\Delta_{fus}H$ for copper and the enthalpy for liquid copper up to the high-temperature limit of the furnace or sample container.

The German chemist Robert Bunsen (1811–1899) developed the ice calorimeter shown in Figure 4-4, which is named after him. The Bunsen ice calorimeter is still used today. A recent example is the measurements by Richet et al. (1982) of the enthalpies of quartz, cristobalite, and amorphous and molten silica. Bunsen's design improved on that of Lavoisier and Laplace because the amount of ice melted is determined from the volume, not the weight, of the water formed. The problem with the latter approach, which was used by Lavoisier and Laplace, is that some of the water always sticks to the ice and does not drain out of the calorimeter. In Bunsen's design, this problem is avoided because the volume change upon melting is measured by the change in the position of mercury in a calibrated capillary tube. Chapter 8 in McCullough and Scott (1968) gives a good description of the design and use of a modern Bunsen ice calorimeter.

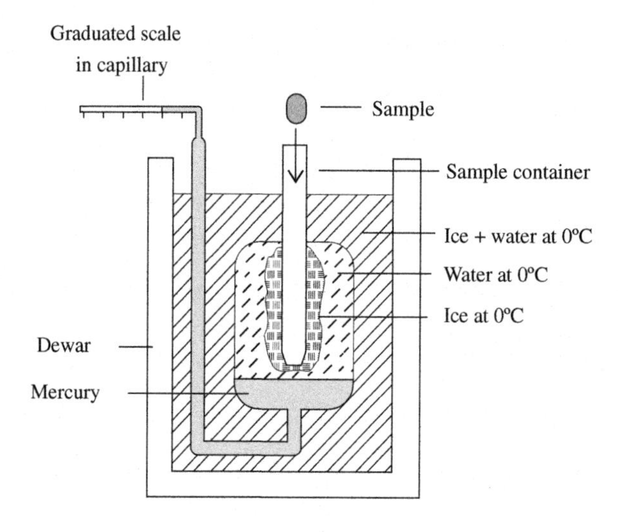

FIGURE 4-4

A schematic diagram of a Bunsen ice calorimeter.

D. Drop calorimeters

Drop calorimeters are used for enthalpy measurements at temperatures above 298 K. Drop calorimeters, like ice calorimeters, consist of two parts: a high-temperature furnace and the calorimetric assembly. The *diphenyl ether drop calorimeter* is an isothermal calorimeter and is similar to a Bunsen ice calorimeter (see Figure 4-4). The diphenyl ether melts at 26.87°C (300.02 K); the large volume increase gives a sensitivity for heat measurement about three times as great as that of the ice calorimeter. Enthalpies measured with a diphenyl ether drop calorimeter are typically referenced to 300 K. Weill et al. (1980) and Stebbins et al. (1983) used this method to measure the enthalpy of fusion of anorthite ($CaAl_2Si_2O_8$), diopside ($CaMgSi_2O_6$), albite ($NaAlSi_3O_8$), sanidine ($KAlSi_3O_8$), and nepheline ($NaAlSiO_4$).

The *copper block calorimeter*, which is shown in Figure 4-5, is another type of drop calorimeter. The copper block is also known as the *receiving calorimeter*. Typically the copper block is massive enough (about 20 kg) that its temperature does not increase by more than 5 K after a hot sample is dropped into it. The copper block is mounted on knife edges inside a sealed brass container that is continually flushed with dry gas (e.g., CO_2). The exterior of the block and the interior of the brass container are polished and gold-plated to minimize heat transfer between them, and the brass container is immersed in a constant-temperature water bath held at 298 K.

In drop calorimetery, the enthalpy of the sample is determined from the temperature increase of the copper block and the known enthalpy ($H_T - H_{298}$) of copper metal—for example, from

$$(H_{T_1} - H_{298})_{sample} \times (T_1 - 298)_{sample} = (H_T - H_{298})_{Cu} \times (298 - T_2)_{Cu} \tag{4-16}$$

in which T_1 is the sample temperature inside the furnace and T_2 is the final temperature of the copper block after the sample is dropped into it. Other metals, such as aluminum, can also be used in *metal*

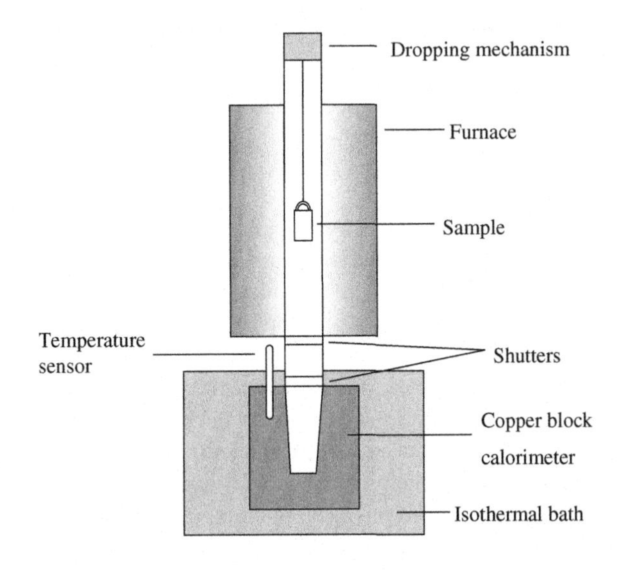

FIGURE 4-5

A schematic diagram of a copper block calorimeter. For simplicity, the brass container around the calorimeter is not shown.

block calorimeters if their thermal properties are very well known and they are good conductors of heat. Enthalpies measured with metal block calorimeters are typically referenced to 298 K. Drop calorimetry is sometimes called the *method of mixtures*. The method of mixtures uses a mixing process (i.e., dropping the hot sample into the metal block calorimeter) to measure the thermal properties of the sample in terms of the calorimeter's known thermal properties. The thermodynamics group at the U.S. Bureau of Mines used a copper block calorimeter described by Southard (1941) to measure enthalpies of a large number of minerals. Many of the fitted enthalpy data are reported by Kelley (1960).

Figure 4-6 shows (a) enthalpy and (b) heat capacity data for corundum (α-Al_2O_3) and molten alumina from 298 K to 3100 K. The curves are smoothed data derived from numerous experimental enthalpy measurements on corundum and molten alumina. There are obviously large discontinuities in both the enthalpy and heat capacity data. These occur at the melting point of corundum, which is 2327 $\pm$ 6 K. The difference between the heat contents of corundum and molten alumina at 2327 K is the enthalpy of fusion of corundum, which is computed in the following example.

Example 4-4. Calculate the enthalpy of fusion of corundum from heat content data:

$$\Delta_{fus}H\ (Al_2O_3) = (H_{2327} - H_{298})_{melt} - (H_{2327} - H_{298})_{solid} = H_{2327}(\text{melt}) - H_{2327}\ (\text{solid}) \quad (4\text{-}17)$$

Maier-Kelley fits to the tabulated enthalpies of corundum and molten alumina are

$$(H_T - H_{298})_{solid} = 118.7170\,T + 4.7517 \times 10^{-3}\,T^2 + 38.757 \times 10^5\,T^{-1} - 48{,}817\ \text{J mol}^{-1} \quad (4\text{-}18)$$

$$(H_T - H_{298})_{melt} = 192.464\,T - 82{,}016\ \text{J mol}^{-1} \quad (4\text{-}19)$$

These two equations give the enthalpy of fusion of corundum by substitution into Eq. (4-17):

$$\Delta_{fus}H\ (Al_2O_3) = (H_{2327} - H_{298})_{melt} - (H_{2327} - H_{298})_{solid} \quad (4\text{-}17)$$

$$= (365.85 - 254.83)\ \text{kJ mol}^{-1} \quad (4\text{-}20)$$

$$\Delta_{fus}H\ (Al_2O_3) = 111.02\ \text{kJ mol}^{-1} \quad (4\text{-}21)$$

which is 111.0 $\pm$ 4.2 kJ mol^{-1}, within the experimental uncertainties. The enthalpy of fusion of corundum is given by Eq. (4-17) because molten alumina crystallizes to corundum upon cooling, instead of forming a glass with different (and variable) thermal properties. Thus, the enthalpies ($H_T - H_{298}$) for corundum and molten alumina are both measured with respect to the same starting material, which is corundum at 298 K.

Figure 4-6(b) also graphically illustrates Eq. (4-3) because the area under the heat capacity curve gives the enthalpy. For example, the area under the heat capacity curve of corundum from 298 K–2327 K is equal to ($H_{2327} - H_{298}$) for corundum. Likewise, the area under the heat capacity line for molten alumina from 2327 K to 3000 K is equal to the enthalpy increment ($H_{3000} - H_{2327}$) for molten aluminum oxide.

We can now see why the error in ($H_T - H_{298}$) calculated from the integrated form of the C_P equation is significantly smaller than that for the C_P fit itself. Relatively large errors in the C_P curves of corundum or molten alumina are needed to produce large changes in the areas under them and in the enthalpies of these phases. Thus, as seen in Example 4-1, the error in ($H_T - H_{298}$) is about 10 times smaller than the error in the C_P fit. However, we can also see that defining the C_P curve by measuring

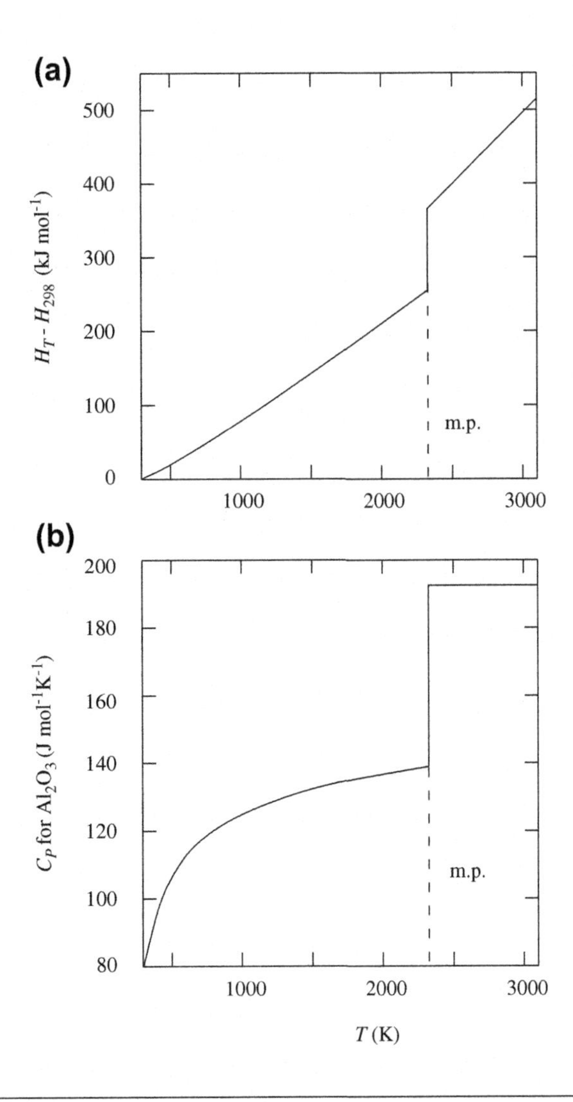

FIGURE 4-6

(a) The molar enthalpy ($H_T - H_{298}$) for corundum (α-Al_2O_3) and molten alumina. (b) The corresponding molar heat capacity ($C_{P,m}$) for corundum and molten alumina. These are smoothed data from the NIST-JANAF tables.

the area under it, that is, using drop calorimetry to measure ($H_T - H_{298}$), is very difficult to do accurately because small errors in the area (enthalpy) measurement are magnified when we use Eq. (4-4) to calculate the heat capacity.

Third, Figure 4-6(b) shows that the heat capacity of molten alumina is apparently independent of temperature. The heat capacities of other molten oxides and molten silicates are also apparently constant with temperature. However, it is quite possible that more accurate and precise measurements over larger temperature ranges will show some dependence of the heat capacity of molten oxides and silicates on temperature. As discussed in Section I-B, more accurate and precise heat capacity measurements for water, liquefied gases, and liquid metals show temperature-dependent C_P values.

Finally, Figure 4-6 shows that the heat capacity curve of corundum, like that of other solids, is steeper at lower temperatures than at higher temperatures. That is, the derivative $(\partial C_P/\partial T)_P$, which is the slope of the curve, is larger at lower temperatures. As C_P becomes a stronger function of temperature, it is much harder to get accurate C_P data from enthalpy measurements.

Drop calorimetry and ice calorimetry are very useful at high temperatures where adiabatic calorimetry (discussed below) is experimentally difficult. Drop calorimetry is versatile and gives accurate enthalpy data to very high temperatures in favorable cases (~3200 K for molten alumina, as shown). Furthermore, drop calorimetry can be done when the temperature of the receiving calorimeter is higher than that of the sample. In this variant, known as *transposed temperature-drop calorimetry,* the enthalpy change when a cooler sample is dropped into a hotter calorimeter is measured. As shown in the following example, this is a very useful method.

Example 4-5. Ziegler and Navrotsky (1986) used transposed temperature-drop calorimetry to measure the enthalpy of fusion of diopside and derived $\Delta_{fus}H = 138.5$ kJ mol^{-1}. Figure 4-7 shows their data and fits to it. Their data for the enthalpy of molten diopside give the equation

$$(H_T - H_{298}) = 332.77(\pm 8.5)T - 88{,}349(\pm 13{,}640) \text{ J mol}^{-1} \tag{4-22}$$

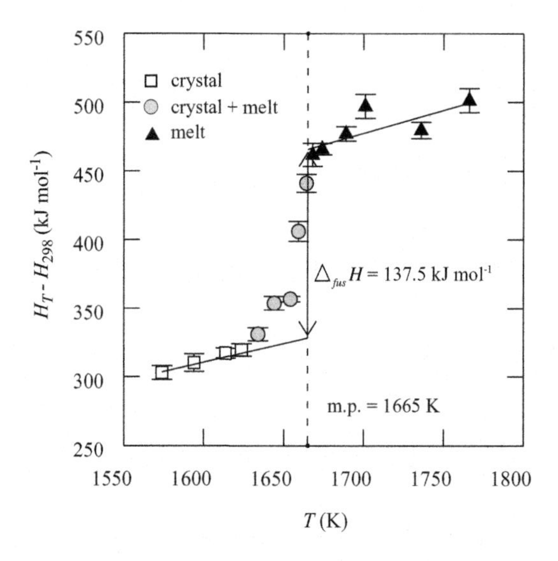

FIGURE 4-7

Enthalpy data for solid and molten diopside ($CaMgSi_2O_6$) measured by Ziegler and Navrotsky (1986) using transposed temperature-drop calorimetry. The points show the experimental data. The lines are fits to the enthalpies of the solid and melt and are from Kelley (1960) and Ziegler and Navrotsky (1986), respectively. The enthalpy of fusion of diopside is 137.5 kJ mol^{-1}, which is the difference between the enthalpy of the solid and enthalpy of the melt at the melting point of diopside (1665 K).

Modified from Ziegler and Navrotsky (1986).

Equation (4-22) gives 465.71 kJ mol^{-1} for the enthalpy of molten diopside at its melting point (1665 K). We use an equation from Kelley (1960) for the enthalpy of solid diopside:

$$(H_T - H_{298}) = 221.208T + 16.401 \times 10^{-3}T^2 + 65.856 \times 10^5T^{-1} - 89,500 \text{ J mol}^{-1} \quad (4\text{-}23)$$

that also fits the data points in Figure 4-7. At 1665 K, Eq. (4-23) gives 328.23 kJ mol^{-1} for the enthalpy of solid diopside. The difference of the two enthalpy values (465.71 – 328.23) gives $\Delta_{fus}H$ (diopside) = 137.5 kJ mol^{-1}, with an uncertainty of a few kJ mol^{-1}, like the experimental data. The arrow in Figure 4-7 shows our computation graphically. Our calculated $\Delta_{fus}H$ of diopside is very similar to a value of 137.3 kJ mol^{-1} reported by Lange et al. (1991) from a differential scanning calorimetry study of diopside melting.

Molten diopside cools to a glass and does not crystallize to solid diopside. If either ice calorimetry or drop calorimetry is used (i.e., a hot sample is dropped into a cold calorimeter), the enthalpy difference between solid diopside and glass formed from the melt during cooling must then be determined to calculate the enthalpy of fusion. We illustrate this type of calculation in Chapter 5, after we describe the calorimetric methods used to measure the ΔH difference between crystalline and glassy materials. This is not necessary in transposed temperature-drop calorimetry, where the enthalpy difference between solid diopside (at its melting point) and molten diopside (at the melting point) can be measured.

There are several disadvantages with drop calorimetric methods. Several of the problems to be avoided have already been mentioned. Perhaps the most significant disadvantage of drop calorimetry arises from the requirement that the sample must be in thermodynamically reproducible states when it is hot and cold. However, a number of silicates, including anorthite, alkali feldspars, nepheline, wollastonite, and diopside, form glasses upon cooling. Drop calorimetric measurements of enthalpies of fusion and of molten silicate enthalpies for these minerals are complicated by glass formation because the quenched melts do not have the same thermodynamic properties as the crystalline silicates. In these cases, it is also necessary to measure the enthalpy differences between the crystalline and amorphous silicates at low temperature (e.g., 298 K if a copper block calorimeter is used). The enthalpy differences are measured using solution calorimetry or another method (see Chapter 5).

A second disadvantage, which was mentioned earlier in Section I-B, is that very accurate enthalpy measurements—for example, to 0.01%—are needed to obtain C_P data to 0.1% accuracy at high temperatures. This accuracy is difficult to obtain, especially if C_P changes significantly with temperature. A third disadvantage is that any solid-state phase transition must be sufficiently rapid so that it goes to completion during the heating time in the furnace. Finally, a fourth disadvantage of drop calorimetry is that the heat effects become harder to measure as the temperatures of the heated sample and of the receiving calorimeter approach one another. This problem arises because typically $(\partial C_P/\partial T)_P$ is large and $(H_T - H_{298})$ is small for a temperature T near 298 K (as shown in Figure 4-6). Despite these disadvantages, ice calorimetry and drop calorimetry remain the best methods for measuring enthalpies of solids and melts at high temperatures. These techniques provided most of the high-temperature enthalpy and heat capacity data in thermodynamic data compilations.

E. Low-temperature adiabatic calorimeters

Adiabatic calorimeters are used to measure heat capacity by the input of a precisely known amount of electrical energy that raises the temperature of a sample by a small temperature increment (see Eq. 3-39). Ideally, there is no heat exchange between an adiabatic calorimeter and its surroundings (hence the

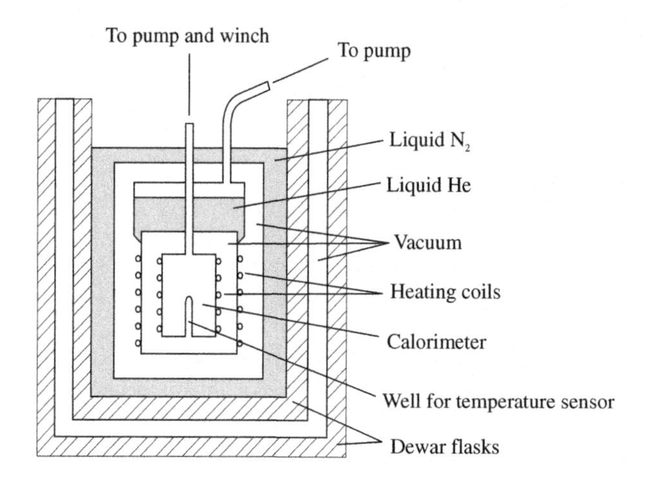

FIGURE 4-8

A schematic diagram of a low-temperature adiabatic calorimeter used to measure the constant-pressure heat capacity of solids and frozen gases.

name). The only heat input is that due to the controlled electrical heating. Traditionally, low-temperature *adiabatic calorimeters* have been used for heat capacity measurements of solids, cryogenic liquids, and frozen gases from a few K or a few tens of K to about room temperature. In addition to heat capacities, adiabatic calorimeters can measure enthalpies of solid-state phase transitions, fusion, and vaporization. The operation of an adiabatic calorimeter is best understood by reference to the schematic diagram in Figure 4-8.

The calorimeter is a small container made of a metal, such as gold or copper, with good thermal conductivity. It has a well for a temperature sensor, which always has to be at exactly the same temperature as the calorimeter and its contents. The calorimeter is heated by passing an electrical current through a wire coil wrapped around it. The calorimeter is suspended inside another metal container, which is called the *adiabatic shield*. The adiabatic shield also has an electrically heated wire coil wrapped around it. The adiabatic shield is maintained at a temperature very close to that of the calorimeter because both are heated electrically ($\Delta T \sim 0.001$ K). This minimizes heat transfer between them. There is a high vacuum in the space between the calorimeter and shield to further minimize heat exchange by conduction or convection. Finally, the calorimeter and shield are polished and gold-plated to minimize heat exchange by radiation.

This assembly is suspended inside a set of evacuated dewars (thermos bottles) that are maintained at low temperature by liquid nitrogen and liquid helium. Provisions are made for filling (or emptying) the calorimeter with various samples (solids, gases, etc.), for connections of electrical leads and gas lines to outside equipment, for raising (or lowering) the calorimeter and shield to and from thermal contact with a liquid helium bath, and for pumping on the liquid helium to further reduce its temperature. The purpose of the entire apparatus is to measure ($\delta q/dT$) as the calorimeter is electrically heated and thus to measure the heat capacity of the sample inside the calorimeter. Chapters 5 and 6 in McCullough and Scott (1968) give detailed descriptions of the construction, operation, and data reduction for low-temperature calorimeters.

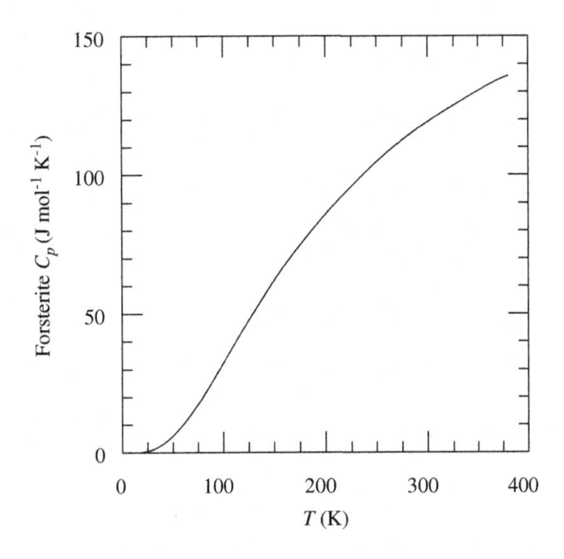

FIGURE 4-9

The low-temperature molar heat capacity ($C_{P,m}$) of forsterite (Mg_2SiO_4) measured by Robie et al. (1982) using a low-temperature adiabatic calorimeter.

Figure 4-9 shows heat capacity (C_P) data for forsterite (Mg_2SiO_4) that were measured by low-temperature adiabatic calorimetry at the U.S. Geological Survey by Robie et al. (1982). The data in Figure 4-9 extend from 6.23 K to 380.31 K. Their data are extrapolated to 0 K by the Debye method, which we describe later. The shape of the forsterite heat capacity curve is typical of that for solids without phase transitions in this temperature region. The accuracy of the forsterite C_P data varies from $\pm 5\%$ in the 5–15 K range, to $\pm 0.1\%$ above 20 K. Robie et al. (1982) compared their data with earlier forsterite C_P measurements made by Kelley (1943) at the U.S. Bureau of Mines. Kelley's heat capacity data extend to only 50 K, and he extrapolated his data to absolute zero from that temperature. He had to do this because cooling to lower temperatures requires liquid helium and/or low-temperature refrigerators of various kinds, which were not readily available until recently. Robie et al. (1982) found that the two data sets for the heat capacity of forsterite agree within 0.3% in the overlapping temperature ranges. We present other examples of low-temperature heat capacity data later in this chapter and when we discuss the third law of thermodynamics and calculation of absolute entropies from C_P data (see Chapter 9).

F. High-temperature adiabatic calorimeters

High-temperature adiabatic calorimetry is useful for measurements on materials that do not cool back to the same initial state that they were in originally—in other words, materials like liquid sulfur or glass-forming silicates. Liquid sulfur does not cool back to crystalline orthorhombic sulfur, which is composed of S_8 rings and is the stable form of sulfur at room temperature. Instead it quenches to plastic sulfur, which is composed of sulfur chains and rings of varying sizes. The thermodynamic properties of plastic sulfur depend on the quench rate and are variable. Once molten, glass-forming silicates quench to glass instead of to the original crystalline silicates. The thermodynamic properties of the glasses also depend on the quench rate and are variable. We have already discussed ways in which the enthalpies of these types of materials can be measured by drop calorimetry. Alternatively, the heat capacities of these types of materials can be directly, and often more easily, measured by high-temperature adiabatic calorimetry.

Two different methods are used in high-temperature adiabatic calorimetry: continuous heating and intermittent heating. In the continuous heating method, electrical energy, temperature, and time are simultaneously measured to obtain ($\delta q/dT$) and the heat capacity of the sample. The continuous heating method is useful for measuring heat capacities but not enthalpies of transitions or heat evolved in slow processes such as order–disorder phenomena. The intermittent heating method is used in these cases. In the intermittent heating method, electrical energy is input to the calorimeter for a known time period, the power is turned off, and the sample temperature is measured after thermal equilibrium is reached. The process is then repeated, and so on. The intermittent heating method is more difficult than the continuous heating method, but it is useful for a wider range of phenomena and heat leak corrections can be made with less error.

High-temperature adiabatic calorimetry becomes harder to do as temperatures increase because of the difficulty in maintaining adiabatic conditions between the calorimeter and adiabatic shield. Much of the difficulty is due to heat transfer by radiation, which becomes increasingly important at higher temperatures. Nevertheless, adiabatic calorimeters operating up to 1750 K have been used to measure the heat capacities of the α, γ, and δ phases of iron metal and the enthalpies of the $\alpha \rightarrow \gamma$ and $\gamma \rightarrow \delta$ phase transitions at 1185 K and 1667 K, respectively. Chapter 9 in McCullough and Scott (1968) and Chapter 2 in Kubaschewski et al. (1993) give good descriptions of high-temperature adiabatic calorimeters and their use.

Figure 4-10 shows an example of heat capacity data obtained by high-temperature adiabatic calorimetry. The data shown are for stoichiometric FeS (troilite), which is ubiquitous in meteorites. About 115 data points are plotted; these were measured by Grønvold and Stølen (1992), who used intermittent heating adiabatic calorimetry. Three different transitions with C_P maxima at about 420 K, 440 K, and 590 K are observed. The intermittent heating method was essential for resolving the three solid-state phase transitions and for calculating their enthalpies.

The 420 K transition is a *structural transition* between two different variants of the nickel arsenide (NiAs)-like crystal structure for FeS. It is characterized by contraction of the *c* axis by ~1% and disappearance of X-ray lines characteristic of the higher-temperature structure. The structural rearrangement is related to breaking up of triangular clusters of Fe atoms in the low-temperature form of FeS. More information about the structural transition, which is still incompletely understood, is given by Grønvold and Stølen (1992) and in the references they cite. The 420 K transition looks like the Greek letter λ and is an example of a *lambda transition*, which is typically observed for order–disorder processes. The maximum C_P value observed by Grønvold and Stølen (1992) for the 420 K transition is 2745.42 J mol^{-1} K^{-1}. There is an abrupt rise and fall of the heat capacity to the observed maximum within a 0.18 K interval. The largest C_P values for this transition are not shown in Figure 4-10; otherwise the rest of the curve would simply appear flat and featureless.

The 440 K and 590 K transitions are *magnetic transitions* due to shifts in the orientation of the electron spins, which we represent schematically as spin up ($\uparrow$) or down ($\downarrow$). Neutron diffraction and magnetic studies show that at about 450 K, the electron spins in FeS shift from being aligned

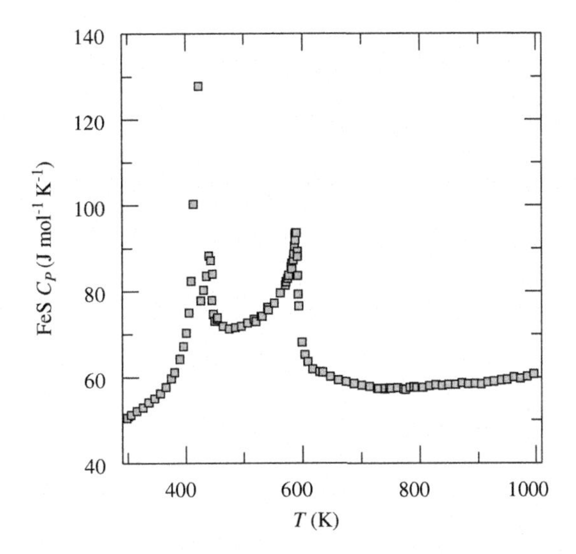

FIGURE 4-10

The high-temperature molar heat capacity ($C_{P,m}$) of troilite (FeS) measured by Grønvold and Stølen (1992) using a high-temperature adiabatic calorimeter.

antiferromagnetically with adjacent electron spins antiparallel (↓↑↓↑↓↑) along the *c* axis to positions perpendicular to it. The transition at 590 K, which is the *Néel temperature*, is a magnetic order–disorder transition. The electron spins shift from being antiferromagnetically aligned at lower temperatures to being paramagnetically disordered (↑↑↓↑↑↓↓↑) at higher temperatures. Although we focused on FeS as an example of high-temperature adiabatic calorimetry used to measure C_P data, all pyrrhotites display complex magnetic behavior.

G. Differential scanning calorimetry

Another technique of measuring high-temperature heat capacities is by *differential scanning calorimetry* (DSC). Figure 4-11 schematically illustrates a heat-flux DSC with a cylinder-type measuring system. In this method, two identical sample containers are heated at the same rate. One container is empty or, as illustrated, contains a reference material, such as corundum, of which the thermal properties are well known. As the two crucibles are heated, the heat flow per unit time is different in each because they have different thermal properties. The different heat flows ($\delta q/t$) translate into a temperature difference ($\Delta T/t$) between the two containers. The heat capacity of the sample is then computed from the ratio of the heat flow and temperature difference:

$$\frac{\delta q/T}{\Delta T/T} = \frac{\delta q}{\Delta T} = \overline{C} \text{ as } \Delta T \rightarrow 0 = C = \left(\frac{\delta q}{dT}\right) \quad (4\text{-}24)$$

The very small temperature differences between the two crucibles are measured by *thermopiles*, which are thermocouples connected together in series. Typically, heat capacities measured by DSC

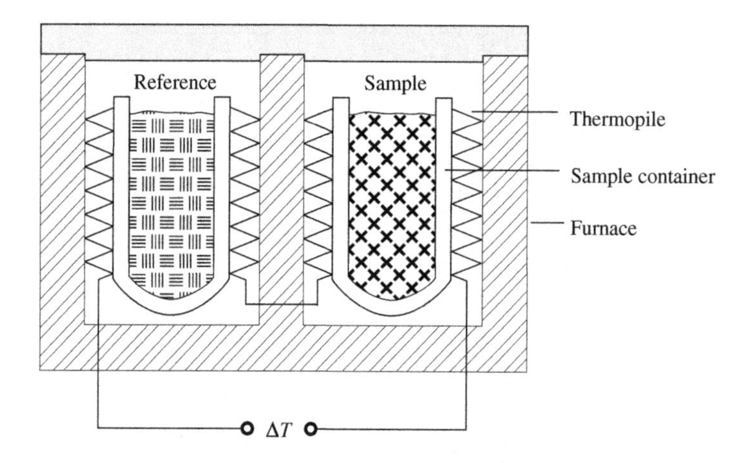

FIGURE 4-11

A schematic diagram of a heat-flux differential scanning calorimeter.

have accuracies of ±1% compared to accuracies of a few tenths of a percent for C_P measurements by high-temperature adiabatic calorimetry or very accurate drop calorimetry.

H. Calorimeters for gases

We only briefly review calorimetric methods used to measure C_P or C_V values of gases because the thermal properties of gases in the ideal state are easily calculated from spectroscopic data and molecular parameters using statistical mechanics. The thermal properties of real gases can be computed or estimated from those of the ideal gas. However, to do this some data on the heat capacities of real gases are needed.

It is virtually impossible to measure C_V values of gases in a static calorimeter because the heat capacity of the calorimeter is generally much larger than that of the gas itself. Thus, the heat capacity of the gas is a tiny difference between two large values (the heat capacity of the calorimeter plus gas and that of the empty calorimeter). The calorimetric methods that have been used to measure C_V values of gases are as follows. Low-temperature adiabatic calorimetry of gases confined inside strong metal containers has been used to study the thermal properties of ortho-H_2, para-H_2, and heavy hydrogen (D_2).

The second method used to measure C_V is via the rapid heating produced by an explosion. The *explosion calorimeter* is a strong steel sphere in which a known amount of heat is liberated by explosion of a gas mixture such as H_2-O_2 or by thermal decomposition of a reactive gas such as ozone (O_3) inside the calorimeter. The rapid temperature rise is measured as a function of time and the C_V of the hot gas can be calculated. However, this method gives uncertain results because the vibrational and electronic state of the hot gas is poorly constrained.

The third method used to measure C_V is a high-pressure constant-volume calorimeter. This is basically a strong steel sphere that has an internal electrical heater. The heat capacity of the gas is calculated from the heat input and temperature increase via Eq. (3-40). This calorimeter

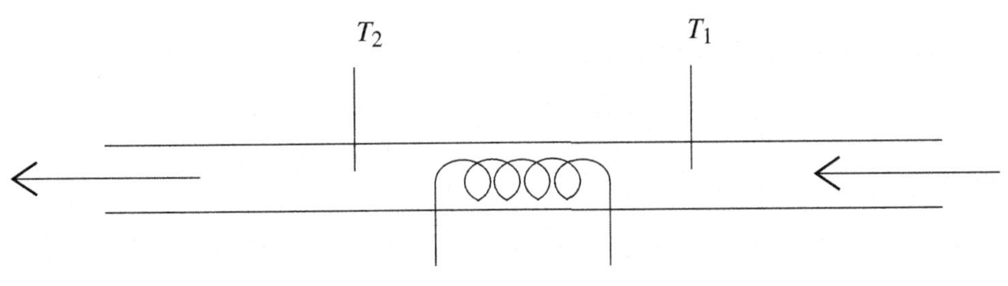

FIGURE 4-12

A schematic diagram of a flow calorimeter used to measure the constant-pressure heat capacity of gases. The arrow indicates the gas flow direction.

is useful for gases at or near their *critical point*, which is the temperature at which the distinction between gas and liquid vanishes; only one fluid phase exists at and above the critical point.

Finally, the fourth method used to measure C_V is the *steam calorimeter* invented by John Joly (1857–1933), professor of geology and mineralogy at Trinity College in Dublin. The steam calorimeter measures the difference of heat capacity between an empty metal bulb and one filled with gas. Both bulbs are suspended from a sensitive balance and are heated by steam. The amount of water condensed on the two metal bulbs is different and gives the C_V of the gas.

The C_P values of gases are easier to measure, because a *flow calorimeter* such as the one schematically illustrated in Figure 4-12 can be used. The gas flows over an electrically heated wire at a constant pressure and flow rate. The temperature of the gas is measured before (T_1) and after (T_2) passing over the heated filament. The heat capacity of the gas is then calculated from

$$C_{P,m} = \frac{(E - e)}{n(T_2 - T_1)} \tag{4-25}$$

where E is the electrical energy per second, e is the energy lost to the surroundings, n is the molar flow rate per second of the gas, and $(T_2 - T_1)$ is the temperature difference. The amount of energy e, which is lost to the surroundings, is determined empirically. The molar flow rate of the gas is simply the volumetric flow rate per second corrected to standard temperature and pressure (STP, 0°C, and 1 bar) divided by the molar volume of an ideal gas at STP.

Example 4-6. A geochemist measures the C_P of carbon dioxide by passing pure CO_2 over an electrically heated filament at one bar pressure and a STP flowrate of 25.3 $cm^3\ s^{-1}$. The temperature of the CO_2 increases from 420°C to 424°C after passing over the filament, which is heated by a current of 0.0245 A at 9 V. What is the molar C_P of CO_2 at the mean $T = 422$°C (695 K)? Assume that corrections for energy lost to the surroundings can be neglected.

Equation (4-25) gives the molar C_P and Eq. (4-7) gives the electrical energy per second:

$$C_{P,m} = \frac{(9\ \text{V})(0.0245\ \text{A})}{(25.3/22{,}711)(424 - 420)} = \frac{0.22050\ \text{J}}{(1.129 \times 10^{-3}\ \text{mol})(4°\text{C})} = 49.48\ \text{J mol}^{-1}\ \text{K}^{-1} \tag{4-26}$$

This is 0.5% lower than the value of 49.73 J mol^{-1} K^{-1} from the equation for CO_2 in Table 3-4.

III. THERMAL PROPERTIES OF SOLIDS

A. Dulong-Petit and Neumann-Kopp heat capacity rules

In 1819, Dulong and Petit (see sidebar) proposed the empirical rule that the atomic heat capacity of all solid elements was equal to three times the gas constant R:

$$3R = 3\,(8.314\text{ J mol}^{-1}\text{ K}^{-1}) = 24.9\text{ J (g atom)}^{-1}\text{ K}^{-1} \tag{4-27}$$

where the abbreviation *g* atom stands for *gram atom*, which is a mole of atoms.

PIERRE LOUIS DULONG (1785-1838)

Dulong began practicing medicine in 1803, when he was still a teenager. At the time, no MD was required. Dulong found that he could not make a living this way, however, because he was so kind hearted that he not only treated his patients free, but also offered to buy their prescriptions for them. Dulong then became a botanist, and finally, a chemist. He worked in Claude Louis Berthollet's (1748-1822) private laboratory. Dulong later took on teaching and administrative positions despite the fact that these took time away from his research because he had to provide for his family.

Dulong contributed to several areas of chemistry. He began to study reaction equilibria when he was able to dissolve barium salts, thought to be insoluble, in heated solutions. With Petit, he developed the Dulong-Petit rule, which states that the product of the specific heat and atomic weight for any solid element is approximately constant.

Dulong faced several tragedies during his lifetime. He was orphaned at four and half years of age. As an adult, he blinded himself in one eye and lost a finger after an accident preparing nitrogen trichloride, a spontaneously explosive oil that he discovered. However, he did not allow this to deter him and he continued to pursue scientific discoveries.

ALEXIS THÉRÈSE PETIT (1791–1820)

Alexis Thérèse Petit was a bright young man and a good scholar. He entered the École Polytechnique as soon as he was 16, which was the minimum age for admission. It is said that he was intellectually prepared at the age of 10. He was at the head of his class and was hired as a teaching assistant upon graduation. He became a physics professor at the Lycée Bonaparte in 1810 and then at the École Polytechnique in 1815.

Petit's most important work was related to temperature measurement. He and Francois Arago (1786–1853) measured changes in refractive index with change in temperature. His work in collaboration with Pierre Louis Dulong included comparing mercury and air thermometers. Together they showed that a gas thermometer is the only reliable standard for temperature measurement. With Dulong, Petit put forth the Dulong-Petit rule, which states that the product of the specific heat and atomic weight for any solid element is approximately constant.

Petit was educated in a very traditional manner and evidently taught the same way, but he was also open to new ideas. His work on refractive indices led him to be one of the early supporters of the wave theory of light. He also supported the atomic theory, which put him among the minority of French scientists at the time, since this idea was vociferously opposed by the respected scientist Claude Louis Berthollet (1748–1822).

Perhaps it was fortunate that Petit was something of a prodigy, for he died young of tuberculosis, which he contracted shortly after the death of his wife.

Although we now know that the *Dulong-Petit rule* only holds at about room temperature and is an approximation, Figure 4-13 shows that it is a remarkably good approximation. There are a few exceptions, notably the light elements (B, Be, C), several of the alkalis (Na, K, Rb, Cs), and plutonium

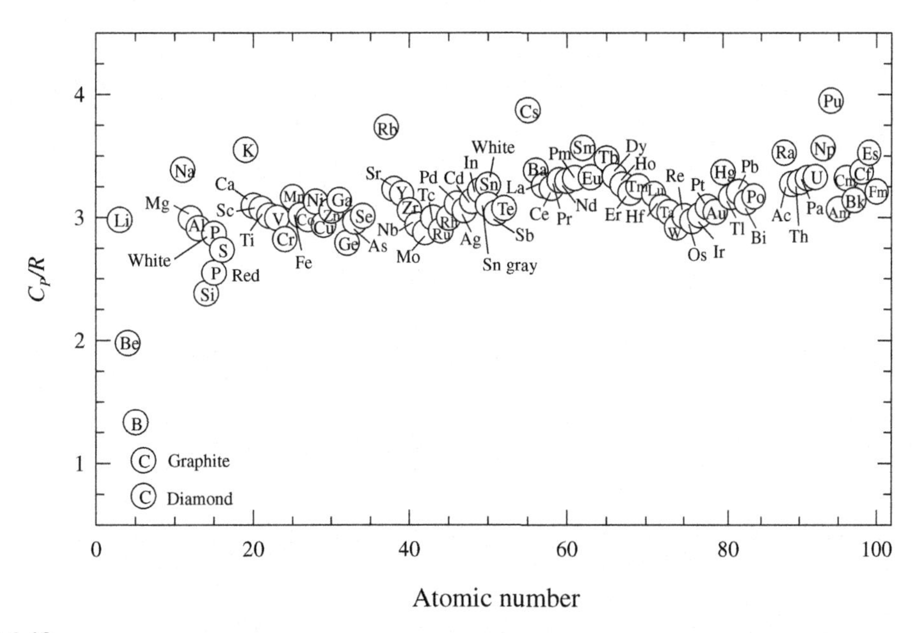

FIGURE 4-13

This plot of $C_{P,m}/R$ for elements that are either solid or liquid at room temperature illustrates the first-order accuracy of the Dulong-Petit rule.

(Pu). The different heat capacities exhibited by different polymorphic forms of the same element, such as for diamond and graphite, gray and white tin, and red and white phosphorus, also show that the Dulong-Petit rule is only approximate. Yet it seems to contain a grain of truth, and this rule is the basis for the Einstein and Debye heat capacity models.

The Dulong-Petit rule was extended to minerals by the German mineralogist and physicist Franz Neumann (1798–1895) in 1831. He noticed that the heat capacities of several carbonates, sulfides, and oxides were equal to the sum of the heat capacities of their constituent elements. The German chemist Hermann Kopp (1817–1892) arrived at a similar conclusion in the 1860s. The *Neumann-Kopp rule* states that the heat capacity per gram atom of all solid compounds is the same. Mathematically this means that $\Delta C_P = 0$ for formation of a solid compound from solid reactants, which can be elements, oxides, sulfides, and so on. The use of the Neumann-Kopp rule is illustrated in Figure 4-14, which compares the measured and estimated heat capacities for ~90 minerals. With a few exceptions such as franklinite ($ZnFe_2O_4$, 25% error), cobalt olivine (Co_2SiO_4, 16% error), and chromite ($FeCr_2O_4$, 15% error), the estimated heat capacities from the Neumann-Kopp rule agree with the measured values within 3.5%. Thus, the Neumann-Kopp rule is very useful, even though it is only an approximation. We illustrate its use in the following example.

Example 4-7. (a) Estimate $C_{P,m}$ of galena (PbS) at 298 K from the C_P values of its constituent elements. The C_P values of Pb metal and orthorhombic sulfur are 26.65 and 22.75 J mol^{-1} K^{-1}, respectively. Their sum is 49.40 J mol^{-1} K^{-1}, which is 0.2% smaller than the actual molar C_P of galena

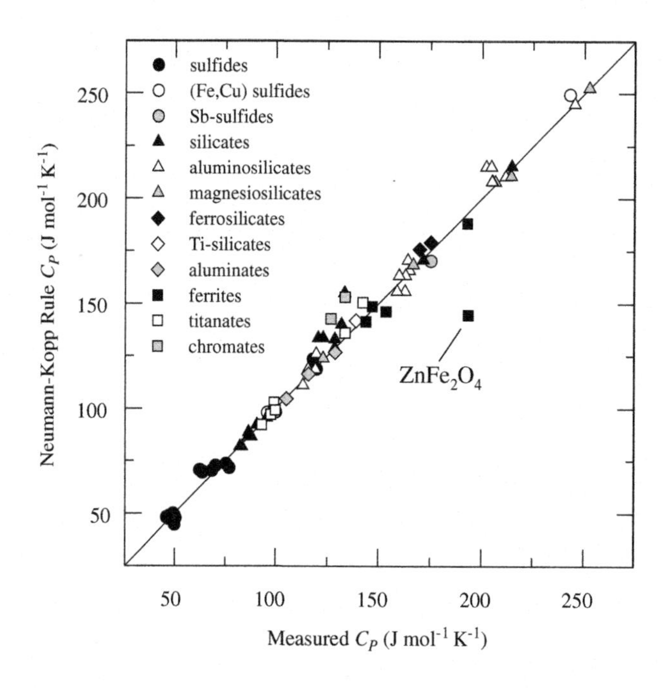

FIGURE 4-14

Heat capacities estimated with the Neumann-Kopp rule are compared to measured values for 89 different minerals. All values are at 298.15 K.

($49.49\ J\ mol^{-1}\ K^{-1}$). (b) Estimate $C_{P,m}$ at 298 K for merwinite ($Ca_3MgSi_2O_8$) from the $C_{P,m}$ values of its constituent oxides. The heat capacities of CaO, MgO, SiO_2 (quartz) are 42.05, 37.24, and 44.60 J $mol^{-1}\ K^{-1}$, respectively. Their sum (i.e., $3CaO + MgO + 2SiO_2$) is 252.59 J $mol^{-1}\ K^{-1}$ and the measured $C_{P,m}$ for merwinite is 252.36 J $mol^{-1}\ K^{-1}$, a 0.1% error.

These examples are at 298 K, but the Neumann-Kopp rule also can be used to estimate heat capacity of solids at higher temperatures. It is a good approximation when the constituent elements of a compound follow the Dulong-Petit rule. Thus, the Neumann-Kopp rule is not a good approximation at low temperatures because it applies only when the heat capacities of solid elements have reached the classic value of $3R$, which occurs at or slightly below room temperature for most elements. Finally, the Neumann-Kopp rule works best for reactions that do not involve gases because ΔC_P is large in these cases, so the approximation fails.

B. Typical shape of heat capacity curves

Figure 4-15 shows the variation with temperature of $C_{P,m}$ and $C_{V,m}$ for forsterite from 5–1800 K. The overall shapes of the heat capacity curves for forsterite are typical for solids with only one crystalline form. The variation of heat capacity with temperature at low temperatures, exemplified by the C_P curve of forsterite shown in Figure 4-9, is primarily due to oscillations of atoms (or molecules) in the crystal lattice about their positions—that is, to the lattice component of the heat capacity. Theoretical models proposed by the Nobel Prize–winning physicists Albert Einstein and Peter Debye (see sidebars) in the early 20th century first explained the lattice heat capacity of solids and are briefly described in the following section.

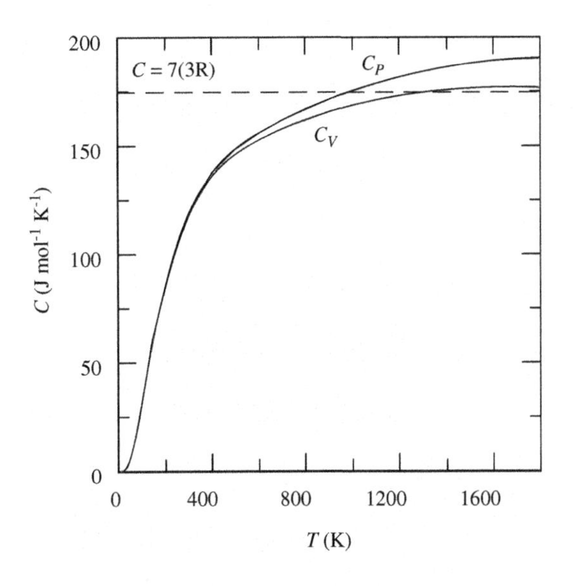

FIGURE 4-15

The molar constant-pressure heat capacity of forsterite from 0 K to 1800 K. The smoothed values given by Robie et al. (1982) and Robie and Hemingway (1995) are plotted.

C. Einstein and Debye models

Einstein published his theory of low-temperature heat capacity in 1907, shortly after Sir James Dewar (1842–1923) measured the heat capacity of diamond at low temperatures. Dewar found that the average heat capacity of diamond in the 20–50 K region was less than 1% of the Dulong-Petit value. This was a perplexing result at the time because thermodynamics could not explain this behavior. Einstein explained Dewar's results by treating a monatomic crystalline solid, such as diamond, as a system of $3N$ independent oscillators (where N is the number of atoms). He also assumed that the N atoms were free to vibrate in three mutually independent perpendicular directions and did so with a single characteristic frequency ν. Einstein derived an expression for the internal energy of such a solid, which, upon differentiation $[(\partial E/\partial T)_V]$, gave the following expression for the C_V per gram atom (taking N as Avogadro's number N_A):

$$C_V = 3R\frac{u^2e^u}{(e^u-1)^2} = 3Rf_E \tag{4-28}$$

In Eq. (4-28) e is an exponential and $u = h\nu/kT = \theta_E/T$, where h is Planck's constant, ν is the characteristic vibrational frequency of the solid, k is Boltzmann's constant, T is the absolute temperature, and θ_E is the *Einstein temperature*. The exponential terms in Eq. (4-28) are called the *Einstein function* and are written as f_E. The values of θ_E and ν are found empirically by matching heat capacity data at some chosen temperature.

ALBERT EINSTEIN (1879-1955)

Albert Einstein is probably the most famous physicist of all time. He was born in Germany and was educated at the ETH in Zurich, Switzerland. After graduation in 1901, he became a patent examiner, and while in this job did some of the most important work of his career. In 1905, he published scientific papers explaining Brownian motion, the photoelectric effect, and proposed his Theory of Special Relativity. Two years later he published his model for the low temperature heat capacity of solids. Einstein's genius then became apparent and he moved from the patent office to professorships in Berne, Zurich, Prague, and Berlin. While professor in Berlin Einstein developed his Theory of General Relativity, which predicted that gravity bends light and other effects. He won the 1921 Nobel Prize in Physics for his work on the photoelectric effect. While visiting the US in 1932-1933, Hitler and the Nazi party came to power in Germany. Einstein resigned his position in Berlin, stayed in the US for the rest of his life, and became an American citizen in 1940.

A key prediction of Special Relativity is the conversion of matter into energy, given by Einstein's famous equation $E = mc^2$. After the discovery of nuclear fission, the practical use of this equation for building atomic bombs became clear. Einstein was a pacifist and firmly opposed to the idea of war, however, he saw the threat posed by Nazi Germany to the rest of the world and was afraid that the Germans might already be developing nuclear weapons. In 1939, he wrote a letter to President Franklin Roosevelt, alerting him to this possibility and urging him to act. This letter spurred the beginning of the Manhattan Project, which developed the first atomic bombs.

PETRUS (PETER) JOSEPHUS WILHELMUS DEBYE (1884-1966)

Peter Debye was born in the Netherlands and studied in Germany. From 1910 to 1935, he was a professor of physics at several European universities. Debye won the Nobel Prize for Chemistry in 1936, shortly after becoming Director of the Kaiser Wilhelm Institute for Physics in Berlin. He left Germany in 1939 when he immigrated to the US. Debye spent the rest of his life as Professor of Chemistry at Cornell University in Ithaca, NY.

Debye's research was relevant to chemistry and physics. Early in his career, he proposed a theory for the heat capacity of solids. The Debye T^3 law is part of this theory and is used to extrapolate heat capacity data from liquid helium temperatures to absolute zero. He developed, with Hückel, a new theory about electrolyte solutions. He imagined that each ion in the solution is surrounded by ions of the opposite charge. The Debye–Hückel equation is widely used to model the properties of ions in electrolyte solutions. With Scherrer, Debye developed powder X-ray diffraction, which uses small samples of powder instead of single crystals.

Two points are worth noting before we continue. First, the C_V per gram atom is the same as the molar C_V only for monatomic solids. The molar C_V must be divided by the number of atoms per molecule (i.e., the number of gram atoms per mole) to give C_V per gram atom. Second, the difference between C_V and C_P of a solid is insignificant at these low temperatures. We discuss the computation of this difference, which is about 5% at 298 K and becomes larger at higher temperatures, later in Section III-H.

Figure 4-16 shows a comparison of Einstein's model with modern data for the temperature-dependent heat capacity of diamond. The qualitative agreement is obvious and is very important, even though the model does not exactly reproduce the measurements. Equation (4-28) predicts that C_V

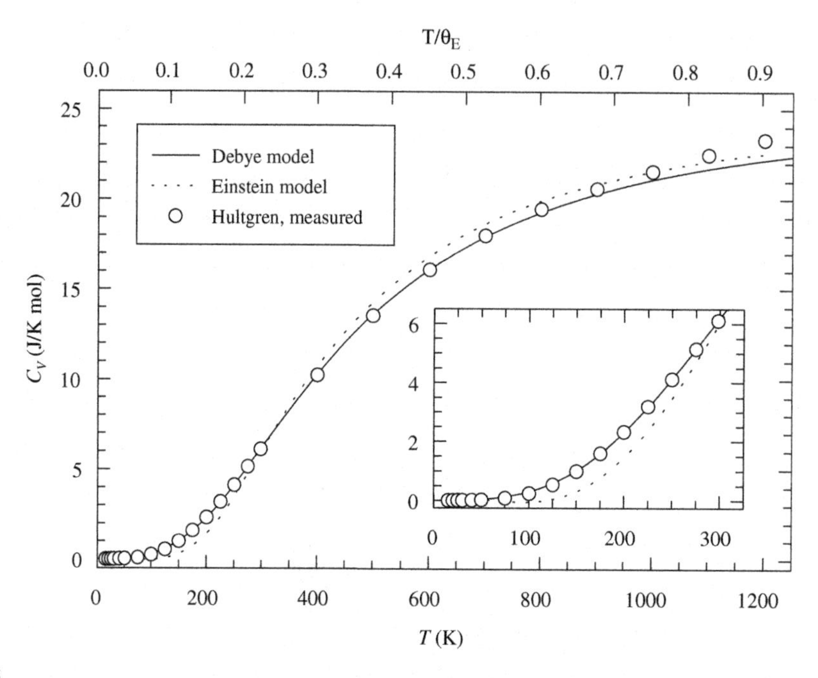

FIGURE 4-16

The molar constant-volume heat capacity of diamond from 0 K to 1,200 K is compared to predictions of the Einstein and Debye models for low-temperature heat capacity.

The $C_{V,m}$ data are from Hultgren et al. (1973).

asymptotically approaches zero as the temperature approaches absolute zero; it also predicts that C_V approaches a limiting value of $3R$ as the temperature approaches infinity. We can see this for ourselves with an example using silver (Ag) metal.

Example 4-8. Use $\theta_E = 210$ K for Ag to compute C_V per gram atom of Ag from 10–500 K. Neglect the difference between C_V and C_P in your work. Equation (4-28) gives C_V and Table 4.1 shows the results with a comparison to C_P data for Ag (Hultgren et al. 1973).

However, as more low-temperature heat capacity data accumulated, it became clear to Einstein and others that his model did not reproduce the data very well. For example,

Table 4-1 Heat capacity of silver

T (K)	$\theta_E/T = u$	f_E	C_V/R (calcd)	C_P/R (obs)
10	21.0	3.3×10^{-7}	9.9×10^{-7}	2.3×10^{-2}
50	4.2	0.27	0.81	1.40
210	1	0.92	2.76	2.94
298.15	0.70	0.96	2.88	3.05
500	0.42	0.99	2.97	3.17
1,000	0.21	0.996	2.99	3.59

Figure 4-16 and Example 4-8 show that Einstein's model underestimates the low-temperature heat capacity. In general, Einstein's model underestimates C_V for $\theta_E/T \geq 10$, that is, at and below 20 K for a material having $\theta_E = 200$ K. The fit of Einstein's model to heat capacity data can be improved using two vibrational frequencies of ν and ν/2, but this makes the model even more empirical.

In 1912 Peter Debye modified Einstein's model. The important changes were that the atomic oscillators in the crystal were assumed to be coupled instead of independent and that a range of vibrational frequencies terminating at a maximum frequency ν_{max} was assumed instead of a single characteristic vibrational frequency. Debye's equation for the C_V per gram atom is

$$C_V = \left[\frac{12}{x^3}\int_0^x \frac{u^3}{(e^u - 1)}du - \frac{3x}{(e^x - 1)}\right] = f_D\left(\frac{T}{\theta_D}\right) \text{ J (g atom)}^{-1}\text{ K}^{-1} \quad (4\text{-}29)$$

where $x = h\nu_{max}/kT = \theta_D/T$, and $u = h\nu/kT$ where θ_D is the *Debye temperature* and the other terms are the same as before. The Debye temperature, like the Einstein temperature, is different for each material, but θ_D and θ_E are also different for the same material. For example, Kieffer (1985) gives $\theta_D = 306$ K and $\theta_E = 210$ K for halite and $\theta_D = 942$ K and $\theta_E = 567$ K for periclase.

As shown in Figure 4-16, Debye's model fits the heat capacity data for diamond better than Einstein's model does. The good agreement of Debye's model with heat capacity data for several types of minerals can be illustrated by plotting C_P per gram atom versus T/θ_D, as done in Figure 4-17. All heat capacity data lie along the same curve in this type of plot.

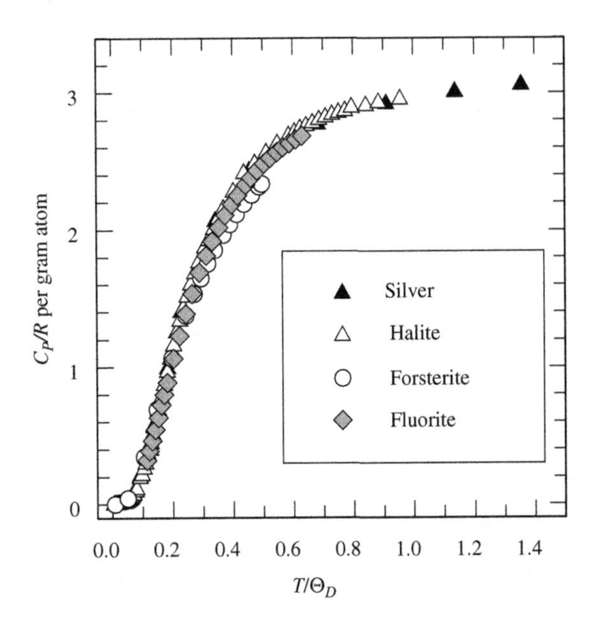

FIGURE 4-17

The heat capacities per gram atom of fluorite (CaF_2) $\theta_D = 470$ K, forsterite (Mg_2SiO_4) $\theta_D = 768$ K, halite (NaCl) $\theta_D = 280$ K, and silver metal (Ag) $\theta_D = 220$ K fall onto the same curve when plotted against (T/θ_D).

Three other factors leading to acceptance of Debye's model are (1) θ_D values can be computed from the elastic properties of a material, (2) the model predicts a dependence of θ_D upon isotopic mass that agrees with the observed dependence, and (3) the model predicts a T^3 dependence for heat capacity of nonmetallic solids at low temperatures. The latter point is of most interest for us, and we discuss it next.

At low temperatures near 0 K, Eq. (4-29) reduces to this simple form:

$$C_V = \frac{12\pi^4}{5}R\left(\frac{T}{\theta_D}\right)^3 = 1943.8\left(\frac{T}{\theta_D}\right)^3 = aT^3 \text{ J (g atom)}^{-1}\text{ K}^{-1} \quad (4\text{-}30)$$

where a is a constant equal to $(1943.8/\theta_D^3)$. Equation (4-30) is known as the *Debye* T^3 law. It is usually valid for $T < \theta_D/50$, which is below 10 K for many materials. The T^3 dependence predicted by Eq. (4-30) is illustrated in Figure 4-18, where C_P/T is plotted versus T^2 for several nonmetallic elements and minerals. The heat capacity data fall on straight lines that have slopes equal to $1943.8/\theta_D^3$ (per gram atom) and intercepts of 0, that is, C_V goes to 0 as T approaches absolute zero. Next we give examples of calculating θ_D from low-temperature C_P data.

Example 4-9. (a) The Debye temperature for forsterite can be calculated from the slope of the line in Figure 4-18 (2.9439×10^{-5}) using the equation

$$\theta_D = \left(\frac{1943.8 \times 7 \text{ g atoms/mol}}{2.9439 \times 10^{-5}}\right)^{\frac{1}{3}} = (4.6220 \times 10^8)^{\frac{1}{3}} = 773 \text{ K} \quad (4\text{-}31)$$

where we have taken account of the seven gram atoms per mole of forsterite. For comparison, Robie et al. (1982) found $\theta_D = 768\pm15$ K from their data. (b) Low-temperature heat capacity measurements for argon give $a = 2.8467\times10^{-3}$. What is the Debye temperature for Ar? Using Eq. (4-30) we calculate

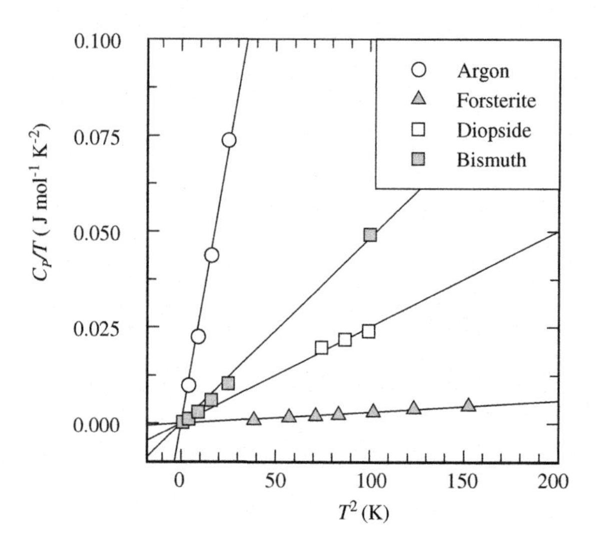

FIGURE 4-18

An illustration of the Debye T^3 law for several nonmetallic solids (argon ice, bismuth, diopside ($CaMgSi_2O_6$), and forsterite). The extrapolated heat capacities are zero at absolute zero.

$\theta_D = 88$ K for Ar, which agrees with the value of 90 K listed in Gopal (1966). (c) Some low-temperature heat capacity data for fluorite (CaF_2) are listed here:

T (K)	17.5	19.9	21.5	25.6	29.1	34.0	36.8	39.8
C_V (J mol^{-1} K^{-1})	0.2803	0.4301	0.5506	0.9121	1.3849	2.2426	2.7740	3.4978
C_V per gram atom	0.0934	0.1434	0.1835	0.3040	0.4616	0.7475	0.9247	1.1659
Debye *T*	481	474	472	475	470	467	471	472

The value of θ_D for each data point is given by

$$\theta_D = T \times \left(\frac{1943.8}{C_V}\right)^{\frac{1}{3}} \tag{4-32}$$

which is a rearranged version of Eq. (4-30). The average Debye temperature calculated from the preceding data is 473 K, in good agreement with the value of 470 K given by Gopal (1966).

In general, the Debye T^3 law is very useful for extrapolation of low-temperature heat capacity data to zero Kelvin. This extrapolation is necessary because it is theoretically impossible to reach 0 K and because it is impractical for most low-temperature calorimeters to operate below liquid helium temperatures (5–10 K). As mentioned earlier in Section II-E, Kelley's heat capacity data on forsterite had to be extrapolated from 50 K to 0 K. This was done empirically using a combination of Debye and Einstein functions that gave the best fit to the heat capacity curve from 50–298 K. Kelley and King (1961) discuss this method, which was used extensively by Kelley and his colleagues at the U.S. Bureau of Mines for extrapolation to absolute zero of low-temperature heat capacities of minerals.

D. Limitations of Debye's model

The Debye model has several limitations that should be understood to use it properly. One shortcoming is that the heat capacities of layered materials such as graphite and boron nitride (BN) show a T^2, instead of a T^3, dependence at low temperatures. Another deficiency is that θ_D is temperature dependent, with typical variations of about 10%, instead of being constant. As noted by Gopal (1966), the value of θ_D at a temperature equal to $2\theta_D$ often gives a good fit to the C_V curve. But this fit is not exact; otherwise, all the points in Figure 4-17 would fall exactly along the same curve. Other limitations of the Debye model are that the C_E, C_M, C_λ, and other components of C_V (see Eq. 4-2) lead to deviations from the predicted T^3 behavior at low temperatures. We now discuss some of these heat capacity contributions in more detail because they are important for thermal properties of metals and minerals.

E. Electronic heat capacity of metals

The heat capacities of metals follow the Debye T^3 law at low temperatures but have a nonzero intercept at 0 K when C_V/T is plotted versus T^2. This behavior is illustrated in Figure 4-19, which shows the low-temperature C_P of manganese metal. The 0 K intercept is nonzero because

$$C_V = \left(\frac{1943.8}{\theta_D^3}\right)T^3 + \gamma T = aT^3 + \gamma T = C_{lat} + C_E \text{ J (g atom)}^{-1} \text{ K}^{-1} \tag{4-33}$$

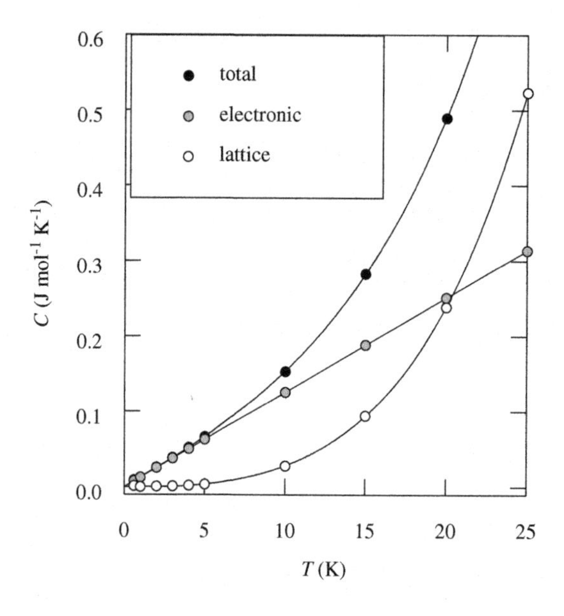

FIGURE 4-19

A plot of $C_{V,m}/T$ versus T^2 for manganese metal (Mn). The intercept gives γ, the electronic heat capacity coefficient for Mn metal.

for Mn and other metals. The linear term is the *electronic heat capacity* (C_E) and the nonzero intercept at 0 K is γ, the *electronic heat capacity coefficient,* which is characteristic for a metal. The electronic heat capacity is the heat capacity of the free electrons in a metal. It is a small fraction of the total heat capacity except at low temperatures, below about 20 K. There C_E, which varies linearly with temperature, becomes larger than C_{lat}, which varies with T^3. The electronic heat capacity also becomes large at high temperatures comparable to the *Fermi temperature*, which is different for each metal but is about 90,000 Kelvins. However, we are interested in C_E at low temperatures because it is part of the measured heat capacities of metals and alloys.

Example 4-10. Calculate the Debye temperature and electronic heat capacity coefficient γ of Mn metal using the data plotted in Figure 4-19. Equation (4-33) can be rearranged to

$$C_V/T = aT^2 + \gamma \tag{4-34}$$

which shows that a plot of C_V/T versus T^2 has a slope equal to a ($1943.8/\theta_D^3$) and an intercept of γ at absolute zero. A subset of the data used to make Figure 4-19 is listed here. The difference between C_P and C_V is insignificant at these low temperatures.

T (K)	1	2	3	4	5	10	15
C_V J K^{-1} (g atom)$^{-1}$	0.01255	0.02552	0.03849	0.05188	0.0661	0.1531	0.28245
C_V/T J K^{-2} (g atom)$^{-1}$	0.01255	0.01276	0.01283	0.01297	0.01322	0.01531	0.01883

The line slope is 2.7805×10^{-5} so $\theta_D = 412$ K and $\gamma = 1.26\times10^{-2}$ J K^{-2} (g atom)$^{-1}$. For reference, Gopal (1966) lists $\theta_D = 420$ K and $\gamma = 1.8\times10^{-2}$ for manganese.

F. Magnetic heat capacity of minerals

We begin by briefly reviewing the magnetic properties of solids and, as done earlier in Section II-F, we use arrows to schematically represent electron spins. After doing this we derive magnetic work and magnetic heat capacity.

Magnetic properties of solids

Diamagnetic materials are nonmagnetic and are feebly repelled by a magnetic field, such as that from a strong permanent magnet or electromagnet. Forsterite (Mg_2SiO_4), quartz (SiO_2), water, halite (NaCl), diamond, and graphite are examples of diamagnetic materials. Electron spins of diamagnetic materials are randomly oriented. All materials are diamagnetic to some extent, but the other types of magnetic behavior dominate diamagnetic behavior in the other types of magnetic materials.

Ferromagnetic materials are the materials that we normally think of as being magnetic and of which magnets are made. They are strongly pulled into a magnetic field. Examples of ferromagnetic materials include Fe, Ni, Co, Gd, Dy, Fe_3C (cementite or cohenite), $FeBe_5$, Cu_2MnAl, Cu_2MnIn, Au_2MnAl, Fe_2B, MnAs, MnBi, MnB, CrTe, CrO_2, $CrBr_3$, EuO, and $GdCl_3$. Ferromagnetic materials have aligned electron spins of equal magnitude (↑↑↑↑↑↑). Many, but not all, ferromagnets are good conductors. Some such as $CrBr_3$ and $GdCl_3$ are insulators. Ferromagnetic materials remain magnetic up to a critical temperature known as the *Curie temperature*, where thermal motions randomize the electron spins. Curie temperatures are different for each material and span a wide range. For example, Curie temperatures are 1395 K for Co, 1043 K for Fe, 633 K for Ni, 77 K for EuO, 37 K for $CrBr_3$, and only 2.2 K for $GdCl_3$.

Ferrimagnetic materials are also magnetic and are pulled into a magnetic field. In fact, strong magnetism was first discovered in magnetite, which is a ferrimagnetic material. Other examples of ferrimagnets include $NiFe_2O_4$ (trevorite), $MgFe_2O_4$ (magnesioferrite), $MnFe_2O_4$, $BaFe_{12}O_{19}$, $Y_3Fe_5O_{12}$ (yttrium iron garnet), and other ferrites and garnets. Ferrimagnetic materials have two or more sublattices that are magnetized in different directions. The most common ferrimagnetic behavior, displayed by spinel ferrites, has adjacent antiparallel electron spins of unequal magnitude (↓↑↓↑↓↑). Most ferrimagnets are poor electrical conductors. Ferrimagnetic materials remain magnetic up to a critical temperature known as the *Néel temperature*, where thermal motions randomize the magnetic spins. Néel temperatures are different for each material and span a wide range. For example, Néel temperatures are 858 K for magnetite and trevorite, 733 K for $BaFe_{12}O_{19}$, 713 K for magnesioferrite, 573 K for $MnFe_2O_4$, and 560 K for $Y_3Fe_5O_{12}$.

Paramagnetic materials are normally nonmagnetic and have disordered adjacent electron spins (↑↑↓↑↑↓↓↑). However, the application of a magnetic field, such as from a strong magnet or electromagnet, to a paramagnetic material induces magnetization M proportional to the strength H of the applied field with the spins oriented parallel to the direction of the applied field. Paramagnetic materials are pulled into a magnetic field with increasing strength as temperature decreases. This behavior follows *Curie's law*:

$$\chi_M = M_m/H = C/T \quad (4\text{-}35)$$

where M_m is molar magnetization (A m^2 mol^{-1}), H is magnetic field strength (A m^{-1}), χ_M is molar magnetic susceptibility (m^3 mol^{-1}), and C is Curie's constant. The molar magnetic susceptibility is reported in cm^3 mol in most data compilations. Paramagnetic materials often have unpaired electrons and include liquid and gaseous O_2, liquid and gaseous NO, chalcanthite ($CuSO_4 \cdot 5H_2O$), ferric ammonium alum ($Fe(NH_4)(SO_4)_2 \cdot 12H_2O$), retgersite ($NiSO_4 \cdot 6H_2O$), gadolinium sulfate octahydrate ($Gd_2(SO_4)_3 \cdot 8H_2O$), and many other transition metal and lanthanide compounds. Ferromagnetic materials are paramagnetic above their Curie temperatures, whereas ferrimagnetic and antiferromagnetic materials are paramagnetic above their Néel temperatures.

Antiferromagnetic materials are highly ordered magnetically with two or more sublattices that are magnetized in different directions. The net effects cancel out, and antiferromagnetic materials behave like paramagnetic materials. There are several ways in which the spins can be oriented in antiferromagnetic materials. The simplest orientation is probably adjacent antiparallel electron spins of equal magnitude (↓↑↓↑↓↑). The magnetic orientation of antiferromagnetic materials is randomized by thermal motions at the Néel temperature. Antiferromagnetic minerals (with their Néel temperatures) include hematite (α-Fe_2O_3, 953 K), troilite (FeS, 590 K), bunsenite (NiO, 520 K), eskolaite (Cr_2O_3, 307 K), and pyrolusite (MnO_2, 84 K). The olivine group minerals cobalt olivine (Co_2SiO_4, 49.8 K), fayalite (Fe_2SiO_4, 64.9 K), tephroite (Mn_2SiO_4, 47.4 K), and liebenbergite (Ni_2SiO_4, 29.2 K) are also examples of antiferromagnets. Many perovskites, including $LaFeO_3$, $NdFeO_3$, and $BiFeO_3$, are also antiferromagnetic.

Magnetic energy, work, and heat capacity

When a magnetic solid is placed into a magnetic field, the strength H of the field changes, as does the inductance B of the magnetic material. The change in the field strength can be seen from the change in contours of constant field strength. For example, these changes can be seen by putting iron filings on a piece of paper above a magnet and then bringing another magnet close to the first one. The change in magnetic inductance is analogous to the alignment of a compass needle in Earth's magnetic field. The magnetic field strength H, the magnetic inductance B, and the magnetization M are all vector quantities.

The differential work done to change the magnetic inductance of a solid by an amount dB is

$$\delta w = H\,dB \int dV \tag{4-36}$$

where V is a volume element inside the magnetic field (in m^3), H is the magnetic field strength in A m^{-1}, and B is the magnetic induction in tesla (1 $T = 1$ N A^{-1} $m^{-1} = 1$ kg s^{-2} A^{-1}). The inductance B and the field strength H are related by μ_0, the magnetic permeability of free space, via

$$B = \mu_0 H \tag{4-37}$$

where $\mu_0 = 4\pi \times 10^{-7}$ N A^{-2} in the SI system and is dimensionless with a value of unity in the cgs system. Equation (4-37) is for a magnetic field in free space without any material inside the field. However, we are interested in what happens when a mineral is put into the magnetic field.

In this case it is useful to rewrite Eq. (4-36) in terms of the magnetization M of the mineral because magnetic work can then be related to magnetic susceptibility via Curie's law. This is done using the definition of M_V, the magnetization per unit volume, which is the change in H from the initial to final field strength when a magnetic solid is placed into it:

$$M_V = \Delta H = H_f - H_i = B/\mu_0 - H_i = J/\mu_0 \tag{4-38}$$

where M_V has units of A m^{-1}, and J is the magnetic polarization of the magnetic solid in tesla. The differential dB in Eq. (4-36) can thus be rewritten as

$$dB = \mu_0 \, dM_V \tag{4-39}$$

by differentiating Eq. (4-38) and rearranging. Substituting Eq. (4-39) into Eq. (4-36) gives

$$\delta w = H \mu_0 \, dM_V \int dV = H \mu_0 \, dM \tag{4-40}$$

as the work done to change the magnetization M (A $m^{-1} \times m^3$ = A m^2) of a magnetic solid by an amount dM. The product ($H \mu_0 \, dM = BdM$) is in joules (1 J = 1 N m = 1 Kg m^2 s^{-2}, as defined in Chapter 3). When work is done by a magnetic field to increase the magnetization of a magnetic mineral, dM is positive. When work is done by a magnetic mineral on the field, dM is negative. The first law for a magnetic material is then rewritten as

$$dE = \delta q - \delta w = \delta q - PdV + H\mu_0 dM = \delta q - PdV + BdM \tag{4-41}$$

and in the case where no PV work is done, Eq. (4-41) becomes

$$dE = \delta q + H\mu_0 dM = \delta q + BdM \tag{4-42}$$

Equation (4-42) is analogous to the first law because H and M are analogous with P and V. The magnetic field strength H and the magnetic inductance B are intensive variables, and the magnetization M of the solid is an extensive variable (= $M_V \times V$). Two magnetic heat capacities that are analogous to C_V and C_P can thus be defined:

$$C_M = \left(\frac{\delta q}{dT}\right)_M \tag{4-43}$$

$$C_H = \left(\frac{\delta q}{dT}\right)_H \tag{4-44}$$

where C_M is the magnetic heat capacity at constant magnetization and C_H is the magnetic heat capacity at constant field strength.

At low temperatures comparable to those at which the electronic heat capacity of metals is important, the magnetic heat capacity of ferromagnetic and ferrimagnetic solids is proportional to $T^{3/2}$ and C_V is given by

$$C_V = aT^3 + \gamma T + \delta T^{3/2} = C_{lat} + C_E + C_M \text{ J (g atom)}^{-1} \text{ K}^{-1} \tag{4-45}$$

where δ is the magnetic heat capacity coefficient. It is difficult to disentangle the electronic and magnetic contributions to the heat capacity of ferromagnetic materials that are electrical conductors. However, it is easier to evaluate Eq. (4-45) for ferrimagnetic materials such as magnetite that are nonconductors and have no C_E contribution (i.e., $\gamma = 0$). An example is given in Figure 4-20 where low-temperature heat capacity data for magnetite are plotted. The intercept on this plot gives $\delta = 3.6 \times 10^{-4}$ J mol^{-1} $K^{-5/2}$. Because C_M varies as $T^{3/2}$ while C_{lat} varies as T^3, C_M can become dominant at low

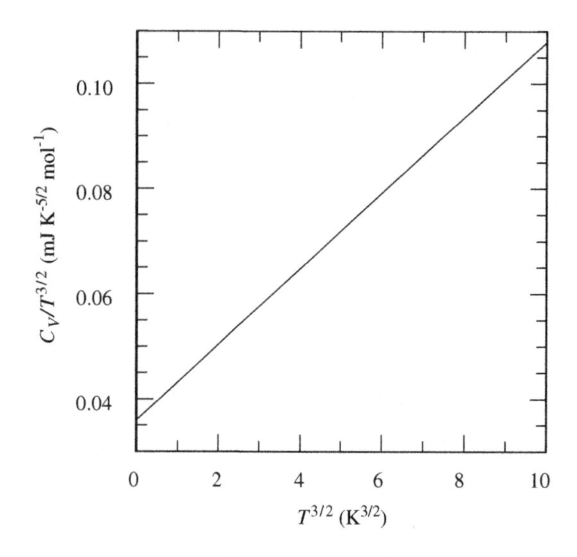

FIGURE 4-20

A plot of $C_V/T^{3/2}$ versus $T^{3/2}$ for magnetite (Fe_3O_4). The intercept gives δ, the magnetic heat capacity coefficient for magnetite.

The line plotted is from Dixon et al. (1965).

temperatures. The comparison of C_M and C_{lat} for magnetite is the topic of one of the problems at the end of this chapter.

Figure 4-21 shows heat capacities of four olivine minerals in the vicinity of their Néel points. The sharp discontinuities in the heat capacity curves, known as *lambda transitions*, are due to transitions from magnetically ordered antiferromagnetic states at lower temperatures to magnetically disordered paramagnetic states at higher temperatures. The large heat capacities in the lambda transitions are due to the magnetic heat capacities C_M of these four minerals. Figure 4-22 illustrates magnetic lambda transitions in metallic Ni (631.5 K) and Fe (1043 K). Two or more polynomials are generally needed to represent C_P in the vicinity of lambda transitions.

G. Lambda transitions of minerals

Lambda transitions occur in the heat capacity curves of many minerals as transitions from ordered to disordered states occur with increasing temperature. At the transition temperature (T_{trans}), the thermal energy is sufficient to overcome the energy barrier (ΔE) between the ordered and disordered states, that is, $\Delta E \sim kT_{trans}$, where k is Boltzmann's constant. A schematic lambda transition is shown in Figure 4-23; other examples appeared earlier in Figures 4-10 and 4-21. The order–disorder transitions are due to changes in crystal structure, in magnetic properties, in molecular orientations in a solid, electric properties, or other effects. For example, β-brass, which is an alloy of copper and zinc with the formula CuZn, undergoes a structural lambda transition at ~742 K whereby it changes from a lower-temperature phase with cubic symmetry to a higher-temperature disordered phase. Structural lambda transitions in troilite (420 K), quartz (844 K), and brass (742 K) are illustrated in Figures 4-10 and 4-23. Magnetic lambda transitions in troilite (440 K and 590 K), fayalite (64.9 K), Ni-olivine (29.2 K),

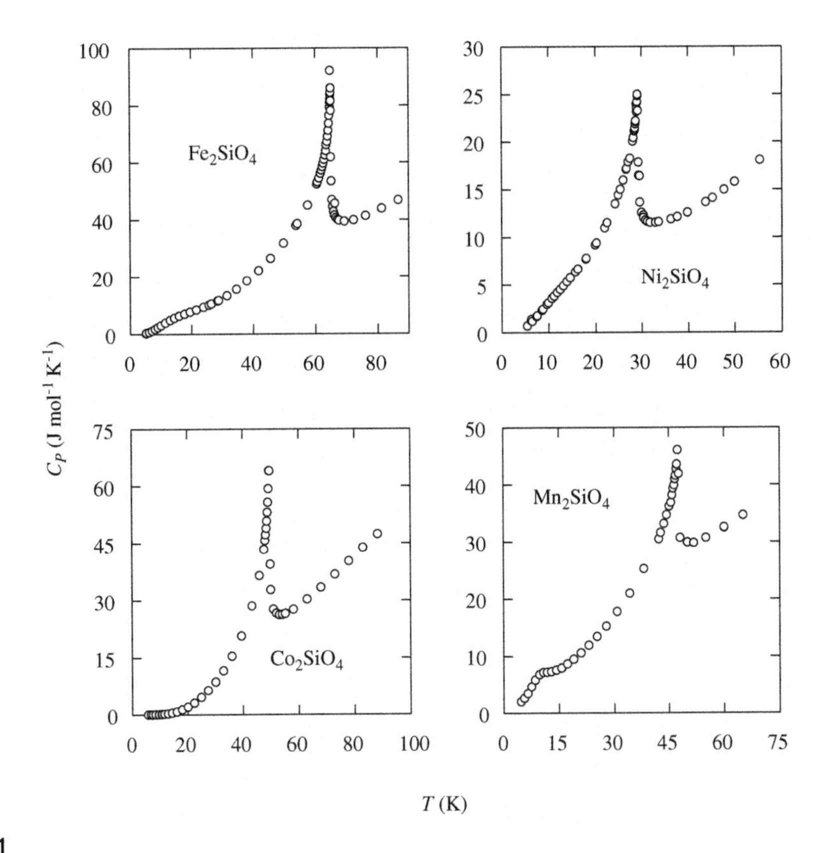

FIGURE 4-21

A plot showing the antiferromagnetic-to-paramagnetic transitions in fayalite (Fe_2SiO_4), Ni olivine (Ni_2SiO_4), Co olivine (Co_2SiO_4), and tephroite (Mn_2SiO_4). The heat capacity data are from Robie et al. (1982, 1984).

Co-olivine (49.8 K), tephroite (47.4 K), Ni metal (631.5 K), and Fe metal (1043 K) are shown in Figures 4-10, 4-21, and 4-22. A lambda transition at 242 K due to changes in the molecular orientation of NH_4^+ ions in solid ammonium chloride (NH_4Cl) is shown in Figure 4-23.

H. The difference between C_P and C_V of solids

The variation of heat capacity with high temperatures where C_P is greater than $3R$ is due to several effects, including the anharmonicity of lattice vibrations and the electronic heat capacity. This behavior is illustrated by the C_P and C_V curves of forsterite in Figure 4-15. As we mentioned earlier, $C_P - C_V$ is about 5% of C_P at room temperature. The molar $C_P - C_V$ difference can be calculated from the equation

$$(C_P - C_V)_m = \frac{\alpha^2 V_m T}{\beta_T} \tag{4-46}$$

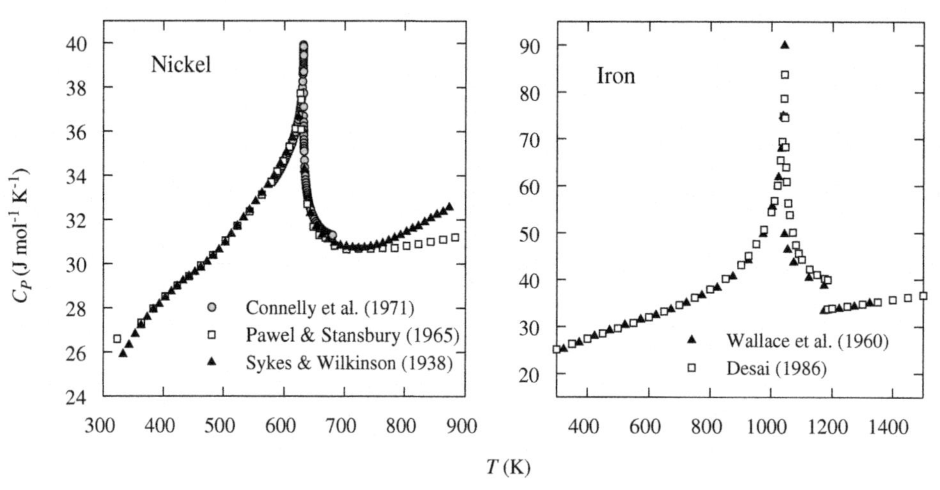

FIGURE 4-22

A plot showing the molar heat capacities of Fe and Ni metal near their Curie points (1043 K for Fe and 631.5 K for Ni).

The data plotted are from Connelly et al. (1971), Desai (1986), Pawel and Stansbury (1964), Sykes and Wilkinson (1938), and Wallace et al. (1960).

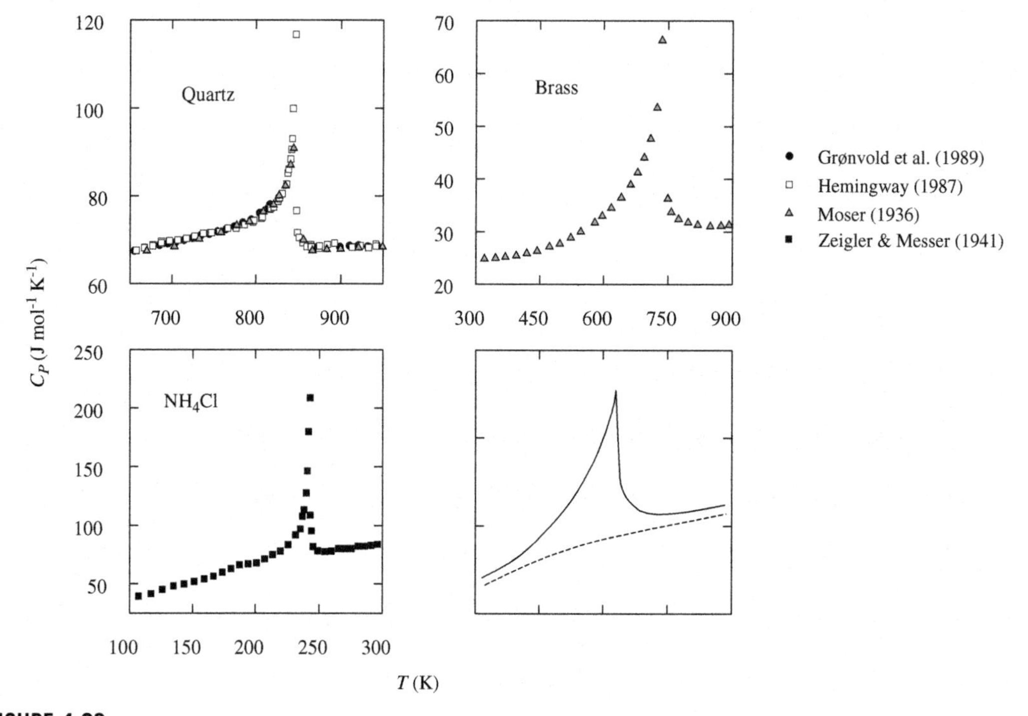

FIGURE 4-23

A plot showing the molar heat capacities of quartz, brass, and NH_4Cl near their lambda anomalies and a schematic lambda transition.

The data plotted are from Moser (1936), Ziegler and Messer (1941), Hemingway (1987), and Grønvold et al. (1989).

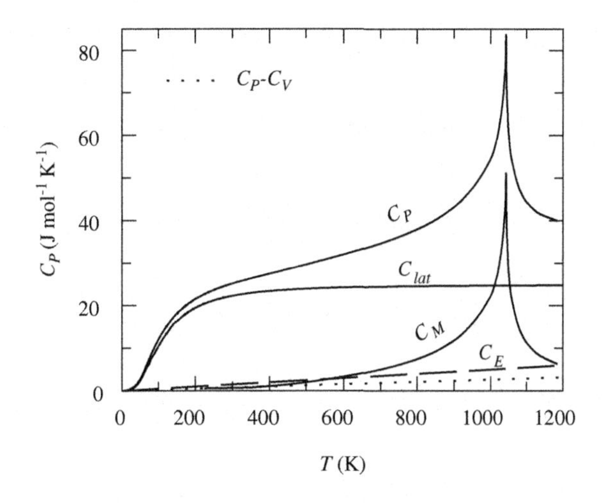

FIGURE 4-24

The molar heat capacity of Fe metal showing measured C_P data, ($C_P - C_V$), and the lattice (C_{lat}), electronic (C_E), and magnetic (C_M) contributions to C_V.

where α is the isobaric thermal expansion coefficient, β_T is the isothermal compressibility, V_m is the molar volume, and T is in Kelvins. We derive Eq. (4-46) in Chapter 8. In principle, Eq. (4-46) gives ($C_P - C_V$), but calculations are difficult in practice because the temperature dependence of α, β, and V_m must all be known. Thus, an approximate relationship, such as

$$(C_P - C_V)_m = A(C_P)^2 T \tag{4-47}$$

which is the *Nernst-Lindemann equation*, is often used instead. The empirical constant A is evaluated from the measured $C_P - C_V$ term at one temperature, usually room temperature, and is remarkably constant with temperature. For example, A for Cu metal varies from $\sim 6.40 \times 10^{-5}$ mol J^{-1} at 100 K to $\sim 6.44 \times 10^{-5}$ mol J^{-1} at 1000 K, which is only a 0.6% difference. Another example is provided by Figure 4-24, which shows C_P, C_{lat}, C_M, C_E, and $C_P - C_V$ for Fe metal over a 1200 K range. The $C_P - C_V$ term is only a small fraction of the total heat capacity.

I. Calculating enthalpies for solids with phase changes

Earlier in this chapter, we showed how to integrate an empirical heat capacity equation to derive an equation for enthalpy as a function of temperature. This is a straightforward procedure unless we need to compute the enthalpy of a material that has one or more solid-state phase changes. As illustrated in Figure 4-25, iron metal is an excellent example of a material that has several different stable solid polymorphs as a function of temperature. How do we calculate ($H_T - H_{298}$) for iron metal and other elements and minerals with similar behavior?

Example 4-11. Iron metal has three solid-state phase transitions: a ferromagnetic-to-paramagnetic transition at the Curie point (1043 K), a transition from the body-centered cubic α phase ferrite to the face-centered cubic γ phase austentite (1185 K), and a transition from the γ phase back to the

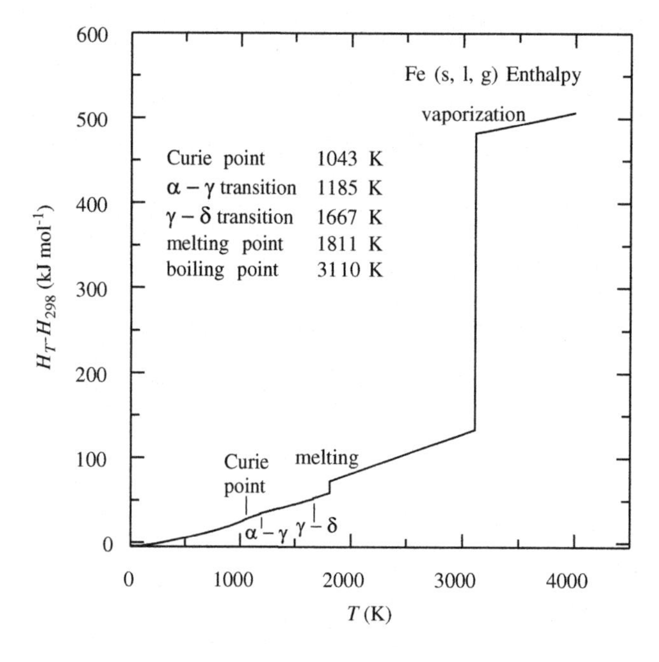

FIGURE 4-25

A graph of the enthalpy of iron metal, liquid, and gas from 0 K to 4000 K. Smoothed data from Desai (1986) and the NIST-JANAF tables are plotted. All enthalpies are shown to scale.

body-centered cubic δ phase (1667 K). The melting point of Fe is 1811 K and the normal boiling point is 3110 K. What is the enthalpy of monatomic Fe (g) at 3200 K?

The enthalpy $(H_{3200} - H_{298})$ of Fe (g) at 3200 K is the sum of the heat contents of each phase (α, γ, δ, liquid, gas) and the enthalpies of the different phase transitions:

$$\begin{aligned}(H_{3200} - H_{298})_{\text{Fe(g)}} = & \int_{298}^{1043} C_P(\alpha - \text{Fe})dT + \Delta_{trans}H_{1043}(\text{Curie}) + \int_{1043}^{1185} C_P(\alpha - \text{Fe})dT \\ & + \Delta_{trans}H_{1185}(\alpha \rightarrow \gamma) + \int_{1185}^{1667} C_P(\gamma - \text{Fe})dT + \Delta_{trans}H_{1667}(\gamma \rightarrow \delta) \\ & + \int_{1667}^{1811} C_P(\delta - \text{Fe})dT + \Delta_{fus}H_{1811}(\delta \rightarrow \text{liq}) + \int_{1811}^{3110} C_P(\text{liq Fe})dT \\ & + \Delta_{vap}H_{3110}(\text{liq} \rightarrow \text{gas}) + \int_{3110}^{3200} C_P(\text{Fe gas})dT \end{aligned} \quad (4\text{-}48)$$

The different terms in Eq. (4-48) can be evaluated using the heat capacity equations given in Appendix 2. Starting with the first term in Eq. (4-48), we then have

$$\int_{298}^{1043} C_P(\alpha - \mathrm{Fe})dT = (H_{1043} - H_{298})_{\alpha-\mathrm{Fe}} = 25{,}916\ \mathrm{J\ mol^{-1}} \tag{4-49}$$

$$\Delta_{trans} H_{1043}\ (\mathrm{Curie}) = 1{,}364\ \mathrm{J\ mol^{-1}} \tag{4-50}$$

$$\int_{1043}^{1185} C_P(\alpha - \mathrm{Fe})dT = (H_{1185} - H_{1043})_{\alpha-\mathrm{Fe}} = 6{,}619\ \mathrm{J\ mol^{-1}} \tag{4-51}$$

$$\Delta_{trans} H_{1185}\ (\alpha \rightarrow \gamma) = 900\ \mathrm{J\ mol^{-1}} \tag{4-52}$$

$$\int_{1185}^{1667} C_P(\gamma - \mathrm{Fe})dT = (H_{1667} - H_{1185})_{\gamma-\mathrm{Fe}} = 17{,}391\ \mathrm{J\ mol^{-1}} \tag{4-53}$$

$$\Delta_{trans} H_{1667}\ (\gamma \rightarrow \delta) = 850\ \mathrm{J\ mol^{-1}} \tag{4-54}$$

$$\int_{1667}^{1811} C_P(\delta - \mathrm{Fe})dT = (H_{1811} - H_{1667})_{\delta-\mathrm{Fe}} = 5{,}737\ \mathrm{J\ mol^{-1}} \tag{4-55}$$

$$\Delta_{fus} H_{1811}\ (\delta \rightarrow \mathrm{liq}) = 13{,}810\ \mathrm{J\ mol^{-1}} \tag{4-56}$$

$$\int_{1811}^{3110} C_P(\mathrm{liq\ Fe})dT = (H_{3110} - H_{1811})_{\mathrm{Fe}(liq)} = 58{,}449\ \mathrm{J\ mol^{-1}} \tag{4-57}$$

$$\Delta_{vap} H_{3110}\ (\mathrm{liq} \rightarrow \mathrm{gas}) = 349{,}213\ \mathrm{J\ mol^{-1}} \tag{4-58}$$

$$\int_{3110}^{3200} C_P(\mathrm{Fe\ gas})dT = (H_{3200} - H_{3110})_{\mathrm{Fe(g)}} = 2{,}406\ \mathrm{J\ mol^{-1}} \tag{4-59}$$

Summing up all these terms, we obtain

$$(H_{3200} - H_{298})_{\mathrm{Fe(g)}} = 482{,}655\ \mathrm{J\ mol^{-1}} \tag{4-60}$$

The Curie point is a heat capacity maximum, but as discussed, there is no isothermal heat effect at the Curie point, which is a lambda transition. Instead, as illustrated in Figure 4-22, there is an elevation of the C_P curve over a wide temperature range in the vicinity of the Curie point. Kelley (1960) modeled the heat effect due to the elevated C_P curve by assigning $\Delta_{trans}H_{1043}$ (Curie) = 1364 J mol^{-1}, which is not strictly correct, but it allows us to use Maier-Kelley equations or other simple polynomials to represent the heat capacity data for iron. More complex polynomials with more terms can be used to represent the lambda transition and heat capacity curve in its vicinity, or we could numerically integrate the heat capacity at and in the vicinity of the lambda transition.

PROBLEMS

1. (a) Integrate the molar C_P equation for N_2 (g) given in Table 3-4 to find an equation for the molar enthalpy, $(H_T - H_{298})$, of N_2 (g). (b) Calculate the molar enthalpy of N_2 (g) at 2000 K.
2. Sinoite (Si_2N_2O) occurs in enstatite chondrite meteorites. Use the Neumann-Kopp rule and the following data to estimate its C_P at 298.15 K.

$$Si_3N_4 \text{ (solid)} + SiO_2 \text{ (quartz)} = 2\ Si_2N_2O\text{(sinoite)}$$

Mineral	$C_{P,m}$ (298.15 K) J mol^{-1} K^{-1}
Si_3N_4	93.01
quartz	44.59

3. Calculate the lattice (C_{lat}) and electronic (C_E) components of the heat capacity of Fe metal at low temperatures and calculate the Debye temperature for Fe from the following data (Desai, 1986).

T (K)	1	2	3	4	5	10	15	20
$C_{P,m} \times 10^3$ (J mol^{-1} K^{-1})	4.961	10.04	15.35	21.01	27.13	69.8	138	256

4. Normal heptane (n-C_7H_{16}) is a constituent of gasoline. Molar heat capacity data for gaseous C_7H_{16} are given here. (a) Fit a C_P equation of your choice to these data. (b) Integrate this equation to get an expression for the molar enthalpy $(H_T - H_{298})$ of C_7H_{16} (g), and calculate $(H_{1000} - H_{298})$ for one mole of n-heptane.

T (K)	298	400	600	800	1000
C_P (cal mol^{-1} K^{-1})	39.67	50.42	68.33	81.43	91.2

5. Estimate the molar C_P of petalite ($LiAlSi_4O_{10}$) at 298.15 K using the Neumann-Kopp rule and the C_P data tabulated here.

Oxide	Li_2O	Al_2O_3	SiO_2
C_P (J mol^{-1} K^{-1})	54.10	79.03	44.6

6. Staveley and Linford (1969) measured the low-temperature heat capacity of calcite and aragonite (polymorphs of $CaCO_3$). They report $C_{P,m} = 83.74$ J mol^{-1} K^{-1} and $(C_P - C_V) = 1.12$ J mol^{-1} K^{-1} at 298.15 K for calcite. (a) Use the Nernst-Lindemann relation to compute $(C_P - C_V)$ at 100-degree intervals from 300 K to 1200 K, and (b) use your results and the C_P equation from Table 3-3 to plot C_V and C_P from 300 K to 1200 K for calcite.
7. Calculate the Debye temperature for pyrite (FeS_2) from the low-temperature specific heat data tabulated here. (The numbers in parenthesis are exponents, so $1.465(-4) = 1.465 \times 10^{-4}$.)

T (K)	8.87	11.37	45.88	53.99
c_p (J g^{-1} K^{-1})	1.465(–4)	2.790(–4)	1.761(–2)	3.004(–2)

8. The normal freezing point of air-saturated water is 273.15 K, where the heat of fusion is 6009.5 J mol^{-1}. However, water can be supercooled 40 degrees below this point. Calculate the heat of fusion of water at –40°C using the C_P data tabulated here.

	$C_{P,m}$ (J mol^{-1} K^{-1})	
T (°C)	**Water**	**Ice**
–40	85.99	32.52
–30	81.46	33.82
–20	78.46	35.13
–10	76.94	36.45
0	76.17	37.78

9. Assume that the volume of a mineral varies with pressure (P) and temperature (T) according to the equation $V = V_0 - AP + BT$, that is, this is the equation of state for the mineral. The internal energy E of the mineral is given by the equation $E = CT + (V - V_0)^2/2A$, where V_0, A, B, and C are constants. Derive expressions for the constant-volume heat capacity C_V and the constant-pressure heat capacity C_P.
10. The Lindemann equation relates the Debye temperature θ_D to the melting point (T_m), molar volume (V_m), and molecular weight (μ) of a solid with a constant (c'):

$$\theta_D = \left(c'/\sqrt[3]{V_m}\right)\left(\sqrt{\frac{T_m}{\mu}}\right)$$

(a) Find the value of the constant for nonmetals by plotting the data tabulated here:

Solid	Chlorargyrite (AgCl)	Sylvite (KCl)	Water Ice	Ge	Si
$(T_m/\mu)^{1/2}/V_m^{1/3}$	0.764	1.117	1.444	1.710	3.378
θ_D	180	230	315	362	630

(b) Estimate the melting point of periclase (MgO), which has a Debye temperature of 820 K, a molar volume of 11.25 cm^3, and a molecular weight of 40.304 g mol^{-1}.

11. Coughlin et al. (1951) measured the heat content ($H_T - H_{298}$) of synthetic hematite (Fe_2O_3) using drop calorimetry. A subset of their data is given here. Use these data to (a) derive a polynomial for $H_T - H_{298}$, (b) derive a polynomial for C_P, (c) tabulate C_P at 298.15 K and from 400 K to 900 K in 100° increments.

Measured Heat Content of Hematite

T (K)	$H_T - H_{298}$ (cal mol^{-1})	T (K)	$H_T - H_{298}$ (cal mol^{-1})
374.8	2005	777.6	15,290
445.4	4170	872.9	18,930
547.7	7300	924.5	21,030
650.1	10,710	948	21,970

12. Very accurate and precise measurements show that the heat capacity of liquid mercury decreases with temperature, from its melting point (234.29 K) to its one atmosphere boiling point (629.84 K). (a) Calculate an equation for $H_T - H_{298}$ (J mol^{-1}) for liquid mercury from the heat capacity equation $C_P = 6.44 + 0.19 \times 10^5\ T^{-2}$ cal mol^{-1} K^{-1} given by Kelley (1960), and (b) tabulate your results at 100-degree increments from 300 K to 600 K.

13. A geochemist wants to measure the molar constant-pressure heat capacity of ammonia (NH_3) and performs an experiment similar to that in Example 4-6. Ammonia at one bar pressure is passed over an electrically heated filament at a STP flow rate of 32.80 cm^3 s^{-1}. The temperature of the gas increases from 419°C to 425°C. The filament is heated by a 0.0468 A current at 9 V. What is the $C_{P,m}$ of NH_3 at 422°C? Neglect corrections for energy lost to the surroundings.

14. Stout and Piwinskii (1982) used drop calorimetry to measure the enthalpy of liquid diopside ($CaMgSi_2O_6$) from 1695 K to 2562 K. Use their data, tabulated here, to calculate the C_P equation.

T (K)	$H_T - H_{298}$ (J g^{-1})	T (K)	$H_T - H_{298}$ (J g^{-1})
1695	1718.8	2126	2412.6
1824	1928	2266	2599.8
1826	1947.4	2422	2845.1
1973	2172.5	2562	3013.6

15. The sound speed (w, in m s^{-1}) in an ideal gas is related to γ, the ratio of C_P/C_V by the equation $w^2 = RT\gamma/\mu$, where μ is the molecular weight of the gas. Calculate the sound speed in dry air at 25°C assuming $\gamma = 1.40$.

16. Calculate C_{lat}, C_M, and θ_D for magnetite using the coefficients $\delta = 0.36$ mJ $K^{-5/2}$ mol^{-1}, $\alpha = 0.072$ mJ K^{-4} mol^{-1}, determined by Dixon et al. (1965). Consider a temperature range of about 1.5 K to 4 K. Also calculate the temperature below which $C_M > C_{lat}$.

17. Show that $(H_T - H_0) = TC_P/4$, where C_P is the heat capacity at T (K) for any solid that obeys the Debye T^3 law.

18. The complex sulfide djerfisherite is found in three types of meteorites: irons, enstatite chondrites, and enstatite achondrites (aubrites). Its ideal chemical formula is $K_3CuFe_{12}S_{13}Cl$. Use Kopp's rule to estimate the molar C_P of djerfisherite at 298.15 K. Do not use elements as reactants; instead, use less complex sulfides and other compounds.

19. The amount of latent heat transferred from Earth's surface into the atmosphere is an important contribution to the overall heat balance of the planet. Use the average rainfall rate of 97.3 g cm^{-2} yr^{-1} to estimate this latent heat flux (W m^{-2}). You also need the enthalpy of vaporization of water at 288 K, Earth's average surface temperature. Compute this from the enthalpy of vaporization at 0°C (597.3 cal g^{-1}) and specific heat difference between water vapor and liquid water ($\Delta c_p = -0.566$ cal g^{-1} K^{-1}).

20. Rhodochrosite ($MnCO_3$) occurs in hydrothermal veins and dissolved in calcite ($CaCO_3$) and siderite ($FeCO_3$). Upon heating, it loses CO_2 and MnO (manganosite) forms:

$$MnCO_3\ (\text{rhodochrosite}) = MnO\ (\text{manganosite}) + CO_2\ (g)$$

$$\Delta C_P = 16.560 - 63.890 \times 10^{-3}T + 32.658 \times 10^{-6}T^2\ \text{J mol}^{-1}\ \text{K}^{-1}$$

Calculate $\Delta(H_T - H_{298})$ for rhodochrosite decomposition from the ΔC_P equation. This will be used in subsequent problems dealing with the same reaction.

CHAPTER

Thermochemistry

5

Wenn eine Verbindung stattfindet, so ist die entwickelte Wärmemenge constant, es mag die Verbindung direct oder indirect und zu wiederholten Malen geschehen.
(The amount of heat developed when a compound forms is always the same whether the combination occurs directly or indirectly.)
—Germain Henri Hess (1840)

The term *enthalpy* (H) was introduced and defined in Chapter 3, and the relationship between enthalpy and the constant-pressure heat capacity (C_P) was described in detail in Chapter 4. In this chapter we discuss *thermochemistry*, which is the measurement and interpretation of enthalpy changes due to chemical reactions, formation of solutions, and changes in state.

This chapter is divided into three sections. In Section I, we introduce several basic concepts and definitions, including the distinction between the ΔH and ΔE of reaction; the definitions of exothermic and endothermic reactions; the thermochemical laws of Lavoisier and Laplace, Hess, and Kirchhoff; the concepts of elemental reference states and the standard state; the enthalpy of solution and dilution; and the enthalpy of ionic reactions in aqueous solutions. In Section II we describe the use of solution and combustion calorimetry for measuring reaction enthalpies. In Section III we discuss several applications of thermochemistry, including adiabatic flame temperatures, bond dissociation energies, ionization potentials, and electron affinities.

As stated in Chapter 4, throughout the book we write 298 K; however, you should interpret this as 298.15 K, and all our computations are done with respect to 298.15 K. The new nomenclature introduced in this chapter follows that recommended by the International Union of Pure and Applied Chemists (IUPAC), with some exceptions.

I. SOME BASIC CONCEPTS

A. The distinction between the ΔH and ΔE of reaction

Many of the chemical reactions and phase changes that occur in our everyday life, such as CO_2 bubbling out of a soda, sugar dissolving in a cup of coffee, milk turning sour, developing and fixing photographic film, hardening of epoxy, or rusting of iron and steel, occur at atmospheric pressure, which is constant at about one bar, within 1–3% (i.e., 10–30 millibars), except under extreme weather conditions. Thus, the heat effects associated with these chemical reactions and phase changes are enthalpy changes because the heat flow q_P for a constant-pressure process

Copyright © 2013 Elsevier Inc. All rights reserved.

$$H_2 - H_1 = \Delta H = q_P \tag{3-38}$$

is an enthalpy change ΔH (see Section II-H of Chapter 3). When Eq. (3-38) is applied to a chemical reaction, the enthalpy change ΔH is the difference between the sum of enthalpies of the products and the sum of enthalpies of the reactants:

$$\Delta H = \sum (\nu_i H_i)_{pr} - \sum (\nu_i H_i)_{re} \tag{5-1}$$

where the enthalpy of each compound (H_i) is multiplied by its stoichiometric coefficient (ν_i) in the chemical equation, and the subscripts *pr* and *re* stand for products and reactants, respectively.

Consider the reaction of hematite (Fe_2O_3) with hydrochloric acid gas to form condensed $FeCl_3$:

$$Fe_2O_3 \text{ (hematite) } + 6\ HCl \text{ (gas)} = 2\ FeCl_3 \text{ (molysite)} + 3\ H_2O \text{ (liquid)} \tag{5-2}$$

which occurs in volcanic vents and fumaroles (e.g., at Mount Saint Augustine in Alaska).

The enthalpy change for reaction (5-2) is

$$\begin{aligned} \Delta H &= (3H_{H_2O} + 2H_{FeCl_3}) - (6H_{HCl} + H_{Fe_2O_3}) \\ &= 3(-285.83) + 2(-399.5) - 6(-92.31) - (-826.2) = -276.43 \text{ kJ mol}^{-1} \end{aligned} \tag{5-3}$$

where the enthalpies of the different compounds were taken from Appendix 1. Likewise, the internal energy change for reaction (5-2) is the difference between the sum of the internal energies of the products and the sum of the internal energies of the reactants, with each E_i value multiplied by the appropriate stoichiometric coefficient, that is, an equation analogous to Eq. (5-1).

The ΔH and ΔE changes for reaction (5-2) are numerically different because H is defined as

$$H = E + PV \tag{3-37}$$

Hence the difference between the ΔH and ΔE of reaction (5-2) is given by

$$\Delta H - \Delta E = \Delta(E + PV) - \Delta E = \Delta(PV) \tag{5-4}$$

which becomes

$$\Delta H - \Delta E = P\Delta V \tag{5-5}$$

when reaction (5-2) occurs at constant pressure.

Equations (5-4) and (5-5) are valid for any chemical reaction or any change of state taking place at constant pressure. In general, the numerical difference between ΔH and ΔE is negligible for reactions involving only condensed phases such as solids, liquids, and melts, because $P\Delta V$ is small relative to ΔE. However, $P\Delta V$ is much larger for reactions involving gases, and it must be taken into account explicitly. The following example illustrates these points.

Example 5-1. (a) Calculate $\Delta H - \Delta E$ for the hydration of lime at a constant pressure of one bar. This reaction is one of several that occurs during hardening of Portland cement:

$$CaO \text{ (lime)} + H_2O \text{ (liquid)} = Ca(OH)_2 \text{ (portlandite)} \tag{5-6}$$

and has $\Delta H = -65{,}400$ J mol^{-1} and $\Delta V = -1.768$ cm^3 mol^{-1} at 298 K. Using Eq. (5-5) and the energy conversion factors in Table 3-2 we compute

$$\Delta H - \Delta E = P\Delta V = -(0.1768 \text{ J bar}^{-1} \text{ mol}^{-1})(1 \text{ bar}) \sim -0.18 \text{ J mol}^{-1} \quad (5\text{-}7)$$

at one bar pressure. In this case $P\Delta V$ is only about 0.0003% of ΔH and is insignificant. In other words, $\Delta H = \Delta E$ for reaction (5-6) for all practical purposes.

(b) Now we consider a reaction involving gases, and we compute the difference between ΔH and ΔE for rust formation on iron metal at a constant pressure of one bar. Rust formation is a complex process that we represent with the simplified chemical reaction

$$4 \text{ Fe (metal)} + 2 \text{ H}_2\text{O (liquid)} + 3 \text{ O}_2 \text{ (gas)} = 2 \text{ Fe}_2\text{O}_3 \cdot \text{H}_2\text{O (goethite)} \quad (5\text{-}8)$$

Reaction (5-8) has $\Delta H = -1678.7$ kJ mol^{-1} and $\Delta V = -67{,}223.3$ cm^3 mol^{-1} at 298 K. We find

$$\Delta H - \Delta E = P\Delta V = -(6722.3 \text{ J bar}^{-1} \text{ mol}^{-1})(1 \text{ bar}) \sim -6.7 \text{ kJ mol}^{-1} \quad (5\text{-}9)$$

so the difference between ΔH and ΔE is larger (~0.4%) than for lime hydration.

Sometimes an equation such as

$$\Delta H - \Delta E = P\Delta V \sim RT\Delta n \quad (5\text{-}10)$$

where Δn is the change in number of moles of gas in the reaction is used, for example, $\Delta n = -3$ for reaction (5-8). However, Eq. (5-10) is an approximation for two reasons. Real gases are nonideal and another equation of state must be used for the $P\Delta V$ correction. Second, the molar volumes of condensed phases are neglected in the $RT\Delta n$ term, which only accounts for $P\Delta V$ of the gas phase.

Equations (5-6) and (5-8) in Example 5-1 are the first of many chemical equations we will use in this chapter and throughout the rest of the book. Instead of writing out designations such as gas, liquid, metal, or solid one or more times in each equation, we will abbreviate using aq for an aqueous solution, c for a crystalline solid, g for a gas, l for a liquid, m for a metal, and s for a noncrystalline solid. However, we will write out mineral names. For simplicity, we may omit the (g) in chemical equations where it is obvious that a compound is a gas.

B. Exothermic and endothermic reactions

Exothermic reactions are chemical reactions that produce heat. In Section II-B of Chapter 3, we defined the heat flow q as negative when heat flows from the system to the surroundings. Thus, exothermic reactions have a negative ΔH of reaction. The word *exothermic* is derived from *exo,* the Greek word for outside, and *therme*, the Greek word for heat. Table 5-1 gives examples of several exothermic reactions. *Endothermic reactions* are chemical reactions that take heat from the surroundings. In Chapter 3, we defined the heat flow q as positive when a system gains heat from the surroundings. Thus, endothermic reactions have a positive ΔH of reaction. The word *endothermic* is derived from *endo*, the Greek word for inside, and *therme*. Table 5-2 gives examples of several endothermic reactions. The description of chemical reactions as exothermic or endothermic dates to 1869, when these terms were introduced by the French thermochemist Marcellin Berthelot (see sidebar).

PIERRE MARCELLIN BERTHELOT (1827–1907)

Berthelot was a French chemist whose work confirmed several important ideas. He used bomb calorimetry to measure the enthalpies of hundreds of reactions. He thereby verified Hess' law, although he was not the first to propose it. He coined the terms endothermic and exothermic. He thought that the heat evolved by reactions drove them; however, we now know that some endothermic reactions proceed spontaneously.

His work in organic chemistry led to some very important conclusions. Berthelot synthesized many organic compounds, thus showing that biological processes are not required to form them. He believed that chemical processes were nothing mystical, but that they were subject to the same physical laws that apply to everything in the universe. While experimenting with the synthesis of organic compounds, Berthelot noticed that his reactions never went to completion. Instead, they stopped at some definite midway point. The products were present in some proportion to the unused reactants which was the same regardless of how much material he started with. This was the beginning of the study of chemical equilibria.

Berthelot had interests beyond his organic chemistry experiments. Active in the government, he was elected as a senator for the first time in 1871, despite the fact that he had not run for the office. He was chosen by voters because of his work as the president of the Comité Scientifique pour la Défense de Paris during the Franco-Prussian war. He also studied the history of chemistry, which is commonly thought to have originated in Egypt. He felt that alchemy arose out of a misinterpretation of the practical knowledge of Egyptian metal smiths. By analyzing metal artifacts from Egypt and Mesopotamia, he created the new field of chemical archaeology. Berthelot intertwined his interests in science and philosophy. He felt there was no limit to possible scientific attainments and that a Utopia could be achieved through scientific principles by the year 2000.

Happily married for 45 years, Berthelot died within an hour after his wife. They had six children.

Home heating with natural gas, which contains >80% methane in most U.S. cities, is an example of an exothermic reaction. When natural gas is burned in air, for example, in a furnace that is heating a house or apartment building, CH_4 combustion occurs by the reaction

$$2\,CH_4 + 4\,O_2 = 2\,CO_2 + 4\,H_2O\,(l) \qquad \Delta H_{298} = -1780.72\text{ kJ} \tag{5-11}$$

where ΔH_{298} stands for the ΔH value at 298.15 K. Of course, the gas is being burned at a much higher temperature than this. Later in this chapter we will describe how ΔH values change with temperature and how to calculate flame temperatures. For now we will discuss and compare ΔH values at 298 K. The combustion of ethane (C_2H_6) or propane (C_3H_8), the next most abundant hydrocarbons in natural gas, is also exothermic, for example,

Table 5-1 Examples of Exothermic Reactions

Reaction	$\Delta_r H^o$ (kJ mol^{-1})
$\frac{1}{2}$ H_2 (g) + $\frac{1}{2}$ Cl_2 (g) = HCl (g) formation of hydrogen chloride gas from the elements	−92.31
C (graphite) + 2 H_2 (g) = CH_4 (g) formation of methane from the elements	−74.81
NH_3 (g) + H_2S (g) = NH_4SH (s) reaction of NH_3 and H_2S to form solid ammonium hydrosulfide	−90.32
KOH (c) + ∞ H_2O (l) = K^+ (aq) + OH^- (aq) + ∞ H_2O (l) potassium hydroxide dissolving in water to make an infinitely dilute solution	−57.46
CaO (lime) + H_2O (l) = $Ca(OH)_2$ (portlandite) hydration of lime	−65.40
H_2O_2 (aq) = H_2O (l) + $\frac{1}{2}$ O_2 (g) decomposition of aqueous hydrogen peroxide to water and O_2 (g)	−94.66
2 NO_2 (g) = N_2O_4 (g) formation of nitrogen tetroxide from nitrogen dioxide	−57.36

Enthalpy values at 298.15 K from Appendix 1.

$$2\,C_2H_6 + 7\,O_2 = 4\,CO_2 + 6\,H_2O\,(1) \quad \Delta H_{298} = -3119.6\,\text{kJ} \tag{5-12}$$

$$C_3H_8 + 5\,O_2 = 3\,CO_2 + 4\,H_2O\,(1) \quad \Delta H_{298} = -2220.0\,\text{kJ} \tag{5-13}$$

The ΔH value for combustion of one mole of a hydrocarbon or any other combustible material in air or any other gas is the molar enthalpy of combustion ($\Delta_c H_m$) or molar heat of combustion for the material and is written with the name of the combusted material, its phase, and the temperature following in parenthesis. Thus, $\Delta_c H_m$ (CH_4, g, 298) = −890.36 kJ mol^{-1} is the molar enthalpy of combustion for methane gas at 298.15 K, and $\Delta_c H_m$ (C_3H_8, g, 298) = −2220.0 kJ mol^{-1} is the molar enthalpy of combustion for propane gas at 298.15 K. Liquid water is a product in reactions (5-11) to (5-13) because enthalpy of combustion values are experimentally measured at 100% relative humidity to ensure its formation. The enthalpy of combustion is important industrially for evaluating the caloric value and the cost of fuels such as coal, gas, and oil. The enthalpy of combustion is also important scientifically for determining enthalpy of formation values (see the following discussion), which cannot be directly measured because of experimental difficulties.

Several examples of endothermic reactions are given in Table 5-2. Two well-known endothermic reactions are the formation of hydrogen iodide gas (HI) and the formation of nitric oxide gas (NO) from the constituent elements:

$$\frac{1}{2}\,H_2\,(g) + \frac{1}{2}\,I_2\,(c) = HI\,(g) \quad \Delta H_{298} = +26.50\,\text{kJ mol}^{-1} \tag{5-14}$$

$$\frac{1}{2}\,N_2\,(g) + \frac{1}{2}\,O_2\,(g) = NO\,(g) \quad \Delta H_{298} = +91.27\,\text{kJ mol}^{-1} \tag{5-15}$$

Table 5-2 Examples of Endothermic Reactions

Reaction	$\Delta_r H^o$ (kJ mol^{-1})
½ H_2 (g) + ½ I_2 (c) = HI (g) formation of hydrogen iodide gas from the elements	26.50
½ N_2 (g) + ½ O_2 (g) = NO (g) formation of nitric oxide from the elements	91.27
NH_4Cl (c) + ∞ H_2O (l) = NH_4^+ (aq) + Cl^- (aq) + ∞ H_2O (l) NH_4Cl dissolving in water to make an infinitely dilute solution	14.06
NaCl (c) + ∞ H_2O (l) = Na^+ (aq) + Cl^- (aq) + ∞ H_2O (l) table salt dissolving in water to make an infinitely dilute solution	3.88
C (graphite) = C (diamond) transformation of graphite into diamond	1.85
S (rhombic) = S (monoclinic) transformation of rhombic to monoclinic sulfur	0.36
H_2O (ice) = H_2O (water) ice melting to water (273.15 K)	6.0095
Ga (metal) = Ga (l) gallium metal melting in your hand (302.15 K)	5.585
Hg (l) = Hg (g) vaporization of liquid mercury to Hg (g)	61.38
H_2O (l) = H_2O (g) water vaporizing to steam (373.15 K)	40.656
C_3H_6O (l) = C_3H_6O (g) acetone vaporization at its boiling point (329.3 K)	29.09
CO (g) = C (g) + O (g) dissociation of carbon monoxide to the atoms	1076.4
$CaCO_3$ (limestone) = CaO (lime) + CO_2 (g) calcination of limestone during cement manufacture	179.0
I_2 (c) = I_2 (g) sublimation of iodine	62.42

Enthalpy values at 298.15 K unless noted otherwise. Data are from Appendix 1.

The ΔH value in each case is the *molar enthalpy of formation* ($\Delta_f H_m$) because the HI or NO are formed from their constituent elements. Likewise, the ΔH values for hydrogen chloride (HCl) gas and CH_4 in Table 5-1 are the molar enthalpies of formation of these two compounds.

C. Some other types of ΔH values

The ΔH for reaction of ammonia (NH_3) and hydrogen sulfide (H_2S) to form solid ammonium hydrosulfide (NH_4SH) in Table 5-1 is a *molar enthalpy of reaction* ($\Delta_r H_m$). Likewise, the ΔH for calcination of limestone in Table 5-2 is also a $\Delta_r H_m$ value. The ΔH for reaction of 1-butene (H_3C-CH_2-CH=CH_2) with hydrogen to form n-butane (H_3C-CH_2-CH_2-CH_3):

$$C_4H_8\ (g) + H_2(g) = C_4H_{10}\ (g) \qquad \Delta H_{355} = -126.95\ \text{kJ mol}^{-1} \qquad (5\text{-}16)$$

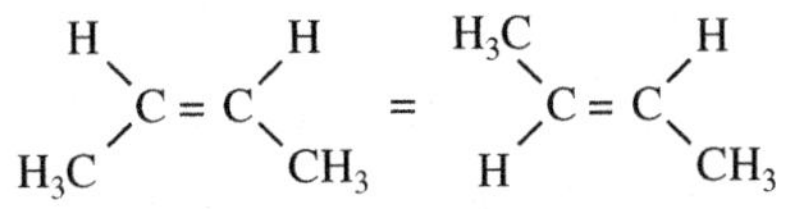

FIGURE 5-1

The isomerization of the cis and trans geometrical isomers of 2-butene. The H and CH_3 groups are on the same side of the double bond in the cis isomer and on opposite sides in the trans isomer.

is the *molar enthalpy of hydrogenation* of 1-butene. The ΔH for conversion of the cis and trans isomers of 2-butene (H_3C-CH=CH-CH3) that is illustrated in Figure 5-1 is the *molar enthalpy of isomerization* of 2-butene. (The hydrocarbons 1-butene, 2-butene, and n-butane are three of the many volatile organic pollutants emitted in car exhaust in large urban areas. Thermodynamic properties of these volatile emissions are important for understanding the generation of air pollution and how to control it.)

Table 5-2 also lists ΔH values for several types of phase changes. The ΔH for one mole of ice melting to water or for one mole of solid gallium melting to the liquid metal is a molar enthalpy of fusion ($\Delta_{fus}H_m$) value. The ΔH for sublimation of one mole of crystalline iodine (I_2) to the vapor is the molar enthalpy of sublimation ($\Delta_{sub}H_m$) of iodine. The ΔH values for vaporization of one mole of water, mercury, or acetone to the corresponding vapor are molar enthalpy of vaporization ($\Delta_{vap}H_m$) values. Fusion, sublimation, and vaporization are endothermic changes in state that take heat from the surroundings. Thus, all $\Delta_{fus}H_m$, $\Delta_{sub}H_m$, and $\Delta_{vap}H_m$ values are positive. The ΔH values for the transformation of one mole of graphite into diamond or of one mole of orthorhombic sulfur into monoclinic sulfur are molar enthalpy of transition ($\Delta_{trans}H_m$) values. Some phase transitions produce heat and are exothermic; other phase transitions take heat from the surroundings and are endothermic. We discuss phase transitions and $\Delta_{trans}H$ values in more detail after we introduce the second law of thermodynamics in Chapter 6.

D. Lavoisier-Laplace Law

In Chapter 4 we described the ice calorimeter developed by the French scientists Lavoisier and Laplace in the late 18th century (see Figure 4-3). They used the calorimeter to make some of the earliest measurements of heats of combustion, including those of charcoal, hydrogen, phosphorus, and sulfur. In 1780, as a result of this work, they developed the *Lavoisier-Laplace law*, which states that the heat required to decompose a compound to its elements is liberated when the compound is formed from its elements. Their law can be generalized to the statement that the heat effect caused by a chemical reaction in one direction is exactly equal in magnitude, but opposite in sign, to that associated with the same reaction in the reverse direction. We now know that the Lavoisier-Laplace law is simply a consequence of the first law of thermodynamics. Otherwise it would be possible to make a perpetual-motion machine that could derive energy from the difference in the enthalpy of a reaction run in both the forward and reverse directions.

Example 5-2. Lavoisier and Laplace observed that combustion of one pound of charcoal to CO_2 melted 96.375 pounds of ice. Converting their values from the old French units of measure into metric units, we find that burning 489.95 grams of charcoal to CO_2 melted 47.218 kg ice. Taking $\Delta_{fus}H_m$ (ice) = 6009.5 J mol^{-1}, we can calculate the ΔH for the reaction

$$C\ (\text{charocal}) + O_2 = CO_2 \qquad \Delta H = -386.1\ \text{kJ mol}^{-1} \tag{5-17}$$

Using the Lavoisier-Laplace law, we conclude that for the reverse process,

$$CO_2 = O_2 + C\ (\text{charcoal}) \qquad \Delta H = +386.1\ \text{kJ mol}^{-1} \tag{5-18}$$

The ΔH value measured by Lavoisier and Laplace for reaction (5-17) is 7.4 kJ mol^{-1} smaller than the accepted value of $\Delta_f H_m$ (CO_2, g, 298) = −393.510 kJ mol^{-1}. The difference is plausibly due to using charcoal instead of pure graphite as well as other factors. But though the difference is very large by today's standards, it is remarkable that Lavoisier and Laplace obtained a result so close to the accepted value in work done over 200 years ago.

E. Hess's Law

The Swiss chemist Germain Henri Hess (see sidebar) formulated his law of heat summation in 1840, at about the same time that James Joule was starting his ingenious experiments on the mechanical equivalent of heat (see Section I-C in Chapter 3). Hess originally applied his law to compound formation. He later generalized it after doing more experiments. *Hess's law* says that the overall enthalpy change for any chemical reaction at constant pressure or at constant volume is the same whether the reaction takes place in one step or in several steps. An important result of Hess's and the Lavoisier-Laplace laws is that chemical equations (and their ΔH values) can be added and subtracted like algebraic equations. Hess's law is valid because enthalpy is a state function (see Section X-A of Chapter 2). Thus, ΔH values are the same regardless of how they are calculated. Hess's law is extremely useful because two or more ΔH values can be combined to give the desired ΔH value for a reaction that for some reason cannot be studied directly.

GERMAIN HENRI HESS (1802–1850)

Germain Hess was born in Switzerland but moved to Russia with his family while he was still young. There he remained for the rest of his life. He was a scholar of many talents: first a geologist, then a medical doctor, and finally a thermochemist. He published a chemistry textbook in 1834 that was widely used in Russia for many years.

Hess's interest in geology shaped the path his life would take. Hess wrote his doctoral thesis on the compositions of Russian mineral waters and their medicinal properties, combining his interests in mineralogy and medicine. He went on to analyze Russian salt samples, river water, and other natural resources. In the process he discovered several new minerals and proved that the composition of cobalt oxide is Co_3O_4. Hess also made contributions to the local government by giving advice on the teaching of chemistry and by helping to develop a water supply for St. Petersburg.

Hess showed that the amount of heat developed by a reaction is independent of the reaction path. His name has been immortalized in the study of thermodynamics with Hess's law, which, in 1840, he was first to clearly state. Hess's law says that the enthalpy of a chemical change is the same regardless of the means by which it is achieved. It doesn't matter how many steps are taken. This law is a consequence of the conservation of energy, but careful and accurate measurements of enthalpy changes had to be made before it was accepted.

Example 5-3. Use Hess's law to compute the molar enthalpy of formation ($\Delta_f H_{m,\,298}$) at 298 K for carbon monoxide (CO) gas. This is the ΔH for the reaction

$$\text{C (graphite)} + \frac{1}{2}\,\text{O}_2\ \text{(g)} = \text{CO (g)} \tag{5-19}$$

We calculate this ΔH by subtracting ΔH for burning CO from ΔH for burning graphite to CO_2. Our calculation uses data (in kilocalories, kcal) from Rossini (1939):

$$\text{C (graphite)} + \text{O}_2 = \text{CO}_2 \qquad \Delta_f H_{m,298} = -94.0518 \pm 0.0108\ \text{kcal mol}^{-1} \tag{5-20}$$

$$\text{CO} + \frac{1}{2}\,\text{O}_2 = \text{CO}_2 \qquad -\Delta_c H_{m,298} = -(-67.636 \pm 0.029)\ \text{kcal mol}^{-1} \tag{5-21}$$

$$\text{C (graphite)} + \frac{1}{2}\text{O}_2\ \text{(g)} = \text{CO (g)} \qquad \Delta_f H_{m,298} = -26.416 \pm 0.031\ \text{kcal mol}^{-1} \tag{5-19}$$

Converting from cal to J, we get $\Delta_f H_{m,298}$ (CO, g) = – 110.524 ± 0.13 kJ mol^{-1}. The equation

$$\Delta_f H_{m,298}\ (\text{CO}) = \Delta_f H_{m,298}\ (\text{CO}_2) - \Delta_c H_{m,298}\ (\text{CO}) \tag{5-22}$$

is equivalent to the subtraction we did earlier. Rossini's (1939) data for the $\Delta_f H_m$ values of CO and CO_2 are still the accepted values given in thermodynamic data compilations today.

F. Kirchhoff's equation

Up to this point we have not considered the effect of temperature on enthalpies of reaction. However, we can easily do this using concepts developed in Chapter 3. As defined earlier, the ΔH of any reaction is the difference between the enthalpy of the products and the reactants:

$$\Delta H = \sum (\nu_i H_i)_{pr} - \sum (\nu_i H_i)_{re} \tag{5-1}$$

where the enthalpy of each compound (H_i) is multiplied by its stoichiometric coefficient (ν_i) in the chemical equation, and the subscripts *pr* and *re* stand for products and reactants, respectively.

Differentiating Eq. (5-1) with respect to temperature gives us

$$\left[\frac{\partial(\Delta H)}{\partial T}\right]_P = \sum \left[\nu_i \left(\frac{\partial H_i}{\partial T}\right)_P\right]_{pr} - \sum \left[\nu_i \left(\frac{\partial H_i}{\partial T}\right)_P\right]_{re} \tag{5-23}$$

which can be rewritten using Eq. (3-44):

$$C_P = \left(\frac{\delta q}{dT}\right)_P = \left(\frac{\partial H}{\partial T}\right)_P \tag{3-44}$$

to substitute for the terms on the right in Eq. (5-23). This substitution gives us the equation

$$\left[\frac{\partial(\Delta H)}{\partial T}\right]_P = \sum [\nu_i C_{P,i}]_{pr} - \sum [\nu_i C_{P,i}]_{re} \tag{5-24}$$

This can be simplified by replacing the right side of Eq. (5-24) by ΔC_P, giving

$$\left[\frac{\partial(\Delta H)}{\partial T}\right]_P = \Delta C_P \tag{5-25}$$

GUSTAV ROBERT KIRCHHOFF (1824–1887)

Gustav Kirchhoff was a German physicist whose first work was on electricity. He formulated a series of laws governing closed circuits. He demonstrated that electrical current moves at the speed of light through conductors.

Later, along with Robert Bunsen, he pioneered the field of spectroscopy. Their first spectrometer consisted of a cigar box, a prism, and two old telescope lenses. They later developed much more sophisticated pieces of equipment that even allowed them to compare two spectra at the same time. Out of this work came many great discoveries. Kirchhoff stated that the ratio of the light absorbed to the light emitted by a body depends only on its temperature, not its composition. At the same time, every element has a unique pattern of spectral lines. This led Bunsen and Kirchhoff to discover the element cesium in 1860 and the element rubidium in 1861 in samples of mineral water. The use of the spectrometer let them see materials that were present in tiny amounts that were impossible to detect otherwise.

Kirchhoff went on to apply this technology to analyzing the composition of the sun, where he found many of the same elements that he saw on Earth. He also determined that materials absorb light of the same wavelengths that they would emit when heated, which explained the dark lines he saw in the sun's spectrum.

which is known as *Kirchhoff's equation* because the German physicist Gustav Kirchhoff (see sidebar) derived this result in 1858. Equation (5-25) is valid for any reaction without phase changes in reactants or products. If one or more reactants and/or products undergoes a phase transition, the ΔH of transition and the heat capacities of both phases involved in the transition must be included to calculate the temperature dependence of the ΔH of reaction. We illustrate this type of calculation in the next section. Equation (5-25) says that the temperature coefficient of the enthalpy of reaction has units of joules per degree and is equal to the difference in heat capacities between products and reactants. Rearranging Eq. (5-25) and integrating between an upper temperature T_2 and a lower temperature T_1 we obtain

$$\Delta H_{T_2} - \Delta H_{T_1} = \int_{T_1}^{T_2} \Delta C_P dT \tag{5-26}$$

Usually, ΔC_P is a function of temperature and is represented by empirically derived polynomial equations, such as the modified Maier-Kelley equations used in Chapters 3 and 4:

$$C_{P,m} = a + b \times 10^{-3}\, T + c \times 10^5\, T^{-2} + d \times 10^{-6}\, T^2 (\text{J mol}^{-1}\text{ K}^{-1}) \tag{5-27}$$

for molar heat capacities. If a modified Maier-Kelley equation is used, then ΔC_P is given by

$$\Delta C_P = \Delta a + \Delta b \times 10^{-3}\, T + \Delta c \times 10^5\, T^{-2} + \Delta d \times 10^{-6} T^2 (\text{J mol}^{-1}\ K^{-1}) \tag{5-28}$$

Sometimes more complex polynomials may be needed to represent ΔC_P to the required accuracy. Conversely, if heat capacity values are only available at one temperature, for example, 298 K, a constant ΔC_P must be used. In this case the error involved is small as long as the product $\Delta C_P \Delta T$ is much less than the average ΔH of reaction over the temperature interval considered.

Example 5-4. Use Kirchhoff's equation to calculate $\Delta_r H_T$ as a function of temperature for

$$CO_2 + H_2S = OCS + H_2O\ (g) \qquad \Delta_r H_{298} = +33.87\ \text{kJ mol}^{-1} \tag{5-29}$$

from 298 K to 1500 K. Reaction (5-29) forms carbonyl sulfide (OCS) in volcanic gases (e.g., at Nyiragongo in the Congo). We first compute ΔC_P (J mol^{-1} K^{-1}) from the data in Table 3-4:

$$C_{P,m}\ (H_2O,\ g) = 25.8 + 18.025 \times 10^{-3} T + 2.356 \times 10^5\, T^{-2} - 2.554 \times 10^{-6}\, T^2 \tag{5-30}$$

$$+C_{P,m}\ (OCS,\ g) = 44.021 + 17.61 \times 10^{-3} T - 6.514 \times 10^5\, T^{-2} - 4.437 \times 10^{-6}\, T^2 \tag{5-31}$$

$$-C_{P,m}\ (H_2S,\ g) = -[24.605 + 26.856 \times 10^{-3} T + 1.914 \times 10^5\, T^{-2} - 5.82 \times 10^{-6}\, T^2] \tag{5-32}$$

$$-C_{P,m}\ (CO_2,\ g) = -[46.789 + 7.714 \times 10^{-3}\, T - 11.701 \times 10^5\, T^{-2}] \tag{5-33}$$

$$\Delta C_{P,m} = -1.573 + 1.065 \times 10^{-3}\, T + 5.629 \times 10^5\, T^{-2} - 1.171 \times 10^{-6}\, T^2 (\text{J mol}^{-1}\text{ K}^{-1}) \tag{5-34}$$

We now substitute $\Delta C_{P,m}$ from Eq. (5-34) into Eq. (5-26) and evaluate the integral between 298 K and a higher temperature T, which varies up to 1500 K. This gives us

$$\Delta_r H_T = 33870 - 1.573(T - 298.15) + \frac{1}{2}\left(1.065 \times 10^{-3}\right)\left[T^2 - (298.15)^2\right]$$
$$-5.629 \times 10^5\left[T^{-1} - (298.15)^{-1}\right] - \frac{1}{3}\left(1.171 \times 10^{-6}\right)\left(T^3 - (298.15)^3\right] \text{ J mol}^{-1} \quad (5\text{-}35)$$

After evaluating and rearranging Eq. (5-35) we obtain this expression for $\Delta_r H_T$:

$$\Delta_r H_T = 36190 - 1.573T + 5.325 \times 10^{-4}T^2 - 5.629 \times 10^5 T^{-1} - 3.903 \times 10^{-7}T^3 \text{J mol}^{-1} \quad (5\text{-}36)$$

Figure 5-2 shows that the $\Delta_r H$ values calculated from Eq. (5-36) are almost identical to those computed from data tabulated in the NIST-JANAF tables. The uncertainty of ± 1.3 kJ mol^{-1} in the enthalpy of reaction plotted in Figure 5-2 is mainly from the uncertainty in the enthalpy of formation of OCS. The small difference between the $\Delta_r H$ values from Eq. (5-36) and the tabulated values would vanish if six- or seven-term polynomial fits were used for the C_P data.

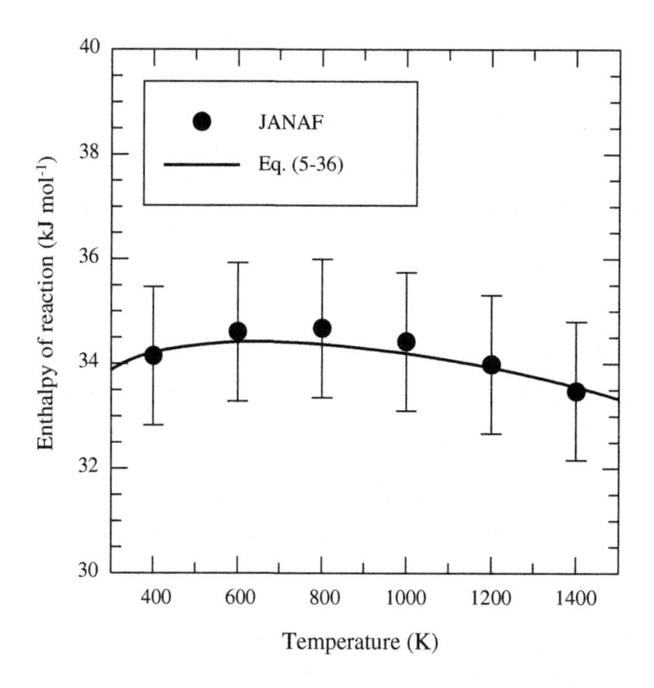

FIGURE 5-2

A comparison of the calculated enthalpy of reaction for $H_2S + CO_2 = OCS + H_2O$ with tabulated data in the NIST-JANAF tables. The error bars show the 2σ uncertainties of ± 1.32 kJ mol^{-1} on the $\Delta_r H$ values.

Kirchhoff's equation can also be written in terms of the enthalpy differences between the products and reactants. We do this by separately integrating the heat capacities of the products and reactants and rewriting the C_P integrals as the heat contents:

$$\Delta H_{T_2} - \Delta H_{T_1} = \int_{T_1}^{T_2} \sum [\nu_i C_{P,i}]_{pr} dT - \int_{T_1}^{T_2} \sum [\nu_i C_{P,i}]_{re} dT \quad (5\text{-}37)$$

$$\Delta H_{T_2} - \Delta H_{T_1} = \sum [\nu_i (H_{T_2} - H_{T_1})]_{\text{products}} - \sum [\nu_i (H_{T_2} - H_{T_1})]_{\text{reactants}} \quad (5\text{-}38)$$

Equation (5-38) can be evaluated using polynomial fits to the enthalpy functions. Many times a ΔH value at 0 K is computed from the ΔH value at 298 K, or vice versa. In these cases tabular values for the enthalpy functions are used as illustrated in the following example.

Example 5-5. Use Kirchhoff's equation in the form of Eq. (5-38) to compute the molar enthalpy of formation of CO (g) at 0 K ($\Delta_f H_{m,0}$) from the $\Delta_f H_{m,298}$ value and enthalpies of the products and reactants. We are asked to find ΔH for reaction (5-19):

$$\text{C (graphite)} + \frac{1}{2}\text{O}_2\text{ (g)} = \text{CO (g)} \quad (5\text{-}19)$$

at absolute zero. Taking $T_2 = 298$ K and $T_1 = 0$ K, we can substitute into Eq. (5-38) to get

$$\begin{aligned} \Delta H_{298} - \Delta H_0 &= \sum [\nu_i (H_{T_2} - H_{T_1})]_{\text{products}} - \sum [\nu_i (H_{T_2} - H_{T_1})]_{\text{reactants}} \\ &= (H_{298} - H_0)_{\text{CO}} - \frac{1}{2}(H_{298} - H_0)_{\text{O}_2} - (H_{298} - H_0)_{\text{graphite}} \\ &= 8.671 - 0.5(8.680) - 1.050 = 3.281 \text{ kJ mol}^{-1} \end{aligned} \quad (5\text{-}39)$$

Rearranging and substituting for ΔH_{298} we obtain the desired answer:

$$\Delta H_0 = \Delta_f H_0(\text{CO}) = -110.524 - 3.281 = -113.805 \text{ kJ mol}^{-1} \quad (5\text{-}40)$$

which is identical to the value given in the NIST-JANAF tables. As discussed later in Section III-B, $\Delta_f H_{m,0}$ for CO is important because it is related to the dissociation energy of the triple bond in CO. In general, enthalpies of reaction at 0 K and 298 K are related by the equations

$$\Delta_r H_0 = \Delta_r H_{298} + \sum \nu_i (H_{298} - H_0)_{\text{reactants}} - \sum \nu_i, (H_{298} - H_0)_{\text{products}} \quad (5\text{-}41)$$

$$\Delta_r H_{298} = \Delta_r H_0 + \sum \nu_i (H_{298} - H_0)_{\text{products}} - \sum \nu_i (H_{298} - H_0)_{\text{reactants}} \quad (5\text{-}42)$$

where the enthalpy ($H_{298} - H_0$) of each reactant and product is multiplied by its stoichiometric coefficient (e.g., 1 for graphite, ½ for O_2, and 1 for CO in this example).

G. Elemental reference states

We introduce the concepts of elemental reference states and the standard enthalpy of formation by considering quartz (SiO_2), which is the second most abundant mineral (after feldspars) in Earth's crust.

Earlier we defined the molar enthalpy of formation as the ΔH of reaction for the formation of one mole of a compound from its constituent elements. However, quartz formation from its constituent elements can occur by any of these reactions:

$$\text{Si (crystal)} + O_2\ (g) = SiO_2(\text{quartz}) \qquad \Delta H_{298} = -910.70\ \text{kJ mol}^{-1} \tag{5-43}$$

$$\text{Si (l)} + O_2\ (g) = SiO_2(\text{quartz}) \qquad \Delta H_{298} = -959.170\ \text{kJ mol}^{-1} \tag{5-44}$$

$$\text{Si (g)} + O_2\ (g) = SiO_2\ (\text{quartz}) \qquad \Delta H_{298} = -1360.70\ \text{kJ mol}^{-1} \tag{5-45}$$

$$\text{Si (c)} + 2\ \text{O (g)} = SiO_2\ (\text{quartz}) \qquad \Delta H_{298} = -1409.04\ \text{kJ mol}^{-1} \tag{5-46}$$

$$\text{Si (g)} + 2\ \text{O (g)} = SiO_2\ (\text{quartz}) \qquad \Delta H_{298} = -1859.04\ \text{kJ mol}^{-1} \tag{5-47}$$

which have different ΔH values. Reactions (5-43) to (5-47) are related to each other because we can write equations such as

$$\text{Si (c)} = \text{Si (l)} \qquad \Delta H_{298} = 48.47\ \text{kJ mol}^{-1} \tag{5-48}$$

$$\text{Si (c)} = \text{Si (g)} \qquad \Delta H_{298} = 450.0\ \text{kJ mol}^{-1} \tag{5-49}$$

$$O_2\ (g) = 2\ \text{O (g)} \qquad \Delta H_{298} = 498.34\ \text{kJ mol}^{-1} \tag{5-50}$$

that can be combined with one of the preceding reactions to give another reaction and its ΔH value. For example, subtracting Eq. (5-48) from Eq. (5-43) gives reaction (5-44) and its ΔH value.

It is convenient to agree on one reaction to define the molar enthalpy of formation of quartz at 298 K, which can then be used for a large number of thermodynamic calculations via application of Hess's law and Kirchhoff's equation. However, which reaction should we use? The answer is reaction (5-43) and the reason is that crystalline silicon and O_2 gas are the stable forms of silicon and oxygen at 298 K. Crystalline Si and O_2 (g), at one bar pressure and behaving as an ideal gas, are the reference states for the elements silicon and oxygen at 298 K.

The *reference state* of an element is generally, but not always, the stable form of the element at the ambient temperature. For gaseous elements we must also specify that the gas is at one bar pressure and behaves as an ideal gas. *By definition, the enthalpy of formation of an element in its reference state is zero at all temperatures.* This definition is necessary because we can measure enthalpy differences but not absolute enthalpies. (Thus, the H_i values in Eqs. (5-1) to (5-3) are actually the enthalpies of formation of each product and reactant and are not absolute values.) Defining $\Delta_f H = 0$ for an element in its reference state at all temperatures provides a consistent base for measuring and calculating enthalpy of reaction values. What are the reference states for some common elements at 298 K, and how do they vary with temperature?

The reference state for silicon is crystalline Si below its melting point (1685 K), liquid Si between its melting point and boiling point (3543 K at one bar pressure), and monatomic Si gas at and above its boiling point. Analogous reference states are chosen for almost all other elements that are solid or liquid at room temperature and have only one stable solid phase, for example, aluminum (Al), bromine (Br),

Table 5-3 Reference States of Iron

Stable Form	Temperature Range (K)
alpha Fe crystal	0–1043
beta Fe crystal	1043–1185
gamma Fe crystal	1185–1667
delta Fe crystal	1667–1811 (m.p. of Fe)
liquid Fe	1811–3110 (b.p. of Fe)
Fe gas	3110–infinite *T*

gallium (Ga), iodine (I), lead (Pb), magnesium (Mg), mercury (Hg), platinum (Pt), potassium (K), silver (Ag), sodium (Na), tin (Sn), and zinc (Zn).

Iron is an example of an element with two or more stable solid phases. Iron metal exists in four different solid forms as a function of temperature. The temperature-dependent reference states for iron are listed in Table 5-3. Calcium (Ca), manganese (Mn), titanium (Ti), and uranium (U) are four other examples of the many elements that have two or more stable solid phases, all of which are elemental reference states depending on the temperature.

There are four exceptions to our definition of an elemental reference state as the most stable form of an element at the ambient temperature. Carbon is the first exception. A particular type of high-purity, synthetic graphite that can be prepared reproducibly is chosen as the reference state for carbon at all temperatures. The choice of Acheson spectroscopic graphite is a practical choice. Neither graphite nor diamond melt at ambient pressure, so liquid carbon is not a good choice for the carbon reference state at high temperatures. The sublimation temperature of graphite is ~3970 K at one bar pressure and is uncertain by about 100 degrees. Furthermore, carbon vapor is composed of C atoms and C_2 to C_8 molecules in proportions that vary with temperature and pressure. Thus, carbon vapor is also a poor choice for a reference state. Diamond, which is unstable with respect to graphite, could be used as the carbon reference state, and in fact diamond was chosen as the reference state for carbon in the pioneering tables by Bichowsky and Rossini (1936). They did this because natural graphite from different localities has different properties, and synthetic graphite with reproducible properties is hard to prepare. However, diamond was abandoned as the carbon reference state and graphite is used at all temperatures for the reasons discussed previously.

Phosphorus is the second exception. Either crystalline white P or the triclinic allotrope of crystalline red P are used as the reference state of phosphorus in different thermodynamic data compilations, with about half of the compilations using white P and half using red P. White phosphorus is used as a reference state because it is easier to prepare, while red phosphorus is used as a reference state because it is the stable form at ambient conditions. This lack of agreement can create problems if data from a compilation using white phosphorus as the reference state are combined with data from a compilation using red phosphorus as the reference state. Thermodynamic calculations involving phosphorus and its compounds are important in biogeochemistry, soil science, and aqueous geochemistry because phosphorus is an essential nutrient element, phosphate minerals such as berlinite ($AlPO_4$) are important for controlling the amount of dissolved phosphorus in natural waters, and orthophosphate ions (PO_4^{3-}) are the major P-bearing species in aqueous solutions.

Tin is the third example of an element in which the choice of the zero enthalpy phase is not obvious. At ambient temperature, tin exists as white and gray tin. The white polymorph is assigned zero enthalpy, but gray tin is more stable ($\Delta_f H^o_{298} = -2.09$ kJ mol^{-1}). However, almost all compilations use white tin as the tin reference state, so the confusion that is found for phosphorus and sulfur (see the following discussion) does not exist for tin. The tin reference state changes to liquid tin at the melting point and then changes to monatomic Sn gas at and above the boiling point of tin.

Sulfur is the fourth exception, and in this case the temperature at which the reference state changes from liquid sulfur to sulfur vapor is the problem. Orthorhombic sulfur is the stable form of sulfur up to 368.3 K, where it transforms to monoclinic sulfur. Then monoclinic sulfur is stable for about 20 degrees until it melts at 388.36 K. Liquid sulfur boils at ~719 K at one bar pressure. However, sulfur vapor, like carbon vapor, is a mixture of many species (S to S_{10}) in proportions that vary with temperature and pressure. Thus, sulfur vapor is a poor choice for the sulfur reference state at high temperatures. Instead, diatomic sulfur vapor (S_2) is chosen as the reference state, but different thermodynamic data compilations use either the one bar boiling point of ~719 K or a fictive boiling point of ~882 K (at which the S_2 gas pressure is one bar) as the transition temperature for the liquid S to S_2 gas reference states.

What are the reference states for elements that are gases at 298 K? These elements turn into liquids and then solids (except helium) as the temperature decreases to absolute zero. Molecular gases such as O_2 dissociate to atoms with increasing temperature. Gaseous H_2, He, Ne, Ar, Kr, Xe, Rn, N_2, O_2, F_2, and Cl_2 are chosen as the reference states, respectively, for the elements hydrogen, helium, neon, argon, krypton, xenon, radon, nitrogen, oxygen, fluorine, and chlorine at all temperatures. This choice is made for convenience, so we do not have to worry about liquefaction, condensation, and dissociation as a function of temperature.

H. The standard enthalpy of formation

We now return to our example of quartz formation from its constituent elements. We agreed that Eq. (5-43) is the reaction defining the enthalpy of formation of quartz. The ΔH for reaction (5-43) is called the *standard enthalpy of formation* of quartz and is written as

$$\Delta_f H^o_{298} \text{ (quartz)} = -910.7 \text{ kJ mol}^{-1} \quad (5\text{-}43)$$

where the subscript 298 indicates that the ΔH is for 298.15 K. The superscript ° indicates that the ΔH is for standard-state conditions. The *standard state* is the baseline for defining changes in enthalpy. The standard-state pressure is one bar. Until the 1970s the standard-state pressure was defined as one atmosphere. The small pressure change of ~1.3% does not affect the ΔH of reactions involving ideal gases, because the enthalpy of an ideal gas is only a function of temperature (see Chapter 3, Section IV). The standard state for pure solids, liquids, and gases is the pure crystalline solid, or pure liquid, or pure gas (behaving ideally) at one bar pressure. Thus, Si (c) and O_2 (g, ideal) at one bar pressure are the standard states for silicon and oxygen. These are also the reference states for silicon and oxygen. Pure crystalline quartz at one bar pressure is the standard state for quartz.

All the ΔH values in Tables 5-1 and 5-2 are standard-state values and should be written with the superscript ^{o}. Likewise, all the ΔH values in the text and examples are standard-state values, and the superscript ^{o} should be used for them also. This was not done because the standard state had not been defined until this section.

The following example illustrates the concepts of elemental reference states and the standard enthalpy of formation. It is lengthy because the elemental reference states change five times and the stable phase of the product changes three times over the temperature range used in the thermodynamic calculations. However, this is exactly the sort of situation often encountered in thermodynamic calculations for everyday problems in geochemistry and related fields.

Example 5-6. Chalcocite (Cu_2S) is an important ore mineral of copper, which is extracted from it by roasting chalcocite in air to form CuO (tenorite) and reducing the CuO to Cu metal by heating with charcoal. The high-temperature thermodynamic properties of chalcocite are thus important for copper production and are also important for understanding the origin of the ore deposits. In this example we use Hess's law and Kirchhoff's equation to calculate ΔH for formation of chalcocite from its constituent elements in their reference states via

$$2\,\text{Cu(ref state)} + \text{S(ref state)} = \text{Cu}_2\text{S(chalcocite)} \qquad \Delta_f H^o_{298} = -83.9 \pm 1.1 \text{ kJ mol}^{-1} \tag{5-51}$$

from 298 K to 1400 K. At 298 K the reference states of copper and sulfur are copper metal and orthorhombic sulfur. However, copper, sulfur, and Cu_2S all undergo phase changes in the 298–1400 K range, so a series of equations is necessary to give ΔH as a function of temperature.

We begin by calculating ΔC_P from 298 K to 368.54 K, which is the temperature at which orthorhombic sulfur transforms to monoclinic sulfur. We use Eq. (5-28) and C_P data from the appendices to do this:

$$C_{P,m}\,(\text{Cu}_2\text{S},\ \alpha) = 49.238 + 92.787 \times 10^{-3}\,T (\text{J mol}^{-1}\,\text{K}^{-1}) \tag{5-52}$$

$$-2\,C_{P,m}\,(\text{Cu, metal}) = -2[20.531 + 8.611 \times 10^{-3}\,T + 1.55 \times 10^{5}\,T^{-2}]\,(\text{J mol}^{-1}\,\text{K}^{-1}) \tag{5-53}$$

$$-C_{P,m}\,(\text{S, ortho}) = -[11.007 + 53.058 \times 10^{-3}\,T - 46.526 \times 10^{-6}\,T^2]\,(\text{J mol}^{-1}\,\text{K}^{-1}) \tag{5-54}$$

$$\Delta C_{P,m} = -2.831 + 22.507 \times 10^{-3}\,T - 3.100 \times 10^{5}\,T^{-2} + 46.526 \times 10^{-6}\,T^2\,(\text{J mol}^{-1}\,\text{K}^{-1}) \tag{5-55}$$

We now substitute Eq. (5-55) for $\Delta C_{P,m}$ into Eq. (5-26) and evaluate the integral between 298 K and $T = 368.54$ K. We then obtain the equation

$$\begin{aligned}\Delta_f H^o_T = {} & -83{,}900 - 2.831(T - 298.15) + \frac{1}{2}(22.507 \times 10^{-3})[T^2 - (298.15)^2] \\ & + 3.100 \times 10^{5}[T^{-1} - (298.15)^{-1}] + \frac{1}{3}(46.526 \times 10^{-6})[T^3 - (298.15)^3]\ \text{J mol}^{-1}\end{aligned} \tag{5-56}$$

After evaluating and rearranging Eq. (5-56), we get this expression:

$$\begin{aligned}\Delta_f H^o_T = {} & -85{,}507.08 - 2.831\,T + 11.2535 \times 10^{-3}\,T^2 + 3.100 \times 10^{5}\,T^{-1} \\ & + 15.509 \times 10^{-6}\,T^3\ \text{J mol}^{-1}\end{aligned} \tag{5-57}$$

for the standard molar enthalpy of formation of Cu_2S, valid from 298.15 K to 368.54 K.

The next temperature range is from 368.54 K, where monoclinic sulfur becomes stable, to 376 K and at which chalcocite undergoes a phase transition from the lower-temperature orthorhombic (α) phase to the higher-temperature hexagonal (β) phase. From 368.54 K to 376 K, ΔC_P is given by

$$C_{P,m}\ (\mathrm{Cu_2S},\ \alpha) = 49.238 + 92.787 \times 10^{-3}\,T\ (\mathrm{J\ mol^{-1}\ K^{-1}}) \tag{5-52}$$

$$-2\,C_{P,m}\ (\mathrm{Cu, metal}) = -2[20.531 + 8.611 \times 10^{-3}\,T + 1.550 \times 10^{5}\,T^{-2}]\ (\mathrm{J\ mol^{-1}\ K^{-1}}) \tag{5-53}$$

$$-C_{P,m}\ (\mathrm{S,\ mono}) = -[17.318 + 20.243 \times 10^{-3}\,T]\ (\mathrm{J\ mol^{-1}\ K^{-1}}) \tag{5-58}$$

$$\Delta C_{P,m} = -9.142 + 55.322 \times 10^{-3}\,T - 3.100 \times 10^{5}\,T^{-2}\ (\mathrm{J\ mol^{-1}\ K^{-1}}) \tag{5-59}$$

The standard molar enthalpy of formation from 368.54 K to 376 K is computed using Eq. (5-26) with the integral evaluated between 368.54 K and 376 K. The ΔC_P is given by Eq. (5-59) and $\Delta_f H^o_{368.54}$ is computed by subtracting the molar enthalpy of transition for orthorhombic to monoclinic sulfur from the ΔH value from Eq. (5-57):

	$\Delta_f H^o_{368.54}$ (kJ mol^{-1})
2 Cu (metal) + S (ortho) = Cu_2S (α)	−83.404
S (mono) = S (ortho)	−0.402
2 Cu (metal) + S (mono) = Cu_2S (α)	−83.806

Substituting into Eq. (5-26) gives the standard molar enthalpy of formation of chalcocite as

$$\Delta_f H^o_T = -9.142\,T + 27.661 \times 10^{-3}\,T^2 + 3.100 \times 10^{5}\,T^{-1} - 85035.42\ \mathrm{J\ mol^{-1}} \tag{5-60}$$

over the 368.54 K to 376 K temperature range.

The next temperature range is from 376 K to 388.36 K, which is the melting point of monoclinic sulfur. The ΔC_P is given by

$$C_{P,m}\ (\mathrm{Cu_2S},\ \beta) = 118.402 - 57.835 \times 10^{-3}\,T + 0.718 \times 10^{5}\,T^{-2} + 21.825 \times 10^{-6}\,T^2 (\mathrm{J\ mol^{-1}\ K^{-1}}) \tag{5-61}$$

$$-2\,C_{P,m}\ (\mathrm{Cu,\ metal}) = -2\left[20.531 + 8.611 \times 10^{-3}\,T + 1.550 \times 10^{5}\,T^{-2}\right]\ (\mathrm{J\ mol^{-1}\ K^{-1}}) \tag{5-53}$$

$$-C_{P,m}\ (\mathrm{S,\ mono}) = -\left[17.318 + 20.243 \times 10^{-3}\,T\right]\ (\mathrm{J\ mol^{-1}\ K^{-1}}) \tag{5-58}$$

$$\Delta C_{P,m} = 60.022 - 95.3 \times 10^{-3}\,T - 2.382 \times 10^{5}\,T^{-2} + 21.825 \times 10^{-6}\,T^2\ (\mathrm{J\ mol^{-1}\ K^{-1}}) \tag{5-62}$$

We calculate $\Delta_f H^o_{376}$ using Eq. (5-60) and the ΔH for the $\alpha \rightarrow \beta$ transition in Cu_2S:

	$\Delta_f H^o_{376}$ (kJ mol^{-1})
2 Cu (metal) + S (mono) = Cu_2S (α)	−83.738
Cu_2S (α) = Cu_2S (β)	3.62
2 Cu (metal) + S (mono) = Cu_2S (β)	−80.118

The standard molar enthalpy of formation over this temperature range is then

$$\Delta_f H^o_T = 60.022\,T - 47.650 \times 10^{-3}\,T^2 + 2.382 \times 10^5\,T^{-1} + 7.275 \times 10^{-6}\,T^3 - 96{,}969.7 \text{ J mol}^{-1} \quad (5\text{-}63)$$

Monoclinic sulfur melts at 388.36 K. From 388.36 K to 720 K, ΔC_P is given by

$$C_{P,m}\,(Cu_2S, \beta) = 118.402 - 57.835 \times 10^{-3}\,T + 0.718 \times 10^5\,T^{-2} + 21.825 \times 10^{-6}\,T^2\ (\text{J mol}^{-1}\,\text{K}^{-1}) \quad (5\text{-}61)$$

$$-2\,C_{P,m}\,(\text{Cu, metal}) = -2[20.531 + 8.611 \times 10^{-3}\,T + 1.550 \times 10^5\,T^{-2}]\ (\text{J mol}^{-1}\,\text{K}^{-1}) \quad (5\text{-}53)$$

$$-C_{P,m}\,(\text{S, liq}) = -[786.186 - 1793.968 \times 10^{-3}\,T - 357.507 \times 10^5\,T^{-2} + 1175.291 \times 10^{-6}\,T^2] \quad (5\text{-}64)$$

$$\Delta C_{P,m} = -708.846 + 1718.911 \times 10^{-3}\,T + 355.125 \times 10^5\,T^{-2} - 1153.466 \times 10^{-6}\,T^2 (\text{J mol}^{-1}\,\text{K}^{-1}) \quad (5\text{-}65)$$

We compute $\Delta_f H^o_{388.36}$ using Eq. (5-63) and the enthalpy of fusion of monoclinic sulfur:

	$\Delta H^o_{388.36}$ (kJ mol^{-1})
2 Cu (metal) + S (mono) = Cu_2S (β)	−79.807
S (liq) = S (mono)	−1.718
2 Cu (metal) + S (liq) = Cu_2S (β)	−81.525

$$\Delta_f H^o_T = -708.846\,T + 859.4555 \times 10^{-3}\,T^2 - 355.125 \times 10^5\,T^{-1} - 384.4887 \times 10^{-6}\,T^3 + 178099.7 \text{ J mol}^{-1} \quad (5\text{-}66)$$

At 720 K, hexagonal (β) Cu_2S transforms to a cubic (γ) phase. From 720 K to 882.117 K, where the vapor pressure of S_2 over liquid sulfur reaches one bar, ΔC_P is given by

$$C_{P,m}\,(Cu_2S, \gamma) = 83.844 - 1.356 \times 10^{-3}\,T + 2.189 \times 10^5\,T^{-2} - 0.600 \times 10^{-6}\,T^2\ (\text{J mol}^{-1}\,\text{K}^{-1}) \quad (5\text{-}67)$$

$$-2\,C_{P,m}\,(\text{Cu, metal}) = -2[20.531 + 8.611 \times 10^{-3}\,T + 1.550 \times 10^{5}\,T^{-2}]\,(\text{J mol}^{-1}\,\text{K}^{-1}) \quad (5\text{-}53)$$

$$-C_{P,m}\,(\text{S, liq}) = -[786.186 - 1793.968 \times 10^{-3}\,T - 357.507 \times 10^{5}\,T^{-2} + 1175.291 \times 10^{-6}\,T^{2}] \quad (5\text{-}64)$$

$$\Delta C_{P,m} = -743.404 + 1775.390 \times 10^{-3}\,T + 356.596 \times 10^{5}\,T^{-2} - 1175.891 \times 10^{-6}\,T^{2}\,(\text{J mol}^{-1}\,\text{K}^{-1}) \quad (5\text{-}68)$$

Combining $\Delta_f H^o_{720}$ calculated from Eq. (5-66) with the enthalpy for the $\beta \rightarrow \gamma$ transition gives:

	$\Delta_f H^o_{720}$ (kJ mol^{-1})
2 Cu (metal) + S (liq) = Cu_2S (β)	−79.560
Cu_2S (β) = Cu_2S (γ)	1.17
2 Cu (metal) + S (liq) = Cu_2S (γ)	−78.390

$$\Delta_f H^o_T = -743.404\,T + 887.695 \times 10^{-3}\,T^2 - 356.596 \times 10^{5}\,T^{-1} - 391.9637 \times 10^{-6}\,T^3 + 192{,}506.5\ \text{J mol}^{-1} \quad (5\text{-}69)$$

for the standard molar enthalpy of formation of chalcocite from 720 K to 882.117 K.

From 882.117 K to 1358 K, which is the melting point of Cu metal, ΔC_P is given by:

$$C_{P,m}\,(\text{Cu}_2\text{S},\ \gamma) = 83.844 - 1.356 \times 10^{-3}\,T + 2.189 \times 10^{5}\,T^{-2} - 0.600 \times 10^{-6}\,T^{2}\,(\text{J mol}^{-1}\,\text{K}^{-1}) \quad (5\text{-}67)$$

$$-2\,C_{P,m}\,(\text{Cu, metal}) = -2[20.531 + 8.611 \times 10^{-3}\,T + 1.550 \times 10^{5}\,T^{-2}]\,(\text{J mol}^{-1}\,\text{K}^{-1}) \quad (5\text{-}53)$$

$$-\frac{1}{2}C_{P,m}\,(\text{S}_2,\ \text{g}) = -\frac{1}{2}[34.658 + 3.251 \times 10^{-3}\,T - 2.795 \times 10^{5}\,T^{-2} - 0.205 \times 10^{-6}\,T^{2}]\,(\text{J mol}^{-1}\,\text{K}^{-1}) \quad (5\text{-}70)$$

$$\Delta C_{P,m} = 25.453 - 20.2035 \times 10^{-3}\,T + 0.4865 \times 10^{5}\,T^{-2} - 0.4975 \times 10^{-6}\,T^{2}\,(\text{J mol}^{-1}\,\text{K}^{-1}) \quad (5\text{-}71)$$

We compute $\Delta_f H^o_{882.117}$ from Eq. (5-69) and the enthalpy of vaporization of liquid sulfur to S_2 (g):

	$\Delta_f H^o_{882.117}$ (kJ mol^{-1})
2 Cu (metal) + S (liq) = Cu_2S (γ)	−81.990
½ S_2 (g) = S (liq)	−½(100.888)
2 Cu (metal) + ½ S_2 (g) = Cu_2S (γ)	−132.434

$$\Delta_f H^o_T = 25.433\,T - 10.10175 \times 10^{-3}\,T^2 - 0.4865 \times 10^5\,T^{-1} - 0.1658 \times 10^{-6}\,T^3 - 146{,}857 \text{ J mol}^{-1} \quad (5\text{-}72)$$

Finally, from 1358 K to 1400 K, ΔC_P is given by:

$$C_{P,m}\,(\text{Cu}_2\text{S},\ \gamma) = 83.844 - 1.356 \times 10^{-3}\,T + 2.189 \times 10^5\,T^{-2} - 0.6 \times 10^{-6}\,T^2\ (\text{J mol}^{-1}\ \text{K}^{-1}) \quad (5\text{-}67)$$

$$-2\,C_{P,m}\,(\text{Cu, liq}) = -2[32.844]\ (\text{J mol}^{-1}\ \text{K}^{-1}) \quad (5\text{-}73)$$

$$-\frac{1}{2}C_{P,m}\,(\text{S}_2,\ \text{g}) = -\frac{1}{2}\,[34.658 + 3.251 \times 10^{-3}\,T - 2.795 \times 10^5\,T^{-2} - 0.205 \times 10^{-6}\,T^2]\ (\text{J mol}^{-1}\ \text{K}^{-1}) \quad (5\text{-}70)$$

$$\Delta C_{P,m} = 0.827 - 2.9815 \times 10^{-3}\,T + 3.5865 \times 10^5\,T^{-2} - 0.4975 \times 10^{-6}\,T^2\ (\text{J mol}^{-1}\ \text{K}^{-1}) \quad (5\text{-}74)$$

We calculate $\Delta_f H^o_{1358}$ by using Eq. (5-72) and the enthalpy of melting for copper:

	$\Delta_f H^o_{1358}$ (kJ mol^{-1})
2 Cu (metal) + S (liq) = Cu_2S (γ)	−131.372
2 [Cu (liq) = Cu (metal)]	−2(13.1)
2 Cu (liq) + ½ S_2 (g) = Cu_2S (γ)	−157.572

$$\Delta_f H^o_T = 0.827\,T - 1.49075 \times 10^{-3}\,T^2 - 3.5865 \times 10^5\,T^{-1} - 0.1658 \times 10^{-6}\,T^3 - 155{,}266.8 \text{ J mol}^{-1} \quad (5\text{-}75)$$

Even more complex calculations are needed if $\Delta_f H^o_T$ for pyrite (FeS_2), chalcopyrite ($CuFeS_2$), or bornite (Cu_5FeS_4) is required, because Fe metal undergoes several solid-state phase changes that must be taken into account (see Table 5-3 and also Figure 4-25 and Section III-I of Chapter 4).

I. Enthalpies of solution and dilution

We now discuss the ΔH values in Tables 5-1 and 5-2 that involve solutions. The dissolution of solid ammonium chloride (NH_4Cl) in liquid water is endothermic, and you would feel your hand getting colder if you held a container in your hand while NH_4Cl (s) was dissolving in the water in it. Conversely, the dissolution of potassium hydroxide (KOH) pellets in liquid water is highly exothermic and you could burn yourself if you held a container in your hand while KOH (s) was dissolving in the water in it. In both cases the heat effect depends on the amount of solute (the NH_4Cl or KOH) and the amount of solvent (water). This point is illustrated in Figure 5-3.

Because enthalpy of solution values vary with the solute-to-solvent ratio, it is customary to list the ΔH for dissolution of one mole of solute as a function of the number of moles of solvent. For example,

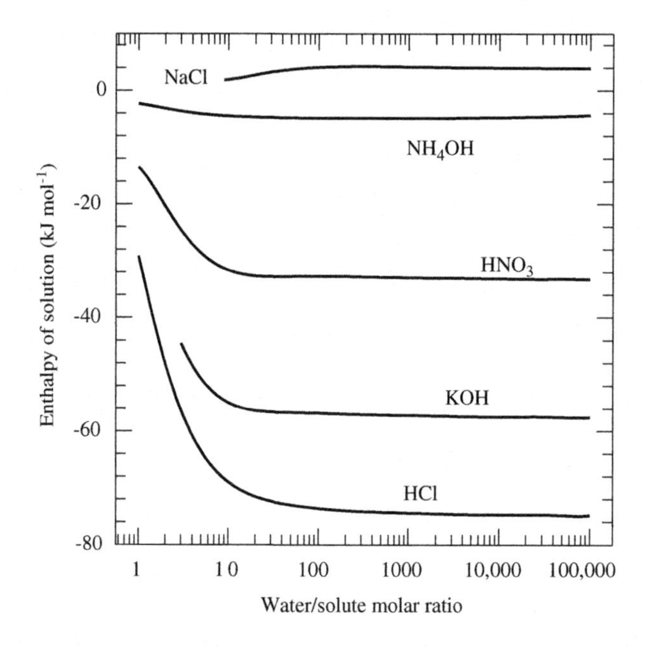

FIGURE 5-3

Enthalpies of solution plotted as a function of the water/solute molar ratio for some common acids, bases, and salts. Each enthalpy of solution approaches a constant value as the water/solute ratio increases.

the *integral molar enthalpy of solution* ($\Delta_{soln}H_m$) is -48.19 kJ for dissolution of one mole of KOH in four moles of water at 298 K. This reaction can be written as

$$\text{KOH (c)} + 4\ \text{H}_2\text{O (l)} = \text{KOH(in 4 H}_2\text{O, l)} \tag{5-76}$$

where (c) stands for crystalline solid, and (l) stands for liquid. Likewise, $\Delta_{soln}H_m = -56.43$ kJ for dissolution of one mole KOH in 20 moles of water at 298 K:

$$\text{KOH (c)} + 20\ \text{H}_2\text{O (l)} = \text{KOH(in 20 H}_2\text{O, l)} \tag{5-77}$$

and $\Delta_{soln}H_m = -57.3$ kJ for dissolution of one mole KOH in 1500 moles of water at 298 K:

$$\text{KOH (c)} + 1500\ \text{H}_2\text{O (l)} = \text{KOH(in 1500 H}_2\text{O, l)} \tag{5-78}$$

However, you will notice that Table 5-1 lists $\Delta H = -57.46$ kJ for KOH dissolving in water, and also that the equation is written as

$$\text{KOH (c)} + \infty\ \text{H}_2\text{O (l)} = \text{K}^+\ \text{(aq)} + \text{OH}^-\ \text{(aq)} + \infty\ \text{H}_2\text{O (l)} \tag{5-79}$$

where the (aq) stands for an infinitely dilute aqueous solution. What does Eq. (5-79) mean?

First, the K^+ and OH^- ions are shown because KOH dissociates into these ions when it dissolves in water. Liquid water appears on both sides of Eq. (5-79) to emphasize that KOH pellets are dissolved in an infinite amount of water, but in general, we would write this reaction without including the water on both sides. The ΔH value for reaction (5-79) is different than that in the three preceding examples, but it is not greatly different than $\Delta_{soln}H_m = -57.30$ kJ for dissolution of one mole KOH in 1500 moles of water. Equation (5-79) represents formation of an *infinitely dilute solution* in which the one mole of KOH is dissolved in a much larger quantity of water, and the KOH concentration in the solution is so small that it is zero for all practical purposes. The ΔH value for KOH dissolved in an infinitely dilute solution is the *standard integral molar enthalpy of solution* ($\Delta_{soln}H^o_m$). It is the limiting value that is approached as the molar water to solute ratio approaches infinity. The asymptotic approach to this limiting value is shown for KOH and other common solutes in Figure 5-3. The infinitely dilute solution, like the ideal gas, is a hypothetical concept that can be approached but never achieved. It is useful because the enthalpy of solution is a function of the solvent-to-solute molar ratio.

The *molar enthalpy of dilution* ($\Delta_{dil}H_m$) is the difference in the integral molar enthalpies of solution for two solutions with different water/solute molar ratios. For example, $\Delta_{dil}H_m = -8.24$ kJ when we add 16 moles of water to the aqueous KOH solution formed in Eq. (5-76) and is calculated from this Hess's law cycle:

$$\text{KOH (c)} + 20\ \text{H}_2\text{O (l)} = \text{KOH(in 20 H}_2\text{O, l)} \qquad \Delta_{soln}H_m = -56.43\ \text{kJ} \qquad (5\text{-}77)$$

$$-[\text{KOH (c)} + 4\ \text{H}_2\text{O (l)} = \text{KOH(in 4 H}_2\text{O, l)}] \qquad \Delta_{soln}H_m = -48.19\ \text{kJ} \qquad (5\text{-}76)$$

$$\text{KOH(in 4 H}_2\text{O, l)} + 16\ \text{H}_2\text{O (l)} = \text{KOH(in 20 H}_2\text{O, l)} \qquad \Delta_{dil}H_m = -8.24\ \text{kJ} \qquad (5\text{-}80)$$

In this illustration the enthalpy of dilution has the same sign as the enthalpy of solution, but this does not occur in all cases.

Example 5-7. The Great Salt Lake in Utah has a salinity of ~25%, which is mainly dissolved NaCl (halite) with smaller amounts of other Na and Mg salts. Calculate the heat effect when halite is dissolved in water to give Great Salt Lake water. We begin by converting the salinity into the water/NaCl molar ratio. Two hundred fifty grams NaCl is 4.278 moles and 750 grams water is 41.632 moles, so the water/NaCl molar ratio is ~10 in the Great Salt Lake. The heat effect is the difference in the enthalpies of formation of halite and NaCl (in 10 H_2O l) and is the integral molar enthalpy of solution for NaCl dissolved in 10 moles of water (see Figure 5-3):

$$\Delta_{soln}H_m = \Delta_f H^o_{298}\ (\text{NaCl, in 10 H}_2\text{O, l}) - \Delta_f H^o_{298}\ (\text{NaCl, halite}) = 1.92\ \text{kJ mol}^{-1} \qquad (5\text{-}81)$$

Note that we are careful to distinguish the solution of NaCl in 10 moles water from an infinitely dilute aqueous solution of NaCl, which would be written as NaCl (aq) and would have all NaCl dissociated into Na^+ and Cl^- aqueous ions.

J. Standard enthalpies of formation for aqueous ions

Experimental measurements show that the enthalpy of reaction for neutralization of a dilute solution of a strong acid by a dilute solution of a strong base at 298 K is independent of the acid (e.g., hydrochloric, nitric, sulfuric) or base (e.g., sodium hydroxide, potassium hydroxide, ammonium hydroxide) used. The only reaction taking place is

$$\text{H}^+\ (\text{aq}) + \text{OH}^-\ (\text{aq}) = \text{H}_2\text{O (l)} \qquad (5\text{-}82)$$

and the standard molar enthalpy of neutralization is given by

$$\Delta_r H^o_{298} = \Delta_f H^o_{298}\,(H_2O,\ l) - \Delta_f H^o_{298}\,(OH^-,\ aq) - \Delta_f H^o_{298}\,(H^+,\ aq) = -55.815\ \text{kJ mol}^{-1} \tag{5-83}$$

In principle, we can calculate the standard enthalpies of formation of the H^+ and OH^- aqueous ions using Eq. (5-83), but how do we separately measure the $\Delta_f H^o_{298}$ values for the two ions?

This is done by defining the standard enthalpy of formation of the H^+ aqueous ion to be zero. The formation of H^+ (aq) from hydrogen in its reference state, which is H_2 gas at one bar pressure behaving as an ideal gas, is

$$\frac{1}{2}\,H_2\,(g) = H^+\,(aq) + e^- \tag{5-84}$$

which occurs when H_2 (g) at one bar pressure is bubbled over a platinum electrode in a dilute aqueous solution. As stated previously, the ΔH for Eq. (5-84) is set to zero by definition,

$$\Delta_f H^o_{298}\,(H^+,\ aq) = 0 \tag{5-85}$$

Equation (5-85) is the basis for measurements of the standard enthalpies of formation of all aqueous ions. We can now determine the standard enthalpy of formation of the OH^- aqueous ion by rearranging Eq. (5-83) to give

$$\Delta_f H^o_{298}\,(OH^-,\ aq) = \Delta_f H^o_{298}\,(H_2O,\ l) - \Delta_r H^o_{298} - \Delta_f H^o_{298}\,(H^+,\ aq) \tag{5-86}$$

$$= -285.830 - (-55.815) - 0 = -230.015\ \text{kJ mol}^{-1} \tag{5-87}$$

The standard enthalpies of formation of other aqueous ions are derived in a similar manner.

For example, when nitric acid is dissolved in a very large amount of water, that is, in an infinitely dilute solution, the reaction taking place is

$$HNO_3\,(l) + \infty\ H_2O\,(l) = H^+\,(aq) + NO^-{}_3\,(aq) + \infty\,H_2O\,(l) \tag{5-88}$$

which can also be written as

$$HNO_3\,(l) + \infty\,H_2O\,(l) = HNO_3\,(aq) + \infty\ H_2O\,(l) \tag{5-89}$$

The observed enthalpy change is $-$ 32.71 kJ mol^{-1} and is the standard integral molar enthalpy of solution of nitric acid. We can write two equivalent equations for this enthalpy change:

$$\Delta_{soln} H^o_m(HNO_3) = \Delta_f H^o_{298}\,(H^+,\ aq) + \Delta_f H^o_{298}\ NO_3^-\,(aq) - \Delta_f H^o_{298}\,(HNO_3,\ l) \tag{5-90}$$

$$\Delta_{soln} H^o_m(HNO_3) = \Delta_f H^o_{298}\,(HNO_3,\ aq) - \Delta_f H^o_{298}\,(HNO_3, l) \tag{5-91}$$

Equations (5-90) and (5-91) are equal to one another. We use this fact and the known $\Delta_f H^o_{298}$ of liquid nitric acid to compute $\Delta_f H^o_{298}$ of the nitrate ($NO^-{}_3$) aqueous ion:

$$\Delta_f H^o_{298}\ NO^-{}_3\,(aq) = \Delta_{soln} H^o_m(HNO_3) + \Delta_f H^o_{298}\,(HNO_3, l) - \Delta_f H^o_{298}(H^+, aq) \tag{5-92}$$

$$\Delta_f H^o_{298}\ NO^-{}_3\,(aq) = -32.71 + (-174.14) - 0 = -206.85\ \text{kJ mol}^{-1} \tag{5-93}$$

Nitric acid is a strong acid that is fully dissociated into ions in dilute solutions. Also recall that we have $\Delta_f H^o_{298}$ (H^+, aq) = 0. Thus, the standard enthalpies of formation of an infinitely dilute solution of aqueous nitric acid and of the nitrate ion are the same. Likewise, the standard enthalpies of formation of infinitely dilute hydrochloric acid (HCl), sulfuric acid (H_2SO_4), and perchloric acid ($HClO_4$) are identical to the $\Delta_f H^o_{298}$ values for the Cl^-, $SO^{2-}{}_4$, and $ClO^-{}_4$ ions.

Example 5-8. Cadmium is widely used in nickel-cadmium rechargeable batteries to power portable electronic devices. When Cd is leached from discarded Ni-Cd batteries in landfills, it dissolves as Cd^{2+}, which is a toxic environmental pollutant. Calculate the standard enthalpy of formation for the Cd^{2+} aqueous ion from enthalpy of solution data for cadmium nitrate.

We need to use the standard integral molar enthalpy of solution ($\Delta_{soln}H^o_m$). This is the limiting value that is approached asymptotically as the water-to-$Cd(NO_3)_2$ molar ratio approaches infinity in a very dilute aqueous solution as shown by the equation

$$Cd(NO_3)_2\ (c) + \infty H_2O\ (l) = Cd^{2+}\ (aq) + 2\ NO^-{}_3\ (aq) + \infty H_2O\ (l) \tag{5-94}$$

The $\Delta_{soln}H^o_m = -33.72$ kJ mol^{-1} for Eq. (5-94), which can also be written as

$$\Delta_{soln}H^o_m = \Delta_f H^o_{298}\ (Cd^{2+}, aq) + 2\ \Delta_f H^o_{298}\ (NO^-{}_3\ (aq) - \Delta_f H^o_{298}\ (Cd(NO_3)_2,\ c) \tag{5-95}$$

We computed the standard enthalpy of formation for the nitrate ion in Eq. (5-93). Substituting this value and the tabulated $\Delta_f H^o_{298}$ value for crystalline Cd nitrate into Eq. (5-95) and rearranging gives the standard enthalpy of formation for the Cd^{2+} aqueous ion as

$$\Delta_f H^o_{298}\ (Cd^{2+},\ aq) = -33.72 - 2(-206.85) + (-456.39) = -76.41\ \text{kJ mol}^{-1} \tag{5-96}$$

II. CALORIMETRIC MEASUREMENT OF REACTION ENTHALPIES

We gave an overview of calorimetry and discussed the basic types of calorimeters that are used to measure the heat capacity and enthalpy of solids, liquids, and gases in Section II of Chapter 4. In this section we briefly discuss calorimetric methods for measuring enthalpies of chemical reactions and of physical changes. The latter enthalpy effects include those due to polymorphic phase transitions (e.g., in the Al_2SiO_5 polymorphs), to vitrification of silicate glasses, to order–disorder transitions, to mixing in silicate melts, and to mixing in minerals such as olivines and pyroxenes that are solutions of two or more components. The two major methods that are used are solution calorimetry and combustion calorimetry. Articles and books that describe the design, construction, and operation of calorimeters used to measure reaction enthalpies include Marsh and O'Hare (1994), Rossini (1956), Skinner (1962), and Sunner and Månsson (1979).

A. Solution calorimetry

Aqueous acid solution calorimetry

Aqueous acid solution calorimetry was developed by Julius Thomsen in Denmark (see sidebar) and Marcellin Berthelot in France (see sidebar) during the latter part of the 19th century. Solution calorimeters measure the heat effect due to dissolution of a solute in some solvent. The heat effect can be exothermic or endothermic. The solute can be a natural or synthetic mineral or the oxides that make up the mineral. The solvent can be an aqueous acid such as hydrochloric acid or hydrofluoric acid, or a molten oxide such as lead borate or sodium molybdate, or a molten metal such as liquid bismuth, cadmium, indium, or tin. Oxide melt calorimetry is used in geochemistry and is described in Section II; molten metal calorimetry is used in metallurgy and is not discussed here. A Hess's law cycle and/or Kirchhoff's equation are generally used to determine the enthalpy of reaction of the sample being studied from the calorimetric data.

HANS PETER JÖRGEN JULIUS THOMSEN (1826–1909)

Julius Thomsen was a professor of chemistry at the University of Copenhagen and head of the chemical laboratory there. He was simultaneously a chemistry professor at the Polytekniske Laereanstalt. He also delved into business. At the time, the European alkali industry required large amounts of soda (Na_2CO_3), which was imported from Egyptian natron lakes. However, Denmark owned the world's largest deposits of cryolite (Na_3AlF_6), located in Ivittuut, Greenland. Taking advantage of this resource, Thomsen developed a method for making soda from cryolite and opened a string of factories. Subsequently, the Belgian engineer Ernest Solvay discovered a more efficient method for making sodium carbonate. However, other uses for the cryolite Thomsen was purifying included making aluminum, enamel, and opaque glass. Thomsen's most important contributions were made in the laboratory, however.

Thomsen expressed the idea that atoms are composed of smaller particles. After the discovery of the first noble gas, he predicted the existence of five others, along with their positions on the periodic table. Within the next few years, these gases were all discovered, and it was found that Thomsen's predictions of their atomic weights were quite accurate.

Thomsen made numerous calorimetric measurements. Like Berthelot, he developed the idea that chemical affinity correlated with the amount of heat produced by a reaction. He gave up the notion, however, after observing that some endothermic reactions occur spontaneously. We now know that the Gibbs free energy of reaction, not the enthalpy of reaction, is the key factor determining whether a reaction occurs spontaneously. Thomsen performed his experiments himself and would not take on students in the laboratory because he feared that the results obtained by different researchers could not be reliably compared.

Suppose we are interested in measuring the standard enthalpy of formation of forsterite (Mg_2SiO_4) from the elements, which is $\Delta_f H^o_{298}$ (Mg_2SiO_4, forsterite) $= -2173.0$ kJ mol^{-1}, and is the enthalpy change at 298.15 K for the reaction

$$2\ \mathrm{Mg\ (m)} + \mathrm{Si\ (c)} + 2\ \mathrm{O_2\ (g)} = \mathrm{Mg_2SiO_4\ (forsterite)} \quad (5\text{-}97)$$

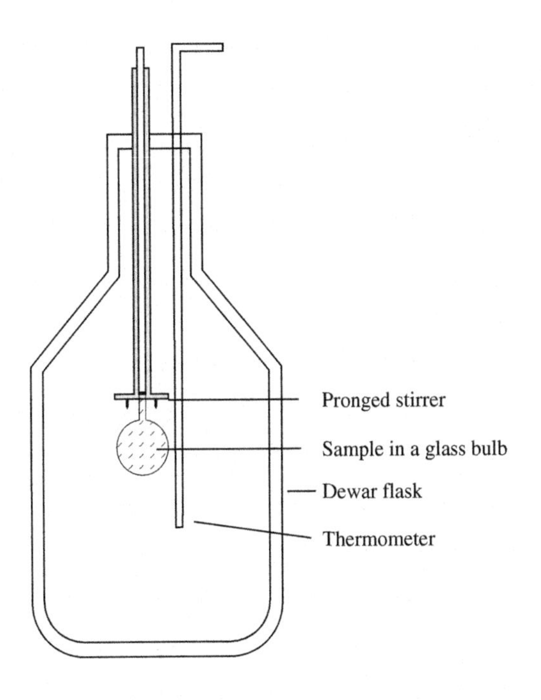

FIGURE 5-4

A schematic diagram of a solution calorimeter that uses an aqueous acid as the solvent.

How could we do this? We could put a stoichiometric mixture of Mg metal, crystalline Si, and O_2 gas into a sealed tube and wait for something to happen. However, after a few days we would conclude that measuring the enthalpy change for Eq. (5-97) is not practical at room temperature because no reaction occurs. Instead, we need to measure the enthalpy changes for several other reactions, which rapidly proceed to completion, and then use Hess's law to calculate the ΔH value for reaction (5-97). This can be done using solution calorimetry.

Figure 5-4 is a schematic diagram of an aqueous acid solution calorimeter. The diagram is much simpler than the actual calorimeter, which is described by Southard (1940), but it illustrates the basic principles. The calorimeter is an insulated container such as a dewar flask that is filled with aqueous hydrochloric acid. The sample is put into a glass bulb that can be lowered into the calorimeter, where it can be broken by rapidly pulling it up against the pronged stirrer. A thermometer inside the calorimeter measures the temperature of the calorimeter (and all of its contents) before, during, and after the sample is dropped into it. Figure 5-5 is a schematic plot of temperature versus time during an imaginary calorimeter experiment. The enthalpy of reaction is calculated from the temperature rise (for an exothermic reaction) or from the temperature decrease (for an endothermic reaction) of the water bath. Figure 5-5 illustrates an exothermic reaction that increases the calorimeter temperature by nine degrees. The enthalpy of solution is calculated from the observed temperature increase and the carefully measured thermal properties of the calorimeter. This calculation is illustrated by the simplified equation

$$\Delta_{soln}H = C_P\Delta T \tag{5-98}$$

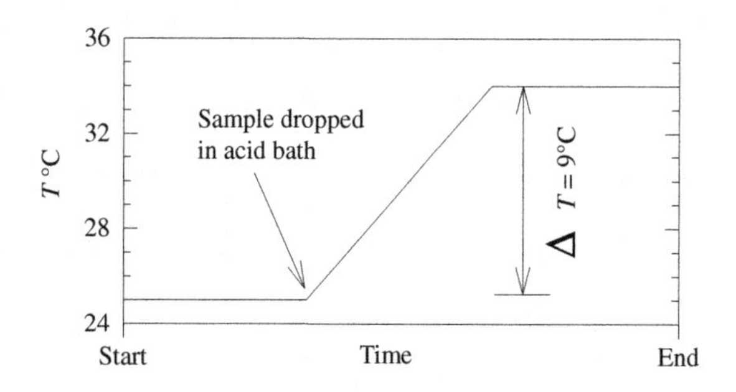

FIGURE 5-5

A schematic temperature-versus-time plot for an exothermic reaction in an aqueous acid solution calorimeter.

where C_P is the constant-pressure heat capacity of the calorimeter and all of its contents, ΔT is the temperature rise measured by the thermometer, and $\Delta_{soln}H$ is the enthalpy of solution of the sample. The $\Delta_{soln}H$ value is then recalculated on a molar basis. The heat capacity of the calorimeter plus its contents is carefully determined beforehand by calibration measurements. The calculation of the enthalpy of solution is more complex than indicated by Eq. (5-98), but this description explains the basic points. An aqueous acid solution calorimeter using hydrofluoric acid is described by Torgeson and Sahama (1948). It operates on the same principle but is made of platinum, which is not corroded by the HF acid.

Modern aqueous acid solution calorimeters typically operate at 25–90°C, have a minimum sample weight of 5–25 mg, and give $\Delta_{soln}H$ values with uncertainties of less than 0.1%. A typical $\Delta_{soln}H$ value for quartz dissolving in 20% (by mass) aqueous hydrofluoric acid at ~346 K is about – 140 kJ mol^{-1}. The uncertainty in $\Delta_f H$ values for formation of a mineral from its constituent oxides is ~0.4 kJ mol^{-1}. However, aqueous acid solution calorimetry has several disadvantages. One is the incomplete dissolution of oxides such as periclase (MgO), corundum (Al_2O_3), and eskolaite (Cr_2O_3) under some conditions. Another is difficulty in measuring $\Delta_f H$ values for Fe^{2+}-bearing compounds and minerals (Navrotsky, 1979; Hovis et al., 1998). Nevertheless, aqueous acid solution calorimetry has been and continues to be very useful in determining $\Delta_f H$ values of minerals, as demonstrated by the following example.

Example 5-9. King et al. (1967) measured the enthalpy of formation of forsterite from the oxides by measuring the heat of solution for MgO (periclase), SiO_2 (quartz), and Mg_2SiO_4 (forsterite) in aqueous acid solution (10% HF and 10% HCl) at 73.7°C (~346 K). In the following equations the acid is given as HF for simplicity. Also, in each case the mineral samples were at 298 K before being dissolved in the acid (346 K), giving an acid solution at 346 K. King et al. (1967) began by dissolving SiO_2 (quartz) in a fresh batch of the aqueous acid:

$$SiO_2\ (qtz) + 6\ HF\ (aq) = H_2SiF_6\ (aq) + 2\ H_2O\ \ (aq) \tag{5-99}$$

$$\Delta_{soln}H^o\ (\text{quartz}) = -143.72\ \text{kJ mol}^{-1} \tag{5-100}$$

Then the stoichiometric amount of MgO (periclase) was dissolved in the same solution:

$$2\ MgO\ (\text{periclase}) + 4\ HF\ (aq) = 2\ MgF_2\ (p) + 2\ H_2O\ (aq) \tag{5-101}$$

$$\Delta_{soln}H^o\ (\text{periclase}) = -296.60\ \text{kJ mol}^{-1} \tag{5-102}$$

Reaction (5-101) gave a precipitate of MgF_2, which is denoted by the *p*. Then forsterite was dissolved in a new batch of the acid solution with the same concentration and volume as that used for dissolving quartz and periclase:

$$Mg_2SiO_4 \text{ (forsterite)} + 10 \text{ HF (aq)} = 2 \text{ MgF}_2 \text{ (p)} + H_2SiF_6 \text{ (aq)} + 4 \text{ H}_2O \text{ (aq)} \quad (5\text{-}103)$$

$$\Delta_{soln}H^o \text{ (forsterite)} = -382.33 \text{ kJ mol}^{-1} \quad (5\text{-}104)$$

The amount of forsterite used in Eq. (5-103) gives a solution with the same composition as the acid used to dissolve quartz and periclase. A Hess's law calculation gives $\Delta_{ox}H^o_{298}$ for the formation of forsterite from its constituent oxides. Equations (5-99) and (5-101) are added together and Eq. (5-103) is subtracted from their sum. When this is done all the solution terms on the right sides of the equations cancel out and we get this reaction (at 298 K):

$$2 \text{ MgO (periclase)} + SiO_2 \text{ (quartz)} = Mg_2SiO_4 \text{ (forsterite)} \quad (5\text{-}105)$$

The enthalpy of formation for reaction (5-105) is given by the two equations:

$$\Delta_{ox}H^o \text{ (Mg}_2\text{SiO}_4) = \Delta_{soln}H^o \text{ (quartz)} + \Delta_{soln}H^o \text{ (periclase)} - \Delta_{soln}H^o \text{ (forsterite)} \quad (5\text{-}106)$$

$$\Delta_{ox}H^o \text{ (Mg}_2\text{SiO}_4) = -143.72 + (-296.6) - (-382.33) = -57.99 \text{ kJ mol}^{-1} \quad (5\text{-}107)$$

This Hess's law calculation is graphically illustrated in Figure 5-6.

A second Hess's law calculation is used to find ΔH for reaction (5-97) by adding the ΔH value for Eq. (5-105) with the ΔH values for the formation of periclase and quartz from their constituent elements at 298 K:

$$2 \text{ MgO (per)} + SiO_2 \text{ (qtz)} = Mg_2SiO_4 \text{ (for)} \qquad \Delta_{ox}H^o = -57.99 \text{ kJ mol}^{-1} \quad (5\text{-}105)$$

$$2 \text{ Mg (m)} + O_2 \text{ (g)} = 2 \text{ MgO (periclase)} \qquad 2\Delta_f H^o_{298} = -1203.20 \text{ kJ mol}^{-1} \quad (5\text{-}108)$$

$$\text{Si (c)} + O_2 \text{ (g)} = SiO_2 \text{ (quartz)} \qquad \Delta_f H^o_{298} = -910.7 \text{ kJ mol}^{-1} \quad (5\text{-}109)$$

$$2 \text{ Mg(m)} + \text{Si (c)} + 2 \text{ O}_2 \text{ (g)} = Mg_2SiO_4 \text{ (for)} \qquad \Delta_f H^o_{298} = -2171.89 \text{ kJ mol}^{-1} \quad (5\text{-}97)$$

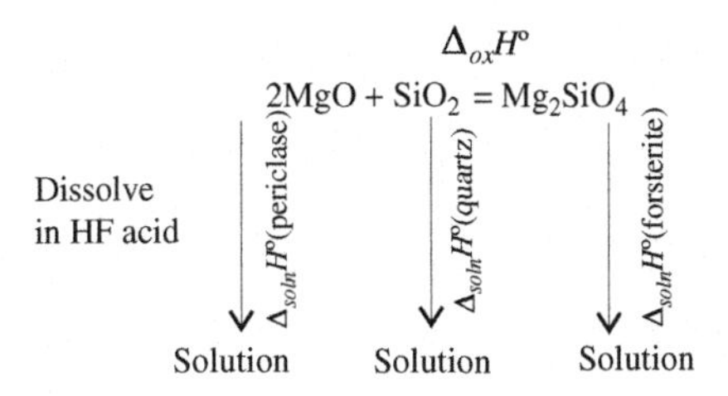

FIGURE 5-6

A graphical illustration of the Hess's law calculation for the enthalpy of formation of forsterite (Mg_2SiO_4) from its constituent oxides in Example 5-9.

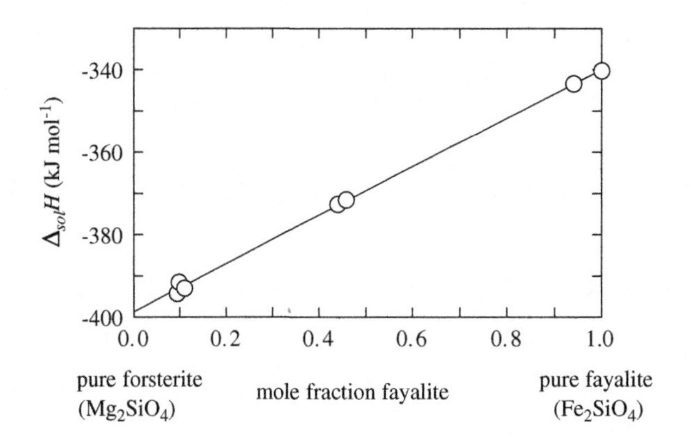

FIGURE 5-7

The results of Torgeson and Sahama (1948) for the enthalpies of solution of olivines dissolved in aqueous HF acid.

This example illustrates the great utility of solution calorimetry for indirectly measuring ΔH values that cannot be obtained via direct measurements.

However, you may have realized that the final result of $-$ 2171.89 kJ mol^{-1} is 1.1 kJ different than the value we originally gave for the standard enthalpy of formation of forsterite. The recommended $\Delta_f H^o_{298}$ value is -2173.0 ± 2.0 kJ mol^{-1}, which is taken from the thermodynamic data compilation by Robie and Hemingway (1995). Their recommended value is based on several sets of measurements, not just the work of King et al. (1967). The 1.1 kJ difference is well within the uncertainty of ± 2.0 kJ. It is not unusual for multiple determinations of an enthalpy value to give different results. For example, Colinet and Pasturel (1994) list eight solution calorimetric values of $\Delta_{ox} H^o_{298}$ (Mg_2SiO_4) for Eq. (5-105) that range from -57.82 kJ mol^{-1} to -63.26 kJ mol^{-1} with an unweighted mean value of -60.50 kJ mol^{-1}.

Aqueous acid solution calorimetry is a versatile method that is not restricted to enthalpy of formation measurements. Torgeson and Sahama (1948) measured the enthalpies of solution in aqueous

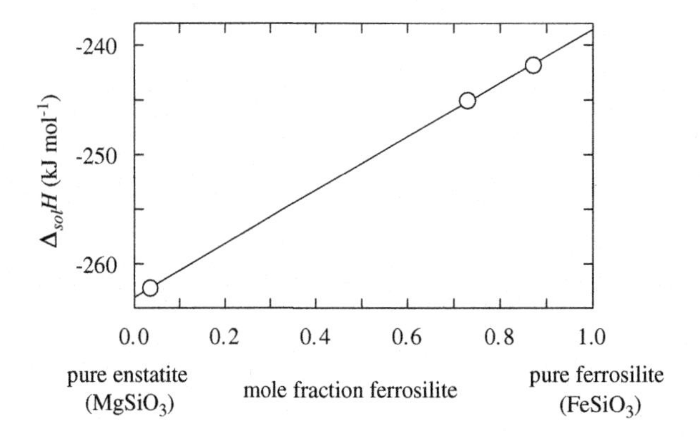

FIGURE 5-8

The results of Sahama and Torgeson (1949) for the enthalpies of solution of orthopyroxenes dissolved in aqueous HF acid.

hydrofluoric acid of seven olivines ranging in composition from 90.6% Mg_2SiO_4 (forsterite, Fo) to pure Fe_2SiO_4 (fayalite, Fa). Sahama and Torgeson (1949) presented similar data for three orthopyroxenes ranging in composition from 96.5% $MgSiO_3$ (enstatite, En) to 87.1% $FeSiO_3$ (ferrosilite, Fs). Their results, which are illustrated in Figures 5-7 and 5-8, show linear variations of the ΔH of solution of the olivines and orthopyroxenes in the HF acid solution. Their results indicate that the substitution of Fe^{2+} for Mg^{2+} in olivine and orthopyroxene does not significantly disturb the mineral structures, leading to ideal mixing of the fayalite and forsterite end members in olivine and of the enstatite and ferrosilite end members in orthopyroxene. (More recent work indicates that the trends are not exactly linear, but we will come back to this when we discuss the thermodynamic properties of solutions.)

Oxide melt calorimetry

Solution calorimetry in an oxide melt is a relatively new method that was developed by Kleppa and colleagues in the 1960s (Colinet and Pasturel, 1994; Kleppa, 2001). In this method samples weighing ~5–15 mg are dissolved in oxide melts at temperatures of about 960–1180 K and measurements of the area under a plot of heat flow versus time give the enthalpy of solution. A typical $\Delta_{soln}H$ value for quartz dissolving in molten lead borate at ~970 K is about −5 kJ mol^{-1}. The uncertainty in $\Delta_{soln}H$ values is ~0.5–1% and uncertainties in $\Delta_{ox}H$ values are ~0.4 kJ mol^{-1}, similar to uncertainties in $\Delta_{ox}H$ values measured by aqueous acid calorimetry. Oxide melt calorimeters are dual or twin calorimeters, analogous to the differential scanning calorimeter illustrated in Figure 4-11. The use of twin calorimeters (one for the samples and one as a control) cancels out unwanted thermal effects. The molten oxides that are used as solvents include lead borate ($2PbO \cdot B_2O_3$) at ~970 K, which is the most widely used oxide melt, lead cadmium borate ($9PbO \cdot 3CdO \cdot 4B_2O_3$) at ~970–1170 K, sodium molybdate ($3Na_2O \cdot 4MoO_3$) at ~970 K, and a mixture of lithium and sodium borates ($LiBO_2 + NaBO_2$) at ~1180 K. These oxides are used because they dissolve acidic oxides (e.g., SiO_2) and basic oxides (e.g., MgO, CaO) with reproducible enthalpies of solution that are independent of the solute concentration, presence of other solutes, and slight variations in the composition of the molten oxide solvent. Oxide melt calorimetry is a versatile technique that has been applied to a wide range of materials (e.g., carbonates, fluorides, glasses, high-pressure phases, hydrous minerals, nitrides). An important point is that oxide melt calorimetry provides experimentally measured ΔH values at high temperatures, which can be compared to ΔH values derived from emf measurements made in the same temperature range (see Chapter 10). However, like aqueous acid solution calorimetry, oxide melt calorimetry has a few disadvantages. One is that the same batch of solvent should be used for the most accurate work. Another is that molten lead borate cannot be used at low oxygen pressures where the Pb^{2+} ion is reduced (Colinet and Pasturel, 1994; Navrotsky, 1979; 1997). However, the advantages outweigh the disadvantages, as illustrated by the following example.

Example 5-10. Charlu et al. (1975) report the enthalpies of solution at 970 K of periclase, corundum, and spinel ($MgAl_2O_4$) in molten lead borate as 1.18, 7.73, and 14.29 kcal mol^{-1}, respectively. The changes in state that they studied are exemplified by the equation

$$\text{sample}(970\text{ K}) + \text{oxide melt}(970\text{ K}) = \text{sample dissolved in oxide melt}(970\text{ K}) \quad (5\text{-}110)$$

Using their results, what is the enthalpy of formation of spinel from its constituent oxides at 298 K? To answer this question we first use a Hess's law cycle to compute

$$\Delta_{ox}H^o_{970} = \Delta_{soln}H^o_{970}(MgO) + \Delta_{soln}H^o_{970}(Al_2O_3) - \Delta_{soln}H^o_{970}(MgAl_2O_4) \quad (5\text{-}111)$$

$$\Delta_{ox}H^o_{970} = 1.18 + 7.73 - 14.29 = -5.38\text{ kcal mol}^{-1} = -22.51\text{ kJ mol}^{-1} \quad (5\text{-}112)$$

from the enthalpy of solution data. We now have to correct the result to 298 K using Kirchhoff's equation. We have several options to do this, including Eq. (5-26), or Eq. (5-37), or Eq. (5-38). We will use enthalpy functions and Eq. (5-38). The enthalpies ($H_{970} - H_{298}$) for MgO, Al_2O_3, and spinel are 31.40, 74.34, and 105.36 kJ mol^{-1}, respectively. The enthalpies can be computed using the heat capacity equations given in Appendix 2 or interpolated from the values listed by Robie and Hemingway (1995). Substituting the enthalpy values and our result from Eq. (5-112) into Eq. (5-38) we obtain $\Delta_{ox}H^o_{298} = -22.13$ kJ mol^{-1}, which agrees with the value of -21.8 kJ mol^{-1} calculated from the heat of formation data in Appendix 1.

B. Combustion calorimetry

Combustion calorimetry dates back to the pioneering work of Lavoisier and Laplace, which we discussed in Section I-D and Example 5-2. From 1800 to about 1840, a number of scientists, including Sir Humphry Davy, Count Rumford, and Pierre Dulong, measured heats of combustion of charcoal, iron, oils, sulfur, various gases (CH_4, CO, H_2), and woods. However, their results, like those of Lavoisier and Laplace, were limited by their rudimentary equipment.

Subsequent work by Pierre Favre (1813–1880) and Johann Silbermann (1806–1865) in France (during the period 1842–1852), by Hess in Russia at about the same time, by Julius Thomsen in Copenhagen (from 1851–1885), and by Marcellin Berthelot in France (from 1864–1905) began the development of modern combustion calorimetry. Favre introduced the word *calorie* as a unit of heat, and Hess coined the term *thermochemistry*. The work done by Berthelot and Thomsen is especially important because they provided an enormous amount of thermochemical data that are still used today. They also developed the precursors of the modern bomb calorimeter (Bertholet) and flame calorimeter (Thomsen). The Royal Society of Chemistry jointly awarded the Davy Medal to Berthelot and Thomsen in 1883 for their thermochemical research.

Combustion in oxygen

Figure 5-9 is a schematic diagram of an oxygen bomb calorimeter, which is given this name because samples are combusted in pure oxygen at an elevated pressure of about 25 bars. A carefully weighed sample is put into the platinum crucible and then the bomb is sealed and placed into an insulated water bath. The bomb is repeatedly flushed with O_2 (g) to remove all air. After all air has been removed, the oxygen pressure is increased to ~25 bars. The temperature of the water bath is monitored with a sensitive thermometer. Once the temperature is stable, the electrical circuit is turned on, which heats up the filament and ignites a small paper fuse, which helps ignite the combustion reaction. The temperature of the water bath rises because all combustions are exothermic. The temperature is monitored until it drops back to the initial value, and the experiment is then over. The enthalpy of reaction is calculated in the same fashion as described for aqueous acid solution calorimetry, using an equation such as Eq. (5-98). The thermal properties of the combustion calorimeter are determined by passing a known amount of electrical energy through it or by combusting a standard material, such as benzoic acid ($C_7H_6O_2$), which has a precisely known enthalpy of combustion.

Oxygen bomb calorimetry can be used to measure $\Delta_c H$ values for any oxidation reaction that gives complete combustion to one or more well-defined products. The enthalpies of formation of many geologically important oxides, including calcia, periclase, corundum, hematite (Fe_2O_3), eskolaite (Cr_2O_3), and rutile, were measured successfully by oxygen bomb calorimetry. Likewise, the $\Delta_f H^o_{298}$ values for many of the rare earth sesquioxides R_2O_3, where R is one of the rare earth metals, have also

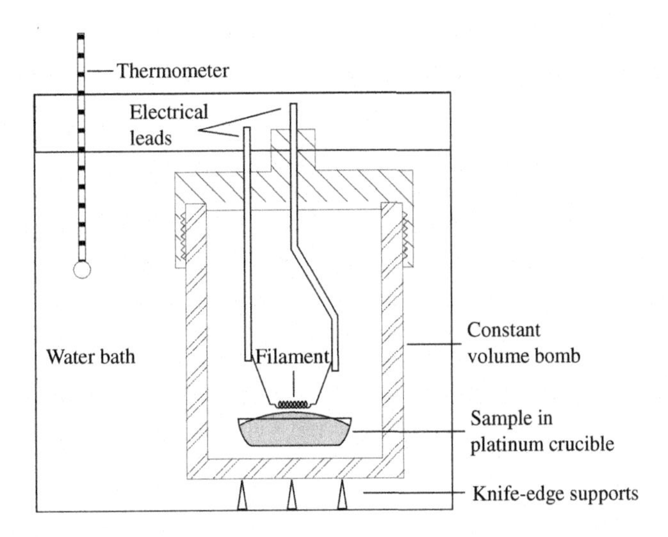

FIGURE 5-9

A schematic diagram of an oxygen bomb combustion calorimeter.

been measured by oxygen bomb calorimetry. However, some elements cannot be studied because they are too radioactive or they decay too rapidly. The radioactive elements technetium Tc, promethium Pm, radium Ra, francium Fr, and elements 99–114 fall into this category. Other elements such as boron do not burn easily in oxygen. A few elements (gold, platinum) do not burn at all. Other elements such as the alkali elements burn but give poorly defined mixtures of one or more oxides, or, like terbium, give an oxide with variable composition. The enthalpies of formation of oxides of 50 elements have been measured successfully by oxygen bomb calorimetry. The enthalpies of formation of many borides, carbides, nitrides, and lower valent oxides (e.g., burning wüstite to magnetite + hematite) have also been measured by this technique. Finally, oxygen bomb calorimetry has been widely used to measure enthalpies of formation of organic compounds (e.g., the constituents of gasoline), drugs (e.g., aspirin), and organometallic compounds (e.g., the gasoline additive tetraethyl lead). The following example illustrates one of these many applications.

Example 5-11. Holley and Huber (1951) measured the standard energy of combustion ($\Delta_c E^o$) of Mg metal to crystalline MgO in pure oxygen at ~25 atmospheres pressure in a bomb calorimeter at 297.8 K. They obtained a value for $\Delta_c E^o = -24{,}667 \pm 20\ \text{J g}^{-1}$ of Mg metal. They note that correction of $\Delta_c E^o$ to 298.15 K is less than the 20 J uncertainty and was neglected. Calculate the standard molar enthalpy of combustion and standard molar enthalpy of formation of MgO (periclase) from their data.

We first multiply the $\Delta_c E^o$ value by the atomic weight of Mg (24.305 g mol^{-1}) and calculate the standard molar energy of combustion $\Delta_c E_m^o = -599{,}531\ \text{J mol}^{-1}$. We convert this to the standard molar enthalpy of combustion using Eq. (5-10):

$$\Delta H - \Delta E = P\Delta V \sim RT\Delta n = (8.314)(298.15)(-0.5) = -1239\ \text{J mol}^{-1} \qquad (5\text{-}10)$$

and obtain $\Delta_c H_m^o = \Delta_f H_{298}^o = -600.8 \pm 0.5\ \text{kJ mol}^{-1}$, which is also the standard molar enthalpy of formation of MgO because it is the ΔH for the formation of MgO from its constituent elements in their reference states. Our value is identical to that of Holley and Huber (1951) when it is recalculated for the 0.6% change in the atomic weight of Mg metal since 1951. However, it is 0.8 kJ different than the recommended value of -601.6 ± 0.3 kJ mol^{-1}, that is, an average of the $\Delta_f H^o_{298}$ from Huber and Holley (1951) and three other $\Delta_f H^o_{298}$ values, two from aqueous acid solution calorimetry and one from another combustion calorimetry study.

Combustion in fluorine and other halogens

Fluorine combustion calorimetry dates back to work in the 1890s by Berthelot and Henri Moissan (1852–1907), the French chemist who first isolated elemental fluorine and won the Nobel Prize in Chemistry in 1906. During World War II workers in the Manhattan Project made major advances in producing and handling F_2 for UF_6 production. This expertise led to development of modern fluorine bomb calorimetry, which is analogous to oxygen bomb calorimetry. However, instead of burning in O_2 (g) to form oxides, elements and compounds burn in high pressure F_2 (g) to form fluorides.

Elemental fluorine is extremely reactive, so fluorine bomb calorimetry is very useful for studying elements and compounds that cannot be burned in oxygen or that do not oxidize completely to one or more well-defined products. Enthalpies of formation have been measured for borides, carbides, fluorides, nitrides, oxides (including SiO_2), phosphides, sulfides (e.g., orpiment, realgar, stibnite, molybdenite), selenides, and tellurides. The next example discusses the measurement of $\Delta_f H^o_{298}$ for quartz, which resolved a longstanding discrepancy.

Example 5-12. Wise et al. (1963) combusted Si and quartz in a fluorine bomb calorimeter:

$$\mathrm{Si\ (c)} + 2\,\mathrm{F_2\ (g)} = \mathrm{SiF_4\ (g)} \tag{5-113}$$

$$\mathrm{SiO_2\ (quartz)} + 2\,\mathrm{F_2\ (g)} = \mathrm{SiF_4\ (g)} + \mathrm{O_2\ (g)} \tag{5-114}$$

and reported standard combustion energies ($\Delta_c E^o$) of $-13{,}721.7$ cal g^{-1} for Si and -2800.3 cal g^{-1} for quartz. Calculate $\Delta_f H^o_{298}$ of SiF_4 (g) and of quartz from their data.

We convert calories to joules and multiply by the atomic weight of Si (28.0855) and the gram formula weight of quartz (60.0843) to get -1612.43 ± 0.80 kJ mol^{-1} and -703.98 ± 1.17 kJ mol^{-1} for the standard molar energies of combustion of crystalline Si and quartz, respectively. The uncertainties are those given by Wise et al. (1963). The $\Delta_c E^o_m$ value for Si is converted to the $\Delta_c H^o_m$ value using Eq. (5-10):

$$\Delta H - \Delta E = P\Delta V \sim RT\Delta n = (8.314)(298.15)(-1.0) = -2479\ \mathrm{J\ mol^{-1}} \tag{5-10}$$

and we obtain $\Delta_c H^o_m = \Delta_f H^o_{298} = -1614.91 \pm 0.80$ kJ mol^{-1}, which is also the standard molar enthalpy of formation of SiF_4 (g) because it is the ΔH for the formation of SiF_4 from its constituent elements in their reference states. This result is 0.03 kJ different than that of Wise et al. (1963) because of a small change in the accepted atomic weight of silicon, and it is 0.09 kJ different than the recommended value of -1615.0 ± 0.8 kJ mol^{-1} given in Appendix 1.

The $\Delta_c E^o_m$ value for quartz is identical to the $\Delta_c H^o_m$ value because there is no change in the number of moles of gas in reaction (5-114). However, in this case the $\Delta_c H^o_m$ value is not the same as the $\Delta_f H^o_{298}$ value. The latter value is that for reaction (5-43) and can be calculated with a Hess's law cycle by subtracting Eq. (5-114) from Eq. (5-113). This yields

$$\Delta_f H^o_{298}\ (\mathrm{SiO_2, quartz}) = \Delta_c H^o_m(\mathrm{Si}) - \Delta_c H^o_m(\mathrm{SiO_2, quartz}) = -910.9 \pm 1.4\ \mathrm{kJ\ mol^{-1}} \tag{5-115}$$

which is virtually identical to the recommended value of -910.7 ± 1.0 kJ mol^{-1} in Appendix 1.

Finally, we note that combustion calorimetry using chlorine or bromine has also been used to determine heat of formation values for some compounds, including LiCl, $BeCl_2$, $CaCl_2$ (hydrophilite), $BaCl_2$, $AlCl_3$, $TiBr_4$, $ZrCl_4$, $NbBr_5$, and $TaBr_5$. Most of this work involves direct reaction of an element with Cl_2 or Br_2, and much less work has been done on combustion of compounds in chlorine or bromine because they are less reactive than fluorine.

III. SOME APPLICATIONS OF THERMOCHEMISTRY

A. Calculations of adiabatic flame temperatures

The adiabatic flame temperature is the temperature to which a reactive mixture, such as rocket fuel, may rise after combustion. This calculation assumes that the enthalpy of reaction heats up the combustion gases (and any unburned or unreactive starting materials) without any heat losses due to dissociation of the combustion gases or heat losses to the surroundings ($\delta q = 0$). Adiabatic flame temperatures are upper limits to actual flame temperatures because heat losses due to conduction and radiation occur, the combustion of the starting materials may be incomplete, and dissociation of combustion gases to atoms and radicals consumes energy that would otherwise be used for heating. Calculations for air-fuel flames should include heating of nonreactive gases, such as nitrogen and argon; otherwise the results will be much too high. Adiabatic flame temperatures are of interest for designing incinerators for burning toxic waste, for rocket fuels, and for designing emission control systems for vehicles, factories, and power plants.

Example 5-13. During the 1950s scientists studied mixtures of reactive chemicals for use as rocket fuels, and it was discovered that cyanogen (C_2N_2) burning in an equimolar amount of pure oxygen gives very high temperatures. Burning in a C_2N_2–O_2 flame occurs via

$$C_2N_2 + O_2 = 2\,CO + N_2 \tag{5-116}$$

with very little dissociation of the very stable combustion products. Calculate the adiabatic flame temperature for a C_2N_2–O_2 flame.

We begin by calculating the enthalpy of reaction (5-116) at 298 K from the heat of formation data given in Appendix 1:

$$\Delta_r H^o_{298} = 2\Delta_f H^o_{298}(CO) - \Delta_f H^o_{298}(C_2N_2) = 2(-110.53) - 309.10 = -530.16 \text{ kJ} \tag{5-117}$$

Reaction (5-116) is very exothermic and releases a large amount of energy, which heats the CO and N_2 well above 2000 K, the upper range of the C_P data for these gases in Table 3-4. Instead of extrapolating the data in Table 3-4, we computed linear least squares fits to enthalpy data for the 298–6000 K range in Chase (1999) and obtained the equations

$$(H_T - H_{298})_{CO} = -11.4087 + 0.0332\,T + 5.5412 \times 10^{-7} T^2 \text{ kJ mol}^{-1} \tag{5-118}$$

$$(H_T - H_{298})_{N2} = -11.2801 + 0.0329\,T + 5.7985 \times 10^{-7}\,T^2 \text{ kJ mol}^{-1}. \tag{5-119}$$

The 530.16 kJ released by Eq. (5-116) is now equated to the heat contents of two moles of CO and one mole of nitrogen. After rearranging the result we obtain the quadratic equation

$$1.6881 \times 10^{-6}\,T^2 + 0.0993\,T - 564.2575 = 0 \tag{5-120}$$

which has the solution $T = 5219$ K, the calculated adiabatic flame temperature of an equimolar $C_2N_2-O_2$ flame. The actual flame temperature is about 4610 K (Conway et al., 1953). The 600-degree difference is mainly due to the neglect of heat losses, because dissociation of CO and N_2 to atoms and radicals is negligible.

B. Dissociation and atomization energies

Dissociation and atomization energies of molecules are important for chemistry in Earth's atmosphere, in the atmospheres of other planets, in the atmospheres of brown dwarfs and cool stars, and for combustion chemistry in cars, trucks, factories, incinerators, and power plants.

Dissociation energy

Simply speaking, the *dissociation energy* (D) is the energy needed (per mole) to break a chemical bond in a molecule. More precisely, the dissociation energy of a diatomic molecule such as H_2 is the energy for molecular dissociation to the constituent atoms

$$H_2\ (g) = 2\ H\ (g) \tag{5-121}$$

with the H_2 molecule and the H atoms in their lowest energy levels and behaving as ideal gases. The dissociation energy of the H_2 molecule is

$$D_0^o = 432.070\ \text{kJ mol}^{-1} \tag{5-122}$$

which corresponds to the value at absolute zero. The superscript o means that the H atoms are in their lowest energy level, and the subscript 0 means that the H_2 molecule is in its lowest energy level. This more precise definition of the dissociation energy is necessary for two reasons. First, many dissociation energies are determined by spectroscopy and are for molecules and atoms in their lowest energy levels, that is, a 0 K dissociation energy. Second, sometimes the dissociation energy of a molecule is measured from the minimum of the potential energy curve that describes bonding in the molecule. This dissociation energy is the D_e value and is slightly larger than the D_0^o value we have just defined. The D_e value is important for understanding chemical bonding in molecules and is used by chemists and physicists studying this topic. However, throughout the rest of this book we use and discuss D_0^o values, which we will just write as D unless we need to specify a temperature, which will be written as a subscript.

The books by Cottrell (1958) and Gaydon (1968) give good descriptions of the underlying theory and techniques used to measure atomiziation and dissociation energies. These two books and Lide (2000) compile values of atomization and dissociation energies.

The dissociation energy and the enthalpies of formation of a molecule and its constituent atoms are related via the general equation

$$D = \Delta_r H^o = \sum \Delta_f H^o(\text{atoms}) - \Delta_f H^o(\text{molecule}) \tag{5-123}$$

where the dissociation energy and the enthalpies of formation are measured at the same temperature. For example, the dissociation energy for H_2 (g) at 298 K is the ΔH for the reaction

$$H_2\ (g) = 2\ H\ (g) \tag{5-121}$$

$$\Delta_r H^o_{298} = 2[\Delta_f H^o_{298}(\mathrm{H,g})] - \Delta_f H^o_{298}(\mathrm{H_2,g}) = 2\Delta_f H^o_{298}(\mathrm{H,g}) = D^o_{298}(\mathrm{H_2,g}) \quad (5\text{-}124)$$

$$D(\mathrm{H_2,g}) = 2(217.998) = 435.996 \text{ kJ mol}^{-1} \quad (5\text{-}125)$$

This value is 3.926 kJ larger than the D^o_0 value given in Eq. (5-122) because of the slightly different enthalpies $(H_{298} - H_0)$ for the H_2 molecule and the two H atoms. The dissociation energies at 0 K and 298 K can be related to one another by using Kirchhoff's equation

$$D^o_{298} - D^o_0 = \sum \nu_i (H_{298} - H_0)_{\text{products}} - \sum \nu_i (H_{298} - H_0)_{\text{reactants}} \quad (5\text{-}38)$$

Eq. (5-38) is a general equation, which we rewrite as

$$\Delta H_{298} - \Delta H_0 = 2(H_{298} - H_0)_H - (H_{298} - H_0)_{H_2} \quad (5\text{-}38)$$

for H_2 dissociation via reaction (5-121). Substituting numerical values into this yields

$$D^o_{298} - D^o_0 = 2(6.197) - 8.468 = 3.926 \text{ kJ mol}^{-1} \quad (5\text{-}126)$$

We use the temperatures as subscripts on the dissociation energies in Eq. (5-126).

Example 5-14. The $\Delta_f H^o_{298}$ value for CO (g) from Example 5-3 is -110.524 kJ mol^{-1}. Calculate D^o_0 (in electron volts) using this value and auxiliary data. The D^o_0 value of CO is the ΔH change for dissociation of CO to its constituent atoms

$$\mathrm{CO(g)} = \mathrm{C(g)} + \mathrm{O(g)} \quad (5\text{-}127)$$

at absolute zero. The ΔH change for reaction (5-127) can be computed as follows. First, we use Eq. (5-123) to calculate D^o_{298} for CO, which is $\Delta_r H^o_{298}$ for reaction (5-127).

$$D^o_{298}(\mathrm{CO}) = 716.68 + 249.17 - (-110.524) = 1076.37 \text{ kJ mol}^{-1} \quad (5\text{-}128)$$

The second step is to use D^o_{298} to calculate D^o_0. This is done using Kirchhoff's equation (5-38), and after rearranging and substituting enthalpy data from Appendix 1 we obtain

$$D^o_0(\mathrm{CO,g}) = D^o_{298}(\mathrm{CO,g}) + (H^o_{298} - H^o_0)_{\mathrm{CO}} - [(H^o_{298} - H^o_0)_{\mathrm{C(g)}} + (H^o_{298} - H^o_0)_{\mathrm{O(g)}}] \quad (5\text{-}129)$$

$$D^o_0(\mathrm{CO,g}) = 1076.37 + 8.671 - (6.536 + 6.725) = 1071.78 \text{ kJ mol}^{-1} \quad (5\text{-}130)$$

From Table 3-2 we see that one eV is equal to about 96.4853 kilojoules per mole, thus

$$D^o_0(\mathrm{CO,g}) = 11.11 \text{ eV} \quad (5\text{-}131)$$

which agrees with the value determined by molecular spectroscopy,

Atomization energy

The *atomization energy* is the energy needed (per mole) to dissociate a molecule to atoms. The atomization and dissociation energies are identical for a diatomic molecule, but they are different for all polyatomic molecules. The dissociation energy of each chemical bond is different even in molecules such as H_2O, NH_3, CH_4, and SF_6 (sulfur hexafluoride). Example 5-15 illustrates the relationships between individual bond dissociation energies, the atomization energy, and the enthalpy of formation of a molecule.

Example 5-15. Perfluoromethane (CF_4) is a powerful greenhouse gas that is present at about 74 parts per trillion (i.e., a mole fraction of 7.4×10^{-11}) in Earth's troposphere. One molecule of CF_4 causes as much greenhouse warming as 10,000 CO_2 molecules. Perfluoromethane is emitted into Earth's troposphere as a by-product of aluminum smelting and apparently also is formed in natural fluorites (CaF_2) via unknown processes (Khalil et al., 2003; Harnisch and Eisenhauer, 1998; Harnisch et al., 2000). The atomization energy, enthalpy of formation, and bond dissociation energies of CF_4 are of interest for modeling its sources and sinks. Compute the atomization energy and $\Delta_f H^o_{298}$ of CF_4 from the C-F bond dissociation energies given in Dixon et al. (1999): D (F_3C–F) = 128.4, D (F_2C-F) = 85.9, D (FC-F) = 122.9, and D (C-F) = 128.4 kcal mol^{-1}. The atomization energy is the sum of the individual bond energies and is 465.6 kcal mol^{-1} (1948.07 kJ mol^{-1}) at absolute zero. The $\Delta_f H^o_{298}$ of CF_4 can be calculated from the 0 K atomization energy via Kirchhoff's equation (5-38) and a Hess's law cycle. The first computation yields the atomization energy at 298 K:

$$\Delta_{at}H_{298}(CF_4) = \Delta_{at}H_0(CF_4) + (H_{298} - H_0)_{C\ (g)} + 4(H_{298} - H_0)_{F\ (g)} - (H_{298} - H_0)_{CF4\ (g)} \tag{5-132}$$

$$\Delta_{at}H_{298}(CF_4) = 1948.07 + 6.536 + 4(6.518) - 12.730 = 1967.95 \text{ kJ mol}^{-1} \tag{5-133}$$

The second computation done by rearranging and substituting into Eq. (5-123) gives the standard enthalpy of formation:

$$\Delta_f H^o_{298}\ (CF_4) = 4(79.38) + 716.68 - 1967.95 = -933.75 \text{ kJ mol}^{-1} \tag{5-134}$$

This result agrees with the tabulated heat of formation value of -933.20 ± 0.75 kJ mol^{-1} listed in Appendix 1 within the stated uncertainty.

We can also use bond energies to estimate enthalpies of formation, and vice versa. If we divide the CF_4 atomization energy of 1948.07 kJ mol^{-1} by four, we obtain the mean C–F bond energy in CF_4, which is 487 kJ mol^{-1}. In the absence of any other information we could use this value in conjunction with other bond energies to estimate the heat of formation of other related compounds. For example, we can estimate the heat of formation of perfluoroethane (C_2F_6), which is another potent greenhouse gas released during aluminum smelting, using the mean C–F bond energy of 487 kJ mol^{-1} and the mean C–C bond energy of 346 kJ mol^{-1} (Cottrell, 1958). The estimated atomization energy of C_2F_6 is 3268 kJ at 0 K versus the actual value of 3222 kJ. The atomization energy at 298 K calculated using $(H_{298} - H_0) = 20.27$, 6.536, and 6.518 kJ mol^{-1} for C_2F_6, C, and F, respectively, is 3300 kJ. The estimated $\Delta_f H^o_{298}$ value for C_2F_6 is calculated using the $\Delta_f H^o_{298}$ values for C (g) and F (g) given previously and is -1390 kJ mol^{-1} versus the actual value of -1344 kJ mol^{-1} from Appendix 1. Another example of using bond energies to estimate $\Delta_f H$ values is given in the problem about CF_3SF_5 (trifluoromethyl sulfur pentafluoride), which is 18,000 times stronger than CO_2 for greenhouse warming and is increasing at 6% per year.

C. Ionization energy and electron affinity

Positive ions (cations) and negative ions (anions) are important for chemistry in the ionospheres of Earth and other planets as well as for chemistry in the highly ionized gas surrounding spacecraft, meteors, and meteorites during hypersonic entry in planetary atmospheres. In some cases such as CF_4 and other perfluorocarbons, reactions with ions in Earth's upper atmosphere are the only known destruction processes for greenhouse gases.

Ionization energy

The *ionization energy* (*IE*) is the energy required to remove an electron from a gaseous atom or molecule at 0 K and is the ΔH for the reaction

$$A = A^+ + e^- \tag{5-135}$$

where A stands for a gaseous atom of any element or for a gaseous molecule. The ionization energy is also called the *ionization potential* (*IP*). The ionization energy is different from the standard enthalpy of formation of a gas phase cation, which is the ΔH for forming the cation from its constituent elements in their reference states. The following example illustrates this point.

Example 5-16. The first ionization potential of potassium is 4.34066 electron volts. This is the enthalpy change for reaction (5-136) at absolute zero. Calculate the standard heat of formation of K^+ (g) at 298.15 K, which is the ΔH change for reaction (5-137):

$$K(g) = K^+(g) + e^-(g) \tag{5-136}$$

$$K(s) = K^+(g) + e^-(g) \tag{5-137}$$

Conversion of the ionization potential to $\Delta_f H^o_{298}$ for K^+ (g) involves both Hess's law and Kirchhoff's law. The first step is computation of ΔH for reaction (5-137) at 0 K:

$$K(g) = K^+(g) + e^-(g) \qquad \Delta H^o_0 = 418.807 \text{ kJ mol}^{-1}$$
$$K(s) = K(g) \qquad \Delta_f H^o_0 = 89.883 \text{ kJ mol}^{-1}$$
$$K(s) = K^+(g) + e^-(g) \qquad \Delta_f H^o_0 = 508.690 \text{ kJ mol}^{-1}$$

The ionization potential was converted to kJ mol^{-1} using the energy conversion factors from Table 3-2. The heat of vaporization of K at 0 K was computed from data in Appendix 1. We next calculate ΔH for reaction (5-137) at 298 K. This computation involves the cycle

$$\begin{array}{ccccl} & & \Delta_f H^o_{298} & & \\ & K(s) & = & K^+(g) + e^-(g) & 298.15\text{ K} \\ (H^o_{298} - H^o_0)_{K(s)} & \uparrow & & \downarrow & -(H^o_{298} - H^o_0)_{K^+(g)} \\ & & & & -(H^o_{298} - H^o_0)_{e^-(g)} \\ & K(s) & = & K^+(g) + e^-(g) & 0\text{ K} \\ & & \Delta_f H^o_0 & & \end{array}$$

The enthalpy increments for K (s), K^+ (g), and the electron gas are from Appendix 1.

$$\Delta_f H^o_{298} = \Delta_f H^o_0 - (H^o_{298} - H^o_0)_{K(s)} + (H^o_{298} - H^o_0)_{K^+(g)} + (H^o_{298} - H^o_0)_{e^-(g)} \tag{5-138}$$

$$= 508.690 - 7.080 + 6.197 + 6.197 \text{ kJ mol}^{-1} = 514.004 \text{ kJ mol}^{-1} \quad (5\text{-}139)$$

for the heat of formation of K^+ gas at 298 K, which is the same as the value in Appendix 1.

Electron affinity

The *electron affinity* is the energy required to remove an electron from a gaseous atom or molecule and is the ΔH for the reaction

$$A^- (g) = A (g) + e^- (g) \quad (5\text{-}140)$$

which is different than the heat of formation of the anion, as illustrated next.

Example 5-17. The electron affinity of F (g) is 3.401190 electron volts. Calculate the standard heat of formation of F^- (g) at 298 K. The electron affinity of fluorine is the enthalpy change for reaction (5-141) at 0 K. The $\Delta_f H^o_{298}$ for F^- (g) is the ΔH for Eq. (5-142) at 298 K:

$$F^-(g) = F(g) + e^- (g) \quad (5\text{-}141)$$

$$\frac{1}{2}F_2 (g) + e^- (g) = F^- (g) \quad (5\text{-}142)$$

Calculate $\Delta_f H^o_{298}$ for F^- (g) from the electron affinity using Kirchhoff's and Hess's laws. The $\Delta_r H$ for reaction (5-141) at 298 K can be calculated from the cycle:

$$\begin{array}{lcccr} & & \Delta_r H^o_{298} & & \\ & F(g) + e^-(g) & = & F^-(g) & 298.15 \text{ K} \\ (H^o_{298} - H^o_0)_{F(g)} \; (H^o_{298} - H^o_0)_{e^-(g)} & \uparrow & & \downarrow \; -(H^o_{298} - H^o_0)_{F^-(g)} & \\ & F(g) + e^-(g) & = & F^-(g) & 0 \text{ K} \\ & & \Delta_r H^o_0 & & \end{array}$$

where the enthalpy increments for F (g), e^- (g), and F^- (g) are taken from Appendix 1. Note that reaction (5-141) has been reversed so that F^- (g) is the product. The $\Delta_r H^o_0 = -3.401190$ electron volts for this reaction. Thus, formation of F^- (g) is exothermic. The $\Delta_r H$ at 298 K is

$$\Delta_r H^o_{298} = \Delta_r H^o_0 - (H^o_{298} - H^o_0)_{F(g)} - (H^o_{298} - H^o_0)_{e^-(g)} + (H^o_{298} - H^o_0)_{F^-(g)} \quad (5\text{-}143)$$

$$= -328.162 - 6.518 - 6.197 + 6.197 \text{ kJ mol}^{-1} = -334.680 \text{ kJ mol}^{-1} \quad (5\text{-}144)$$

The ΔH change for reaction (5-142) is obtained using Hess's law:

$$F(g) + e^- (g) = F^- (g) \qquad \Delta_r H^o_{298} = -334.680 \text{ kJ mol}^{-1}$$

$$\frac{1}{2} F_2 (g) = F(g) \qquad \Delta_f H^o_{298} = 79.390 \text{ kJ mol}^{-1}$$

$$\frac{1}{2} F_2 (g) + e^- (g) = F^- (g) \qquad \Delta_f H^o_{298} = -255.290 \text{ kJ mol}^{-1}$$

The formation of F^- as well as Cl^-, Br^-, and I^- is exothermic because halogens are more stable after gaining an electron, which closes their outer electron shells.

PROBLEMS

1. The standard enthalpy of formation ($\Delta_f H^o_{298}$) of CO_2 (g) is the ΔH change at 298.15 K for

$$\text{C (graphite)} + \text{O}_2\text{(g)} = \text{CO}_2\text{ (g)}$$

Use the $\Delta_f H^o_{298}$ values tabulated here to calculate the ΔH changes at 298.15 K for CO_2 formation from (a) diamond + O_2 (g); (b) graphite + 2 O (g); (c) diamond + 2 O (g); (d) C (g) + O_2 (g); and (e) C (g) + 2 O (g).

Substance	C (graphite)	C (diamond)	C (g)	CO_2 (g)	O_2 (g)	O (g)
$\Delta_f H^o_{298}$ (kJ mol^{-1})	0	1.85	716.68	−393.52	0	249.17

2. The Cs^+/Cs gas ratio is a temperature indicator in the atmospheres of brown dwarfs (failed stars). Calculate the standard enthalpy of formation ($\Delta_f H^o_{298}$) of Cs^+ (g)—that is, the ΔH for

$$\text{Cs(metal)} = \text{Cs}^+\text{ (g)} + \text{e}^-\text{ (g)}$$

from the following data: the first IP for Cs (g) = 375.704 kJ mol^{-1}, $\Delta_f H^o_{298}$ (Cs, g) = 76.500 kJ mol^{-1}, and the ($H^o_{298} - H^o_0$) values for Cs (g), Cs^+ (g), and e^- (g), which are each 6.197 kJ mol^{-1}.

3. Calculate the heat evolved when casting how plaster (α-$CaSO_4 \cdot ½ H_2O$) reacts with water and hardens to gypsum ($CaSO_4 \cdot 2 H_2O$). Use the $\Delta_f H^o_{298}$ values (kJ mol^{-1}) below (Kelley et al., 1941).

H_2O (liq)	α-$CaSO_4 \cdot ½ H_2O$ (c)	$CaSO_4 \cdot 2 H_2O$ (gypsum)
−285.83	−1559.88	−2005.81

4. Enstatite ($MgSiO_3$) with the ilmenite structure becomes stable at about 20 GPa (200 kilobar) in the MgO-SiO_2 system. Calculate the $\Delta_r H^o_{298}$ between the orthopyroxene and ilmenite phases

$$\text{MgSiO}_3\text{ (orthopyroxene)} = \text{MgSiO}_3\text{ (ilmenite)}$$

using the following data from Ashida et al. (1988) for enthalpy of solution in lead borate melts:

$$\text{MgSiO}_3\text{ (orthopyroxene)} = \text{MgSiO}_3\text{ (dissolved in Pb borate melt)}\ \Delta H = 111.23 \pm 1.11\text{ kJ mol}^{-1}$$

$$\text{MgSiO}_3\text{ (ilmenite)} = \text{MgSiO}_3\text{ (dissolved in Pb borate melt)} \qquad \Delta H = 52.00 \pm 4.11\text{ kJ mol}^{-1}$$

5. Calculate the enthalpy of solution ($\Delta_{soln}H^o_{298}$) for table salt dissolved in water in an infinitely dilute solution from the tabulated $\Delta_f H^o_{298}$ values (kJ mol^{-1}) here:

NaCl (halite)	Na^+ (aq)	Cl^- (aq)
−411.3	−240.3	−167.1

6. Precipitation of malachite ($Cu_2(OH)_2CO_3$) from groundwater in the oxidized region of an ore deposit occurs via the net reaction

$$2\,Cu^{2+}\,(aq) + CO_3^{2-}\,(aq) + 2\,OH^-\,(aq) = Cu_2(OH)_2CO_3\ (\text{malachite})\ \Delta_r H^o_{298} = -48.6\ \text{kJ mol}^{-1}$$

Calculate the enthalpy of formation of the Cu^{2+} ion using this reaction and the data here:

Substance	$\Delta_f H^o_{298}$ (kJ mol^{-1})
CO_3^{2-} (aq)	−675.2
OH^- (aq)	−230.0
malachite	−1054.0

7. Nitric oxide (NO) forms when N_2 burns in air-rich flames. It is easily ionized and is important for the chemistry of ions in flames. The ionization potential of nitric oxide (NO) is 9.26436 eV. Calculate the heat of formation of NO^+ (g) at 0 K.
8. Reactions of ozone (O_3) with ions are important in the air surrounding meteors and spacecraft during hypersonic entry into Earth's upper atmosphere. The reaction

$$O^- + O_3 \rightarrow O + O_3^-$$

has $\Delta H^o_{300} = -63$ kJ mol^{-1} at 300 K. Estimate $\Delta_f H^o_{298}$ for O_3^-(g) from this value by using the electron affinity of O gas (1.4611 eV) and the enthalpy of formation of ozone (141.8 kJ mol^{-1}). There are two approximations you have to make in solving this problem. What are they?
9. Nitric oxide (NO) is a pollutant formed by high temperature combustion. The spectroscopically measured dissociation energies (D^o_0) for N_2, O_2, and NO are 78,715 cm^{-1}, 41,260 cm^{-1}, and 52,400 cm^{-1}, respectively. Calculate $\Delta_f H^o_{298}$ for NO, that is ΔH for the reaction ½ N_2 (g) + ½ O_2 (g) = NO (g) from these data.
10. The hydroperoxyl (HO_2) radical is important for atmospheric chemistry on Earth, including the HO_x catalytic cycle for ozone destruction. Litorja and Ruscic (1998) used photoionization mass spectroscopy to determine D^o_0 (H-OOH) = 362.8 ± 3.3 kJ mol^{-1}. Use this value and auxiliary data from Appendix 1 to compute $\Delta_f H^o_0$ of HO_2 (g). *Hint:* Consider the products formed by breaking the H-OOH bond in H_2O_2 (g).

11. Kiseleva et al. (1979) reported the results tabulated here for heats of solution of different phases in molten lead borate ($2\ PbO \cdot B_2O_3$) at 1170 K.

(a) Use these data to calculate ΔH for formation of clinoenstatite and forsterite from their constituent oxides at this temperature.

Phase	$\Delta_{soln}H^o_{1170}$ (kJ mol^{-1})
MgO (periclase)	8.66
SiO_2 (quartz)	3.77
$MgSiO_3$ (clinoenstatite)	47.2
Mg_2SiO_4 (forsterite)	80.5

(b) Use the data that follows to correct your answer from part (a) to 298.15 K.

Phase	$(H^o_{1170} - H^o_{298})$ (kJ mol^{-1})
MgO (periclase)	41.67
SiO_2 (quartz)	60.72
$MgSiO_3$ (clinoenstatite)	96.19
Mg_2SiO_4 (forsterite)	139.54

12. Tephroite (Mn_2SiO_4) and rhodonite ($MnSiO_3$) are the most common manganese silicates and are important sources of this metal. Navrotsky and Coons (1976) measured heats of solution for several phases in molten lead borate at 986 K. Use their data that follow to calculate ΔH for these reactions: (a) MnO (manganosite) + $MnSiO_3$ (rhodonite) = Mn_2SiO_4, (b) 2 $MnSiO_3$ (rhodonite) = Mn_2SiO_4 + SiO_2 (quartz), and (c) $MnSiO_3$ (rhodonite) = $MnSiO_3$ (pyroxmangite).

Phase	$\Delta_{soln}H^o_{986}$ (kJ mol^{-1})
SiO_2 (quartz)	−3.18
MnO (manganosite)	5.82
$MnSiO_3$ (rhodonite)	29.2
$MnSiO_3$ (pyroxmangite)	28.95
Mn_2SiO_4 (tephroite)	59.62

13. Zinc sulfide has two polymorphs, sphalerite and wurtzite, with different thermodynamic properties. Adami and King (1964) used HCl solution calorimetry and Hess's law to derive $\Delta H^o_{298} = -109.58$ kJ mol^{-1} for the reaction

$$ZnO\ (zincite) + H_2S\ (g) = ZnS\ (wurtzite) + H_2O\ (l)$$

Use their results and data from Appendix 1 to calculate $\Delta_f H^o_{298}$ for ZnS (wurtzite).

14. Large impacts apparently took place on Earth early in its history. Calculate (a) the energy required to heat up the entire Earth to 3000 K and (b) the size of a rocky impactor with a 20 km s^{-1} velocity and a density of 3.3 g cm^{-3} required to do this. Assume Earth has a mass of 5.97×10^{24} kg and is 67% forsterite and 33% Fe (by mass) and start at an initial temperature of 500 K. Use the following data for Fe and Mg_2SiO_4 in your calculations, and show your work.

Species	$(H^o_{3000} - H^o_{298})$ (kJ mol^{-1})	$(H^o_{500} - H^o_{298})$ (kJ mol^{-1})
Mg_2SiO_4 (s, liq)	572.69	27.49
Fe (reference state)	127.27	5.53

15. Nitrogen tetroxide (N_2O_4) and unsymmetrical dimethyl hydrazine ($N_2H_8C_2$) spontaneously react when mixed together:

$$H_2NN(CH_3)_2\ (l) + N_2O_4\ (l) = 2\ N_2\ (g) + 2\ CO_2\ (g) + 4\ H_2O(g)$$
$$\Delta_r H^o_{298} = -1784.12\ \text{kJ mol}^{-1}$$

and were used as self-igniting fuel in the Apollo lunar landers. Calculate the adiabatic flame temperature of this rocket engine. Necessary C_P data are in Table 3-4.

16. Ammonium chloride (NH_4Cl) sublimes to a 1:1 mixture of NH_3 (g) and HCl (g):

$$NH_4Cl\ (s) = NH_3\ (g) + HCl\ (g) \qquad \Delta_r H^o_{298} = 42{,}160\ \text{cal mol}^{-1}\ (\text{of } NH_4Cl)$$

The ΔH given is from sublimation pressure measurements. Devise a thermochemical cycle and calculate $\Delta_r H^o{}_{298}$ for NH_4Cl sublimation using data from Stephenson (1944).

Reaction	$\Delta_r H^o_{298}$ (cal mol^{-1})
NH_3 (g) + H_2O (l) = NH_4OH (aq)	−8285
NH_4^+ (aq) + OH^- (aq) = NH_4OH (aq)	−865
H_2O (l) = H^+ (aq) + OH^- (aq)	13,385
HCl (g) = H^+ (aq) + Cl^- (aq)	−17,880
NH_4Cl (s) = NH_4^+ (aq) + Cl^- (aq)	3600

This problem is of interest for interpretation of Galileo probe mass spectrometer data and for microwave observations of Cl-bearing gases in Jupiter's atmosphere.

17. In a pioneering study, Haber and Tamaru (1915) used flow calorimetry to measure the heat of decomposition of ammonia to its constituent elements in their reference states. Use their data and the heat contents ($H^o_T - H^o_{298}$) for N_2, H_2, and NH_3 given here to calculate the standard enthalpy of formation of ammonia.

temperature (°C)	466	503	554	659
ΔH^o (calories mol^{-1})	12,670	12,700	12,860	13,100

	($H^o_T - H^o_{298}$) kJ mol^{-1}		
T (K)	NH_3	N_2	H_2
700	16.872	11.937	11.749
800	21.853	15.046	14.702
900	27.113	18.223	17.676
1000	32.637	21.463	20.680

18. Isopentane and n-pentane are anthropogenic air pollutants. Although they have the same chemical formula (C_5H_{12}), isopentane has an atmospheric lifetime of 7.5 days, whereas n-pentane has a lifetime of only 6.2 days. This suggests that isopentane has a more negative molar enthalpy of formation. Use the data given here to show whether or not this suggestion is true.

$$\text{n-}C_5H_{12}\ (g) + 8\ O_2\ (g) = 5\ CO_2\ (g) + 6\ H_2O\ (l) \qquad \Delta_c H_m = -3536.15\ \text{kJ mol}^{-1}$$
$$\text{iso-}C_5H_{12}\ (g) + 8\ O_2\ (g) = 5\ CO_2\ (g) + 6\ H_2O\ (l) \qquad \Delta_c H_m = -3528.12\ \text{kJ mol}^{-1}$$

19. The standard enthalpy of formation of the Zn^{2+} aqueous ion (-153.39 kJ mol^{-1}) is equal to the standard integral molar enthalpy of solution of zinc metal in acid, that is, the ΔH for the reaction

$$Zn\ (\text{metal}) + 2\ H^+\ (aq) = Zn^{2+}\ (aq) + H_2\ (g)$$

The standard enthalpy of formation of infinitely dilute H_2SO_4 (aq) is -909.27 kJ mol^{-1}. Use these data to calculate the standard integral molar enthalpy of solution of zinkosite ($ZnSO_4$).

20. The data of Sahama and Torgeson (1949) for $\Delta_{soln}H^o$ of orthopyroxenes in aqueous HF acid are represented by the equation $\Delta_{soln}H^o = -263.045 + 24.483(X_{Fs})$, where X_{Fs} is the ferrosilite mole fraction. Calculate ΔH^o for the reaction

$$Fe_2SiO_4\ (\text{fayalite}) + SiO_2\ (\text{quartz}) = 2\ FeSiO_3\ (\text{ferrosilite})$$

using data from this problem and heats of solution of -263.045 and -138.072 kJ mol^{-1}, respectively, for fayalite and quartz. Is formation of ferrosilite from fayalite + quartz exothermic or endothermic?

21. Boerio-Goates et al. (2001) measured the standard molar enthalpies of combustion ($\Delta_c H^o_m$) and formation ($\Delta_f H^o_{298}$) of crystalline adenosine ($C_{10}H_{13}N_5O_4$). Their data are important for calculating thermodynamic properties of adenosine 5′-triphosphate (ATP), which transfers energy inside living cells via its breakdown and reformation. Calculate $\Delta_f H^o_{298}$ of adenosine (c) from $\Delta_c H^o_m = -5139.4$ kJ mol^{-1} at 298 K for the combustion reaction

$$C_{10}H_{13}N_5O_4\ (c) + 11.25\ O_2\ (g) = 10\ CO_2\ (g) + 6.5\ H_2O\ (l) + 2.5\ N_2\ (g)$$

22. Sturges et al. (2000) discovered the potent greenhouse gas CF_3SF_5 (trifluoromethyl sulfur pentafluoride) in Earth's atmosphere. This gas has an abundance of ~0.12 parts per trillion (in 1999), is increasing at 6% per year, and is apparently produced when SF_6 decomposes in high-voltage equipment where it is used as a gaseous insulator. Estimate the heat of formation of CF_3SF_5 using these rounded off $\Delta_f H^o_{298}$ values from Appendix 1: CF_3 (-472 ± 5), SF_5 (-903 ± 10), CS_2 (117 ± 1), CS (280 ± 1), and S (277 ± 0.1). Also estimate the uncertainty on your answer using quadratic combination of errors.

CHAPTER

The Second Law of Thermodynamics and Entropy

6

Die Entropie der Welt strebt einem Maximum zu. *(The entropy of the universe increases to a maximum.)*
—Rudolf Clausius (1865)

Es ist unmöglich, eine Maschine zu konstruieren, die fortdauernd die Wärme der Umgebung in äußere Arbeit verwandelt. *(It is impossible to construct a machine that continuously converts the heat of its surroundings into external work.)*
—Walther Nernst (1924)

The law that entropy always increases—the second law of thermodynamics—holds, I think, the supreme position among the laws of Nature.
—Sir Arthur Eddington (1929)

Chapters 3 through 5 described the first law of thermodynamics and several of its applications. We can summarize this discussion by stating that the first law expresses the equivalence of the various forms of energy in general and of thermal energy (heat) and mechanical energy (work) in particular. The first law and the equations derived from it are general in their application. (The relationships derived for an ideal gas can be modified for real gases, as we will discuss later.)

In this chapter we introduce the second law of thermodynamics. The second law is fundamentally different from the first law because it predicts the direction of spontaneous change for all chemical and physical processes. It also quantifies the amount of waste heat that is produced when heat is transformed into electrical or mechanical work. The first law does not consider either of these two important questions. These predictions depend on *entropy*, which is a new state function that is defined by the second law.

This chapter is divided into nine sections. Section I reviews key definitions that were introduced earlier and are important for this chapter's discussion. In particular, we review the definitions of reversible and irreversible processes. In Section II we discuss the historical development of the second law, which took 30 years. We give several verbal statements of the second law in Section III. These statements are easy to understand and convey the meaning of the second law. Section IV describes the Carnot cycle, which is a reversible thermodynamic cycle of an ideal machine that can be used as a heat engine (i.e., for converting heat into work) or as a heat pump (i.e., for cooling and refrigeration). The Carnot cycle is one of the most important concepts in thermodynamics and is the foundation of the second law. We use the Carnot cycle to calculate the maximum possible efficiency of any real heat engine and to show it is impossible to reach absolute zero. In Section V we use the Carnot cycle to define the thermodynamic (or absolute) temperature scale. We define entropy, discuss entropy changes

Copyright © 2013 Elsevier Inc. All rights reserved.

for a system and its surroundings during reversible and irreversible (or spontaneous) processes, and give a mathematical statement of the second law in Section VI. In Section VII we discuss some applications of the second law. We first derive the combined first and second law, also known as the *fundamental equation*. Then we use the fundamental equation to calculate entropy changes for chemical and physical processes. We also describe entropy calculations for solids, liquids, and gases. We close Section VII with a description of entropy estimation methods for minerals. We introduce two new state functions, the Gibbs and Helmholtz free energies, in Section VIII, and use them to quantify the maximum work and net work that are obtained when heat is transformed into work. We then use the Gibbs free energy to calculate entropies for aqueous ions. Section IX discusses entropy and probability. A brief review of basic concepts about probability is followed by a discussion of the Boltzmann-Planck equation. We derive equations for the entropy of mixing for gases, liquids, and solids. We then present and use equations for the configurational entropy of minerals with mixing on more than one crystallographic site. Next we use the Boltzmann-Planck relation to calculate the probability of impossible events, such as boulders rolling uphill at the expense of the thermal energy of their surroundings. Finally, Maxwell's demon, an imaginary being that may be able to violate the second law, is also discussed.

LUDWIG EDWARD BOLTZMANN (1844–1906)

Ludwig Boltzmann was an Austrian theoretical physicist who believed that the observable behavior of materials was linked to occurrences on an atomic level. Furthermore, he believed that changes on the atomic level were governed by probability and chance. He used statistical mechanics to show that entropy is related to disorder or randomness. However, the well-known Boltzmann-Planck equation $S = k \log W$, which quantitatively relates entropy and the disorder of a system, was first used by Max Planck and not by Boltzmann himself.

During his lifetime, few accepted Boltzmann's belief in atoms as real physical particles, thinking instead that they were simply a useful model for mathematical calculations. Wilhelm Ostwald, a respected and influential scientist at that time and a colleague of Boltzmann, did not think Boltzmann's kinetic theory was useful even as a hypothesis. An intelligent and sensitive man, Boltzmann regrettably suffered from depression and health problems, made worse by the rejection of his ideas by established scientists. He ultimately committed suicide after several unsuccessful attempts. The Boltzmann-Planck equation $S = k \log W$ is engraved on his tomb in the main cemetery in Vienna, Austria.

I. A REVIEW OF SOME IMPORTANT DEFINITIONS

Several key concepts that were introduced in Chapters 2 and 3 are essential for understanding the Carnot cycle, entropy, the second law, and the other new ideas presented in this chapter. We briefly review these concepts here, but you may want to reread the relevant sections in Chapters 2 and 3. A *system* is the part of the universe that we are studying, and we can define it in such a way as to make it easier to solve the problem at hand. We define the *surroundings* as everything outside the system. *Closed systems* exchange energy but not matter with the surroundings, whereas *isolated systems* do not exchange energy or matter with the surroundings. A *state variable* is a property such as internal energy or enthalpy that depends only on the difference between the initial and final states of the system and not on the path between the initial and final states.

The distinction between reversible and irreversible processes is very important. A *reversible process* is a process that passes through a succession of equilibrium states so that all the changes in the system and the surroundings can be restored by a reversal of the direction of the process without producing any other changes in the system or surroundings. To meet these conditions, a reversible process has to be carried out infinitely slowly, with infinitesimally small steps, and without any friction, hysterisis, or electrical resistance. This is impossible to do in any real process. A reversible process is the ideal limiting case of a real process.

An *irreversible process* does not pass through a succession of equilibrium states and cannot be reversed without making some changes in the system and/or in the surroundings. All natural processes are irreversible to a greater or lesser extent. The extent of irreversibility can be quantified using the entropy change for each process because the entropy change increases as the degree of irreversibility increases. Examples of irreversible processes include clocks and watches slowing down as their mainsprings unwind, explosions, gas expanding into a vacuum, hot food cooling down as it sits on a table, ice cubes melting in a glass of soda, logs burning in a fire, mixing of two gases or solutions, salt dissolving in water, and sugar dissolving in coffee. The cooling of Earth's interior over geologic time as radioactive potassium, thorium, and uranium slowly decay away is an example of an irreversible process on a much longer time scale. All these processes are moving toward a final state in which no further spontaneous changes occur. The common denominator for all irreversible processes is an entropy increase. In many cases energy is lost or dissipated. However, this is not the essential feature of an irreversible process, because mixing of gases or liquids or solids dissolving in liquids (examples of diffusion down a concentration gradient) does not involve the dissipation of energy.

II. HISTORICAL DEVELOPMENT OF THE SECOND LAW

The development of the second law, the concept of entropy, and of thermodynamics in general, is closely connected with the development of the steam engine. The chemist and physiologist Lawrence J. Henderson (1878–1942) summarized this situation in 1917 with his famous statement that "Science owes more to the steam engine than the steam engine owes to science."

James Watt developed the first practical steam engine in 1769. This led to the emergence of Great Britain as the major industrial power in the world and was one of the seminal events that marked the start of the Industrial Revolution. It also intensified the rivalry between Great Britain and France, which needed to develop efficient steam engines to strengthen its economy.

As the Industrial Revolution progressed, Watt and others continued to improve the design and operation of steam engines. Each engine had some ratio of work done to coal burned. This ratio is called the *thermal efficiency* and is represented by the Greek letter eta (η):

$$\eta = \frac{\text{work obtained}}{\text{heat input}} = \frac{w}{q_h} \tag{6-1}$$

where w represents the work done and q_h the heat taken by the steam from the boiler. The goal of Watt and other engineers was to increase the efficiency of steam engines to get the most work from the amount of coal burned. This would decrease costs (for coal) and increase output (from work done). Improvements, such as better insulation on the boiler and steam lines, better bearings, and better lubrication, reduced heat losses and friction. But engineers wondered if there was a theoretical limit to the efficiency of steam engines that could not be exceeded. This is the question that French engineer Sadi Carnot (pronounced *sadie car-no*; see sidebar) addressed and answered in a small book entitled *Reflections on the Motive Power of Fire*, published in 1824, when he was only 28 years old.

NICOLAS LÉONARD SADI CARNOT (1796–1832)

Sadi Carnot was born in Paris and spent most of his short life there. As a child, he was educated at home by his father, who was a well-known general under Napoleon. He completed his education at the prestigious École Polytechnique. Carnot also joined the army but left military service in 1828 to devote himself to study.

Carnot's contribution to thermodynamics was his idealized steam engine. He was the first to imagine the functioning of a heat engine in terms of a series of reversible processes. His model was used to show that there is a limit to the efficiency of any heat engine, which depends only on the difference in temperature between its hot and cold reservoirs, not on the size of the engine, friction, or any other mechanical aspect. Carnot's work anticipated the second law of thermodynamics and would later influence Kelvin and Clausius.

Carnot published only one book, which remained obscure for several reasons. First, he wrote it in a style that was too casual for scientists but too detailed for laymen. The book received only one citation in the body of literature of the time (in a paper by B. P. E. Clapeyron). Also, Carnot died of cholera at a young age. The custom at that time was to burn the possessions of victims of contagious diseases, and so only one copy of his manuscript survived.

Carnot discussed the operation of steam engines and other heat engines that used air or another gas instead of steam. He considered an ideal heat engine operating in a closed cycle with reversible processes. Carnot proved that a reversible heat engine has the theoretical maximum efficiency and that all reversible heat engines working between the same two temperatures are equally efficient. His arguments laid the foundation for the second law of thermodynamics. Carnot was also the first person to use the concept of a closed cycle in thermodynamics, but he did not use indicator diagrams to illustrate the operation of his ideal heat engine. Carnot's book was written for the general public and received a good review. However, it did not make a big impression at the time, perhaps due to the political instability in France. Ironically, because Carnot wrote his book for the general public, scientists remained unaware of his work and its importance for the field of thermodynamics. Carnot died of cholera eight years later, and most of his papers and possessions were burned, as was customary with victims of epidemics. By this time, his book had been forgotten by almost everyone. Lord Kelvin later recounted that when he was looking for a copy in Paris in the late 1840s, virtually none of the booksellers had heard of it, although they knew of Carnot's illustrious father, who was one of Napoleon's generals.

Fortunately, Émile Clapeyron, a French mining and railroad engineer (see sidebar in Chapter 7), read Carnot's book and recognized its significance. In 1834, Clapeyron published his only paper in thermodynamics, entitled "Memoir on the Motive Power of Heat," in a French engineering journal. In this paper, he derived the Clapeyron equation (see Chapter 7), used indicator diagrams to illustrate the Carnot cycle, and presented Carnot's results mathematically.

But Clapeyron's paper, like Carnot's book, did not make a big impact on French scientists. Clapeyron's work remained relatively unknown until his paper was translated into German and republished in 1843. Over 20 years after Carnot's book had appeared, his work and that of Clapeyron was finally appreciated and recognized by the German physicist Rudolf Clausius (see sidebar) and by William Thomson (later Lord Kelvin; see sidebar in Chapter 2). Both of these scientists proceeded to develop the second law of thermodynamics. Clausius presented his results at a scientific meeting in early 1850 and published them shortly thereafter. Thomson presented his work at another scientific meeting in late 1851 and published his results in 1853. Both scientists continued working on questions related to the second law for the next several years and made important contributions to thermodynamics, which we discuss later.

RUDOLF JULIUS EMMANUEL CLAUSIUS (1822–1888)

Rudolf Clausius was a well-respected German physicist who built on Sadi Carnot's work through careful analysis. He postulated that heat could not pass by itself from one body to another at higher temperature, which is one way of stating the second law of thermodynamics.

Clausius valued clarity and simplicity and strove to write the laws of thermodynamics succinctly. He expressed them in the following manner: (1) the energy of the universe is constant, and (2) the entropy of the universe tends toward a maximum. His version of the second law expands on Carnot's in that it uses the concept of entropy, a term that Clausius coined.

Clausius faced hardships during his life that may have reduced the amount of scientific work he was able to do later. First, he was wounded in 1870 during the Franco-Prussian War, during which he led a student ambulance corps. Second, he was faced with the task of raising his six children alone after his wife died giving birth in 1875.

Clausius was always exacting in his work, yet was often unaware of or ambivalent toward the work of his contemporaries. For example, he seemed reticent to make use of statistical methods. In his writings on the kinetic theory of gases, he acknowledges that the molecules in a gas will travel at various speeds, but he uses average velocities instead of adopting Maxwell's distribution equations, which estimate the number of particles at any given speed. Clausius never recognized the statistical nature of the second law, either.

III. THE SECOND LAW OF THERMODYNAMICS

Before getting into mathematical equations, we present several verbal statements of the second law of thermodynamics. Clausius stated the second law as: "Heat cannot pass of its own accord from a colder to a hotter body." This statement is easy to understand because your cup of coffee sitting on your desk *never* gets warmer by extracting heat from the desk, your hand, or the surrounding air. You can heat up the coffee using a hot plate or a microwave oven, but the heating is not spontaneous and is being done by an external machine.

Another similar statement of the second law from Lord Kelvin is: "Two bodies which are at different temperatures exchange heat in such a manner that the heat flows naturally from the hotter to the colder body." Refrigeration reverses this natural transfer of heat, but a machine is required to do this, which means that the heat flow is not spontaneous.

In his book *Theory of Heat*, published in 1871, the Scottish physicist James Clerk Maxwell (see sidebar) wrote: "Admitting heat to be a form of energy, the second law asserts that it is impossible, by the unaided action of natural processes, to transform any part of the heat of a body into mechanical work, except by allowing heat to pass from that body into another at a lower temperature." Maxwell's statement, like those of Clausius and Kelvin, says that it is impossible for us to extract energy from our surroundings and sell it to the local electric or gas company at a profit. This is impossible because it would cost us more money to get the energy from our surroundings than we would receive by selling it to these utilities.

The Nobel Prize–winning German physicist Max Planck (see sidebar in Chapter 9) stated the second law in this form: "It is impossible to construct an engine which will work in a complete cycle, and produce no effect except the raising of a weight and the cooling of a heat-reservoir." In other words, we cannot completely convert heat into the potential energy of a raised weight without expending more energy than we will get back when the weight falls down. There is always some waste heat that cannot be completely converted into useful work. Otherwise, we could run a refrigerator by extracting heat from the surrounding room, use this heat to raise a weight, and then use the falling weight to run an electrical generator to run the refrigerator. Planck's statement and that of Nernst's at the start of this chapter tells us that a perpetual-motion machine that runs by extracting heat from its surroundings is impossible.

JAMES CLERK MAXWELL (1831–1879)

James Clerk Maxwell was a Scottish physicist with wide-ranging interests that included adult education, the physics of Saturn's rings, electricity and magnetism, the kinetic theory of gases, and thermodynamics. Maxwell is famous for his equations in electricity and magnetism that are based on Michael Faraday's pioneering experimental studies. His thermodynamics work is less well known, although his book *Theory of Heat* is one of the best thermodynamics texts written. Another one of Maxwell's important contributions is his ingenious work on the statistical distribution of molecular velocities in a gas. Boltzmann subsequently gave a mathematical proof of Maxwell's hypothesis. The equation for molecular velocities in a gas is now known as the *Maxwell-Boltzmann distribution*.

Maxwell also raised a question that continues to challenge scientists to this day: He wondered if the second law of thermodynamics could be defeated by an intelligent being who could obtain work from the heat of his environment. Maxwell's demon was supposed to open and shut a door between two vessels of gas, to let only fast-moving molecules in and slow ones out. This would make one part of the gas hot and the other part cold, and a heat engine could be run from the temperature difference. Lord Kelvin named this being *Maxwell's Demon*. It has since been shown that several effects, including Brownian motion, would prevent the demon from doing his job. Nevertheless, new incarnations of the demon pop up from time to time, although none has yet defeated the second law.

IV. CARNOT CYCLE

A. Steam engines and heat engines

A steam engine converts heat into work. Water is heated in a boiler and converted to steam. The steam expands through a valve into a cylinder with a piston. The expansion moves the piston outward and produces work via coupling the piston movement to other hardware. The cooled steam is withdrawn from the cylinder through another valve, the piston is returned to its original position by the steam engine, and the next expansion cycle begins. In the simplest terms, the steam engine takes heat from the boiler (a high-temperature reservoir), converts some of the heat into work, and discards the leftover steam (to a low-temperature reservoir). The basic operating principle of the steam engine is shown in Figure 6-1. A gasoline-powered internal combustion engine, a diesel engine, a jet engine, and a steam- or gas-powered turbine operate according to the same basic principle. Steam engines and these other engines are called *heat engines*.

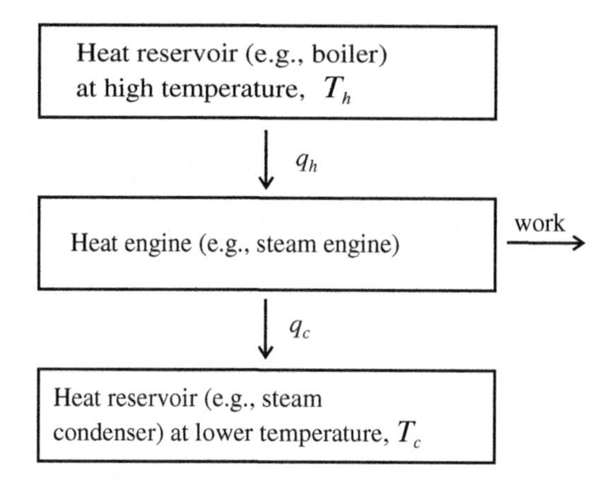

FIGURE 6-1

A summary of the basic operating principles of a steam engine or another heat engine.

B. Carnot heat engine

The *Carnot cycle* is the name given to the heat engine that Carnot considered in his book. Carnot considered the efficiency of a hypothetical, ideal heat engine that operated reversibly and had no losses due to friction or imperfect insulation. No real engine operates reversibly because a reversible process is an ideal process that can be approached but never achieved.

However, we can build an engine that *approximates* a Carnot cycle to do work or generate power, and we can study a Carnot cycle mathematically using indicator (*PV*) diagrams, such as those we used in Chapter 3. The operation of the Carnot cycle is schematically illustrated with the indicator diagram in Figure 6-2 and is described here. We use one mole of steam, behaving as an ideal gas that is inside the cylinder of a steam engine, in our discussion of the Carnot cycle. However, we will show that the efficiency of the Carnot cycle does not depend on whether we use steam, air, or another gas.

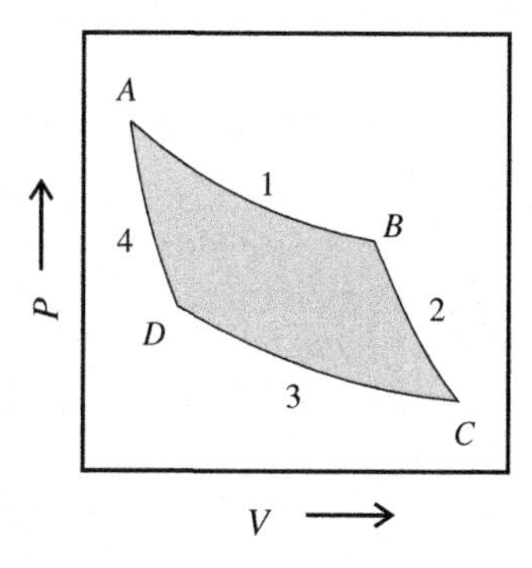

FIGURE 6-2

An indicator diagram for the Carnot cycle. The four steps in the Carnot cycle are (1) a reversible isothermal expansion, (2) a reversible adiabatic expansion, (3) a reversible isothermal compression, and (4) a reversible adiabatic compression. The shaded area is the work done by the engine after going through one cycle.

The first step in the Carnot cycle is a reversible isothermal expansion of the steam inside the cylinder. During this step, the volume of the steam increases from V_A to V_B, and the pressure of the steam decreases from P_A to P_B. The steam absorbs an amount of heat q_h from a heat reservoir at temperature T_h (h = high or hot). According to Eq. (3-57),

$$dE = nC_V dT \tag{3-57}$$

the internal energy of the steam does not change because there is no temperature change during an isothermal process. Thus, ΔE_1 for the first step of the Carnot cycle is

$$\Delta E_1 = nC_V \Delta T = (1)C_V(0) = 0 \tag{6-2}$$

The C_V in Eq. (3-57) and Eq. (6-2) is the constant-volume heat capacity of the steam. The amount of work w_1 done by the steam is

$$w_1 = \int_{V_A}^{V_B} PdV = \int_{V_A}^{V_B} nRT_h \frac{dV}{V} = RT_h \ln\left(\frac{V_B}{V_A}\right) \tag{3-55}$$

which is equal to the area under curve AB. The heat q_h absorbed by the steam is equal to the work w_1 that the steam does against the piston during expansion,

$$q_h = w_1 = RT_h \ln\left(\frac{V_B}{V_A}\right) \tag{6-3}$$

because of the conservation of energy:

$$\Delta E_1 = 0 = q_h - w_1 \tag{3-15}$$

Thus, w_1 and q_h are nonzero, but ΔE_1 is zero for the first step.

The second step in the cycle is a reversible adiabatic expansion from point B to point C as the piston moves outward in the cylinder. By definition, there is no heat transfer between a system and its surroundings during an adiabatic process. Thus, the heat flow q_2 during this step is zero. As a result, the steam has to use part of its internal energy to push the piston outward during this step. The temperature of the steam decreases to a lower temperature T_c (c = cold) as it does work w_2 equal to the area under curve BC. The internal energy change ΔE_2 and the work w_2 done during this step are related via the equation

$$\Delta E_2 = -w_2 \tag{6-4}$$

which follows from the first law (Eq. 3-15) because $q_2 = 0$. The work w_2 done is also given by

$$w_2 = -nC_V \Delta T = -C_V(T_c - T_h) \tag{6-5}$$

C_V in Eq. (6-5) is the average constant-volume heat capacity of the steam over the temperature interval from T_h to T_c. Thus, ΔE_2 and w_2 are nonzero, but q_2 is zero during the second step.

The third step in the cycle is a reversible isothermal compression from point C to point D as the piston moves inward in the cylinder. During this step, the volume of the steam decreases from V_C to V_D and the pressure of the steam increases from P_C to P_D. The steam has to transfer an amount of heat q_c to a heat reservoir at temperature T_c to maintain isothermal conditions. The internal energy of the steam does not change because this is an isothermal step ($\Delta E_3 = 0$). The work w_3 done to compress the steam is equal to the area under the curve *CD*. The heat q_c that is transferred to the heat reservoir at T_c and the work done w_3 are given by the equation

$$q_c = w_3 = RT_c \ln\left(\frac{V_D}{V_C}\right) \tag{6-6}$$

Both the work done to compress the steam and the heat transferred from the steam to the heat reservoir are negative because V_C is larger than V_D. This agrees with the definitions in Section II-C of Chapter 3 that work done on the system and heat transferred from the system to the surroundings are negative. Thus, w_3 and q_c are nonzero, but ΔE_3 is zero for the third step.

The fourth step, which closes the cycle, is a reversible adiabatic compression from point *D* to point *A*. By definition, the heat flow q_4 is zero for this step. The temperature of the steam increases from T_c to T_h as work w_4 equal to the area under curve *DA* is done to compress the steam. The internal energy change ΔE_4 and the work w_4 done on the steam during this step are related via the equation

$$w_4 = -\Delta E_4 = -nC_V\Delta T = -C_V(T_h - T_c) \tag{6-7}$$

Thus, ΔE_4 and w_4 are nonzero, but q_4 is zero during this step.

We now analyze the work, heat, and internal energy change for the entire Carnot cycle. The work done by the steam in going through the cycle once is

$$w = w_1 + w_2 + w_3 + w_4 \tag{6-8}$$

equal to the shaded area *ABCD* in Figure 6-2. Equation (6-8) can be rewritten as

$$w = \left[RT_h \ln\left(\frac{V_B}{V_A}\right)\right] + [-C_V(T_c - T_h)] + \left[RT_c \ln\left(\frac{V_D}{V_C}\right)\right] + [-C_V(T_h - T_c)] \tag{6-9}$$

Looking at Eq. (6-9) we see that the work w_2 done by the steam during the reversible adiabatic expansion and the work w_4 done on the steam during the reversible adiabatic compression cancel each other out. Thus, Eq. (6-9) reduces to

$$w = w_1 + w_3 = \left[RT_h \ln\left(\frac{V_B}{V_A}\right)\right] + \left[RT_c \ln\left(\frac{V_D}{V_C}\right)\right] \tag{6-10}$$

We can evaluate Eq. (6-10) using Eq. (3-70), which relates the temperature and volume of an ideal gas undergoing a reversible adiabatic process. Points *A* and *D* lie on one adiabatic curve and are related to the ratio (T_h/T_c) via the equation

$$\left(\frac{T_h}{T_c}\right) = \left(\frac{V_D}{V_A}\right)^{\gamma-1} \tag{6-11}$$

while points *B* and *C* lie on another adiabatic curve and are also related to the ratio (T_h/T_c) via

$$\left(\frac{T_h}{T_c}\right) = \left(\frac{V_C}{V_B}\right)^{\gamma-1} \tag{6-12}$$

Setting Eq. (6-11) equal to Eq. (6-12) we find that

$$\frac{V_D}{V_A} = \frac{V_C}{V_B} \tag{6-13}$$

which is equivalent to $V_A V_C = V_B V_D$. Rearranging we then obtain

$$\frac{V_B}{V_A} = \frac{V_C}{V_D} \tag{6-14}$$

which we can substitute into Eq. (6-10) to yield

$$w = \left[RT_h \ln\left(\frac{V_B}{V_A}\right)\right] + \left[RT_c \ln\left(\frac{V_D}{V_C}\right)\right] = R(T_h - T_c)\ln\left(\frac{V_B}{V_A}\right) \tag{6-15}$$

The net heat absorbed by the steam in going through the Carnot cycle once is

$$q = (q_1 + q_2 + q_3 + q_4) = (q_h + 0 - q_c + 0) = q_h - q_c \tag{6-16}$$

because $q_2 = q_4 = 0$ for the reversible adiabatic steps. The internal energy change ΔE of the steam in going through the Carnot cycle once is zero because

$$\Delta E = \Delta E_1 + \Delta E_2 + \Delta E_3 + \Delta E_4 = 0 + nC_V(T_c - T_h) + 0 + nC_V(T_h - T_c) = 0 \tag{6-17}$$

Thus, from the first law (Eq. 3-15), the work done by the steam in going through the Carnot cycle once is equal to the net heat absorbed by the steam in the cycle

$$w = q_h - q_c \tag{6-18}$$

The $(q_h - q_c)$ term in Eq. (6-18) is the same as the $(w_1 + w_3)$ term in Eq. (6-10) because Eqs. (6-3) and (6-6) show that the heat flow and work during steps 1 and 3 are equal. We can confirm Eq. (6-18) using Eq. (6-15), and Eqs. (6-3), (6-6), and (6-13) to substitute for w and $q_h - q_c$, respectively:

$$R(T_h - T_c)\ln\left(\frac{V_B}{V_A}\right) = RT_h \ln\left(\frac{V_B}{V_A}\right) - RT_c \ln\left(\frac{V_D}{V_C}\right) = R(T_h - T_c)\ln\left(\frac{V_B}{V_A}\right) \tag{6-19}$$

Thus, we have shown that the two sides of Eq. (6-18) are identical.

We now consider the thermal efficiency of the Carnot cycle, which is given by Eq. (6-1):

$$\eta = \frac{\text{work obtained}}{\text{heat input}} = \frac{w}{q_h} \tag{6-1}$$

Rewriting Eq. (6-1) using Eq. (6-15) to substitute for w and Eq. (6-3) to substitute for q_h gives

$$\eta = \left[R(T_h - T_c)\ln\left(\frac{V_B}{V_A}\right)\right] \Big/ \left[RT_h \ln\left(\frac{V_B}{V_A}\right)\right] = \frac{T_h - T_c}{T_h} = 1 - \frac{T_c}{T_h} \tag{6-20}$$

We can also derive Eq. (6-20) from Eq. (6-1) by substituting for w using Eq. (6-18), replacing q_h and q_c by Eqs. (6-3) and (6-6), and using Eq. (6-14) to equate the V_B/V_A and V_D/V_C terms.

Equation (6-20) shows that the thermal efficiency of the Carnot cycle is proportional to the temperature difference between the hot and cold reservoirs and is inversely proportional to the temperature of the high-temperature reservoir. The thermal efficiency is the same whether steam, air, NH_3, another gas, a water + steam mixture, or an NH_3 liquid + vapor mixture is used in the Carnot cycle. The only way to increase the thermal efficiency is to increase the temperature difference. No variables other than T_h and T_c are in Eq. (6-20), and hence no other variables play a role in determining the thermal efficiency of the Carnot cycle.

Example 6-1. (a) Calculate the thermal efficiency of a Carnot cycle that has $T_h = 766$ K and $T_c =$ 316 K. These temperatures correspond to those in high-pressure steam power plants. We can find the efficiency using Eq. (6-20):

$$\eta = \frac{T_h - T_c}{T_h} = \frac{766 - 316}{766} = 0.59 \tag{6-21}$$

In this case we have a thermal efficiency of 59%, which means that 59% of the heat absorbed from the high-temperature reservoir is converted into work. (b) Calculate the thermal efficiency of a Carnot cycle that operates between a hot reservoir at 458 K and a cold reservoir at 316 K. These temperatures correspond to those in moderate-pressure steam boilers. The efficiency

$$\eta = 1 - \frac{T_c}{T_h} = 1 - \frac{316}{458} = 0.31 \tag{6-22}$$

of 31% is lower than in the first case because $\Delta T/T_h$ is smaller.

It is important to understand that the thermal efficiencies calculated for the Carnot cycle are theoretical maximum values that are larger than the thermal efficiencies of any real engine. As we noted at the start of this section, all real engines operate irreversibly. Heat leaks and frictional losses can be made smaller and smaller as technology improves, but we can never make an engine that operates reversibly (i.e., goes through a succession of equilibrium states) because a real engine cannot work infinitely slowly with infinitely small steps without any dissipative effects. Thus, the thermal efficiency of the Carnot cycle is the upper limit to the thermal efficiency that can be reached by any real engine.

C. Carnot refrigerator

A refrigerator is a heat engine working in reverse and is a form of heat pump. This is illustrated in Figure 6-3. A refrigerator takes heat q_c from a cold reservoir at temperature T_c with input of work w and rejects heat q_h to a hot reservoir at temperature T_h. A refrigerator runs on electric current, removes heat from the items inside it, and rejects the heat into the surrounding room.

Now let's look at a refrigerator in more practical terms. Figure 6-4 shows the basics of how a refrigerator operates. Liquefied refrigerant from the liquid storage reservoir flows through a throttling valve into an evaporator. The evaporator receives an amount of heat q_c from items at T_c inside the refrigerator. This heat flow vaporizes the liquid refrigerant inside the evaporator. The vaporized refrigerant flows into the compressor, where it is compressed adiabatically (or nearly so). This gives a denser, hotter gas that flows into the condenser. The hot, high-pressure gas in the condenser is cooled by air cooling or water cooling. The condenser rejects an amount of heat q_h into the surrounding room at T_h. The cooling in the condenser reliquefies the refrigerant, which flows into the liquid storage vessel, and the cycle starts over.

Without going into the engineering details, good refrigerants are gases that can be liquefied under moderate pressure at a temperature slightly higher than room temperature, have a freezing point below the lowest temperature in the refrigeration cycle, have a critical point above the highest temperature in the cycle, are chemically stable and nonreactive with the materials in the refrigeration unit, and have a moderate vapor pressure at the lowest temperature in the cycle. (The critical point is the highest temperature at which liquid + gas coexist. The two phases merge into one phase above the critical point. We discuss this in Chapters 7 and 8.) Good refrigerants should also give a satisfactory coefficient

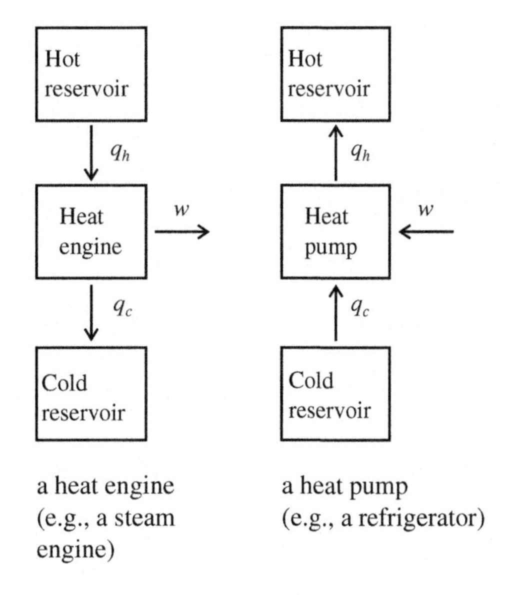

FIGURE 6-3

A schematic comparison of the basic operation of a heat engine and a heat pump.

of performance, which we define in the next paragraph. In addition, refrigerants for home use should be nontoxic. Gases that are used as refrigerants include those listed in Table 6-1. The chlorofluorocarbon (CFC) gases are being phased out as refrigerants because they are greenhouse gases and also cause ozone depletion when they are photochemically decomposed in Earth's stratosphere. The hydrofluorocarbon (HFC) gases are replacements for CFC refrigerants. However, HFC gases are also greenhouse gases. Ironically, propane and isobutane, which are flammable and were replaced by the CFC gases in the 1930s, are now the preferred alternatives to the CFC and HFC refrigerants.

The efficiency of any refrigerator is described by its coefficient of performance (β), which is the amount of heat (q_c) removed from the low-temperature reservoir to keep it at a specified low temperature divided by the amount of work (w) needed to remove the heat:

$$\beta = -\frac{q_c}{w} \tag{6-23}$$

Equation (6-23) has a minus sign because we defined work done on the system as negative in Chapter 3. Performance coefficients are greater than unity, with larger values indicating greater efficiency. For example, a typical temperature inside a freezer is about 5°F (258 K), and a typical condenser temperature on a freezer is 86°F (303 K). In this case, the observed coefficients of performance using ammonia or CO_2 as refrigerants are 4.76 or 2.56, respectively. In contrast, the coefficient of performance for a Carnot refrigerator (i.e., a reverse Carnot heat engine) operating between the same temperatures is

$$\beta_{\text{Carnot}} = -\frac{q_c}{w} = \frac{T_c}{T_h - T_c} = \frac{258}{303 - 258} = 5.73 \tag{6-24}$$

which is the theoretical maximum coefficient of performance under these conditions.

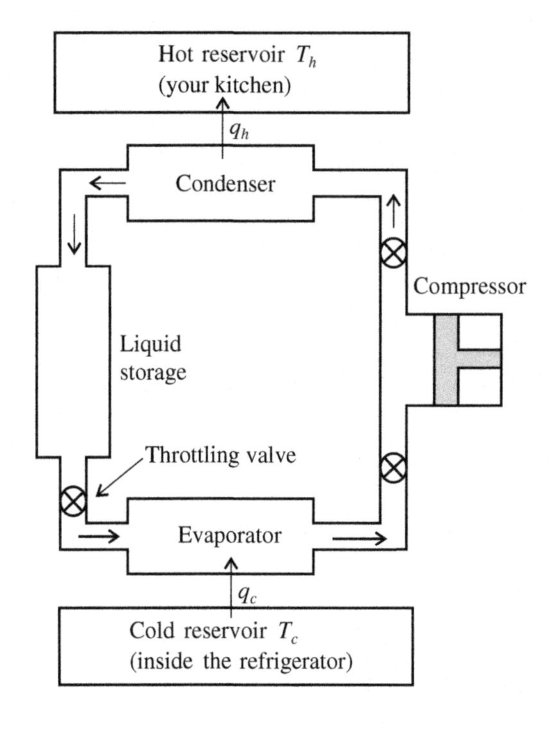

FIGURE 6-4

A schematic diagram of a refrigerator and its operation.

Table 6-1 Properties of Selected Refrigerants[a]

Name or Code	Formula	f.p. (K)	b.p. (K)	T_{crit} (K)	P_{crit} (bar)
ammonia	NH_3	195.48	239.82	405.5	113.5
carbon dioxide[b]	CO_2	194.8 (s)	216.55	304.1	73.75
CFC-11	CCl_3F	162.04	296.8	471.2	44.1
CFC-12	CF_2Cl_2	115	243.3	384.95	41.36
ethyl ether	$(C_2H_5)_2O$	157.0	307.58	466.74	36.38
isobutane	$i\text{-}C_4H_{10}$	134.8	261.42	408.15	36.48
methyl chloride	CH_3Cl	175.44	249.06	416.25	66.79
HFC-23	CHF_3	117.97	191.0	299.3	48.58
HFC-32	CH_2F_2	137	221.6	351.6	58.30
HFC-134	CH_2FCF_3	184	253.2	391.8	–
propane	C_3H_8	85.46	231.1	369.82	42.5
sulfur dioxide	SO_2	197.70	263.1	430.8	78.84

[a]*Properties are from the* CRC Handbook of Chemistry and Physics.
[b]*CO_2 sublimes at 194.8 K. Its triple point is at 216.55 K and 5.18 bars pressure.*

Example 6-2. One kilogram of liquid water at 25°C is put into a freezer at −15°C to make ice cubes for a party. Calculate the amount of electrical work (in kilowatt hours) needed to make the ice cubes using a Carnot refrigerator or using a refrigerator with ammonia refrigerant. Assume that both refrigerators reject heat at a condenser temperature of 30°C. The enthalpy of fusion of water is 6009.5 J mol^{-1} at 0°C, and the average C_P values for water and ice are 75.54 J mol^{-1} K^{-1} (0°C to 25°C) and 36.45 J mol^{-1} K^{-1} (−15°C to 0°C), respectively.

The amount of heat that has to be removed is equal to the enthalpy of liquid water from 0°C to 25°C, the enthalpy of fusion, and the enthalpy of ice from 0°C to −15°C per kilogram. One kilogram of water is equal to 55.51 moles of water, so the amount of heat (q_c) is

$$q_c = 55.51[75.54(25) + 6009.5 + 36.45(15)] = 468{,}768 \text{ J} \tag{6-25}$$

The coefficient of performance for the Carnot refrigerator is

$$\beta_{\text{Carnot}} = -\frac{q_c}{w} = \frac{T_c}{T_h - T_c} = \frac{258}{303 - 258} = 5.73 \tag{6-26}$$

and the amount of work done in Joules is

$$-w = \frac{q_c}{\beta_{\text{Carnot}}} = \frac{468{,}768}{5.73} = 81{,}809 \text{ J} \tag{6-27}$$

The coefficient of performance for the ammonia refrigerator is 4.76, the value given previously. The work done by the ammonia refrigerator is

$$-w = \frac{q_c}{\beta_{NH_3}} = \frac{468{,}768}{4.76} = 98{,}481 \text{ J} \tag{6-28}$$

Using the conversion factors in Table 3-2, we calculate that 0.02 kilowatt hours of electrical work are required for cooling with the Carnot refrigerator, and that 0.03 kilowatt hours are required with the ammonia refrigerator. However, our calculation does not consider the work that has to be done to maintain the kilogram of ice cubes at −15°C, given imperfect insulation, because we do not know the size of the heat leak into the freezer from the surrounding room.

D. Unattainability of absolute zero

We now use the Carnot refrigerator to demonstrate that it is impossible to reach 0 K (absolute zero) because an infinite amount of work has to be done to remove all the heat from a material.

Example 6-3. One mole (63.55 grams) of copper metal is cooled from 300 K in an attempt to reach 0 K. How much work has to done to cool the copper to absolute zero?

We first use Eq. (6-24) to compute the coefficient of performance of the Carnot refrigerator from 300 K to 0.1 K, the lowest temperature at which Hultgren et al. (1973) present thermal data for copper metal:

$$\beta_{\text{Carnot}} = -\frac{q_c}{w} = \frac{T_c}{T_h - T_c} = \frac{0.1}{300 - 0.1} = 3.33 \times 10^{-4} \tag{6-29}$$

This is a much smaller value than that for either the Carnot or the ammonia refrigerators in the last example. Based on the data tabulated by Hultgren et al. (1973), the heat q_c that must be removed from the low-temperature reservoir at T_c is

$$q_c = (H^o_{300} - H^o_{0.1}) = 5052.0 \text{ J mol}^{-1} \tag{6-30}$$

The work that has to be done is found by rearranging Eq. (6-24) and is

$$-w = \frac{q_c}{\beta_{\text{Carnot}}} = \frac{5052.0}{3.33 \times 10^{-4}} = 1.52 \times 10^7 \text{ J mol}^{-1} \tag{6-31}$$

The remaining heat that has to be extracted between 0.1 K and 0 K can be calculated by integrating Eq. (4-33):

$$(H^o_{0.1} - H^o_0) = \int_0^{0.1} (aT^3 + \gamma T)dT = \frac{1}{4}aT^4 + \frac{1}{2}\gamma T^2 \tag{6-32}$$

Equation (6-32) gives the enthalpy of Cu metal at these low temperatures, which is due to Debye (lattice vibrations) and electronic contributions. The method given in Example 4-10 is used to calculate the Debye constant $a = 4.878 \times 10^{-5}$ J K^{-4} (g atom)$^{-1}$ and the electronic heat capacity coefficient $\gamma = 6.817 \times 10^{-4}$ J K^{-2} (g atom)$^{-1}$ from the low-temperature heat capacity data tabulated by Hultgren et al. (1973). The remaining enthalpy is 3.41×10^{-6} J mol^{-1} and is mainly due to the electronic heat capacity of the Cu metal. This heat is to be extracted from the copper at 0 K and rejected to the hot reservoir at 300 K. This seems like a trivial problem because the amount of heat left in the copper is so small. However, the coefficient of performance for the Carnot refrigerator is now zero:

$$\beta_{\text{Carnot}} = \frac{T_c}{T_h - T_c} = \frac{0}{300 - 0} = 0 \tag{6-33}$$

Consequently, the remaining amount of work to be done is infinite:

$$-w = \frac{q_c}{\beta_{\text{Carnot}}} = \frac{3.42 \times 10^{-6}}{0} = \infty \tag{6-34}$$

and we cannot cool the copper to absolute zero.

This conclusion does not depend on the material that we are cooling. We can show this by going back to Eq. (6-24):

$$-\frac{q_c}{w} = \frac{T_c}{T_h - T_c} \tag{6-35}$$

and rearranging it to solve for the work needed to reach T_c, the temperature of the cold reservoir:

$$w = -q_c \frac{T_h - T_c}{T_c} = -q_c \frac{T_h - 0}{0} = \infty \tag{6-36}$$

Equation (6-36) immediately shows us that an infinite amount of work is needed to reach 0 K, independent of our starting temperature (T_h) or the amount of heat that must be removed (q_c).

V. THERMODYNAMIC TEMPERATURE SCALE

The thermodynamic, or Kelvin, temperature scale was first proposed by William Thomson in 1848. This is the temperature scale that we use throughout this book, and we discussed its development in Section II-C of Chapter 2. In our earlier discussion we said the thermodynamic temperature scale is based on the Carnot cycle, and that it does not depend on the physical properties of any substance. We now show how the thermal efficiency of the Carnot cycle defines the thermodynamic temperature scale.

Equation (6-1) for the thermal efficiency of the Carnot cycle can be written in different ways that are equivalent to each other. For example, Eq. (6-20) gives the efficiency (η) in terms of the temperatures of the hot and cold reservoirs:

$$\eta = \frac{T_h - T_c}{T_h} = 1 - \frac{T_c}{T_h} \tag{6-20}$$

Alternatively, we can rewrite Eq. (6-1) using Eq. (6-18) to substitute for the work w and get an expression for the efficiency in terms of the heat (q_h) absorbed from the hot reservoir and the heat (q_c) rejected to the cold reservoir:

$$\eta = \frac{w}{q_h} = \frac{q_h - q_c}{q_h} = 1 - \frac{q_c}{q_h} \tag{6-37}$$

Equations (6-20) and (6-37) are equal to one another, so we know that

$$1 - \frac{T_c}{T_h} = 1 - \frac{q_c}{q_h} \tag{6-38}$$

We can rearrange Eq. (6-38) to give the expression

$$\frac{q_c}{q_h} = \frac{T_c}{T_h} \tag{6-39}$$

which shows that the temperature of each heat reservoir is proportional to the amount of heat transferred to it or from it during operation of the Carnot cycle.

Equation (6-39) is a general equation that defines the thermodynamic temperature scale. This temperature scale is also called the *absolute temperature scale* because thermodynamic temperatures are absolute and completely independent of the physical properties of any material. The zero point, or absolute zero, on the thermodynamic scale is the temperature at which thermal efficiency of the Carnot cycle is unity, in other words, $T_c = 0$ in Eq. (6-38) gives $\eta = 1$. The size of the degree on the thermodynamic scale is fixed as one kelvin, which is 1/273.16 of the thermodynamic (Kelvin) temperature of the triple point of water (where ice, liquid water, and water vapor are in equilibrium). Temperature measurements on the Kelvin scale are made using different types of thermometers, thermocouples, and pyrometers that are calibrated with respect to the fixed temperature points on the international temperature scale. Chapter 2 describes basic principles of temperature measurement and development of the international temperature scale.

VI. ENTROPY

A. Definition of entropy

In the last section we defined the thermodynamic temperature scale using the Carnot cycle. We now use the Carnot cycle to define entropy. We start with Eq. (6-39):

$$\frac{q_c}{q_h} = \frac{T_c}{T_h} \tag{6-39}$$

and rearrange it by dividing both sides by T_c and multiplying both sides by q_h to give

$$\frac{q_c}{T_c} = \frac{q_h}{T_h}$$

or, equivalently,

$$\frac{q_c}{T_c} - \frac{q_h}{T_h} = 0 \tag{6-41}$$

Equations (6-40) and (6-41) show that the quantity q/T is conserved in the Carnot cycle. That is, there is no change in q/T around the entire cycle. The quantity q/T thus behaves like the internal energy and enthalpy, which are unchanged when we go through one complete Carnot cycle. As we showed in Section II-B of this chapter, the net heat flow $q_h - q_c$ for the Carnot cycle is not zero. However, when we divide the heat flow by the temperature of the heat reservoir, the net change in the quantity q/T is zero.

Clausius realized that the quantity q/T has the characteristics of a state function, and he demonstrated that it is conserved for any reversible cycle as follows. We can divide the Carnot cycle shown in Figure 6-2 or indeed any reversible cycle, such as the air standard Otto cycle in Figure 3-5 or the cycle in Figure 3-9, into a much larger number of tiny Carnot cycles. This is possible because any PV path can be approximated by a combination of adiabatic and isothermal steps (i.e., the zigzag line in Figure 6-5a). As we increase the number of adiabatic and isothermal steps in the zigzag line, it becomes smoother and smoother, until it is impossible to distinguish from the smooth PV path.

We can thus represent any cycle on a PV diagram (indicator diagram) by an infinite number of infinitesimally small Carnot cycles (see Figure 6-5b). When we do this, all the parts of the infinitesimally small Carnot cycles inside the larger cycle cancel each other out. We are left with the larger cycle itself, which is built up from an infinite number of Carnot cycles. Equations (6-40) and (6-41) apply to each of the infinitesimally small Carnot cycles. The amounts of heat transferred to and from heat reservoirs during each of these cycles are also infinitesimally small, so we now can write the integral

$$\int_{cycle} \frac{\delta q_{rev}}{T} = 0 \tag{6-42}$$

for any cycle that is conducted reversibly.

Clausius introduced the name *entropy* for the quantity $\delta q_{rev}/T$. *Entropy* is from the Greek word τροπη for transformation or change (a literal translation is *turning*). Entropy has the symbol S. It is a state function, and thus the entropy change during a process is path independent and depends only on

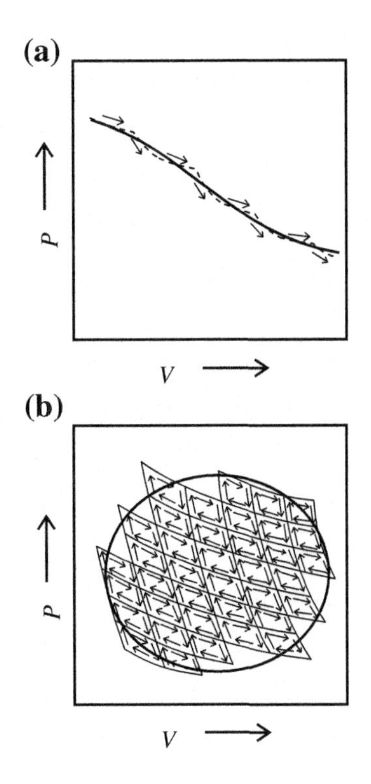

FIGURE 6-5

(a) Schematic illustration representing an arbitrary PV path by a combination of isothermal and adiabatic steps. (b) Schematic illustration representing any reversible cycle by an infinite number of infinitesimal Carnot cycles.

the initial and final values. Entropy is also an extensive variable, and thus the entropy of a system depends on the size of the system. The differential dS is defined as

$$dS = \frac{\delta q_{rev}}{T} \tag{6-43}$$

where δq_{rev} is the heat flow for reversible changes in a closed system.

B. Entropy changes in reversible and irreversible processes

The differential entropy change dS for reversible processes in a closed system is given by Eq. (6-43). The corresponding entropy change for the surroundings has the same value but an opposite sign to that for the system because heat absorbed by the system is lost from the surroundings, and vice versa. Thus, the two entropy changes cancel out for a reversible process:

$$dS_{sys} + dS_{surr} = 0 \tag{6-44}$$

During a reversible process, entropy is transferred between a system and its surroundings, but entropy is not produced. The entropy change due to transport of entropy between a system and its surroundings is denoted by d_eS.

On the other hand, entropy always increases (dS is positive) for irreversible changes in a closed system or in an isolated system. The entropy created inside a system by irreversible processes is denoted d_iS. The entropy change of a closed or isolated system during reversible and irreversible processes is given by the equations

$$dS = (d_eS + d_iS) \tag{6-45}$$

$$dS = \frac{\delta q_{rev}}{T} + d_iS \tag{6-46}$$

with $d_iS = 0$ for reversible processes and $d_iS > 0$ for irreversible processes. Equations (6-43) to (6-46) are mathematical statements of the second law. They show that entropy always increases for irreversible processes. The second law is sometimes called the *law of increasing entropy.*

VII. SOME APPLICATIONS OF THE SECOND LAW

A. The combined first and second law

We obtain a very useful equation, called the *combined first and second law,* or the *fundamental equation of thermodynamics*, by combining the first law of thermodynamics:

$$dE = \delta q - \delta w \tag{3-11}$$

with the definition of entropy given in Eq. (6-43) to obtain

$$dE = TdS - \delta w \tag{6-47}$$

If we are considering only pressure-volume work, then we can substitute for δw using

$$\delta w = PdV \tag{3-18}$$

to obtain the *combined equation for the first and second laws*:

$$dE = TdS - PdV \tag{6-48}$$

We now use the *fundamental equation* to compute entropy changes for reversible processes. The equations that we derive are also used to compute entropy changes for real processes, which are irreversible to a greater or lesser extent. We can do this because entropy is a state function and its value is path independent. At least in principle we can always devise a reversible process to restore a system and its surroundings to their original states. Because entropy is transferred but not produced in a reversible process, the ΔS change that we calculate is that due to the irreversible process itself.

B. Entropy change during reversible adiabatic processes

By definition, $\delta q = 0$ for an adiabatic process because the system does not exchange heat with the surroundings. So, from Eq. (6-43) we have

$$dS = \frac{\delta q}{T} = \frac{0}{T} = 0 \tag{6-49}$$

for a reversible adiabatic process. Note that the entropy *change* is zero for a reversible adiabatic process but that the system undergoing the process does not have zero entropy. A reversible adiabatic process is also an *isentropic process* because it is a constant entropy process. However, an irreversible adiabatic process is not isentropic.

C. Entropy change during reversible isothermal processes

The internal energy change is zero for a reversible isothermal process because

$$dE = nC_V dT = nC_V(0) \tag{3-57}$$

Thus, the fundamental equation reduces to

$$TdS = PdV \tag{6-50}$$

which can be integrated between the initial state (A) and final state (B) of the system, giving

$$T\Delta S = T(S_B - S_A) = \int_A^B PdV \tag{6-51}$$

Equation (6-51) is a general equation that can be evaluated once we know volume as a function of pressure for a material. Using the ideal gas law to evaluate the integral and rearranging gives

$$\Delta S = nR \ln\left(\frac{V_B}{V_A}\right) \tag{6-52}$$

which can be rewritten as

$$\Delta S = nR \ln\left(\frac{P_A}{P_B}\right) \tag{6-53}$$

for the entropy change of an ideal gas during a reversible isothermal process.

Example 6-4. Figure 3-7 is the indicator diagram for the reversible isothermal expansion of an ideal gas from 0.05 liter and 8 bars to 0.40 liter and 1 bar at 298.15 K. We calculated the work done by the gas, the internal energy change, the enthalpy change, and the heat absorbed by the gas in Example 3-4. Now we want to calculate the entropy change during this process. We will use Eqs. (6-52) and (6-53) to do this so that we can demonstrate that the two equations give the same result. Starting with Eq. (6-52) we have

$$\Delta S = nR \ln\left(\frac{V_B}{V_A}\right) = (1.614 \times 10^{-2}\ \text{mol})(8.314\ \text{J mol}^{-1}\text{K}^{-1}) \ln\left(\frac{0.40}{0.05}\right) = 0.279\ \text{J K}^{-1} \tag{6-54}$$

using the number of moles of gas that we had calculated in Example 3-4. Using Eq. (6-53) gives

$$\Delta S = nR \ln\left(\frac{P_A}{P_B}\right) = (1.614 \times 10^{-2}\ \text{mol})(8.314\ \text{J mol}^{-1}\text{K}^{-1}) \ln\left(\frac{8.0}{1.0}\right) = 0.279\ \text{J K}^{-1} \tag{6-55}$$

which agrees exactly with the result computed using Eq. (6-52).

D. Entropy change during reversible isochoric processes

No pressure-volume work is done during a reversible isochoric process because the volume is constant. The fundamental equation (6-48) then becomes

$$dE = TdS \tag{6-56}$$

By rewriting Eq. (6-56) using Eq. (3-57) we obtain

$$\left(\frac{\partial S}{\partial T}\right)_V = \frac{nC_V}{T} \tag{6-57}$$

where n is the number of moles. The integrated form of Eq. (6-57) is

$$\Delta S = n\int \frac{C_V}{T}dT \tag{6-58}$$

Equations (6-57) and (6-58) are general equations for the variation of entropy with temperature at constant volume. They apply to all materials and are not limited to ideal gases.

At low temperatures where the Debye T^3 law Eq. (4-30) is valid ($T < \theta_D/50$, which is below 10 K for many materials), Eq. (6-58) can be rewritten as

$$\Delta S = \int_0^T \frac{C_V}{T}dT = \int_0^T \frac{aT^3 + \gamma T}{T}dT = \frac{1}{3}aT^3 + \gamma T = \frac{1}{3}C_V \cong \frac{1}{3}C_P \tag{6-59}$$

where a is a constant equal to $1943.8/\theta_D^3$ and γ is the electronic heat capacity coefficient, which were both defined in Section III of Chapter 4. The electronic heat capacity coefficient is needed for metals and other conductive materials but is zero for most minerals, which are insulators. Equation (6-59) is integrated between absolute zero and a higher temperature T, and the entropy at 0 K is taken as zero. The latter assumption is justified by the third law of thermodynamics, which we discuss in Chapter 9.

Example 6-5. Compute the entropy change for forsterite (Mg_2SiO_4) between 0 and 5 K, the lowest temperature at which heat capacity measurements were made by Robie et al. (1982). The molar heat capacity data given in Table 5 of Robie et al. (1982) and Eq. (6-59) yield

$$\Delta S = (S_5^o - S_0^o) = \frac{1}{3}C_V \cong \frac{1}{3}C_P = \frac{0.0032}{3} = 0.011 \text{ J mol}^{-1}\text{K}^{-1} \tag{6-60}$$

in exact agreement with the entropy tabulated in their paper. As discussed in Section III of Chapter 4, $(C_P - C_V) \sim 0$ at these low temperatures and using C_P does not introduce any error.

E. Entropy change during reversible isobaric processes

In Chapter 3, we defined the heat flow at constant pressure as the change in enthalpy (Eq. 3-43). We can thus rearrange and substitute dH into Eq. (6-49) and obtain

$$dH = TdS \tag{6-61}$$

Substituting for dH using the definition of the constant-pressure heat capacity from Eq. (3-44) and rearranging, we get

$$\left(\frac{\partial S}{\partial T}\right)_P = \frac{C_P}{T} \tag{6-62}$$

which is equivalent to the integrated form:

$$\Delta S = \int \frac{C_P}{T} dT \tag{6-63}$$

Equations (6-62) and (6-63) are general equations for the variation of entropy with temperature at constant pressure. They apply to all materials and are not limited to ideal gases. In many cases Maier-Kelley equations (or the modified four-term Maier-Kelley equations) are used for heat capacity data above 298 K. In these cases Eq. (6-63) becomes

$$\Delta S = \int \frac{C_P}{T} dT = \int \frac{a + bT + cT^{-2} + dT^2}{T} dT = a \ln T + bT - \frac{1}{2} cT^{-2} + \frac{1}{2} dT^2 \tag{6-64}$$

Example 6-6. Calculate the entropy change for corundum (α-Al_2O_3) heated from 298 K to its melting point of 2327 K. Equation (4-18) gives a Maier-Kelley fit to enthalpy of corundum:

$$(H_T - H_{298})_{solid} = 118.7170\ T + 4.7517 \times 10^{-3}\ T^2 + 38.757 \times 10^5\ T^{-1} - 48{,}817 \text{ J mol}^{-1} \tag{4-18}$$

from 298 to 2327 K. The corresponding heat capacity equation is

$$C_{P,m} = 118.717 + 9.5034 \times 10^{-3}\ T - 38.757 \times 10^5\ T^{-2} \text{ J mol}^{-1}\text{ K}^{-1} \tag{6-65}$$

Substituting Eq. (6-65) into Eq. (6-64) and evaluating gives

$$\Delta S = 118.717 \ln (T/298.15) + 9.5034 \times 10^{-3}(T - 298.15) + 19.3785 \times 10^5(T^{-2} - 298.15^{-2}) \tag{6-66}$$

The calculated ΔS values ($S_T - S_{298}$) and C_P/T curve for corundum are plotted in Figure 6-6. The ΔS between any two temperatures is the area under the C_P/T curve for the same temperature range. This figure can be compared to Figure 4-6, showing the enthalpy and heat capacity of corundum, which are related in an analogous manner, that is, enthalpy is the area under the C_P curve between two specified temperatures.

F. Entropy change during a phase transition

If we are concerned with the entropy change at constant pressure and constant temperature, for example, for a phase transition such as melting, sublimation, vaporization, or between two polymorphs of a solid, we can integrate Eq. (6-61) between the initial and final states (i.e., the two phases involved in the transition) to get

$$\Delta_{trans} S = \frac{\Delta_{trans} H}{T_{trans}} \tag{6-67}$$

where $\Delta_{trans}H$ is the enthalpy of transition and T_{trans} is the transition temperature. Equation (6-67) is also a general equation.

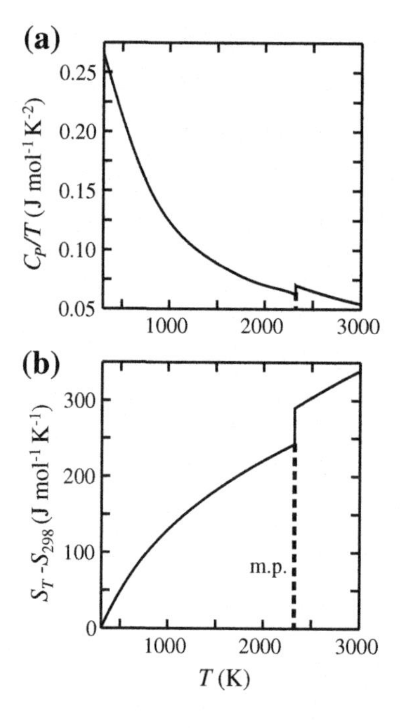

FIGURE 6-6

(a) The C_p/T curve for corundum (α-Al_2O_3) and molten alumina. (b) The corresponding molar entropy difference ($S_T - S_{298}$) for corundum and molten alumina. Compare to the heat capacity and enthalpy curves for corundum and molten alumina in Figure 4-6.

Example 6-7. Corundum melts at 2327 K and the enthalpy of fusion is 111.02 kJ mol^{-1}. The enthalpy of molten Al_2O_3 is given by Eq. (4-19),

$$(H_T - H_{298})_{melt} = 192.464\,T - 82{,}016 \text{ J mol}^{-1} \tag{4-19}$$

from 2327 K to 3000 K. Calculate the entropy of fusion for corundum and the entropy change for molten alumina heated from the melting point to 3000 K. The entropy of fusion is computed by substituting numerical values into Eq. (6-67):

$$\Delta_{trans}S = \frac{\Delta_{trans}H}{T_{trans}} = \frac{111{,}020}{2327} = 47.71 \text{ J mol}^{-1}\text{K}^{-1} = 9.54 \text{ J (g atom)}^{-1}\text{K}^{-1} \tag{6-68}$$

Differentiation of Eq. (4-19) gives a constant heat capacity of 192.464 J mol^{-1} K^{-1} for molten alumina. The ΔS values ($S_T - S_{2327}$) for the melt are computed using Eq. (6-63) and are

$$(S_T - S_{2327}) = 192.464 \ln(T/2327) \tag{6-69}$$

The ΔS values ($S_T - S_{298}$) for the melt are obtained by combining Eqs. (6-68) and (6-69) with the ΔS value for corundum at the melting point (2327 K) and are given by the equation

$$\Delta S = (S_T - S_{298})_{melt} = 289.48 + 192.464 \ln(T/2327) \tag{6-70}$$

The ΔS values and C_P/T curve for molten alumina are also plotted in Figure 6-6.

Table 6-2 lists enthalpies and entropies of fusion for common minerals and ices that are important for terrestrial geology and planetary science. The entropy of fusion is the difference of the entropy of the solid and liquid at the melting point and is large when there is a large difference in bonding and structure between the two phases. The entropy of each phase can be viewed as the sum of a vibrational component (S_{vib}) and a configurational component (S_{config}), each of which is larger for the liquid than the solid. The $\Delta_{fus}S$ values in Table 6-2 are given per gram atom and span a wide range from 1.5 J g atom^{-1} K^{-1} (cristobalite) to 14.2 J g atom^{-1} K^{-1} (solid Ar). An empirical correlation known as *Richard's rule* states that

Table 6-2 Enthalpy and Entropy of Fusion of Selected Minerals and Ices

Material	T_m (K)	$\Delta_{fus}H^o$ (kJ mol^{-1})	$\Delta_{fus}S^o$ (J g atom^{-1}K^{-1})
H_2	13.80	0.117	4.24
Ne	24.56	0.332	13.52
N_2	63.14	0.721	5.71
CO	68.09	0.8355	6.13
Ar	83.81	1.188	14.2
C_2H_6	90.35	2.857	3.95
CH_4	90.69	0.937	2.07
H_2S	187.61	2.377	4.22
NH_3	195.48	5.655	7.23
CO_2	216.55	8.326	12.8
H_2O	273.15	6.0095	7.33
$NaAlSi_3O_8$ (albite)	1393	64.5 ± 3.0	3.6 ± 0.2
FeS (troilite)	1463 ± 3	31.46 ± 2.09	10.8 ± 0.7
Fe_2SiO_4 (fayalite)	1490	89.3 ± 1.1	8.6 ± 0.1
Mn_2SiO_4 (tephroite)	1620	89.0 ± 0.5	7.85 ± 0.04
$FeTiO_3$ (ilmenite)	1640	90.8 ± 0.4	11.07 ± 0.05
$CaMgSi_2O_6$ (diopside)	1670 ± 5	137.7 ± 2.0	8.2 ± 0.1
$CaTiSiO_5$ (titanite)	1670	123.8 ± 0.4	9.27 ± 0.03
Co_2SiO_4	1688	103 ± 15	8.7 ± 1.3
$Ca_2MgSi_2O_7$ (akermanite)	1731	123.6 ± 3.2	6.0 ± 0.2
Fe	1811 ± 5	13.81 ± 0.84	7.6 ± 0.5
$CaSiO_3$ (pseudowollast.)	1821	57.3 ± 2.8	6.3 ± 0.3
$CaAl_2Si_2O_8$ (anorthite)	1830	133.0 ± 4.0	5.6 ± 0.2
$MgSiO_3$ (enstatite)	1834	73.2 ± 6.0	8.0 ± 0.6
Ni_2SiO_4	1920	221 ± 26	16.4 ± 1.9
SiO_2 (cristobalite)	1999	8.92 ± 1.0	1.5 ± 0.2
Mg_2SiO_4 (forsterite)	2163	102.8 ± 5.0	6.8 ± 0.3
Mg_2SiO_4 (forsterite)	2174	142 ± 14	9.3 ± 0.9
$MgAl_2O_4$ (spinel)	2408	107 ± 11	6.4 ± 0.6

A gram atom is a mole of atoms. There are two gram atoms per mole of N_2, three per mole of H_2O, four per mole of NH_3, five per mole of CH_4, and so on. The values for diopside and enstatite are for metastable congruent melting.

$$\Delta_{fus}S = \frac{\Delta_{fus}H}{T_m} \sim 8.4 \text{ J (g atom)}^{-1}\text{K}^{-1} \tag{6-71}$$

where T_m is the melting point. This correlation is only approximate, which is demonstrated by the fact that data for 150 elements, minerals, and inorganic compounds give an average $\Delta_{fus}S$ value of 8.8 ± 4.4 J g atom^{-1} K^{-1}.

In the absence of other data, Richard's rule gives a zeroth-order approximation for the enthalpy of fusion. Kubaschewski and Alcock (1979) note that better estimates of $\Delta_{fus}H$ values are obtained from correlations with the entropies of fusion of related compounds with similar structures. However, other factors such as the cation charge, ionic radius, coordination number, and crystal field stabilization energy (CFSE arising from the d electrons of transition metal cations) also affect the entropy of fusion. For example, the $\Delta_{fus}S = 16.4 \pm 1.9$ J g atom^{-1} K^{-1} for Ni_2SiO_4 is significantly larger (by ~6 J g atom^{-1} K^{-1}) than expected from the trend of $\Delta_{fus}S$ versus ionic radius for Mn, Fe, Co, Mg, and Ni olivines. The difference is attributed to the large CFSE of Ni^{2+} in octahedral sites in olivine, which Ni^{2+} prefers to the molten olivine (Sugarawa and Akaogi, 2003). Nevertheless, Richard's rule is a quick way to estimate $\Delta_{fus}H$ values, which are not easily measured for many refractory phases.

Example 6-8. Periclase (MgO) melts at 3141 ± 13 K (Yamada et al., 1998) but its enthalpy of fusion is unknown. An estimate of 77 ± 6 kJ mol^{-1} was derived from the MgO-ZrO_2 phase diagram by Kelley (1936). Periclase has the NaCl structure and the mean $\Delta_{fus}S = 11.3 \pm 2.1$ J g atom^{-1} K^{-1} for 13 halides of Li, Na, K, and Ag with the NaCl structure. This $\Delta_{fus}S$ value gives an estimated $\Delta_{fus}H = 71 \pm 13$ kJ mol^{-1}, which agrees with Kelley's estimate.

However, there is a much better correlation between the entropy of vaporization and the boiling points of compounds. This correlation, known as *Trouton's rule*, is

$$\Delta_{vap}S = \frac{\Delta_{vap}H}{T_b} \sim 88 \text{ J mol}^{-1}\text{K}^{-1} \tag{6-72}$$

where T_b is the normal boiling point. Table 6-3 lists enthalpies and entropies of vaporization for liquids with boiling points ranging from 77 K (liquid nitrogen) to 4120 K (molten platinum). The $\Delta_{vap}S$ values vary from ~72 to ~122 J mol^{-1} K^{-1} with a tendency for higher values at higher temperatures. Figure 6-7 shows data for ~150 elements, halides, and oxides with boiling points ranging up to ~5900 K. The linear least-squares fit to the data has a slope equal to $\Delta_{vap}S = 109$ J mol^{-1} K^{-1}. There is an apparent tendency for refractory elements to have larger $\Delta_{vap}S$ values, ranging up to ~140 J mol^{-1} K^{-1} for tungsten. However, uncertainties up to a few hundred degrees exist for the extrapolated boiling points of several refractory metals (W, Hf, Nb, Os, Re, Rh, Ru, Ta, Tc, and U), leading to large uncertainties in the $\Delta_{vap}S$ values. Thus, we recommend using $\Delta_{vap}S = 109 \pm 10$ J mol^{-1} K^{-1} for estimating $\Delta_{vap}H$ values, where the uncertainty is estimated excluding the most refractory and most volatile elements.

Example 6-9. The enthalpy of vaporization of silicate rock is important for modeling ablation of rocky meteors in Earth's upper atmosphere. Using data from the JANAF tables, the boiling point of pure molten silica is calculated as 3160 K (Schaefer and Fegley, 2004). The measured boiling point of tektite glass, which is ~70% SiO_2 by weight, is ~3103 K (Centolanzi and Chapman, 1966). Estimate the enthalpy of vaporization of molten silica using an average boiling point of 3130 K. We take $\Delta_{vap}S = 109 \pm 10$ J mol^{-1} K^{-1} and compute $\Delta_{vap}H = 341 \pm 31$ kJ mol^{-1} (~5680 ± 520 J g^{-1}). Our estimate is similar to but ~6% lower than the value of 6050 J g^{-1} used for the heat of vaporization of rocky meteors (Bronshten, 1983; Öpik, 1958).

Table 6-3 Enthalpy and Entropy of Vaporization of Selected Liquids

Material	b.p. (K)	$\Delta_{vap}H^o$ (kJ mol^{-1})	$\Delta_{vap}S^o$ (J mol^{-1}K^{-1})
N_2	77.36	5.577	72.09
CO	81.61	6.040	74.01
Ar	87.19	6.448	73.95
CH_4	111.67	8.180	73.25
C_2H_6	184.5	14.703	79.69
NH_3	239.7	23.351	97.42
H_2S	212.8	18.673	87.75
H_2O	373.15	40.6563	108.95
Hg	629.8	59.2	94.00
Re_2O_7	635	75 ± 15	118 ± 24
S	719	65.22	90.7
As_4O_6 (claudetite, l)	732	59.4 ± 4.2	81 ± 6
Na (l)	1158	96.947	83.72
$PbCl_2$ (cotunnite, l)	1225	126.8 ± 5.0	104 ± 4
PbS (galena, l)	1615	153.3 ± 6.3	95 ± 4
NaCl (halite, l)	1738	170.3 ± 0.8	98.0 ± 0.5
Ag	2437	251.0 ± 0.8	103.0 ± 0.3
CaF_2 (fluorite, l)	2783	312 ± 17	112 ± 6
Fe	3110 ± 20	349.6 ± 1.3	112 ± 1
Pt	4122 ± 170	504.1 ± 2.0	122 ± 5

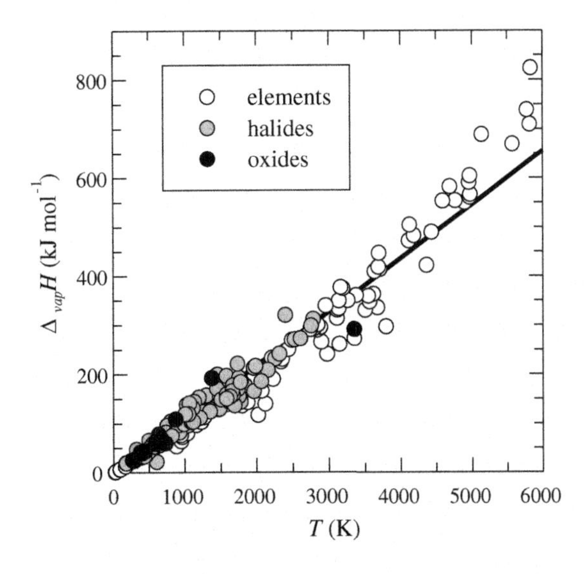

FIGURE 6-7

An illustration of the correlation between the normal boiling point and enthalpy of vaporization (Trouton's rule). The slope of the line, which is 109 J mol^{-1} K^{-1}, is the average entropy of vaporization.

G. Absolute entropy calculations for solids, liquids, and gases

The third law of thermodynamics, which we discuss in Chapter 9, states that entropy is an absolute quantity and that

$$S^o(0\text{ K}) = 0\text{ J mol}^{-1}\text{ K}^{-1} \tag{6-73}$$

for a pure, perfectly ordered crystalline solid. Thus, if we have a solid such as Fe metal and we measure its heat capacity from 0 K to 298.15 K, its absolute entropy at 298.15 K is given by

$$S^o_{298} = \int_0^{298.15} \frac{C_P}{T} dT \tag{6-74}$$

which can be calculated numerically, graphically, or analytically.

We use iron as an example to demonstrate calculations of the absolute entropy as a function of temperature for materials that undergo a number of phase transitions such as solid-state polymorphic transformations, melting, and vaporization.

Example 6-10. Iron metal has three solid-state phase transitions: a ferromagnetic-to-paramagnetic transition at the Curie point (1043 K), a transition from the body-centered cubic α phase ferrite to the face-centered cubic γ phase austentite (1185 K), and a transition from the γ phase back to the body-centered cubic δ phase (1667 K). The melting point of Fe is 1811 ± 5 K and the normal boiling point is 3110 ± 20 K. What is the entropy of monatomic Fe (g) at 3200 K?

This example is a companion to Example 4-11 in Chapter 4, where the enthalpy (heat content) of monatomic Fe (g) at 3200 K was calculated. The absolute entropy of Fe (g) at 3200 K is the sum of the entropy contributions of each phase (α, γ, δ, liquid, gas) and the entropies of the different phase transitions:

$$\begin{aligned} S^o_T(\text{Fe, g}) = S^o_{298} &+ \int_{298}^{1043} \frac{C_P(\alpha - Fe)}{T} dT + \frac{\Delta_{trans}H^o_{1043}}{1043}(\text{Curie point}) + \int_{1043}^{1185} \frac{C_P(\alpha - \text{Fe})}{T} dT \\ &+ \frac{\Delta_{trans}H^o_{1185}}{1185}(\alpha \rightarrow \gamma) + \int_{1185}^{1667} \frac{C_P(\gamma - \text{Fe})}{T} dT + \frac{\Delta_{trans}H^o_{1667}}{1667}(\gamma \rightarrow \delta) \\ &+ \int_{1667}^{1811} \frac{C_P(\delta - \text{Fe})}{T} dT + \frac{\Delta_{fus}H^o_{1811}}{1811}(\delta \rightarrow \text{liq}) + \int_{1811}^{3110} \frac{C_P(\text{liq Fe})}{T} dT \\ &+ \frac{\Delta_{vap}H^o_{3110}}{3110}(\text{liq} \rightarrow \text{gas}) + \int_{3110}^{3200} \frac{C_P(\text{Fe, g})}{T} dT \end{aligned} \tag{6-75}$$

The different terms in Eq. (6-75) can be evaluated using the heat capacity equations given in Appendix 2. The first term in Eq. (6-75) is the absolute entropy of α-Fe metal at 298 K:

$$S^o_{298} = 27.085 \text{ J mol}^{-1} \text{ K}^{-1} \tag{6-76}$$

$$\int_{298}^{1043} \frac{C_P(\alpha - \text{Fe})}{T} dT = (S^o_{1043} - S^o_{298})_{\alpha-\text{Fe}} = 40.847 \text{ J mol}^{-1} \text{ K}^{-1} \tag{6-77}$$

$$\frac{\Delta_{trans} H^o_{1043}}{1043} = \frac{1364}{1043} = \Delta_{trans} S^o_{1043} = 1.308 \text{ J mol}^{-1} \text{ K}^{-1} \tag{6-78}$$

$$\int_{1043}^{1185} \frac{C_P(\alpha - \text{Fe})}{T} dT = (S^o_{1185} - S^o_{1043})_{\alpha-\text{Fe}} = 5.944 \text{ J mol}^{-1} \text{ K}^{-1} \tag{6-79}$$

$$\frac{\Delta_{trans} H^o_{1185}}{1185} = \frac{900}{1185} = \Delta_{trans} S^o_{1185} = 0.759 \text{ J mol}^{-1} \text{ K}^{-1} \tag{6-80}$$

$$\int_{1185}^{1667} \frac{C_P(\gamma - \text{Fe})}{T} dT = (S^o_{1667} - S^o_{1185})_{\gamma-\text{Fe}} = 12.275 \text{ J mol}^{-1} \text{ K}^{-1} \tag{6-81}$$

$$\frac{\Delta_{trans} H^o_{1667}}{1667} = \frac{850}{1667} = \Delta_{trans} S^o_{1667} = 0.510 \text{ J mol}^{-1} \text{ K}^{-1} \tag{6-82}$$

$$\int_{1667}^{1811} \frac{C_P(\delta - \text{Fe})}{T} dT = (S^o_{1811} - S^o_{1667})_{\delta-\text{Fe}} = 3.300 \text{ J mol}^{-1} \text{ K}^{-1} \tag{6-83}$$

$$\frac{\Delta_{fus} H^o_{1811}}{1811} = \frac{13,810}{1811} = \Delta_{fus} S^o_{1811} = 7.626 \text{ J mol}^{-1} \text{ K}^{-1} \tag{6-84}$$

$$\int_{1811}^{3110} \frac{C_P(\text{liq Fe})}{T} dT = (S^o_{3110} - S^o_{1811})_{\text{Fe(liq)}} = 24.279 \text{ J mol}^{-1} \text{ K}^{-1} \tag{6-85}$$

$$\frac{\Delta_{vap} H^o_{3110}}{3110} = \frac{349,213}{3110} = \Delta_{vap} S^o_{3110} = 112.287 \text{ J mol}^{-1} \text{ K}^{-1} \tag{6-86}$$

$$\int_{3110}^{3200} \frac{C_P(\text{Fe, g})}{T} dT = (S^o_{3200} - S^o_{3110})_{\text{Fe(g)}} = 0.761 \text{ J mol}^{-1} \text{ K}^{-1} \tag{6-87}$$

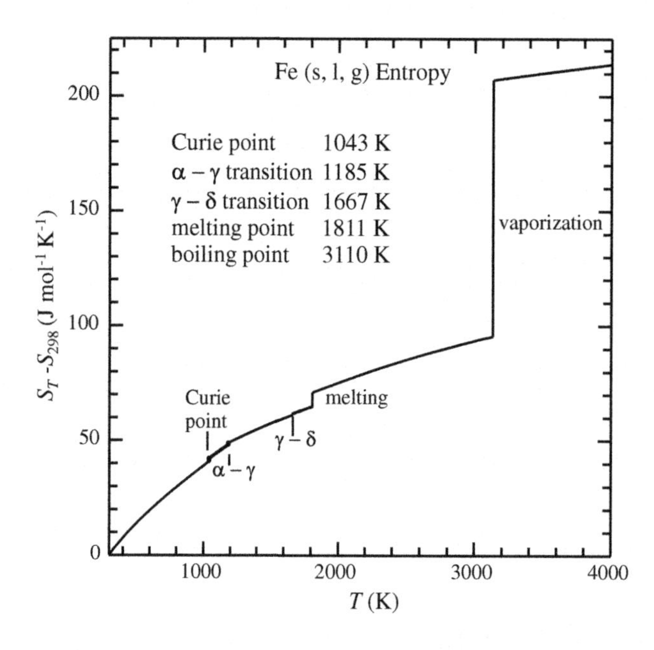

FIGURE 6-8

A graph of the entropy difference ($S_T - S_{298}$) for iron metal, liquid, and gas from 298 K to 4000 K. The entropy differences are shown to scale. Compare to the enthalpy curve for iron metal, liquid, and gas in Figure 4-25.

Summing up all these terms, we obtain

$$S^o_{3200}(\text{Fe, g}) = 236.981 \text{ J mol}^{-1}\text{K}^{-1} \tag{6-88}$$

for the absolute entropy of monatomic Fe (g) at 3200 K. These calculations are illustrated in Figure 6-8, which can be compared to Figure 4-25, showing the contributions to the enthalpy of monatomic Fe (g) at the same temperature.

H. Entropy changes for chemical reactions without phase changes

We first consider entropy of reaction calculations for chemical reactions without phase changes in either the products or reactants. Entropy changes for chemical reactions are calculated analogously to enthalpy changes (see Eq. 5-1). The standard entropy of reaction $\Delta_r S^o$ is the difference between the sum of entropies of the products and the sum of entropies of the reactants:

$$\Delta_r S^o = \sum (\nu_i S^o_i)_{pr} - \sum (\nu_i S^o_i)_{re} \tag{6-89}$$

where the entropy of each compound (S^o_i) is multiplied by its stoichiometric coefficient (ν_i) in the chemical equation, and the subscripts *pr* and *re* stand for products and reactants, respectively. In general, the superscript o denotes entropy values at standard-state conditions and a subscript such as 298 denotes the temperature. Thermodynamic data tabulations typically list S^o values for elements and

Table 6-4 Thermodynamic Properties of Forsterite (Mg_2SiO_4) from 298K to 2000 K

T	C_p^o	S_T^o	$-\left[\frac{G_T^o - H_{298}^o}{T}\right]$	$H_T^o - H_{298}^o$	$\Delta_f H_T^o$	$\Delta_f G_T^o$	$\log_{10} K_f$
(K)		(J mol^{-1} K^{-1})			kJ mol^{-1}		
298.15	118.61	94.11	94.11	0.00	−2173.91	−2054.54	359.94
300	119.10	94.84	94.11	0.22	−2173.93	−2053.80	357.59
400	137.71	131.96	99.05	13.16	−2174.15	−2013.70	262.96
500	148.28	163.91	108.90	27.50	−2173.56	−1973.64	206.18
600	155.77	191.63	120.43	42.72	−2172.62	−1933.74	168.34
700	163.57	216.20	132.39	58.67	−2171.43	−1894.02	141.33
800	167.80	238.33	144.27	75.24	−2170.07	−1854.48	121.08
900	171.41	258.30	155.85	92.21	−2168.72	−1815.11	105.34
1000	174.64	276.53	167.02	109.51	−2184.62	−1774.46	92.69
1100	177.62	293.32	177.75	127.13	−2183.13	−1733.52	82.32
1200	180.43	308.90	188.04	145.03	−2181.20	−1692.72	73.68
1300	183.11	323.44	197.90	163.21	−2178.89	−1652.11	66.38
1400	185.69	337.11	207.36	181.65	−2434.16	−1604.94	59.88
1500	188.17	350.01	216.44	200.34	−2429.72	−1545.86	53.83
1600	190.58	362.23	225.18	219.28	−2425.12	−1487.09	48.55
1700	192.92	373.85	233.58	238.46	−2470.57	−1428.31	43.89
1800	195.19	384.94	241.69	257.86	−2465.48	−1367.15	39.67
1900	197.40	395.56	249.51	277.49	−2460.21	−1306.27	35.91
2000	199.55	405.74	257.07	297.34	−2454.78	−1245.68	32.53

compounds, for example, at 298 K (as in Appendix 1) or at 298 K and at higher temperatures, as shown in Table 6-4 for forsterite.

Example 6-11. Calculate the entropy of formation of forsterite from its constituent elements in their reference states at 298 K. This is the entropy change for the reaction

$$2\ \mathrm{Mg(metal)} + \mathrm{Si(crystal)} + 2\ \mathrm{O_2\ (g)} = \mathrm{Mg_2SiO_4(forsterite)} \tag{5-97}$$

at 298 K and is given by

$$\Delta_f S_{298}^o(\mathrm{Mg_2SiO_4}) = S_{298}^o(\mathrm{Mg_2SiO_4}) - 2S_{298}^o(\mathrm{O_2,\ g}) - S_{298}^o(\mathrm{Si,\ c}) - 2S_{298}^o(\mathrm{Mg,\ m}) \tag{6-90}$$

$$\Delta_f S_{298}^o(\mathrm{Mg_2SiO_4}) = 94.1 - 2(205.15) - 18.81 - 2(32.67) = -400.35\ \mathrm{J\ mol^{-1}\ K^{-1}} \tag{6-91}$$

Formation of forsterite from its constituent elements at 298 K has a large negative ΔS^o change because two moles of O_2 gas are consumed.

We now illustrate $\Delta_r S$ calculations at temperatures other than 298 K. The variation of the entropy of reaction $\Delta_r S$ with temperature is given by the equations

$$\left[\frac{\partial(\Delta S)}{\partial T}\right]_P = \frac{\Delta C_P}{T} \tag{6-92}$$

$$\Delta S = \int \frac{\Delta C_P}{T} dT \tag{6-93}$$

which are analogous to Eqs. (6-62) and (6-63) for a single element or compound. The ΔC_P is

$$\Delta C_P = \sum \left[\nu_i C_{P,i}\right]_{pr} - \sum \left[\nu_i C_{P,i}\right]_{re} \tag{6-94}$$

the difference between the sum of heat capacities of the products and the sum of heat capacities of the reactants, with each term multiplied by its appropriate stoichiometric coefficient. The entropy of reaction $\Delta_r S$ at any temperature T relative to a reference temperature T_r is given by

$$\Delta_r S_T = \Delta_r S_{T_r} + \int_{T_r}^{T} \frac{\Delta C_P}{T} dT \tag{6-95}$$

Convenient reference temperatures are 298 K for high-temperature reactions and 0 K for low-temperature reactions. If modified Maier-Kelley equations are used for the heat capacities of products and reactants, the entropy of reaction is given by

$$\Delta_r S_T = \Delta_r S_{T_r} + \Delta a \ln\left(\frac{T}{T_r}\right) + \Delta b(T - T_r) - \frac{1}{2}\Delta c\left(\frac{1}{T^2} - \frac{1}{T_r^2}\right) + \frac{1}{2}\Delta d(T^2 - T_r^2) \tag{6-96}$$

Equations (6-95) and (6-96) apply to chemical reactions where no phase changes occur for either the products or the reactants.

Example 6-12. Use the heat capacity equations in Table 3-3 and Appendix 2 and the data from Example 6-11 to calculate the entropy of formation of forsterite from its constituent elements in their reference states at 900 K. The ΔC_P for reaction (5-97) is

$$\Delta C_P = C_P(Mg_2SiO_4) - 2C_P(O_2) - C_P(Si) - 2C_P(Mg) \tag{6-97}$$

Numerical evaluation of Eq. (6-97) using the data sources listed gives

$$\Delta C_P = 31.975 - 25.985 \times 10^{-3}T - 30.952 \times 10^5 T^{-2} + 9.666 \times 10^{-6} T^2 \text{ J mol}^{-1}\text{ K}^{-1} \tag{6-98}$$

Substituting this expression into Eq. (6-95), integrating between 298 K and 900 K, and using our result from Example 6-11 for the entropy of reaction at 298 K yields

$$\begin{aligned}\Delta_r S^o_{900} &= -400.35 + 31.975 \ln\left(\frac{900}{298.15}\right) - 25.985 \times 10^{-3}(900 - 298.15) + \frac{1}{2} \times 30.952 \\ &\quad \times 10^5\left(\frac{1}{900^2} - \frac{1}{298.15^2}\right) + \frac{1}{2} \times 9.666 \times 10^{-6}(900^2 - 298.15^2) \\ &= -400.35 + 7.67 = -392.68 \text{ J mol}^{-1}\text{K}^{-1}\end{aligned} \tag{6-99}$$

This $\Delta_r S^o$ value is about the same as that at 298 K because the additional entropy change due to heating Mg metal, Si crystal, O_2 gas, and forsterite is much smaller than that due to reaction of two moles of oxygen gas.

I. Entropy changes for chemical reactions with phase changes

If either the products or reactants (or both) in a chemical reaction undergo one or more phase changes over the temperature range being studied, we have to take the $\Delta_{trans}S$ values into account when we calculate the $\Delta_r S$ value. Transition entropies for the reactants are subtracted from the $\Delta_r S$ value and transition entropies for the products are added to the $\Delta_r S$ value. In general, we rewrite Eq. (6-95) as

$$\Delta_r S_T = \Delta_r S_{T_r} + \int_{T_r}^{T_{tr1}} \frac{\Delta C_P}{T} dT - \frac{\Delta_{tr1} H}{T_{tr1}} + \int_{T_{tr1}}^{T_{tr2}} \frac{\Delta C'_P}{T} dT + \frac{\Delta_{tr2} H}{T_{tr2}} + \int_{T_{tr2}}^{T} \frac{\Delta C''_P}{T} dT \qquad (6\text{-}100)$$

where T_{tr1} is the temperature of a phase transition in a reactant, T_{tr2} is the temperature of a phase transition in a product, and ΔC_P, $\Delta C'_P$, and $\Delta C''_P$ denote the three different ΔC_P values in the three temperature ranges below the first phase transition, between the two phase transitions, and above the second phase transition, respectively. The order of these phase transitions may be reversed from that shown in Eq. (6-100), and a phase transition in a product may occur at a lower temperature than a phase transition in a reactant. The following example uses forsterite formation to illustrate how to do calculations with Eq. (6-100).

Example 6-13. Calculate the entropy of reaction for forsterite formation from its constituent elements at 1000 K. This is above the melting point of Mg metal and below the melting point of silicon. The reaction involved is

$$2\ \text{Mg (liquid)} + \text{Si (crystal)} + 2\ \text{O}_2\ \text{(g)} = \text{Mg}_2\text{SiO}_4\ \text{(forsterite)} \qquad (6\text{-}101)$$

and the $\Delta_r S$ is given by

$$\Delta_r S^o_{1000} = \Delta_r S^o_{298} + \int_{298}^{923} \frac{\Delta C_P}{T} dT + \int_{923}^{1000} \frac{\Delta C'_P}{T} dT - 2\Delta_{fus} S^o_{923}(\text{Mg}) \qquad (6\text{-}102)$$

The first integral from 298 K to 923 K is evaluated using the ΔC_P from Example 6-12. The $\Delta C'_P$ term in the second integral is different because the heat capacity of molten Mg is used instead of the C_P of Mg metal. This equation for $\Delta C'_P$ (J mol^{-1} K^{-1}) is

$$\Delta C'_P = -48.173 + 79.227 \times 10^{-3} T - 31.356 \times 10^5 T^{-2} - 15.734 \times 10^{-6} T^2 \qquad (6\text{-}103)$$

The entropy of fusion of Mg metal is 9.18 J mol^{-1} K^{-1} (taking $\Delta_{fus}H = 8477$ J mol^{-1} at 923 K). Substituting the numerical values into Eq. (6-102) gives

$$\Delta_r S^o_{1000} = -400.35 + 7.99 + 0.80 - 2(9.18) = -409.92\ \text{J mol}^{-1}\ \text{K}^{-1} \qquad (6\text{-}104)$$

for the entropy of formation of forsterite at 1000 K.

J. Entropy estimation methods

The $\Delta_r S$ values for reactions of two or more solids to form another solid are typically smaller than the $\Delta_r S$ values for reactions in which gases are consumed or produced. In many cases the $\Delta_r S$ values are zero plus or minus a few J mol^{-1} K^{-1}. Thus, entropies of solids can be estimated using a Neumann-Kopp approach as done for heat capacity in Example 4-7 of Chapter 4.

Example 6-14. (a) Proustite (Ag_3AsS_3) is a silver ore mineral. Estimate its entropy at 298 K from the entropies of acanthite (Ag_2S) and orpiment (As_2S_3). We do this using the equation

$$S^o_{298}(Ag_3AsS_3) = \frac{3}{2}S^o_{298}(Ag_2S) + \frac{1}{2}S^o_{298}(As_2S_3) = \frac{3}{2}(142.89) + \frac{1}{2}(163.6) = 296.1 \text{ J mol}^{-1}\text{K}^{-1} \tag{6-105}$$

The measured S^o_{298} of proustite from low-temperature calorimetry is 303.7 J mol^{-1} K^{-1}, which is about 2.5% larger. (b) Estimate the 298 K entropy of ferrosilite, which is unstable at ambient pressure, from the entropies of quartz and fayalite. We do this using the chemical reaction

$$2\ FeSiO_3\ (\text{ferrosilite}) = Fe_2SiO_4\ (\text{fayalite}) + SiO_2\ (\text{quartz}) \tag{6-106}$$

Assuming that $\Delta_r S = 0$ we compute

$$S^o_{298}(FeSiO_3) = \frac{1}{2}\left[S^o_{298}(Fe_2SiO_4) + S^o_{298}(\text{quartz})\right] = \frac{1}{2}\left[151.0 + 41.46\right] = 96.2 \text{ J mol}^{-1}\text{K}^{-1} \tag{6-107}$$

versus the tabulated value of 94.6 J mol^{-1} K^{-1}, so the estimate is 1.7% too large.

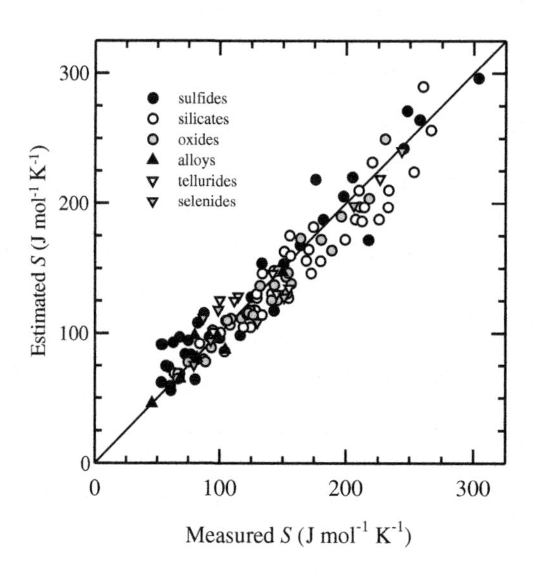

FIGURE 6-9

Entropies estimated with a Neumann-Kopp additive method are compared to measured values for about 140 different minerals. All values are at 298.15 K.

Figure 6-9 compares the estimated and measured entropies for about 140 minerals; 90 of these minerals were also included in Figure 4-14. The estimated entropies can be higher or lower than the observed entropies, and the average error is $-1 \pm 15\%$. This average is comparable to the average error of 3.5% on the estimated heat capacities in Figure 4-14, but the standard deviation is larger than for the heat capacity estimates. However, a Neumann-Kopp approach for entropy estimation is very easy and gives a value in the absence of experimental data.

Alternatively, Latimer (1951) devised a method for estimating the entropy of inorganic compounds from the entropies of their constituent ions. His method takes the effects of ionic charge, mass, and size into account. In some cases, his method works better than an additive Neumann-Kopp approach. For example, S^o_{298} of Ag_2S (acanthite) is 142.89 J mol^{-1} K^{-1}, Latimer's method gives 141.42 J mol^{-1} K^{-1}, and an additive approach gives 117.41 J mol^{-1} K^{-1}. However, the situation is reversed for As_2S_3 (orpiment), which has $S^o_{298} = 163.6$ J mol^{-1} K^{-1}. In this case the additive approach gives 167.5 J mol^{-1} K^{-1}, whereas Latimer's method gives 112.1 J mol^{-1} K^{-1}. Tables of entropy values for cations and anions are given in several references such as Kubaschewski and Alcock (1979) and Stull and Prophet (1967), so they do not have to be reproduced again here.

VIII. GIBBS AND HELMHOLTZ FREE ENERGY

It is convenient to define two new state functions, the Helmholtz free energy A and the Gibbs free energy G, to calculate the maximum work (A) and the net work (G) that can be obtained when heat is transformed into work. We discuss each of these functions here.

A. Helmholtz free energy

The German physicist Helmholtz (see Chapter 3) defined *Helmholtz free energy* as

$$A = E - TS \tag{6-108}$$

in 1882. The Helmholtz free energy is also known as the *work function*, the *Helmholtz function*, or *free energy*. The symbol A stands for *arbeit*, which is the German word for *work*. However, European texts commonly use the symbol F for the Helmholtz free energy.

The Helmholtz free energy is an extensive function, like internal energy and entropy. If we consider a reversible isothermal change of state, the change in A is given by

$$\Delta A = \Delta E - T\Delta S \tag{6-109}$$

The $T\Delta S$ term is the amount of heat that cannot be used to do work. Equation (6-43) shows that we can replace $T\Delta S$ in Eq. (6-109) by q_{rev} for this process. This gives

$$\Delta A = \Delta E - q_{rev} \tag{6-110}$$

Substituting for ΔE using the first law (Eq. 3-15), we find that

$$\Delta A = (q_{rev} - w) - q_{rev} = -w \tag{6-111}$$

Thus, the decrease in the Helmholtz free energy ($-\Delta A$) for a reversible, isothermal process is the work done by the system. Because the process is reversible, this is the maximum amount of work that can be done by the system (see Example 3-6 in Chapter 3). The work that can be obtained from an irreversible process is less than the maximum amount of work that is obtained when the same process is conducted reversibly.

Example 6-15. Calculate ΔA for the reversible, isothermal expansion of an ideal gas from 0.05 liter, 8 bars to 0.40 liter, 1 bar at 298.15 K. Figure 3-7 illustrates this process, which we also used in Examples 3-4 and 6-4. The internal energy change (see Example 3-4) is zero because the gas is expanded isothermally:

$$dE = C_V dT = 0 \tag{3-57}$$

The entropy change (see Example 6-4) is 0.279 J K^{-1}. The change in the Helmholtz free energy ΔA is given by Eq. (6-109) and is

$$\Delta A = \Delta E - T\Delta S = 0 - (298.15)(0.279) = -83.2 \text{ J} \tag{6-112}$$

Earlier in Example 3-4 we calculated that the gas did 83.2 J of work during its expansion. Equation (6-112) shows that $-\Delta A$ is equal to the work done by the gas.

The Helmholtz free energy is a criterion of equilibrium for reactions at constant volume and temperature. If ΔA is zero, a reaction is at equilibrium. If ΔA is less than zero, a reaction is spontaneous; if ΔA is greater than zero, a reaction is not spontaneous. However, it is usually more convenient to use the Gibbs free energy (defined in the next section) as a criterion of equilibrium because many reactions take place at constant pressure. The Helmholtz free energy is more widely used to calculate equations of state of geological materials such as iron, olivine, and pyroxene. We briefly discuss this topic in Chapter 8.

B. Gibbs free energy

The *Gibbs free energy G* is a state function that was invented by the American physicist J. W. Gibbs in 1875 (see sidebar in Chapter 7). It is defined by the equations

$$G = E + PV - TS \tag{6-113}$$

$$G = H - TS \tag{6-114}$$

Equation (6-114) is analogous to Eq. (6-108) for the Helmholtz free energy. The two functions are related to one another by the equation

$$G = A + PV \tag{6-115}$$

The Gibbs free energy is also known as the *free energy*, the *Gibbs function*, or *available energy*. It is an extensive function like the internal energy *E*, entropy *S*, and Helmholtz free energy *A*.

The change in *G* for a reversible, isothermal change of state is given by

$$\Delta G = \Delta H - T\Delta S \tag{6-116}$$

This is a very important equation that we will use throughout the rest of the book. It shows that the Gibbs free energy change for a chemical reaction can be calculated from the enthalpy and entropy changes for the reaction.

For now, we use Eq. (6-116) to illustrate the relationship of the two free energies. In general, for a reversible isothermal change in state, we have

$$\Delta G = \Delta E + \Delta(PV) - T\Delta S = \Delta A + \Delta(PV) \quad (6\text{-}117)$$

If we consider a reversible change in state that is both isothermal and isobaric (e.g., ice melting to water), we can rewrite Eq. (6-117) as

$$\Delta G = \Delta E + P\Delta V - T\Delta S = \Delta A + P\Delta V \quad (6\text{-}118)$$

Equation (6-117) shows that the Gibbs free energy change for a reversible process at constant temperature is identical (or nearly so) to the Helmholtz free energy change because PV is constant for an ideal gas, or nearly so for a real gas, along an isotherm (see the discussion of Boyle's law in Chapter 2). If the reversible process is both isothermal and isobaric, Eq. (6-118) shows that ΔG and ΔA differ by an amount equal to the work ($P\Delta V$) that has to be done against the constant external pressure. The ΔG change is thus the net work that can be obtained from a reversible process under isothermal and isobaric conditions.

Example 6-16. (a) Calculate ΔG for the reversible, isothermal expansion of an ideal gas that was considered in Example 6-15 and some earlier examples. We can compute ΔG using Eq. (6-116):

$$\Delta G = \Delta H - T\Delta S = 0 - (298.15)(0.279) = -83.2\text{ J} \quad (6\text{-}119)$$

or by using Eq. (6-117):

$$\Delta G = \Delta A + \Delta(PV) = -83.2 + 0 = -83.2\text{ J} \quad (6\text{-}120)$$

Both equations show that the changes in the Gibbs and Helmholtz free energies are equal in this case. If necessary, you should review Chapter 3 to remind yourself that ΔH is zero because the temperature is constant, and the $\Delta(PV)$ term is zero because PV is constant along an isotherm.

(b) Calculate ΔG and ΔA for ice melting to water in air at 0 Celsius and one atmosphere pressure using the data in Problem 3-18 and $\Delta_{fus}S$ for ice from Table 6-2. The PV work done is

$$P\Delta V = 10^5\text{ Pa} \times (1.000160\text{ cm}^3\text{ g}^{-1} - 1.091\text{ cm}^3\text{ g}^{-1}) \times (1\text{ m}^3/10^6\text{ cm}^3) \times 18.015\text{ g mol}^{-1} \quad (6\text{-}121)$$

$$P\Delta V = -0.1636\text{ J mol}^{-1} \sim -0.16\text{ J mol}^{-1}. \quad (6\text{-}122)$$

The internal energy change is

$$\Delta E = \Delta H - P\Delta V = 6009.5\text{ J mol}^{-1} + 0.16\text{ J mol}^{-1} \sim 6009.7\text{ J mol}^{-1} \quad (6\text{-}123)$$

The Gibbs free energy change is

$$\Delta G = \Delta H - T\Delta S = 6009.5 - (273.15)(6009.5/273.15) = 0\text{ J mol}^{-1} \quad (6\text{-}124)$$

because $\Delta_{fus}S$ is equal to $\Delta_{fus}H$ divided by the melting point (see Eq. 6-67). The ΔA change is

$$\Delta A = \Delta E - T\Delta S = \Delta H - P\Delta V - T\Delta S = -P\Delta V = 0.1636\text{ J mol}^{-1} \sim 0.16\text{ J mol}^{-1} \quad (6\text{-}125)$$

and is slightly positive because water is slightly denser than ice. Thus, the surroundings do PV work on the system (ice + water), which has a smaller volume after melting. The ΔG change is ~0.16 J mol^{-1}

smaller than the ΔA change because $P\Delta V$ is negative. The two equations (6-116) and (6-118) have to give the same result for ΔG, but it is important to understand why each equation gives the result that it does.

In this example, ice and water coexist at equilibrium at 0°C in air at one atmosphere pressure and ΔG is zero. In general, the ΔG change for a chemical reaction or a phase change at constant temperature and pressure determines whether or not the reaction or phase change is favorable. A process with a negative ΔG value is favorable, one with a positive ΔG value is unfavorable, and a process with a ΔG value equal to zero is at equilibrium.

Finally, we note that Gibbs called free energy the thermodynamic potential and used the Greek letter zeta (ζ) to represent it. Today, most books use G for the Gibbs free energy. However, the symbols used for the Helmholtz and Gibbs free energies in American and European books are inconsistent. Some old books use the symbol Z for the Gibbs free energy. In their famous 1923 book, Lewis and Randall used the symbol F for the Gibbs free energy. Their notation was subsequently used in many American thermodynamic books. But as mentioned earlier, F stands for Helmholtz free energy in European books, which use G for the Gibbs energy.

C. Gibbs free energy changes for chemical reactions

The ΔG for a reaction can be calculated analogously to the ΔH change, namely from the equation

$$\Delta_r G = \sum (\nu_i \Delta_f G_i)_{pr} - \sum (\nu_i \Delta_f G_i)_{re} \quad (6\text{-}126)$$

where $\Delta_r G$ is the Gibbs free energy change for a chemical reaction, $\Delta_f G$, the Gibbs free energy of formation of a material is multiplied by its stoichiometric coefficient (ν_i) in the chemical equation, and the subscripts *pr* and *re* stand for products and reactants, respectively. As with enthalpy, we can only measure differences in the Gibbs free energy but not its absolute value. We can also calculate $\Delta_r G$ for a reaction from Eq. (6-116) using $\Delta_r H$ and $\Delta_r S$, the enthalpy and entropy changes for the reaction. *The Gibbs free energy of formation of an element in its reference state is zero at all temperatures by definition.* The standard Gibbs free energy is denoted with the superscript *o* and refers to the pure crystalline solid, or pure liquid, or pure gas (behaving ideally) at one bar pressure.

Example 6-17. (a) Calculate the standard Gibbs free energy of formation of forsterite at 298 K. Table 6-4 and Example 6-11 give $\Delta_f H^o$ and $\Delta_f S^o$ at 298 K, which we substitute into Eq. (6-116):

$$\Delta_f G^o_{298} = \Delta_f H^o_{298} - T\Delta_f S^o_{298} = -2173.91 \times 10^3 - (298.15)(-400.35) = -2054.55 \text{ kJ mol}^{-1} \quad (6\text{-}127)$$

The result we get is identical to the value listed in Table 6-4. (b) Calculate the standard Gibbs free energy of reaction $\Delta_r G^o$ at 298 K for

$$2\,\text{MgO (periclase)} + \text{SiO}_2\text{ (quartz)} = \text{Mg}_2\text{SiO}_4\text{ (forsterite)} \quad (6\text{-}128)$$

We can do this using the $\Delta_f G^o$ values listed in Appendix 1:

$$\begin{aligned}\Delta_r G^o_{298} &= \Delta_f G^o_{298}\text{ (forsterite)} - \Delta_f G^o_{298}\text{ (quartz)} - 2\Delta_f G^o_{298}\text{ (periclase)} \\ &= -2054.5 - (-856.3) - 2(-569.3) = -59.6 \text{ kJ mol}^{-1}\end{aligned} \quad (6\text{-}129)$$

We now do the same calculation using Eq. (6-116). We first evaluate the $\Delta_r H^o$ term,

$$\begin{aligned}\Delta_r H^o_{298} &= \Delta_f H^o_{298}\,(\text{forsterite}) - \Delta_f H^o_{298}\,(\text{quartz}) - 2\Delta_f H^o_{298}\,(\text{periclase}) \\ &= -2173.9 - (-910.7) - 2(601.6) = -60.0 \text{ kJ mol}^{-1}\end{aligned} \tag{6-130}$$

then we evaluate the $\Delta_r S^o$ term,

$$\begin{aligned}\Delta_r S^o_{298} &= S^o_{298}\,(\text{forsterite}) - S^o_{298}\,(\text{quartz}) - 2S^o_{298}\,(\text{periclase}) = 94.1 - 41.46 - 2(26.95) \\ &= -1.26 \text{ J mol}^{-1}\text{K}^{-1}\end{aligned} \tag{6-131}$$

and finally we calculate $\Delta_r G^o$,

$$\Delta_r G^o_{298} = \Delta_r H^o_{298} - T\Delta_r S^o_{298} = -60.0 \times 10^3 - (298.15)(-1.26) = -59.6 \text{ kJ mol}^{-1} \tag{6-132}$$

This result agrees with the value from Eq. (6-129) and shows that the two computational methods give the same result.

D. The entropy of aqueous ions

We now use the Gibbs free energy to calculate entropy values for aqueous ions. These calculations use the standard integral molar enthalpy of solution (defined in Chapter 5 as the ΔH for formation of an infinitely dilute solution) and the analogous standard integral molar Gibbs free energy of solution. As discussed in Chapter 5, enthalpy of formation values for aqueous ions are calculated using the convention that the standard enthalpy of formation of the H^+ aqueous ion is defined to be zero. The formation of H^+ (aq) from hydrogen in its reference state, which is H_2 gas at one bar pressure behaving as an ideal gas, is

$$\frac{1}{2}H_2\,(g) = H^+(aq) + e^-. \tag{5-84}$$

Reaction (5-84) occurs when H_2 (g) at one bar pressure is bubbled over a platinum electrode in a dilute aqueous solution. The standard Gibbs free energy of formation for Eq. (5-84) and the standard entropy of the H^+ aqueous ion are also defined as zero. Because both $\Delta_r H^o_{298}$ and $\Delta_r G^o_{298}$ are zero, $\Delta_r S^o_{298}$ must also be zero. However, the entropy of H_2 gas is not zero, so formally we must also have a nonzero value for the entropy of the electron in Eq. (5-84):

$$S^o_{298}(e^-) = \Delta_r S^o_{298} - S^o_{298}(H^+) + \frac{1}{2}S^o_{298}(H_2, g) = 65.34 \text{ J mol}^{-1}\text{ K}^{-1} \tag{6-133}$$

The entropy values for other aqueous ions are determined from reaction networks in a manner analogous to that which we used for computing $\Delta_f H$ values for aqueous ions.

Example 6-18. (a) The chloride aqueous ion (Cl^-) is the major anion (i.e., negatively charged ion) in sea water. Calculate its entropy using the standard integral molar enthalpy and standard integral Gibbs free energy of solution for HCl (g) in water. This reaction is

$$HCl\,(g) + \infty H_2O\,(l) = H^+\,(aq) + Cl^-\,(aq) + \infty H_2O\,(l) \tag{6-134}$$

The $\Delta_{soln}H^o_m$ and $\Delta_{soln}G^o_m$ values at 298 K for reaction (6-134) are calculated from enthalpy and Gibbs free energy of formation data in Appendix 1:

$$\Delta_{soln}H^o_m = \Delta_f H^o_{298}(\mathrm{HCl, ai}) - \Delta_f H^o_{298}(\mathrm{HCl, g}) = -167.08 - (-92.31) = -74.77 \text{ kJ mol}^{-1} \quad (6\text{-}135)$$

$$\Delta_{soln}G^o_m = \Delta_f G^o_{298}(\mathrm{HCl, ai}) - \Delta_f G^o_{298}(\mathrm{HCl, g}) = -131.22 - (-95.30) = -35.92 \text{ kJ mol}^{-1} \quad (6\text{-}136)$$

The abbreviation *ai* in the preceding equations stands for an aqueous ionized solution of HCl at infinite dilution. In other words, the HCl is fully ionized into H^+ and Cl^- ions in the aqueous solution. The $\Delta_{soln}S^o_m$ value at 298 K for Eq. (6-134) is computed by rearranging Eq. (6-116):

$$\Delta_{soln}S^o_m = (\Delta_{soln}H^o_m - \Delta_{soln}G^o_m)/298.15 = -130.30 \text{ J mol}^{-1} \text{ K}^{-1} \quad (6\text{-}137)$$

The entropy of the chloride ion is then given by

$$\begin{aligned} S^o_{298}(\mathrm{Cl^-}) &= \Delta_{soln}S^o_{298} - S^o_{298}(\mathrm{H^+}) + S^o_{298}(\mathrm{HCl, g}) = -130.3 - 0 + 186.902 \\ &= 56.60 \text{ J mol}^{-1}\text{K}^{-1} \end{aligned} \quad (6\text{-}138)$$

(b) The Mg^{2+} aqueous ion is the second most abundant cation (i.e., positively charged ion) after Na^+ in sea water. Calculate its entropy from $S^o_{298} = 366.0 \text{ J mol}^{-1} \text{ K}^{-1}$ for $MgCl_2 \cdot 6\,H_2O$ (bischofite), the standard integral molar enthalpy of solution ($\Delta_{soln}H^o_m = -16.10 \text{ kJ mol}^{-1}$), and the standard integral molar Gibbs free energy of solution ($\Delta_{soln}G^o_m = -25.75 \text{ kJ mol}^{-1}$) for $MgCl_2 \cdot 6\,H_2O$ (bischofite) dissolution in water. This reaction is

$$\mathrm{MgCl_2 \cdot 6\,H_2O}\ (\text{bischofite}) = \mathrm{Mg^{2+}}\ (\mathrm{aq}) + 2\,\mathrm{Cl^-}\ (\mathrm{aq}) + 6\,\mathrm{H_2O}\ (\mathrm{l}) \quad (6\text{-}139)$$

The six water molecules in bischofite have to be considered in reaction (6-139) to maintain mass balance. The water molecules in other hydrated salts also have to be considered in the dissolution reactions of these salts. The $\Delta_{soln}S^o_m$ value for Eq. (6-139) is given by

$$\Delta_{soln}S^o_m = (\Delta_{soln}H^o_m - \Delta_{soln}G^o_m)/298.15 = 32.37 \text{ J mol}^{-1} \text{ K}^{-1} \quad (6\text{-}140)$$

The entropy of the Mg^{2+} ion is then given by

$$\begin{aligned} S^o_{298}(\mathrm{Mg^{2+}}) &= \Delta_{soln}S^o_m - 2S^o_{298}(\mathrm{Cl^-}) - 6S^o_{298}(\mathrm{H_2O, l}) + S^o_{298}(\text{bischofite}) \\ &= 32.37 - 2(56.60) - 6(69.95) + 366.0 = -134.53 \text{ J mol}^{-1}\text{K}^{-1} \end{aligned} \quad (6\text{-}141)$$

The entropy of the Mg^{2+} ion is negative, as is the entropy of many other ions. This is just a consequence of defining ionic entropies relative to an arbitrary value of zero for the aqueous H^+ ion. The negative entropy value means that the water molecules around Mg^{2+} are more ordered than the water molecules around H^+. Several other common ions in river water and in sea water also have negative entropies ($\text{J mol}^{-1} \text{ K}^{-1}$), including carbonate ($CO_3^{2-}$, $S^o_{298} = -50.0$), ferric iron (Fe^{3+}, $S^o_{298} = -280$), phosphate (PO_4^{3-}, $S^o_{298} = -222$), and Al^{3+} ($S^o_{298} = -340$).

IX. ENTROPY AND PROBABILITY

A. Some basic concepts about probability

If something is certain to happen, it has a probability of one. Conversely, if something is impossible, it has a probability of zero. The probability of an event is defined as the number of ways that event can happen divided by the total number of possible outcomes. For example, when a six-sided die is rolled, there are six possible ways the die could land, or six possible outcomes. There is only one way to roll a three, so the probability of rolling a three is

$$p(3) = \frac{1}{6} = 0.167 \tag{6-142}$$

However, there are three ways to roll an even number, because two, four, or six would count. So:

$$p(even) = \frac{3}{6} = \frac{1}{2} = 0.50 \tag{6-143}$$

If the total number of possible outcomes is known, the probability of an event can be determined from the equation

$$p(event) = \frac{\text{number of times the event happens}}{\text{number of times tried}} \tag{6-144}$$

How can we determine the probability of two events happening? For example, if we roll a six-sided die twice, what is the probability of rolling a three both times? The probability of two events happening is given by

$$p(x,y) = p(x) \times p(y) \tag{6-145}$$

if x and y are independent results. So the probability of rolling two threes in a row is given by

$$p(3,3) = \frac{1}{6} \times \frac{1}{6} = \frac{1}{36} = 0.0278 \tag{6-146}$$

which has about a 2.8% chance of happening. The probability of rolling a three three times in a row is given by

$$p(3,3,3) = \frac{1}{6} \times \frac{1}{6} \times \frac{1}{6} = \frac{1}{216} = 0.00463 \tag{6-147}$$

which has about a 0.5% chance of happening, that is, about 5 chances out of 1000. Equation (6-145) can be generalized to the equation

$$p\left(\sum n_i\right) = p(n_1) \times p(n_2) \times p(n_3) \times \cdots \times p(n_i) = \prod p(n_i) \tag{6-148}$$

which says that the probability of i independent events taking place is the product of the individual probabilities. Equation (6-148) is valid for calculating the probability of i events taking place sequentially or simultaneously.

Example 6-19. (a) Calculate the probability of rolling a total of eight on two six-sided dice. There are five different ways to get a total of eight. We could get two on the first die and six on the other, six on the first and two on the other, three on the first and five on the other, five on the first and three on the other, or four on both. The total number of possible outcomes when rolling two dice is 36. So,

$$p(8) = \frac{5}{36} = 0.139 = 13.9\% \tag{6-149}$$

The probability of rolling a total of two, which can only happen if we get ones on both dice, is 1/36, or about 3%, so we can see that the chances of getting eight are much better.

(b) Calculate the probability of rolling a total of seven on two six-sided dice. There are six different ways to get a total of seven, which are $1 + 6$, $2 + 5$, $3 + 4$, $4 + 3$, $5 + 2$, and $6 + 1$. Thus,

$$p(7) = \frac{6}{36} = 0.167 = 16.7\% \tag{6-150}$$

Now let's consider a bigger system. Imagine we took a deck of playing cards and spread them all over the floor, face-down. Then we pick up one card at a time. What is the probability of picking up the cards in a certain order, say ace of hearts, two of hearts, three of hearts, and so on, through all four suits, say hearts, then spades, then diamonds, and then clubs? Obviously, there is only one way this can happen. However, we also need to figure out the total number of possible outcomes to find the probability.

There are 52 options on the floor for the first card to pick up, so we have a 1 in 52 chance of getting the ace of hearts. After that card is removed, 51 remain, so there is a 1 in 51 chance of picking up the two of hearts. We can see that the probability of picking up the cards in the order we want will be determined by

$$\frac{1}{52} \times \frac{1}{51} \times \frac{1}{50} \times \cdots \times \frac{1}{2} \times \frac{1}{1} \tag{6-151}$$

This can also be written as

$$\frac{1}{52!} = \frac{1}{8.07 \times 10^{67}} = 1.24 \times 10^{-68} \tag{6-152}$$

The exclamation point ! is the mathematical symbol for a factorial, and it refers to

$$N! = N \times (N-1) \times (N-2) \times \cdots \times (N-(N-1)) \tag{6-153}$$

where N is an integer.

Example 6-20. Calculate the factorials for each of the integers 3 to 7.

$$3! = 3 \times 2 \times 1 = 6 \tag{6-154}$$

$$4! = 4 \times 3 \times 2 \times 1 = 24 \tag{6-155}$$

$$5! = 5 \times 4 \times 3 \times 2 \times 1 = 120 \tag{6-156}$$

$$6! = 6 \times 5 \times 4 \times 3 \times 2 \times 1 = 720 \tag{6-157}$$

$$7! = 7 \times 6 \times 5 \times 4 \times 3 \times 2 \times 1 = 5040 \tag{6-158}$$

It would be tedious to calculate factorials for slightly larger numbers, and almost impossible for integers comparable to Avogadro's number (~6×10^{23}). However, the factorial of a number $N > 10$ can be fairly accurately found using Stirling's formula

$$N! = (N^N/e^N)\sqrt{2\pi N} \tag{6-159}$$

which can be simplified to

$$\ln (N!) = N \ln N - N \tag{6-160}$$

for numbers comparable to Avogadro's number.

Because 52! is a very large number, we can see that the chances of picking up the cards in the specified order are very small. If it took about five minutes to spread the cards out and pick them up and we had to do this 8.07×10^{67} times, it could take as long as 7.7×10^{62} years to pick them up in the order we want. This is enormously longer than the age of the universe, which is about 1.5×10^{10} years.

Example 6-21. Calculate 52! using a pocket electronic calculator and Eq. (6-159). My pocket calculator gives 8.0658×10^{67} for 52!. Substituting into Eq. (6-159) gives

$$52! = (52^{52}/e^{52})\sqrt{2\pi 52} = (1.7068 \times 10^{89}/3.8310 \times 10^{22})(18.08) = 8.0529 \times 10^{67} \tag{6-161}$$

In this case Eq. (6-159) is a good approximation and is accurate within 0.16%.

The probability of picking up our cards in any specified order is the same, regardless of what that order is or how random it appears to be. We could pick up the cards at random, record the order in which we collected them, and then try to repeat the experiment. It is just as unlikely that we would get the same random sequence again as that we would get the orderly sequence we first mentioned. It is much more likely that we would get a new random sequence. Our deck of cards example shows that for a larger system, the probability of complete order becomes negligible. If you play around with Stirling's formula, you will see that $N!$ quickly becomes astronomically large. Therefore, $1/N!$ becomes very tiny. Probability shows us that disordered states are the most probable ones. Ordered states are so unlikely that they can be considered never to occur spontaneously.

B. Boltzmann-Planck equation

All of us probably associate entropy with the concept of randomness and disorder or disorganization. These associations are all correct and can be put onto a firm foundation and quantified using probability and the *Boltzmann-Planck equation:*

$$S = k \ln \Omega = k \ln W \tag{6-162}$$

where k is Boltzmann's constant, which is equal to the ideal gas constant divided by Avogadro's number ($k = R/N_A = 1.3805 \times 10^{-23}$ J K^{-1}) and Ω (omega) or W is the number of distinguishably different arrangements a system can have. This equation relates entropy to physical properties of the constituents of a system such as the atoms or molecules making up a gas.

We can illustrate the concept of distinguishably different arrangements by thinking about the number of different ways that tumbled stones made of blue sodalite ($Na_4(AlSiO_4)_3Cl$) and green olivine can be made into necklaces. Suppose that we select three different sodalites (denoted S1, S2, and S3), with each being a slightly different size, shape, or color. In this case we could make six different necklaces: S1-S2-S3, S1-S3-S2, S2-S3-S1, S2-S1-S3, S3-S1-S2, S3-S2-S1. We have three possible ways of making the necklace before putting any of the stones onto a string, two alternative ways of stringing the necklace after putting the first stone on it, and only one way of finishing the necklace. The six arrangements are equal to $3 \times 2 \times 1 = 3!$. Now suppose we had six sodalites with slightly different size, shape, or color. We could arrange the six stones into $6! = 720$ different necklaces. Likewise, if we used three sodalites and three olivines, each slightly different from the other, we could make $6! = 720$ different necklaces.

However, if we carefully select three identical sodalites and three identical olivines, we can make the six stones into only 20 different-looking necklaces. For example, B-B-B-G-G-G, or B-G-B-G-B-G, or B-G-B-B-G-G, and so on. We decreased the number of distinguishably different arrangements by picking two sets of three identical stones. In this case, we calculate

$$\Omega = \frac{(3+3)!}{3!3!} = \frac{6!}{6 \times 6} = \frac{720}{36} = 20 \tag{6-163}$$

Because we selected three identical sodalites, we do not have six different arrangements that exist for three different sodalites. Likewise, because we selected three identical olivines, we do not have six different arrangements that exist for three different olivines. Thus, we have to divide the 720 possible arrangements for six different stones by $36 = (3! \times 3!)$. We can generalize Eq. (6-163) for a binary mixture of any two different kinds of objects:

$$\Omega = \frac{(N_1 + N_2)!}{(N_1!)(N_2!)} \tag{6-164}$$

where N_1 is the number of objects of one kind and N_2 is the number of objects of another kind. If we are mixing three different kinds of objects, we can write

$$\Omega = \frac{(N_1 + N_2 + N_3)!}{(N_1!)(N_2!)(N_3!)} \tag{6-165}$$

and so on for more complicated mixtures.

C. Entropy of mixing and probability

We now derive a formula for the entropy of mixing in a binary mixture as a function of the mole fractions of the components of the mixture. We use the entropy of mixing for N_2 (79%) and O_2 (21%) molecules in air as an example in our derivation, but the result we obtain applies for any binary mixture. We start with Eq. (6-164) and rewrite it as

$$\Omega = \frac{(N_{N_2} + N_{O_2})!}{(N_{N_2}!)(N_{O_2}!)} \tag{6-166}$$

The number of N_2 molecules is N_{N2} and the number of O_2 molecules is N_{O2}. If we consider mixing all the N_2 and O_2 molecules in one mole of air, then

$$N_{N_2} + N_{O_2} = N_A \tag{6-167}$$

we have Avogadro's number (N_A) of molecules. The mole fractions of N_2 and O_2 molecules are X_{N2} and X_{O2}, respectively. We can then write the number of molecules as the mole fraction times the total number of molecules:

$$N_{N_2} = X_{N_2} N_A \tag{6-168a}$$

$$N_{O_2} = X_{O_2} N_A \tag{6-168b}$$

Substituting these values into Eq. (6-166) we get

$$\Omega = \frac{N_A!}{(X_{N_2} N_A)! \times (X_{O_2} N_A)!} \tag{6-169}$$

Equation (6-169) gives the total number of different distinguishable arrangements of N_2 and O_2 molecules in one mole of air. This is an extremely large number that we could compute using Stirling's formula, Eq. (6-160). However, we will not do this because we can simplify things first. We substitute Eq. (6-169) into the Boltzmann-Planck equation (6-162) to get

$$S = \frac{R}{N_A} \ln \frac{N_A!}{(X_{N_2} N_A)! \times (X_{O_2} N_A)!} \tag{6-170}$$

We now apply Stirling's formula Eq. (6-160) and rewrite Eq. (6-170) as

$$S = \frac{R}{N_A} [(N_A \ln N_A - N_A) - X_{N_2} N_A \ln (X_{N_2} N_A) + X_{N_2} N_A - X_{O_2} N_A \ln (X_{O_2} N_A) + X_{O_2} N_A] \tag{6-171}$$

Avogadro's number N_A is a common term that can be factored out to give

$$S = R[(\ln N_A - 1) - X_{N_2} \ln (X_{N_2} N_A) + X_{N_2} - X_{O_2} \ln (X_{O_2} N_A) + X_{O_2}] \tag{6-172}$$

which is further simplified using the fact that the two mole fractions sum to unity:

$$S = R[\ln N_A - X_{N_2} \ln (X_{N_2} N_A) - X_{O_2} \ln (X_{O_2} N_A)] \tag{6-173}$$

We remove the remaining N_A terms with the next three steps:

$$S = R[\ln N_A - X_{N_2} \ln X_{N_2} - X_{N_2} \ln N_A - X_{O_2} \ln X_{O_2} - X_{O_2} \ln N_A] \tag{6-174}$$

$$S = R[N_A \ln (1 - X_{N_2} - X_{O_2}) - X_{N_2} \ln X_{N_2} - X_{O_2} \ln X_{O_2}] \tag{6-175}$$

$$S_{mix} = -R[X_{N_2} \ln X_{N_2} + X_{O_2} \ln X_{O_2}] \tag{6-176}$$

Equation (6-176) gives the entropy of mixing for N_2 and O_2 molecules in air. Using mole fractions of 0.79 for N_2 and 0.21 for O_2 we find the entropy of mixing is

$$\begin{aligned} S_{mix} &= -R[0.79 \ln 0.79 + 0.21 \ln 0.21] = -R[-0.186 - 0.328] \\ &= 0.514R = 4.273 \text{ J mol}^{-1}\text{K}^{-1} \end{aligned} \tag{6-177}$$

In general, the entropy of mixing for a binary mixture is given by

$$S_{mix} = -R[X_1 \ln X_1 + X_2 \ln X_2] \tag{6-178}$$

where X_1 and X_2 are the mole fractions of the two components in the mixture. The entropy of mixing for a mixture of any number i of different components is given by

$$S_{mix} = -R\Sigma(X_i \ln X_i) \tag{6-179}$$

Example 6-22. Although dry air is mainly N_2 and O_2, it also contains many other gases. We use the average composition of dry air to calculate the entropy of mixing for making air from the nine most abundant gases in it: N_2 ($X_1 = 0.78084$), O_2 ($X_2 = 0.20946$), Ar ($X_3 = 9.34 \times 10^{-3}$), CO_2 ($X_4 = 3.6 \times 10^{-4}$), Ne ($X_5 = 1.818 \times 10^{-5}$), He ($X_6 = 5.24 \times 10^{-6}$), CH_4 ($X_7 = 1.7 \times 10^{-6}$), Kr ($X_8 = 1.14 \times 10^{-6}$), and H_2 ($X_9 = 5.5 \times 10^{-7}$). Equation (6-179) becomes

$$\begin{aligned} S_{mix} = & -R[0.78084 \ln 0.78084 + 0.20946 \ln 0.20946 + 9.34 \times 10^{-3} \ln 9.34 \times 10^{-3} \\ & +3.6 \times 10^{-4} \ln 3.6 \times 10^{-4} + 1.818 \times 10^{-5} \ln 1.818 \times 10^{-5} + 5.24 \times 10^{-6} \ln 5.24 \times 10^{-6} \\ & +1.7 \times 10^{-6} \ln 1.7 \times 10^{-6} + 1.14 \times 10^{-6} \ln 1.14 \times 10^{-6} + 5.5 \times 10^{-7} \ln 5.5 \times 10^{-7}] \end{aligned} \tag{6-180}$$

After evaluating each term, we have

$$\begin{aligned} S_{mix} = & -R[-0.1932 - 0.3274 - 0.0437 - 0.0029 - 1.98 \times 10^{-4} - 6.37 \times 10^{-5} \\ & -2.26 \times 10^{-5} - 1.56 \times 10^{-5} - 7.93 \times 10^{-6}] \\ = & \ 0.5675R = 4.718 \text{ J mol}^{-1}\text{K}^{-1} \end{aligned} \tag{6-181}$$

Almost all the difference between this value and that computed previously for the entropy of mixing of N_2 and O_2 (4.273 J $mol^{-1}K^{-1}$) is due to the inclusion of Ar and CO_2, which are the third and fourth most abundant gases in air.

Equations (6-178) and (6-179) are general equations that are valid for any ideal solution of gases, liquids, and solids where mixing is completely random and only one substitution is possible. We have to modify Eqs. (6-178) and (6-179) if more than one substitution is possible, as in minerals with several different types of crystallographic sites. We describe these modifications in Section E below. If the solutions are not ideal, then mixing is not completely random. In these cases, which we describe when we consider solution thermodynamics, we have to consider the effects of nonideality on the entropy of mixing.

D. The entropy of mixing of gases

We now derive Eqs. (6-178) and (6-179) without using the Boltzmann-Planck relation to show that the arguments based on probability give exactly the same results as the equations we derived earlier. Equations (6-52) and (6-53) give the entropy change of an ideal gas during a reversible isothermal process. We use these equations to compute the entropy of mixing (ΔS_{mix}) for reversible isothermal mixing of 0.79 moles N_2 and 0.21 moles O_2 to make one mole of air. We have to assume that N_2 and O_2 behave ideally because Eqs. (6-52) and (6-53) were derived for ideal gases. This assumption is not a problem because at ambient temperature and pressure, both gases show $<0.1\%$ deviation from ideality, and we could change the temperature and pressure to make N_2 and O_2 behave even more like ideal gases if this were absolutely necessary.

The N_2 and O_2 are in different compartments separated by a removable wall inside a gas-tight container and are at the same pressure and temperature. The sizes of the two compartments are in proportion to the number of moles of both gases, so the ratio is about 4:1. We remove the wall and allow each gas to expand to fill the entire container. The entropy change for the N_2 is

$$\Delta S_{N_2} = nR \ln\left(\frac{V_B}{V_A}\right) = n_{N_2} R \ln\left(\frac{V_{N_2} + V_{O_2}}{V_{N_2}}\right) = n_{N_2} R \ln \frac{1}{X_{N_2}} \quad (6\text{-}182)$$

The entropy change for the O_2 is

$$\Delta S_{O_2} = nR \ln\left(\frac{V_B}{V_A}\right) = n_{O_2} R \ln\left(\frac{V_{N_2} + V_{O_2}}{V_{O_2}}\right) = n_{O_2} R \ln \frac{1}{X_{O_2}} \quad (6\text{-}183)$$

The entropy of mixing is the sum of the two entropy changes and is

$$\begin{aligned} \Delta S_{mix} &= \Delta S_{N_2} + \Delta S_{O_2} = n_{N_2} R \ln \frac{1}{X_{N_2}} + n_{O_2} R \ln \frac{1}{X_{O_2}} \\ &= -R[n_{N_2} \ln X_{N_2} + n_{O_2} \ln X_{O_2}] \end{aligned} \quad (6\text{-}184)$$

The moles of N_2 and O_2 sum up to the total number of moles in the gas mixture, so we can replace the moles of each gas by its mole fraction and rewrite Eq. (6-184) as

$$\begin{aligned} \Delta S_{mix} &= -R[X_{N_2} \ln X_{N_2} + X_{O_2} \ln X_{O_2}] \\ &= -R[-0.186 - 0.328] \\ &= 0.514R = 4.273 \text{ J mol}^{-1}\text{K}^{-1} \end{aligned} \quad (6\text{-}185)$$

Equation (6-185) gives exactly the same result as Eq. (6-177) because it is the same equation. We could have obtained Eq. (6-185) using any two or more ideal gases, so Eq. (6-185) is also the derivation for the general forms of Eqs. (6-178) and (6-179).

E. Configurational entropy of minerals

Configurational entropy is the entropy arising from mixing of different atoms in metal alloys or from the mixing of different cations or anions in minerals and mineral solid solutions. Examples of configurational entropy include the entropy arising from mixing of Fe, Ni, and Co in meteoritic metal, of Al and Si in high-albite $NaAlSi_3O_8$, of Mg and Fe cations in olivine $(Mg,Fe)_2SiO_4$, and of F, Cl, and OH anions in apatite $Ca_5(PO_4)_3(OH,F,Cl)$. As stated earlier, we have to modify Eqs. (6-178) and (6-179) for minerals with mixing on more than one crystallographic site. In these cases the configurational entropy is given by the equation

$$S_{config} = -R \sum_j \sum_i X_{i,j} \ln X_{i,j} \quad (6\text{-}186)$$

The first summation is for all sites j in a mineral, and the second summation is for all atoms i on all sites in the mineral. We illustrate Eq. (6-186) using melilites, which are generally solid solutions between gehlenite ($Ca_2SiAl_2O_7$) and åkermanite ($Ca_2MgSi_2O_7$). Melilite is common in several types of Ca,Al-rich inclusions in CV3 carbonaceous chondrite meteorites. It is also found in thermally metamorphosed impure limestones, and some types of terrestrial alkaline rocks (nepheline basalts, leucitites, and alnöites).

There are two different tetrahedral sites in melilites where cationmixing occurs. There is one T_1 site where Si, Mg, and Al substitutions are possible, and two T_2 sites where Si and Al substitutions are possible. Thus, for melilites, Eq. (6-186) is written as

$$S_{config} = -R\left[\left(\sum X_i \ln X_i\right)_{T_1} + 2\left(\sum X_i \ln X_i\right)_{T_2}\right] \tag{6-187}$$

The formula of gehlenite can be written as $Ca_2Al(Al_{0.5}Si_{0.5})_2O_7$ to emphasize that only Al atoms are in the T_1 site and that Al and Si atoms mix on the two T_2 sites. The configurational entropy of gehlenite is thus given by

$$S_{config} = -R\left[\left(1 \ln 1\right) + 2\left(\frac{1}{2} \ln \frac{1}{2} + \frac{1}{2} \ln \frac{1}{2}\right)\right] = 2R \ln 2 = 11.526 \text{ J mol}^{-1}\text{K}^{-1} \tag{6-188}$$

However, åkermanite does not have any configurational entropy because only Mg atoms are in the T_1 site and only Si atoms are on the two T_2 sites.

Several other minerals, including inverse spinels, also have two crystallographic sites in a 1:2 ratio on which cations can mix. Normal spinels have the general formula $A^{2+}B^{3+}{}_2O_4$ with all divalent ions on the four coordinated A site and all trivalent ions on the six coordinated B site. Inverse spinels have the general formula $A^{3+}B^{2+}B^{3+}O_4$ with all divalent ions on the B site, half of the trivalent ions on the A site, and the other half of the trivalent ions on the B site. Spinel ($MgAl_2O_4$), hercynite ($FeAl_2O_4$), gahnite ($ZnAl_2O_4$), and galaxite ($MnAl_2O_4$) are examples of normal spinels. Magnetite (Fe_3O_4), magnesioferrite ($MgFe_2O_4$), trevorite ($NiFe_2O_4$), and ülvospinel (Fe_2TiO_4) are examples of inverse spinels. However, spinels do not have to be either normal or inverse and can be a mixture of the two distributions, depending upon their thermal history.

There is no configurational entropy for a normal spinel because there is no cation mixing on the two sites: all cations on the A site are the same, and all cations on the B site are the same. Inverse spinels have configurational entropy because of the cation mixing on the B site. The general equation for configurational entropy of inverse spinels is a rewritten form of Eq. (6-187):

$$S_{config} = -R\left[\left(\sum X_i \ln X_i\right)_{A} + 2\left(\sum X_i \ln X_i\right)_{B}\right] \tag{6-189}$$

Equation (6-189) gives a configurational entropy of 2Rln2 (11.526 J mol^{-1} K^{-1}). Later we show that the configurational entropy can be verified using high-temperature chemical equilibria.

F. Probability of impossible events

Earlier in this chapter we gave examples of irreversible processes, such as hot food cooling as it sits on a table, ice cubes melting in a glass of soda, logs burning in a fire, and hot coffee cooling as it sits in a cup. All of us know that such events are the natural course of things and that cold food and cold coffee never get warmer by extracting heat from the surroundings, that a glass of soda does not spontaneously generate ice cubes, and that fires do not generate wood logs from ashes. The Boltzmann-Planck equation shows that such impossible events involve a decrease in entropy and have an almost infinitely small probability of ever taking place.

Example 6-23. What is the probability that a 100-gram meteorite will spontaneously spring 10 cm into the air at the expense of the thermal energy in its surroundings, which are at 15°C? We have to compute the entropy change of the system for this process. To do this we have to identify the system and analyze the process taking place. The system is the meteorite and its surroundings from which heat is extracted. The process taking place is the same as that in a Carnot engine that extracts heat from the surroundings and uses it to do work. In this case, all of the heat is converted into work that is used to raise the meteorite. There is no cold reservoir and no heat could be rejected to it because all heat is converted into work. The heat q taken from the surroundings is equal to the work done to lift the meteorite vertically against the force of gravity:

$$q = mg\Delta h = (0.1\ \text{kg})(9.80\ \text{m s}^{-2})(0.1\ \text{m}) = -0.098\ \text{J} \tag{6-190}$$

The heat q has a negative sign because we defined heat flow out of a system as negative (if necessary, go back to Chapter 3 to review). The ΔS of the system is given by

$$\Delta S = \frac{q}{T} = \frac{-0.098}{288} = -3.4 \times 10^{-4}\ \text{J K}^{-1} \tag{6-191}$$

The entropy of the surroundings decreases by 0.00034 J K^{-1} because of the heat that is converted into work. The entropy of the meteorite does not change in this process. The probability (Ω) of this event occurring is related to the entropy change via the equation

$$\Delta S = S_{final} - S_{initial} = k \ln \frac{\Omega_{final}}{\Omega_{initial}} \tag{6-192}$$

Rearranging Eq. (6-192) to solve for the probability gives

$$\frac{\Omega_{final}}{\Omega_{initial}} = \exp\left(\frac{\Delta S}{k}\right) = \exp\left(\frac{-3.4 \times 10^{-4}}{1.38 \times 10^{-23}}\right) = \exp(-2.5 \times 10^{19}) \cong \frac{1}{10^{19}} \tag{6-193}$$

Thus, there is about a chance in 10 million trillion that the meteorite will spring into the air at the expense of the thermal energy of the surroundings. One chance in 10 million trillion is about the same ratio as 4 millimeters is to the distance to Proxima Centauri, the nearest star (4.21 light years away from our solar system).

G. Maxwell's demon

In 1871 James Clerk Maxwell postulated the existence of a tiny being, later called a *demon* by Lord Kelvin. The demon was stationed at a trap door in a wall dividing two parts of an isothermal and isobaric container of air. The demon's job was to separate fast (hot) from slow (cold) gas molecules because there is normally a statistical distribution of temperatures among gas molecules, with some being hotter than average and some being colder than average. The demon operated the trap door to let all fast molecules from the other side come into his side of the vessel and to let all slow molecules from his side go into the other side of the vessel. By doing his job the demon raised the temperature of his side of the container and simultaneously decreased the temperature of the other side of the container. Theoretically, Maxwell's demon could now use the temperature difference to do work such as running a television set.

The demon's separation of air into hot and cold molecules violates the second law and should be impossible. You may say Maxwell's demon is impossible in any case because there are no small demons who could operate trap doors to separate air molecules. But the key question about the demon, which is really an ingenious thought experiment, is whether there are any limitations on the second law of thermodynamics. As far as we know, the answer is no. There are no limitations to the second law, and Maxwell's demon cannot separate hot and cold air molecules. But why not?

Maxwell concluded that the demon could not distinguish between fast and slow gas molecules because "no one has yet discovered any practical method of tracing the path of a molecule, or of identifying it at different times." However, isotopes were unknown in 1871 and Maxwell did not know it is possible in principle for the demon to distinguish chemically identical molecules or atoms, for example, $^{15}N^{14}N$ from $^{14}N^{14}N$, or $^{16}O^{18}O$ from $^{16}O^{16}O$, or ^{40}Ar from ^{36}Ar.

In 1912, the Polish physicist Marian Smoluchowski (1872–1917) concluded that Brownian motion (small random motions caused by ambient thermal energy) would cause the trap door to open and close randomly. As a result of the random opening and closing of the trap door, some fast molecules could escape and some slow molecules could enter the demon's side of the container. Thus, the demon cannot do his job unless he also suppresses Brownian motions.

In the late 1940s the French physicist Leon Brillouin (1889–1969) pointed out that the demon, his container, and the air in it were all at the same temperature. As a consequence everything inside the container is emitting photons of the same energy (i.e., at the same wavelength) and nothing can be distinguished. The demon has to use a flashlight to see the air molecules against an otherwise uniform background. Brillouin showed that entropy increases when the demon uses his flashlight, and consequently the demon's separation of hot and cold air molecules obeys the second law of thermodynamics.

PROBLEMS

1. Three sprinters of practically equal ability are training for the Olympics by racing each other. They race a total of four times. Assuming that the most likely outcome occurs, what is the probability that a particular runner will win four races, three races, two races, one race, or no races?
2. The ancient Egyptians made jewelry from electrum, a natural alloy of gold and silver. Compute the configurational entropy of an electrum alloy that is 70 mole percent gold and 30 mole percent silver.
3. What is the probability of 1 ml of water initially at 316 K (outdoor temperature on a very hot day in Texas) spontaneously coming to a boil at the expense of the thermal energy of its surroundings? Use these data to solve the problem: C_P (H_2O, l) = 75.48 J mol^{-1} K^{-1} and ρ (H_2O, l, 316 K) $\cong$ 1 g cm^{-3}.
4. During combustion of gasoline in a car engine, some of the N_2 in air is converted to nitric oxide (NO) via the reaction

$$\frac{1}{2}\ N_2\ (g) + \frac{1}{2}\ O_2\ (g) = NO\ (g)$$

because of the high temperatures inside the cylinder of the engine. Calculate $\Delta_f G^o$ for NO (g) at 2300 K using the data tabulated here and in Table 3-4.

	Thermodynamic Properties at 298.15 K	
Gas	$\Delta_f H^o$ (kJ mol^{-1})	S^o (J mol^{-1} K^{-1})
N_2	0	191.609
O_2	0	205.147
NO	90.291	210.758

5. The lifetime of space probes operating on Venus's surface (average $T = 740$ K) is limited by the high temperature on the surface and internal heat generation by the electronics. A long-term (1-year) lander mission has been proposed that requires refrigeration of the lander using a plutonium-powered radioisotope thermoelectric generator (RTG). Calculate the coefficient of performance of and the work done by a Carnot refrigerator used to cool the space probe. Assume $q_c = 4180$ kJ. This is the enthalpy $(H_{740} - H_{298})$ of 16 kilograms of titanium metal, which is used to make the pressure shells of the probes.
6. Calculate the changes in pressure, internal energy E, enthalpy H, Helmholtz free energy A, Gibbs free energy G, and entropy S during convective descent of one mole of atmospheric gas on Jupiter from the 300 K, 6.75 bars level to the 400 K level. Assume that Jovian atmospheric gas is pure H_2 behaving ideally with an entropy $S^o{}_{300} = 130.68$ J mol^{-1} K^{-1} and a constant heat capacity $C_P =$ 29.08 J mol^{-1} K^{-1} (3.50R).
7. IR reflection spectra indicate that solid N_2 is the major ice on the surface of Triton, Neptune's largest satellite. At Triton's average surface temperature of 38 K, hexagonal β-N_2 (s) is the stable form of N_2 (s). However, this is close to the α-β transition temperature of 35.61 K, below which the cubic α-N_2 is more stable. Use the C_P, entropy, and enthalpy data for α and β-N_2 ices to calculate the ΔG difference between α and β-N_2 at 34 K. You may be interested to know that several groups have discussed the possible conversion of β and α-N_2 on Triton due to seasonal and geographical temperature changes.

$$(\alpha - N_2)S^o_{35.61} = 25.25 \text{ J mol}^{-1}\text{ K}^{-1}$$
$$C_P(\alpha - N_2) = 2.1184\,T - 29.752 \text{ J mol}^{-1}\text{ K}^{-1} \quad \text{from} \quad T = 30 - 35.61 \text{ K}$$
$$\Delta_{trans}H^o(\alpha \rightarrow \beta \text{ at } 35.61 \text{ K}) = +228.91 \text{ J mol}^{-1}$$
$$C_P(\beta - N_2) = 0.3958\,T + 21.8196 \text{ J mol}^{-1}\text{ K}^{-1} \quad \text{from} \quad T = 35.61 - 63.14 \text{ K}$$

8. Write down the other 17 ways in which three identical blue sodalites and three identical green olivines can be made into necklaces.
9. Calculate the probability of rolling a total of nine with two six-sided dice.
10. Calculate the configurational entropy of an alloy made of 0.6 moles Au and 0.4 moles Cu.
11. Draw qualitatively correct diagrams of the Carnot cycle on plots of (a) temperature-entropy, (b) temperature-volume, (c) temperature-pressure, (d) pressure-entropy, and (e) entropy-volume. Use any one of these plots to determine the total change in entropy around the cycle.

12. The change from one atmosphere to one bar in the reference state for thermodynamic data tabulations caused the same change in S_T^o for all ideal gases. (a) What is this change (i.e., $S_{1\ atm}^o - S_{1\ bar}^o$) to five decimal places? (b) Why were the entropies of solids and liquids unaffected by changing the reference state pressure? Assume $\beta V \sim 10^{-4}$ for either solids or liquids.

13. Calculate the lattice (S_{lat}) and electronic (S_E) components of the entropy of Fe metal at low temperatures from the data tabulated in Problem 4-3. Plot and tabulate your results.

14. Calculate the lattice (S_{lat}) and magnetic (S_M) components of the entropy of magnetite using the coefficients in Problem 4-16. Consider a temperature range of about 1.5 K to 4 K. Also calculate the temperature below which $S_M > S_{lat}$.

15. Yamada and colleagues have used a solar furnace to measure the melting points of several refractory oxides, including lime (CaO, m.p. 3174 ± 7 K), hafnia (HfO_2, m.p. 3076 ± 3 K), urania (UO_2, m.p. 3143 ± 28 K), zirconia (ZrO_2, m.p. 2980 ± 25 K), and yttria (Y_2O_3, m.p. 2706 ± 4 K). Lime has the NaCl crystal structure, yttria and urania have a distorted CaF_2 structure, and hafnia and zirconia are cubic at high temperatures. Estimate the enthalpy of fusion for these refractory oxides using data from Example 6-8.

16. Using data from the JANAF tables, the boiling point of pure molten alumina is calculated as 4045 K (Schaefer and Fegley, 2004). Estimate the enthalpy of vaporization of molten alumina using the recommended Trouton's rule constant.

17. Djerfisherite is a rare potassium-bearing sulfide mineral found in enstatite chondrite and enstatite achondrite (aubrite) meteorites. It is occasionally found in terrestrial kimberlites and alkaline rocks. Estimate S_{298}^o for djerfisherite ($K_6NaFe_{24}S_{26}Cl$) using a Neumann-Kopp approach.

18. It has been claimed that technological advances give 80% efficiency for a heat engine operating between 1350 F and 150 F. Are these claims true or false?

19. The Ca^{2+} ion is the major cation in river water. Calculate its entropy at 298 K from $S_{298}^o = 83.39$ J mol^{-1} K^{-1} for $Ca(OH)_2$ (portlandite), $S_{298}^o = -10.9$ J mol^{-1} K^{-1} for OH^- (aq), the standard integral molar enthalpy ($\Delta_{soln}H_m^o = -18.24$ kJ mol^{-1}), and the standard integral molar Gibbs free energy of solution ($\Delta_{soln}G_m^o = +29.71$ kJ mol^{-1}) for $Ca(OH)_2$ (portlandite) dissolution in water (Cox et al., 1989). This reaction is

$$\mathrm{Ca(OH)_2(portlandite)} = \mathrm{Ca^{2+}(aq)} + 2\,\mathrm{OH^-(aq)}$$

20. Calculate the configurational entropy due to (a) mixing of Al and Si atoms on one site in high albite (analbite, $NaAlSi_3O_8$) and (b) mixing of Al and Si atoms on one site in high leucite ($KAlSi_2O_6$).

CHAPTER

7 Phase Equilibria of Pure Materials

Ice-nine was the last gift Felix Hoenikker created for mankind before going to his just reward. He had made a chip of ice-nine. It was blue-white. It had a melting point of one-hundred-fourteen-point-four-degrees Fahrenheit.

—Kurt Vonnegut, *Cats Cradle* (1963)

The Gibbs free energy (G) was introduced and defined in Chapter 6. In this chapter we use the Gibbs free energy to discuss *phase equilibria* of pure materials, which are changes in state between different forms of a pure material. Ice melting to water, water boiling to steam, graphite transforming to diamond, aragonite ($CaCO_3$) changing to calcite ($CaCO_3$), or dry ice (CO_2) subliming to CO_2 (gas) are all examples of phase equilibria. This chapter also shows how to represent phase equilibria using phase diagrams. These concepts and the other new material introduced in this chapter are used throughout the rest of the book.

This chapter is divided into five sections. The first defines the basic concepts we need to discuss phase equilibria. Metastable phases and methods to determine phase transition temperatures from the properties of metastable phases are also described. Section II discusses phase equilibria and the Clapeyron equation. Topics covered in this section include graphical representation of phase equilibria, derivations of the Clapeyron and Clausius-Clapeyron equations, applications of the Clapeyron equation to different types of phase equilibria at ambient pressure and relatively low pressures, and the Gibbs phase rule. Section III describes the effects of total applied pressure on vapor pressure of solids and liquids. Section IV describes Ehrenfest's classification of phase transitions and lambda transitions. Section V discusses melting curves at high pressures, empirical melting laws, computation of phase boundaries using the Gibbs free energy, retrieval of Gibbs free energy data from phase boundaries, and polymorphic phase transitions. Other topics related to phase equilibria occur in subsequent chapters. Thermodynamic properties of gases near their critical points are in Chapter 8, high-pressure phase equilibria of minerals are described in Chapter 10, and binary phase equilibria are discussed in Chapter 12.

I. BASIC CONCEPTS AND DEFINITIONS

A. Phases

We give several examples of phases before giving a thermodynamic definition because the definition is easier to understand with the examples in mind.

Elements

Graphite and diamond are two phases of the element carbon. Liquid oxygen and O_2 (gas) are two phases of the element oxygen. Orthorhombic sulfur, monoclinic sulfur, molten sulfur, and sulfur

Copyright © 2013 Elsevier Inc. All rights reserved.

vapor are four phases of the element sulfur. However, a mixture of O_2 (g) and ozone (O_3) is only one phase. The two gases mix together in all proportions without separating. The properties of the O_2 and ozone gas mixture are uniform throughout. Sulfur vapor contains S_1-S_{10} species in proportions that vary with temperature and the total pressure. The sulfur gases mix together and there is only one gas phase with uniform properties throughout. Carbon vapor, which is even more complex, is also single phase. In general, gases are single phase. Some gas mixtures (e.g., $N_2 + NH_3$, $CH_4 + NH_3$, $Ar + H_2O$, and $He + CO_2$) do separate into different phases, but these effects occur at conditions where the gases have densities like liquids and behave like liquids in many respects (see Section 6.5 of Rowlinson, 1969). The unmixing of gases has practical consequences in geochemistry and planetary science. For example, $He + CO_2$ mixtures are used in some high-pressure liquid chromatography (HPLC) instruments for the analysis and separation of organic compounds. In these cases it is important to avoid phase separation of the $He + CO_2$ gas mixtures (Wells et al., 2003). Gas unmixing is also important for the recovery of oil from depleted oil fields. Phase separation of $CO_2 + H_2O$ mixtures occurs at pressures and temperatures found in Earth's lower crust and upper mantle (e.g., Blencoe et al., 2001) and phase separation of $H_2 + He$ mixtures occurs at conditions found in the interiors of Jupiter and Saturn, the two largest gas giant planets in our solar system (Streett, 1973).

Chemical compounds and minerals

Liquid water, ice, and steam are three phases of H_2O. Quartz, coesite, cristobalite, stishovite, tridymite, and molten silica are six phases of silicon dioxide (SiO_2). Calcite and aragonite are two phases of calcium carbonate ($CaCO_3$). Sphalerite and wurtzite are two phases of zinc sulfide (ZnS). Andalusite, kyanite, and sillimanite are three phases of Al_2SiO_5. Arsenolite and claudetite are two phases of arsenic sesquioxide (As_2O_3). Cinnabar and metacinnabar are two phases of mercury sulfide (HgS).

Mineral solid solutions between two or more end-members are single phase, particularly at higher temperatures and pressures. Some examples of mineral solid solutions that are binary solutions (at least to a first approximation) are olivine (forsterite Mg_2SiO_4–fayalite Fe_2SiO_4), melilite (gehlenite $Ca_2Al_2SiO_7$–åkermanite $Ca_2MgSi_2O_7$), and plagioclase feldspar (anorthite $CaAl_2Si_2O_8$–albite $NaAlSi_3O_8$). However, a number of mineral solid solutions separate into two different phases at lower temperatures. The alkali feldspars (microcline $KAlSi_3O_8$–albite $NaAlSi_3O_8$) and pyroxenes (enstatite $Mg_2Si_2O_6$–diopside $CaMgSi_2O_6$) show this behavior.

The calcination of limestone to make cement involves the chemical reaction

$$CaCO_3\ (s) = CaO\ (lime) + CO_2\ (g) \tag{7-1}$$

which contributes about 3% of global anthropogenic CO_2 emissions. The three phases involved in Eq. (7-1) are CO_2 (g), $CaCO_3$, and lime (CaO). Lime and $CaCO_3$ are both solids, but they are two different phases. A mixture of anorthite ($CaAlSi_2O_8$) and diopside ($CaMgSi_2O_6$), which is an analog for basaltic rocks, contains two solid phases. At sufficiently high temperatures where the two minerals are partially molten, three phases are present (anorthite, diopside, and their melt).

Finally, we can give several examples from everyday life. Coffee with sugar dissolved in it is a single phase. If we add milk to the coffee, we still have a single-phase system. Air is also a single-phase system. An oil-and-vinegar salad dressing has two phases, but there are only two phases even if the oil and vinegar are dispersed into many small droplets.

All the different phases in these examples have several characteristics in common. In each case the different phases have different physical properties. For example, diamond and graphite look different and have different densities and hardness. Water is liquid, ice is solid, and steam is gaseous. The five different solid phases of SiO_2 have different densities and crystal structures. The different solid phases of $CaCO_3$, ZnS, Al_2SiO_5, HgS, and As_2O_3 have different physical properties. Anorthite is a framework silicate, whereas diopside is a chain silicate with different density and optical properties. Their melt has properties different than either mineral.

Second, the different phases are also physically different from one another. Finally, in each case the different phases can be mechanically separated from one another. Vinegar can be separated from the oil in a salad dressing, mineral grains can be separated from a melt, lime can be separated from $CaCO_3$ and CO_2 (g), and liquid oxygen can be separated from O_2 (g). On the other hand, we cannot mechanically separate a mixture O_2 and ozone, or air, into pure gases.

We can now give a thermodynamic definition that can be understood in the context of our examples. A *phase* is a homogeneous, mechanically separable, physically distinct part of a system that is being studied. As our examples illustrate, a phase does not have to be a pure material. A gaseous mixture, a liquid solution, or a solid solution can each be a single phase. However, two different solids, such as lime and calcite, comprise two different phases. But as long as both solids are present, their amount, their grain size, and their shape are irrelevant. A mixture of lime and calcite is a two-phase system without regard to these other factors. Likewise, we have three phases of the element sulfur as long as monoclinic sulfur, molten sulfur, and sulfur vapor are all present, independent of the amount of each phase.

As stated earlier, gases generally mix together and form a single phase. However, many liquids, including oxide and silicate melts, are completely or partially insoluble in one another. Insoluble liquids are also known as *immiscible* liquids. A physical mixture of insoluble liquids is a mixture of different phases. A dramatic illustration of a multiphase mixture of liquids is given by Hildebrand and Scott (1964), who show a photograph of a test tube containing seven different liquids (heptane, aniline, water, perfluorokerosene, phosphorus, gallium, and mercury).

On Earth, immiscibility commonly occurs in three types of igneous systems (McBirney, 1984). Iron-rich tholeiitic magmas can unmix into a felsic, SiO_2-rich melt and a mafic, Fe-rich melt. The Skaergaard intrusion in East Greenland provides examples of this behavior (see Figures 6-25 and 6-26 in McBirney, 1984). Alkaline carbonatite magmas can unmix into a carbonate-rich melt and a silica-rich, alkali-rich melt. The natrocarbonatite lavas erupted by the Oldavai L'Engai volcano in Tanzania, East Africa, are good examples of this behavior. Third, sulfide-rich melts can separate from basaltic magmas. The sulfide ores in the Sudbury region of Ontario, Canada, exemplify this type of immiscibility. Also, the high-Ti, high-Fe mare basalts returned by Apollo 11 exhibit immiscibility between a Fe-rich basaltic melt and a K-rich granitic melt (Roedder, 1979). These examples are for multicomponent systems. We return to the topic of unmixing in Chapter 12 when we consider phase equilibria of binary and more complex systems.

However, liquid crystals, such as those in the display screens of pocket calculators, laptop computers, flat televisions, watches, and postcard thermometers, are examples of pure liquids that form two or more phases. Liquid crystals are also examples of the way thermodynamics makes possible some of the modern technology that we have come to take for granted.

A *liquid crystal* is an anisotropic liquid formed when certain organic compounds melt. Upon further heating the liquid crystal transforms to a normal liquid. These changes in state are illustrated by the behavior of para-azoxyanisole:

$$\text{p-azoxyanisole (solid)} \underset{391.2\text{ K}}{=} C_{14}H_{14}N_2O_3\text{ (nematic)} \underset{408.3\text{ K}}{=} C_{14}H_{14}N_2O_3\text{ (isotropic)} \quad (7\text{-}2)$$

at one atmosphere pressure. Para-azoxyanisole, cholesteryl acetate ($C_{29}H_{48}O_2$), cholesteryl benzoate ($C_{34}H_{50}O_2$), sodium stearate ($C_{17}H_{35}COONa$), and the other organic compounds that form liquid crystals are rod-shaped or plate-shaped molecules about 3–5 nanometers long.

Liquid crystals flow like liquids but have electrical, magnetic, and optical properties that depend on their phase (smectic, cholesteric, or nematic). Different liquid crystals generally exist as one phase, but in some cases liquid crystals change into another liquid crystal phase at an intermediate temperature before their transition to a normal liquid. Smectic (or soapy) liquid crystals are oily and form a series of terraces when they flow over a flat surface. The terrace or step structure forms because the liquid crystal molecules are arranged in layers within which they can move. Nematic (or fibrous) liquid crystals have molecules oriented in one direction but otherwise free floating. Nematic phases formed by optically active molecules form twisted structures that transmit plane polarized light, depending on their molecular orientation. This twisted nematic phase is also called the *cholesteric phase* after cholesteryl acetate and benzoate, which were the first liquid crystals discovered in 1888 by Friedrich Reinitzer.

In liquid crystal displays plane polarized light goes through liquid crystal layers sandwiched between transparent electrically conductive films. After going through the liquid crystals, the light goes through a polarizer at 90° to the first one and a color filter before being displayed on the computer screen. Without an electrical field, molecules in the cholesteric phase twist the polarized light 90°, and it goes through the second crossed polarizer. However, an applied electrical field aligns the molecules parallel to the direction of the field (and the polarized light). No light is transmitted through the second polarizer in this case, and the screen appears black. Millions of liquid crystal pixels are needed to display images on a computer or television screen.

B. Polymorphs, isomorphs, and pseudomorphs

A few pages earlier we said that diamond and graphite are two phases of the element carbon and that calcite and aragonite are two phases of calcium carbonate. The two minerals in each pair are also polymorphs. A *polymorph* is a particular crystalline form of an element or pure compound that crystallizes in different forms. Different polymorphs have different chemical and physical properties but yield the same gas, liquid, or solid upon a change in state. Calcite (rhombohedral) and aragonite (orthorhombic) are polymorphs of calcium carbonate. They both yield lime and CO_2 upon heating. Table 7-1 lists selected polymorphic phases and some of their properties.

It is important to distinguish the term *polymorph* from the terms *isomorph* and *pseudomorph*, which are terms used in mineralogy but have completely different meanings. The German chemist and mineralogist Eilhardt Misterlich (1794–1863) discovered isomorphism and polymorphism (1818–1822). *Isomorphs* are different crystalline substances with essentially identical crystalline forms. For example, calcite ($CaCO_3$) is isomorphic with rhodochrosite ($MnCO_3$) because both minerals are rhombohedral. Several orthorhombic carbonates and sulfates also form isomorphic pairs: barite ($BaSO_4$)

Table 7-1 Polymorphs of Selected Elements and Minerals

Formula	Name	Structure	V_m (cm^3)	S^o_{298} (J)	C^o_P (J)	$\Delta_f H^o_{298}$ (kJ)
C	diamond	cubic	3.417	2.377	6.109	1.895
	graphite	hexagonal	5.299	5.732	8.517	0
S	orthorhombic		15.53	32.056	22.750	0
	monoclinic		16.50	33.028	23.225	0.360
Sn[a]	gray	cubic	20.592	42.204	25.656	−2.25
	white	tetragonal	16.295	50.070	26.694	0
Al_2SiO_5	andalusite	rhombic	51.54	91.4	122.60	−2589.9
	sillimanite	rhombic	49.91	95.4	123.72	−2586.1
	kyanite	triclinic	44.11	82.8	121.58	−2593.8
$CaCO_3$	calcite	trigonal	36.94	91.713	83.471	−1207.621
	aragonite	orthorhombic	34.16	87.990	82.320	−1207.747
FeS_2	pyrite	cubic	23.94	52.90	62.17	−171.5
	marcasite	orthorhombic	24.58	53.90	62.43	−169.5
HgS	cinnabar	hexagonal	28.419	82.42	48.10	−53.35
	metacinnabar	cubic	30.173	88.85	48.54	−49.38
ZnS	sphalerite	cubic	23.834	58.655	45.756	−204.13
	wurtzite	hexagonal	23.85	58.844	45.878	−194.81
$MgSiO_3$	enstatite	monoclinic	3.209	67.86	82.12	−1545.0
	ilmenite	rhombohedral	3.808	60.40	102.64	−1486.6
	perovskite	orthorhombic	4.098	63.60	79.88	−1445.1
SiO_2	α-tridymite	hexagonal	26.53	45.12	44.25	−906.9
	α-cristobalite	cubic	25.74	46.06	44.30	−906.0
	α-quartz	trigonal	22.69	41.46	44.59	−910.7
	coesite	monoclinic	20.64	40.3	44.91	−907.6
	stishovite	tetragonal	14.016	29.50	42.16	−864.0

[a]*Properties for tin are given at 286.2 K, the gray-white transition temperature, and are from Kaufman (1963) and Hultgren et al. (1973).*

Sources: Density data, Robie et al. (1966); thermodynamic data, Appendix 1.

and witherite ($BaCO_3$), celestite ($SrSO_4$) and strontianite ($SrCO_3$), anglesite ($PbSO_4$) and cérussite ($PbCO_3$), and aragonite ($CaCO_3$) and anhydrite ($CaSO_4$). These isomorphic pairs are not polymorphs. Minerals that are isomorphic and polymorphic are *isodimorphic*. Pyrite is isodimorphic because it is isomorphic with pyrite group minerals (hauerite MnS_2, sperrylite $PtAs_2$, smaltite $CoAs_2$) and polymorphic with marcasite (orthorhombic FeS_2). Marcasite is isodimorphic because it is isomorphic with marcasite group minerals (loellingite $FeAs_2$, rammelsbergite $NiAs_2$, arsenopyrite FeAsS) and polymorphic with pyrite. A *pseudomorph* is a newly formed mineral with the crystal structure of the mineral that it replaced. Limonite ($2Fe_2O_3 \cdot 3H_2O$) in the form of pyrite (cubic FeS_2), and malachite (monoclinic $CuCO_3 \cdot Cu(OH)_2$) in the form of cuprite (octahedral Cu_2O) are two examples of pseudomorphs.

C. Allotropes

The Swedish chemist J. J. Berzelius (1779–1848) coined the word *allotrope* in 1841 to describe the red and yellow forms of mercuric iodide (HgI_2; used as a topical antiseptic) and orthorhombic and monoclinic sulfur. The word *allotrope* literally means "other turn" or "other behavior" (Jensen, 2006), but the common usage is a different form of an element that exists in several forms. All polymorphic forms of elements are allotropes, but not all allotropes are polymorphs. For example, O, O_2, and O_3 gases are three allotropes of oxygen, but they are not separate phases. However, O_2 gas, liquid O_2, and the three different crystalline forms of solid O_2 are five different phases of oxygen and are also five different allotropic forms of oxygen.

D. Enantiomorphs and enantiomers

An *enantiomorph* (from *enantios*, the Greek word for opposite, and *morphos*, the Greek word for shape) is one of a pair of objects that are nonsuperposable mirror images of each other. Enantiomorphic crystals are optically active and rotate plane polarized light either clockwise or counterclockwise. Quartz crystals are enantiomorphs. The dextrorotatory (d-) crystals are right-handed and rotate light clockwise. The levorotatory (l-) crystals are left-handed and rotate light counterclockwise. However, the enantiomorphic forms of quartz are not separate phases. The optical activity of quartz crystals is due to their crystal structure, in contrast to the optical activity of organic compounds, which is due to one or more asymmetric carbon atoms in a molecule.

An *enantiomer* (from *enantios*, the Greek word for opposite, and *meros*, the Greek word for part) is an optically active molecule that rotates plane polarized light either clockwise (dextrorotatory, d-) or counterclockwise (levorotatory, l-). All the proteins in our bodies are levorotatory enantiomers that rotate light counterclockwise. We would starve to death if we were fed food composed of dextrorotatory proteins. A *racemic mixture* is an equimolar mixture of d- and l-enantiomers. Optical isomers (i.e., the right-and left-handed forms) and geometric isomers (e.g., cis- and trans- forms, see Figure 5-1) of a compound are not separate phases of the compound.

E. Enantiotropes and monotropes

An *enantiotrope* is a polymorph that undergoes a reversible transformation into another polymorph at atmospheric pressure. All polymorphs are stable over some temperature range at atmospheric pressure, and their transformation into one another is reversible. Enantiotropy exists in systems where the transition temperature between the two solids is below the melting point of both polymorphs. Heating converts the lower-temperature polymorph into the higher-temperature one, which melts at a higher temperature. Monoclinic and orthorhombic sulfur are enantiotropes with a transition temperature of 368.54 K. This is below the melting point of monoclinic sulfur (388.36 K). Thus, heating orthorhombic sulfur at ambient pressure leads to the two reactions

$$\text{sulfur (orthorhombic)} = \text{sulfur (monoclinic)} \qquad T = 368.54\ \text{K} \qquad P = 1\ \text{bar} \qquad (7\text{-}3)$$

$$\text{sulfur (monoclinic)} = \text{sulfur (liquid)} \qquad T = 388.36\ \text{K} \qquad P = 1\ \text{bar} \qquad (7\text{-}4)$$

A *monotrope* is a polymorph that does not have a reversible transformation into another polymorph at atmospheric pressure. One polymorph is always stable and the other is metastable. Monotropy exists in systems where the transition temperature between solid polymorphs is above the melting point of the lower-temperature polymorph. In many cases, upon heating the metastable monotrope melts, then resolidifies to the stable monotrope, which melts at a higher temperature. Iodine monochloride (ICl), which is a topical anti-infective medicine, has two monotropes that melt at 287.0 K and 300.4 K; the latter is the stable form. Benzophenone [$(C_6H_5)_2CO$], which is used to make antihistamines, has two monotropes that melt at 299.6 K and 321.6 K; the latter is the stable form. The concepts of enantiotropy and monotropy have practical applications in the pharmaceutical industry because the more stable of the two polymorphic forms of a solid compound is preferable for use as a drug.

F. Phase transitions and phase equilibria

A *phase transition* is a change in state from one phase to another. The defining characteristic of a phase transition is the abrupt change in one or more physical properties with an infinitesimal change in temperature. Dry ice subliming to CO_2 (g), ice melting to water, water boiling to steam, orthorhombic sulfur transforming to monoclinic sulfur, calcite transforming to aragonite, and graphite transforming to diamond are all examples of phase transitions. Each of these phase transitions can be represented by a reaction. For example, Eq. (7-3) represents the transition of orthorhombic to monoclinic sulfur at 368.54 K at one bar total pressure. Likewise, Eq. (7-4) represents monoclinic sulfur melting to liquid sulfur at 388.36 K at one bar total pressure.

Water boils to steam at 100°C at one atmosphere pressure. (We use a pressure of one atmosphere because the normal boiling point is defined as the temperature at which a liquid boils to vapor with one atmosphere pressure.) However, if we are at high altitudes, water boils at a lower temperature because atmospheric pressure is lower.

In general, the temperature at which a phase transition occurs depends on the total pressure. We could rephrase this to say that the pressure at which a phase transition occurs depends on the temperature. The two statements are equivalent. The set of *P*-*T* points at which a phase transition occurs can thus be represented by a curve on a plot of temperature versus pressure. Figure 7-1 shows an example for water boiling to steam, or more scientifically, for the *vaporization curve* of liquid water. The curve in Figure 7-1 extends from 0°C (where air-saturated liquid water and ice stably coexist at one atmosphere pressure) to 373.946°C (647.096 K, the *critical point* of water where liquid water and steam merge into one phase). The shape of the vaporization curve of liquid water is characteristic of all vaporization curves: convex toward the temperature axis. Figure 7-2 shows the curve for dry ice (solid CO_2) subliming to CO_2 (g). A comparison of Figures 7-1 and 7-2 shows that the vaporization curve for

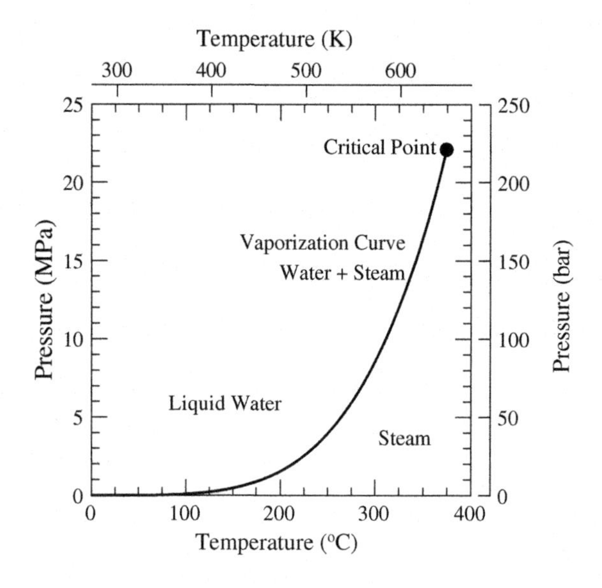

FIGURE 7-1

The vaporization curve for liquid water from 0°C to 374°C.

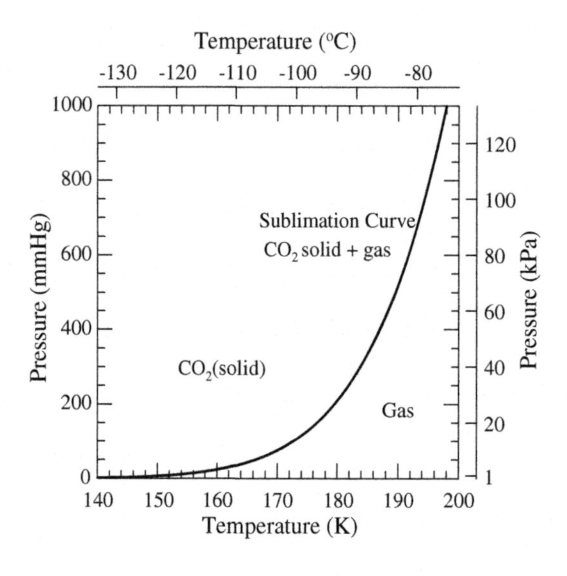

FIGURE 7-2

The sublimation curve for solid CO_2 (dry ice) from 140 K to 200 K.

water and the *sublimation curve* for CO_2 have the same characteristic shape. The vaporization curve for a liquid (e.g., water) and the sublimation curve for a solid (e.g., solid CO_2) are the same as the vapor pressure curves for the liquid and solid.

Equations (7-1) to (7-4) are also examples of *phase equilibria* in which the different phases stably coexist with one another. Thus, orthorhombic and monoclinic sulfur stably coexist at one bar pressure if we maintain the temperature at 368.54 K. *Equilibrium between two or more phases is independent of the amount, size, and shape of the phases, except for very small droplets or particles, for which the surface energy becomes important.* For example, at 273 K the vapor pressure of water droplets with radii of 0.1, 0.01, and 0.001 μm are 1.012, 1.128, and 3.326 times higher, respectively, than that over pure water (Dufour and Defay, 1963). These effects are important for the thermodynamics of clouds in planetary atmospheres.

G. Metastable phases

A *metastable* phase would not be present in a system at equilibrium under the existing conditions (*P*, *T*, composition), and it is present only because the system is not approaching equilibrium at an observable rate. The monoclinic sulfur found in fumaroles at Vesuvius and Vulcano, diamonds, aragonite, and marcasite are examples of metastable minerals (Palache et al., 1944).

Supercooled water is another common example. In 1724 Fahrenheit discovered that liquid water can be cooled below 0°C without crystallizing to ice. Supercooled water coexists in equilibrium with water vapor and is metastable to about −40°C (at ~1 bar pressure) as long as it is free of dust, ice crystals, and other particles. However, if a tiny piece of ice is added to the water, all of it immediately freezes. Below about −40°C supercooled water ceases to be metastable and becomes unstable with respect to ice. When water is cooled to this temperature, freezing spontaneously occurs without the addition of ice or dust. The lower temperature limit for supercooled water depends on pressure and is about −92°C at about 2100 bars pressure.

Figure 7-3 shows the water vapor pressure over water, supercooled water, and ice. The vapor pressure curve for supercooled water is a continuous, smooth extension of the vapor pressure (or vaporization) curve for liquid water (e.g., see Figure 7-1) to temperatures below the triple point of water (0.01°C). In general, the vapor pressure curves for supercooled liquids are continuous, smooth extensions of the vaporization curves of the stable liquids.

The *normal freezing point* (0°C) of water is the freezing point of air-saturated water at one atmosphere total pressure. The freezing point of pure water (without any dissolved air) is 0.0024°C higher (Dorsey, 1940; Waring, 1943). In other words, 0°C is the freezing point for a solution of air dissolved in water and the *freezing point depression* for air dissolved in water is −0.0024°C. As we discuss later in this chapter, pressure changes the freezing/melting points of materials. In the case of water, a pressure of one atmosphere decreases its freezing/melting point by about 0.0075°C from its true value. Pure, air-free water freezes (or ice melts) at 0.0099°C (= 0.0024°C + 0.0075°C). This temperature is the triple-point temperature, which was defined as 0.01°C in 1954 (see Section II-C of Chapter 2). The *triple point* of water is the temperature at which the three phases of ice, liquid water, and water vapor are in equilibrium at a total pressure equal to the water vapor pressure (611.657 Pa = 4.588 mmHg = 6.1166×10^{-3} bars).

The vapor pressure curve for water (and supercooled water) intersects the vapor pressure curve for ice at the triple point (0.01°C). This intersection is emphasized by extrapolating the vapor pressure

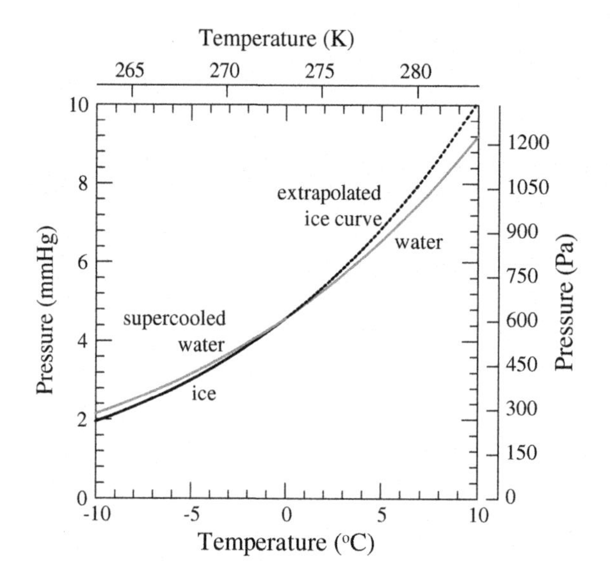

FIGURE 7-3

The vapor pressure curves for water, supercooled water, and ice in the vicinity of the triple point (0.01°C, 273.16 K).

curve for ice above 0.01°C (shown as the dashed line in Figure 7-3). Liquids such as water can be supercooled below their freezing points and superheated above their boiling points. In contrast, solids such as ice cannot be superheated above their melting points. Thus, the vapor pressure over superheated ice is a mathematical extrapolation of the vapor pressure curve for ice, but it does not represent measured data points. (However, in many cases metastable solids can exist inside the stability field of another stable solid polymorph for long or short periods, and we discuss several such examples in this chapter.)

Figure 7-3 shows that the vapor pressure over supercooled water is larger than that over ice. This is a general characteristic of metastable phases they have larger vapor pressures than the corresponding stable phases. The difference between the vapor pressure of supercooled water and ice is initially very small, near the triple point, but it becomes larger with decreasing temperature. The difference in the vapor pressures of supercooled water and ice is related to the difference in their Gibbs free energies (ΔG_T) via the equation

$$\Delta G_T = -RT\ln(P_{meta}/P_{stab}) \tag{7-5}$$

where T is the absolute temperature (K), R is the ideal gas constant, and P_{meta} and P_{stab} are the vapor pressures over the metastable (supercooled water) and the stable (ice) phases. At the transition temperature between the two phases, the triple point of water in this case, the two phases are in equilibrium, their vapor pressures are identical, their Gibbs energies are identical, and the Gibbs energy difference between them is zero. At lower temperatures ice is more stable.

Example 7-1. Some vapor pressure data for ice and supercooled liquid water are tabulated here. Use these data to compute ΔG between the two phases as a function of temperature.

Temp (°C)	0.01	−5	−10	−15
Temp (K)	273.16	268.15	263.15	258.15
P_w (mmHg)	4.588	3.163	2.149	1.436
P_i (mmHg)	4.588	3.013	1.950	1.241
ΔG (J mol^{-1})	0	−108.3	−212.6	−313.2

We first convert from the Celsius to the Kelvin temperature scale and then compute ΔG using Eq. (7-5). Our results (rounded to the nearest 0.1 J mol^{-1}) are in the last row of the preceding table. A least-squares fit to our results gives the equation

$$\Delta G_T = -424.864 - 18.877T + 0.0748T^2 \text{ J mol}^{-1} \quad (7\text{-}6)$$

for the ΔG between supercooled water (less stable) and ice (more stable) from 258 K to 273 K.

A metastable phase also has a higher solubility than the corresponding stable phase. The difference in the solubilities of a metastable phase and a stable phase are related to the difference in their Gibbs free energies (ΔG_T) via the equation

$$\Delta G_T = -RT\frac{1}{n}\ln(x_{meta}/x_{stab}) \quad (7\text{-}7)$$

where x_{meta} and x_{stab} are the solubilities of the metastable and stable phases, respectively, n is the number of atoms in a homonuclear molecule such as I_2 ($n = 2$), P_4 ($n = 4$), and S_8 ($n = 8$), and the other symbols are the same as before. At the transition temperature between the two phases, they coexist in equilibrium, their solubilities are identical, their Gibbs energies are identical, and the Gibbs energy difference between them is zero. Equation (7-7) can be applied when the rate of transition of the metastable phase to the stable phase is slow enough that the metastable phase can equilibrate with the solution before it is transformed to the stable polymorph. Jamieson (1953) used the different solubilities of calcite and aragonite in aqueous solution at high pressures to determine the ΔG between them and to locate the calcite-aragonite phase boundary.

Example 7-2. Brønsted (1906) measured the solubility of prismatic (monoclinic) and octahedral (orthorhombic) sulfur in different organic solvents as a function of temperature. He found both polymorphs dissolve as S_8 molecules and that monoclinic sulfur is 1.276 times more soluble than orthorhombic sulfur at 25.3°C (298.45 K). Thus, the Gibbs free energy difference between the two polymorphs (per mole of S) is

$$\Delta G_T = -(8.314)(298.45)\frac{1}{8}\ln(1.276) = -75.6 \text{ J mol}^{-1} \quad (7\text{-}8)$$

Brønsted's data agree with thermal data (see Chapter 9), showing that orthorhombic sulfur is more stable than monoclinic sulfur by 72.2 J mol^{-1} at 298.15 K.

Finally, a metastable phase has a lower melting point than the corresponding stable phase. We gave several examples of this behavior in Section E above while discussing monotropes. Menthol ($C_{10}H_{20}O$), which smells and tastes like peppermint and is used in the production of cigarettes, cough drops, nasal inhalers, and perfumes, has several metastable polymorphs with lower melting points (δ 31.5°C, γ 33.5–34°C, β 35.5°C) than the stable phase (α 42.5°C).

H. Components

A *component* is a chemically distinct, independently variable constituent used to write algebraic equations describing the chemical composition of all phases involved in phase equilibria in a system. Often either elements or compounds can be chosen as components, and the alternative giving the minimum number of components is usually chosen.

For example, a three-phase mixture of liquid water, ice, and water vapor at the triple point is a system made of one component (H_2O) because the chemical composition of all three phases is H_2O. The algebraic equation giving the composition of each phase is simply

$$H_2O = 1 \times H_2O \tag{7-9}$$

We could also use the elements hydrogen and oxygen as components, but this choice gives two components instead of one. Furthermore, neither H_2 nor O_2 is involved in phase equilibria with ice, water, or water vapor at 273.16 K. However, if we were working at a much higher temperature where thermal dissociation of water vapor is significant, then H_2 and O_2 could be used as components because they are involved in the equilibria taking place. In general, the choice of components for a single-component system such as water, sulfur, silica, or iron is self-evident.

However, the choice of components for binary, ternary, quaternary, and higher-order systems is more difficult. Reaction (7-1) involves the three phases CO_2 (g), $CaCO_3$ (s), and CaO (lime). We can choose the components of this system in several different ways. We could use the elements C, Ca, and O, but none of these elements is involved in phase equilibria with CO_2 (g), $CaCO_3$ (s), and CaO (lime) until temperatures much higher than those at which Eq. (7-1) is important industrially or geologically. Looking at Eq. (7-1) it seems logical to pick CaO (lime) and CO_2 (g) as the two components, which is often done. In this case the three algebraic equations

$$CaO = 1 \times CaO + 0 \times CO_2 \tag{7-10}$$

$$CO_2 = 0 \times CaO + 1 \times CO_2 \tag{7-11}$$

$$CaCO_3 = 1 \times CaO + 1 \times CO_2 \tag{7-12}$$

describe the chemical compositions of the three phases in the CaO-CO_2 system. However, we could also choose CaO and $CaCO_3$ as the two components. In this case we use the equations

$$CaO = 1 \times CaO + 0 \times CaCO_3 \tag{7-13}$$

$$CO_2 = -1 \times CaO + 1 \times CaCO_3 \tag{7-14}$$

$$CaCO_3 = 0 \times CaO + 1 \times CaCO_3 \tag{7-15}$$

to describe the chemical compositions of the three phases in the CaO-$CaCO_3$ system. Finally, we could choose CO_2 and $CaCO_3$ as the two components, which yields

$$CaO = -1 \times CO_2 + 1 \times CaCO_3 \tag{7-16}$$

$$CO_2 = 1 \times CO_2 + 0 \times CaCO_3 \tag{7-17}$$

$$CaCO_3 = 0 \times CO_2 + 1 \times CaCO_3 \quad (7\text{-}18)$$

to describe the chemical compositions of the three phases in the CO_2-$CaCO_3$ system.

As mentioned, CaO (lime) and CO_2 (g) are generally picked as the two components. This is done to avoid negative components that are unnecessary to describe this simple two-component system. But in some cases, such as metamorphic reactions involving complex minerals such as micas and amphiboles, it is advantageous to use negative components to describe mineral compositions. These applications are discussed by Korzhinskii (1959), Thompson (1981), and in several papers in the volume edited by Ferry (1982).

I. Degrees of freedom

The *degrees of freedom* of a system are the number of intensive state variables (e.g., pressure, temperature, concentration) of the components that can be arbitrarily and independently varied without altering the number of phases in the system. For example, there is only one degree of freedom at any *P-T* point along the vaporization curve of water in Figure 7-1. If we are originally on the vaporization curve of water and increase the temperature of the system (liquid water + steam) from 400 K to 450 K, more water vaporizes to steam and the system moves back onto the vaporization curve at a higher temperature and higher vapor pressure of steam. Likewise, if we are originally on the water vaporization curve at a pressure of 150 bars and decrease the pressure of the system to 50 bars, more steam condenses to water and the system moves back onto the vaporization curve at a lower temperature and lower vapor pressure of steam. In each case we independently varied one intensive parameter (*T* or *P*) and the other intensive parameter responded to maintain phase equilibrium.

However, if we are originally on the vaporization curve and change both the temperature and pressure of the system (e.g., decrease *T* and increase *P*), we move off the vaporization curve and change the number of phases in the system (from water + steam to steam in this example).

II. PHASE EQUILIBRIA, THE CLAPEYRON EQUATION, AND THE CLAUSIUS-CLAPEYRON EQUATION

A. Phase diagram for water

Figure 7-4 shows the *P-T* phase diagram for water at relatively low pressures (< 250 MPa). The triple point is the *only P-T* point where the three phases (ice I + water + water vapor) coexist in equilibrium (see Section I-G). The triple-point temperature is defined as 273.16 K and the triple-point pressure is 611.657 Pa. Ice I is the ice that forms in the freezer in your kitchen and outside during winter. It has a hexagonal crystal structure. Other polymorphic forms of ice with different crystal structures are stable at higher pressures and are designated by roman numerals (ice II, ice III, ice V). Ice I and liquid water coexist in equilibrium along the melting curve

$$H_2O \text{ (ice I)} = H_2O \text{ (water)} \quad (7\text{-}19)$$

The Gibbs free energy change for Eq. (7-19) is zero along the melting curve. Hence,

$$G_{ice} = G_{water} \tag{7-20}$$

along this curve. The low-pressure end of the ice I melting curve is the solid-liquid-vapor (*S-L-V*) triple point (273.16 K, 611.657 Pa). The high-pressure end of the ice I melting curve is the ice I–ice III–liquid triple point (251.165 K, 209.9 MPa). Ice III and the other high-pressure forms of water ice are shown in the high-pressure phase diagram for water in Figure 7-5. P. W. Bridgman (1882–1961) discovered many of the high-pressure phases of water ice and determined the high-pressure phase diagram of water. Bridgman (see sidebar) pioneered many areas of high-pressure research in

PERCY WILLIAMS BRIDGMAN (1882–1961)

Percy Bridgman was an American scientist who was born in Cambridge, Massachusetts, where he spent most of his life. He was educated at Harvard and joined the Harvard faculty as a professor in the physics department immediately after getting his PhD in 1905. He held this position until his death in 1961.

Bridgman pioneered high-pressure research. Prior to his work, the French physicist Emile Amagat had studied compressibility of gases at pressures up to 3000 atmospheres. Bridgman greatly extended this range of pressures by developing a seal that became tighter under pressure, ultimately allowing him to reach pressures greater than 100 kilobars. He also developed new methods of measuring high pressures. Over a 50-year period, Bridgman then proceeded to study the effects of pressure on many chemical and physical properties of different gases, liquids, and solids. He determined the high-pressure phase diagram of water and discovered several new crystallographic forms of ice that are only stable at high pressures. This is probably Bridgman's most famous work and led the writer Kurt Vonnegut to include a fictional high-pressure ice called ice-nine in his 1963 novel *Cats Cradle*. Subsequently, ice IX was discovered, and fortunately it does not have the properties Vonnegut ascribed to it.

In addition to his high-pressure research, Bridgman developed a means of growing high-purity single crystals. His method is used to grow gallium arsenide (GaAs) and other semiconductors formed by elements in groups III and V of the periodic table. Bridgman also wrote prolifically on the philosophy of physics, mostly in support of the idea that the field of physics should concern itself only with what is measurable.

Bridgman was an independent perfectionist who seldom collaborated with others and who had few students, yet he was well respected by his colleagues for the quality of his work. He was honored with the Nobel Prize for Physics in 1946. He later developed an incurable cancer so painful that it led him to commit suicide. He viewed this action as a form of euthanasia.

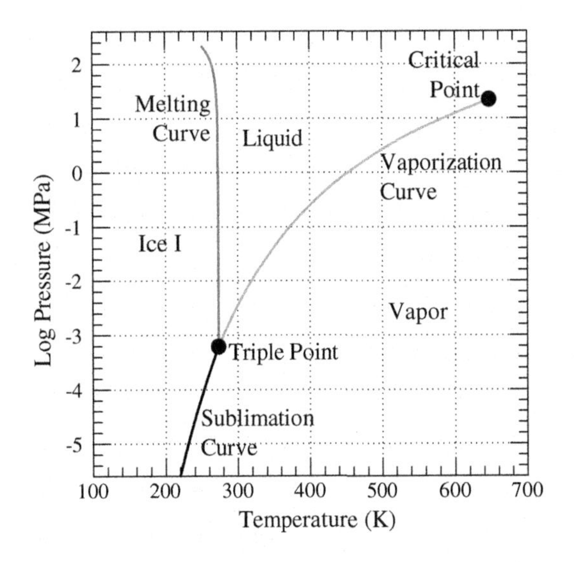

FIGURE 7-4

The *P-T* phase diagram for H_2O at pressures less than 250 MPa.

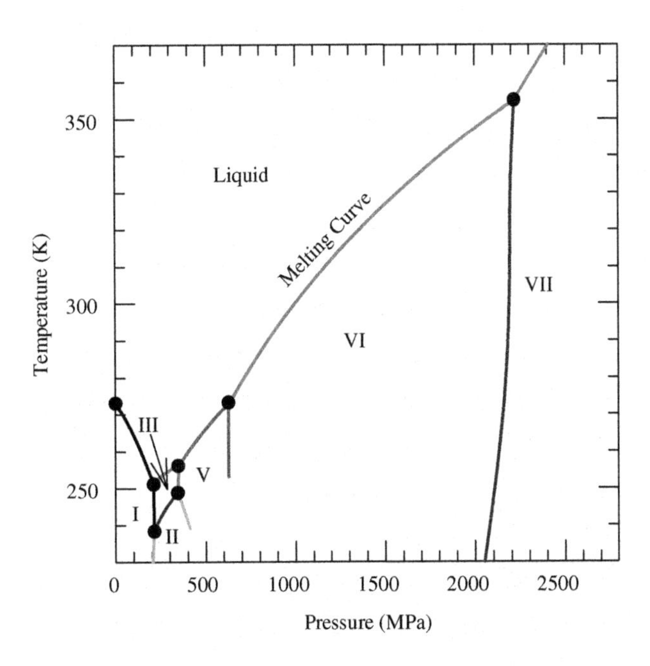

FIGURE 7-5

Part of the *P-T* phase diagram for H_2O at higher pressures. All ice phases along the melting curve are shown.

chemistry, geology, and physics and won the 1946 Nobel Prize in Physics for his research. However, to his dismay, he was never able to convert graphite into diamond and called graphite "nature's best spring." As we discuss later, this modern alchemy was achieved by a research team at the General Electric Company in 1954.

Liquid water and vapor coexist in equilibrium along the vaporization curve

$$H_2O\ (\text{water}) = H_2O\ (\text{vapor}) \quad (7\text{-}21)$$

$$G_{water} = G_{vapor} \quad (7\text{-}22)$$

The low-pressure end of the vaporization curve is the *S-L-V* triple point. The high-pressure end of the vaporization curve is the critical point of water (647.096 K, 22.064 MPa). As far as is known, all vaporization curves for pure one-component systems (e.g., H_2O, CO_2, SO_2, Ar, S, As, Na, Hg) extend between the *S-L-V* triple point and the critical point (where the distinction between vapor and liquid disappears). However, the critical temperatures (T_{crit}) and pressures (P_{crit}) have not been measured for elements with normal boiling points (b.p.) higher than that of cesium (b.p. 944 K, T_{crit} 1938 K, P_{crit} 9.4 MPa). Lithium, Na, and K apparently have critical temperatures higher than Cs. However, the critical parameters for Li (T_{crit} 3223 K, P_{crit} 67 MPa), Na (T_{crit} 2573 K, P_{crit} 35 MPa), and K (T_{crit} 2223 K, P_{crit} 16 MPa) are based on extrapolations of *PVT* data (e.g., see Ambrose, 1980) and have not been measured directly.

Melting curves are different than vaporization curves because there is no known *S-L* critical point. The melting curves of several gases (e.g., H_2, He, Ar, N_2, CO_2, CH_4, NH_3) have been extended far beyond the temperature and pressure of the *L-V* critical point, but no *S-L* critical point has been observed. For example, the melting curve of Ar has been measured to 3300 K and 80 GPa without evidence for a *S-L* critical point (Boehler et al., 2001). These *P-T* conditions are far beyond those of the *L-V* critical point at 150.8 K and 4.87 MPa (Ambrose, 1980).

Ice I and water vapor coexist in equilibrium along the sublimation curve where we have

$$H_2O\ (\text{ice}) = H_2O\ (\text{vapor}) \quad (7\text{-}23)$$

$$G_{ice} = G_{vapor} \quad (7\text{-}24)$$

In general, the high-pressure end of the sublimation curve is the *S-L-V* triple point. In some cases, such as the ices of HCl, CO, and N_2, the low-pressure end of the sublimation curve is a triple point with another solid polymorph that is stable at lower temperatures (S_α-S_β-V). The sublimation curve of O_2 ice is interrupted by two different triple points of this type. However, this is not the case for water and the low-pressure end of the sublimation curve is absolute zero.

B. Derivation of the Clapeyron equation

The Clapeyron equation is useful for calculating *P-T* curves such as those in Figure 7-4. It was first derived by the French engineer Émile Clapeyron (see sidebar) in his only paper about thermodynamics ("Memoir on the Motive Power of Heat"). This paper is notable because of the Clapeyron equation

ÉMILE CLAPEYRON (1799–1864)

Émile Clapeyron was a French engineer born during tumultuous times. He studied at the *École polytechnique*, an institution known both as a great place of learning and as an enclave of leftist politics. After the defeat of Napoleon, Clapeyron and several other graduates of the school found it convenient to relocate to Russia. In 1820, Clapeyron and his good friend and collaborator Gabriel Lamé accepted teaching positions at the *École des Travaux de Saint-Petersburg*. It would not be long, however, before the political climate in Russia would change as well. The Decembrists, a group of young, educated Russians who wanted changes in the judicial system and a more democratic government, staged an unsuccessful revolt in December 1825. This act brought greater scrutiny of all academics, particularly foreign ones, who were seen as a source of dangerous ideas. Clapeyron was placed under surveillance and was even briefly exiled, under armed guard, to a remote village. The French Revolution of 1830 made France safe for intellectuals again, and Clapeyron and Lamé returned in 1831.

Clapeyron's specialty was the design of railroads and locomotives. He designed engines to climb especially long, steep grades, and he designed metal bridges. He developed a scientific method of determining where a station should be placed to serve several cities in the most efficient manner. At the time, this interest in the development of railroads was revolutionary. Trains were seen as a potentially dangerous influence because they would provide transportation to the poor. Furthermore, in train cars, the poor might have a chance to encounter the wealthy. In Russia, there was also concern about the negative impact of encouraging tourism. Visitors from the West might bring their liberal notions in with them. Ultimately, however, the military and economic benefits of trains were recognized and the railroads built.

Through his work as an engineer, Clapeyron became interested in the writings of Sadi Carnot, a fellow student of the *École polytechnique*. He might well be the reason Carnot's ideas were perpetuated, because Carnot had been virtually forgotten. Clapeyron gave Carnot's heat engine cycle a mathematical treatment and illustrated the cycle with a Watt diagram. Later, Kelvin and Clausius would use Clapeyron's paper as the basis for their own determinations of the second law of thermodynamics.

Clapeyron, along with Lamé and Carnot, received the honor of having their names engraved among the 72 names of French scientists on the Eiffel Tower. Other scientists on the list who are of interest to students of thermodynamics are Dulong, Gay-Lussac, and Le Chatelier.

but also because it revived Carnot's ideas and presented them in a mathematical form using indicator diagrams. We use the water phase diagram in Figure 7-4 to help explain the derivation of the Clapeyron equation and to illustrate its use.

Imagine that we have a mixture of ice I and water coexisting in equilibrium at some *P-T* point on the melting curve. We can start at any point on the melting curve, but let's pick an initial *P-T* point near the *S-L-V* triple point. The ice I + water mixture could be contained inside a diamond anvil cell (see Figure 2-15), or another type of high-pressure press, that we can heat or cool by placing it inside a constant-temperature bath. The ice I + water mixture is a closed system because the total amount of H_2O inside the diamond anvil cell remains constant. However, the amount of H_2O present as ice I or as liquid water can vary depending on the temperature and pressure of the diamond anvil cell.

We now make small changes in temperature and pressure, which move the ice I + water system along the melting curve away from the initial *P-T* point and allow the system to return to equilibrium. The exact temperature and pressure changes are unimportant. The two key points are that we stay on the melting curve and that we make infinitesimally small changes in temperature (dT) and pressure (dP) so reversible phase changes occur in response to our actions.

The P and T changes cause a variation in the ratio of ice I to water. The Gibbs free energy of each phase also varies because G is an extensive variable that depends on the amount of material. However, the overall change in Gibbs free energy must equal zero because the ice I + water system reequilibrated at another point on the melting curve. Therefore, the change in Gibbs free energy for ice I (dG_I) must equal the change in Gibbs free energy for the water (dG_W):

$$dG_I = dG_W \tag{7-25}$$

We can evaluate Eq. (7-25) by going back to the definition of Gibbs free energy given in Chapter 6 and writing down the differential of G:

$$G = E + PV - TS \tag{6-113}$$

$$dG = dE + PdV + VdP - TdS - SdT \tag{7-26}$$

We simplify Eq. (7-26) using the combined equation for the first and second laws, also known as the *fundamental equation* (Section VII-A, Chapter 6), to substitute for dE:

$$dE = TdS - PdV \tag{6-48}$$

$$dG = (TdS - PdV) + PdV + VdP - TdS - SdT \tag{7-27}$$

$$dG = VdP - SdT \tag{7-28}$$

Equation (7-28) is the *second fundamental equation* and it expresses the dependence of G on P and T for a closed system, that is, with the molar amounts of all components in the system being held constant. The variables P and T are called the *natural variables* for the Gibbs free energy. The natural variables are generally the most convenient choice for solving problems involving the particular thermodynamic function.

We can now rewrite Eq. (7-25) in terms of the volume of each phase, the entropy of each phase, temperature, and pressure by using Eq. (7-28):

$$dG_I = dG_W \tag{7-25}$$

$$V_I dP - S_I dT = V_W dP - S_W dT \tag{7-29}$$

The subscripts I and W stand for ice I and water, as before. Finally, we rearrange Eq. (7-29) to get an equation expressing (dP/dT) (or vice versa) in terms of the properties of the two phases:

$$\left(\frac{dP}{dT}\right)_{eq} = \frac{S_I - S_W}{V_I - V_W} = \frac{\Delta S_{I-W}}{\Delta V_{I-W}} = \frac{\Delta H_{I-W}}{T\Delta V_{I-W}} \tag{7-30}$$

Equation (7-30) is the *Clapeyron equation.* It can be thought of as a quantitative statement of *LeChatelier's principle*, which heads Chapter 10. We can paraphrase LeChatelier's principle as follows: If a system in equilibrium is subjected to a change that modifies the equilibrium, the system responds to counteract the change and return to equilibrium. Equation (7-30) tells us how the ice I + water system, originally in phase equilibrium along the melting curve, responds to a change in temperature or pressure.

In general, the Clapeyron equation applies to all first-order phase transitions such as melting and freezing, sublimation, vaporization, and polymorphic transitions. As we discuss in more detail later, a first-order phase transition has discontinuities in entropy, volume, enthalpy, and heat capacity (i.e., $\Delta_{trans}S \neq 0$, $\Delta_{trans}V \neq 0$, $\Delta_{trans}H \neq 0$, $\Delta_{trans}C_P \neq 0$). The subscript *eq* on the derivative in Eq. (7-30) specifies that the Clapeyron equation applies along equilibrium lines such as melting, sublimation, solid-state phase transition, and vaporization curves. More precisely, (dP/dT) is evaluated with the total Gibbs free energy of the system held constant. In other words, $\Delta G = 0$ for the system, which can only happen if the Gibbs free energies of the two phases change correspondingly, as indicated by Eq. (7-25). We derived Eq. (7-30) for the ice I + water system, but because the Clapeyron equation applies to a range of phase transitions, we can rewrite Eq. (7-30) in a more general form:

$$\left(\frac{dP}{dT}\right)_{eq} = \frac{\Delta S}{\Delta V} = \frac{\Delta H}{T\Delta V} \tag{7-31}$$

C. Water ice melting curve and the Clapeyron equation

Figure 7-6 shows the melting curve of ice I, which was computed using the recommended equation of state for water (Wagner and Pruß, 2002). This is the same as the melting curve shown in the water phase diagram in Figure 7-4. However, the curvature is now apparent because of the much smaller temperature range in Figure 7-6. The curve is compared with data points from W. Thomson (Lord Kelvin, 1850) and Bridgman (1911), which are in excellent agreement with it.

Water is one of a few substances (including Bi, Sb, Ga, Ge) that have a negative (dT/dP) slope along the melting curve. The enthalpy of melting (i.e., $\Delta H = H_{liquid} - H_{solid}$) is positive for all materials. Likewise, the volume of melting (i.e., $\Delta V = V_{liquid} - V_{solid}$) is positive for most but not all substances. Thus, for most materials the melting point increases with increasing pressure. However, liquid water is denser than ice, which floats in it, so the ΔV of melting is negative, making both the

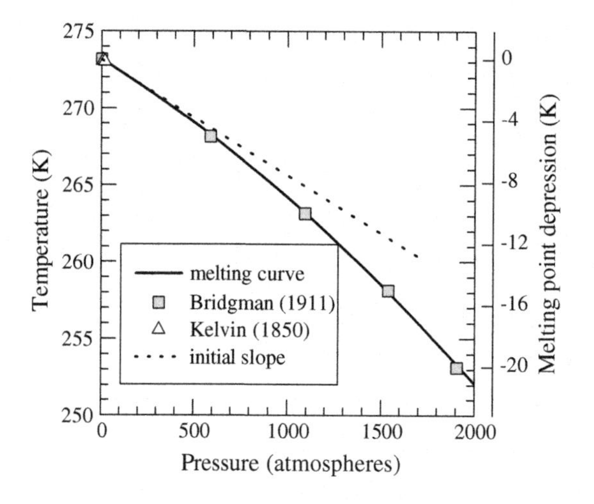

FIGURE 7-6

The Ice I—water melting curve from the *S-L-V* triple point (0.01°C) to the ice I—ice III—water triple point (−22°C).

quotient ($T\Delta V/\Delta H$) and the (dT/dP) slope negative. Figure 7-6 shows that the melting point of ice is decreased under pressure, and we compute this in the following example.

Example 7-3. The melting point of ice is 0.0024°C (273.1524 K) at one atmosphere pressure. Calculate the variation in melting point with increasing pressure and the melting curve of ice I under pressure, assuming that the specific volumes of liquid water (1.00016 $cm^3\ g^{-1}$), ice I (1.09089 $cm^3\ g^{-1}$), and the enthalpy of fusion (333.58 J g^{-1}) remain constant. We are asked to find (dT/dP), so we flip Eq. (7-30) upside down and substitute the numerical values into it, giving

$$\left(\frac{dT}{dP}\right)_{eq} = \frac{T\Delta V_{I-W}}{\Delta H_{I-W}} = \frac{(273.1524\ \text{K})(1.000160 - 1.09089\ \text{cm}^3\text{g}^{-1})}{(333.58\ \text{J g}^{-1})} \tag{7-32}$$

$$\left(\frac{dT}{dP}\right)_{eq} = (-0.074294\ \text{K cm}^3\text{J}^{-1})(0.101325\ \text{J cm}^{-3}\text{atm}^{-1}) \tag{7-33}$$

$$\left(\frac{dT}{dP}\right)_{eq} = -0.00753\ \text{K atm}^{-1} \tag{7-34}$$

The Clapeyron equation predicts that the melting point of ice I decreases by 0.00753 degrees for a pressure increase of one atmosphere. We used the specific volume and enthalpy of fusion in our calculation. However, molar properties can also be used and the answer will be the same.

In 1850, James Thomson predicted that pressure would decrease the melting point of ice, and his brother William Thomson (Lord Kelvin) verified this by measuring the melting point of ice as a function of pressure. Kelvin found that ice melted at −0.059°C at 8.1 atmospheres pressure ($dT/dP = -0.0073$ deg atm^{-1}) and at −0.129°C at 16.8 atmospheres pressure ($dT/dP = -0.0077$ deg atm^{-1}). The equation of the melting curve is found by rearranging and integrating Eq. (7-31):

$$\int_{273.1524}^{T} dT = \int_{1}^{P} \frac{\Delta V}{\Delta S} dP \tag{7-35}$$

Assuming that the specific volumes of ice I and water and the entropy of fusion remain constant, the quotient $\Delta V/\Delta S$ can be moved outside the integral, giving

$$\begin{aligned} T_{melt} &= \frac{\Delta V}{\Delta S}(P - 1) + 273.1524 \\ &= 273.1524 - 0.00753(P - 1) \end{aligned} \tag{7-36}$$

The melting curve calculated from Eq. (7-36) is shown as the dotted line (labeled initial slope) in Figure 7-6. It deviates noticeably from the actual melting curve starting at $P \sim 250$ atmospheres.

The deviations arise from our neglect of the variations with temperature and pressure of the specific volumes of ice I and water, and the enthalpy of melting. Bridgman (1911) measured the melting temperature and ΔV as a function of pressure along the entire ice I—water melting curve. He calculated (dP/dT) from his P-T measurements. Then he calculated ΔH using Eq. (7-31) and his data for ΔV. Finally, he computed ΔE from ΔH and $P\Delta V$ using Eq. (5-5). Bridgman's results are summarized in Table 7-2 and displayed in Figure 7-7. The ΔV for ice melting to water increases with increasing pressure because ice has a lower density than water and is more easily compressed than water. The increased pressure and larger ΔV also increase $P\Delta V$ along the melting curve. However, the ΔH of melting decreases with increasing pressure. The net effect is that (dP/dT) gets smaller, as shown in Figure 7-7. Exactly the same effect is displayed in Figure 7-6, where the (dT/dP) slope gets steeper (larger) with increasing pressure.

Table 7-2 Ice I—Water Melting Curve Parameters

T (°C)	P (bars)	ΔV (cm^3 g^{-1})	(dP/dT)	$P\Delta V$ (cal g^{-1})	ΔH (cal g^{-1})	ΔE (cal g^{-1})
0	0	−0.0900	135.8	−0.00	79.8	79.8
−5	622	−0.1016	113.3	−1.45	73.7	75.1
−10	1152	−0.1122	96.5	−2.96	68.0	71.0
−15	1621	−0.1218	83.1	−4.52	62.5	67.0
−20	2009	−0.1313	72.6	−6.06	57.7	63.6
−22[a]	2157	−0.1352	70.0	−6.7	56.1	62.8

[a]*Ice I—ice III—water triple point.*

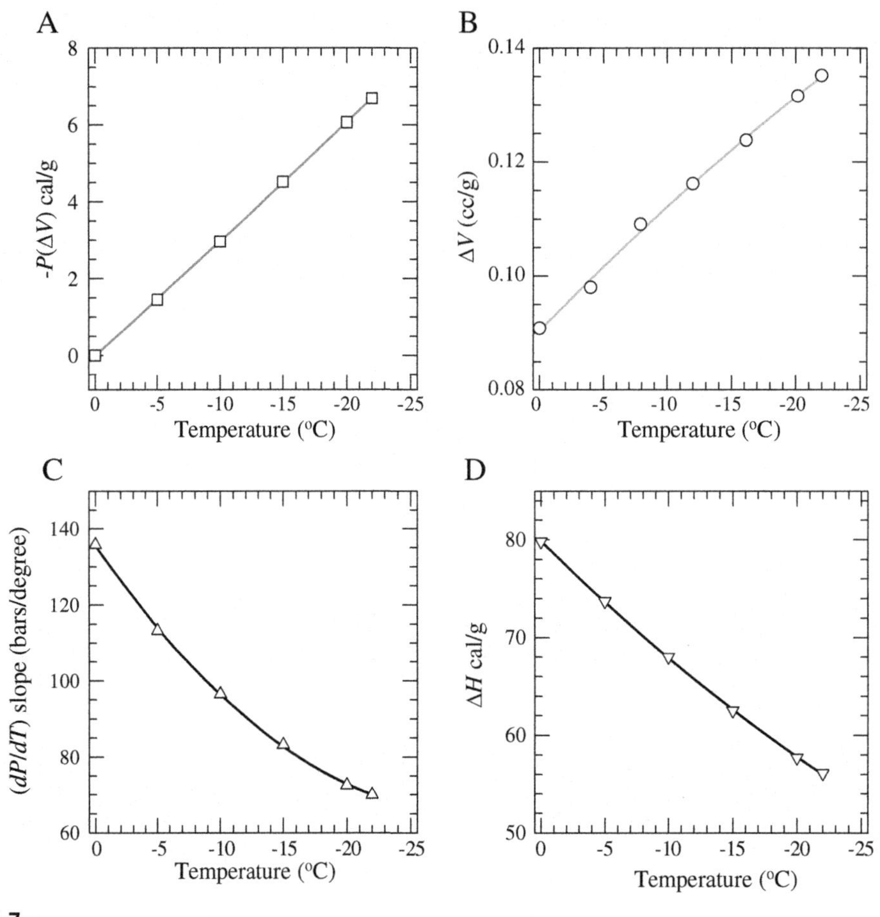

FIGURE 7-7

A four-panel plot of $P\Delta V$, ΔV, ΔH, and (dP/dT) along the ice I—water melting curve based on the data of Bridgman (1911).

D. Vaporization and sublimation curves of water and the clapeyron equation

We already stated that the Clapeyron equation also applies to vaporization and sublimation curves. This is easily illustrated using water because its properties are precisely known.

Example 7-4. Calculate the (dP/dT) slope and enthalpy of vaporization for water at 372 K in the vicinity of the normal boiling point. Using data from Wagner and Pruß (2002) we calculate

$$\left(\frac{dP}{dT}\right)_{eq} = \frac{\Delta S}{\Delta V} = \frac{6.0738 \text{ kJ kg}^{-1}\text{K}^{-1}}{1737.4071 \text{ m}^3\text{kg}^{-1}(0.101325 \text{ kJ m}^{-3}\text{atm}^{-1})} = 0.0345 \text{ atm K}^{-1} \quad (7\text{-}37)$$

This result is easily verified by directly computing the slope from tabulated *P-T* data in Wagner and Pruß (2002). This gives 0.0345 atm K^{-1} in exact agreement with Eq. (7-37). We compute the ΔH of vaporization by rearranging the Clapeyron equation to give

$$\Delta H = T\Delta V\left(\frac{dP}{dT}\right)_{eq}$$

$$= \frac{(372\ \text{K})(1737.4071\ \text{m}^3\text{kg}^{-1})(0.0345\ \text{atm K}^{-1})}{9.86923\ \text{m}^3\text{atm kJ}^{-1}} \quad (7\text{-}38)$$

$$\Delta H = 2259.3\ \text{kJ kg}^{-1}$$

The ΔH value computed from the difference in enthalpy of water and steam at 372 is virtually identical to the preceding result (2259.4 kJ kg^{-1}).

You probably noticed the large ΔV term in the two previous equations. At 372 K, the specific volume of steam is 1737.4081 m^3 kg^{-1}. This is almost the same as the ΔV term. In contrast, the specific volume of water is orders of magnitude smaller (0.00104 m^3 kg^{-1}) and is negligible compared to V_{steam} or to the ΔV term.

The Clapeyron equation can be applied to any vaporization curve. In the previous example we computed the Clapeyron slope and enthalpy of vaporization at one point. However, we could use an equation to represent the variation of pressure with temperature (see Section II-F, which follows). Differentiating this equation with respect to temperature would then yield (dP/dT) along the vaporization curve. This could be combined with data for the ΔV term to compute the enthalpy of vaporization as a function of temperature, or ΔV could be calculated if ΔH were known.

Example 7-5. Calculate ΔV for the sublimation of ice I to water vapor at the triple point (273.16 K). The (dP/dT) slope of the sublimation curve is 4.75×10^{-4} atm K^{-1} and the ΔH for sublimation is 2834.5 J g^{-1} at the triple point. We rearrange Eq. (7-31) and solve for ΔV:

$$\Delta V = \left(\frac{dT}{dP}\right)_{eq}\frac{\Delta H}{T}$$

$$= 2105.3\ \text{K atm}{-1}\frac{2834.5\ \text{J g}^{-1}}{(273.16\ \text{K})(0.101325\ \text{J cm}^{-3}\text{atm}^{-1})} \quad (7\text{-}39)$$

$$= 215{,}604.0\ \text{cm}^3\text{g}^{-1}$$

The ΔV for sublimation is essentially the specific volume of water vapor at the triple point because the specific volume of ice I is only 1.09089 cm^3 g^{-1}.

As we stated earlier, ice I, liquid water, and water vapor coexist in equilibrium at the triple point. Thus, the ΔH for sublimation (2834.5 J g^{-1} for Eq. 7-23) is simply the sum of the ΔH values for fusion (333.58 J g^{-1} for Eq. 7-19) and vaporization (2500.9 J g^{-1} for Eq. 7-21):

$$\Delta_{sub}H = \Delta_{vap}H + \Delta_{fus}H \quad (7\text{-}40)$$

Equation (7-40) is a general equation that applies to all materials, but only at their *S-L-V* triple points. The three ΔH values have different variations with temperature because the ΔC_P values for sublimation, vaporization, and melting are different. However, Eq. (7-40) can be used for approximate computations. For example, it can be used to calculate $\Delta_{fus}H$ from vapor pressure points on the vaporization and sublimation curves. Ideally, this calculation would use vapor pressure data near the triple point, but this is not always possible. Equation (7-40) also can be used to compute a vapor pressure equation from $\Delta_{trans}H$ values and the vapor pressure at the triple point.

An equation analogous to Eq. (7-40) can be written for the ΔH values at *any* triple point. That is, the largest ΔH value is the sum of the two smaller ΔH values. This is true for any type of triple point (e.g., *S-L-V*, S_α-S_β-*L*, S_α-S_β-*V*, S_α-S_β-S_γ). Furthermore, the ΔV changes must obey a similar relation, that is, the largest ΔV change is the sum of the two smaller ΔV changes. Third, the pressure and temperature values for the three separate equilibrium curves that intersect at the triple point must agree. Last, the metastable extension of each equilibrium line is within the angle subtended by the other two equilibrium lines. *These four conditions must be satisfied at any triple point without exception.*

Finally, it is interesting to compare the Clapeyron slopes for melting (-132.8 atm K^{-1}), vaporization (4.51×10^{-4} atm K^{-1}), and sublimation (4.75×10^{-4} atm K^{-1}) of water at the triple point. Figure 7-4 shows that the melting curve is much steeper than either the vaporization or sublimation curves, but it is hard to see the different slopes of the sublimation and vaporization curves. However, the differences between these curves are illustrated in Figure 7-3, where it is apparent that the ice curve (i.e., sublimation curve) is steeper than that for water (the vaporization curve). An important consequence of the difference between the slopes of vaporization and sublimation curves is that the vapor pressure of a low-temperature solid will be overestimated if the break in slope at the triple point is neglected. For example, the vapor pressure curve for supercooled water in Figure 7-3 illustrates how extrapolation of the vaporization curve below the triple-point temperature gives incorrect results for the vapor pressure over ice. Similar errors result if the vapor pressure over liquefied gases is extrapolated below their triple points without taking the difference between $\Delta_{vap}H$ and $\Delta_{sub}H$ into account.

E. Derivation of the Clausius-Clapeyron equation

Examples 7-4 and 7-5 show that to first approximation we can neglect the volume of the condensed phase when we apply the Clapeyron equation to vaporization or sublimation curves:

$$\Delta V \sim V_{vapor} \tag{7-41}$$

Using Eq. (7-41) to substitute for the ΔV term in the Clapeyron equation (7-31) gives us

$$\left(\frac{dP}{dT}\right)_{eq} = \frac{\Delta H}{TV_{vapor}} \tag{7-42}$$

where V_{vapor} is volume of the vapor produced by sublimation or vaporization. If we now assume that the vapor behaves ideally, we can substitute for V_{vapor} using the ideal gas law:

$$\left(\frac{dP}{dT}\right)_{eq} = \frac{\Delta H}{TV_{vapor}} = \frac{P_{vapor}\Delta H}{nRT^2} \tag{7-43}$$

where n is the number of moles of vapor. Grouping like terms on the same side of Eq. (7-43),

$$\frac{dP}{P_{vapor}} = \frac{\Delta H}{nRT^2}dT \tag{7-44}$$

which is equivalent to

$$d\ln P = \frac{\Delta H}{nRT^2}dT \quad \text{or to} \quad d\log P = \frac{\Delta H}{(\ln 10)nRT^2}dT \tag{7-45}$$

Equation (7-45) is the *Clausius-Clapeyron equation*. The approximations that we made for its derivation are (1) the ΔV of vaporization or sublimation is equal to V_{vapor}, (2) the vapor behaves ideally, and (3) the total pressure of the system is equal to the vapor pressure over the condensed phase (solid or liquid). The first approximation is justified for P-T conditions where the molar (or specific) volume of the vapor is much larger than that of the condensed phase. The second approximation is generally good under the same P-T conditions. The third approximation is good for pure systems. It also holds if other gases are present at low pressure and do not dissolve in or react with the material being studied. As we show later in Section III, the vapor pressure of a pure material is affected by total pressure (from a second gas or applied in a press).

Finally, the interpretation of Eq. (7-45) is simple. A plot of $\ln P$ versus temperature (Kelvin) is a curve with a slope given by $(\Delta H/RT^2)$. We illustrate its use in the next example.

Example 7-6. Epsomite ($MgSO_4 \cdot 7H_2O$) and hexahydrite ($MgSO_4 \cdot 6H_2O$) are common evaporite minerals on Earth. Epsomite is also found in carbonaceous chondrites, and there is spectroscopic and geochemical evidence that it may be present on Europa and Mars. However, published data for epsomite are contradictory and fragmentary. Carpenter and Jette (1923) measured the water vapor dissociation pressures from 25°C to 70°C for the reaction

$$MgSO_4 \cdot 7H_2O\ (\text{epsomite}) = MgSO_4 \cdot 6H_2O\ (\text{hexahydrite}) + H_2O\ (g) \tag{7-46}$$

Use the Clausius-Clapeyron equation (7-45) to calculate ΔH values as a function of temperature from a subset of their data, which is summarized here along with our calculations.

The data are plotted as $\ln P$ (mmHg) versus T (Kelvin) according to Eq. (7-45). A least-squares fit to the resulting curve gives the equation

$$\ln P = -78.0393 + 0.453087T - 6.133535 \times 10^{-4}T^2 \tag{7-47}$$

The ΔH of reaction (7-46) is calculated from Eq. (7-45) by rearranging it to get

$$\Delta H = RT^2\left(\frac{d\ln P}{dT}\right) \tag{7-48}$$

We need to find the derivative of $\ln P$ with respect to temperature to solve Eq. (7-48). This is

$$\begin{aligned}\left(\frac{d\ln P}{dT}\right) &= \frac{d}{dT}[-78.0393 + 0.453087T - 6.133535 \times 10^{-4}T^2] \\ &= 0.453087 - 12.26707 \times 10^{-4}T\end{aligned} \tag{7-49}$$

Equation (7-49) gives the slope of the curve at each temperature. The values computed from Eq. (7-49) are listed in the "slope" column of the table. Equation (7-48) gives the ΔH of reaction values that are listed in the "ΔH" column of the table. The ΔH of reaction decreases with increasing temperature, and our results for ΔH agree with those of Carpenter and Jette (1923).

T(°C)	P_W[a]	T(K)	ln P_W	Slope	ΔH[b]
25.00	12.7	298.15	2.5416	0.0873	15.43
32.40	22.8	305.55	3.1268	0.0783	14.52
36.65	31.5	309.80	3.4500	0.0731	13.93
40.12	40.1	313.27	3.6914	0.0688	13.42
45.07	57.2	318.22	4.0466	0.0627	12.62
50.29	79.0	323.44	3.9178	0.0563	11.71
55.33	99.6	328.48	4.0133	0.0501	10.75
60.18	123.5	333.30	4.0973	0.0442	9.76
65.11	153.9	338.26	4.1761	0.0381	8.67
69.74	185.7	342.89	4.2448	0.0325	7.58

[a]*Water vapor pressure in mmHg.*
[b]*ΔH in kcal mol^{-1}.*

As discussed in Chapters 10 and 11, the relative stabilities of epsomite and hexahydrite can be determined from their solubilities in water, from the water vapor pressure over their saturated solutions, and from electromotive force measurements.

The great utility of the Clausius-Clapeyron equation becomes more apparent when it is rewritten making use of the identity

$$d\left(\frac{1}{T}\right) = -\frac{1}{T^2}dT \tag{7-50}$$

Equation (7-45) can now be written in the two equivalent forms:

$$\frac{\partial \ln P}{\partial(1/T)} = -\frac{\Delta H}{nR} \tag{7-51}$$

$$\frac{\partial \log P}{\partial(1/T)} = -\frac{\Delta H}{(\ln 10)nR} \sim -\frac{\Delta H}{2.303nR} \tag{7-52}$$

Equations (7-51) and (7-52) show that a logarithmic plot of vapor pressure data versus inverse temperature will be a straight line with a slope equal to $-\Delta H/nR$ (for a plot of ln P versus $1/T$) or equal to $-\Delta H/(\ln 10)nR \sim -\Delta H/2.303nR$ for a plot of $\log_{10} P$ versus $1/T$. (The approximation arises because 2.303 is not exactly equal to ln 10.) The ΔH is the average enthalpy of sublimation (or vaporization) over the temperature range considered.

F. Applications of the Clausius-Clapeyron equation

We now discuss the relationship between the Clausius-Clapeyron equation and vapor pressure data using the vaporization curve of water as an illustration. Figure 7-8 shows the vaporization curve of

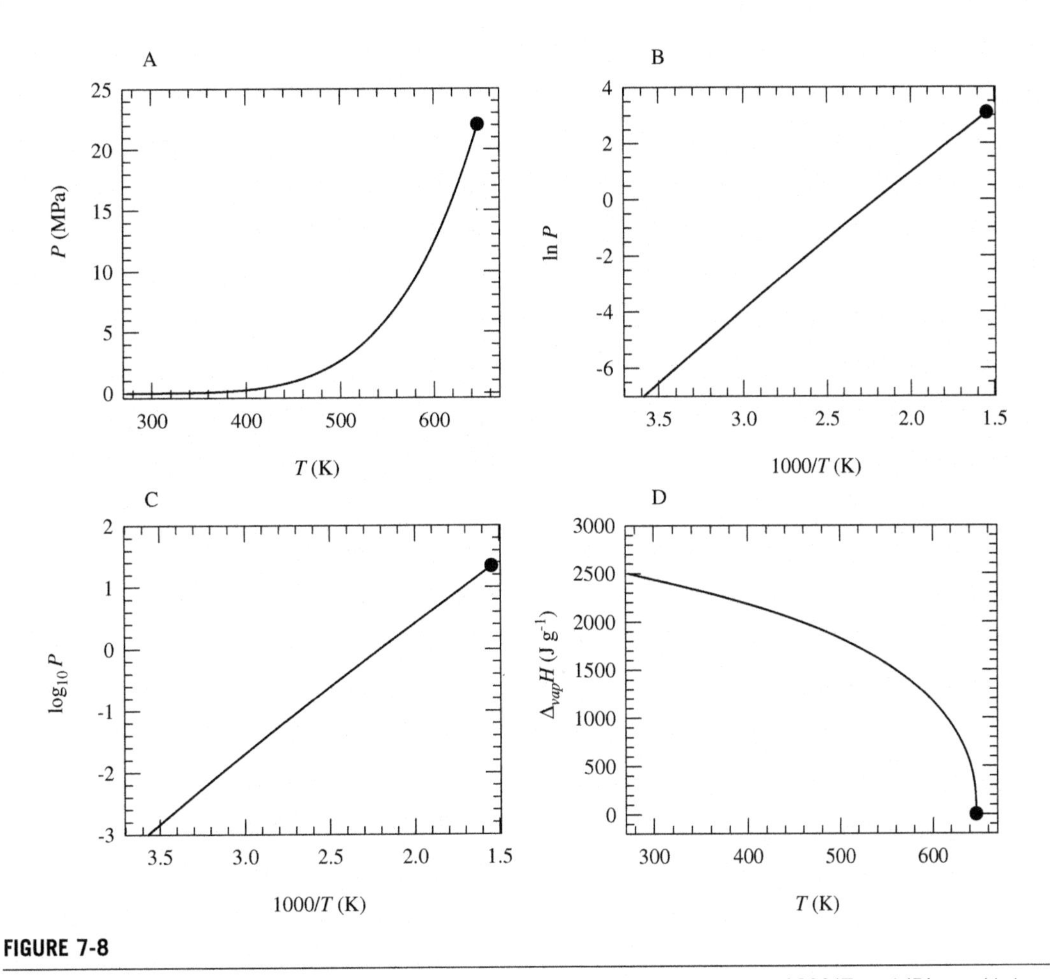

FIGURE 7-8

A four-panel plot showing (A) *P* versus *T*, (B) ln *P* versus 1000/*T*, (C) log *P* versus 1000/*T*, and (D) $\Delta_{vap}H$ along the water vaporization curve from 0°C to 374°C.

water plotted in three different ways and a plot of the enthalpy of vaporization. The black dot in graphs A–D is the critical point at ~ 647 K, 22 MPa. Graph A is a smaller version of Figure 7-1 and shows the steep dependence of the vapor pressure on temperature. Graph B shows the same data plotted as ln*P* versus 1000/*T*. The line in graph B is not perfectly straight but is slightly curved. Nevertheless, the linear least-squares fit

$$\ln P = 10.755 - \frac{4897.7}{T} \tag{7-53}$$

reproduces this line to first approximation. The worst agreement between the curve in graph B and Eq. (7-53) is at about 400 K (1000/*T* ~ 2.5). At 400 K the vapor pressure of water is 0.246 MPa versus a calculated value of 0.226 MPa from Eq. (7-53). The slope of 4897.7 (= $-\Delta H/R$) corresponds to an average enthalpy of vaporization of about 40,720 J mol^{-1} over the entire temperature range from the triple point to the critical point (273.16 − 647.096 K). This is slightly different than the enthalpy of vaporization (40,656.3 J mol^{-1}) at the normal boiling point of 100°C. Graph C shows the same data (as in graph B) but plotted as $\log_{10}P$ versus 1000/*T*. The line in graph C is not perfectly straight but is slightly curved. The linear least-squares fit to this line is

$$\log_{10}P = 4.671 - \frac{2127.05}{T} \tag{7-54}$$

In this case the slope is 2127.05 [$= -\Delta H/(\ln 10)R$], which gives the same ΔH value as it must.

Graph D shows $\Delta_{vap}H$ over the same temperature range. The heat of vaporization is not constant but varies with temperature. The curvature of the lines in graphs B and C results because $\Delta_{vap}H$ is variable. In principle the ΔH value calculated using the Clausius-Clapeyron equation can be assigned to the midpoint temperature. Then the ΔH value can be corrected to a specific temperature (e.g., 298 K, or the boiling point, or the triple point, etc.) using Kirchhoff's equation (Section I-F, Chapter 5). However, this correction becomes less meaningful when the temperature range is large, as in this case.

Another interesting point in graph D is the steep decrease in $\Delta_{vap}H$ in the vicinity of the critical point, where the distinction between vapor and liquid disappears. The steep decrease and eventual disappearance of $\Delta_{vap}H$ at the critical point occurs for all liquids and liquefied gases. Likewise, the entropy of vaporization ($\Delta_{vap}S = \Delta_{vap}H/T$) and $\Delta_{vap}V$ ($= (dT/dP)\cdot\Delta_{vap}H/T$) also disappear at the critical point of all liquids and liquefied gases.

We deliberately used the entire temperature range from the *S-L-V* triple point to the critical point in Figure 7-8 to illustrate the curvature of log *P* versus 1/*T* plots and the variation in ΔH. In many experimental studies the sublimation pressure or vapor pressure data are measured over smaller temperature ranges and/or the ΔH of sublimation (or vaporization) is much larger than the $\Delta C_P \Delta T$ product, which is the variation of ΔH over the temperature range studied. The log *P* versus 1/*T* (or ln *P* versus 1/*T*) plots are generally straight lines in these cases because the variation of ΔH with temperature is much smaller.

Example 7-7. Use the Clausius-Clapeyron equation (7-52) to calculate the enthalpy of sublimation $\Delta_{sub}H$ of Fe metal from the vapor pressure data of Jones et al. (1927) given shortly. These data are from the research group of Irving Langmuir (1881–1957), who won the 1932 Nobel Prize in Chemistry. Langmuir developed the free vaporization (*Langmuir vaporization*) method for vapor pressure measurements during his studies of electric light bulbs at the GE research laboratory (Langmuir, 1913). This method allowed him to measure the vapor pressure of refractory metals such as tungsten by measuring the weight loss from light bulb filaments as a function of time and temperature.

T(K)	**1270**	**1438**	**1562**	**1580**
P_{obs}(bar)	1.417×10^{-9}	9.884×10^{-8}	1.168×10^{-6}	1.630×10^{-6}
P_{calcd}(bar)	1.437×10^{-9}	9.528×10^{-8}	1.181×10^{-6}	1.647×10^{-6}

The data are plotted as log *P* versus 1/*T* according to Eq. (7-52). A linear least-squares fit to the four data points gives this equation for the straight line

$$\log_{10}P \text{ (bars)} = 6.750 - \frac{19{,}802.75}{T} \tag{7-55}$$

The sublimation pressures calculated from Eq. (7-55) are in the bottom row of the table. The calculated values agree with the data within 1.4% except for the point at 1438 K, which is 3.6% different. The average $\Delta_{sub}H$ value over this temperature range is calculated from the slope

$$\Delta_{sub}H = -(\ln 10)(8.314)(-19{,}802.70) = 379.1 \text{ kJ mol}^{-1} \tag{7-56}$$

A good, straight line was obtained in this example because the sublimation pressure data were measured over a relatively small temperature range of 300 degrees and because the enthalpy of sublimation of Fe is much larger than the $\Delta C_P \Delta T$ product (~4 kJ mol^{-1}) over this T range.

We can integrate the Clausius-Clapeyron equation (7-45) making increasingly sophisticated assumptions. The simplest assumption, often made out of necessity, is that ΔH is independent of temperature. This implies $\Delta C_P = 0$ for vaporization or sublimation (i.e., C_P (gas) $= C_P$ (condensed phase)). In this case,

$$\int d\ln P = \int \frac{\Delta H}{nRT^2} dT = \frac{\Delta H}{nR} \int \frac{1}{T^2} dT \tag{7-57}$$

$$\ln P = -\frac{\Delta H}{nRT} + c \tag{7-58}$$

where c is the integration constant. Equation (7-45) is usually integrated between two temperatures, which are the low (T_1) and high (T_2) temperatures of an experimental study. Then we have

$$\ln\left(\frac{P_2}{P_1}\right) = -\frac{\Delta H}{nR}\left(\frac{1}{T_2} - \frac{1}{T_1}\right) \tag{7-59}$$

The next example illustrates the use of Eq. (7-59).

Example 7-8. The S-L-V triple point of cristobalite (SiO_2) is 1999 K and 2.6 Pa. Calculate the one-bar boiling point of molten silica from these values and $\Delta_{vap}H = 485$ kJ mol^{-1}. Solid, liquid, and vapor are in equilibrium at the triple point (review Section II-A, if necessary). Thus, we know the vapor pressure over molten silica at one P-T point (T_1 1999 K, $P_1 = 2.6$ Pa). We can rearrange Eq. (7-59) to solve for the temperature T_2 at which $P_2 = 1$ bar:

$$\begin{aligned} \frac{1}{T_2} &= -\frac{R}{\Delta_{vap}H}\ln\left(\frac{P_2}{P_1}\right) + \frac{1}{T_1} \\ &= -\frac{8.314}{485{,}000}\ln\left(\frac{10^5}{2.6}\right) + \frac{1}{1999} \end{aligned} \tag{7-60}$$

Equation (7-60) gives $T_2 = 3132$ K for the one-bar boiling point of molten silica. For comparison, Schaefer and Fegley (2004) computed 3160 K for the one-bar boiling point of molten silica. Centolanzi and Chapman (1966) measured 3103 K for the one-bar boiling point of tektite glass, which is 70–80 mass % SiO_2. The name *tektites* is derived from *tektos*, the Greek word for molten rock. Tektites are centimeter-size, silica-rich, glassy rounded objects produced from terrestrial surface material during asteroidal or cometary impacts. They occur in strewn fields that are often far away from the original impact sites.

Example 7-8 also illustrates that the Clausius-Clapeyron equation is still valid when the vapor over a solid or liquid is composed of several gaseous species with concentrations that vary with temperature and the total vapor pressure. The vapor pressure of materials that are relatively refractory (solid elements and their compounds, solid and molten alloys, oxides, silicates, and their melts) is often measured by a technique known as *Knudsen effusion mass spectroscopy* (KEMS). In this method the condensed phase and vapor are equilibrated inside a crucible with a pinhole orifice. The vapor streaming out of the orifice is then analyzed using a mass spectrometer. Excellent descriptions of the KEMS technique, Langmuir vaporization, and other methods used to measure the vapor pressure of refractory materials are given in Margrave (1967). KEMS measurements show that silica vaporizes congruently (i.e., the condensed phase and vapor have the same bulk composition). However, the vapor is not pure SiO_2 gas but is composed of SiO, O_2, O, SiO_2, and Si gases in amounts that vary with temperature and the total vapor pressure. Figure 7-9 shows the composition of the vapor over SiO_2 (*s, liq*) plotted as log *P* versus 10,000/*T* for the KEMS data of Kazenas et al. (1985). Notice that their data are for a limited temperature range, which is slightly different for each gas. They did not measure monatomic Si gas and have only three data points (not shown) for monatomic O gas, which lies between SiO_2 and O_2. We could analyze the partial pressure of each gas using the Clausius-Clapeyron equation and derive an enthalpy of sublimation (or vaporization) for each species. This is often done and is one of the principal ways in which the standard enthalpies of formation of high-temperature gases observed in cool stars and protoplanetary disks (e.g., SiO, SiS, AlF, SiC, PN, SO, NaCl) are determined.

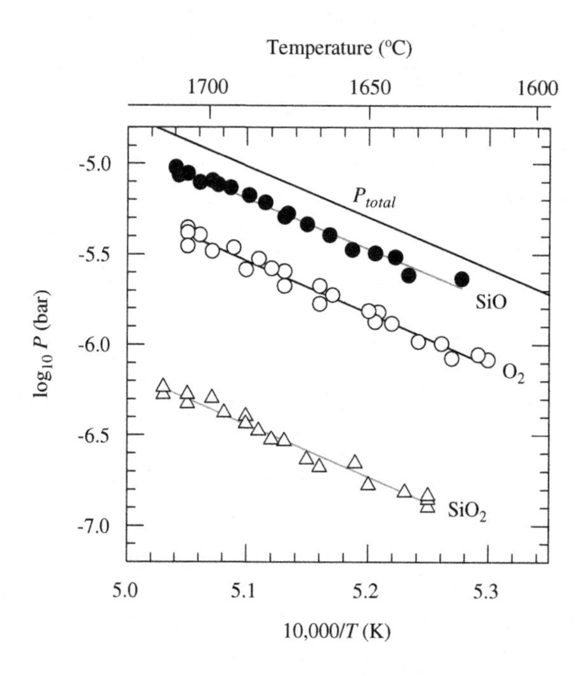

FIGURE 7-9

The results of Kazenas et al. (1985) for the total vapor pressure and the partial pressures of O_2, SiO, and SiO_2 over silica.

If ΔC_P for the evaporation or sublimation is not zero but is independent of temperature, we can use Kirchhoff's equation (5-26) to calculate ΔH as a function of temperature:

$$\Delta H_{T_2} = \Delta H_{T_1} + \int_{T_1}^{T_2} \Delta C_p dT \qquad (5\text{-}26)$$

$$\begin{aligned} \Delta H_{T_2} &= \Delta C_p(T_2 - T_1) + \Delta H_{T_1} \\ &= (\Delta H_{T_1} - T_1 \Delta C_P) + \Delta C_p T_2 \end{aligned} \qquad (7\text{-}61)$$

The temperature T_1 is the reference temperature and is generally 298.15 K if T_2 is above room temperature and is absolute zero if T_2 is below room temperature. Sometimes it is convenient to pick another temperature as T_1. Here we use $T_1 = 298.15$ K and $T_2 = T$. Equation (7-61) is then substituted into the Clausius-Clapeyron equation (7-45), giving

$$d\ln P = \frac{(\Delta H_{298} - 298.15\Delta C_p) + \Delta C_p T}{nRT^2} dT \qquad (7\text{-}62)$$

Integration of Eq. (7-62) yields

$$\ln P = \left(\frac{298.15\Delta C_p - \Delta H_{298}}{nR}\right)\frac{1}{T} + \frac{\Delta C_p}{nR}\ln T + c' \qquad (7\text{-}63)$$

where c' is an integration constant. This is expressed more easily in the form

$$\ln P = A + \frac{B}{T} + C\ln T \qquad (7\text{-}64)$$

where A is the intercept (the integration constant c'), and B and C are defined as

$$B = \frac{298.15\Delta C_p - \Delta H_{298}}{nR} \qquad (7\text{-}65)$$

$$C = \frac{\Delta C_p}{nR} \qquad (7\text{-}66)$$

Equation (7-64) or a variant thereof is often used to represent vapor pressure and dissociation pressure data. Smyth and Adams (1923) studied reaction (7-1) and they give the equation

$$\log_{10} P = 29.119 - \frac{11{,}355}{T} - 5.388\log_{10} T \qquad (7\text{-}67)$$

valid from 587°C 1389°C and 1–779,000 mmHg pressure. The ΔC_P for CO_2 loss from $CaCO_3$ varies and is not constant over this temperature range. However, Smyth and Adams (1923) found that Eq. (7-67) represented the data within their experimental accuracy. Finally, more complicated integrations of the Clausius-Clapeyron equation can be done, for example, by assuming that ΔC_P is given by Maier-Kelley equations. An alternative method is to use polynomial equations to fit log P versus T. The ΔH as a function of temperature can be found by differentiation. If the equation of state of the gas is known, the difference between ΔH for the real gas and an ideal gas can be found, and ΔV can also be computed as a function of temperature.

G. Polymorphic phase transitions

The transition between orthorhombic and monoclinic sulfur is a classic example of a polymorphic phase transition. This transition is important for sulfur geochemistry on Io, the volcanically active satellite of Jupiter. The German physical chemist Gustav Tammann (1861–1938) studied this phase transition and his results are shown in Figure 7-10 (Roozeboom, 1901; Tammann, 1925). The pressure dependence of the transition temperature in Figure 7-10 is

$$T_{trans} = 368.50 + 0.038P + 2.6480 \times 10^{-6}P^2 \tag{7-68}$$

The calorimetric measurements of West (1959) give $\Delta_{trans}H^o = 401.7 \pm 2$ J mol^{-1} at the transition temperature. However, the $\Delta_{trans}V^o$ is much harder to measure. Combining West's result for $\Delta_{trans}H^o$ with the Clapeyron slope from Eq. (7-68) gives

$$\Delta_{trans}V^o = \frac{\Delta_{trans}H^o}{T_{trans}}\left(\frac{dT}{dP}\right)_{eq} = \frac{401.7 \text{ J mol}^{-1}}{368.54 \text{ K}(0.1 \text{ J cm}^{-3} \text{ bar}^{-1})}(0.0380 \text{ K bar}^{-1}) = 0.414 \text{ cm}^3 \text{ mol}^{-1} \tag{7-69}$$

This ΔV value is probably uncertain by $\pm 1\%$ due to the uncertainties in the ΔH of transition and in the Clapeyron slope. Several problems at the end of the chapter are about this transition.

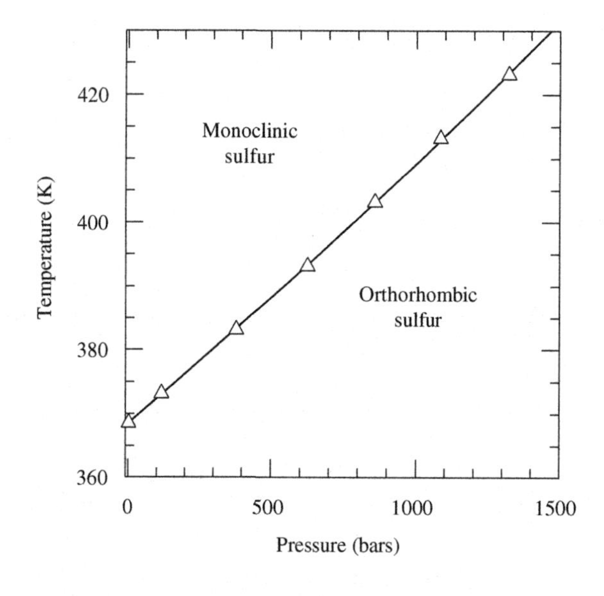

FIGURE 7-10

Pressure dependence of the orthorhombic-monoclinic sulfur transition temperature.

H. The Gibbs phase rule and some of its applications

J. W. Gibbs (1839–1903) is probably the greatest scientist the United States has produced (see sidebar). Gibbs derived the phase rule in his monumental paper "On the Equilibrium of Heterogeneous Substances." This paper, over 320 pages long, was published in two parts in the *Transactions of the Connecticut Academy*, an obscure journal that virtually no one read. Fortunately, Gibbs sent reprints to a number of famous European scientists, including the Dutch physicist J. D. van der Waals (see biographical sidebar in Chapter 8). Van der Waals realized the importance of Gibbs's paper and advised his younger colleague Bakhuis Roozeboom to read it. Roozeboom (1854–1907) then devoted the rest of his life to studying the phase rule and its applications to many problems in chemistry (see sidebar). He stated the phase rule in its present form and began a six-volume treatise, *Die Heterogenen Gleichgewichte vom Standpunkte der Phasenlehre* (*Heterogeneous Equilibria from the Standpoint of the Phase Rule*), that was finished by his students (Aten, Buchner, and Schreinemakers) after his early death.

JOSIAH WILLARD GIBBS (1839–1903)

Willard Gibbs was a Connecticut Yankee who spent almost all his life in and around Yale University in New Haven, Connecticut. He was educated at Yale, where his father was a professor in the Divinity School. Gibbs is best known for his theoretical work, but his doctoral thesis (1863) was about the design of gears, and he patented an improved railway brake (1866). After receiving his doctorate, Gibbs traveled in Europe (1867–1869), where he attended lectures by several of the great chemists and physicists of his time. In 1871, he was appointed professor of mathematical physics in Yale College and held this position until his death. Gibbs never married and lived with his sister and her family. His main activities were walking to and from Yale College, horseback riding, and driving the family carriage.

Gibbs published most of his papers about thermodynamics in *The Transactions of the Connecticut Academy*, an obscure journal. Fortunately, he sent reprints of his work to many scientists in Europe, and two of them, Maxwell and van der Waals, recognized his genius. Maxwell made a plaster cast of the Entropy–Volume–Internal Energy surface of water that he sent to Gibbs as a gift. Van der Waals adopted Gibbs's methods, used them in his own phase equilibria studies, and introduced Bakhuis Roozeboom to Gibbs's work. Thanks to these scientists and to the German physical chemist Wilhelm Ostwald (1853–1932), Gibbs was recognized as a great scientist. He received the Copley Medal of the Royal Society in 1901 and the Rumford Prize of the American Academy of Arts and Sciences in 1881 and was a member of the U.S. National Academy of Sciences and an honorary or a corresponding member of many foreign academies of science.

HENDRIK WILLEM BAKHUIS ROOZEBOOM (1854–1907)

Bakhuis Roozeboom was a physical chemist from The Netherlands. He came from a working-class family (his father was a bookkeeper), and he could not afford a university education. He took a job as an analytical chemist at the Hague rather than pursue a graduate degree until a fire destroyed the factory where he worked. His supervisor's brother-in-law, J. M. van Bemmelen, the professor of chemistry at the University of Leiden, was able to offer him a position as a research assistant. This allowed Roozeboom to continue his studies, and he received his doctorate in 1884. His dissertation was on the phase relations of hydrates of sulfur dioxide, chlorine, bromine, and hydrogen chloride. In 1879, Roozeboom married Catharina Wins, with whom he had six children. To meet the needs of his growing family, Roozeboom took a job teaching at a Leiden girls' school, where he worked from 1881 to 1896.

Roozeboom worked at the University of Leiden at the same time that van der Waals was professor of physics there. Van der Waals introduced Roozeboom to the work of Gibbs, which would shape Roozeboom's research. Roozeboom was the first to apply the phase rule to heterogeneous equilibria, and he was the first to represent equilibria graphically, developing the earliest phase diagrams. He systematically studied the effects of pressure, temperature, and concentration on equilibria and then processed the results into an easily understood picture. In 1896, he replaced van't Hoff as professor of chemistry at the University of Amsterdam. There he applied the phase rule to alloys, which proved to be very important for the manufacture of steel, an iron-carbon alloy. In 1901 he began a six-volume treatise on the phase rule and its applications to heterogeneous equilibria, which is a classic work. However, Roozeboom died from pneumonia in 1907 and three of his students (Aten, Buchner, and Schreinemakers) completed this treatise after his death.

The *Gibbs phase rule* tells us the number of intensive variables (the degrees of freedom F) that we can arbitrarily fix when a system of P phases and C components is in equilibrium at constant temperature and pressure (or constant entropy and volume or constant entropy and pressure or constant temperature and volume):

$$F = C - P + 2 \tag{7-70}$$

We illustrate the use of the phase rule (Eq. 7-70) with several examples.

We first consider the H_2O system and its low-pressure phase diagram in Figure 7-4. The regions labeled ice I, liquid, and vapor each contain one phase (S, L, V) and one component (H_2O). According to the Gibbs phase rule,

$$F = (C - P + 2) = (1 - 1 + 2) = 2 \tag{7-71}$$

we have two degrees of freedom in each *one-phase field*. Thus, within some rather wide limits we can independently and arbitrarily vary pressure and temperature without altering the number of phases present. The ice I, liquid, and vapor regions are called *bivariant fields* (or *divariant fields*) because there are two degrees of freedom in each region.

Two different phases of the one component (H_2O) coexist in equilibrium along the sublimation curve ($S = V$), the vaporization curve ($L = V$), and the melting curve ($S = L$). Thus,

$$F = (C - P + 2) = (1 - 2 + 2) = 1 \tag{7-72}$$

we have one degree of freedom along each *univariant curve*. Within rather wide limits we can arbitrarily and independently vary either pressure or temperature and remain on a univariant curve. This was discussed earlier in Section I-I.

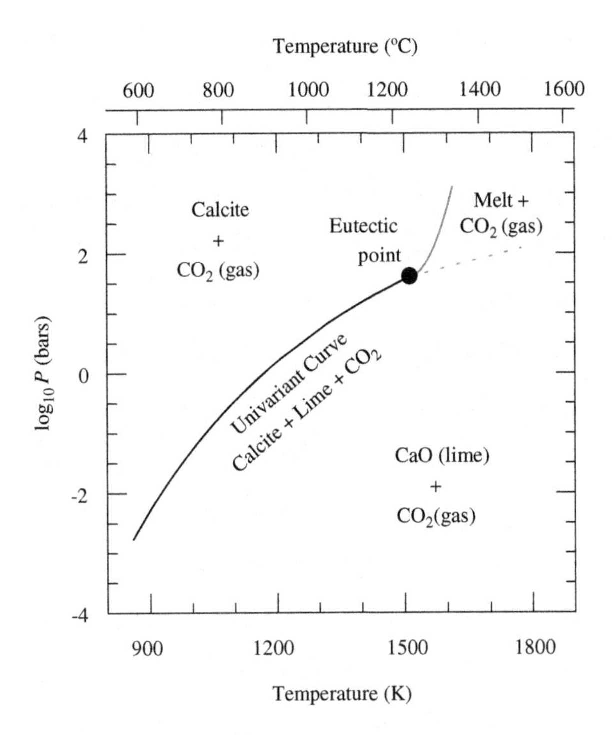

FIGURE 7-11

Pressure-temperature phase diagram for the CaO-CO_2 system based on the data of Smyth and Adams (1923) and Baker (1962).

The three univariant curves intersect one another at the *S-L-V* triple point where all three phases (ice I, liquid, vapor) coexist in equilibrium. In this case we have

$$F = (C - P + 2) = (1 - 3 + 2) = 0 \tag{7-73}$$

There are no degrees of freedom at the triple point, which is an *invariant point*. Neither pressure nor temperature can vary from the triple point *P* and *T* without losing at least one phase.

Figure 7-11 is the *P-T* phase diagram for the $CaO-CO_2$ system based on the work of Smyth and Adams (1923) and Baker (1962). The black curve is the CO_2 dissociation pressure for reaction (7-1) as a function of temperature and is a plot of Eq. (7-67). The two curves at the upper end of the diagram are the CO_2 dissociation pressures for the reactions

$$CaCO_3 \text{ (calcite)} = CaO\text{(in solution in } CaCO_3\text{-rich melt)} + CO_2 \text{ (g)} \tag{7-74}$$

$$CaCO_3 \text{ (in solution in } CaCO_3\text{-poor melt)} = CaO \text{ (lime)} + CO_2 \text{ (g)} \tag{7-75}$$

These two curves intersect the curve for reaction (7-1) at 1513 K and 40.0 bars (the eutectic point). Each of the CO_2 dissociation pressure curves in Figure 7-11 is a univariant curve because three phases coexist along each line. Calcite, lime, and CO_2 gas coexist along the dissociation pressure curve for reaction (7-1). Calcite, CaO (dissolved in $CaCO_3$-rich melt), and CO_2 gas coexist along the dissociation pressure curve for reaction (7-74). Calcium carbonate (dissolved in $CaCO_3$-poor melt), lime, and CO_2 gas coexist along the dissociation pressure curve for reaction (7-75). The Gibbs phase rule shows that all curves are univariant because two components (e.g., CaO and CO_2) and three phases give one degree of freedom.

The three univariant curves divide Figure 7-11 into three different phase fields, which are CaO (lime) + CO_2 gas, $CaCO_3$ (calcite) + CO_2 gas, and melt + CO_2 gas. Each field is bivariant:

$$F = (C - P + 2) = (2 - 2 + 2) = 2 \tag{7-76}$$

There is no triple point in this system. However, there is another type of invariant point called a *eutectic point* at 1513 ± 1 K and 40.0 ± 0.4 bars, which is the intersection point of the three univariant curves in Figure 7-11. The eutectic point is the lowest melting point in the CaO-CO_2 system and has a composition of about 53.1 mole % CaO + 46.9 mole % CO_2 (Baker, 1962). It is a *quadruple point* where calcite + lime + eutectic melt + CO_2 gas coexist. Thus, it is an invariant point according to the Gibbs phase rule:

$$F = (C - P + 2) = (2 - 4 + 2) = 0 \tag{7-77}$$

In general, the Gibbs phase rule (Eq. 7-70) shows that $(c + 1)$ phases coexist along a univariant curve in a system containing c components. The pressure can be expressed as a function of temperature, or vice versa, along the univariant curve. Likewise, $(c + 2)$ phases coexist at an invariant point in a system containing c components. Temperature and pressure are both uniquely specified at the invariant point.

III. EFFECT OF TOTAL PRESSURE ON VAPOR PRESSURE

So far we have implicitly assumed that both the vapor and the solid (or liquid) are under the same pressure, namely the vapor pressure. In other words, we were considering systems where the total pressure was equal to the vapor pressure. However, this is not always the case.

For example, suppose we are considering the vapor pressure of liquid water in a glass sitting on a table at room temperature. In this case, the vapor pressure of the liquid water is about 25 mmHg, while the total pressure on the liquid water is about 785 mmHg. This situation is illustrated in Figure 7-12. The vapor pressure (p) of the liquid water plus the ambient atmospheric pressure equals the total pressure (P) on the liquid water. In this specific example, p is about 25 mmHg and ambient atmospheric pressure is about 760 mmHg, but the derivation does not depend on these exact values.

Despite the presence of a second gas (air), we are still considering the reaction

$$H_2O \text{ (water)} = H_2O \text{ (vapor)} \tag{7-21}$$

At equilibrium, $\Delta G = 0$ and the Gibbs free energies of water vapor and liquid water are equal:

$$G_{liquid} = G_{vapor} \tag{7-22}$$

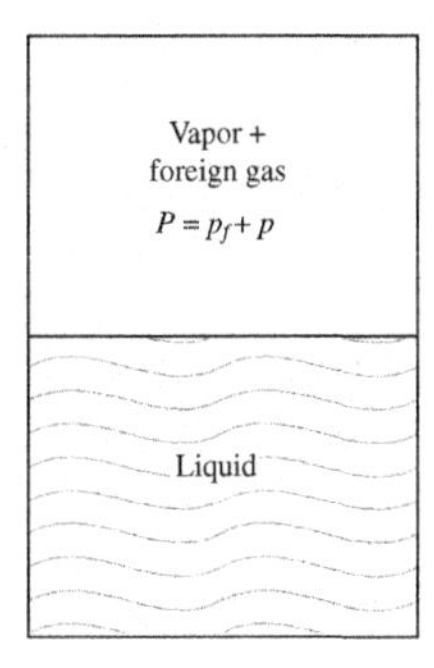

FIGURE 7-12

A cartoon illustrating the effect of applied pressure on the vapor pressure of a liquid or solid.

The Gibbs free energies of the vapor and liquid water can be written as functions of T and P:

$$G_{vapor}(T,\ p) = G_{liquid}(T,\ P) \tag{7-78}$$

where the pressure of the vapor (p) and the total pressure on the liquid water (P) are used. Since we are interested in the effect of total pressure on the vapor pressure, we will differentiate Eq. (7-78) with respect to pressure P:

$$\left(\frac{\partial G_{vapor}}{\partial P}\right)_T = \left(\frac{\partial G_{vapor}}{\partial p}\right)_T \left(\frac{\partial p}{\partial P}\right)_T = \left(\frac{\partial G_{liquid}}{\partial P}\right)_T \tag{7-79}$$

Evaluation of Eq. (7-79) requires the derivative of G with respect to P at constant T. We can derive this by evaluating Eq. (7-28), the second fundamental equation, at constant temperature:

$$dG = VdP - SdT \tag{7-28}$$

$$\left(\frac{\partial G}{\partial P}\right)_T = V \tag{7-80}$$

We substitute Eq. (7-80) into Eq. (7-79) and obtain

$$V_{vapor}\left(\frac{\partial p}{\partial P}\right)_T = V_{liquid} \tag{7-81}$$

We can rearrange Eq. (7-81) to solve for the pressure derivative term:

$$\left(\frac{\partial p}{\partial P}\right)_T = \frac{V_{liquid}}{V_{vapor}} \tag{7-82}$$

Equation (7-82) expresses the variation of vapor pressure (p) with total pressure (P). We can write an analogous equation for sublimation reactions such as

$$H_2O\ (\text{ice}) = H_2O\ (\text{vapor}) \tag{7-23}$$

This analogous equation is

$$\left(\frac{\partial p}{\partial P}\right)_T = \frac{V_{solid}}{V_{vapor}} \tag{7-83}$$

Now we can examine Eqs. (7-82) and (7-83) to see what they tell us about the change in vapor pressure (Δp) with increased total pressure (ΔP):

$$\Delta p = \Delta P \frac{V_{liquid}}{V_{vapor}} \tag{7-84}$$

$$\Delta p = \Delta P \frac{V_{solid}}{V_{vapor}} \tag{7-85}$$

In both cases, we see that vapor pressure is increased as the total pressure on the liquid (or solid) is increased. However, the fractions (V_{liquid}/V_{vapor}) and (V_{solid}/V_{vapor}) are generally very small because the molar volume (or specific volume) of vapor is much larger than that of the liquid or solid. Thus, a very large ΔP is needed to significantly increase the vapor pressure of a solid or liquid in most cases. Assuming that we have an ideal gas, we can replace V_{vapor} by

$$V_{vapor} = \frac{RT}{p} \tag{7-86}$$

and then Eq. (7-82) can be rewritten as follows:

$$V_{vapor}\left(\frac{\partial p}{\partial P}\right)_T = V_{liquid} \tag{7-82}$$

$$\frac{RT}{p}\left(\frac{\partial p}{\partial P}\right)_T = V_{liquid} \tag{7-87}$$

$$RT\frac{dp}{p} = V_{liquid}dP \tag{7-88}$$

Integrating both sides from p_0 (total $p =$ vapor p) to p (vapor p at total pressure P), we get

$$RT\int_{p_0}^{p}\frac{dp}{p} = V_{liquid}\int_{p_0}^{P} dP \tag{7-89}$$

$$RT\ln\left(\frac{p}{p_0}\right) = V_{liquid}(P - p_0) \tag{7-90}$$

Equation (7-90) was derived by the British physicist J. H. Poynting (1852–1914), who also discovered radiation pressure, which can be used to propel "solar sails." The *Poynting equation* gives the variation of vapor pressure (p) from p_0 (vapor $p =$ total p) to a higher pressure P. The two assumptions made in deriving this equation are an ideal vapor and an incompressible liquid. Both assumptions are generally justified at low vapor pressures. Equation (7-90) can be modified for use with nonideal vapors and compressible liquids. The analogous equation for a solid is

$$RT\ln\left(\frac{p}{p_0}\right) = V_{solid}(P - p_0) \tag{7-91}$$

Example 7-9. At 25°C, the vapor pressure and density of water are 23.756 mmHg and 0.99707 g cm^{-3}. Calculate the vapor pressure of water under a total pressure of 50 bars at 25°C.

We first convert the water vapor pressure into bars. This is the p_0 value in Eq. (7-90):

$$p_0 = \frac{23.756 \text{ mmHg}}{750.062 \text{ mmHg bar}-1} = 0.03167 \text{ bars} \tag{7-92}$$

Next, we need to calculate the molar volume of water from its density and molecular weight:

$$V_{liquid} = \frac{18.015 \text{ g mol}^{-1}}{0.99707 \text{ g cm}^{-3}} = 18.0679 \text{ cm}^3 \text{ mol}^{-1} \tag{7-93}$$

Now we substitute the numerical values into Eq. (7-90) and rearrange it:

$$\ln\left(\frac{p}{p_0}\right) = \frac{1}{RT}V_{liquid}(P - p_0)$$

$$= \frac{18.0679}{(83.145)(298.15)}(50 - 0.03167) \tag{7-93}$$

$$= 0.03642$$

$$\left(\frac{p}{p_0}\right) = \exp(0.03642) = 1.0371 \tag{7-94}$$

$$p = p_0\,(1.0371) = (0.03167 \text{ bars})(1.0371) \tag{7-95}$$

$$p = 0.03284 \text{ bars} = 24.636 \text{ mmHg} \tag{7-96}$$

The increased vapor pressure is only 0.88 mmHg larger.

However, the effects of total pressure on the vapor pressures of liquids and solids may be much greater than this in the deep atmospheres and in the interiors of the Jovian planets Jupiter, Saturn, Uranus, and Neptune. Thermochemical equilibrium models predict that aqueous water clouds condense at total pressures of 1000 to 2000 bars in the atmospheres of Uranus and Neptune (Fegley and Prinn, 1986). Deeper inside the Jovian planets, at pressures of hundreds to thousands of kilobars, the vapor pressures of solids may be increased so much that any infalling planetesimals or the planetary cores "dissolve" in the surrounding gas (Fegley and Prinn, 1986). The solubility of solids in gases is a well-known effect (e.g., Morey, 1957) and silica solubility in high-pressure steam occurs industrially and in geological environments on Earth.

IV. TYPES OF PHASE TRANSITIONS

Earlier we stated that the Clapeyron equation applies to all first-order phase transitions. We now discuss what this means in some more detail. In 1933 the Austrian physicist Paul Ehrenfest (1880–1933) classified phase transitions as first-order, second-order, and third-order according to the lowest derivatives of the Gibbs free energy with respect to T and P that show discontinuities. His work is the first attempt to classify phase transitions. It is interesting because it applies to some transitions and it stimulated other attempts to classify and explain phase transitions.

A. First-order phase transitions

According to Ehrenfest's classification, a first-order phase transition has a continuous variation of the Gibbs free energy G with temperature but discontinuous variations of the first derivatives of G with respect to T and P. Melting, vaporization, many solid-state transitions, and liquid crystal transitions are all first-order transitions. All of the phase transitions that we have discussed to this point are first-order transitions, for example,

$$\text{p-azoxyanisole (solid)} = C_{14}H_{14}N_2O_3\ \text{(nematic)} = C_{14}H_{14}N_2O_3\ \text{(isotropic)} \tag{7-2}$$

$$\text{sulfur (orthorhombic)} = \text{sulfur (monoclinic)} \quad T = 368.54\ \text{K} \quad P = 1\ \text{bar} \tag{7-3}$$

$$\text{sulfur (monoclinic)} = \text{sulfur (liquid)} \quad T = 388.36\ \text{K} \quad P = 1\ \text{bar} \tag{7-4}$$

$$H_2O\ \text{(ice I)} = H_2O\ \text{(water)} \tag{7-19}$$

$$H_2O\ \text{(water)} = H_2O\ \text{(vapor)} \tag{7-21}$$

$$H_2O\ \text{(ice)} = H_2O\ \text{(vapor)} \tag{7-23}$$

$$CO_2\ \text{(ice)} = CO_2\ \text{(gas)} \tag{7-97}$$

These first derivatives have discontinuities at the temperature of a first-order phase transition:

$$-\left(\frac{\partial G}{\partial T}\right)_P = S \tag{7-98}$$

$$\left(\frac{\partial G}{\partial P}\right)_T = V \tag{7-80}$$

$$\left[\frac{\partial(G/T)}{\partial(1/T)}\right]_P = H \tag{7-99}$$

The heat capacity, which is the second derivative of G with respect to T,

$$-T\left(\frac{\partial^2 G}{\partial T^2}\right)_P = T\left(\frac{\partial S}{\partial T}\right)_P = C_P \tag{7-100}$$

also shows a discontinuity at the transition temperature. Equations (7-80) and (7-98) are derived from the second fundamental equation (7-28). Equation (7-99) can be derived from Eq. (6-114):

$$G = H - TS \tag{6-114}$$

$$\frac{G}{T} = \frac{H}{T} - S \tag{7-101}$$

$$\left[\frac{\partial(G/T)}{\partial(1/T)}\right]_P = \left[\frac{\partial(H/T - S)}{\partial(1/T)}\right]_P = H \tag{7-102}$$

Equation (7-100) follows from Eqs. (6-62) and (7-98). The continuous nature of the Gibbs free energy and the discontinuities in entropy, volume, and enthalpy are illustrated schematically in Figure 7-13. The actual variations in enthalpy, heat capacity, and entropy at the melting point of

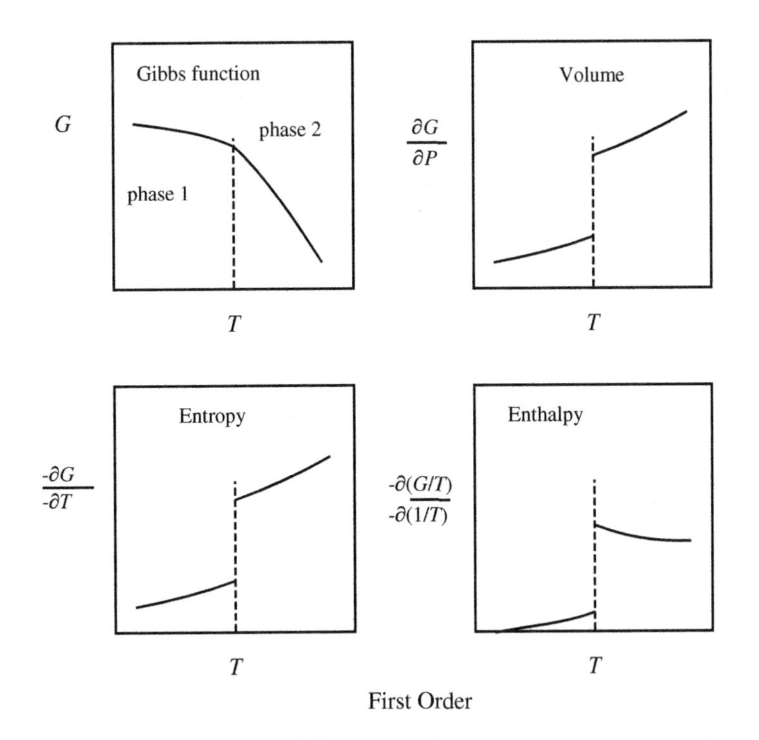

FIGURE 7-13

A schematic diagram of the discontinuities in the volume, entropy, and enthalpy for a first-order phase transition.

corundum (α-Al_2O_3) are shown in Figures 4-6 and 6-4, respectively. Figures 4-25 and 6-8 show the actual variations in enthalpy and entropy at solid-state phase transitions, the melting point, and the boiling point of Fe metal. The discontinuities in S, V, H, and C_P simply mean that for a first-order phase transition, $\Delta S \neq 0$, $\Delta V \neq 0$, $\Delta H \neq 0$, $\Delta C_P \neq 0$, which is what we normally associate with phase transitions, for example, see the data in Table 7-1.

B. Second-order phase transitions

According to Ehrenfest's classification scheme, a second-order phase transition is a phase transition that is continuous in the Gibbs free energy and its first derivatives but discontinuous in the second derivatives of the Gibbs free energy with respect to temperature and pressure. Thus, a second-order phase transition simultaneously has no discontinuities in the entropy, volume, and enthalpy and $\Delta S = 0$, $\Delta V = 0$, and $\Delta H = 0$. However, there are discontinuities simultaneously in the heat capacity C_P, the isothermal compressibility β_T, and the isobaric thermal expansion coefficient α: $\Delta C_P \neq 0$, $\Delta\beta_T \neq 0$, and $\Delta\alpha \neq 0$. The second derivatives of the Gibbs free energy with respect to temperature and pressure are

$$-T\left(\frac{\partial^2 G}{\partial T^2}\right)_P = T\left(\frac{\partial S}{\partial T}\right)_P = C_P \tag{7-100}$$

$$\left(\frac{\partial^2 G}{\partial P^2}\right)_T = \left(\frac{\partial V}{\partial P}\right)_T = -\beta_T V \tag{7-103}$$

$$\frac{\partial}{\partial T}\left[\frac{\partial(G/T)}{\partial(1/T)}\right]_P = \left(\frac{\partial H}{\partial P}\right)_T = C_P \tag{7-104}$$

$$\frac{\partial}{\partial T}\left[\left(\frac{\partial G}{\partial P}\right)_T\right]_P = \frac{\partial^2 G}{\partial T \partial P} = \left(\frac{\partial V}{\partial T}\right)_P = \alpha V \tag{7-105}$$

Equations (7-100) and (7-104) are two different derivations that show the heat capacity C_P is the second derivative of the Gibbs free energy. The variations of G, V, S, and C_P for a second-order phase transition are displayed schematically in Figure 7-14.

We cannot use the Clapeyron equation for second-order phase transitions because they do *not* have discontinuities in entropy, volume, and enthalpy. Equation (7-31) is indeterminate (0/0).

Instead, we need to use analogous equations, known as *Ehrenfest's equations*, to calculate the (dP/dT) slope. These two equations are

$$\left(\frac{dP}{dT}\right)_{eq} = \frac{\Delta C_P}{TV\Delta\alpha_P} \tag{7-106}$$

$$\left(\frac{dP}{dT}\right)_{eq} = \frac{\Delta\alpha_P}{\Delta\beta_T} \tag{7-107}$$

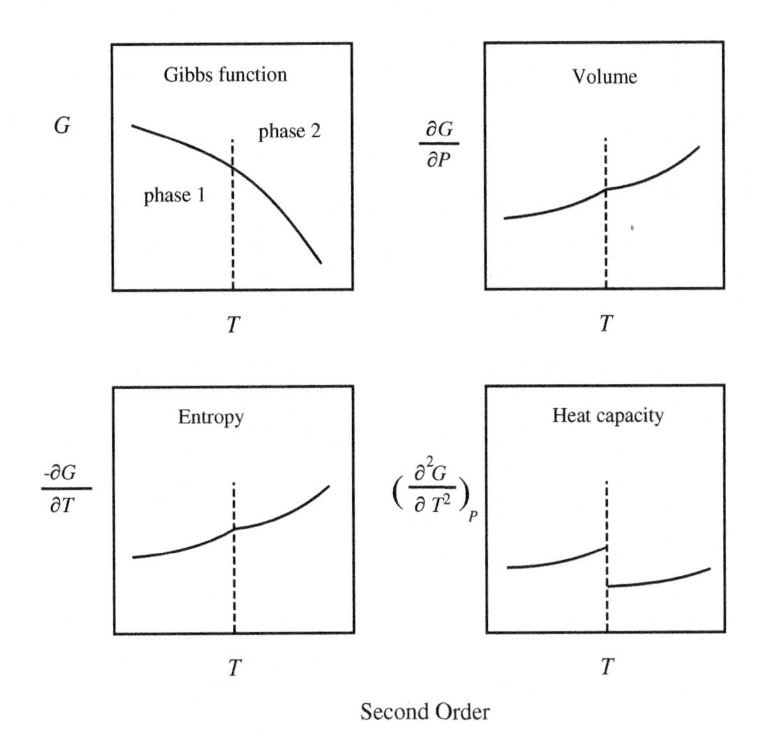

FIGURE 7-14

A schematic diagram of the variations in Gibbs free energy, volume, entropy, and heat capacity for a second-order phase transition.

where α_P is the isobaric thermal expansion coefficient:

$$\alpha_P = \frac{1}{V}\left(\frac{\partial V}{\partial T}\right)_P = \left(\frac{\partial \ln V}{\partial T}\right)_P \tag{2-33}$$

and β_T is the isothermal compressibility:

$$\beta_T = -\frac{1}{V}\left(\frac{\partial V}{\partial P}\right)_T = -\left(\frac{\partial \ln V}{\partial P}\right)_T \tag{2-34}$$

The onset of superconductivity in the absence of a magnetic field is apparently the only second-order phase transition, according to Ehrenfest's classification.

C. Lambda transitions

Lambda transitions occur in the heat capacity curves of many minerals as transitions from ordered to disordered states occur with increasing temperature (Section III-G, Chapter 4). The heat capacity curve looks like the Greek letter λ (lambda) in the vicinity of the phase transition. Figures 4-21 to 4-24 illustrate a schematic lambda transition and observed lambda transitions in the C_P curves of olivines, Ni and Fe metal, quartz, and crystalline NH_4Cl. Examples of lambda transitions include the low-to-high quartz transition (846.5 K), order–disorder transitions in metallic alloys, the onset of ferromagnetism in Fe and Ni metal at their Curie points (1043 K and 631 K, respectively), and the onset of ferroelectricity in $BaTiO_3$ at its Curie point (393 K).

It is unclear for practical and theoretical reasons whether all or some lambda transitions fall into Ehrenfest's category of second-order transitions. The practical reasons for this uncertainty are that it is very difficult, if not impossible, to distinguish between an extremely large but finite value of C_P and an infinite value of C_P at the transition temperature. The values of other properties (e.g., thermal expansivity, isobaric compressibility, entropy) also vary sharply with temperature in the vicinity of a lambda transition and have singularities at the transition. The theoretical reasons are that higher-order phase transitions are apparently more complex than envisioned in Ehrenfest's classification scheme. Lambda transitions and other higher-order phase transitions are the result of order–disorder reactions taking place over a range of temperatures and sometimes involving an intermediate phase at the transition temperature. This is the case for the low-to-high quartz transition. Several articles and books provide good reviews of this topic (Pippard, 1957; Landau and Lifshitz, 1958).

The best-studied lambda transition is probably the lambda curve between "higher-temperature" liquid ^{4}He I and superfluid liquid ^{4}He II (the lower-temperature phase). The heat capacity of liquid ^{4}He II is about 10 times larger than that of liquid ^{4}He I, the thermal conductivity of ^{4}He II is about 10^6 larger than that of ^{4}He I, and ^{4}He II has zero viscosity. Heat capacity data extending within one microdegree of the transition temperature indicate that C_P rises to an infinite value at the transition temperature (Buckingham and Fairbanks, 1961; Keller, 1969).

Helium also has a lighter stable isotope ^{3}He, which is much less abundant than ^{4}He; the ^{3}He/^{4}He atomic ratio in Earth's atmosphere is about 1.4×10^{-6}. The phase diagram and properties of ^{3}He are different than those of ^{4}He (see Keller, 1969) and our discussion is for ^{4}He. However, several points are worth mentioning. First, ^{3}He and ^{4}He are the only substances that remain liquid down to absolute zero. Second, ^{3}He and ^{4}He are the only substances that do *not* have a *S-L-V* triple point. Third, the properties of ^{3}He and ^{4}He are different because they have different masses, they are composed of an odd number (two protons and one neutron in ^{3}He) versus an even number (two protons and two neutrons in ^{4}He) of elementary particles, and they have different nuclear spins (½ spin for ^{3}He versus 0 spin for ^{4}He).

A schematic phase diagram for ^{4}He is shown in Figure 7-15. This plots the stability field of gaseous, liquid, and solid (hexagonal close packed, hcp) ^{4}He at low temperatures (below 10 K) and up to several tens of bar pressure. The small stability field of body-centered cubic (bcc) solid ^{4}He is too small to be shown on the diagram. This oval-shaped field extends from the vicinity of the lambda curve to 1.44 K, 26.3 bars. The critical point of ^{4}He is 5.19 K, 2.27 bars. The one-bar boiling point of liquid ^{4}He I is 4.22 K. Liquid ^{4}He I transforms to superfluid ^{4}He II at 2.172 K, 0.0508 bars. The transition temperature varies with pressure (the lambda curve) and extends to the *S-I-II* triple point at 1.76 K, 30.13 bars. The solid ^{4}He–liquid ^{4}He II melting curve approaches absolute zero with a (dP/dT) slope of zero. In the vicinity of the lambda line, the solid ^{4}He–liquid ^{4}He II and ^{4}He–liquid ^{4}He I melting curves have positive dP/dT slopes.

Experimental measurements have failed to detect an enthalpy change for the lambda transition from liquid ^{4}He I to liquid ^{4}He II. Furthermore, the densities of the two liquid phases are the same at

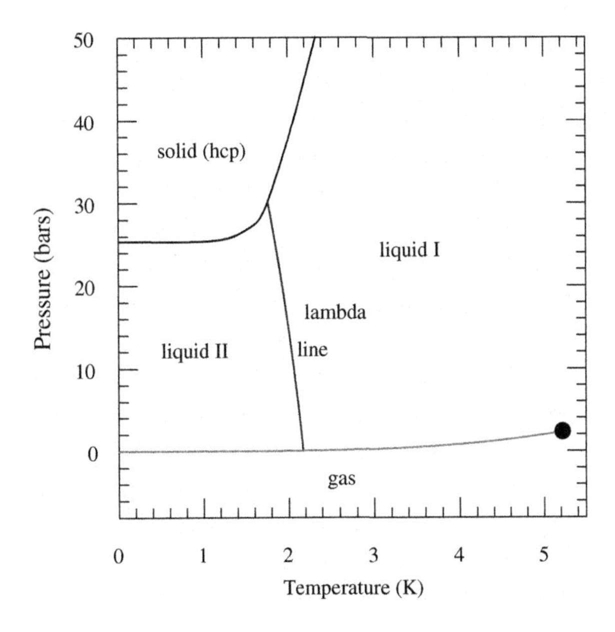

FIGURE 7-15

Schematic phase diagram for ^{4}He.

the lambda point (2.172 K) where ^{4}He vapor, liquid I, and liquid II all coexist. As can be seen from the phase diagram in Figure 7-15, the slopes of the two curves,

$$^4\text{He (liquid II)} = {}^4\text{He (vapor)} \tag{7-108}$$

$$^4\text{He (liquid I)} = {}^4\text{He (vapor)} \tag{7-109}$$

are the same because there is no discontinuity at the lambda point (2.172 K, 0.0508 atm). Thus, the Clapeyron slopes for vaporization of liquid II and liquid I are equal:

$$\left(\frac{dP}{dT}\right)_{eq} = \frac{\Delta S(II - vapor)}{\Delta V(II - vapor)} = \frac{\Delta S(I - vapor)}{\Delta V(I - vapor)} \tag{7-110}$$

Likewise, there is no discontinuity in the slopes of the two freezing curves:

$$^4\text{He (liquid II)} = {}^4\text{He (solid)} \tag{7-111}$$

$$^4\text{He (liquid I)} = {}^4\text{He (solid)} \tag{7-112}$$

Thus, we cannot use the Clapeyron equation to calculate the (dP/dT) slope for the lambda line but must use one of the Ehrenfest equations instead.

Example 7-10. Liquid ^{4}He I and II have the following properties in the vicinity of the lambda point 2.172 K: $\rho = 0.14615$ g cm^{-3}, c_{sat} (liquid I) $= 20.35$ J g^{-1} K^{-1}, c_{sat} (liquid II) $= 25.55$ J g^{-1} K^{-1}, α (liquid I) $= -0.02$ K^{-1}, α (liquid II) $= -0.06$ K^{-1}. Then, applying the first Ehrenfest equation,

$$\left(\frac{dP}{dT}\right)_{eq} = \frac{\Delta C_P}{TV\Delta\alpha} = \frac{(20.35 - 25.55)\ \text{J g}^{-1}\text{K}^{-1}}{(2.172\ \text{K})(6.8423\ \text{cm}^3\text{g}^{-1})(-0.02 + 0.06\ \text{K}^{-1})} \tag{7-113}$$

$$\left(\frac{dP}{dT}\right)_{eq} = -\frac{5.0\ \text{J g}^{-1}\ \text{K}^{-1}}{0.594\ \text{cm}^3\ \text{g}^{-1}}(9.869\ \text{cm}^3\text{atm J}^{-1}) = -86.4\ \text{atm K}^{-1} \tag{7-114}$$

This result is only in fair agreement with the measured value of -110 atm K^{-1} (Keller, 1969). However, the slope of the lambda line varies steeply near the lambda point and is -90 atm K^{-1}, about 0.05 K lower. Also part of the disagreement may result from the difficulty in measuring the specific heat and thermal expansion coefficient so near the lambda point. We cannot use the second Ehrenfest equation (7-107) in this case because the isothermal compressibilities of liquid ^4He I and II are the same, within experimental error.

V. MELTING CURVES AND POLYMORPHIC PHASE TRANSITIONS AT HIGH PRESSURES

A. Effect of high pressure on melting points

Earlier in this chapter we compared the calculated and observed melting curves of ice I and found that the calculated (dT/dP) curve started to deviate from the observed curve at ~250 atmospheres pressure (Example 7-3 and Figure 7-6). In general, the agreement between calculated Clapeyron slopes and observed slopes is only good at temperatures near the one-bar melting point and at "low" pressures because the dependence of ΔS and ΔV (or equivalently of ΔH and ΔV) on pressure and temperature along the melting curve must be considered. The differential volume change of the solid and melt along the melting curve is given by the complete differentials dV_{solid} and dV_{melt} with respect to P and T. We can rewrite Eq. (2-21) for the solid and melt:

$$dV_{solid} = \left(\frac{\partial V_{solid}}{\partial T}\right)_P dT + \left(\frac{\partial V_{solid}}{\partial P}\right)_T dP \tag{7-115}$$

$$dV_{melt} = \left(\frac{\partial V_{melt}}{\partial T}\right)_P dT + \left(\frac{\partial V_{melt}}{\partial P}\right)_T dP \tag{7-116}$$

Along the melting curve, the volumes of the solid and melt are the sum of their volume at the one-bar melting point (V^o) and their volume change (ΔV) along the curve:

$$V_{solid} = V^o_{solid} + \Delta V_{solid} = V^o_{solid} + \int_{m.p.}^{T} \left(\frac{\partial V_{solid}}{\partial T}\right)_P dT + \int_{1}^{P} \left(\frac{\partial V_{solid}}{\partial P}\right)_T dP \tag{7-117}$$

$$V_{melt} = V^o_{melt} + \Delta V_{melt} = V^o_{melt} + \int_{m.p.}^{T} \left(\frac{\partial V_{melt}}{\partial T}\right)_P dT + \int_{1}^{P} \left(\frac{\partial V_{melt}}{\partial P}\right)_T dP \tag{7-118}$$

The ΔV of melting ($V_{melt} - V_{solid}$) at any point along the melting curve is thus given by

$$(\Delta V)_T^P = (\Delta V)_{m.p.}^o + \int_{m.p.}^{T} \left[\left(\frac{\partial V_{melt}}{\partial T}\right)_P - \left(\frac{\partial V_{solid}}{\partial T}\right)_P\right] dT + \int_{1}^{P} \left[\left(\frac{\partial V_{melt}}{\partial P}\right)_T - \left(\frac{\partial V_{solid}}{\partial P}\right)_T\right] dP \quad (7\text{-}119)$$

$$(\Delta V)_T^P = (\Delta V)_{m.p.}^o + \int_{m.p.}^{T} \Delta\left(\frac{\partial V_i}{\partial T}\right)_P dT + \int_{1}^{P} \Delta\left(\frac{\partial V_i}{\partial P}\right)_T dP \quad (7\text{-}120)$$

We can rewrite Eq. (7-120) using the isobaric thermal expansion coefficient (α_P) from Eq. (2-33) and the isothermal compressibility (β_T) from Eq. (2-34):

$$(\Delta V)_T^P = (\Delta V)_{m.p.}^o + \int_{m.p.}^{T} \Delta(\alpha_i V_i) dT - \int_{1}^{P} \Delta(\beta_i V_i) dP \quad (7\text{-}121)$$

The minus sign in Eq. (7-121) is from the definition of β_T in Eq. (2-34). Assuming that $\Delta(\alpha_i V_i)$ and $\Delta(\beta_i V_i)$ are constant along the melting curve, we then obtain

$$(\Delta V)_T^P = (\Delta V)_{m.p.}^o + \Delta(\alpha_i V_i)(T - T_{m.p.}) - \Delta(\beta_i V_i)(P - 1) \quad (7\text{-}122)$$

The differential entropy change of the solid and melt along the melting curve is given by the complete differentials dS_{solid} and dS_{melt} with respect to P and T

$$dS_{solid} = \left(\frac{\partial S_{solid}}{\partial T}\right)_P dT + \left(\frac{\partial S_{solid}}{\partial P}\right)_T dP \quad (7\text{-}123)$$

$$dS_{melt} = \left(\frac{\partial S_{melt}}{\partial T}\right)_P dT + \left(\frac{\partial S_{melt}}{\partial P}\right)_T dP \quad (7\text{-}124)$$

Along the melting curve, the entropies of the solid and melt are the sum of their entropy at the one-bar melting point (S^o) and their entropy change (ΔS) along the curve:

$$S_{solid} = S_{solid}^o + \Delta S_{solid} = S_{solid}^o + \int_{m.p.}^{T} \left(\frac{\partial S_{solid}}{\partial T}\right)_P dT + \int_{1}^{P} \left(\frac{\partial S_{solid}}{\partial P}\right)_T dP \quad (7\text{-}125)$$

$$S_{melt} = S_{melt}^o + \Delta S_{melt} = S_{melt}^o + \int_{m.p.}^{T} \left(\frac{\partial S_{melt}}{\partial T}\right)_P dT + \int_{1}^{P} \left(\frac{\partial S_{melt}}{\partial P}\right)_T dP \quad (7\text{-}126)$$

The ΔS of melting ($S_{melt} - S_{solid}$) at any point along the melting curve is

$$(\Delta S)_T^P = (\Delta S)_{m.p.}^o + \int_{m.p.}^{T} \left[\left(\frac{\partial S_{melt}}{\partial T}\right)_P - \left(\frac{\partial S_{solid}}{\partial T}\right)_P\right] dT + \int_{1}^{P} \left[\left(\frac{\partial S_{melt}}{\partial P}\right)_T - \left(\frac{\partial S_{solid}}{\partial P}\right)_T\right] dP \quad (7\text{-}127)$$

We can rewrite Eq. (7-127) using Eq. (6-62) to substitute for $(\partial S/\partial T)_P$ and one of the Maxwell relations from Chapter 8 $[(\partial S/\partial P)_T = -(\partial V/\partial T)_P]$. This gives us

$$(\Delta S)_T^P = (\Delta S)^o_{m.p.} + \int_{m.p.}^{T} \frac{\Delta C_P}{T} dT - \int_1^P \Delta(\alpha_i V_i) dP \tag{7-128}$$

Finally, assuming that ΔC_P and $\Delta(\alpha_i V_i)$ are constant along the melting curve, we get

$$(\Delta S)_T^P = (\Delta S)^o_{m.p.} + \Delta C_P \ln\left(\frac{T}{T_{m.p.}}\right) - \Delta(\alpha_i V_i)(P-1) \tag{7-129}$$

The (dT/dP) slope along the melting curve is thus given by

$$\left(\frac{dT}{dP}\right)_{eq} = \frac{\Delta V}{\Delta S} = \frac{(\Delta V)^o_{m.p.} + \Delta(\alpha_i V_i)(T - T_{m.p.}) - \Delta(\beta_i V_i)(P-1)}{(\Delta S)^o_{m.p.} + \Delta C_P \ln(T/T_{m.p.}) - \Delta(\alpha_i V_i)(P-1)} \tag{7-130}$$

It should be apparent that Eq. (7-130) reduces to the Clapeyron equation at standard-state conditions (i.e., $T = T_{m.p.}$ and $P = 1$ bar). Equation (7-130) was derived assuming that $\Delta(\alpha_i V_i)$, $\Delta(\beta_i V_i)$, and ΔC_P are constant along the melting curve. This is only true over a limited range, which is larger for two solid polymorphs than it is for a solid and melt. Nevertheless, Eq. (7-130) shows why the initial Clapeyron slope is different than that at higher pressures.

Bridgman (1915) derived Eq. (7-130) and also showed that the curvature of the Clapeyron slope (dT/dP) for the phase boundary between a solid and melt is given by

$$\left(\frac{d^2T}{dP^2}\right)_{eq} = -\frac{1}{\Delta V}\left(\frac{dT}{dP}\right)_{eq}\left[\frac{\Delta C_P}{T}\left(\frac{dT}{dP}\right)^2_{eq} - 2\Delta(\alpha_i V_i)\left(\frac{dT}{dP}\right)_{eq} + \Delta(\beta_i V_i)\right] \tag{7-131}$$

Most melting curves curve downward when melting temperature is plotted as a function of pressure, which means that the second derivative $(d^2T/dP^2) < 0$. This behavior is followed by melting curves with negative Clapeyron slopes (water; see Figure 7-6) and materials with positive Clapeyron slopes (Fe metal, SiO_2, silicates; see Figure 7-16) and other ices (CO, CH_4, N_2, NH_3). Some melting curves, such as those of cesium, rubidium, and tellurium, increase to a maximum temperature, flatten out, and then decrease again. The melting curve of the lead-platinum (Pb-Pt) eutectic is unusual because it curves upward, that is, increasing (dT/dP) slope with increasing pressure (see Figure 7-14 in Kennedy and Newton, 1963). Inequalities involving the difference of heat capacity ΔC_P, thermal expansivity $(\Delta\alpha V)$, and compressibility $(\Delta\beta V)$ between melt and solid can be derived from Eq. (7-131). These inequalities provide qualitative or semiquantitative explanations for different types of melting curves (curving downward, increasing to a maximum, or curving upward) and their observed behavior, for example, larger curvature of the albite and diopside curves versus the linearity of the iron and forsterite melting curves. However, quantitative modeling is difficult or impossible because in most cases at least some of the necessary parameters and/or their variations with temperature and pressure are unknown.

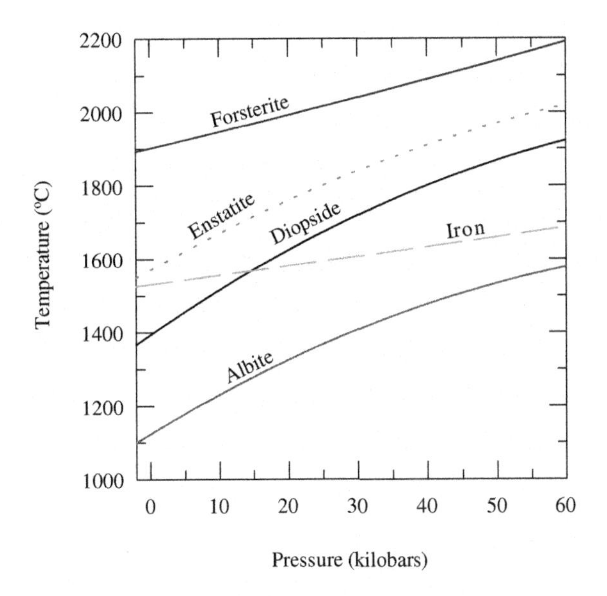

FIGURE 7-16

Melting curves of albite, diopside, enstatite, forsterite, and iron metal.

Thus, various empirical relationships have been devised to describe high-pressure melting curves. One of these is the Simon-Glatzel equation:

$$\frac{P - P_0}{a} = \left(\frac{T_m}{T_0}\right)^c - 1 \tag{7-132}$$

where T_0 and P_0 are the triple-point temperature and pressure, a and c are empirically derived constants that are different for each material, and T_m is the melting point at pressure P. In many cases P_0 is at sufficiently low pressure that it is taken as zero. For example, the constants for Fe metal are $a = 107$ GPa, $c = 1.76$, and $T_0 = 1811$ K.

The Kraut-Kennedy equation is another empirical relationship. This is based on the observation that melting curves are linear if they are plotted against compression ($\Delta V/V_0$) instead of pressure. The Kraut-Kennedy equation is

$$T_m = T_m^0[1 + C(\Delta V/V_0)] \tag{7-133}$$

where T_m^0 and V_0 are the melting point and volume at "zero" pressure (one bar), $\Delta V/V_0$ is the compression, and C is an empirical constant fit to the melting curve.

The Lindemann melting law is a third empirical relationship. It was already introduced in Problem 4-14 of Chapter 4 and is a correlation between the Debye temperature and melting point of a solid. However, by rearranging this correlation we obtain

$$T_m = \theta_D^2 \mu V_m^{2/3}(c')^2 \tag{7-134}$$

where T_m is the melting temperature, θ_D is the Debye temperature, μ is the atomic (or molecular weight), V_m is the molar volume, and c' is a constant for each material. The Lindemann melting law can be used by assuming a value of c' based on other solids with the same crystal structure.

B. Calculations of high-pressure phase boundaries

The synthesis of diamond from graphite is a good example to illustrate the calculation of high-pressure phase boundaries from thermodynamic data. Our discussion is based on the classic paper by Berman and Simon (1955). Their computed phase boundary is essentially the same as that determined experimentally by Bundy et al. (1961) and subsequently by Kennedy and Kennedy (1976). An excellent history of man's attempts to make diamond, the first successful synthesis by Bundy and colleagues at the GE research laboratory in 1954, and subsequent events is given by Hazen (1999). The reaction to make diamonds from graphite is

$$C\ (\text{graphite}) = C\ (\text{diamond}) \tag{7-135}$$

Along the graphite-diamond phase boundary, we have

$$G_T^P\ (\text{graphite}) = G_T^P\ (\text{diamond}) \tag{7-136}$$

because both phases coexist at equilibrium. The superscript ^{o} is not on the Gibbs energies because we are at high pressure and not at standard-state conditions (1 atm or 1 bar pressure, depending on the definition used in the data compilation). Computation of the graphite-diamond phase boundary requires finding the series of P-T points along which $\Delta G = 0$ for Eq. (7-135). Thus, we need to explicitly consider the temperature and pressure dependence of the ΔG of reaction. The temperature dependence of ΔG involves integrals for computing enthalpy and entropy changes as a function of temperature. This has been described in Chapters 4 and 6. The pressure dependence is given by rewriting Eq. (7-80) in terms of ΔG:

$$\left[\frac{\partial(\Delta G)}{\partial P}\right]_T = \Delta V \tag{7-137}$$

The ΔV is the temperature-and pressure-dependent volume change for reaction (7-135). The general equation for finding $\Delta G = 0$ along the phase boundary is

$$(\Delta G)_T^P = \Delta H_{298}^o + \int_{298}^{T} \Delta C_P dT - T\Delta S_{298}^o - T\int_{298}^{T} \frac{\Delta C_P}{T} dT + \int_{P=1}^{P=P} \Delta V dP \tag{7-138}$$

$$(\Delta G)_T^P = (\Delta G)_T^o + \int_{P=1}^{P=P} \Delta V dP \tag{7-139}$$

Equations (7-138) and (7-139) are general equations that apply to any high-pressure phase boundary (polymorphs and melting curves), not just to the graphite-diamond boundary.

We first evaluate the standard Gibbs free energy change. Thermodynamic data at 298 K are given in Table 7-1. Extension of these data to higher temperatures involves heat capacity equations for graphite and diamond that we fit to data from 298 K to 1200 K. The two equations are

$$C_P\ (\text{graphite}) = 3.6691 + 30.2948 \times 10^{-3}T - 2.8238 \times 10^{5}T^{-2} - 11.9816 \times 10^{-6}T^{2}\ \text{J mol}^{-1}\ \text{K}^{-1} \quad (7\text{-}140)$$

$$C_P\ (\text{diamond}) = 2.936 + 30.2563 \times 10^{-3}T - 4.4115 \times 10^{5}T^{-2} - 10.9529 \times 10^{-6}T^{2}\ \text{J mol}^{-1}\ \text{K}^{-1} \quad (7\text{-}141)$$

The ΔC_P is C_P (diamond) $-$ C_P (graphite) and is given by

$$\Delta C_P = -0.7331 - 0.0385 \times 10^{-3}T - 1.5877 \times 10^{5}T^{-2} + 1.0287 \times 10^{-6}T^{2}\ \ \text{J mol}^{-1}\ \text{K}^{-1} \quad (7\text{-}142)$$

We now evaluate the ΔC_P integrals in Eq. (7-138). The enthalpy integral (J mol^{-1}) is

$$\int_{298}^{T} \Delta C_P dT = -321.3 - 0.7331T - 1.925 \times 10^{-5}T^{2} + 1.5877 \times 10^{5}T^{-1} + 3.429 \times 10^{-7}T^{3} \quad (7\text{-}143)$$

The entire enthalpy term (J mol^{-1}) is

$$\Delta H_T^o = \Delta H_{298}^o + \int_{298}^{T} \Delta C_P dT \quad (7\text{-}144)$$

$$\Delta H_T^o = 1573.7 - 0.7331T - 1.925 \times 10^{-5}T^{2} + 1.5877 \times 10^{5}T^{-1} + 3.429 \times 10^{-7}T^{3}$$

Table 7-3 gives ΔH_T^o from 298 K to 1200 K and compares our results to Berman and Simon (1955).

The entropy integral (J mol^{-1} K^{-1}) is given by

$$\int_{298}^{T} \frac{\Delta C_P}{T} dT = 3.250 - 0.7331\ \ln T - 3.85 \times 10^{-5}T + 79{,}385T^{-2} + 5.1435 \times 10^{-7}T^{2} \quad (7\text{-}145)$$

The $\Delta S_{298}^o = -3.355$ J mol^{-1} K^{-1} and the entire entropy term (J mol^{-1} K^{-1}) is

$$\Delta S_T^o = \Delta S_{298}^o + \int_{298}^{T} \frac{\Delta C_P}{T} dT \quad (7\text{-}146)$$

$$\Delta S_T^o = -0.105 - 0.7331\ \ln T - 3.85 \times 10^{-5}T + 79{,}385T^{-2} + 5.1435 \times 10^{-7}T^{2}$$

Table 7-4 gives ΔS_T^o from 298 K to 1200 K and compares our results to Berman and Simon (1955). Now we can combine ΔH_T^o and $-T\Delta S_T^o$ to get ΔG_T^o. Table 7-5 gives our results and a comparison to those of Berman and Simon (1955). The two computations are identical at 298.15 K and diverge with

Table 7-3 Enthalpy Values (J mol^{-1}) for Graphite-Diamond Conversion

Temperature (K)	ΔC_P Integral	ΔH_T^o	Berman (ΔH_T^o)
298.15	0	1895	1895
400	−199	1696	1686
500	−332	1563	1464
600	−429	1466	1318
700	−500	1396	1297
800	−546	1349	1255
900	−570	1325	1234
1000	−572	1323	1213
1100	−550	1345	1213
1200	−504	1391	1172

Table 7-4 ΔS^o Values (J mol^{-1} K^{-1}) for Graphite-Diamond Conversion

Temperature (K)	$-(\Delta C_P/T$ Integral)	$-\Delta S^o$	Berman ($-\Delta S^o$)
298.15	0	3.355	3.356
400	0.580	3.935	3.975
500	0.879	4.234	4.477
600	1.057	4.412	4.728
700	1.166	4.521	4.770
800	1.228	4.583	4.812
900	1.257	4.612	4.853
1000	1.259	4.614	4.853
1100	1.238	4.593	4.895
1200	1.198	4.553	4.895

increasing temperature. The disagreement is 195 J mol^{-1} at 1200 K, and arises from the different C_p equations used. A linear least-squares fit to our ΔG^o data in Table 7-5 gives

$$\Delta G_T^o = 1489.81 + 4.4412T \text{ J mol}^{-1} \tag{7-147}$$

We now compute the temperature and pressure dependence of ΔV for reaction (7-135). Average values of the isobaric thermal expansion coefficients and isothermal compressibilities for diamond (α_D) and graphite (α_G) are given in Tables 2-1 and Table 2-12, respectively. The volume of each phase as a function of T and P relative to its volume at 298.15 K is given by

$$\begin{aligned} V_T^P &= V_{298}^o + \alpha_{av} V_{298}^o (T - 298) - \beta_{av} V_{298}^o (P - 1) \\ &= V_{298}^o [1 + \alpha_{av}(T - 298) - \beta_{av}(P - 1)] \end{aligned} \tag{7-148}$$

Table 7-5 ΔG^o Values (J mol^{-1}) for Graphite-Diamond Conversion

Temperature (K)	ΔG^o	Berman (ΔG^o)
298.15	2895	2895
400	3270	3276
500	3680	3682
600	4113	4142
700	4561	4644
800	5015	5104
900	5476	5607
1000	5937	6067
1100	6397	6569
1200	6855	7050

The volume change for reaction (7-135) as a function of T and P is then given by

$$\begin{aligned}(\Delta V)_T^P = V_D - V_G &= \Delta V_{298}^o + (\alpha_D V_{D,298}^o - \alpha_G V_{G,298}^o)\Delta T - (\beta_D V_{D,298}^o - \beta_G V_{G,298}^o)\Delta P \\ &= \Delta V_{298}^o + \Delta(\alpha_i V_{i,298}^o)\Delta T - \Delta(\beta_i V_{i,298}^o)\Delta P\end{aligned} \quad (7\text{-}149)$$

Equation (7-149) assumes that α and β are constant and independent of temperature and pressure. Neither assumption is strictly correct, but they are usually made because insufficient data are available to do otherwise. Using numerical data from Tables 2-1, 2-12, and 7-1 we obtain

$$\begin{aligned}(\Delta V)_T^P &= -1.882 - 1.368 \times 10^{-4}(T - 298) + 1.493 \times 10^{-5}\Delta P \text{ cm}^3 \text{ mol}^{-1} \\ &= -0.1882 - 1.368 \times 10^{-5}(T - 298) + 1.493 \times 10^{-6}\Delta P \text{ J bar}^{-1}\end{aligned} \quad (7\text{-}150)$$

The calculated values from Eq. (7-150) at one bar pressure are compared to those of Berman and Simon in Table 7-6. The two computations agree within 1% at all temperatures.

Table 7-6 ΔV_T Values (cm^3 mol^{-1}) for Graphite-Diamond Conversion

T (K)	$-\Delta V_T$	Berman ($-\Delta V_T$)
298.15	1.882	1.883
400	1.896	1.895
500	1.910	1.907
600	1.923	1.919
700	1.937	1.929
800	1.951	1.941
900	1.964	1.953
1000	1.978	1.966
1100	1.992	1.978
1200	2.005	1.991

We now use Eq. (7-150) to evaluate the ΔV integral in Eq. (7-139). The pressure along the graphite-diamond phase boundary is much larger than unity, so we can neglect the difference between one bar pressure and zero pressure:

$$\begin{aligned}\int_0^P (\Delta V)_T^P dP &= \int_0^P (-0.1882 - 1.368 \times 10^{-5}(T - 298) + 1.493 \times 10^{-6}P)dP \\ &= [-0.1882 - 1.368 \times 10^{-5}(T - 298)]P + 7.465 \times 10^{-7}P^2 \\ &= (\Delta V_T^o)P + 7.465 \times 10^{-7}P^2\end{aligned} \tag{7-151}$$

The final equation to be solved for the pressure of the graphite-diamond phase boundary as a function of temperature is

$$1489.81 + 4.4412T + (\Delta V_T^o)P + 7.465 \times 10^{-7}P^2 = 0 \tag{7-152}$$

Equation (7-152) is a quadratic equation. The *P-T* points along the graphite-diamond phase-boundary are listed in Table 7-7 and plotted in Figure 7-17. Our phase boundary and that computed by Berman and Simon (1955) are virtually indistinguishable. They intersect at 1280 K, and our boundary is at higher pressure at a given temperature. Experimentally determined *P-T* points from the GE researchers (e.g., Bundy et al., 1961; Strong and Chrenko, 1971) fall on or between the two computed equilibrium lines. The results of Kennedy and Kennedy (1976) are at slightly higher pressures. However, the synthetic diamonds are not made by the direct conversion of graphite to diamond. This requires higher pressures and temperatures because of kinetic barriers. The synthetic diamonds are made by dissolving carbon in molten Fe, Ni alloy. Any form of carbon can be used as the precursor, and in one case, peanut butter was used.

Table 7-7 The Graphite-Diamond Phase Boundary

T (K)	*P* (10^3 atm)	Berman (10^3 atm)
298.15	15.75	16.15
400	18.34	18.25
500	20.90	20.5
600	23.50	23.0
700	26.09	26.0
800	28.70	28.5
900	31.36	31.5
1000	34.02	34.0
1100	36.70	37.0
1200	39.44	39.5
1500	47.75	47.5
2000	62.27	61.0

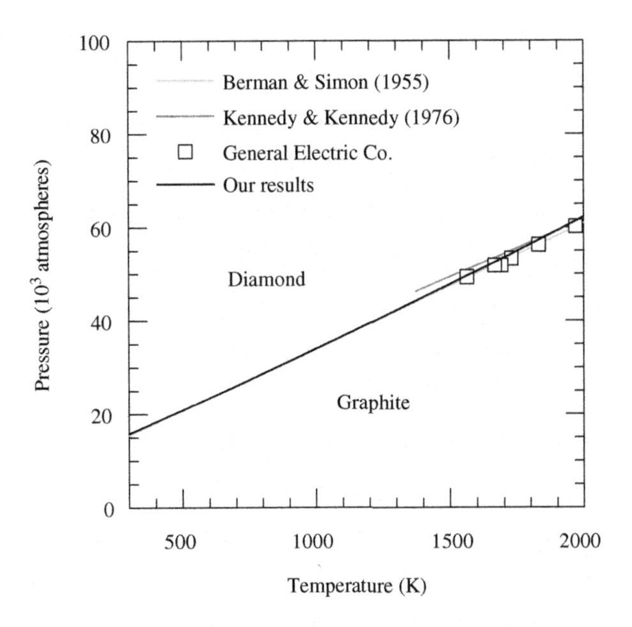

FIGURE 7-17

Comparison of theoretical and experimental data for the graphite-diamond phase boundary.

C. Retrieval of thermodynamic data from phase equilibria

The equations and methods used to compute the graphite-diamond phase boundary can also be used to extract thermodynamic data from high-pressure experiments. Holland and Powell (1990, 1998) describe retrieval of thermodynamic data from petrological phase equilibria. We give two examples here and return to this topic in Chapters 10–12.

Example 7-11. Figure 7-10 shows the pressure dependence of the orthorhombic-to-monoclinic sulfur transition. The equation for the univariant curve along which both phases coexist is

$$T_{trans} = 368.50 + 0.0380P + 2.648 \times 10^{-6}P^2 \quad (7\text{-}68)$$

The Gibbs free energy change for the orthorhombic to monoclinic sulfur phase transition is zero at any P-T pair along the univariant curve because the two sulfur allotropes are in equilibrium:

$$(\Delta G)_T^P = 0 = (\Delta G)_T^o + \int_{P=1}^{P=P} \Delta V dP \quad (7\text{-}139)$$

Tammann measured the ΔV of transition along the univariant curve and his results are given by

$$\begin{aligned} \Delta_{trans} V &= 0.01379 + 2.08 \times 10^{-7} P \text{ cm}^3 \text{ g}^{-1} \\ &= 0.04422 + 6.67 \times 10^{-7} P \text{ J bar}^{-1} \end{aligned} \quad (7\text{-}153)$$

Thus, the standard Gibbs free energy change between orthorhombic and monoclinic sulfur is

$$(\Delta G)^o_T = -\int_{P=1}^{P=P} \Delta V dP = -3.33 \times 10^{-7}P^2 - 0.04422P + 0.04422 \text{ J mol}^{-1} \quad (7\text{-}154)$$

We solve Eq. (7-154) for $\Delta G^o{}_T$ at a given pressure. Then we use Eq. (7-68) to find T_{trans} at the same pressure. Our results are summarized in Table 7-8, which also gives a comparison to the standard Gibbs free energy change derived from thermal data (see Problem 21 in this chapter):

$$\Delta G^o_T = 151.63 + 4.34T + 4.11 \times 10^{-4}T^2 - 0.830T \ln T \text{ J mol}^{-1} \quad (7\text{-}155)$$

The two sets of calculations agree within 5.5% over the pressure ranged studied by Tammann.

The ice I—liquid water melting curve is another good example (see Figures 7-4 and 7-6). Table 7-2 gives Bridgman's data for this univariant curve, which is described by the equation

$$T = 273.16 - 6.9087 \times 10^{-3}P - 1.5124 \times 10^{-6}P^2 \quad (7\text{-}156)$$

The ΔV for ice I melting to liquid water varies along the melting curve and is given by

$$\Delta V(\text{cm}^3\ \text{g}^{-1}) = -0.09013 - 1.695 \times 10^{-5}P - 1.777 \times 10^{-9}P^2 \quad (7\text{-}157)$$

As discussed earlier, ice I and liquid water coexist in equilibrium along the melting curve

$$H_2O \text{ (ice I)} = H_2O \text{ (water)} \quad (7\text{-}19)$$

The Gibbs free energy change for Eq. (7-19) is zero along the melting curve. Hence,

$$G_{ice} = G_{water} \quad (7\text{-}20)$$

along this curve. Thus, the standard free energy change between ice I and liquid water is

$$(\Delta G)^o_T = -\int_1^P \Delta V dP = -0.09014 + 0.09013P + 8.475 \times 10^{-6}P^2 + 5.923 \times 10^{-10}P^3 \text{ cm}^3 \text{ bar g}^{-1}$$

$$= 0.16237P + 1.527 \times 10^{-5}P^2 + 1.067 \times 10^{-9}P^3 \text{ J mol}^{-1} \quad (7\text{-}158)$$

Table 7-8 ΔG^o Values for Orthorhombic-Monoclinic Sulfur Phase Transition

		ΔG^o_T (J mol^{-1})	
***P* (bars)**	***T* (K)**	**Univariant Curve**	**Thermal Data**
300	380.1	−13.3	−13.5
600	392.2	−26.6	−27.0
900	404.8	−40.0	−41.2
1200	417.9	−53.5	−56.3

Table 7-9 ΔG^o Values for the Ice I—Water Melting Curve			
		ΔG^o_T (J mol^{-1})	
P (bars)	T (K)	Univariant Curve[a]	Vapor Pressure Data[a]
0	273.16	0.0	0.0
500	269.33	85.0	83.1
1000	264.74	178.7	179.8
1500	259.39	281.5	288.6
2000	253.29	394.4	407.4

[a]*For reaction (7-19) ice I melting to liquid water.*

Equation (7-158) is constrained to give $\Delta G^o_T = 0$ at zero pressure corresponding to the triple point where ice I, water, and water vapor are in equilibrium. Table 7-9 summarizes our results for the ice I—water melting curve. We solved Eq. (7-158) for the standard Gibbs free energy change as a function of pressure and used Eq. (7-156) to find the corresponding temperature. For a comparison we used Eq. (7-6) to calculate ΔG^o_T from the vapor pressures over supercooled liquid water and ice. The two sets of calculations, which use two different types of data obtained with different experimental methods, agree within 3.5%.

These two examples show that Gibbs energy data can be retrieved from phase equilibrium data, but they also show that the retrieved data are very sensitive to uncertainties in the *P-T* measurements along the phase boundary.

PROBLEMS

1. Calcite and aragonite are two polymorphic forms of $CaCO_3$. Their enthalpies, entropies, and densities are given in Table 7-1. Calculate which polymorph is stable at Earth's surface, calculate the phase boundary pressure at 298.15 K, and derive an equation for pressure along the phase boundary as a function of temperature (T), assuming that $\Delta C_P = 0$ for the phase transition.
2. Jadeite ($NaAlSi_2O_6$) is a common mineral in Alpine-type metamorphic terrains and forms via

$$NaAlSi_3O_8 \text{ (high albite)} = NaAlSi_2O_6 \text{ (jadeite)} + SiO_2 \text{ (quartz)}$$

Use the data tabulated here to find the following: (a) the pressure of the equilibrium phase boundary between high albite and jadeite + quartz at 298.15 K, (b) the slope (dP/dT) of the phase boundary, (c) the pressure of the phase boundary at 1000°C assuming constant slope.

Mineral	V_m (cm^3)	$\Delta_f H^o_{298}$ (kJ mol^{-1})	S^o_{298} (J mol^{-1} K^{-1})
high albite	100.57	−3924.84	223.40
jadeite	60.40	−3027.83	133.50
quartz	22.69	−910.7	41.46

3. Condensation and vaporization of solid and liquid forsterite were probably important processes for chondrule formation in the solar nebula. Nagahara et al. (1994) report that the *S-L-V* triple point of forsterite (Mg_2SiO_4) is 2163 K and 5.2×10^{-5} bar and that molten Mg_2SiO_4 has an enthalpy of vaporization of 443 kJ mol^{-1}. Compute the one-bar boiling point of molten Mg_2SiO_4.
4. The "best fit line" to the calcite-aragonite phase boundary determinations by Jamieson (1953), MacDonald (1956), Griggs et al. (1960), Crawford and Fyfe (1964), Boettcher and Wyllie (1968), and Johannes and Puhan (1971) is given in this table:

T (K)	350	400	500	600	700
P (bar)	4355	5075	6520	7965	9410

(a) What is the (dP/dT) slope (K bar^{-1})? **(b)** Is the extrapolated phase boundary at 298.15 K at a lower pressure, the same pressure, or a higher pressure than that calculated from the data in Table 7-1? **(c)** What is the calculated ΔG^o_{298} value if $\Delta C_P = 0$?

5. Cinnabar (red HgS) is stable at lower temperatures and transforms to metacinnabar (black HgS) at higher temperatures. Use data from Table 7-1 to calculate (a) the transition temperature at ambient pressure (i.e., one bar) and (b) the initial Clapeyron slope (dT/dP). You can assume that $\Delta C_P = 0$ to a good first approximation.
6. Figure 7-5 shows the *P-T* phase diagram for H_2O at pressures relevant for icy bodies in the outer solar system, for example, the icy Galilean satellites of Jupiter, Titan, Triton, Pluto, Charon, and Edgeworth-Kuiper Belt objects. Consider the change in state ice II → ice V. (a) Is the Clapeyron slope positive or negative? (b) Which phase is denser? (c) Which phase has the higher entropy? (d) Is ΔH exothermic or endothermic?
7. Yoder (1952) measured the effect of pressure on the melting point of diopside ($CaMgSi_2O_6$). The one-bar melting point of diopside is 1670 K and $\Delta T = T_P - 1670$, the difference between the melting point at some pressure *P* and at one bar. Use the subset of Yoder's results given here and $\Delta_{fus}H$ in Table 6-2 to calculate the Clapeyron slope (dT/dP) in K bar^{-1} and the ΔV of melting.

P(bars)	498	1014	1514	2587	3248	4212	4987
ΔT (K)	6.5	13.1	19.8	33.4	42.4	54.3	64.9

8. The triple point of N_2 ice is 63.15 K. The triple-point pressure, that is, the saturated N_2 vapor pressure, is 93.96 mmHg. Calculate the normal melting point of N_2 using these data and $\Delta_{fus}V = 0.072$ cm^3 g^{-1}, $\Delta_{fus}H = 720.9$ J mol^{-1}.
9. The two possible geometries for a triple point involving one vapor phase and two condensed phases (e.g., a solid and melt) in a one-component system are shown in Figure 7-18. You should state why a third configuration is impossible as long as one phase is vapor.

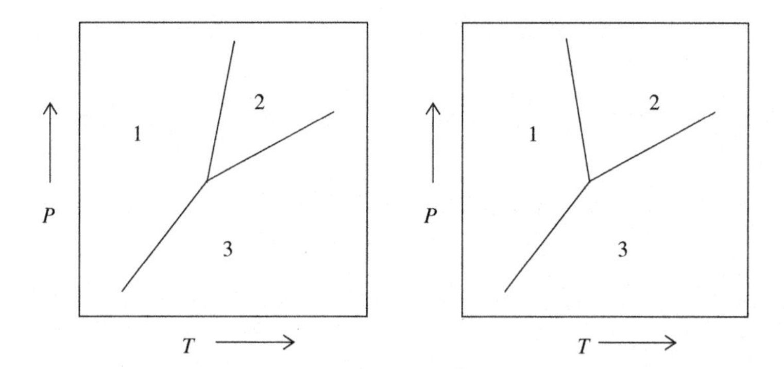

FIGURE 7-18

Schematic diagram of two configurations for *S-L-V* triple point.

10. Figure 7-19 shows the eight possible arrangements of univariant curves around the triple point of three solids S_1-S_2-S_3. In this case the three solids are the Al_2SiO_5 polymorphs. Use the data given in Table 7-1 to quantitatively argue which one is correct and why the other diagrams shown must be incorrect. *Hints:* Compute the Clapeyron slopes. From Le Chatelier's principle we deduce higher T favors higher entropy phases and higher P favors denser phases.
11. Some experimental measurements of the CO_2 pressure as a function of temperature for the calcination of limestone to make cement:

$$CaCO_3\ (s) = CaO\ (lime) + CO_2\ (g) \tag{7-1}$$

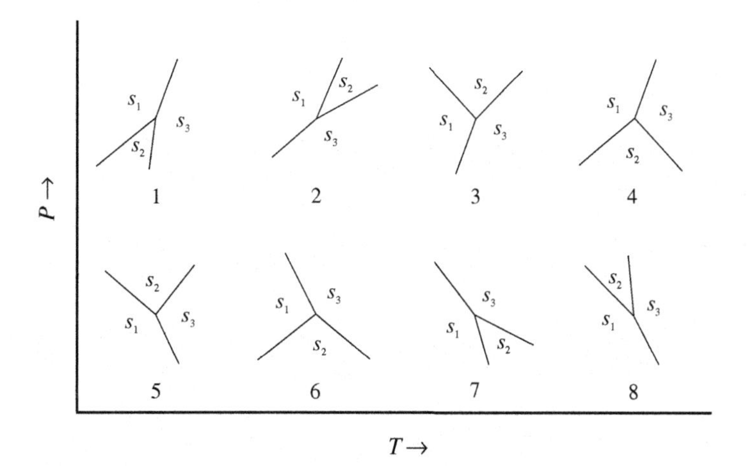

FIGURE 7-19

Schematic diagram of eight possible configurations for the triple point among three polymorphic solids (S_1-S_2-S_3).

are tabulated here (Johnston, 1910; Hill and Winter, 1956). (a) Use the Clausius-Clapeyron equation to derive the dissociation pressure equation, (b) compute the temperature at which the CO_2 pressure is one bar, and (c) find the average ΔH of reaction (to the nearest 0.1 kJ mol^{-1}).

T(°C)	449	466	540	656	673	711	749	800
P (mm-Hg)	0.0170	0.0334	0.432	9.47	14.5	32.7	72.0	183.0

12. The permanent southern polar cap on Mars is CO_2 ice and has a temperature of 148 K during southern winter. The heat of sublimation of CO_2 ice is 25,230 J/mol and the vapor pressure over CO_2 ice at its "sublimation point" of 194.67 K is one atmosphere. Use these data to compute the CO_2 vapor pressure over the Martian southern polar cap in winter.
13. The normal boiling point of tungsten (W) is almost 6000 K and has not been measured experimentally. Estimate its boiling point from the vapor pressure over W metal (Jones et al., 1927; Langmuir, 1913) and the heat of melting (35.40 kJ mol^{-1}) at its melting point of 3680 K.

T (K)	2518	2610	2852	2913	2950	3000	3060	3066
P (10^{-9} bar)	0.1692	0.5081	13.59	35.48	50.81	99.25	189.3	200.2

14. Old Faithful geyser in Yellowstone National Park is at an altitude of 7365 feet (2245 meters) where the average atmospheric pressure is 766 millibars. Calculate the boiling point of water (in degrees Celsius) erupted by Old Faithful. The vapor pressure of water (0—100°C) is given by the equation: $\log_{10} P$ (Mpa) $= 4.082 - 1623.019/T - 101{,}083.15/T^2$.
15. State the number of components, phases, and degrees of freedom for each system: (a) magnesite + periclase + CO_2 (g) in any proportions, (b) thenardite Na_2SO_4 + mirabilite $Na_2SO_4 \cdot 10H_2O$ + H_2O (l) + H_2O (g) in any proportions, (c) ammonium carbonate $(NH_4)_2CO_3$ + NH_3 (g) + CO_2 (g) + H_2O (g) + H_2O (l) in any proportions, (d) N_2 (g) + H_2 (g) + NH_3 (g) in any proportions at room temperature.
16. Anhydrite $CaSO_4$ + gypsum $CaSO_4 \cdot 2H_2O$ commonly occur together in evaporate deposits such as those in Stassfurt, Germany. (a) Is it theoretically possible for anhydrite + gypsum + water to stably coexist? If so, specify the phases and components present and the number of degrees of freedom. (b) Is the system of anhydrite + gypsum + water + water vapor possible? If so, specify the phases and components and the number of degrees of freedom.
17. Thenardite (Na_2SO_4), the solid heptahydrate ($Na_2SO_4 \cdot 7H_2O$), mirabilite ($Na_2SO_4 \cdot 10H_2O$), and water (H_2O) are four phases in the two component Na_2SO_4-H_2O system. List all possible sets of two components (e.g., Na_2SO_4 and H_2O) that can be used in the phase rule to describe this system. Pick any one of these component sets (except Na_2SO_4 and H_2O) and write down algebraic equations for the chemical composition of each of the four phases listed above, for example,

$$Na_2SO_4 = 1 \times Na_2SO_4 + 0 \times H_2O$$

and so on. Thenardite occurs at salt lakes, for example, at Borax Lake, California. Mirabilite occurs associated with halite and gypsum in salt deposits. The heptahydrate does not occur naturally.

18. In the 1950s, Harold Urey suggested that the reaction

$$CaCO_3 \text{ (calcite)} + SiO_2 \text{ (quartz)} = CaSiO_3 \text{ (wollastonite)} + CO_2 \text{ (gas)}$$

buffered, or regulated, the CO_2 pressure on Venus. (a) What are the components in this system and how many are there? (b) Sketch a phase diagram (log P versus $1/T$) for this reaction. (c) Use the Gibbs phase rule to specify the phases and degrees of freedom in the two fields and along the univariant line.

19. Io is the innermost Galilean satellite of Jupiter. Its average surface temperature is about 120 ± 10 K and much of its surface is covered by SO_2 ice. However, there are no data for the vapor pressure of SO_2 ice. Estimate its vapor pressure on Io's surface from the data of Giauque and Stephenson (1938) for the vapor pressure and thermal properties of SO_2 at the triple point: $T = 197.64$ K, $P = 12.56$ mmHg, $\Delta_{fus}H = 1769.1$ cal mol^{-1}, $\Delta_{vap}H = 6768.2$ cal mol^{-1}. Also assume ΔC_P (vapor − solid) $= -3.0$ cal mol^{-1} K^{-1}.

20. Roozeboom gives the following data for the melting curve of monoclinic sulfur as a function of pressure. Use these results to compute the equation for the melting curve, the (dT/dP) slope along the melting curve, and the ΔV for melting at one bar pressure. Use $\Delta_{fus}H = 1718$ J mol^{-1}.

P (kg cm^{-2})	199	534	914	1318	1588	1838	2149	2650	3143
T (°C)	120.0	129.9	141.1	151.1	158.1	163.1	170.1	180.1	190.1

21. Derive Eq. (7-155) for $\Delta G^o{}_T$ from the enthalpy of transition (401.7 J mol^{-1}) and the ΔC_P between orthorhombic and monoclinic sulfur

$$\Delta C_P = C_P \text{ (monoclinic)} - C_P \text{ (orthorhombic)} = 0.830 - 8.22 \times 10^{-4} T \text{ J mol}^{-1} \text{ K}^{-1}$$

22. Halstead and Moore (1957) measured the temperature and equilibrium water vapor pressure (in mmHg) along the univariant curve where portlandite $Ca(OH)_2$ and lime CaO coexist. (a) Use the Clausius-Clapeyron equation to calculate the temperature where the water vapor pressure is one bar. (b) Calculate the average enthalpy of reaction for thermal decomposition of portlandite to lime plus water vapor. (c) Use heat capacity data from Appendix 2 to calculate the enthalpy of reaction at 298.15 K, assuming that $\Delta_r H^o$ from part (b) is for an average temperature of 706 K.

T(K)	635	665	680	707	731	754	767.5	776.5
P (mm)	21.5	50.5	75.6	141.75	243.9	418.0	548.0	671.0

23. Goldsmith and Graf (1957) measured the univariant curve for the reaction

$$MnCO_3 \text{ (rhodochrosite)} = MnO \text{ (manganosite)} + CO_2 \text{ (g)}$$

as a function of pressure and temperature from 376°C to 750°C.

T(°C)	376	400	450	500	550	600	650	700	750
P (bars)	1.01	2.10	8.28	26.6	75.9	190	428	855	1552

Use the following data to calculate $\Delta V = V_{MnO} - V_{MnCO3}$ along the univariant curve.

Mineral	V^o_{298} (J bar^{-1})	$10^5 \times \alpha V$ (K)	$10^6 \times \beta V$ (bar)
MnO	1.322	5.4	0.8
$MnCO_3$	3.107	7.6	3.1

CHAPTER

Equations of State

8

The ordinary gaseous and ordinary liquid states are, in short, only widely separated forms of the same condition of matter, and may be made to pass into one another by a series of gradations so gentle that the passage shall nowhere present any interruption or breach of continuity. From carbonic acid as a perfect gas to carbonic acid as a perfect liquid, the transition we have seen may be accomplished by a continuous process, and the gas and liquid are only distant stages of a long series of physical changes.

—Thomas Andrews (1869), discussing his experiments on CO_2

This chapter describes equations of state for gases, liquids, and solids. An equation of state (EOS) describes the relationship among pressure, temperature, and volume for a material. Equations of state are essential for modeling the following: (1) the variation of pressure and temperature with altitude in planetary, satellite, and stellar atmospheres; (2) the variation of pressure and temperature with depth in the interiors of asteroids, planets, and satellites; and (3) chemical equilibria between fluids and solids at high pressures and temperatures in planetary interiors (e.g., metamorphic reactions in Earth's crust and mantle). Up to this point, we have mainly used only one equation of state, the ideal gas law, and have not considered equations of state for liquids and solids in any detail. We now present other equations of state that describe the behavior of nonideal (i.e., real) gases. We also introduce equations of state for condensed phases.

There are six sections in this chapter. Section I deals with thermodynamic formulas, such as the Maxwell relations, that are useful for deriving equations of state from *PVT* measurements and for thermodynamic calculations involving equations of state. Section II reviews the behavior of ideal gases, describes the deviations from ideality displayed by real gases, introduces the important concepts of reduced variables and the law of corresponding states, and discusses equations of state for real gases. The van der Waals equation, the Redlich-Kwong and related cubic equations of state, and the virial equation are discussed. Section III summarizes thermodynamic properties of real gases. The major topics are the Joule-Thomson effect, fugacity, and fugacity coefficients. Section IV discusses thermodynamic properties of gas mixtures. Section V describes critical phenomena of pure gases and gas mixtures. Section VI is a brief introduction to equations of state of solids and liquids. We review thermal expansion coefficients, the isothermal compressibility and bulk modulus, define the adiabatic compressibility and bulk modulus, introduce Grüneisen's parameter, and present the Birch-Murnaghan and other equations of state for solids.

Copyright © 2013 Elsevier Inc. All rights reserved.

I. THE MAXWELL RELATIONS AND THERMODYNAMIC FORMULAS

A. The four fundamental equations

The four fundamental equations are expressions for dE, dH, dG, and dA in terms of their natural variables. We derived the first fundamental equation, also known as the combined first and second law, in Chapter 6. This equation is

$$dE = TdS - PdV \tag{6-48}$$

Equation (6-48) expresses the dependence of the internal energy E on its natural variables entropy S and volume V. We derived the second fundamental equation (for dG) in Chapter 7:

$$dG = VdP - SdT \tag{7-28}$$

Equation (7-28) expresses the dependence of the Gibbs free energy G on its natural variables pressure P and temperature T. We now derive the fundamental equations for enthalpy H and the Helmholtz free energy A. Enthalpy is defined as

$$H = E + PV \tag{3-37}$$

We differentiate Eq. (3-37) and substitute for dE using the first fundamental equation Eq. (6-48):

$$dH = dE + PdV + VdP \tag{8-1}$$

$$dH = (TdS - PdV) + PdV + VdP \tag{8-2}$$

$$dH = TdS + VdP \tag{8-3}$$

Equation (8-3) is the third fundamental equation. It expresses the dependence of enthalpy H on its natural variables entropy S and pressure P. The Helmholtz free energy is defined as

$$A = E - TS \tag{6-108}$$

Differentiating Eq. (6-108) and substituting for dE gives the fourth fundamental equation:

$$dA = dE - TdS - SdT \tag{8-4}$$

$$dA = (TdS - PdV) - TdS - SdT \tag{8-5}$$

$$dA = -PdV - SdT \tag{8-6}$$

Equation (8-6) is the fourth fundamental equation. It gives the dependence of the Helmholtz free energy A on its natural variables V and T. We summarize the four fundamental equations here:

$$dE = TdS - PdV \tag{6-48}$$

$$dH = TdS + VdP \tag{8-3}$$

$$dG = VdP - SdT \tag{7-28}$$

$$dA = -PdV - SdT \tag{8-6}$$

The four fundamental equations are very powerful because we can express the thermodynamic properties of a system as partial derivatives of one of the four potentials (E, H, G, A) with respect to its natural variables. Thus, partial derivatives of E with respect to S and V give all other functions, partial derivatives of H with respect to S and P give all other functions, and so on.

Example 8-1. Use the fundamental equation for dH to derive an equation for the variation of entropy with pressure at constant enthalpy, which is important for gas liquefaction.

We are asked to find the derivative $(\partial S/\partial P)_H$, which is easily done by setting Eq. (8-3) equal to zero ($dH = 0$ for constant enthalpy) and rearranging:

$$dH = TdS + VdP = 0 \tag{8-3}$$

$$TdS = -VdP \tag{8-7}$$

$$\left(\frac{\partial S}{\partial P}\right)_H = -\frac{V}{T} \tag{8-8}$$

We can also differentiate each of the four fundamental equations with respect to the coefficient terms while holding the natural variables constant. This gives eight new equations, two from each of the four fundamental equations. The two relationships so derived from Eq. (6-48) describe the variation of internal energy E with temperature along an isochore and the variation of E with pressure along an adiabat (or isentrope):

$$\left(\frac{\partial E}{\partial T}\right)_V = T\left(\frac{\partial S}{\partial T}\right)_V \tag{8-9}$$

$$\left(\frac{\partial E}{\partial P}\right)_S = -P\left(\frac{\partial V}{\partial P}\right)_S \tag{8-10}$$

The two relationships derived from Eq. (8-3) describe the variation of enthalpy with temperature along an isobar and the variation of H with V along an adiabat:

$$\left(\frac{\partial H}{\partial T}\right)_P = T\left(\frac{\partial S}{\partial T}\right)_P \tag{8-11}$$

$$\left(\frac{\partial H}{\partial V}\right)_S = V\left(\frac{\partial P}{\partial V}\right)_S \tag{8-12}$$

The two relationships derived from Eq. (7-28) describe the variation of the Gibbs free energy with volume along an isotherm and the variation of G with entropy along an isobar:

$$\left(\frac{\partial G}{\partial V}\right)_T = V\left(\frac{\partial P}{\partial V}\right)_T \tag{8-13}$$

$$\left(\frac{\partial G}{\partial S}\right)_P = -S\left(\frac{\partial T}{\partial S}\right)_P \tag{8-14}$$

The relationships derived from Eq. (8-6) describe the variation of the Helmholtz free energy with pressure along an isotherm and the variation of A with entropy along an isochore:

$$\left(\frac{\partial A}{\partial P}\right)_T = -P\left(\frac{\partial V}{\partial P}\right)_T \tag{8-15}$$

$$\left(\frac{\partial A}{\partial S}\right)_V = -S\left(\frac{\partial T}{\partial S}\right)_V \tag{8-16}$$

Several of the preceding equations are alternate formulations of relationships we derived earlier. For example, we discussed entropy changes during reversible isochoric and reversible isobaric processes in Section VII of Chapter 6 and derived the two expressions:

$$\left(\frac{\partial S}{\partial T}\right)_V = \frac{C_V}{T} \tag{6-57}$$

$$\left(\frac{\partial S}{\partial T}\right)_P = \frac{C_P}{T} \tag{6-62}$$

The combination of Eqs. (6-57) and (8-9) gives Eq. (3-42), the definition of the constant-volume heat capacity, whereas the combination of Eqs. (6-62) and (8-11) gives Eq. (3-44), the definition of the constant-pressure heat capacity. Another example is that the right side of Eq. (8-13) is equal to -1 times the inverse of the isothermal compressibility $(\beta_T)^{-1}$, which is also equivalent to -1 times the isothermal bulk modulus K_T:

$$V\left(\frac{\partial P}{\partial V}\right)_T = -\left[-\frac{1}{V}\left(\frac{\partial V}{\partial P}\right)_T\right]^{-1} = -\frac{1}{\beta_T} = -K_T \tag{8-17}$$

Two other examples are that the right side of Eq. (8-16) is equal to $-ST/C_V$, and that the right side of Eq. (8-15) is equal to $P\beta_T V$. Equations (8-9) to (8-16) are useful for calculations involving phase equilibria and *PVT* properties of materials. We give several examples when we discuss equations of state for gases and condensed phases.

B. Extensive and intensive variables and energy function (*E, H, G, A*) derivatives

We introduced the concepts of extensive and intensive variables in Chapter 2. An *extensive variable* is a property that depends on the size of a system or on the quantity of material in a system. The volume V and entropy S are extensive variables. An *intensive variable* is a characteristic property of a system and is independent of the quantity of material in a system. The temperature T and pressure P are intensive variables. The product of an intensive variable (pressure, temperature, force, voltage, magnetic field strength) and an extensive variable (volume, entropy, length, electrical charge, magnetization of a solid) has the dimensions of energy and work (look back at Table 3-6, which lists several of these conjugate pairs). The four fundamental equations contain conjugate pairs of intensive and extensive variables (*PdV, TdS, SdT, VdP*). Equations (3-37), (6-108), and (6-113), defining *H, A,* and *G*, are examples of Legrendre transforms. A Legrendre transform is a transformation of natural variables accomplished by subtracting a conjugate pair from a thermodynamic potential such as internal energy E. Alberty (1997) reviews the use of Legrendre transforms in thermodynamics. We now derive equations that express the extensive variables *V* and *S* and the intensive variables *T* and *P* in terms of the derivatives of the four thermodynamic potentials *E, H, G,* and *A*.

The complete differential of *E* written in terms of its natural variables *S* and *V* is

$$dE = \left(\frac{\partial E}{\partial S}\right)_V dS + \left(\frac{\partial E}{\partial V}\right)_S dV \tag{8-18}$$

The first fundamental equation (6-48) and Eq. (8-18) are equal to one another. Thus, we find that

$$\left(\frac{\partial E}{\partial S}\right)_V = T \tag{8-19}$$

$$-\left(\frac{\partial E}{\partial V}\right)_S = P \tag{8-20}$$

The complete differential of *H* written in terms of its natural variables *S* and *P* is

$$dH = \left(\frac{\partial H}{\partial S}\right)_P dS + \left(\frac{\partial H}{\partial P}\right)_S dP \tag{8-21}$$

Equations (8-3) and (8-21) are equal to one another, so we have

$$\left(\frac{\partial H}{\partial S}\right)_P = T \tag{8-22}$$

$$\left(\frac{\partial H}{\partial P}\right)_S = V \tag{8-23}$$

The complete differential of *G* written in terms of its natural variables *P* and *T* is

$$dG = \left(\frac{\partial G}{\partial P}\right)_T dP + \left(\frac{\partial G}{\partial T}\right)_P dT \tag{8-24}$$

A comparison of Eq. (7-28) and Eq. (8-24) shows that

$$\left(\frac{\partial G}{\partial P}\right)_T = V \tag{8-25}$$

$$-\left(\frac{\partial G}{\partial T}\right)_P = S \tag{8-26}$$

Finally, the complete differential of A written in terms of its natural variables T and V is

$$dA = \left(\frac{\partial A}{\partial T}\right)_V dT + \left(\frac{\partial A}{\partial V}\right)_T dV \tag{8-27}$$

Equations (8-6) and (8-27) are equal to one another, so we have the two identities

$$-\left(\frac{\partial A}{\partial T}\right)_V = S \tag{8-28}$$

$$-\left(\frac{\partial A}{\partial V}\right)_T = P \tag{8-29}$$

We summarize these relationships next. The partial derivatives of internal energy E and enthalpy H with respect to entropy S are equal to temperature T, an intensive variable:

$$\left(\frac{\partial E}{\partial S}\right)_V = \left(\frac{\partial H}{\partial S}\right)_P = T \tag{8-30}$$

The partial derivatives of E and the Helmholtz free energy A with respect to volume V are equal to pressure, which is also an intensive variable:

$$-\left(\frac{\partial E}{\partial V}\right)_S = -\left(\frac{\partial A}{\partial V}\right)_T = P \tag{8-31}$$

The partial derivatives of H and G with respect to P are equal to V:

$$\left(\frac{\partial H}{\partial P}\right)_S = \left(\frac{\partial G}{\partial P}\right)_T = V \tag{8-32}$$

Finally, the partial derivatives of G and A with respect to T are equal to S:

$$-\left(\frac{\partial G}{\partial T}\right)_P = -\left(\frac{\partial A}{\partial T}\right)_V = S \tag{8-33}$$

C. Maxwell relations

The four equations known as the Maxwell relations first appeared in a footnote in Maxwell's book *Theory of Heat.* The Maxwell relations are obtained from the preceding energy function derivatives. The basis of their derivation is that the order of differentiation is irrelevant for perfect differentials, that is the second derivatives of perfect differentials are equal to one another. Internal energy, enthalpy, Gibbs free energy, and the Helmholtz free energy are state functions and their differentials are perfect, or exact, differentials. You should review Section XII of Chapter 2 if you need to refresh your memory about these important concepts.

Equation (8-18) is the complete differential of E with respect to S and V:

$$dE = \left(\frac{\partial E}{\partial S}\right)_V dS + \left(\frac{\partial E}{\partial V}\right)_S dV \tag{8-18}$$

The first derivatives are equal to temperature and pressure (see Section II below):

$$\left(\frac{\partial E}{\partial S}\right)_V = T \tag{8-19}$$

$$-\left(\frac{\partial E}{\partial V}\right)_S = P \tag{8-20}$$

The second derivative of Eq. (8-19) is

$$\frac{\partial^2 E}{\partial V \partial S} = \frac{\partial}{\partial V}\left(\frac{\partial E}{\partial S}\right)_V\Bigg]_S = \frac{\partial}{\partial V}T\Bigg]_S = \left(\frac{\partial T}{\partial V}\right)_S \tag{8-34}$$

The second derivative of Eq. (8-20) is

$$\frac{\partial^2 E}{\partial S \partial V} = \frac{\partial}{\partial S}\left(\frac{\partial E}{\partial V}\right)_S\Bigg]_V = \frac{\partial}{\partial S}(-P)\Bigg]_V = -\left(\frac{\partial P}{\partial S}\right)_V \tag{8-35}$$

These two second derivatives are equal because dE is a perfect differential. Thus, we can write

$$\frac{\partial^2 E}{\partial V \partial S} = \frac{\partial^2 E}{\partial S \partial V} \tag{8-36}$$

Using Eqs. (8-34) and (8-35) to substitute for the second derivatives in Eq. (8-36) gives

$$\left(\frac{\partial T}{\partial V}\right)_S = -\left(\frac{\partial P}{\partial S}\right)_V \tag{8-37}$$

Equation (8-37) is the first Maxwell relation.

The complete differential of H with respect to its natural variables S and P is

$$dH = \left(\frac{\partial H}{\partial S}\right)_P dS + \left(\frac{\partial H}{\partial P}\right)_S dP \tag{8-21}$$

Equation (8-22) shows that the first derivative of H with respect to S is equal to temperature:

$$\left(\frac{\partial H}{\partial S}\right)_P = T \tag{8-22}$$

The second derivative of Eq. (8-22) is

$$\frac{\partial^2 H}{\partial P \partial S} = \frac{\partial}{\partial P}\left(\frac{\partial H}{\partial S}\right)_P\Bigg]_S = \frac{\partial}{\partial P}T\Bigg]_S = \left(\frac{\partial T}{\partial P}\right)_S \tag{8-38}$$

Equation (8-23) shows that the first derivative of H with respect to pressure is equal to volume:

$$\left(\frac{\partial H}{\partial P}\right)_S = V \tag{8-23}$$

The second derivative of Eq. (8-23) is

$$\frac{\partial^2 H}{\partial S \partial P} = \frac{\partial}{\partial S}\left(\frac{\partial H}{\partial P}\right)_S\bigg]_P = \frac{\partial}{\partial S}V\bigg]_P = \left(\frac{\partial V}{\partial S}\right)_P \tag{8-39}$$

The two second derivatives in Eqs. (8-38) and (8-39) are equal to one another because dH is a perfect differential. Thus, we can write

$$\frac{\partial^2 H}{\partial P \partial S} = \frac{\partial^2 H}{\partial S \partial P} \tag{8-40}$$

$$\left(\frac{\partial T}{\partial P}\right)_S = \left(\frac{\partial V}{\partial S}\right)_P \tag{8-41}$$

Equation (8-41) is the second Maxwell relation.

The third Maxwell relation is derived from the complete differential of the Gibbs free energy G with respect to its natural variables P and T:

$$dG = \left(\frac{\partial G}{\partial P}\right)_T dP + \left(\frac{\partial G}{\partial T}\right)_P dT \tag{8-24}$$

The first derivatives in Eq. (8-24) were evaluated earlier and are given by the equations

$$\left(\frac{\partial G}{\partial P}\right)_T = V \tag{8-25}$$

$$\left(\frac{\partial G}{\partial T}\right)_P = -S \tag{8-26}$$

The second derivatives of Eqs. (8-25) and (8-26) are given by the two equations

$$\frac{\partial^2 G}{\partial T \partial P} = \frac{\partial}{\partial T}\left(\frac{\partial G}{\partial P}\right)_T\bigg]_P = \frac{\partial}{\partial T}V\bigg]_P = \left(\frac{\partial V}{\partial T}\right)_P \tag{8-42}$$

$$\frac{\partial^2 G}{\partial P \partial T} = \frac{\partial}{\partial P}\left(\frac{\partial G}{\partial T}\right)_P\bigg]_T = \frac{\partial}{\partial P}(-S)\bigg]_T = -\left(\frac{\partial S}{\partial P}\right)_T \tag{8-43}$$

The two second derivatives are equal to one another, which gives the third Maxwell relation:

$$\frac{\partial^2 G}{\partial T \partial P} = \frac{\partial^2 G}{\partial P \partial T} \tag{8-44}$$

$$\left(\frac{\partial V}{\partial T}\right)_P = -\left(\frac{\partial S}{\partial P}\right)_T \tag{8-45}$$

The fourth Maxwell relation is derived in a similar manner by starting with the complete differential dA of the Helmholtz free energy with respect to its natural variables T and V:

$$dA = \left(\frac{\partial A}{\partial T}\right)_V dT + \left(\frac{\partial A}{\partial V}\right)_T dV \tag{8-27}$$

The two first derivatives are given by the equations

$$\left(\frac{\partial A}{\partial T}\right)_V = -S \tag{8-28}$$

$$\left(\frac{\partial A}{\partial V}\right)_T = -P \tag{8-29}$$

The second derivatives, which are equal to one another, are given by the equations

$$\frac{\partial^2 A}{\partial V \partial T} = \frac{\partial}{\partial V}\left(\frac{\partial A}{\partial T}\right)_V\Bigg]_T = \frac{\partial}{\partial V}(-S)\Bigg]_T = -\left(\frac{\partial S}{\partial V}\right)_T \tag{8-46}$$

$$\frac{\partial^2 A}{\partial T \partial V} = \frac{\partial}{\partial T}\left(\frac{\partial A}{\partial V}\right)_T\Bigg]_V = \frac{\partial}{\partial T}(-P)\Bigg]_V = -\left(\frac{\partial P}{\partial T}\right)_V \tag{8-47}$$

Thus, the fourth Maxwell relation is

$$\left(\frac{\partial S}{\partial V}\right)_T = \left(\frac{\partial P}{\partial T}\right)_V \tag{8-48}$$

A summary of the four Maxwell relations for a closed system is

$$\left(\frac{\partial T}{\partial V}\right)_S = -\left(\frac{\partial P}{\partial S}\right)_V \tag{8-37}$$

$$\left(\frac{\partial T}{\partial P}\right)_S = \left(\frac{\partial V}{\partial S}\right)_P \tag{8-41}$$

$$\left(\frac{\partial V}{\partial T}\right)_P = -\left(\frac{\partial S}{\partial P}\right)_T \tag{8-45}$$

$$\left(\frac{\partial S}{\partial V}\right)_T = \left(\frac{\partial P}{\partial T}\right)_V \tag{8-48}$$

An analogous set of Maxwell relations can also be derived for an open system.

Maxwell's relations are valuable because they relate quantities that are easy to measure experimentally to functions that are not easy to measure. They greatly extend our ability to apply thermodynamics to practical problems in chemistry, geology, materials science, and physics.

For example, we can use one of the Maxwell relations to derive a very useful expression for the difference between the constant-pressure and constant-volume heat capacities of any material. In Chapter 3 we showed that this difference can be written as

$$C_P - C_V = \left(\frac{\partial V}{\partial T}\right)_P \left[P + \left(\frac{\partial E}{\partial V}\right)_T\right] \tag{3-49}$$

We used a rewritten version of Eq. (3-49) in Chapter 4. This equation is much more useful and is

$$(C_P - C_V)_m = \frac{\alpha_P^2 V_m T}{\beta_T} \tag{4-46}$$

where α_P is the isobaric thermal expansion coefficient, β_T is the isothermal compressibility, V_m is the molar volume, and T is in kelvins. One way to derive Eq. (4-46) from Eq. (3-49) is to start with a thermodynamic equation of state

$$\left(\frac{\partial E}{\partial V}\right)_T = T\left(\frac{\partial S}{\partial V}\right)_T - P \tag{8-49}$$

and use this to substitute into Eq. (3-49). This gives us

$$\begin{aligned} C_P - C_V &= \left(\frac{\partial V}{\partial T}\right)_P \left[P + \left(\frac{\partial E}{\partial V}\right)_T\right] = \left(\frac{\partial V}{\partial T}\right)_P \left[P + \left(-P + T\left(\frac{\partial S}{\partial V}\right)_T\right)\right] \\ &= T\left(\frac{\partial V}{\partial T}\right)_P \left(\frac{\partial S}{\partial V}\right)_T \end{aligned} \tag{8-50}$$

The change in entropy due to volume changes at constant temperature is hard to measure, so we want to recast Eq. (8-50) into a different form. We do this using the Maxwell relation

$$\left(\frac{\partial S}{\partial V}\right)_T = \left(\frac{\partial P}{\partial T}\right)_V \tag{8-48}$$

Equation (8-50) becomes

$$C_P - C_V = T\left(\frac{\partial V}{\partial T}\right)_P \left(\frac{\partial P}{\partial T}\right)_V \tag{8-51}$$

Equation (8-51) is used to calculate $(C_P - C_V)$ of real gases from PVT data or with an equation of state. The right side of Eq. (8-51) contains the thermal pressure coefficient, which is equal to

$$\left(\frac{\partial P}{\partial T}\right)_V = -\frac{(\partial V/\partial T)_P}{(\partial V/\partial P)_T} \tag{2-32}$$

Substituting Eq. (2-32) back into Eq. (8-51) gives the useful expression

$$C_P - C_V = -\frac{T(\partial V/\partial T)_P^2}{(\partial V/\partial P)_T} \tag{8-52}$$

Equation (8-52) shows that $(C_P - C_V)$ can be evaluated from measurements of thermal expansivity $(\partial V/\partial T)_P$ and compressibility $(\partial V/\partial P)_T$. Bridgman made extensive use of Eq. (8-52) in his high-pressure studies and used it to derive $(C_P - C_V)$ for mercury and many other materials. Finally, we can use the definitions of the isobaric thermal expansion coefficient (α_P) and the isothermal compressibility (β_T) to rewrite Eq. (8-52) to get the desired result:

$$(C_P - C_V)_m = \frac{\alpha_P^2 V_m T}{\beta_T} \tag{4-46}$$

Equation (4-46) is a general equation that applies to solids, liquids, and gases. As shown in Section III in Chapter 2, Eq. (3-49), and thus also Eq. (4-46), reduces to $(C_P - C_V) = R$ in the case of an ideal gas. Furthermore, Eq. (4-46) has the great advantage that only easily measured experimental quantities are required to evaluate $(C_P - C_V)$. This is illustrated in the next example.

Example 8-2. Calculate C_V at 298 K for forsterite from the C_P (118.61 J mol^{-1} K^{-1}), isobaric thermal expansion coefficient α_P (4.36×10^{-5} K^{-1}), isothermal compressibility β_T (7.8×10^{-7} bar^{-1}), and its molar volume (43.79 cm^3 mol^{-1}) at the same temperature. Using Eq. (4-46) we get

$$\begin{aligned} C_V &= C_P - \frac{\alpha_P^2 V_m T}{\beta_T} \\ &= 118.61 - \frac{(4.36 \times 10^{-5}\ \text{K}^{-1})^2 (43.79\ \text{cm}^3\ \text{mol}^{-1})(0.1\ \text{J cm}^{-3}\ \text{bar}^{-1})(298.15\ \text{K})}{7.8 \times 10^{-7}\ \text{bar}^{-1}} \\ &= 115.43\ \text{J mol}^{-1}\ \text{K}^{-1} \end{aligned} \tag{8-53}$$

D. Bridgman's thermodynamic formulas

We defined 10 thermodynamic variables (P, V, T, q, w, E, H, S, A, and G) in Chapters 2–6. A large number of thermodynamic formulas can be written using these 10 variables and their derivatives. Each partial first derivative involves three variables with one in the numerator, one in the denominator, and one held constant, for example, $(\partial V/\partial T)_P$, for a total of $10 \times 9 \times 8 = 720$ first derivatives. The number of thermodynamic formulas involving any four of these derivatives is

$$\frac{720 \times 719 \times 718 \times 717}{4!} \cong 11 \times 10^9 \tag{8-54}$$

(In other words, there are about 11×10^9 different ways to choose four objects out of a total of 720.)

In 1914 the physicist P. W. Bridgman developed a method for systematically tabulating the minimum number of equations necessary to write all of these formulas. We briefly review his approach here. More details about the Bridgman thermodynamic formulas are given in his paper (Bridgman, 1914) and in the revised version published as a small book (Bridgman, 1961).

Bridgman divided the 720 partial first derivatives into 10 groups on the basis of the variable held constant during differentiation. Thus, the 72 derivatives evaluated at constant pressure are in one group, the 72 derivatives evaluated at constant temperature are in another group, and so on. Equation

(2-65), the chain rule, is used to write any of the 72 derivatives in each of the 10 groups in terms of the two variables in the derivative and an arbitrary variable x:

$$\left(\frac{dy}{dt}\right) = \left(\frac{dy}{dx}\right)\left(\frac{dx}{dt}\right) \tag{2-65}$$

For example, the derivative $(\partial G/\partial P)_T$ can be written as

$$\left(\frac{\partial G}{\partial P}\right)_T = \left(\frac{\partial G}{\partial x}\right)_T \Big/ \left(\frac{\partial P}{\partial x}\right)_T = \left(\frac{\partial G}{\partial x}\right)_T \left(\frac{\partial x}{\partial P}\right)_T \tag{8-55}$$

The derivatives with respect to x are then abbreviated as the differential in the numerator. The abbreviated notation for the derivatives with respect to x in Eq. (8-55) is

$$\left(\frac{\partial G}{\partial x}\right)_T = (\partial G)_T \tag{8-56}$$

$$\left(\frac{\partial P}{\partial x}\right)_T = (\partial P)_T \tag{8-57}$$

Each group of 72 derivatives can be expressed as ratios of the nine derivatives with respect to the arbitrary variable x. As shown by Eq. (8-55), the dx terms drop out of these ratios, and we are left with ratios of nine differentials. For example using the notation introduced previously, the nine differentials at constant temperature are

$$\left(\frac{\partial P}{\partial x}\right)_T = (\partial P)_T \tag{8-57}$$

$$\left(\frac{\partial V}{\partial x}\right)_T = (\partial V)_T \tag{8-58}$$

$$\left(\frac{\delta q}{\partial x}\right)_T = (\delta q)_T \tag{8-59}$$

$$\left(\frac{\delta w}{\partial x}\right)_T = (\delta w)_T \tag{8-60}$$

$$\left(\frac{\partial E}{\partial x}\right)_T = (\partial E)_T \tag{8-61}$$

$$\left(\frac{\partial H}{\partial x}\right)_T = (\partial H)_T \tag{8-62}$$

$$\left(\frac{\partial S}{\partial x}\right)_T = (\partial S)_T \tag{8-63}$$

$$\left(\frac{\partial A}{\partial x}\right)_T = (\partial A)_T \tag{8-64}$$

$$\left(\frac{\partial G}{\partial x}\right)_T = (\partial G)_T \tag{8-56}$$

Note that we continue to use the notation δq and δw for the differentials of heat and work because these two variables are path dependent.

A different arbitrary variable can be chosen for each group, and we can write nine analogous differentials for each of the remaining nine groups (constant pressure, constant volume, and so on) for a total of 90 differentials. The total number of equations is cut in half to 45 if the arbitrary variables are related to one another by the cyclic rule (Eq. 2-70). Then relationships such as

$$(\partial G)_P = -(\partial P)_G \tag{8-65}$$

are valid. Finally, the value of one differential has to be fixed and three fundamental derivatives have to be specified. Bridgman chose to set $(\partial T)_P = 1$ and to specify thermal expansivity at constant pressure $(\partial V/\partial T)_P$, isothermal compressibility $(\partial V/\partial P)_T$, and constant-pressure heat capacity $C_P = (\delta q/dT)_P$ as the three fundamental derivatives because these quantities can be measured experimentally. The 90 differentials that Bridgman defined in this manner are listed in Table 8-1. Bridgman's thermodynamic formulas can also be derived using mathematical functions known as Jacobians. This approach is described in a paper by Shaw (1935), but we do not discuss it here. The versatility of Bridgman's method for summarizing thermodynamic formulas and the use of Table 8-1 is illustrated with several worked examples.

Example 8-3. Use Bridgman's method to evaluate the derivative $(\partial G/\partial P)_T$. First we rewrite the derivative using the abbreviated notation described previously. Then we find the values of the two differentials from Table 8-1 and evaluate the derivative:

$$\left(\frac{\partial G}{\partial P}\right)_T = \frac{(\partial G)_T}{(\partial P)_T} = \frac{-V}{-1} = V \tag{8-66}$$

Example 8-4. Use Bridgman's thermodynamic formulas to derive the Maxwell relation:

$$\left(\frac{\partial S}{\partial V}\right)_T = \left(\frac{\partial P}{\partial T}\right)_V \tag{8-48}$$

Using the differentials in Table 8-1 we obtain

$$\left(\frac{\partial S}{\partial V}\right)_T = \frac{(\partial S)_T}{(\partial V)_T} = \left(\frac{\partial V}{\partial T}\right)_P \Big/ \left[-\left(\frac{\partial V}{\partial P}\right)_T\right] = -\left(\frac{\partial V}{\partial T}\right)_P \left(\frac{\partial P}{\partial V}\right)_T \tag{8-67}$$

The right side of Eq. (8-67) is evaluated using the cyclic rule (Eq. 2-70):

$$-\left(\frac{\partial V}{\partial T}\right)_P \left(\frac{\partial P}{\partial V}\right)_T = \left(\frac{\partial P}{\partial T}\right)_V \tag{8-68}$$

Combining Eqs. (8-67) and (8-68) gives the desired result, namely Eq. (8-48).

Table 8-1 Bridgman's Thermodynamic Formulas

The derivatives are related via equations such as $(\partial T)_P = -(\partial P)_T$

constant P ($dP = 0$)—Isobaric process

$$(\partial T)_P = 1$$

$$(\partial V)_P = \left(\frac{\partial V}{\partial T}\right)_P$$

$$(\delta q)_P = C_P$$

$$(\delta w)_P = P\left(\frac{\partial V}{\partial T}\right)_P$$

$$(\partial E)_P = C_P - P\left(\frac{\partial V}{\partial T}\right)_P$$

$$(\partial H)_P = C_P$$

$$(\partial S)_P = \frac{C_P}{T}$$

$$(\partial A)_P = -S - P\left(\frac{\partial V}{\partial T}\right)_P$$

$$(\partial G)_P = -S$$

constant T ($dT = 0$)—Isothermal process

$$(\partial P)_T = -1$$

$$(\partial V)_T = -\left(\frac{\partial V}{\partial P}\right)_T$$

$$(\delta q)_T = T\left(\frac{\partial V}{\partial T}\right)_P$$

$$(\delta w)_T = -P\left(\frac{\partial V}{\partial P}\right)_T$$

$$(\partial E)_T = T\left(\frac{\partial V}{\partial T}\right)_P + P\left(\frac{\partial V}{\partial P}\right)_T$$

$$(\partial H)_T = -V + T\left(\frac{\partial V}{\partial T}\right)_P$$

$$(\partial S)_T = \left(\frac{\partial V}{\partial T}\right)_P$$

$$(\partial A)_T = P\left(\frac{\partial V}{\partial P}\right)_T$$

$$(\partial G)_T = -V$$

Table 8-1 Bridgman's Thermodynamic Formulas *(continued)*

constant V ($dV = 0$)—Isochoric (constant-volume) process

$$(\partial P)_V = -\left(\frac{\partial V}{\partial T}\right)_P$$

$$(\partial T)_V = \left(\frac{\partial V}{\partial P}\right)_T$$

$$(\delta q)_V = C_P\left(\frac{\partial V}{\partial P}\right)_T + T\left(\frac{\partial V}{\partial T}\right)_P^2$$

$$(\delta w)_V = 0$$

$$(\partial E)_V = C_P\left(\frac{\partial V}{\partial P}\right)_T + T\left(\frac{\partial V}{\partial T}\right)_P^2$$

$$(\partial H)_V = C_P\left(\frac{\partial V}{\partial P}\right)_T + T\left(\frac{\partial V}{\partial T}\right)_P^2 - V\left(\frac{\partial V}{\partial T}\right)_P$$

$$(\partial S)_V = \frac{1}{T}\left[C_P\left(\frac{\partial V}{\partial P}\right)_T + T\left(\frac{\partial V}{\partial T}\right)_P^2\right]$$

$$(\partial A)_V = -S\left(\frac{\partial V}{\partial P}\right)_T$$

$$(\partial G)_V = -V\left(\frac{\partial V}{\partial T}\right)_P - S\left(\frac{\partial V}{\partial P}\right)_T$$

constant q ($\delta q = 0$)—Adiabatic process

$$(\partial P)_q = -C_P$$

$$(\partial T)_q = -T\left(\frac{\partial V}{\partial T}\right)_P$$

$$(\partial V)_q = -C_P\left(\frac{\partial V}{\partial P}\right)_T - T\left(\frac{\partial V}{\partial T}\right)_P^2$$

$$(\delta w)_q = -P\left[C_P\left(\frac{\partial V}{\partial P}\right)_T + T\left(\frac{\partial V}{\partial T}\right)_P^2\right]$$

$$(\partial E)_q = P\left[C_P\left(\frac{\partial V}{\partial P}\right)_T + T\left(\frac{\partial V}{\partial T}\right)_P^2\right]$$

$$(\partial H)_q = -VC_P$$

$$(\partial S)_q = 0$$

$$(\partial A)_q = P\left[C_P\left(\frac{\partial V}{\partial P}\right)_T + T\left(\frac{\partial V}{\partial T}\right)_P^2\right] + TS\left(\frac{\partial V}{\partial T}\right)_P$$

$$(\partial G)_q = TS\left(\frac{\partial V}{\partial T}\right)_P - VC_P$$

(Continued)

Table 8-1 Bridgman's Thermodynamic Formulas *(continued)*

constant w ($\delta w = PdV = 0$)—No PV work is done for derivatives with w held constant.

$$(\partial P)_w = -P\left(\frac{\partial V}{\partial T}\right)_P$$

$$(\partial T)_w = P\left(\frac{\partial V}{\partial P}\right)_T$$

$$(\partial V)_w = 0$$

$$(\delta q)_w = P\left[C_P\left(\frac{\partial V}{\partial P}\right)_T + T\left(\frac{\partial V}{\partial T}\right)_P^2\right]$$

$$(\partial E)_w = P\left[C_P\left(\frac{\partial V}{\partial P}\right)_T + T\left(\frac{\partial V}{\partial T}\right)_P^2\right]$$

$$(\partial H)_w = P\left[C_P\left(\frac{\partial V}{\partial P}\right)_T + T\left(\frac{\partial V}{\partial T}\right)_P^2 - V\left(\frac{\partial V}{\partial T}\right)_P\right]$$

$$(\partial S)_w = \frac{P}{T}\left[C_P\left(\frac{\partial V}{\partial P}\right)_T + T\left(\frac{\partial V}{\partial T}\right)_P^2\right]$$

$$(\partial A)_w = -PS\left(\frac{\partial V}{\partial P}\right)_T$$

$$(\partial G)_w = -P\left[V\left(\frac{\partial V}{\partial T}\right)_P + S\left(\frac{\partial V}{\partial P}\right)_T\right]$$

constant E ($dE = 0$)—The internal energy of the system remains constant.

$$(\partial P)_E = -C_P + P\left(\frac{\partial V}{\partial T}\right)_P$$

$$(\partial T)_E = -T\left(\frac{\partial V}{\partial T}\right)_P - P\left(\frac{\partial V}{\partial P}\right)_T$$

$$(\partial V)_E = -C_P\left(\frac{\partial V}{\partial P}\right)_T - T\left(\frac{\partial V}{\partial T}\right)_P^2$$

$$(\delta q)_E = -P\left[C_P\left(\frac{\partial V}{\partial P}\right)_T + T\left(\frac{\partial V}{\partial T}\right)_P^2\right]$$

$$(\delta w)_E = -P\left[C_P\left(\frac{\partial V}{\partial P}\right)_T + T\left(\frac{\partial V}{\partial T}\right)_P^2\right]$$

$$(\partial H)_E = -V\left[C_P - P\left(\frac{\partial V}{\partial T}\right)_P\right] - P\left[C_P\left(\frac{\partial V}{\partial P}\right)_T + T\left(\frac{\partial V}{\partial T}\right)_P^2\right]$$

$$(\partial S)_E = -\frac{P}{T}\left[C_P\left(\frac{\partial V}{\partial P}\right)_T + T\left(\frac{\partial V}{\partial T}\right)_P^2\right]$$

$$(\partial A)_E = P\left[C_P\left(\frac{\partial V}{\partial P}\right)_T + T\left(\frac{\partial V}{\partial T}\right)_P^2\right] + S\left[T\left(\frac{\partial V}{\partial T}\right)_P + P\left(\frac{\partial V}{\partial P}\right)_T\right]$$

$$(\partial G)_E = -V\left[C_P - P\left(\frac{\partial V}{\partial T}\right)_P\right] + S\left[T\left(\frac{\partial V}{\partial T}\right)_P + P\left(\frac{\partial V}{\partial P}\right)_T\right]$$

Table 8-1 Bridgman's Thermodynamic Formulas *(continued)*

Constant H ($dH = 0$)—Isenthalpic process

$$(\partial P)_H = -C_P$$

$$(\partial T)_H = V - T\left(\frac{\partial V}{\partial T}\right)_P$$

$$(\partial V)_H = -C_P\left(\frac{\partial V}{\partial P}\right)_T - T\left(\frac{\partial V}{\partial T}\right)_P^2 + V\left(\frac{\partial V}{\partial T}\right)_P$$

$$(\delta q)_H = VC_P$$

$$(\delta w)_H = -P\left[C_P\left(\frac{\partial V}{\partial P}\right)_T + T\left(\frac{\partial V}{\partial T}\right)_P^2 - V\left(\frac{\partial V}{\partial T}\right)_P\right]$$

$$(\partial E)_H = V\left[C_P - P\left(\frac{\partial V}{\partial T}\right)_P\right] + P\left[C_P\left(\frac{\partial V}{\partial P}\right)_T + T\left(\frac{\partial V}{\partial T}\right)_P^2\right]$$

$$(\partial S)_H = \frac{VC_P}{T}$$

$$(\partial A)_H = -\left[S + P\left(\frac{\partial V}{\partial T}\right)_P\right]\left[V - T\left(\frac{\partial V}{\partial T}\right)_P\right] + PC_P\left(\frac{\partial V}{\partial P}\right)_T$$

$$(\partial G)_H = -VC_P - VS + TS\left(\frac{\partial V}{\partial T}\right)_P$$

Constant S ($dS = 0$)—Adiabatic or isentropic process

$$(\partial P)_S = -\frac{C_P}{T}$$

$$(\partial T)_S = -\left(\frac{\partial V}{\partial T}\right)_P$$

$$(\partial V)_S = -\frac{1}{T}\left[C_P\left(\frac{\partial V}{\partial P}\right)_T + T\left(\frac{\partial V}{\partial T}\right)_P^2\right]$$

$$(\delta q)_S = 0$$

$$(\delta w)_S = -\frac{P}{T}\left[C_P\left(\frac{\partial V}{\partial P}\right)_T + T\left(\frac{\partial V}{\partial T}\right)_P^2\right]$$

$$(\partial E)_S = \frac{P}{T}\left[C_P\left(\frac{\partial V}{\partial P}\right)_T + T\left(\frac{\partial V}{\partial T}\right)_P^2\right]$$

$$(\partial H)_S = -\frac{VC_P}{T}$$

$$(\partial A)_S = \frac{1}{T}\left[PC_P\left(\frac{\partial V}{\partial P}\right)_T + PT\left(\frac{\partial V}{\partial T}\right)_P^2 + TS\left(\frac{\partial V}{\partial T}\right)_P\right]$$

$$(\partial G)_S = -\frac{1}{T}\left[VC_P - TS\left(\frac{\partial V}{\partial T}\right)_P\right]$$

(Continued)

Table 8-1 Bridgman's Thermodynamic Formulas *(continued)*

constant A ($dA = 0$)—The Helmholtz free energy of the system remains constant

$$(\partial P)_A = S + P\left(\frac{\partial V}{\partial T}\right)_P$$

$$(\partial T)_A = -P\left(\frac{\partial V}{\partial P}\right)_T$$

$$(\partial V)_A = S\left(\frac{\partial V}{\partial P}\right)_T$$

$$(\delta q)_A = -P\left[C_P\left(\frac{\partial V}{\partial P}\right)_T + T\left(\frac{\partial V}{\partial T}\right)_P^2\right] - TS\left(\frac{\partial V}{\partial T}\right)_P$$

$$(\delta w)_A = PS\left(\frac{\partial V}{\partial P}\right)_T$$

$$(\partial E)_A = -P\left[C_P\left(\frac{\partial V}{\partial P}\right)_T + T\left(\frac{\partial V}{\partial T}\right)_P^2\right] - S\left[T\left(\frac{\partial V}{\partial T}\right)_P + P\left(\frac{\partial V}{\partial P}\right)_T\right]$$

$$(\partial H)_A = \left[S + P\left(\frac{\partial V}{\partial T}\right)_P\right]\left[V - T\left(\frac{\partial V}{\partial T}\right)_P\right] - PC_P\left(\frac{\partial V}{\partial P}\right)_T$$

$$(\partial S)_A = -\frac{1}{T}\left[PC_P\left(\frac{\partial V}{\partial P}\right)_T + PT\left(\frac{\partial V}{\partial T}\right)_P^2 + TS\left(\frac{\partial V}{\partial T}\right)_P\right]$$

$$(\partial G)_A = S\left[V + P\left(\frac{\partial V}{\partial P}\right)_T\right] + PV\left(\frac{\partial V}{\partial T}\right)_P$$

constant G ($dG = 0$)—Derivatives taken along equilibrium (eq) or saturation (σ) curves

$$(\partial P)_G = S$$

$$(\partial T)_G = V$$

$$(\partial V)_G = V\left(\frac{\partial V}{\partial T}\right)_P + S\left(\frac{\partial V}{\partial P}\right)_T$$

$$(\delta q)_G = -TS\left(\frac{\partial V}{\partial T}\right)_P + VC_P$$

$$(\delta w)_G = P\left[V\left(\frac{\partial V}{\partial T}\right)_P + S\left(\frac{\partial V}{\partial P}\right)_T\right]$$

$$(\partial E)_G = V\left[C_P - P\left(\frac{\partial V}{\partial T}\right)_P\right] - S\left[T\left(\frac{\partial V}{\partial T}\right)_P + P\left(\frac{\partial V}{\partial P}\right)_T\right]$$

$$(\partial H)_G = VC_P + VS - TS\left(\frac{\partial V}{\partial T}\right)_P$$

$$(\partial S)_G = \frac{1}{T}\left[VC_P - TS\left(\frac{\partial V}{\partial T}\right)_P\right]$$

$$(\partial A)_G = -S\left[V + P\left(\frac{\partial V}{\partial P}\right)_T\right] - PV\left(\frac{\partial V}{\partial T}\right)_P$$

Bridgman's formulas are useful for many other applications. For example, they can be used to derive properties along the saturation curve (i.e., the vaporization curve) of a liquid. In Chapter 4 we noted that measurements of the heat capacity of volatile liquids do not give C_V or C_P but give C_σ, the heat capacity along the saturation curve. The liquid and vapor coexist in equilibrium along the saturation curve. Thus, the Gibbs free energy G is constant along this curve. Equations for C_σ and other properties along the saturation curve involve derivatives at constant G, which can be evaluated using Bridgman's formulas.

II. REAL GAS EQUATIONS OF STATE AND THERMODYNAMICS

A. *PVT* behavior of ideal gases

We discussed the *PVT* behavior of ideal gases in Chapter 2, and it is worth reviewing this behavior before describing how real gases differ. Ideal gases obey Boyle's law and Charles's law. Boyle's law states that the product of pressure times volume (for a fixed amount of gas, such as the molar volume V_m) has a constant value (C) along an isotherm:

$$PV_m = C \tag{2-9}$$

For example, Table 2-6 shows that the *PV* product for one mole of air is constant within 0.03% of the one bar value for pressures of 1–10 bars along the 350 K isotherm. Furthermore, all ideal gases have the same molar *PV* product at the same temperature because the ideal gas law shows

$$PV_m = RT \tag{2-20}$$

Thus, a plot of the molar *PV* product as a function of pressure at constant temperature is a horizontal line for an ideal gas. You can see this for yourself by plotting the data in Table 2-6 or the data in Figure 2-13(a) as *PV* versus pressure.

Charles's law (or Gay-Lussac's law) states that the volume of a fixed amount of gas is proportional to temperature (t °C) at constant pressure:

$$V = V_0(1 + \alpha_0 t) \tag{2-12}$$

where V is the volume of a fixed amount of gas at any temperature t, V_0 is the volume of the same amount of gas at 0°C, and α_0 is the proportionality constant. The value of α_0 is determined by extrapolating measurements at several different low pressures to the limiting value of zero pressure (e.g., see Table 2-8). The value of α_0 is the same for all ideal gases and is

$$\alpha_0 = \frac{(V - V_0)}{V_0 t} = \frac{1}{273.15} = 36.610 \times 10^{-4}\ {}^\circ\mathrm{C}^{-1} \tag{2-13}$$

An equivalent statement of Charles's law is that the quotient of volume divided by temperature (for a fixed amount of gas, such as the molar volume V_m) is constant along an isobar:

$$\frac{V_m}{T} = \frac{V_{m,0}}{T_0} \tag{2-17}$$

No gas behaves ideally under all conditions, but many gases approach ideality at normal temperatures as their pressure is reduced to lower and lower values. Table 2-7 illustrates this point. It shows the zero pressure values of the molar *PV* product for 22 different gases at 273.15 K. These values were obtained by extrapolating the PV_m product measured at several different low pressures to the limiting value of zero pressure. The mean *PV* product for the 22 gases is 22.41400 atm L mol^{-1}. This value is the molar volume of an ideal gas at one atmosphere pressure and 273.15 K. These conditions are standard temperature and pressure (STP). The molar volume of an ideal gas at one bar pressure and 273.15 K is 22.71098 bar L mol^{-1}, about 1% larger because one bar pressure is about 1% smaller than one atmosphere pressure.

The combination of Boyle's law and Charles's law gives the ideal gas equation of state (Eq. 2-20). Figure 2-12 shows the *PVT* surface for an ideal gas. This is smooth without any discontinuities. The two-dimensional projections of this surface illustrated in Figure 2-13 are also smooth without any discontinuities. To first approximation, these diagrams show the actual behavior of real gases at temperatures much less than or much greater than the critical temperature for pressures much less than the critical pressure. However, they do not show either the vapor-liquid phase transition or the vapor-liquid critical point (see Chapter 7). These are two of the important deviations from ideality that are displayed by real gases and that should be included in an equation of state for real gases.

B. Deviations from ideality

We start by using CO_2 to illustrate the deviations from ideality that are displayed by real gases. We picked CO_2 because it plays an important role in the chemistry of Earth and other planets, for example, CO_2 is the major gas in the atmospheres of Mars and Venus, is generally the second most abundant gas (after steam) in terrestrial volcanic gases, is an important constituent of high-pressure fluids in Earth's crust and mantle, and is an important greenhouse gas in the atmospheres of Earth, Mars, and Venus. Solid CO_2 (dry ice) is also important because it occurs in the polar caps on Mars, on Triton (Neptune's largest satellite), and in comets.

Figure 8-1 shows isotherms for CO_2 plotted as the molar *PV* product versus pressure. If CO_2 behaved ideally and followed Boyle's law, all these isotherms would be horizontal lines. This is not the case. Instead the shapes of the isotherms depend on temperature. For example, consider the 0°C isotherm, which is representative of isotherms between the *S-L-V* triple point of CO_2 at 216.55 K, 5.18 bars and the critical point at 304.13 K, 73.77 bars. The 0°C isotherm initially curves downward toward the pressure axis because CO_2 is more compressible than an ideal gas at these pressures and temperatures. The 0°C isotherm becomes a straight line vertically downward when CO_2 is compressed to its saturation vapor pressure (34.85 bars) and liquid CO_2 forms. The liquid/gas ratio increases as the straight-line segment of the isotherm traverses the two-phase field (inside the critical dome) where liquid and gaseous CO_2 coexist. All the CO_2 is liquefied at the bottom of the vertical line where the isotherm (for liquid CO_2) curves sharply upward and continues to higher pressures as a line with a nearly constant slope. Other isotherms behave similarly until the critical point is reached (indicated by the hollow point on the dome at 73.77 bars).

The characteristic shape of the isotherms changes above the critical temperature because CO_2 cannot exist as two separate phases at temperatures above its critical temperature. Thus, the vertically downward straight-line segment that occurs inside the two-phase field is missing. The 50°C isotherm is representative of those between the critical temperature and the 448.5°C isotherm. The 50°C isotherm

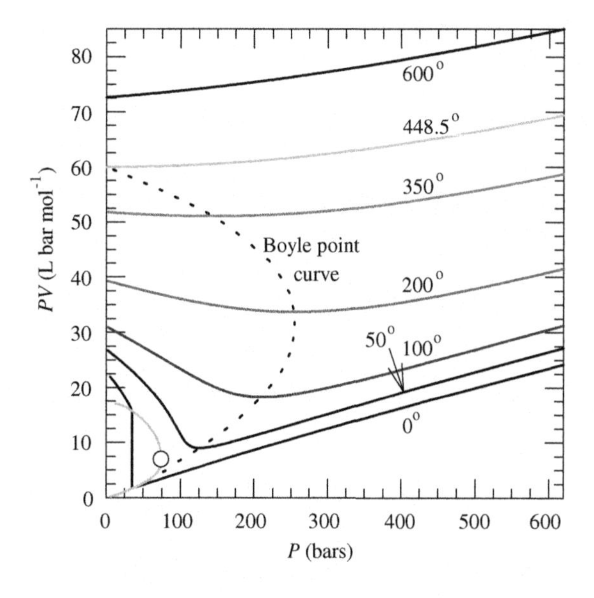

FIGURE 8-1

Isotherms for CO_2 from 0°C to 600°C.

initially curves downward toward the pressure axis because CO_2 is more compressible than an ideal gas at these pressures and temperatures. The 50°C isotherm reaches a minimum at 126 bars ($PV = 8.97$), where its slope goes through zero. It then curves upward away from the pressure axis and continues to higher pressures as a nearly straight line of constant slope. The minimum point on the 50°C isotherm is the *Boyle point* because Boyle's law is obeyed where the isotherm has zero slope and is horizontal. The Boyle point is defined by

$$\left[\frac{\partial(PV)}{\partial P}\right]_T = \left[\frac{\partial(PV)}{\partial(1/V)}\right]_T = \left[\frac{\partial(PV)}{\partial\rho}\right]_T = 0 \qquad \text{(8-69)}$$

The Boyle point curve on Figure 8-1 shows the temperature dependence of the CO_2 Boyle point. The minimum points on the isotherms become broader and shallower as temperature increases. The high-temperature end of the Boyle point curve, 448.5°C for CO_2, is the *Boyle temperature*. The Boyle temperature is the high-temperature end of the Boyle point curve where the zero slope (horizontal) part of an isotherm is at zero pressure.

The characteristic shape of the isotherms changes above the Boyle temperature because there is no longer any minimum point. The 600°C isotherm for CO_2 is representative of isotherms above its Boyle temperature (448.5°C). This isotherm immediately curves upward away from the pressure axis and does not go through a minimum. Carbon dioxide is less compressible than an ideal gas everywhere along the 600°C isotherm and along all other isotherms above its Boyle temperature. The higher-pressure branches of the lower-temperature isotherms eventually intersect their initial PV values and CO_2 is then less compressible than an ideal gas. For example, the crossovers on the 0, 100, and 350°C isotherms occur at 578, 614, and 281 bars, respectively.

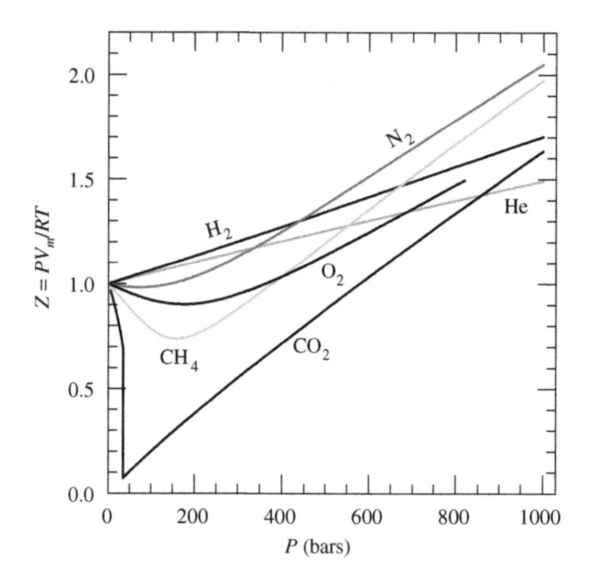

FIGURE 8-2

Compression factors for H_2, He, N_2, O_2, CH_4, and CO_2 at 0°C.

Figure 8-2 shows isotherms that are plotted as the molar PV product (PV_m) divided by RT, which is a new quantity defined as the *compression factor* (Z):

$$Z = \frac{PV_m}{RT} \tag{8-70}$$

The compression factor is always unity for ideal gases because $PV_m = RT$, the ideal gas law. But this is not the case for real gases. The compression factors plotted in Figure 8-2 vary as a function of pressure from 0 to 1000 bars along the 273.15 K (0°C) isotherm. The compression factor is useful because it immediately shows deviations from ideality ($Z \neq 1$), the size of deviations, and whether or not a gas is more compressible ($Z < 1$) or less compressible ($Z > 1$) than an ideal gas under the same P-T conditions.

The gases included in Figure 8-2 are chosen because they are important gases in different planetary atmospheres. Hydrogen and helium are the two most abundant gases in the atmospheres of Jupiter, Saturn, Uranus, and Neptune. Nitrogen is the most abundant gas on Earth and on Titan, the largest satellite of Saturn. Oxygen is the second most abundant gas on Earth and is a trace gas (~ 0.13%) in the Martian atmosphere. Methane is the second most abundant gas in Titan's atmosphere and is an important greenhouse gas on Earth, Titan, and the Jovian planets. Finally, CO_2 is included to facilitate comparison with Figure 8-1.

The 0°C (273.15 K) isotherms for H_2, He, N_2, O_2, CH_4, and CO_2 plotted in Figure 8-2 show the same characteristic behavior as that exhibited at different temperatures by the CO_2 isotherms in Figure 8-1. This similarity arises because the gases plotted in Figure 8-2 have different Boyle temperatures: 23 K (He), 110 K (H_2), 327 K (N_2), 406 K (O_2), 510 K (CH_4), and 722 K (CO_2). Thus, the He and H_2 isotherms are nearly straight lines, the N_2 isotherm has a shallow, broad minimum at about 72.5 bars, the O_2 isotherm has a deeper, broad minimum at about 175 bars, the CH_4 isotherm has a deeper, more well-defined minimum at about 160 bars, and the CO_2 isotherm has a sharp, clean

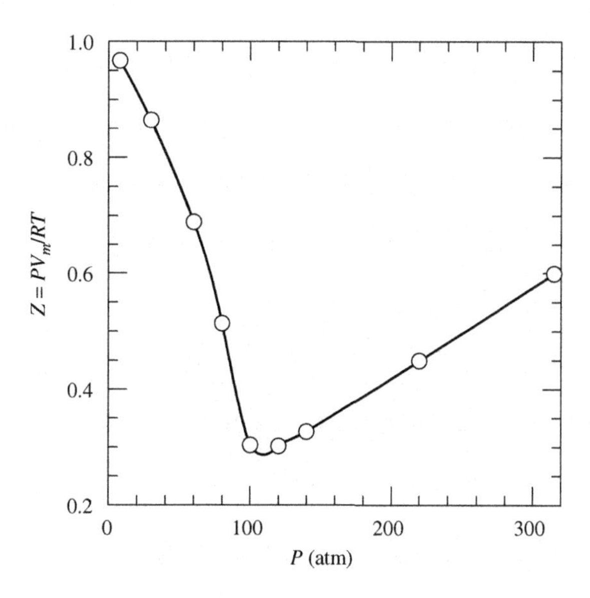

FIGURE 8-3

Compression factor for N_2O at 50°C.

minimum at 34.85 bars, its saturation vapor pressure at 273.15 K. The minima in the N_2, O_2, CH_4, and CO_2 isotherms correspond to the Boyle points of these gases at 273.15 K.

Example 8-5. Nitrous oxide (N_2O) is widely used as a replacement for CFC gases in refrigeration, an antibacterial foaming agent in the dairy industry, a surgical anesthetic, and a fuel oxidant in racing cars. These uses and natural sources emit N_2O into Earth's atmosphere where it is a greenhouse gas and is also involved in ozone destruction. Calculate its compression factor Z (and plot Z versus P) at 323 K (50°C) from a subset of the *PVT* data of Couch et al. (1961), who measured the specific volume ($cm^3\ g^{-1}$) from 8 to 315 atmospheres pressure.

P (atm)	8	30	60	80	100	120	140	220	315
V	72.797	17.344	6.905	3.862	1.824	1.513	1.404	1.230	1.145
Z	0.967	0.864	0.688	0.513	0.303	0.301	0.326	0.449	0.599

We multiplied V by the molecular weight of N_2O (44.013 g mol^{-1}) to convert from specific to molar volume and used $R = 82.0578\ cm^3\ atm\ K^{-1}\ mol^{-1}$ to compute Z. Figure 8-3 shows our calculated compression factor curve, which is similar to those for CO_2 and C_2H_4.

C. Law of corresponding states

The law of corresponding states was proposed by the Dutch physicist Johannes Diederik van der Waals (1837–1923), who won the 1910 Nobel Prize in Physics for his work on the thermodynamics of gases and liquids (see sidebar). The law of corresponding states says that the properties of gases and liquids (e.g., vapor pressure, enthalpy of vaporization, compression factor) are the same and fall onto the same

JOHANNES DIDERIK VAN DER WAALS (1837–1923)

JOHANNES DIDERIK VAN DER WAALS

J. D. van der Waals was the son of a carpenter and the oldest of 10 children. In his childhood, primary (or elementary) school was not free and there was no secondary school in The Netherlands. His parents could only afford to pay for his education in primary school and he was self-taught thereafter. At the age of 14 he became a pupil-teacher, an apprentice who helps the elementary school teacher with large classes. Over the next few years, van der Waals passed a series of exams that allowed him to become a primary school teacher. Secondary schools known as HBS schools were founded in The Netherlands in 1863. After passing further examinations, van der Waals became a physics teacher in an HBS school in the Hague, the Dutch capital city, in 1865. The HBS schools were more advanced than today's high schools; some of the best Dutch scientists of the time, including two Nobel Prize winners (van der Waals and H. A. Lorentz), began their careers as HBS teachers.

Van der Waals wrote his doctoral dissertation (*On the Continuity of the Gaseous and Liquid States*) in 1873 when he was 36 years old and still a physics teacher at the HBS school in the Hague. Van der Waals became deputy director of his HBS school in 1874. One year later he became a member of the Dutch Royal Academy of Sciences, and in 1877 he was appointed professor of physics at the University of Amsterdam. He held this position until he retired in 1908. Then his former position was split into two professorships, with his son, J. D. van der Waals, Jr. (1873–1971), the new professor of physics and his former student, Philipp Kohnstamm (1875–1951), the professor of thermodynamics. Two years after his retirement, van der Waals was awarded the 1910 Nobel Prize in Physics for "his studies of the physical state of liquids and gases." Van der Waals's work influenced several generations of Dutch chemists and physicists, including J. P. Kuenen, Philipp Kohnstamm, A. Michels, Kammerlingh Onnes, and Bakhuis Roozeboom.

curve when they are expressed in terms of reduced variables. The Van der Waals equation and other equations of state with two parameters (discussed in the next sections) can be written in terms of reduced variables. These are ratios of pressure, temperature, volume (or density) to the critical pressure, critical temperature, critical volume (or density):

$$P_R = \frac{P}{P_C} = \pi \tag{8-71}$$

$$T_R = \frac{T}{T_C} = \theta \tag{8-72}$$

$$V_R = \frac{V}{V_C} = \phi \tag{8-73}$$

$$\rho_R = \frac{\rho}{\rho_C} = \frac{V_C}{V} \tag{8-74}$$

When this is done, we obtain a reduced equation of state that expresses one reduced variable as a function of the other two reduced variables, for example,

$$V_R = f(T_R, P_R) \tag{8-75}$$

The exact nature of the function relating the reduced variables is different for each equation of state. However, the general principle is the same. Two gases with the same values of reduced temperature (T_R), reduced pressure (P_R), and reduced volume (V_R) (or reduced density ρ_R) are in corresponding states. Equation (8-75) shows this statement is true if any two of the three reduced variables are the same because the third variable is related to the other two by the reduced equation of state. The pressures, temperatures, and volumes (or densities) of two gases in corresponding states are related via the equations

$$\frac{P}{P_C} = \frac{P'}{P'_C} \tag{8-76}$$

$$\frac{T}{T_C} = \frac{T'}{T'_C} \tag{8-77}$$

$$\frac{V}{V_C} = \frac{V'}{V'_C} \tag{8-78}$$

$$\frac{\rho}{\rho_C} = \frac{\rho'}{\rho'_C} \tag{8-79}$$

The properties of the two different gases in Eqs. (8-76) to (8-79) are denoted by symbols with and without primes. Specific examples of corresponding behavior are given below.

The law of corresponding states is related to the forces between the atoms or molecules in a gas and it has been derived from the properties of gas atoms and molecules using statistical mechanics (Pitzer, 1939). The heavier noble gases (Ar, Kr, Xe) and other molecules with spherical or nearly spherical force fields (N_2, O_2, CO, CH_4) follow the law of corresponding states almost exactly (Guggenheim, 1945). Other molecules show larger deviations from the law of corresponding states because their interatomic or intermolecular forces are more complex. However, even in these cases, the law of corresponding states is an extremely useful approximation that can be used to estimate unknown properties such as vapor pressure, enthalpy of vaporization, critical density, pressure, and temperature from those properties that have been measured for other chemically similar atoms or molecules.

The simplest way to illustrate the law of corresponding states is to compare the critical compression factor (Z_C) for different gases. This is defined as

$$Z_C = \frac{P_C V_C}{RT_C} \tag{8-80}$$

If the law of corresponding states was completely correct, all gases would have the same value of the critical compression factor Z_C. Table 8-2 lists Z_C values for 31 gases, including all the major gases and many important trace gases in the atmospheres of Earth, Venus, Mars, Jupiter, Saturn, Uranus, Neptune, Pluto, and Titan, the largest satellite of Saturn. Table 8-2 also includes most of the refrigerants in Table 6-1 and sulfur vapor, which is erupted from volcanoes on Jupiter's satellite Io. To first approximation, the mean Z_C value (0.28 ± 0.02) is fairly constant. However, closer scrutiny of Table 8-2 reveals trends due to molecular mass, polarity, shape, and size.

The mean Z_C value for H_2 and its isotopic variants (HD, D_2), He, and Ne is 0.31 ± 0.01 (D_2 0.313, Ne 0.311, HD 0.310, n-H_2 0.306, e-H_2 0.303, He 0.302). The high Z_C values for these gases are due to quantum effects. (Normal or n-H_2, has a 3:1 ortho/para ratio like H_2 gas at room temperature and higher temperatures. However, equilibrium or e-H_2 has a temperature-dependent ortho/para ratio that decreases as the temperature approaches absolute zero, where all ortho H_2 is converted into para H_2. The ortho/para ratio of H_2 is important for the thermal structure of the upper atmospheres of the gas giant planets in our solar system.)

The noble gases Ar, Kr, Xe, and small nonpolar molecules (e.g., N_2, CO, O_2, and CH_4) with spherical or nearly spherical force fields have mean Z_C values of 0.289 ± 0.003. Hydrocarbons are generally ellipsoidal molecules and their mean Z_C value is also fairly constant. This is one expression of Kammerlingh Onnes's *principle of mechanical equivalence,* which states that molecules of the same general shape have similar properties. The other molecules in Table 8-2 have Z_C values ranging from 0.231 (H_2O) to 0.298 (CS_2). These differences are related to polarity, shape, and size.

The trends in the values of the critical compression factor Z_C led to the use of an additional parameter to improve correlations based on the law of corresponding states. Hougen et al. (1960) take the value of Z_C as the third parameter. Another approach discussed by Pitzer and Brewer (1961) is to use the acentric factor ω, which is determined from reduced vapor-pressure curves (explained in a moment) and is a dimensionless parameter defined as

$$\omega = -\log \frac{P_{sat}}{P_C} \quad \text{at} \quad T_R = 0.70 \tag{8-81}$$

The saturated vapor pressure (P_{sat}) at a temperature equivalent to $T_R = 0.70$ is used in Eq. (8-81). Pitzer and Brewer (1961) describe calculation of the compression factor and other thermodynamic properties using the reduced temperature, reduced pressure, and acentric factor. The acentric factor and other parameterized correlations are extensively used by industry and are discussed in Chao and Robinson (1979, 1986), Prausnitz et al. (1999), and Reid et al. (1977).

The reduced vapor-pressure curves in Figure 8-4 illustrate another application of the law of corresponding states. The vapor-pressure curves in Figure 8-4 are plotted in terms of the reduced temperature (T_R) and reduced vapor pressure (P_{sat}/P_C). If the law of corresponding states was completely correct, Eqs. (8-76) and (8-77) would be obeyed and the reduced vapor-pressure curves would be identical for all gases. This is not the case. Figure 8-4 shows that the reduced vapor pressures

Table 8-2 Physical Properties of Selected Gases

Gas	T_b (K)	$V_{l,m}$	$V_{g,m}$	$\Delta_v H$	T_c (K)	P_c (bar)	$V_{c,m}$	Z_C
^3He	3.19	51.46	125.7	33.0	3.31	1.14	72	0.302
^4He	4.2	32.02	242.2	83.4	5.19	2.27	57	0.302
n-H_2	20.28	28.48	1504	898	33.2	12.97	65	0.306
Ne	27.14	16.72	2110	1730	44.49	26.79	42	0.304
N_2	77.36	34.75	6077	5577	126.2	33.9	89	0.290
Air	78.90	33.11	8764	5867	132.53	37.86	84.52	0.290
CO	81.61	35.37	6443	6040	132.91	34.99	93	0.295
Ar	87.30	28.63	6921	6448	150.8	48.7	75	0.291
O_2	90.19	28.04	7162	6818	154.58	50.43	73	0.288
CH_4	111.67	37.98	8837	8180	190.58	46.04	99	0.288
Kr	119.78	34.73	9508	8991	209.41	55.0	91	0.287
C_2H_4	169.38	49.42	13,436	13,534	282.34	50.39	130	0.280
Xe	165.03	44.69	13,121	12,543	289.77	58.41	118	0.286
CHF_3	233	55.40	2337	13,415	299.30	48.58	133	0.259
CO_2	216.6	37.35	3197	15,420	304.1	73.75	94	0.274
C_2H_6	184.5	55.28	14,645	14,703	305.42	48.80	148	0.285
PH_3	185.38	45.57	14,782	14,598	324.5	65.4		
CH_2F_2	221.5	42.70	18,690	19,908	351.3	57.78	122	0.241
C_3H_8	231.06	75.84	18,243	18,778	369.82	42.50	203	0.281
H_2S	212.8	37.23	17,656	18,673	373.2	89.4	98	0.284
OCS	222.87	51.0		18,506	378.8	63.49	135	0.272
CF_2Cl_2	244.26	81.37	18,439	19,920	384.95	41.36	217	0.280
NH_3	239.7	24.99	18,905	23,351	405.5	113.5	72	0.244
i-C_4H_{10}	261.48	97.87	20,560	21,269	408.15	36.48	263	0.283
CH_3Cl	248.9	50.39	19,471	21,497	416.25	66.79	139	0.268
SO_2	263.08	43.82	20,683	24,937	430.8	78.84	122	0.269
$(C_2H_5)_2O$	307.7	106.48	23,443	26,520	466.74	36.38	280	0.262
CCl_3F	296.8	93.31	21,347	25,140	471.2	44.1	248	0.279
CS_2	319.4	62.20		26,740	552	79.0	173	0.298
H_2O	373.13	18.80	30,138	40,656	647.14	220.5	56	0.231
$C_{10}H_8$	490				748.4	40.51	413	0.269
sulfur	719	73.88	50,195	65,220	1313	182.1	158	0.264
								Mean $\pm 1\sigma = 0.279 \pm 0.018$.

Notes: T_b is the normal boiling point, $V_{g,m}$, $V_{l,m}$, and $V_{c,m}$ are molar volumes (cm^3/mol) of gas and liquid at the boiling point and of fluid at the critical temperature T_C, $\Delta_v H$ is the enthalpy of vaporization at T_b (J/mol), $Z_C = P_C V_C / RT_C$, n-H_2 is normal H_2 with an ortho/para ratio of 3:1 (see Chapter 4, Section I); CHF_3 HFC-23, CH_2F_2 HFC-32, C_3H_8 propane, CF_2Cl_2 CFC-12, i-C_4H_{10} isobutane, CH_3Cl methyl chloride, $(C_2H_5)_2O$ diethyl ether, and CCl_3F CFC-11 are refrigerants (see Table 6-1); $C_{10}H_8$ naphthalene is found in moth balls. Data for CO_2 are at its triple point.

Sources: Ambrose (1980), Cragoe (1928), Lemmon et al. (2000), NIST Chemistry Web Book, *Rau et al. (1973), Sifner (1999), Timmermans (1950).*

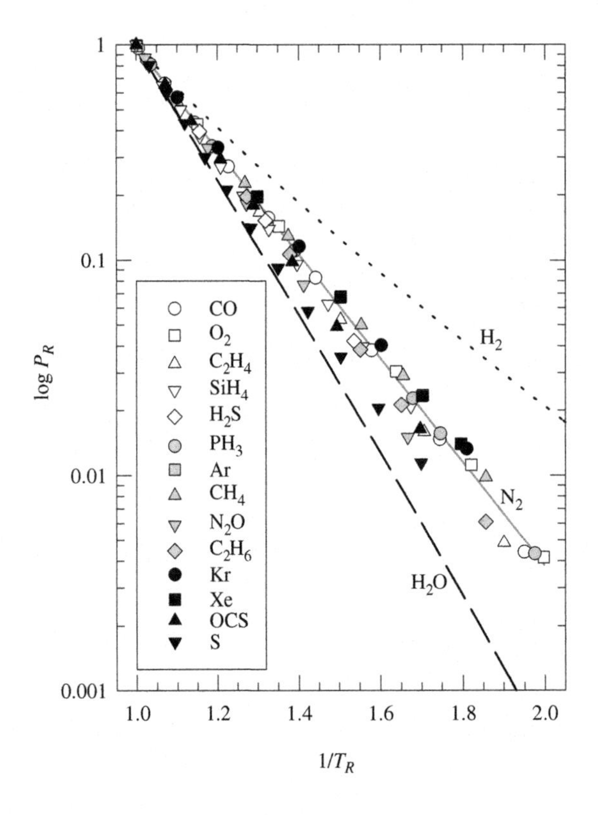

FIGURE 8-4

Reduced vapor-pressure plot for N_2, CO, O_2, C_2H_4, SiH_4, H_2S, PH_3, Ar, CH_4, N_2O, C_2H_6, Kr, Xe, OCS, S, H_2, and H_2O.

of a number of gases (CO, O_2, C_2H_4, SiH_4, H_2S, PH_3, Ar, CH_4, N_2O, C_2H_6, Kr, Xe, OCS, S) are similar, but not identical, to that of N_2, which is plotted as the solid line. A linear least-squares fit of P_R versus T_R for N_2 gives the equation of the line, which is

$$\log_{10} P_R = 2.4109 - \frac{2.4164}{T_R} \tag{8-82}$$

Equation (8-82) can be used to estimate the vapor pressure of a material from its critical constants or to estimate one of the critical constants from knowledge of the others and vapor-pressure data. Nitrogen was chosen for comparison because it has a Z_C value of 0.290, almost identical to the mean value of 0.289 for noble gases and small, nonpolar molecules.

Example 8-6. Carbon dioxide is a gas at room temperature and is an important constituent of volcanic gases on Earth. Sulfur is solid at room temperature but is emitted in volcanic gases on Io. Show that their reduced vapor pressures are nearly identical. We do this using the three pairs of vapor-pressure data in the following table. Each pair shows the vapor pressure of CO_2 and S at three different temperatures. The reduced temperatures of CO_2 and S are the same in each pair, and their reduced vapor pressures are almost identical.

	T (K)	P (bar)	$T_R = T/T_C$	$P_R = P/P_C$
CO_2	225.39	7.47	0.741	0.10
sulfur	973	16.7	0.741	0.09
CO_2	260.12	24.27	0.855	0.33
sulfur	1123	54.5	0.855	0.30
CO_2	294.85	59.62	0.970	0.81
sulfur	1273	146.4	0.970	0.80

Figure 8-4 shows that in general the agreement between reduced vapor pressures of different materials becomes worse at lower temperatures (lower T_R values). Also, the vapor pressure of other gases, notably H_2, He, Ne, and H_2O, deviate significantly from the line. These deviations are due to quantum mechanical effects for the light gases (H_2, He, Ne) and to hydrogen bonding for water. The reduced vapor-pressure curves for H_2 and H_2O are shown in Figure 8-4.

Hougen et al. (1960) give an empirical equation that uses the reduced temperature and the critical compression factor Z_C to estimate the reduced vapor pressure. This equation is

$$\log_{10}P_R = \frac{A(T_R - 1)}{T_R} - 10^{-8.68(T_R - b)^2} \tag{8-83}$$

The A and b terms are constants that depend on the critical compression factor Z_C of a gas:

$$A = 16.25 - 73.85Z_C + 90Z_C^2 \tag{8-84}$$

$$b = 1.80 - 6.20Z_C \tag{8-85}$$

Equation (8-83) gives a better match to the reduced vapor-pressures curves of hydrocarbons and some other gases than provided by Eq. (8-82). Equation (8-83) can be used to find Z_C by empirically fitting vapor-pressure data until the best match between the calculated and observed curves is obtained. This was done for elemental sulfur, selenium, arsenic, and phosphorus vapors (Rau, 1975a,b and references therein).

A third illustration of the law of corresponding states is to plot the compression factors for different gases along reduced-temperature isotherms as a function of reduced pressure. The reduced-temperature isotherms in Figure 8-5 are plotted as a function of the reduced pressure. The isotherms near the two-phase region ($T_R \leq 1$) have steep slopes. Isotherms inside the two-phase region are vertical lines of infinite slope (see Figure 8-1). Rowlinson (1958) showed that isotherms (T_R) plotted against the reduced density (ρ_R) are much smoother, with slopes that are always finite, even inside the two-phase region. However, in most cases one is interested in properties of gases as a function of temperature and pressure, so the reduced-temperature and-pressure plots are more useful.

Example 8-7. Carbon dioxide is the major gas (96.5%) in the atmosphere of Venus. Estimate its compression factor Z (and hence its deviation from ideality) for average surface conditions on Venus (735.3 K, 92.1 bars) by using critical parameters from Table 8-2 and the reduced-variable plot in Figure 8-5. The critical T and P for CO_2 are 304.1 K and 73.75 bars. Average surface conditions on Venus correspond to $T_R = (735.3/304.1) = 2.4$ and $P_R = (0.965 \times 92.1/73.75) = 1.2$. These points correspond to $Z = 1$ on Figure 8-5. The measured PVT data for CO_2 give $Z = 1.005$, in excellent agreement with the Z value estimated using reduced variables. Carbon dioxide is predicted to be ideal with $Z = 1$ within 0.5% on Venus.

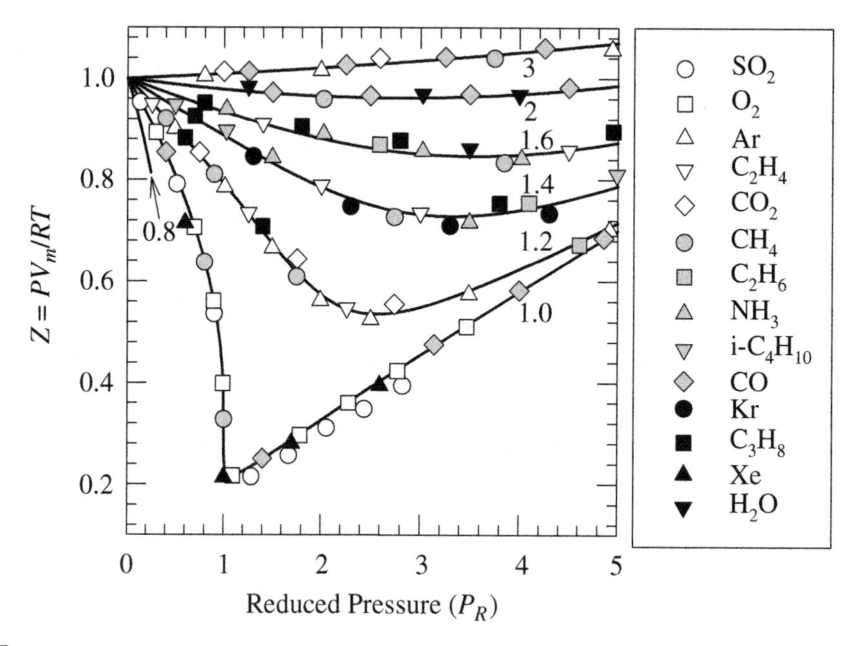

FIGURE 8-5

Reduced-temperature and-pressure plot for compression factors.

Correlations using only reduced temperature and pressure without additional parameters are estimated to have errors of 15–30% (Dodge, 1944; Hougen et al., 1960). The errors are largest for the light gases influenced by quantum mechanical effects (H_2, He, Ne) and for water, which is affected by hydrogen bonding. The errors for H_2, He, and Ne are reduced if the equations suggested by Newton (1935) are used to compute their reduced temperatures and pressures:

$$T_R = \frac{T}{T_C + 8} \tag{8-86}$$

$$P_R = \frac{P}{P_C + 8} \tag{8-87}$$

Example 8-8. Hydrogen is the major gas in the cold, dense atmospheres of Uranus and Neptune. Experimental data give $Z = 1.00$ for H_2 at 100 K and 31 bars. Use Figure 8-5 to estimate Z for H_2 at the same P-T point and then use Newton's equations to compute reduced temperature and pressure and make a second estimate. Table 8-2 gives $T_C = 33.2$ K and $P_C = 12.97$ bars for H_2. The reduced temperature $T_R = 100/33.2 = 3.01$ and the reduced pressure $P_R = 31/12.97 = 2.39$. Figure 8-5 gives $Z = 1.03$ at this point. This is 3% larger than the actual value. The second estimate uses Eqs. (8-86) and (8-87) to calculate reduced variables. In this case $T_R = 2.43$ and $P_R = 1.48$. This point is between the lines on Figure 8-5, with $Z = 1.015$ for $T_R = 3$ and $Z = 0.97$ for $T_R = 2$. A linear interpolation gives $Z = 0.99$ at $T_R = 2.43$, which is 1% smaller than the actual value.

Figure 8-5 covers only a small range of reduced temperatures and pressures. More extensive charts of Z as a function of reduced temperature and pressure are in Dodge (1944) and Hirschfelder et al.

(1954). As done with the reduced vapor pressure, the critical compression factor Z_C or the acentric factor ω can be used with T_R and P_R for estimating compression factors. Hougen et al. (1960) give a set of tables that use Z_C as the third parameter; Pitzer and Brewer (1961) give tables using the accentric factor. Finally, as we discuss later, reduced-variable plots and tables are also used to estimate thermal properties and fugacity coefficients of gases under P-T conditions where there are no experimental data. In general, these empirical correlations are very useful in cases where the equation of state is unknown.

D. Van der Waals equation

J. D. van der Waals derived the van der Waals equation in his doctoral thesis, written in 1873 when he was 36 years old (see sidebar). An annotated English translation of his thesis (Rowlinson, 2004) was recently republished and is worth reading. The van der Waals equation is a major improvement over the ideal gas equation because van der Waals considered the finite volumes of gas atoms or molecules and their interatomic or intermolecular forces. For simplicity, he assumed that atomic or molecular volumes and the forces between the atoms or molecules in a gas are constant. This assumption is only an approximation because the atomic or molecular volumes and the interatomic or intermolecular forces actually depend on the pressure and temperature of a gas. Consequently, the van der Waals equation is useful at P-T conditions where the molar volume of a gas is large compared to the volume occupied by the gas atoms or molecules and where the interatomic or intermolecular forces between the gas atoms or molecules are small compared to the thermal energy at the ambient temperature. The van der Waals equation is the first real gas equation of state to qualitatively explain the continuity of states, vapor-liquid condensation, and the vapor-liquid critical point of gases (e.g., the explanation given by van der Waals of Andrews [1869] results for CO_2). It is also the prototype for other (more complex) cubic equations of state including the Redlich-Kwong and Peng-Robinson equations and their various modifications (Redlich and Kwong 1949; Holland and Powell 1991; Prausnitz et al. 1999). The van der Waals equation can be written in several equivalent forms:

$$\left(P + \frac{a}{V^2}\right)(V - b) = nRT \tag{8-88}$$

$$P = \frac{RT}{V_m - b} - \frac{a}{V_m^2} \tag{8-89}$$

$$V_m^3 - \left[b + \frac{RT}{P}\right]V_m^2 + \left(\frac{a}{P}\right)V_m - \frac{ab}{P} = 0 \tag{8-90}$$

The a and b are the van der Waals constants for a gas. The a term is related to the attractive forces between gas atoms or molecules and a/V_m is the effective energy of attraction between the gas particles (atoms or molecules). Excluding ^3He, ^4He, and H_2 because of quantum mechanical effects, the ratio a/V_m is about one-third of the enthalpy of vaporization of the liquefied gas at the normal boiling point ($a/V_m = 0.33 \pm 0.32$ for the gases in Tables 8-2 and 8-3). The quotient a/V_m has the dimensions of energy or force. Thus, a/V_m^2 has the same dimensions as pressure and is the internal pressure of the gas particles. The co-volume b is a repulsive term and is related to the finite volumes of the atoms or molecules in the gas. The ratio $b/V_{liq} = 1.16 \pm 0.19$ for the 31 gases in Tables 8-2 and 8-3 for which

Table 8-3 Constants for Cubic Equations of State

	van der Waals[a]		Redlich-Kwong[b]		Pitzer
Gas	$10^{-4}a$ cm^6 bar mol^{-2}	b cm^3 mol^{-1}	$10^{-4}a'$ cm^6 bar K$^{1/2}$ mol^{-2}	b' cm^3 mol^{-1}	ω
^3He	2.803	30.18	5.2	20.92	−0.473
^4He	3.460	23.76	8.0	16.47	−0.365
n-H_2	24.78	26.60	144.7	18.44	−0.218
Ne	21.55	17.26	145.6	11.96	0
N_2	137.0	38.69	1560	26.82	0.037
Air	135.3	36.38	1578	25.22	
CO	147.2	39.48	1720	27.36	0.049
Ar	136.2	32.18	1695	22.31	−0.004
O_2	138.2	31.86	1741	22.08	0.021
CH_4	230.1	43.02	3218	29.82	0.011
Kr	232.5	39.57	3410	27.43	−0.002
C_2H_4	461.4	58.23	7856	40.36	0.087
Xe	419.2	51.56	7232	35.74	0.002
CHF_3	537.8	64.03	9428	44.38	0.260
CO_2	365.7	42.86	6462	29.70	0.223
C_2H_6	557.5	65.05	9872	45.09	0.100
CH_2F_2	622.9	63.19	11,831	43.80	0.271
C_3H_8	938.5	90.44	18,288	62.68	0.153
H_2S	454.4	43.39	8894	30.07	0.100
OCS	659.1	62.01	12,999	42.98	0.099
CF_2Cl_2	1045	96.73	20,774	67.05	0.176
NH_3	422.5	37.13	8621	25.74	0.250
i-C_4H_{10}	1332	116.3	27,264	80.60	0.176
CH_3Cl	756.6	64.77	15,641	44.89	0.156
SO_2	686.5	56.79	14,439	39.36	0.251
$(C_2H_5)_2O$	1746	133.3	38,231	92.42	0.281
CCl_3F	1468	111.0	32,297	76.97	0.188
CS_2	1125	72.62	26,780	50.33	0.115
H_2O	553.9	30.50	14,278	21.14	0.344
$C_{10}H_8$	4032	192.0	111,779	133.08	0.302
S	2761	74.94	101,377	51.94	0.171

[a]*The van der Waals a for $^3He = 2.803 \times 10^4$ cm^6 bar mol^{-2}, and so on.*

data are available. This ratio is more constant than a/V_m and is only slightly larger (1.20 ± 0.14) if ^3He, ^4He, and H_2 are excluded.

The a and b terms in the van der Waals equation can be calculated from three different properties of a gas: (1) its deviations from the ideal gas law, (2) its critical temperature and pressure, and (3) its cooling upon expansion. However, the values of a and b calculated from these three different data sets are slightly different. The differences arise because the forces between atoms and molecules and their volumes depend on the pressure and temperature of the gas. In his thesis, van der Waals used the deviations of CO_2 and other gases from the ideal gas law to calculate their values of a and b. These deviations are given by the constant-pressure and constant-volume thermal expansion coefficients determined from gas thermometry. The thermal expansion coefficient for an ideal gas is the same whether heating is done at constant pressure or constant volume (see Section II-A). However, this is not the case for real gases, and the two thermal expansion coefficients have slightly different values.

Example 8-9. Calculate the van der Waals coefficients a and b for N_2 using the constant-volume ($\kappa_v = 3.6744 \times 10^{-3}\ K^{-1}$) and constant-pressure ($\kappa_p = 3.6732 \times 10^{-3}\ K^{-1}$) thermal expansion coefficients determined by gas thermometry (with $P_0 = 1001.9$ mmHg, $T_0 = 273.15$ K, $T_1 = 373.15$ K; Jeans, 1954). The constant-volume thermal expansion coefficient is defined by the equation

$$P_1 = P_0[1 + \kappa_v(T_1 - T_0)] \tag{8-91}$$

Equation (8-91) can be rewritten in terms of the van der Waals equation, giving (Jeans, 1954)

$$a = P_0 V_m^2 [\kappa_v T_0 - 1] \tag{8-92}$$

The molar volume of N_2 is 16,992.1 $cm^3\ mol^{-1}$ at 1001.9 mmHg pressure (1.318 atm = 1.336 bars) and 273.15 K. Substituting numerical values into Eq. (8-92) yields

$$\begin{aligned} a &= (1.318\ \text{atm})(16{,}992.1\ \text{cm}^3\text{mol}^{-1})^2[(3.6744 \times 10^{-3}\ \text{K}^{-1})(273.15\ \text{K}) - 1] \\ &= 139.4 \times 10^4\ \text{cm}^6\text{atm mol}^{-2} = 141.2 \times 10^4\text{cm}^6\text{bar mol}^{-2} \end{aligned} \tag{8-93}$$

The constant-pressure thermal expansion coefficient is defined by the equation

$$V_1 = V_0[1 + \kappa_p(T_1 - T_0)] \tag{8-94}$$

The molar volume of N_2 is 19,490.1 $cm^3\ mol^{-1}$ at $T_1 = 373.15$ K and $P = 1001.9$ mmHg pressure.

Equation (8-94) can also be rewritten in terms of the van der Waals equation (Jeans, 1954), giving

$$b = -V_0\left[T_0\kappa_p - 1 - \frac{a}{PV_0}\left(\frac{1}{V_0} + \frac{1}{V_1}\right)\right] \tag{8-95}$$

Substituting numerical values into Eq. (8-95) yields

$$\begin{aligned} b &= -(16{,}992.1)\left[(273.15\ )(3.6732 \times 10^{-3}\ ) - 1 - \frac{141.2 \times 10^4}{(1.336\ )(16992.1)}\left(\frac{1}{16992.1} + \frac{1}{19490.1}\right)\right] \\ &= 59.76\ \text{cm}^3\text{mol}^{-1} \end{aligned} \tag{8-96}$$

Example 8-9 used thermal expansion coefficients derived from gas thermometry, but the required values for computing a and b can be derived from tabular values of volume (or density) as a function of pressure and temperature for a gas. For example, Rau (1975a,b) used *PVT* data for arsenic and phosphorus vapor to compute his van der Waals constants.

Making the approximation that gas atoms or molecules are hard spheres of radius r, it can be shown that b is equal to four times the molar volume occupied by gas molecules (Jeans, 1954):

$$b = 4N_A \frac{4}{3} \pi r^3 \quad (8\text{-}97)$$

Thus, the diameter of a "hard sphere" gas atom or molecule can be derived from b via the formula

$$d \text{ (nm)} = 0.0926 b^{1/3} \quad (8\text{-}98)$$

Example 8-10. Calculate the size of a N_2 molecule using the "hard sphere" approximation and the van der Waals equation. Using $b = 59.76$ cm^3 mol^{-1} from Example 8-9 and Eq. (8-98) we get

$$d \, (N_2) = 0.0926(59.76)^{1/3} = 0.362 \text{ nm} \quad (8\text{-}99)$$

This is slightly smaller than the accepted value of 0.370 nm for the diameter of a "hard sphere" N_2 molecule (Hirschfelder et al., 1954, p. 1111).

Table 8-3 lists values of the van der Waals constants a and b for several gases found in planetary and satellite atmospheres. These values were calculated from the critical parameters in Table 8-2 using two equations, which we will derive later:

$$a = \frac{27bRT_C}{8} \quad (8\text{-}100)$$

$$b = \frac{RT_C}{8P_C} \quad (8\text{-}101)$$

Example 8-11. Calculate the van der Waals constants for ethane (C_2H_6) from its critical parameters and then compute the molar volume of ethane at its normal boiling point (184.5 K, 1 atm). Table 8-2 lists $T_C = 305.42$ K and $P_C = 48.80$ bars. We put these values into Eq. (8-101) to find b and then use the result and Eq. (8-100) to compute a:

$$b = \frac{(83.145 \text{ cm}^3 \text{ bar K}^{-1} \text{ mol}^{-1})(305.42 \text{ K})}{8(48.80 \text{ bars})} = 65.05 \text{ cm}^3 \text{ mol}^{-1} \quad (8\text{-}102)$$

$$a = \frac{27(65.05 \text{ cm}^3 \text{ mol}^{-1})(83.145 \text{ cm}^3 \text{ bar K}^{-1} \text{ mol}^{-1})(305.42 \text{ K})}{8} = 5.58 \times 10^6 \text{ cm}^6 \text{ bar mol}^{-2} \quad (8\text{-}103)$$

We now compute V_m using Eq. (8-90). This can be done iteratively or via the general solution to a cubic equation (e.g., Lodders and Fegley, 1998, p. 11). The result is $V_m = 14{,}832$ cm^3 mol^{-1}, which is 1.3% larger than the actual value of 14,644 cm^3 mol^{-1}.

Figure 8-6 shows a comparison of the observed compression factor for CO_2 at 735.3 K (the mean surface temperature on Venus) with values calculated from the ideal gas, van der Waals, Peng-Robinson, and virial equations. The Peng-Robinson equation is a cubic equation and the virial

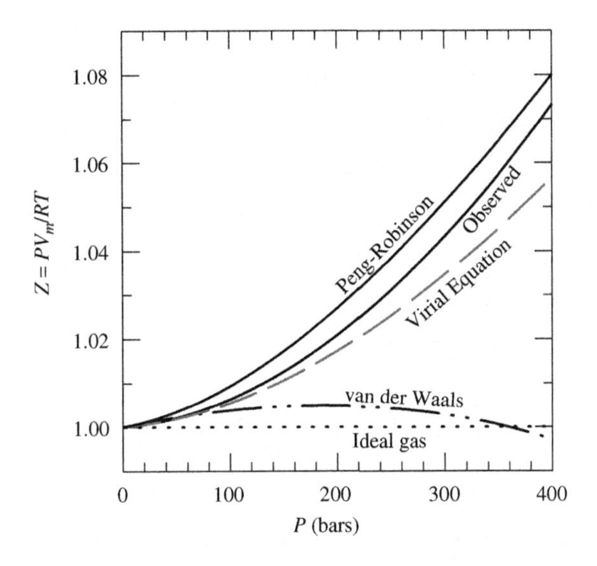

FIGURE 8-6

Comparison of observed and calculated values for the CO_2 compression factor at 735 K. Van der Waals coefficients ($a = 131.2 \times 10^4$ cm^6 bar mol^{-2} and $b = 24.53$ cm^3 mol^{-1}) determined from *PVT* data for CO_2 were used in the calculations. The virial coefficients for CO_2 at 735 K are $B = 1.3$ cm^3 mol^{-1} and $C = 1266$ cm^6 mol^{-2}.

equation is an infinite power series. They are described in Sections D and E next. The pressure scale in Figure 8-6 is extended well beyond the average surface pressure of 92.1 bars on Venus. Example 8-7 showed that CO_2 is ideal within 0.5% at Venus surface conditions and this is also seen in Figure 8-6 (at 92.1 bars the observed $Z = 1.0055$ versus 1 for ideality). This good agreement arises because Venus's average surface temperature (735.3 K) is close to the Boyle temperature of CO_2 (721.6 K) where Boyle's law is obeyed.

Figure 8-6 shows that the Z values computed from the van der Waals, Peng-Robinson, and virial equations generally agree better with the observed values than do the Z values computed from the ideal gas law. In general, the deviations between the observed and calculated values increase with increasing pressure. At the maximum pressure shown of 400 bars the ideal gas law is 7.3% lower, the van der Waals equation is 7.6% lower, the virial equation is 1.6% lower, and the Peng-Robinson equation is 0.65% higher than the observed value ($Z = 1.0732$). At pressures of 0–110 bars the van der Waals equation fits the data better than the Peng-Robinson equation. The reason for this is that the van der Waals coefficients were calculated from high-temperature *PVT* data for CO_2, so the agreement is better than if a and b values from Table 8-3 were used.

E. Other empirical equations of state

The deviations between the observed *PVT* behavior of real gases and the behavior predicted by the van der Waals equation led to the development of many other empirical equations of state for real gases. Well over 100 different equations of state have been proposed for real gases, for example, see these books (Chao and Robinson, 1979, 1986; Hirschfelder et al., 1954; Pitzer, 1995; Prausnitz et al., 1999;

Rowlinson, 1958). However, none of these equations is completely satisfactory for representing *PVT* properties of gases and liquids over the full range of *P-T* conditions of interest in Earth and planetary sciences. The van der Waals equation remained the prototype for other, more complex cubic equations, such as the Redlich-Kwong and the Peng-Robinson equations, and their various modifications. Redlich and Kwong (1949) proposed the equation

$$P = \frac{RT}{V_m - b'} - \frac{a'}{\sqrt[2]{T}(V_m(V_m + b'))} \tag{8-104}$$

At the critical point the Redlich-Kwong constants a′ and b′ are given by

$$a' = \frac{0.42748R^2(T_C)^{2.5}}{P_C} \tag{8-105}$$

$$b' = 0.08664\frac{RT_C}{P_C} \tag{8-106}$$

These constants are listed in Table 8-3. The Redlich-Kwong equation is generally an improvement over the van der Waals equation. However, as done in Example 8-9 and in Figure 8-6 with the van der Waals equation, when greater accuracy is desired the Redlich-Kwong constants should be evaluated by fitting them to *PVT* data over the range of interest instead of computing them from the critical parameters.

In fact, the constants for the van der Waals equation, the Redlich-Kwong equation, and any other cubic equation of state vary with temperature. Recognizing this fact, Soave (1972) rewrote the Redlich-Kwong equation and introduced a temperature-dependent expression for a':

$$P = \frac{RT}{V_m - b'} - \frac{a'(T)}{V_m(V_m + b')} \tag{8-107}$$

$$a'(T) = a'(T_C)\alpha(T) \tag{8-108}$$

$$a'(T_C) = 0.42748\frac{R^2T_C^2}{P_C} \tag{8-109}$$

$$\alpha(T) = \left[1 + (0.480 + 1.574\omega - 0.176\omega^2)(1 - \sqrt{T_R})\right]^2 \tag{8-110}$$

The b' constant is the same as in the Redlich-Kwong equation and is given by Eq. (8-106). The T_R is reduced temperature and ω is Pitzer's acentric factor. This is defined by Eq. (8-81) and is also listed in Table 8-3. Equation (8-110) is different for H_2. In this case $\alpha(T)$ is given by

$$\alpha(T) = 1.202\exp(-0.30288T_R) \tag{8-111}$$

Peng and Robinson (1976) developed an equation similar to the Soave equation. It is

$$P = \frac{RT}{V_m - b'''} - \frac{a'''(T)}{V_m(V_m + b''') + b'''(V_m - b''')} \tag{8-112}$$

$$b''' = 0.07780 \frac{RT_C}{P_C} \tag{8-113}$$

$$a''' = a'''(T_C)\alpha(T_R, \omega) \tag{8-114}$$

$$a'''(T_C) = 0.45724 \frac{R^2 T_C^2}{P_C} \tag{8-115}$$

$$\alpha = \left[1 + \beta(1 - T_R^{1/2})\right]^2 \tag{8-116}$$

$$\beta = 0.3764 + 1.54226\omega - 0.26992\omega^2 \tag{8-117}$$

The last cubic equation of state we discuss is the one proposed by Kerrick and Jacobs (1981). This is a modified version of the Redlich-Kwong equation that is designed to give a good fit to *PVT* data for H_2O and CO_2, and their mixtures up to 10 kilobars. Their equation uses a constant co-volume b'' (equal to 29 and 58 $cm^3\ mol^{-1}$ for H_2O and CO_2, respectively), whereas a'' is a function of temperature and pressure:

$$P = \frac{RT(1 + y + y^2 - y^3)}{V_m(1 - y)^3} - \frac{a''}{\sqrt[2]{T}\left[V_m(V_m + b'')\right]} \tag{8-118}$$

$$y = \frac{b''}{4V_m} \tag{8-119}$$

$$a'' = c + \frac{d}{V_m} + \frac{e}{V_m^2} \tag{8-120}$$

The values of the c, d, and e coefficients for pure H_2O and CO_2 are given by the equations

$$c\ (H_2O) = \left[290.78 - 0.30276T + 1.4774 \times 10^{-4}\ T^2\right] \times 10^6 \tag{8-121}$$

$$d\ (H_2O) = \left[-8374 + 19.437T - 8.148 \times 10^{-3}\ T^2\right] \times 10^6 \tag{8-122}$$

$$e\ (H_2O) = \left[76{,}600 - 133.9T + 0.1071\ T^2\right] \times 10^6 \tag{8-123}$$

$$c\ (CO_2) = \left[28.31 + 0.10721T - 8.81 \times 10^{-6}\ T^2\right] \times 10^6 \tag{8-124}$$

$$d\ (CO_2) = \left[9380 - 8.53T + 1.189 \times 10^{-3}\ T^2\right] \times 10^6 \tag{8-125}$$

$$e\ (CO_2) = \left[-368{,}654 + 715.9T + 0.1534T^2\right] \times 10^6 \tag{8-126}$$

Figures 8-6 to 8-8 give comparisons of the different cubic equations of state. Figure 8-6 is relevant to surface conditions on Venus and was discussed earlier. Figure 8-7 shows the 400 K isotherm for H_2 and compares the observed compression factor to that calculated from the Redlich-Kwong, Soave, and virial equations of state. The 400 K isotherm is a straight line because it is significantly above the Boyle temperature of H_2 (110 K). It is relevant to Jupiter's atmosphere, which is 86% H_2 and has an adiabatic temperature-pressure profile below the 166 K, 1 bar level. The Galileo probe measured P and T down to the 427 K, 22 bar level in Jupiter's atmosphere (Seiff et al., 1998) and Jupiter's atmospheric pressure is about 20 bars at 400 K. All of the equations of state shown in Figure 8-7 give values that are virtually identical to the observed values up to much higher pressures. Figure 8-8 is relevant to aqueous fluids in Earth's crust. It shows the compression factor from 1 bar to 10 kilobars for H_2O at 1000 K. This isotherm has a minimum because it is below the Boyle temperature of water (1538 K). The Kerrick-Jacobs equation gives a good fit to the observed values over the entire range, although its results tend to be slightly higher above 8 kilobars. The Peng-Robinson equation is valid from 1 to 400 bars but gives Z values that are too large at higher pressures. The virial equation fits the data up to the Boyle point at about 1200 bars but then gives values that are significantly lower than the observed compression factor. These figures show that the van der Waals and Redlich-Kwong equations can be used up to pressures of a few hundred bars. Other cubic equations of state with temperature-dependent coefficients (e.g., the Soave, Peng-Robinson, and Kerrick-Jacobs equations) may give increased accuracy in some cases. However, this entails increased complexity. Finally, the virial equation of state, which we

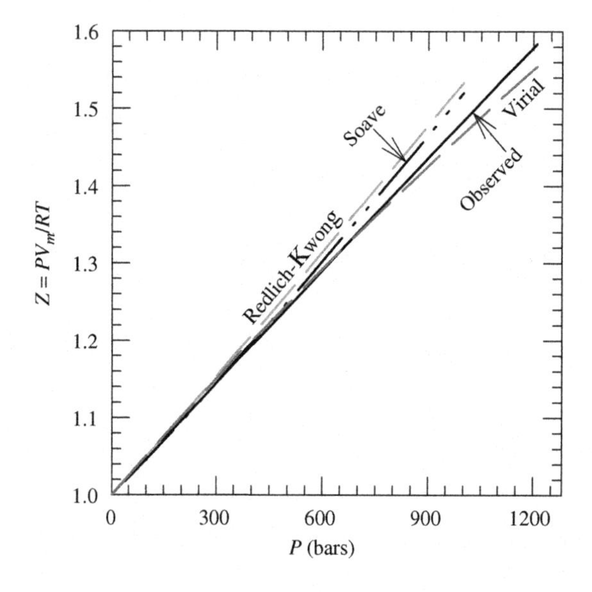

FIGURE 8-7

Comparison of observed and computed values for the H_2 compression factor at 400 K. The virial coefficients for H_2 at 400 K are $B = 16.9$ cm^3 mol^{-1} and $C = 290$ cm^6 mol^{-2}.

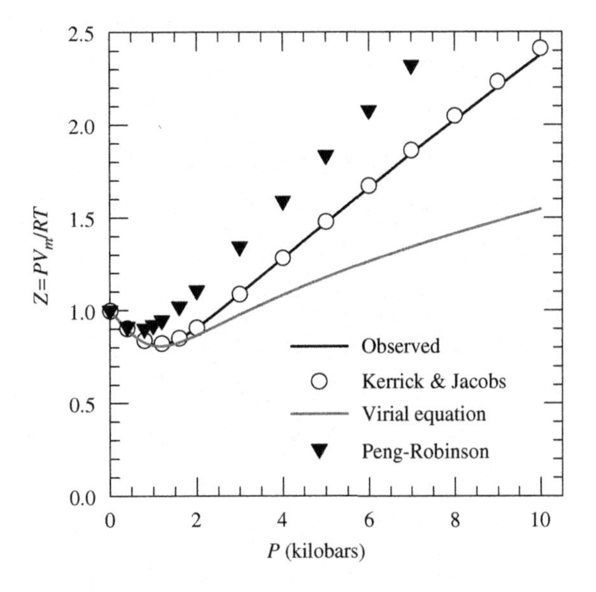

FIGURE 8-8

Comparison of observed and computed values of the H_2O compression factor at 1000 K. The H_2O virial coefficents at 1000 K are $B = -21.8$ cm^3 mol^{-1} and $C = 657$ cm^6 mol^{-2}.

discuss next, generally gives good agreement with *PVT* data up to pressures where molar volume is about the same as that at the critical point.

F. Virial equation of state

A *virial equation* uses an infinite power series for the equation of state of a gas. The word *virial* is derived from *vis* (plural *vires*), the Latin word for force or strength. In 1870, Rudolf Clausius (see sidebar in Chapter 6), who developed the second law of thermodynamics and the concept of entropy, also developed the *virial theorem.* This is a general theorem in physics that gives the ratio of the time-averaged kinetic energy to the time-averaged potential energy of a mechanical system, such as the atoms (or molecules) in a gas. The virial (Ξ_i) of particle i is defined by

$$\Xi_i = \overline{-\frac{1}{2}(\mathbf{r_i}\bullet\mathbf{F_i})} = \overline{K_i} \tag{8-127}$$

The time-averaged kinetic energy of the particle is $\overline{K_i}$ and the term $\overline{-(1/2)(\mathbf{r_i}\bullet\mathbf{F_i})}$ is the time-averaged vector product of the forces acting on the particle as it moves in three-dimensional space. The vector product takes into account the force component acting in the direction of motion of the particle and is given by a cosine term such as in Eq. (3-6). The virial coefficients thus give the deviations from ideality in terms of the forces between gas particles (atoms or molecules).

The power series format of the virial equation is general and can be used to represent the van der Waals equation and the other empirical equations of state discussed in the last section. In 1901 the

HEIKE KAMERLINGH ONNES (1853–1926)

Kamerlingh Onnes was born in Groningen, The Netherlands, where his father owned a tile factory. He graduated from an HBS school in 1870 and entered the University of Groningen. At the end of his freshman year he won the gold medal for his essay on vapor density measurements and their interpretation. These types of measurements are still done today and provide information on the molecules present in the saturated vapor of arsenic, phosphorus, selenium, and sulfur as a function of temperature.

Kamerlingh Onnes spent his second year in college at the University of Heidelberg, where he was an assistant to Robert Kirchhoff (see sidebar in Chapter 5). Onnes received his doctorate in 1879 and spent a few years as an assistant to one of the chemistry professors at Leiden. During this time he developed his principle of mechanical equivalence and became acquainted with van der Waals. They remained close friends for the rest of their lives. In 1882 Onnes became the first professor of experimental physics at the University of Leiden. He began a 40-year research program in experimental low-temperature physics that culminated in the 1913 Nobel Prize for Physics.

Kamerlingh Onnes was world-famous; the press called him "le gentleman du absolute zero." In 1908 Onnes became the first person to liquefy helium. His student and successor, W. H. Keesom, was the first person to solidify helium, in 1926. The helium gas that Onnes used when he first liquefied it cost about $2500 per cubic foot and was obtained by heating large amounts of monazite sand from North Carolina. Uranium and thorium are found in monazite and produce He by their radioactive decay. Shortly before Onnes liquefied helium, it was discovered in natural gas in Kansas. Subsequently, other natural gas wells were also found to contain helium, and today it is extracted from natural gas. This helium is predominantly ^4He. The lighter isotope ^3He is produced from decay of tritium and did not become available for study until after World War II.

Dutch physicist Kammerlingh Onnes (see sidebar) proposed using a virial expansion in terms of the inverse volume (i.e., density) to write the equation of state for gases and liquids:

$$\frac{PV_m}{RT} = Z = 1 + \frac{B}{V_m} + \frac{C}{V_m^2} + \frac{D}{V_m^3} + \cdots \tag{8-128}$$

$$\frac{PV_m}{RT} = Z = 1 + B\rho_m + C\rho_m^2 + D\rho_m^3 + \cdots \tag{8-129}$$

The $B, C, D \ldots$ terms in Eqs. (8-128) and (8-129) are the second, third, fourth … virial coefficients. The first coefficient is the ideal gas term, which is equal to unity in the equations above. The higher virial coefficients give the deviation of a real gas from ideality. As density decreases to zero ($\rho_m \rightarrow 0$), the volume increases to infinity ($V_m \rightarrow \infty$), and the virial equations (8-128) and (8-129) reduce to the ideal gas equation $PV_m/RT = 1$.

The virial coefficients for pure gases depend only on temperature and are independent of pressure or density. The virial coefficients for gas mixtures depend on temperature and the composition of the gas mixture. We describe their derivation later. The values of the virial coefficients are determined from *PVT* data for gases. The second virial coefficient *B* is defined by

$$B = \left[\frac{\partial Z}{\partial(1/V)}\right]_T = \left(\frac{\partial Z}{\partial \rho}\right)_T \quad \text{as } \rho \rightarrow 0 \tag{8-130}$$

Likewise, the third virial coefficient *C* is defined by the equation

$$C = \frac{1}{2}\left[\frac{\partial^2 Z}{\partial(1/V)^2}\right]_T = \frac{1}{2}\left(\frac{\partial^2 Z}{\partial \rho^2}\right)_T \quad \text{as } \rho \rightarrow 0 \tag{8-131}$$

The practical use of these two definitions is illustrated by rewriting Eq. (8-128) as

$$V_m\left[\frac{PV_m}{RT} - 1\right] = V_m(Z-1) = \Delta = B + \frac{C}{V_m} + \cdots \tag{8-132}$$

The quantity $V_m(Z-1)$ is also called Δ because it represents the deviation from the molar volume of an ideal gas (Deming and Shupe, 1932). Ideally a graph of $V_m(Z-1)$ versus the inverse molar volume ($1/V_m$ = molar density ρ_m), gives a straight line with an intercept equal to *B* and a slope equal to *C*. In many cases *PVT* data give a curve instead of a straight line, and a polynomial equation is used to represent the curve. The simplest polynomial that gives constant values of the second virial coefficient *B* is used as the correct equation. The value of *C* is more sensitive to the number of coefficients in the polynomial, and in some cases it may not be possible to obtain an essentially constant value of *C*. The intercept of the curve is equal to *B* and the limiting value of the slope at zero density is equal to *C*. Alternatively, a plot of Δ versus ρ_m can be graphically extrapolated to zero density as done by Deming and Shupe (1932).

Example 8-12. Methane (CH_4) is the major C-bearing gas in the atmospheres of Jupiter, Saturn, Uranus, and Neptune. Calculate the second and third virial coefficients for CH_4 at 273.15 K from the *PVT* data of Douslin et al. (1964). A subset of their data for molar densities (in mol L^{-1}) and compression factors is given here, along with our calculations of $V_m(Z-1)$ (in cm^3 mol^{-1}).

ρ_m	0.75	2.5	4	7.5	9.5	11	12
Z	0.961416	0.883086	0.828977	0.750615	0.738373	0.747658	0.763987
$V_m(Z-1)$	−51.4453	−46.7656	−42.7558	−33.2513	−27.5397	−22.9402	−19.6678

Figure 8-9 shows the plot of these data, which are curved instead of being a straight line. Thus, a polynomial equation is required to fit the data. Trial and error shows that the values of the second and

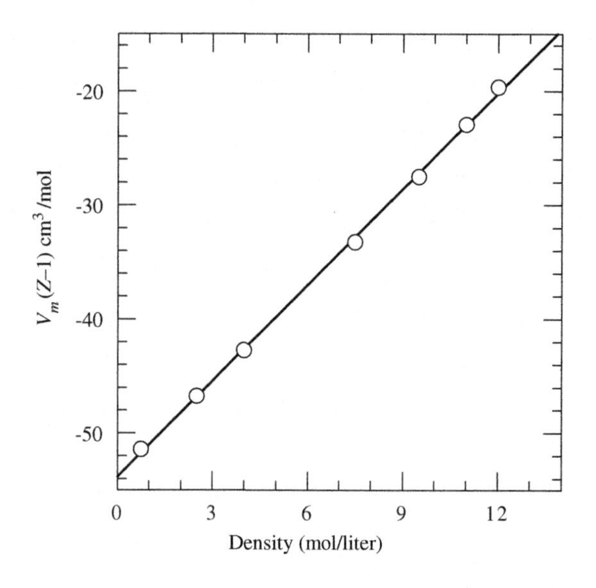

FIGURE 8-9

A graph illustrating calculation of the second (B) and third (C) virial coefficients for methane at 273.15 K.

third virial coefficients are essentially constant for polynomials with five or six terms. The simpler equation is preferable and is

$$V_m(Z-1) = -53.442 + 2.657\rho_m + 9.076 \times 10^{-3}\rho_m^2 - 2.1530 \times 10^{-3}\rho_m^3 + 2.0749 \times 10^{-4}\rho_m^4 \quad (8\text{-}133)$$

The intercept of Eq. (8-133) gives $B = -53.442$ cm^3 mol^{-1}. For comparison, Douslin et al. (1964) list $B = -53.35$ cm^3 mol^{-1} from analysis of their entire data set at 273.15 K. The first derivative of Eq. (8-133) is the slope at any point along the curve. This equation is

$$\frac{\partial}{\partial \rho}[V_m(Z-1)] = 2.657 + 1.8152 \times 10^{-2}\rho_m - 6.459 \times 10^{-3}\rho_m^2 + 8.2996 \times 10^{-4}\rho_m^3 \quad (8\text{-}134)$$

The slope at zero density gives $C = 2657$ cm^6 mol^{-2} versus the value of 2620 cm^6 mol^{-2} given by Douslin et al. (1964). Note that the value of C was multiplied by 1000 because the molar density was given in mol L^{-1}. If we had used mol cm^{-3} for ρ_m, this would not have been necessary.

Figure 8-10 shows the temperature dependence of the second virial coefficient B for several common gases. The Boyle temperature, which we defined earlier in terms of the slope of PV isotherms, is the temperature where the second virial coefficient B goes through zero as it changes sign from negative (at low temperatures) to positive at higher temperatures. It is often convenient to represent the temperature dependence of B using polynomials in terms of inverse temperature:

$$B\ (\text{air}) = 39.410 - 10{,}794/T - 979{,}238/T^2 \quad (8\text{-}135)$$

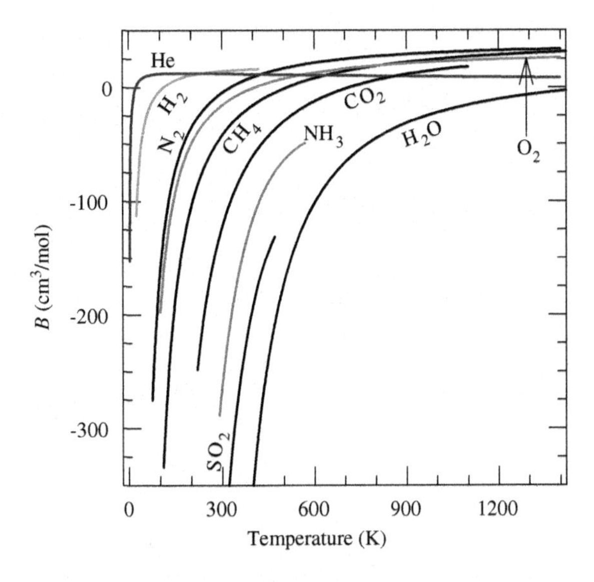

FIGURE 8-10

A plot of the second virial coefficients (B) for several gases.

Equation (8-135) is valid from 100 K to 1400 K and reproduces the tabular data of Levelt Sengers et al. (1971) within ± 0.8 cm^3 mol^{-1}. This is within experimental accuracy that is generally several cm^3 mol^{-1}. Table 8-4 gives polynomial fit coefficients for the second virial coefficients of the major gases found in planetary atmospheres. In some cases, such as He and H_2O, polynomials that are more complicated are needed to represent the B values. Table 8-5 gives the Boyle temperatures for several of these gases.

Table 8-4 Temperature Variation of Second Virial Coefficients[a]

Gas	b_0	b_1	b_2	T range
^4He[b]	12.276	−183.366	−2417.6	24–1400
H_2	22.956	−2383.7	−21,511	24–420
N_2	40.962	−10,230	−996,606	75–1400
Air	39.41	−10,794	−979,238	100–1400
O_2	33.280	−10,297	−1,260,749	100–1400
Ar	30.605	−9183.6	−1,248,231	80–1000
CH_4	41.042	−15,091	−2,823,547	110–1500
CO_2	39.748	−14,327	−10,230,967	250–1000
NH_3	−133.981	145,219	−55,120,126	270–580
SO_2	−266.0	257,900	−91,471,332	290–470
H_2O[c]	−31.801	203,141	−316,706,588	300–3000

[a] $B = b_0 + b_1/T + b_2/T^2 + (b_3/T^3 + b_4/T^4)$.
[b] Also $b_3 = 9071.5$ and $b_4 = -11,190$.
[c] Also $b_3 = 1.56614 \times 10^{11}$ and $b_4 = -3.30642 \times 10^{13}$.

Table 8-5 Joule-Thomson Coefficients for Some Gases[a]

Gas	Z	μ_{JT} (K bar^{-1})	T_B (K)	T_{JT} (K)	C_P/R
H_2O	1.232	6.664	1540	2690	4.241
NH_3	0.9896	2.811		1990	4.421
Xe	0.9947	1.818	768	1456	2.500
CO_2	0.9950	1.093	715	1355	4.466
Kr	0.9979	0.803	575	1090	2.500
CH_4	0.9983	0.438	510	968	4.286
Ar	0.9994	0.366	412	780	2.500
O_2	0.9993	0.268	406	764	3.533
CO	0.9997	0.250	342.5	624	3.505
Air	0.9997	0.237	347	639	3.506
N_2	0.9998	0.215	327	621	3.503
^4He	1.0004	−0.062	23	47	2.500
H_2	1.0006	−0.034	110	195.5	3.468
Ne	1.0006	−0.030	122	231	2.500

[a] *Compression factors (Z), Joule-Thomson coefficients (μ_{JT}), and C_P/R values at 298.15 K, 1 bar except data for H_2O at 373.15 K, 1 bar.*

The temperature dependence of the third virial coefficient C is more complicated than that of the second virial coefficient. Figure 8-11 shows the third virial coefficient of hydrogen. Generally, only the second and perhaps the third virial coefficients can be determined as a function of temperature with any accuracy and precision. Usually, even the sign of the fourth virial coefficient (D) cannot be reliably determined.

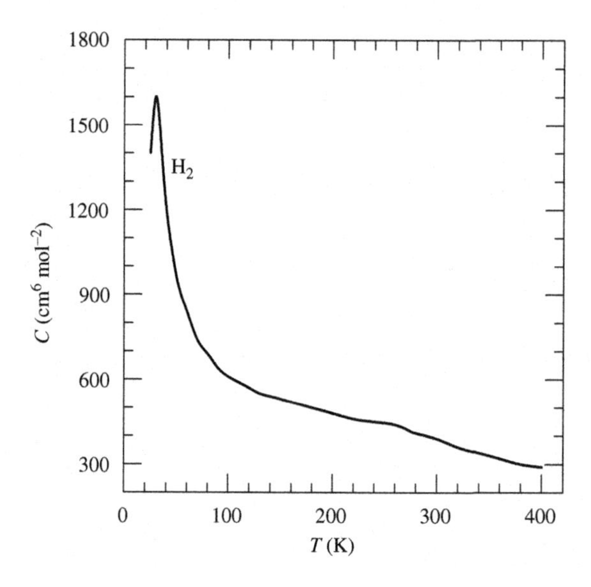

FIGURE 8-11

A plot of the third virial coefficient (C) for hydrogen.

An important property of the virial coefficients is that they can be related to the forces between the atoms (or molecules) in a gas by statistical mechanics (statistical computations of the thermodynamic properties of systems on the atomic and molecular scale). Thus, the second virial coefficient arises from interactions between pairs of gas particles (atoms or molecules), the third virial coefficient arises from interactions between three gas particles, the fourth virial coefficient from interactions between four gas particles, and so on. The useful pressure range of the virial equation is generally limited to densities equal to the critical density at any temperature. This limitation is due to insufficient knowledge about the values of the higher virial coefficients.

Example 8-13. Use the virial equation to calculate PV_m (atm cm^3 mol^{-1}) for air at 1, 120, and 200 atmospheres pressure at 300 K, compare calculated and observed values (Din, 1962), and compare the size of the second (B/V) and third (C/V^2) terms in the virial equation. Values of the virial coefficients at 300 K are $B = -7.45$ cm^3 mol^{-1} and $C = 1500$ cm^6 mol^{-2}. Results are given here.

P (atm)	1	120	200
obs V_m	24,610	204.80	127.50
obs PV_m	24,610	24,576	25,500
virial PV_m	24,610	24,601	25,456
1st term	1	1	1
$-B/V$	3×10^{-4}	0.036	0.06
C/V^2	2.5×10^{-6}	0.036	0.09

The virial equation gives a good fit to the observed values with <0.2% error up to 200 atmospheres pressure. At this point the molar volume of air is 127.5 cm^3, which is comparable to that at the critical point (84.5 cm^3). Over the 1–200 atmosphere pressure range the relative contributions of the three terms in the virial equation changed from 1, 3×10^{-4}, and 2.5×10^{-6} (1 atm), to 1, 0.036, 0.036 (120 atm), to 1, 0.06, and 0.09 (200 atm). The fourth virial coefficient D, which is unknown, is needed to extend the calculated values to higher pressures.

It is common to use a power series to fit PVT data as a function of molar density and to identify the polynomial coefficients with the coefficients in the infinite power series of the virial equation. This has to be done carefully because the derived values of the virial coefficients depend on the number of terms in the polynomial and may have no underlying theoretical significance. In general, the errors in the second virial coefficient (B) may be small, but the errors in the third (C) and higher (D, E, ...) virial coefficients become increasingly larger. For example, the compression factor ($Z = PV_m/RT$) for the 1000 K isotherm of water in Figure 8-8 can be represented by a fifth-order polynomial equation:

$$Z = \frac{PV_m}{RT} = 1.0016 - \frac{22.832}{V_m} + \frac{687.718}{V_m^2} + \frac{1.0891 \times 10^{-3}}{V_m^3} - \frac{7591.51}{V_m^4} + \frac{2,470,047}{V_m^5} \tag{8-136}$$

In this case, both the second and third coefficients in Eq. (8-136) are within 5% of the actual values ($B = -21.8$ cm^3 mol^{-1} and $C = 657$ cm^6 mol^{-2}) of the second and third virial coefficients for water at 1000 K. The fourth and higher virial coefficients of water are unknown and no comparisons are possible for the higher coefficients in Eq. (8-136).

In many cases polynomial equations are written using pressure, instead of density, as the independent variable. The 1000 K isotherm for water in Figure 8-8 can also be fit using a fifth-order polynomial equation in terms of pressure:

$$Z = \frac{PV_m}{RT} = 1.0027 - 3.6238 \times 10^{-4}P + 2.28446 \times 10^{-7}P^2 - 4.32824 \times 10^{-11}P^3 + 3.82248 \times 10^{-15}P^4 - 1.28058 \times 10^{-19}P^5 \quad (8\text{-}137)$$

However, this fifth-order equation does not fit the data as well as Eq. (8-136). Strictly speaking, the coefficients in Eq. (8-137) are not virial coefficients because they are not directly related to the intermolecular forces between gas molecules. It is possible to convert the coefficients of the power series in pressure to those of the power series in density ($1/V$), and this is the subject of one of the problems at the end of the chapter. Values of virial coefficients for pure gases and gas mixtures are given by Levelt-Sengers et al. (1971) and Dymond and Smith (1980). More comprehensive discussions of the theoretical foundations of the virial equation are given by Hirschfelder et al. (1954) and Mason and Spurling (1969).

III. THERMODYNAMIC PROPERTIES OF REAL GASES

A. The Joule-Thomson effect

The Joule-Thomson effect is the cooling or heating observed during the adiabatic and isenthalpic expansion of fluids (e.g., gases, liquids, magmas). It is important for the liquefaction of gases and refrigeration. Some geologists have suggested that Joule-Thomson cooling may also be important for magmas but special conditions are required (Waldbaum, 1971; Ramberg, 1971). It is unlikely that silicate magmas can be cooled by Joule-Thomson expansion, although the effect may be important for cryogenic volcanism in the outer solar system. Joule and Thomson (later Lord Kelvin) discovered this effect during a series of experiments on gas expansion that they conducted from 1852 to 1862 (Joule and Thomson, 1854, 1862; Thomson and Joule, 1853). Their work was done to improve on Joule's free expansion experiment (see Chapter 3), which showed

$$\left(\frac{\partial E}{\partial V}\right)_T = 0 \quad (3\text{-}51)$$

within his experimental error for a real gas. Another consequence of Eq. (3-51) is that

$$\left(\frac{\partial E}{\partial P}\right)_T = \left(\frac{\partial E}{\partial V}\right)_T \left(\frac{\partial V}{\partial P}\right)_T = 0 \quad (8\text{-}138)$$

However, Eqs. (3-51) and (8-138) are only equal to zero for an ideal gas because the internal energy of a real gas is a function of its volume and pressure.

We can easily show this for a van der Waals gas. Equation (8-49) describes the variation of internal energy E with volume along an isotherm and is a general equation,

$$\left(\frac{\partial E}{\partial V}\right)_T = T\left(\frac{\partial S}{\partial V}\right)_T - P \quad (8\text{-}49)$$

The entropy derivative in this equation can be replaced using the Maxwell relation

$$\left(\frac{\partial S}{\partial V}\right)_T = \left(\frac{\partial P}{\partial T}\right)_V \tag{8-48}$$

We then get an expression for *(dE/dV)* in terms of pressure, volume, and temperature:

$$\left(\frac{\partial E}{\partial V}\right)_T = T\left(\frac{\partial P}{\partial T}\right)_V - P \tag{8-139}$$

Writing the van der Waals equation (8-89) for n moles of gas, we see that

$$\left(\frac{\partial P}{\partial T}\right)_V = \frac{\partial}{\partial T}\left(\frac{nRT}{V-b} - \frac{a}{V^2}\right)_V = \frac{nR}{V-b} \tag{8-140}$$

Using Eq. (8-140) and the van der Waals equation to substitute into Eq. (8-139), we find

$$\left(\frac{\partial E}{\partial V}\right)_T = T\frac{nR}{V-b} - \left[\frac{nRT}{V-b} - \frac{a}{V^2}\right] = \frac{a}{V^2} \tag{8-141}$$

Equation (8-141) shows that the internal energy of a van der Waals gas depends on its volume and is related to the internal pressure term. Thomson realized that the internal energy of a real gas should depend on its volume and pressure, and he suggested new experiments to Joule.

Figure 8-12 is a schematic of the apparatus used in the Joule-Thomson experiments. Gas at higher pressure is allowed to escape through a porous plug or a throttling valve into a lower-pressure container. Joule and Thomson initially used a ball of cotton or a bunched-up silk handkerchief as the porous plug. Subsequently, they and other scientists used throttling valves or porous plugs made of other materials. The whole apparatus is insulated, so there is no heat flow into or out of it; thus, the expansion is adiabatic. Thermometers and pressure gauges on the two sides of the plug measure the temperature (ΔT) and pressure differences (ΔP).

Initially, the gas (e.g., air) is at pressure P_1, temperature T_1, and volume V_1, and it is expanded to pressure P_2 at temperature T_2 with volume V_2. Because the expansion is done adiabatically, there is no heat flow between the system and the surroundings ($q = 0$). The difference between the work done in

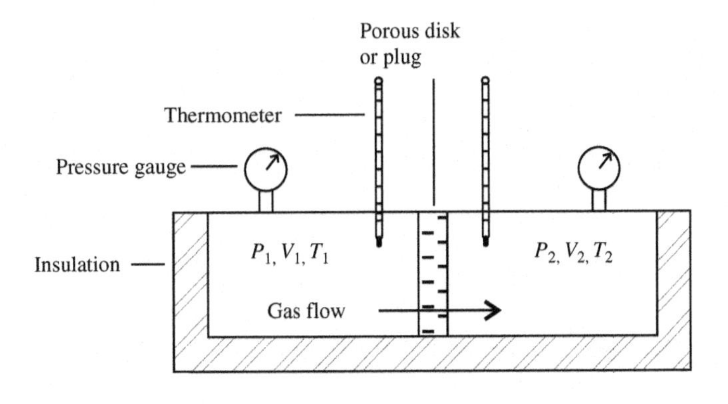

FIGURE 8-12

A schematic diagram of the Joule-Thomson experiment.

forcing the gas through the porous plug (w_1) and the work done during expansion on the other side of the plug (w_2) is

$$w_2 - w_1 = P_2V_2 - P_1V_1 \tag{8-142}$$

The first law ($E = q - w$) shows that the internal energy change of the gas is given by

$$E_2 - E_1 = (q_2 - w_2) - (q_1 - w_1) = w_1 - w_2 \tag{8-143}$$

Combining these two equations and rearranging, we find that

$$E_2 + P_2V_2 = E_1 + P_1V_1 \tag{8-144}$$

Each side of Eq (8-144) is equal to enthalpy H. Thus we can rewrite it as

$$H_2 = H_1 \tag{8-145}$$

Equation (8-145) shows that the Joule-Thomson experiment is done at constant enthalpy. If necessary, you can go back to Chapter 3 to review the first law, enthalpy, and related concepts.

The data obtained in the Joule-Thomson experiment are ΔT as a function of ΔP at constant enthalpy, and as ΔT and ΔP become smaller and smaller the derivative being measured is

$$\left(\frac{\partial T}{\partial P}\right)_H = \mu \tag{8-146}$$

Equation (8-146) defines the adiabatic Joule-Thomson (JT) coefficient (μ). It is the temperature change (K) with pressure (bar) when a gas expands adiabatically at constant enthalpy.

Example 8-14. Calculate μ for air and CO_2 using some of the data of Joule and Thomson (1854). The calculated μ values are equal to ($\Delta T/\Delta P$) $\times$ 14.5038 psi/bar and are given in the following table. The values measured by Joule and Thomson are close to the actual values of 0.246 K/bar for air and 1.219 K/bar for CO_2 under the same conditions.

Gas	P_1 (psi)	P_2 (psi)	ΔP (psi)	T_1 (°C)	T_2 (°C)	ΔT (°C)	μ (K/bar)
air	20.969	14.624	−6.345	17.006	16.898	−0.108	0.247
CO_2	75.324	14.723	−60.601	12.844	7.974	−4.870[a]	1.208

[a]*Observed for an air-CO_2 mixture with 95.51% CO_2. Their corrected ΔT for pure CO_2 is −5.049°C.*

Thus, expansion of air from 100 bars to one bar pressure would cool it by about 24.6 degrees.

In their experiments Joule and Thomson found that μ for air and CO_2 decreases with increasing temperature. Their equation for the cooling of air (ΔT in Kelvin) as a function of the pressure change (ΔP in atmospheres) during isenthalpic expansion is

$$\Delta T = 0.276\left(\frac{273.15}{T}\right)^2 \Delta P \tag{8-147}$$

Subsequent experimental work on the Joule-Thomson expansion of H_2 by the famous Polish scientist Karol Olszewski (1846–1915) showed that μ is also a function of pressure.

The JT coefficient μ is zero for ideal gases and nonzero for real gases. The pressure change ($P_2 - P_1$) in the Joule-Thomson experiment is always negative, and thus μ is positive for cooling and negative for heating. Table 8-5 lists the compression factor Z and μ for several gases found in planetary atmospheres. The data in Table 8-5 show that the size of the Joule-Thomson effect is correlated with the nonideality of the gas. Steam has the largest deviation from ideality ($Z = 1.232$) and gives the largest cooling effect (6.66 degrees per bar). Conversely, N_2 has the smallest deviation from ideality ($Z = 0.9998$) and gives the smallest cooling effect (0.215 degrees per bar). The light gases H_2, He, and Ne are also nearly ideal but are warmed by expansion.

The Joule-Thomson coefficient is related to the constant pressure heat capacity C_P via

$$\mu = \left(\frac{\partial T}{\partial P}\right)_H = -\frac{(\partial H/\partial P)_T}{(\partial H/\partial T)_P} = -\frac{(\partial H/\partial P)_T}{C_P} \tag{8-148}$$

Equation (8-148) is an application of Eq. (2-70), the cyclic rule. If the Joule-Thomson experiment is done isothermally instead of adiabatically, the enthalpy change (J mol^{-1}) of the gas per unit pressure (bar) is given by the derivative

$$\left(\frac{\partial H}{\partial P}\right)_T = -\mu C_P = \varphi \tag{8-149}$$

Equation (8-149) defines the isothermal Joule-Thomson coefficient (φ).

The temperature and pressure dependence of μ is illustrated in Figure 8-13. This has four parts showing (a) enthalpy H, (b) heat capacity C_P, (c) the derivative $(\partial H/\partial P)_T$, which is the slope of the enthalpy curve, and (d) the adiabatic JT coefficient μ of N_2 as a function of pressure at 0°C. The same isotherm is shown in Figure 8-2 where the molar PV product for N_2 is plotted as a function of pressure. The adiabatic JT coefficient for N_2 is 0.253 at 0°C and one bar pressure. It decreases with increasing pressure and μ goes through zero at about 400 bars (see Figure 8-13d). This is also the minimum enthalpy point along this isotherm (Figure 8-13a), and the derivative $(\partial H/\partial P)_{TX} = 0$ at this point (Figure 8-13c). Equations (8-148) and (8-149) require that the adiabatic JT coefficient μ and the isothermal JT coefficient φ are both zero at this point ($P \sim 400$ bars), which is the *Joule-Thomson inversion point* for N_2 at 0°C (273.15 K). The JT inversion point is the pressure at a given temperature where the adiabatic and isothermal JT coefficients μ and φ switch sign and go through zero. The set of all these inversion points (for μ or φ) is a parabolic curve known as the *Joule-Thomson inversion curve*. The high-temperature end of the inversion curve, where $\mu = 0$ and $\varphi = 0$ at zero pressure, is the *Joule-Thomson inversion temperature*. Table 8-5 also lists Joule-Thomson inversion temperatures. Most values are accurate to a few degrees but the values for H_2O, NH_3, and CO_2 are probably uncertain by ± 100 K.

We can understand the origin, magnitude, and sign of the Joule-Thomson effect by rearranging Eq. (8-148) and rewriting it using the definition of enthalpy ($H = E + PV$):

$$\mu = \left(\frac{\partial T}{\partial P}\right)_H = -\frac{1}{C_P}\left[\left(\frac{\partial E}{\partial P}\right)_T + \left(\frac{\partial (PV)}{\partial P}\right)_T\right] \tag{8-150}$$

Equation (8-150) shows that $\mu = 0$ for an ideal gas because the terms inside the brackets are zero:

$$\left(\frac{\partial E}{\partial P}\right)_T = 0 \tag{8-151}$$

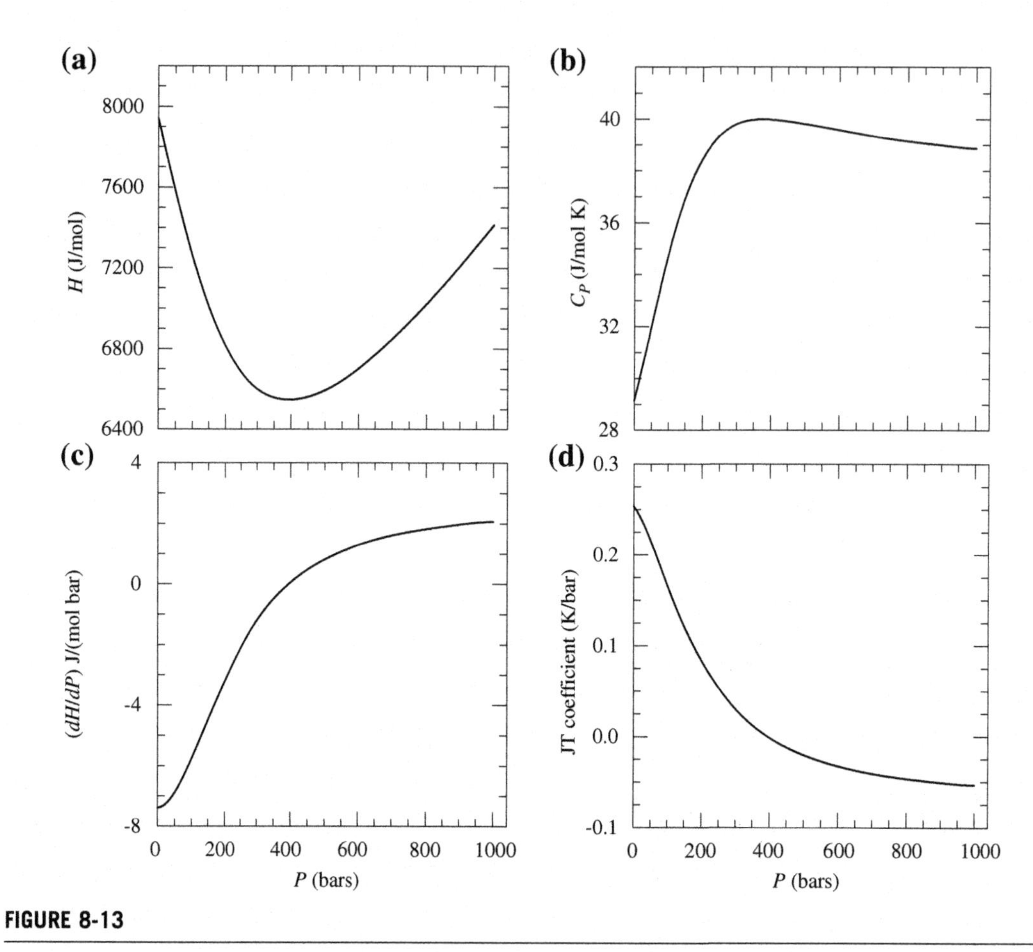

FIGURE 8-13

(a) Enthalpy, (b) heat capacity, (c) $(\partial H/\partial P)_T$, and (d) JT coefficient μ for N_2 at 273.15 K as a function of pressure. Note that the inversion point for μ corresponds to the minimum in the enthalpy curve.

$$\left(\frac{\partial(PV)}{\partial P}\right)_T = \left(\frac{\partial(nRT)}{\partial P}\right)_T = 0 \tag{8-152}$$

Equation (8-150) also shows that the dependence of internal energy on pressure and deviations from Boyle's law are the two factors causing nonzero μ values for real gases. The derivative $(dE/dP)_T$ is always negative for real gases. For example, the values for N_2 and H_2 at 0°C and one bar are about –6.38 and –0.62 J mol^{-1} bar^{-1}, respectively. The derivative $[d(PV)/dP]$ gives deviations from Boyle's law. It is zero at a Boyle point along an isotherm and either negative (below) or positive (above) the Boyle point. The values for N_2 and H_2 at 0°C and one bar are about –1.00 and 1.45 J mol^{-1} bar^{-1}. The value for N_2 is negative because 0°C is slightly below its Boyle temperature (54°C, 327 K). The value for H_2 is positive because 0°C is well above its Boyle temperature (–163°C, 110 K). The sum of the two derivatives is

$$\left[\left(\frac{\partial E}{\partial P}\right)_T + \left(\frac{\partial (PV)}{\partial P}\right)_T\right] = \left(\frac{\partial H}{\partial P}\right)_T = -\mu C_P = \varphi \tag{8-153}$$

Hydrogen has a positive (dH/dP) and a negative μ value because of its small negative (dE/dP) and large positive [$d(PV)/dP$]. Nitrogen has a negative (dH/dP) and a positive μ value because both derivatives are negative. The JT coefficients μ and φ have opposite signs.

The adiabatic JT coefficient μ is also given by

$$\mu = \left(\frac{\partial T}{\partial P}\right)_H = \frac{1}{C_P}\left[T\left(\frac{\partial V}{\partial T}\right)_P - V\right] \tag{8-154}$$

Equation (8-154) shows that μ is zero for an ideal gas and can be calculated from heat capacity and *PVT* data for a real gas or other fluid. The volume of an ideal gas is given by nRT/P, thus

$$\left(\frac{\partial V}{\partial T}\right)_P = \frac{\partial}{\partial T}\left(\frac{nRT}{P}\right)_P = \frac{nR}{P} \tag{8-155}$$

Consequently, the term inside the square brackets of Eq. (8-154) is zero for an ideal gas:

$$\left[T\left(\frac{\partial V}{\partial T}\right)_P - V\right] = T\frac{nR}{P} - V = V - V = 0 \tag{8-156}$$

The JT coefficients μ and φ are zero at all temperatures and pressures for an ideal gas.

The term inside the square brackets of Eq. (8-154) is generally not equal to zero for a real gas because its volume is not given by the ideal gas law. However, $\mu = 0$ and $\varphi = 0$ for a real gas at pressure and temperature points, where

$$T\left(\frac{\partial V}{\partial T}\right)_P = V \tag{8-157}$$

Equation (8-157) defines the Joule-Thomson inversion curve. Another equivalent definition is

$$\frac{1}{T} = \frac{1}{V}\left(\frac{\partial V}{\partial T}\right)_P = \alpha_P \tag{8-158}$$

Equation (8-158) shows that the Joule-Thomson inversion curve is the set of *P-T* points where the isobaric thermal expansion coefficient (α_P) of the real gas is equal to the inverse temperature.

The dependence of μ on temperature and pressure is qualitatively explained by using the van der Waals equation to evaluate Eq. (8-154). Hoxton (1919) gives this derivation, which yields

$$\mu = \frac{1}{C_P}\left[\frac{2a}{RT} - b - P\left(\frac{6ab}{R^2T^2} - \frac{4a^2}{R^3T^3}\right)\right] \tag{8-159}$$

The a and b in Eq. (8-159) are the van der Waals constants and C_P is the constant-pressure heat capacity for a gas. Equation (8-159) makes several predictions about the behavior of μ. First, it predicts that μ is negative for H_2 and He because $a < b$ for the van der Waals constants of H_2 and He. Second, Eq. (8-159) predicts that μ decreases with increasing temperature because the term $2a/RT$

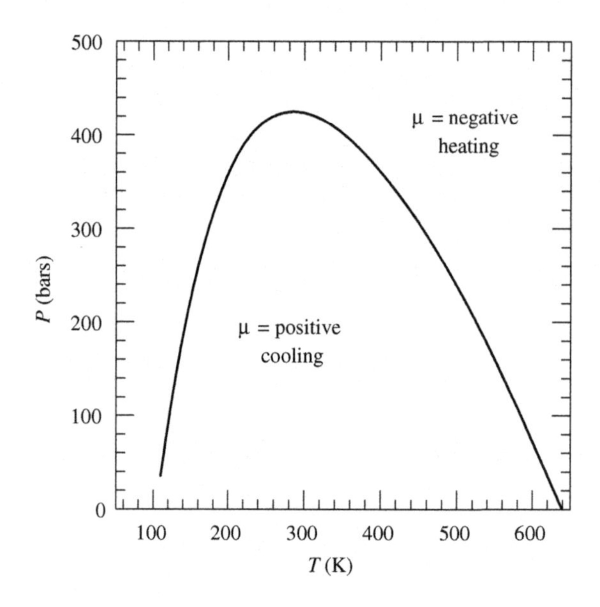

FIGURE 8-14

The Joule-Thomson inversion curve for air.

becomes smaller as T increases. Third, it predicts that μ decreases with increasing pressure and that the pressure dependence is smaller than the temperature dependence because the terms in parentheses are smaller than $2a/RT$. All these predictions agree with the observed behavior of μ, which is illustrated in Figure 8-13(d) for N_2 and in Figure 8-14 for air. The Joule-Thomson inversion curve of air is characteristic of the inversion curve for all gases. The maximum pressure and temperature on the inversion curve, which are about 430 bars and 640 K for air, are the highest pressure and temperature at which isenthalpic cooling of a gas is possible. The Joule-Thomson inversion curve is parabolic like the Boyle temperature curve but lies at higher pressure at the same temperature. Both curves can be calculated from PVT data for a gas. However, uncertainties in the PVT data are magnified in their derivatives and the computed curves have larger uncertainties than the original measurements. This situation is analogous to the computation of C_P values from enthalpy data that we discussed in Chapter 4.

B. Gibbs free energy, pressure, and fugacity

The change in the Gibbs free energy G as the pressure of a gas is isothermally varied from an arbitrary initial pressure P_1 to some final pressure P_2 and can be calculated by rearranging Eq. (7-80):

$$\left(\frac{\partial G}{\partial P}\right)_T = V \tag{7-80}$$

$$dG = VdP \tag{8-160}$$

Equation (8-160) applies to any gas, and it can be evaluated using the equation of state for a gas. This is more or less simple to do depending on the complexity of the equation of state. In the case of an ideal gas, the volume V can be replaced by nRT/P. We then get the equation

$$dG = \frac{nRT}{P}dP = nRT(d\ln P) \tag{8-161}$$

Integration of Eq. (8-161) gives

$$\int_{G_1}^{G_2} dG = nRT \int_{P_1}^{P_2} d\ln P \tag{8-162}$$

$$G_2 - G_1 = nRT \ln\frac{P_2}{P_1} \tag{8-163}$$

If we now take the pressure P_1 as the standard-state pressure of one bar, G_1 is equal to G^o, the standard Gibbs free energy, and Eq. (8-163) can be rearranged to give

$$G = G^o + nRT \ln P \tag{8-164}$$

Equation (8-163) gives the isothermal change in Gibbs free energy of an ideal gas between any two arbitrary pressures. The related Eq. (8-164) gives G for an ideal gas at an arbitrary pressure P relative to the standard-state pressure of one bar. Both equations, but Eq. (8-164) in particular, are important for chemical equilibria of ideal gases.

Example 8-15. Astronomers are currently detecting planets around other stars. These planets are called *extrasolar planets* (*exoplanets*). Imagine that an Earth-like exoplanet has been found with a surface temperature and pressure of 300 K and 10 bars and an atmosphere composed of air. Calculate $G - G^o$ per mole of air at 300 K and 10 bars assuming ideality. Equation (8-164) gives

$$G - G^o = nRT \ln P = (1)(8.314)(300)\ln 10 = 5743.1 \text{ J mol}^{-1} \tag{8-165}$$

By analogy with its major constituents N_2, O_2, and Ar, we could define the standard Gibbs free energy of air (G^o) as zero at the standard-state pressure of one bar. In this case the calculated value of $G - G^o = 5743.1$ J mol^{-1} is the Gibbs free energy of air at 10 bars pressure. If we take a nonzero value of G^o for air at one bar pressure, then G at 10 bars is 5743.1 J mol^{-1} larger.

Equations (8-161) to (8-165) are very useful, but they only apply to ideal gases. We now want to derive analogous equations that apply to real gases. This is done using the concept of fugacity, which was developed in 1901 by the American physical chemist Gilbert N. Lewis (1875–1946).

Fugacity (*f*) has the same dimensions as pressure and it can be regarded as the pressure that a real gas would have if it behaved ideally. Thus, the fugacity and pressure of an ideal gas are equal to one another. In general, the pressure and fugacity of a real gas are related by

$$\gamma = \frac{f}{P} \tag{8-166}$$

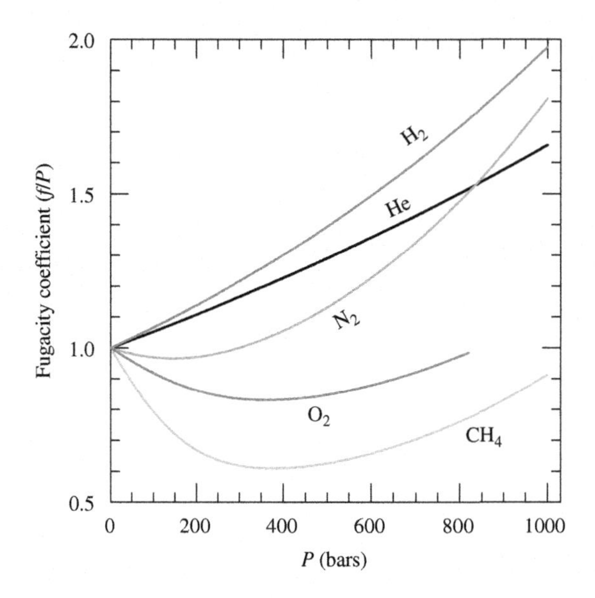

FIGURE 8-15

Fugacity coefficients for H_2, He, N_2, O_2, and CH_4 at 0°C.

The *fugacity coefficient* (γ) is unity for an ideal gas and is a positive number that can be less than one or greater than one for a real gas. Figure 8-15 illustrates fugacity coefficients as a function of pressure at 0°C for H_2, He, N_2, O_2, and CH_4. It is the counterpart to Figure 8-2, which shows compression factors (*Z*) for the same gases under the same conditions. The curves in Figure 8-15 were calculated from the data in Figure 8-2 using the equations described here.

Fugacity is partially defined by the equation

$$dG = nRT(d\ln f) \tag{8-167}$$

which is deliberately chosen to be analogous to the equation

$$dG = nRT(d\ln P) \tag{8-161}$$

for an ideal gas. The definition of fugacity is completed by requiring that fugacity approach pressure as the pressure approaches zero, that is, $f/P \rightarrow 1$ as $P \rightarrow 0$. This requirement is necessary because the fugacity of a gas only becomes equal to the gas pressure at sufficiently low pressures where the real gas approaches ideality. The fugacity and pressure of many real gases are observed to be approximately equal at one bar pressure (i.e., $\gamma \sim 1$ as shown in Figure 8-15). This approximate equality is not observed for all gases and it is not part of the definition of fugacity.

C. Calculating the fugacity of a real gas

The fugacity of a real gas is calculated by combining Eqs. (8-160) and (8-167) to get an equation involving the fugacity, pressure, and volume of a gas:

$$VdP = nRT(d\ln f) \tag{8-168}$$

The difference between the volume predicted by the ideal gas law and the actual volume V of a real gas is the residual volume α. Thus, we can rewrite V as

$$V = \frac{nRT}{P} - \alpha \tag{8-169}$$

Our previous discussion of equations of state for real gases shows that the residual volume α in Eq. (8-169) can be positive or negative and may be a complicated function of temperature and pressure. Equation (8-168) is now rewritten using Eq. (8-169) to substitute for V:

$$\left(\frac{nRT}{P} - \alpha\right)dP = nRT(d\ln f) \tag{8-170}$$

Rearranging Eq. (8-170) and simplifying gives

$$d\ln f = \frac{1}{nRT}\left(\frac{nRT}{P} - \alpha\right)dP \tag{8-171}$$

$$d\ln f = \left(\frac{1}{P} - \frac{\alpha}{nRT}\right)dP \tag{8-172}$$

Equation (8-172) is now integrated between an arbitrary initial pressure P_1 (where fugacity $= f_1$) and a final pressure P_2 (where fugacity $= f_2$):

$$\int_{f_1}^{f_2} d\ln f = \int_{P_1}^{P_2}\left(\frac{1}{P} - \frac{\alpha}{nRT}\right)dP \tag{8-173}$$

The integral on the right side of Eq. (8-173) can be written as the difference between an integral involving the term $1/P$ and an integral involving the term α/nRT. This gives us

$$\ln\frac{f_2}{f_1} = \int_{P_1}^{P_2}\frac{1}{P}dP - \int_{P_1}^{P_2}\frac{\alpha}{nRT}dP \tag{8-174}$$

Equation (8-174) is now simplified by moving the constant terms $1/nRT$ out of the α integral and by setting the initial pressure P_1 equal to zero. As mentioned earlier, fugacity and pressure are equal in the limit of zero pressure where real gases behave ideally. The resulting equation is

$$\ln\frac{f_2}{f_0} = \ln\frac{P_2}{P_0} - \frac{1}{nRT}\int_0^{P_2}\alpha dP \tag{8-175}$$

Dropping the subscripts on P_2 and f_2 and rearranging, we eliminate the terms f_0 and P_0 as we approach the limit of zero pressure where f_0/P_0 approaches unity and obtain the important equation

$$\ln\frac{f}{P} = -\frac{1}{nRT}\int_0^{P}\alpha dP \tag{8-176}$$

Equation (8-176) is the basis for calculating f and f/P for gases. Depending on the particular problem at hand, the residual volume α is evaluated using different methods.

Example 8-16. Carbon monoxide could be the major gas in the atmosphere of a Venus-like exoplanet with a graphite-rich crust. Use the *PVT* data tabulated by Deming and Shupe (1931) to calculate f and $\gamma = f/P$ for CO at a surface $T = 300°C$, and $P = 25 - 200$ atmospheres pressure.

Equation (8-169) was used to compute α values from tabulated molar volumes. The equation

$$\alpha = -23.3844 - 7.8708 \times 10^{-3}P + 3.2297 \times 10^{-6}P^2 \text{ cm}^3 \text{ mol}^{-1} \text{ atm}^{-1} \quad (8\text{-}177)$$

represents the calculated α values. Equation (8-177) was substituted into Eq. (8-176), integrated, and evaluated using $R = 82.0489$ cm^3 atm mol^{-1} K^{-1} and $0°C = 273.18$ K, which were used by Deming and Shupe (1931) to compute their tabular *PVT* data. The resulting equation is

$$\ln\frac{f}{P} = 4.9724 \times 10^{-4}P + 8.3680 \times 10^{-8}P^2 - 2.2891 \times 10^{-11}P^3 \quad (8\text{-}178)$$

The fugacities and fugacity coefficients computed from Eq. (8-178) at $P = 25 - 200$ atmospheres are listed here. The computed values agree with those (lit.) of Deming and Shupe (1931).

P (atm)	25	50	75	100	150	200
ln (f/P)	0.01248	0.02507	0.03775	0.05054	0.07639	0.1026
$\gamma = f/P$	1.0126	1.0254	1.0385	1.0518	1.0794	1.1080
f (atm)	25.32	51.27	77.89	105.2	161.9	221.6
f (lit.)	25.31	51.27	77.88	105.2	161.9	221.6

The compression factor Z is frequently tabulated in books and papers giving *PVT* data for gases. Thus, it is convenient to rewrite Eq. (8-176) in terms of Z by using Eq. (8-169) to substitute for α:

$$-\frac{\alpha}{nRT} = -\frac{1}{nRT}\left(\frac{nRT}{P} - V\right) = -\left(\frac{1}{P} - \frac{V}{nRT}\right) = -\frac{1}{P}\left(1 - \frac{PV_m}{RT}\right) = \frac{Z-1}{P} \quad (8\text{-}179)$$

$$\ln\frac{f}{P} = \int_0^P \frac{Z-1}{P}dP \quad (8\text{-}180)$$

Example 8-17. Air in the atmosphere of the Earth-like exoplanet from Example 8-15 is actually slightly nonideal at 300 K and 10 bars. Calculate its fugacity, f/P, and $G - G^o$. The virial coefficients for air at 300 K ($B = -7.45$ cm^3 mol^{-1} and $C = 1500$ cm^6 mol^{-2}) are given in Example 8-13. We first rewrite the virial equation in terms of pressure:

$$\frac{PV_m}{RT} = Z = 1 + B'P + C'P^2 + D'P^3 + \cdots \quad (8\text{-}181)$$

The second and third virial coefficients in terms of volume are converted to the respective virial coefficients in terms of pressure using the formulas

$$B' = \frac{B}{RT} = \frac{-7.45 \text{ cm}^3 \text{ mol}^{-1}}{(83.145 \text{ cm}^3 \text{ bar mol}^{-1}\text{K}^{-1})(300 \text{ K})} = -2.99 \times 10^{-4} \text{ bar}^{-1} \quad (8\text{-}182)$$

$$C' = \frac{(C - B^2)}{(RT)^2} = \frac{(1500 + 7.45^2) \text{ cm}^6 \text{ mol}^{-2}}{(83.145 \times 300)^2 \text{ cm}^6 \text{ bar}^2 \text{ mol}^{-2}} = 2.50 \times 10^{-6} \text{ bar}^{-2} \quad (8\text{-}183)$$

For reference Eq. (8-181) gives $Z = 0.9973$ with these virial coefficients in good agreement with $Z = 0.9972$ in the tables of Hilsenrath et al. (1955). Combining Eqs. (8-180) and (8-181) gives

$$\ln\frac{f}{P} = \int_0^P \frac{Z-1}{P} dP = \int_0^P (B' + C'P) dP = B'P + \frac{C'P^2}{2} \quad (8\text{-}184)$$

Substituting numerical values into Eq. (8-184) and solving gives $\ln(f/P) = -0.002865$, $f/P = 0.9971$, and $f = 9.971$ bars. The integrated form of Eq. (8-167) is used to compute $G - G^o$:

$$G - G^o = nRT \ln f = (1)(8.314)(300) \ln 9.971 = 5735.9 \text{ J mol}^{-1} \quad (8\text{-}185)$$

Equation (8-167) is integrated from a hypothetical standard state of the real gas behaving ideally at one bar pressure. We can thus relate $\gamma = f/P$ and $G - G^o$ via the equation

$$\gamma = \frac{f}{P} = \exp\{[(G - G^o)_{real} - (G - G^o)_{ideal}]/RT\} \quad (8\text{-}186)$$

The G^o values in Eq. (8-186) are the same as the standard Gibbs energy of formation values $\Delta_f G^o$ discussed in previous chapters and given in thermodynamic data compilations.

Example 8-17 also illustrates that the compression factor Z is equal to the ratio f/P at low pressures. This equality exists because only the second virial coefficient is important at low pressures, as we showed in Example 8-13. Then Eq. (8-184) simplifies to

$$\ln\frac{f}{P} = B'P \quad (8\text{-}187)$$

Example 8-13 also shows that the second virial term is much smaller than unity at low pressures. We approximate the exponential of Eq. (8-187) by an infinite series known as a *Maclaurin series* (described in Mellor [1955] and many basic calculus books):

$$\frac{f}{P} = \exp(B'P) = 1 + B'P + \frac{(B'P)^2}{2} + \frac{(B'P)^3}{3} + \cdots \quad (8\text{-}188)$$

The series in Eq. (8-188) is terminated after the $B'P$ term and using Eq. (8-181) gives

$$\frac{f}{P} \cong 1 + B'P = Z \quad (8\text{-}189)$$

Finally, we can use any equation of state written in terms of density to find fugacity via

$$\ln\frac{f}{P} = Z - 1 - \ln Z + \int_0^{\rho_m} (Z-1)d\ln\rho_m \tag{8-190}$$

If the virial coefficients for Eq. (8-129) are available, then Eq. (8-190) is used in the form

$$\ln\frac{f}{P} = B\rho_m + \frac{(C+B^2)\rho_m^2}{2} + \left(BC + \frac{D}{3}\right)\rho_m^3 + \cdots \tag{8-191}$$

Pitzer (1995) illustrates the derivation of these equations.

Example 8-18. Douslin et al. (1964) give *PVT* data for CH_4 at 0°C. Use Eq. (8-191) to calculate f/P for CH_4 from their density data. We use the values of the second and third virial coefficients that were calculated in Example 8-12 ($B = -53.442$ cm^3 mol^{-1}, $C = 2657$ cm^6 mol^{-2}) and set $D = 0$. The first three terms in Eq. (8-191) were used and the results are given in the following table.

ρ_m (mol L^{-1})	0.75	2.5	4	7.5	9.5
f/P	0.962	0.888	0.836	0.737	0.683
P (bars)	16.4	50.1	75.3	128	159

You should plot these values to convince yourself that the calculated values of f/P fall on the CH_4 curve in Figure 8-15. Equations for calculating fugacity from the van der Waals and related equations of state are described in Section IV about gas mixtures.

D. Calculating fugacity from reduced variables

In some cases, *PVT* data are not available to calculate the fugacity of a real gas. In these situations, we can estimate f/P using the law of corresponding states. This method is illustrated by rewriting Eq. (8-180) in terms of reduced pressure $P_R = P/P_C$

$$\ln\frac{f}{P} = \int_0^{P_R} \frac{Z-1}{P_R} dP_R \tag{8-192}$$

Figure 8-5 shows that the compression factor Z is the same, or nearly so, for many different gases at the same values of reduced temperature and pressure. Newton (1935) evaluated the integral in Eq. (8-192) graphically and prepared generalized charts of f/P as a function of P_R along different T_R isotherms. Figure 8-16 shows contours of f/P as a function of P_R and T_R using data from Hougen et al. (1960). Equations (8-86) and (8-87) are used to compute reduced temperature and pressure for H_2, He, and Ne, while T_R and P_R for all other gases are computed from Eqs. (8-71) and (8-72).

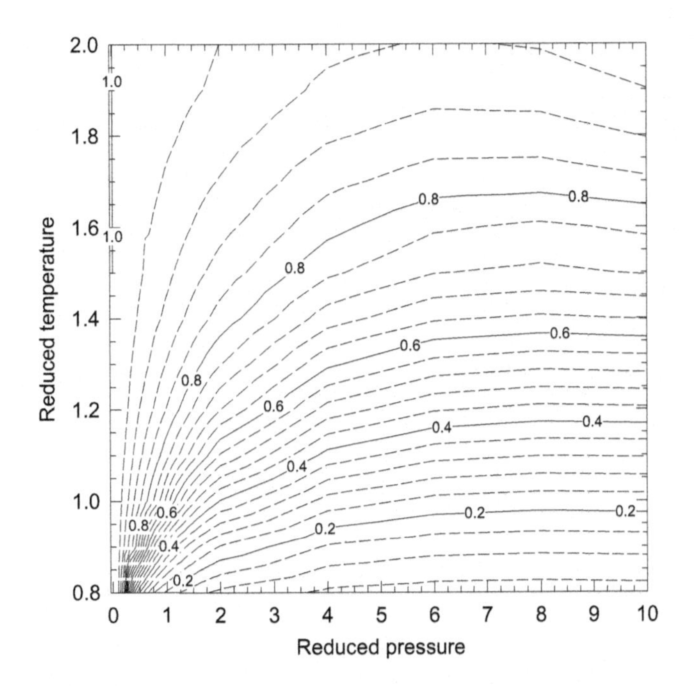

FIGURE 8-16

Reduced-temperature and-pressure plot for fugacity coefficients.

Plotted using data from Hougen et al. (1960).

Example 8-19. Use Figure 8-16 to estimate *f*/*P* for O_2 and CH_4 in Figure 8-15 at 600 bars pressure. The following table summarizes the results and compares the estimated and actual *f*/*P* values. The critical temperatures and pressures used in the computations are from Table 8-2.

gas	O_2	CH_4
$T_R = 273/T_C$	1.8	1.4
$P_R = 600/P_C$	12	13
f/*P* estimated	0.925	0.69
f/*P* actual	0.88	0.66

IV. THERMODYNAMIC PROPERTIES OF GAS MIXTURES

In many cases, we are interested in the fugacity or fugacity coefficient of a gas, which is part of a more or less complex gas mixture. The atmospheres of planets and satellites in our solar system and in other planetary systems, volcanic gases on Venus, Earth, and Io, natural gases on Earth, and metamorphic fluids in Earth's crust and mantle are some examples of important gas mixtures in the earth and planetary sciences (e.g., see Table 8-6). Equation (8-176) is also the basis of fugacity calculations for gas mixtures. However, the residual volume α in Eq. (8-176) now involves the partial molar volume $\overline{V}_i$

Table 8-6 Some Important Natural Gas Mixtures

Planetary Atmospheres[a]	
Venus	CO_2 (96.5%)-N_2 (3.5%)-SO_2 (0.02%)
Earth	N_2 (78.1%)-O_2 (20.9%)-Ar (0.9%)
Mars	CO_2 (95.3%)-N_2 (2.7%)-Ar (1.6%)
Jupiter	H_2 (86.2%)-He (13.6%)-CH_4 (0.2%)
Saturn	H_2 (88%)-He (11.9%)-CH_4 (0.4%)
Uranus	H_2 (82.5%)-He (15.2%)-CH_4 (2.3%)
Neptune	H_2 (80%)-He (19%)-CH_4 (1–2%)
Satellite Atmospheres[a]	
Titan	N_2 (95.1%)-CH_4 (4.9%) at the surface
Io	SO_2 (>90%)-SO (<10%)
Volcanic Gases	
Earth[b]	H_2O (78.7%)-SO_2 (11.5%)-CO_2 (3.2%)-H_2S (3.2%)
Io[c]	SO_2 (81.6%)-S_2 (11.6%)-SO (6.3%)

[a]*From Lodders and Fegley (2011).*
[b]*Kilauea volcano, HI, Table 6.13 in Lodders and Fegley (1998).*
[c]*Pele volcano, nominal gas composition.*

of gas i in a mixture instead of the molar volume $V_m = V/n$ of a pure gas. Equation (8-169) gives α for a pure gas and the equation

$$\alpha_i = \frac{RT}{P} - \overline{V}_i \tag{8-193}$$

gives α for gas i in a mixture. The partial molar volume $\overline{V}_i$ is the change in the total volume V of the gas mixture at constant P and T when the number of moles n_i of gas i is varied, holding the number of moles of all other gases constant (i.e., at otherwise constant composition):

$$\overline{V}_i = \left(\frac{\partial V}{\partial n_i}\right)_{P,T,n \neq n_i} \tag{8-194}$$

Using the definition of $\overline{V}_i$ the analogous form of Eq. (8-176) for a gas mixture is

$$RT \ln \gamma_i = RT \ln \frac{f_i}{P_i} = -\int_0^P \left(\frac{RT}{P} - \overline{V}_i\right) dP \tag{8-195}$$

The methods used to evaluate Eq. (8-195) depend on the particular problem and on the accuracy with which we have to know the fugacity or fugacity coefficient for the gas(es) of interest.

A. Lewis and Randall fugacity rule

The simplest method is the *Lewis and Randall fugacity rule*, also known as the *G. N. Lewis fugacity rule* (Lewis and Randall, 1923, pp. 225–227). This states that the fugacity f_{mix} of a gas in a mixture is equal to its mole fraction (X_i) in the mixture multiplied by its fugacity as a pure gas f_{pure} at the same temperature T and total pressure P as the gas mixture:

$$f_{mix} = X_i f_{pure} \quad (8\text{-}196)$$

Equation (8-166) defines the fugacity coefficient γ. Using this to rewrite Eq. (8-196) we see that γ of a gas in a mixture is the same as that of the pure gas under the same P-T conditions:

$$\gamma_i\ (\text{mixture}) = \gamma_i\ (\text{pure}) \quad (8\text{-}197)$$

Thus, the Lewis and Randall fugacity rule says that the fugacity of an individual gas in a mixture can be determined by finding the fugacity of the pure gas under the same conditions. The fugacity of the pure gas is computed using any of the methods described previously.

Example 8-20. Use the G. N. Lewis fugacity rule to estimate the fugacity and f/P for H_2 and He at the 0°C level of Jupiter's atmosphere where the pressure is 4.92 bars (Seiff et al., 1998). Hydrogen is 86.2% and He is 13.6% of Jupiter's atmosphere (Table 8-6). Figure 8-15 gives $\gamma = 1.0031$ for H_2 and $\gamma = 1.0025$ for He at 4.92 bars pressure. Using Eq. (8-196) we compute

$$f_{H2}(\text{Jupiter}) = 0.862 f_{H2}(\text{pure}) = 0.862(1.0031 \times 4.92) = 4.25 \text{ bars} \quad (8\text{-}198)$$

$$f_{He}\ (\text{Jupiter}) = 0.136 f_{He}\ (\text{pure}) = 0.136(1.0025 \times 4.92) = 0.67 \text{ bars} \quad (8\text{-}199)$$

We find the fugacities are 4.25 and 0.67 bars for H_2 and He, the fugacity coefficients for H_2 and He in Jupiter's atmosphere are the same as the pure gases, according to Eq. (8-197), and that both gases are ideal within 0.31% at this level of Jupiter's atmosphere.

The Lewis and Randall fugacity rule is very convenient and is widely used. However, it is important to understand its limitations. It is exact for gases that mix ideally and is an approximation for gas mixtures that are nonideal. The fugacity f_{mix} of gas i in a gas mixture is related to the fugacity of the pure gas f_{pure} at the same T and total P by the equation (Gillespie, 1925):

$$RT \ln f_{mix} = RT\ln(X_i f_{pure}) + \int_0^P \left(\overline{V}_i - \frac{V_i}{n_i}\right) dP \quad (8\text{-}200)$$

Equation (8-200) reduces to Eq. (8-196) if the value of the integral is zero. Thus, over the entire integration path (from $P = 0$ to $P = P$ of the gas mixture) the Lewis and Randall rule requires that the partial molar volume of a gas in a mixture and the molar volume of the pure gas are identical:

$$\frac{V_i}{n_i} = \overline{V}_i = \left(\frac{\partial V}{\partial n_i}\right)_{P,T,n \neq n_i} \quad (8\text{-}201)$$

When Eq. (8-201) is true the molar volume (V_m) of the gas mixture is the sum of the mole fraction (X_i) times molar volume ($V_{m,i}$) products for each gas in the mixture at all pressures:

$$V_m = \sum_{i=1} X_i V_{m,i} \quad (8\text{-}202)$$

Equation (8-202) is *Amagat's law*, proposed by the French physicist Émile Amagat (1841–1915), who made pioneering studies of gases under high pressures.

The molar and partial molar volumes of an ideal gas are equal at the same P and T, but the molar and partial molar volumes of a real gas are different at the same pressure and temperature. The Lewis and Randall fugacity rule is only a good approximation under conditions where Eq. (8-201) or Eq. (8-202) are satisfied or approximately correct. The next few examples illustrate these conditions. They include low pressures where many gases mix ideally, or nearly so (P up to ~100–300 bars), high pressures where gases behave like liquids ($P > 1000$ bars), and mixtures of gases with similar reduced properties. The Lewis and Randall fugacity rule is a good approximation at all pressures for the overwhelmingly dominant gas in a mixture (e.g., H_2 in the atmospheres of Jupiter, Saturn, Uranus, and Neptune, and CO_2 in the atmospheres of Venus and Mars). Conversely, the Lewis fugacity rule fails for condensable gases that are minor constituents of mixtures and for gases near their critical points.

B. Virial equation for gas mixtures

The virial equation is the only equation of state for gas mixtures that does not involve any arbitrary assumptions. Equations (8-128) and (8-129) are written as

$$\frac{PV_m}{RT} = Z_{mix} = 1 + \frac{B_{mix}}{V_m} + \frac{C_{mix}}{V_m^2} + \frac{D_{mix}}{V_m^3} + \cdots \tag{8-203}$$

$$\frac{PV_m}{RT} = Z_{mix} = 1 + B_{mix}\rho_m + C_{mix}\rho_m^2 + D_{mix}\rho_m^3 + \cdots \tag{8-204}$$

The B_{mix}, C_{mix}, D_{mix},... virial coefficients are functions of temperature and the composition of the gas mixture. The equation for the second virial coefficient B_{mix} of a binary gas mixture composed of gases 1 and 2 with mole fractions X_1 and X_2 that sum to unity is

$$B_{mix} = X_1^2 B_{11} + X_2^2 B_{22} + 2X_1 X_2 B_{12} \tag{8-205}$$

Equation (8-205) is an exact equation without any arbitrary assumptions. It is from the statistical mechanics of real gases (Hirschfelder et al., 1954). The B_{11} coefficient arises from interactions of two particles (atoms or molecules) of gas 1, the B_{22} coefficient is from interactions of two particles of gas 2, and the B_{12} coefficient is from interactions between particles of gas 1 and gas 2. The B_{11} and B_{22} coefficients are the same as the second virial coefficients for pure gases that we discussed earlier. The B_{12} coefficient is the *interaction* or *cross-virial coefficient* for the two gases and is determined experimentally or theoretically for the gas mixture. It is only a function of temperature and is independent of pressure, density, or the composition of the gas mixture. The equation for B_{mix} in a mixture of n different gases is a generalization of Eq. (8-205):

$$B_{mix} = \sum_{i=1}^{n} \sum_{k=1}^{n} X_i X_k B_{ik} \tag{8-206}$$

Likewise, the equation for the third virial coefficient (C_{mix}) of a binary gas mixture composed of gases 1 and 2 with mole fractions X_1 and X_2 that sum to unity is

$$C_{mix} = X_1^3 C_{111} + X_2^3 C_{222} + 3X_1^2 X_2 C_{112} + 3X_1 X_2^3 C_{122} \tag{8-207}$$

Equation (8-207) is also an exact equation without any arbitrary assumptions. The C_{111} coefficient arises from interactions of three particles (atoms or molecules) of gas 1. The C_{222} coefficient arises from interactions of three particles of gas 2. The C_{111} and C_{222} coefficients are the same as the third virial coefficients of the pure gases. The C_{112} and C_{122} coefficients arise from interactions between three particles of the two different gases. They are the interaction or cross-virial coefficients. They only depend on temperature and are independent of pressure, density, or the composition of the gas mixture. The equation for C_{mix} in a mixture of n different gases is

$$C_{mix} = \sum_{i=1}^{n}\sum_{k=1}^{n}\sum_{m=1}^{n} X_i X_k X_m C_{ikm} \tag{8-208}$$

For ternary and more complex mixtures, Eq. (8-208) includes cross terms such as C_{ikm} that involve the interaction of three different gases. Measurement of these cross terms is extremely difficult, and values are known for only a few gas mixtures. Prausnitz et al. (1999) discuss virial coefficients for gas mixtures in more detail and give examples from the literature.

Example 8-21. Gibby et al. (1929) measured second virial coefficients for H_2, He, and their mixture (B_{mix}) from 25°C to 175°C. Find the interaction virial coefficient for H_2-He from their data. The following table gives the data and results. We solved a rearranged version of Eq. (8-205) for B_{12}:

$$B_{\mathrm{H2He}} = \left\{B_{mix} - \left(B_{\mathrm{He}}X_{\mathrm{He}}^2 + B_{\mathrm{H2}}X_{H2}^2\right)\right\}/\left(2X_{\mathrm{H2}}X_{\mathrm{He}}\right) \tag{8-209}$$

T(°C)	25.0	50.0	75.0	100.4	125.2	150.1	175.0
B_{11} (H_2)	14.713	15.049	15.386	15.543	15.744	15.543	15.408
B_{22} (He)	11.438	11.371	11.259	10.900	11.079	10.362	10.922
B_{mix}	14.309	14.444	14.287	14.421	13.905	14.376	13.905
B_{12}	15.55	15.69	15.26	15.63	14.41	15.81	14.65

The virial coefficients are in cm^3 mol^{-1}. The mixture is 49.89% H_2 and 50.11% He. Three decimal places were used to avoid round-off errors in the calculations. However, comparisons with the B values for H_2 and He calculated from Table 8-4 and with the B_{12} values for H_2-He from Dymond and Smith (1980) indicate the virial coefficients have uncertainties of a few cm^3 mol^{-1}.

The fugacity coefficient (γ_1) for component 1 in a binary mixture is

$$\ln\gamma_1 = \frac{2}{V_m}\left[X_1B_{11} + X_2B_{12}\right] + \frac{3}{2}\frac{1}{V_m^2}\left[X_1^2C_{111} + 2X_1X_2C_{112} + X_2^2C_{122}\right] - \ln Z_{mix} \tag{8-210}$$

The fugacity coefficient (γ_2) for component 2 in a binary mixture is

$$\ln\gamma_2 = \frac{2}{V_m}\left[X_2B_{22} + X_1B_{12}\right] + \frac{3}{2}\frac{1}{V_m^2}\left[X_2^2C_{222} + 2X_1X_2C_{122} + X_1^2C_{112}\right] - \ln Z_{mix} \tag{8-211}$$

The general equation for the fugacity coefficient (γ_i) of component i in a gas mixture is

$$\ln\gamma_i = \frac{2}{V_m}\sum_{k=1}^{m} X_kB_{ik} + \frac{3}{2}\frac{1}{V_m^2}\sum_{k=1}^{m}\sum_{o=1}^{m} X_kX_oC_{iko} - \ln Z_{mix} \tag{8-212}$$

If only the second virial coefficients are known, Eq. (8-212) simplifies to

$$\ln \gamma_i = \frac{2}{V_m}\sum_{k=1}^{m} X_k B_{ik} - \ln Z_{mix} \quad (8\text{-}213)$$

This is equivalent to

$$\ln \gamma_i = \left[2\sum_{k=1}^{m} X_k B_{ik} - B_{mix}\right]\frac{P}{RT} \quad (8\text{-}214)$$

However, the two latter equations are only valid for densities less than or equal to about 50% of the critical density of the gas mixture.

Example 8-22. Use the virial equation to compute fugacity coefficients for H_2 and He in Jupiter's atmosphere (see Example 8-20). The second virial coefficients at 273.15 K computed from the data in Table 8-4 are $B_{H2} = 13.94$ and $B_{He} = 11.57$ cm^3 mol^{-1}. The cross-virial coefficient $B_{12} = 16.06$ cm^3 mol^{-1} from the equation

$$B_{12} = 11.436 + 1987.94/T - 197,677/T^2 \text{ cm}^3 \text{ mol}^{-1} \quad (8\text{-}215)$$

This is a least-squares fit to the data of Gibby et al. (1929) and Dymond and Smith (1980) for B_{12} from 123 K to 448 K. The second virial coefficient B_{mix} is calculated from Eq. (8-205):

$$B_{mix} = (0.862)^2 13.94 + (0.136)^2 11.57 + 2(0.862)(0.136)(16.06) = 14.34 \text{ cm}^3 \text{ mol}^{-1} \quad (8\text{-}216)$$

The fugacity coefficients for H_2 ($\gamma = 1.0031$) and He ($\gamma = 1.0036$) are calculated from Eq. (8-213):

$$\ln \gamma_{H2} = [2(0.862 \times 13.94 + 0.136 \times 16.06) - 14.34]\frac{4.92}{(83.144 \times 273.15)} = 3.05 \times 10^{-3} \quad (8\text{-}217)$$

$$\ln \gamma_{He} = [2(0.136 \times 11.57 + 0.862 \times 16.06) - 14.34]\frac{4.92}{(83.144 \times 273.15)} = 3.57 \times 10^{-3} \quad (8\text{-}218)$$

The γ values are close to those from the Lewis and Randall fugacity rule (see Example 8-20).

C. Van der Waals equation

The virial coefficients are unknown for many complex gas mixtures encountered in nature, and a cubic equation of state is useful in these cases. We first consider the van der Waals equation. The fugacity coefficient for a van der Waals gas (e.g., Lewis and Randall, 1923) is

$$\ln \frac{f}{P} = \ln \gamma = \ln\left(\frac{RT}{V_m - b}\right) + \frac{b}{V_m - b} - \frac{2a}{RTV_m} - \ln P \quad (8\text{-}219)$$

The fugacity coefficient for a component in a van der Waals gas mixture is

$$\ln\gamma_i = \ln\left(\frac{V_m}{V_m - b_{mix}}\right) + \frac{b_i}{V_m - b_{mix}} - \frac{2\sqrt{a_i}\sum_{k=1}^{m} X_k (a_k)^{1/2}}{RTV_m} - \ln Z_{mix} \quad (8\text{-}220)$$

The van der Waals constants for the mixture are a_{mix} and b_{mix}. The constants for a binary gas mixture of gas 1 and gas 2 with mole fractions $X_1 + X_2 = 1$ are (Kohnstamm 1926):

$$a_{mix} = X_1^2 a_1 + 2X_1X_2a_{12} + X_2^2 a_2 \quad (8\text{-}221)$$

$$a_{12} = \sqrt{a_1 a_2} \quad (8\text{-}222)$$

$$b_{mix} = X_1^2 b_1 + 2X_1X_2b_{12} + X_2^2 b_2 \quad (8\text{-}223)$$

$$b_{12} = \frac{1}{8}\left(\sqrt[3]{b_1} + \sqrt[3]{b_2}\right)^3 \quad (8\text{-}224)$$

Example 8-23. Calculate the fugacity coefficients (γ) for N_2, O_2, and their mixture (78.85% N_2, 21.15% O_2) at 0°C, 100 bars and compare the results to those from the Lewis fugacity rule.

Equations (8-221) to (8-224) and data in Table 8-3 give the constants for the N_2-O_2 mixture:

$$10^4 a_{12} = \sqrt{(137.0)(138.2)} = 137.6 \quad (8\text{-}225)$$

$$\begin{aligned} a_{mix} &= \left[0.7885^2(137.0) + 2(0.7885)(0.2115)(137.6) + 0.2115^2(138.2)\right] \times 10^4 \\ &= 137.3 \times 10^4 \ \text{cm}^6 \ \text{bar mol}^{-2} \end{aligned} \quad (8\text{-}226)$$

$$b_{12} = \frac{1}{8}\left(\sqrt[3]{38.69} + \sqrt[3]{31.86}\right)^3 = \frac{6.5524^3}{8} = 35.16 \ \text{cm}^3 \ \text{mol}^{-1} \quad (8\text{-}227)$$

$$b_{mix} = 0.7885^2(38.69) + 2(0.7885)(0.2115)(35.16) + 0.2115^2(31.86) = 37.21 \ \text{cm}^3 \ \text{mol}^{-1} \quad (8\text{-}228)$$

The van der Waals equation and Eq. (8-70) give the molar volume $V_m = 210.5$ cm^3 mol^{-1} and compression factor $Z_{mix} = 0.927$ of the gas mixture at 273.15 K and 100 bars. The value of γ_{mix} is

$$\ln \gamma_{mix} = \ln\left(\frac{83.144 \times 273.15}{210.5 - 37.21}\right) + \frac{37.21}{210.5 - 37.21} - \frac{2 \times 137.3 \times 10^4}{(83.144 \times 273.15 \times 210.5)} - \ln(100) \quad (8\text{-}229)$$

$$\gamma_{mix} = \exp(-0.0892) = 0.915 \quad (8\text{-}230)$$

The fugacity coefficients for N_2 and O_2 are

$$\ln \gamma_{N2} = \ln\left(\frac{210.5}{210.5 - 37.21}\right) + \frac{38.69}{210.5 - 37.21} - \frac{200\sqrt{137} \times 21.15 \times \sqrt{138.2}}{(83.144 \times 273.15 \times 210.5)} - \ln(0.927) \quad (8\text{-}231)$$

$$\gamma_{N_2} = \exp(0.3718) = 1.450 \quad (8\text{-}232)$$

$$\ln\gamma_{O2} = \ln\left(\frac{210.5}{210.5 - 37.21}\right) + \frac{31.86}{210.5 - 37.21} - \frac{200\sqrt{138.2} \times 78.85 \times \sqrt{137}}{(83.144 \times 273.15 \times 210.5)} - \ln(0.927) \quad (8\text{-}233)$$

$$\gamma_{O2} = \exp(0.0003) = 1.0003 \quad (8\text{-}234)$$

The γ values from the Lewis fugacity rule are those for the pure gases at the same pressure and temperature, for which Figure 8-15 gives $\gamma_{N2} = 0.970$ and $\gamma_{O2} = 0.918$. The van der Waals equation and the Lewis fugacity rule both predict $\gamma_{N2} > \gamma_{O2}$, but the actual values differ significantly.

D. Other cubic equations of state

Redlich and Kwong (1949) describe fugacity calculations for gas mixtures and give the equation

$$\log\gamma_i = 0.4343(Z_{mix} - 1)\frac{B_i}{B_{mix}} - \log(Z_{mix} - B_{mix}P) - \frac{A_{mix}^2}{B_{mix}}\left(\frac{2A_i}{A_{mix}} - \frac{B_i}{B_{mix}}\right)\log\left(1 + \frac{B_{mix}P}{Z_{mix}}\right) \quad (8\text{-}235)$$

for the fugacity coefficient γ_i of gas i in a mixture. The A, B, and Z_{mix} terms are

$$A_i = \left(\frac{0.42748}{P_C}\right)^{1/2}\left(\frac{T_C}{T}\right)^{5/4} = \left(\frac{a_i'}{R^2T^{5/2}}\right)^{1/2} \quad (8\text{-}236)$$

$$B_i = \frac{0.08664}{P_C}\frac{T_C}{T} = \frac{b_i'}{RT} \quad (8\text{-}237)$$

$$A_{mix} = \sum_i X_iA_i \quad (8\text{-}238)$$

$$B_{mix} = \sum_i X_iB_i \quad (8\text{-}239)$$

$$Z_{mix} = \frac{PV_m}{RT} \quad (8\text{-}240)$$

Example 8-24. Use the Redlich-Kwong equation to calculate fugacity coefficients for H_2 and N_2 in a gas giant exoplanet with an atmosphere of 50% H_2 and 50% N_2, $T_{surface} = 0°C$, and surface $P = 600$ atmospheres. We first compute A and B for each gas and the gas mixture:

$$A_{H2} = \left(\frac{0.42748}{12.97}\right)^{1/2}\left(\frac{33.2}{273.15}\right)^{5/4} = 0.01303 \quad (8\text{-}241)$$

$$A_{N2} = \left(\frac{0.42748}{33.9}\right)^{1/2}\left(\frac{126.2}{273.15}\right)^{5/4} = 0.04277 \quad (8\text{-}242)$$

$$A_{mix} = \sum_i X_i A_i = 0.5(0.01303) + 0.5(0.04277) = 0.02790 \tag{8-243}$$

$$B_{H2} = \frac{0.08664}{12.97}\frac{33.2}{273.15} = 8.12 \times 10^{-4} \tag{8-244}$$

$$B_{N2} = \frac{0.08664}{33.9}\frac{126.2}{273.15} = 1.18 \times 10^{-3} \tag{8-245}$$

$$B_{mix} = \sum_i X_i B_i = 0.5(8.12 \times 10^{-4}) + 0.5(1.18 \times 10^{-3}) = 9.96 \times 10^{-4} \tag{8-246}$$

Next we compute the compression factor for the gas mixture Z_{mix} from the equations

$$Z = \frac{1}{(1-h)} - \frac{(A^2/B)h}{(1+h)} \tag{8-247}$$

$$h = \frac{BP}{Z} = \frac{b'}{V_m} \tag{8-248}$$

Equation (8-247) is another form of the Redlich-Kwong equation (8-104). It is solved using the general solution for a cubic equation or by trial and error starting from an initial guess for Z (or V_m). The molar volume of the gas mixture $V_m = 55.0287\ \text{cm}^3\ \text{mol}^{-1}$ and $Z_{mix} = 1.47308$ for this example. Finally, we substitute numerical values into Eq. (8-235) and find the fugacity coefficients. Pressure in atmospheres was converted into bars in the following equations:

$$\log\gamma_{H2} = \frac{1}{\ln 10}(1.47308 - 1)\frac{8.12}{9.96} - \log(1.47308 - 607.95 \cdot 9.96 \times 10^{-4})$$
$$- \frac{0.0279^2}{9.96 \times 10^{-4}}\left\{\frac{2(0.01303)}{0.0279} - \frac{8.12}{9.96}\right\}\log\left(1 + \frac{607.95 \cdot 9.96 \times 10^{-4}}{1.47308}\right) \tag{8-249}$$
$$= 0.21532$$

$$\log\gamma_{N2} = \frac{1}{\ln 10}(1.47308 - 1)\frac{11.8}{9.96} - \log(1.47308 - 607.95 \cdot 9.96 \times 10^{-4})$$
$$- \frac{0.0279^2}{9.96 \times 10^{-4}}\left\{\frac{2(0.04277)}{0.0279} - \frac{11.8}{9.96}\right\}\log\left(1 + \frac{607.95 \cdot 9.96 \times 10^{-4}}{1.47308}\right) \tag{8-250}$$
$$= 0.08524$$

The calculated fugacity coefficients are 1.64 for H_2 and 1.22 for N_2. These are close to the experimental values of 1.62 for H_2 and 1.30 for N_2 in this gas mixture at 0°C and 600 atmospheres pressure (Merz and Whitaker, 1928).

Fugacity equations based on the modified Redlich-Kwong, Peng-Robinson, and Soave equations of state are more complex than those described above. They are in the original references (Kerrick and Jacobs, 1981; Peng and Robinson, 1976; Soave, 1972).

V. CRITICAL PHENOMENA OF REAL GASES

A. Historical background

Around 1790 the Dutch physician Martinus van Marum (1750–1837) inadvertently liquefied NH_3 by compression to about seven atmospheres pressure during experiments testing whether or not Boyle's law was valid for all gases or only for air. Starting in 1822, the French nobleman Charles, Baron Cagniard de la Tour (1777–1859), experimented with heating ethyl alcohol (C_2H_5OH), diethyl ether [$(C_2H_5)_2O$], and water inside sealed tubes. His results for the critical temperature (188°C) and pressure (37–38 atmospheres) of diethyl ether are close to the actual values of 194°C and 36.1 atmospheres. At about the same time as de la Tour began his work, the great English scientist Michael Faraday (see sidebar) liquefied a number of gases by heating one end of a sealed, V-shaped tube and cooling the other end of the tube in air or in a freezing mixture. The hot end of the tube contained a compound that decomposed when heated to produce the gas he wanted to liquefy. For example, Faraday liquefied Cl_2 by heating chlorine clathrate hydrate ($Cl_2 \cdot 10\ H_2O$) at one end of the V-shaped tube while keeping the other end at room temperature. Sulfur dioxide, H_2S, CO_2, N_2O, NH_3, and cyanogen (C_2N_2) were liquefied in a series of similar experiments over a 20-year period. Faraday distinguished between coercible gases, which he liquefied, and permanent gases, such as N_2, O_2, H_2, CH_4, and CO, which he could not liquefy in this fashion. However, he did not know why this distinction existed.

MICHAEL FARADAY (1791–1867)

Faraday was one of four children of a blacksmith and a farmer's daughter. He only had a few years of schooling and educated himself while apprenticed to a bookbinder by reading books brought in for rebinding. By good fortune, in the spring of 1812 he was given a ticket to attend the last of four public lectures by Sir Humphry Davy, professor of chemistry at the Royal Institution and then at the height of his career. Afterward, Faraday sent Davy his notes and drawings of Davy's lecture and asked for a job. Davy granted his request and Faraday accompanied Davy and his wife on an 18-month tour of continental Europe, during which time they met many famous scientists of the day. On their return Faraday began working as Davy's assistant and published his first scientific paper on the chemical analysis of lime from Tuscany, Italy, in 1816.

During the next 10 years, Faraday began his experiments on gas liquefaction, studied oxidation of meteoritic iron and synthetic iron alloys, discovered benzene and the first chlorinated organic compounds, and produced glass with a high refractive index. Faraday became director of the laboratory at the Royal Institution in 1825 and began his famous weekly lecture series.

Faraday's greatest contributions to science, however, were his discoveries relating to electricity and magnetism. Faraday was the first to map a magnetic field. He found that the lines of the field travel out one pole and loop around to the opposite pole rather than going along a straight, polarized path. He discovered electromagnetic induction in 1831. Faraday did fundamental research in electrochemistry and introduced the terms *anode, anion, cathode, cation, electrode, electrolyte, electrolysis,* and *ion.* Later in his career, Faraday discovered a connection between magnetism and light when he used his high refractive index glass to discover magneto-optical rotation (rotation of plane-polarized light by substances placed in a magnetic field). His experimental research paved the way for Maxwell's theories and the modern electrical industry.

A few years later, in the 1860s, Thomas Andrews (1813–1885), a professor of chemistry at Queens College, Belfast, Ireland, discovered critical phenomena of real gases in his experiments on CO_2, which he called carbonic acid (see the epigraph heading this chapter). His discovery explains that Faraday could not liquefy N_2, O_2, H_2, CH_4, and CO by applying any amount of pressure at room temperature, because it is above the critical temperatures of all these gases (see Table 8-2). We defined the critical point and discussed some aspects of the critical behavior of water and helium in Chapter 7. We now explore these topics in some more detail using CO_2 as an example.

B. Critical behavior of carbon dioxide

Figure 8-17 shows the *P-T* phase diagram for CO_2 at low pressures. The vaporization curve terminates at the *L-V* critical point where the distinction between gas and liquid disappears. Figure 7-4 in Chapter 7 illustrates the same thing for water. At temperatures above the critical temperature only one phase, gas, or, more correctly, supercritical fluid, exists. As shown in Figure 8-18, CO_2 behaves more and more like an ideal gas as temperature increases above the critical temperature, for example, compare the 270, 290, 304.1, and 360 K isotherms for CO_2.

Figure 8-18 illustrates the *L-V* critical point for CO_2 at 304.1 K and 73.75 bars. In general, the *L-V* critical point is an inflection point for isotherms on the *PV* plane and is defined by

$$\left(\frac{\partial P}{\partial V}\right)_T = 0 \quad T = T_C \tag{8-251}$$

$$\left(\frac{\partial^2 P}{\partial V^2}\right)_T = 0 \quad T = T_C \tag{8-252}$$

$$\left(\frac{\partial^3 P}{\partial V^3}\right)_T < 0 \quad T = T_C \tag{8-253}$$

Equations (8-251) to (8-253) apply to any *L-V* critical point and any real gas equation of state.

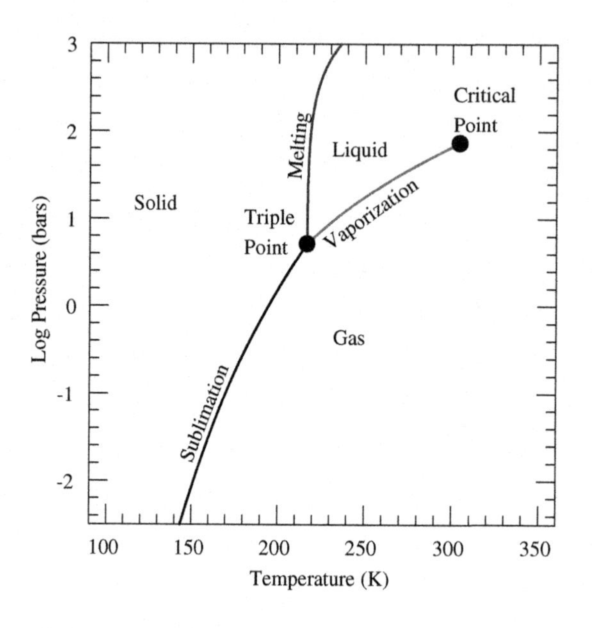

FIGURE 8-17

The *P-T* phase diagram for CO_2 at low pressures.

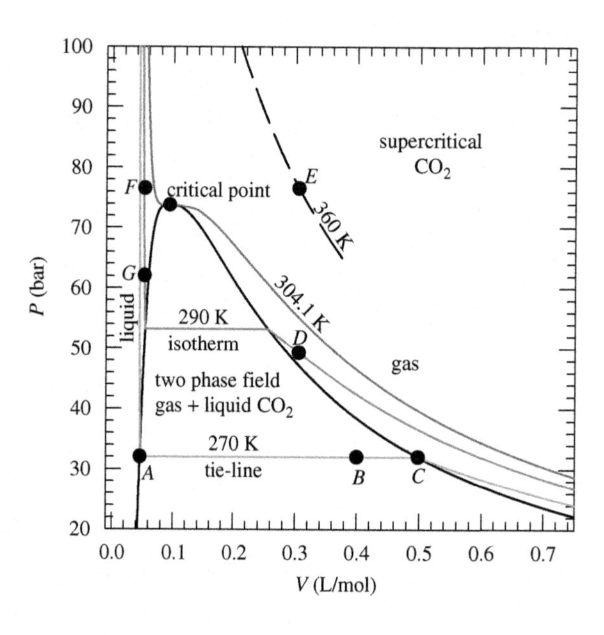

FIGURE 8-18

Isotherms for CO_2 in the vicinity of the critical point.

We now discuss the behavior and shape of isotherms below, at, and above the critical point of real gases using the data for CO_2 in Figure 8-18 as an example. We start with the 270 K isotherm, which is about 34 degrees below the critical point. As CO_2 gas is isothermally compressed, its pressure and volume follow the 270 K curve on the right side of the diagram. The 270 K isotherm deviates from Boyle's law and the *PV* product varies from 17.90 to 15.95 L bar mol^{-1} as the molar volume decreases from 0.726 to 0.498 L mol^{-1} (at point *C*). The slope of this (and any other) *P-V* isotherm is given by the general equation

$$slope = \left(\frac{\partial P}{\partial V}\right)_T = -\frac{1}{V\beta_T} = -K_T V \qquad (8\text{-}254)$$

where β_T and K_T are the isothermal compressibility and isothermal bulk modulus, defined by Eqs. (2-34) and (2-47), respectively. Notice that the 270 K isotherm becomes flat inside the black curve and then becomes much steeper in the liquid field on the left side of Figure 8-18. Gaseous CO_2 is more compressible and has a larger molar volume than liquid CO_2. Thus, (dP/dV) is smaller for the gas and larger for the liquid. The horizontal part of the isotherm is inside the two-phase field, where increasing pressure changes the ratio of gas to liquid to give zero slope (see the following discussion). Figure 8-18 also shows that (dP/dV) increases (i.e., steeper slopes) with increasing temperature for the 270–360 K isotherms for CO_2 (g), but that (dP/dV) is approximately constant with temperature for CO_2 liquid (e.g., compare the 270 K and 290 K isotherms).

The bold black curve in Figure 8-18 is the *coexistence curve.* It separates a two-phase field inside the curve from a one-phase field outside the curve. Liquid and gaseous CO_2 coexist inside the two-phase field and CO_2 (g), or liquid CO_2, or supercritical CO_2 exist in the one-phase field. Different parts of the coexistence curve have different names. The right side of the coexistence curve is the *saturation curve,* also known as the *dewpoint line* or the *orthobaric vapor curve.* The left side is the *bubble-point curve* or the *orthobaric liquid curve.*

The saturation curve intersects the 270 K isotherm at point *C*, where the first drop of liquid CO_2 forms and the 270 K isotherm enters the two-phase region where CO_2 gas and liquid coexist. Visually we would see that a meniscus appears in the container holding the CO_2 sample. The horizontal part of the 270 K isotherm is a tie line. The horizontal part of the 290 K isotherm and of any other isotherm inside the two-phase region is a tie line. A *tie line* is a line that joins two coexisting phases (e.g., CO_2 liquid and gas). The position of a point on the tie line is a measure of the relative amounts of the two phases. As the two-phase (gas + liquid) CO_2 mixture is further compressed and we move from point *C* to *B*, the pressure remains constant, but the proportions of gas and liquid change (more liquid CO_2 forms) and the molar volume of the mixture decreases. We calculate the proportions of gas and liquid from the *lever rule*, illustrated in the next example.

Example 8-25. Use the lever rule to calculate the relative and absolute proportions of CO_2 liquid and gas at point *B* in Figure 8-18. The two ends of the 270 K tie line are at molar volumes of 0.498 L mol^{-1} (point *C*, gas side) and 0.0465 L mol^{-1} (point *A*, liquid side). The molar volume is 0.4 L mol^{-1}at point *B*. The relative proportions of the two phases at point *B* are

$$\frac{\text{gas}}{\text{liquid}} = \left|\frac{(B-A)}{(B-C)}\right| = \left|\frac{(0.400-0.0465)}{(0.400-0.498)}\right| = \frac{0.35350}{0.0980} = 3.6071 \qquad (8\text{-}255)$$

Equation (8-255) is the *lever rule*. The vertical lines around the fraction indicate we use the absolute value of the fraction. In this case, the gas/liquid molar ratio is 3.6071 at point *B*. From mass balance the mole fractions of gas (X_g) and liquid (X_l) are related via

$$X_g + X_l = 1 = 3.6071X_l + X_l = 4.6071X_l \tag{8-256}$$

Thus, $X_l \sim 0.2171$ and $X_g \sim 0.7829$ at point *B*, or in other words, point *B* is a mixture of 21.71% liquid and 78.29% gaseous CO_2. This result agrees with our intuition that point *B* is gas rich because it is closer to point *C* on the saturation curve than to point *A* on the bubble-point curve.

Continued compression moves us across the 270 K tie line to point *A*, where the last bubble of gaseous CO_2 disappears and the 270 K isotherm intersects the bubble-point curve. The nearly vertical segment of the 270 K isotherm on the left side of Figure 8-18 is an isotherm for liquid CO_2. As mentioned earlier, it is steeper than the isotherm for CO_2 gas because liquid CO_2 is less compressible (smaller β_T, larger K_T) than gaseous CO_2.

Figure 8-18 shows that the points bounding the two sides of the coexistence curve get closer to each other with increasing temperature. Eventually, at 304.1 K, the two sides of the curve merge into one point, the critical point, and there is no distinction between gaseous and liquid CO_2. A second phase does not appear, and no distinct meniscus is visible. A different effect, known as critical opalescence, occurs and light scattering by the fluid CO_2 increases at this point.

Above the critical temperature, there is no longer any distinction between gaseous and liquid CO_2. Instead, there is a *continuity of states*. We illustrate this concept using the path *D-E-F-G* in Figure 8-18. We start with CO_2 gas at point *D* on the 290 K isotherm, which is below the critical temperature. We warm the CO_2 gas to point *E* on the 360 K isotherm, which is above the critical temperature. We then compress the CO_2 to point *F*, which is on the continuation of the 290 K isotherm and is also above the critical point of CO_2. Finally, we cool the CO_2 to point *G*, also on the 290 K isotherm but below the critical point of CO_2. At this latter point, we now have liquid CO_2. However, at no point along the path *D-E-F-G* did two phases coexist. Only one phase, fluid CO_2, was present, and the gas → liquid transformation occurred smoothly, without phase separation or the appearance of a meniscus. The behavior of other gases at and near their critical points is analogous to that described here for CO_2.

We now consider some of the relationships between the *PT* and *PV* plots of the CO_2 phase diagram in Figures 8-17 and 8-18. The flat parts of the isotherms inside the coexistence curve in Figure 8-18 correspond to the temperature-dependent pressures along the vaporization curve in Figure 8-17. In other words, if the pressures corresponding to the isotherms in Figure 8-18 were plotted on Figure 8-17, they would fall onto the vaporization curve. In principle, the vapor pressure equation for liquid CO_2 could be determined from the *PVT* data. Figure 8-19 illustrates the excellent agreement between the liquid CO_2 vapor pressure determined from flat parts of isotherms (Michels et al., 1937) with vapor pressures tabulated by NIST.

The enthalpy of vaporization ($\Delta_{vap}H$) of liquid CO_2 can also be determined from the data in Figure 8-18 by using the Clapeyron equation:

$$\left(\frac{dP}{dT}\right)_{eq} = \frac{\Delta S}{\Delta V} = \frac{\Delta H}{T\Delta V} \tag{7-31}$$

The variation of pressure with temperature determined from the isothermal segments inside the coexistence curve gives an equation that is differentiated to find (dP/dT). The volumes of the

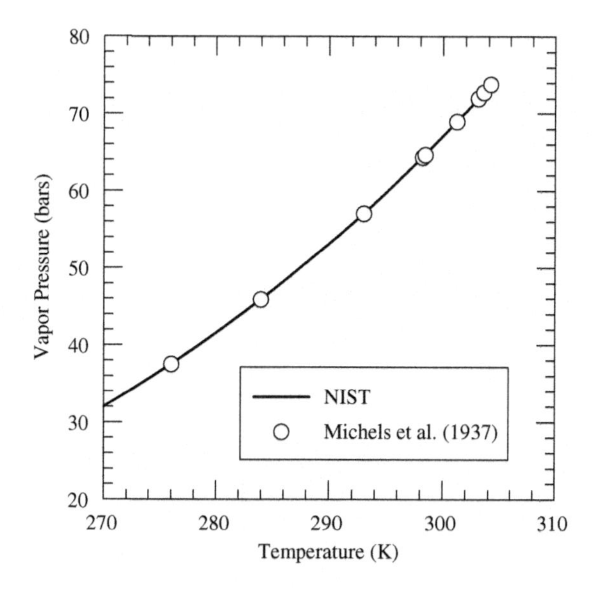

FIGURE 8-19

The vapor pressure of liquid CO_2 from isotherms measured by Michels et al. (1937).

coexisting gas and liquid are given by the intersection of the coexistence curve with the isotherms, so we know $\Delta V = V_g - V_l$ as a function of temperature. We can rearrange Eq. (7-31) and solve for ΔH as a function of temperature:

$$\Delta_{vap}H = T \cdot \Delta V \cdot \frac{dP}{dT} \tag{8-257}$$

The calculations are the subject of Problem 8-27 at the end of the chapter, and Figure 8-20 shows the results. The points are the $\Delta_{vap}H$ values derived from the *PVT* data of Michels et al. (1937) and the curve is $\Delta_{vap}H$ tabulated by NIST. Figure 8-20 also shows that the enthalpy of vaporization for liquid CO_2 decreases with increasing temperature and $\Delta_{vap}H \rightarrow 0$ as $T \rightarrow T_C$. This behavior is also observed for water (see Figure 7-8) and is general behavior displayed by all materials.

Figure 8-21 shows the orthobaric densities (i.e., the densities of the coexisting gas and liquid) along the coexistence curve. These are plotted in *Amagat density units.* One Amagat density unit is the density of one mole of ideal gas at STP (273.15 K, 1 atmosphere) and is equal to the reciprocal of the molar volume at STP (22.41400 L mol^{-1}). Thus, a density of 237.70 Amagats is

$$237.70 \text{ Amagats} = \frac{237.70}{22.41400 \text{ L mol}^{-1}} = 10.605 \text{ mol L}^{-1} \tag{8-258}$$

This is the density of CO_2 at its critical point determined by Michels et al. (1937) and corresponds to a molar volume of 94.3 cm^3 mol^{-1}. Figure 8-21 shows the mean orthobaric density,

$$\rho_{mean} = \frac{(\rho_g + \rho_l)}{2} \tag{8-259}$$

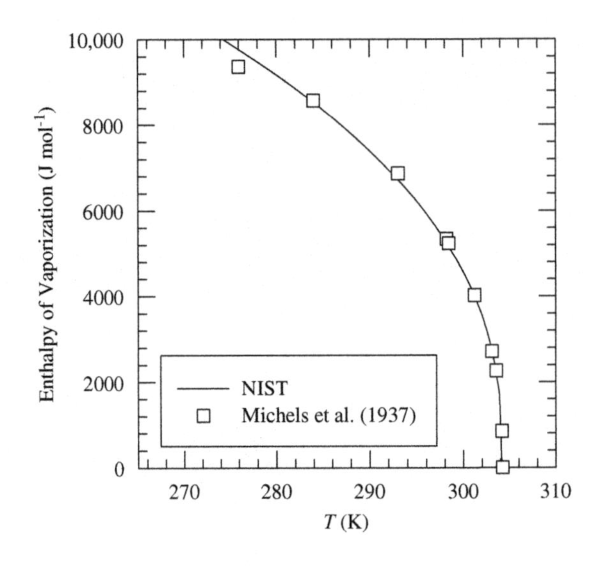

FIGURE 8-20

The enthalpy of vaporization of liquid CO_2 from isotherms measured by Michels et al. (1937).

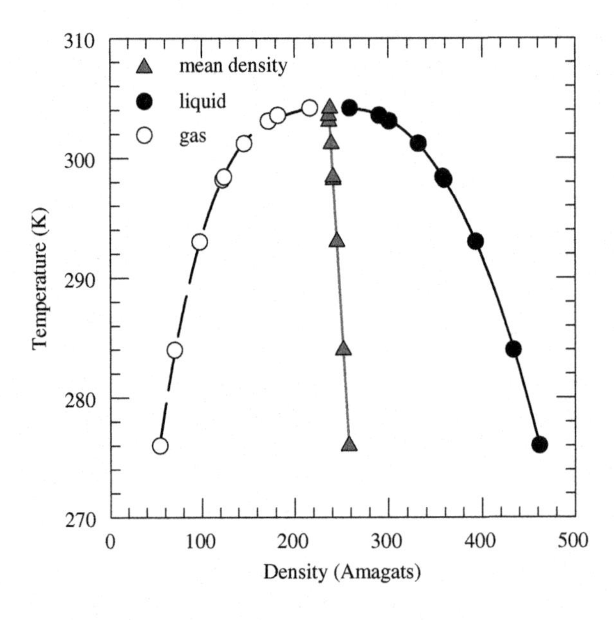

FIGURE 8-21

Orthobaric densities measured by Michels et al. (1937) for CO_2 in the vicinity of the critical point.

The mean orthobaric density passes through the density at the critical point. This is an illustration of the *law of rectilinear diameters* proposed by Cailletet and Mathias. The temperature dependence of the orthobaric densities in the vicinity of the critical temperature, and whether or not the mean density line is straight or curved, are important research topics (e.g., Rowlinson, 2004).

C. Van der Waals equation

Equations (8-100) and (8-101) give the van der Waals constants a and b from the critical temperature and pressure of a gas. We now derive these two equations.

The first and second derivatives of the van der Waals equation (8-88) at the critical point are

$$\left(\frac{\partial P}{\partial V}\right)_T = \frac{\partial}{\partial V}\left[\frac{RT_C}{V_C - b} - \frac{a}{V_C^2}\right]_T = -\frac{RT_C}{(V_C - b)^2} + \frac{2a}{V_C^3} = 0 \qquad (8\text{-}260)$$

$$\left(\frac{\partial^2 P}{\partial V^2}\right)_T = \frac{2RT_C}{(V_C - b)^3} - \frac{6a}{V_C^4} = 0 \qquad (8\text{-}261)$$

Equations (8-260) and (8-261) are used to find values of a and b at the critical point. These are

$$a_C = \frac{9RT_C V_C}{8} \qquad (8\text{-}262)$$

$$b_C = \frac{V_C}{3} \qquad (8\text{-}263)$$

Substituting these values back into the van der Waals equation (8-88) gives

$$P_C = \frac{RT_C}{V_C - b_C} - \frac{a_C}{V_C^2} = \frac{RT_C}{\left(V_C - \frac{V_C}{3}\right)} - \left[\frac{\frac{9RT_C V_C}{8}}{V_C^2}\right] \qquad (8\text{-}264)$$

Equation (8-264) is solved for V_C, which is given by

$$V_C = \frac{3RT_C}{8P_C} \qquad (8\text{-}265)$$

Finally, Eq. (8-265) is substituted back into Eqs. (8-262) and (8-263), giving

$$a = \frac{27bRT_C}{8} \qquad (8\text{-}100)$$

$$b = \frac{RT_C}{8P_C} \qquad (8\text{-}101)$$

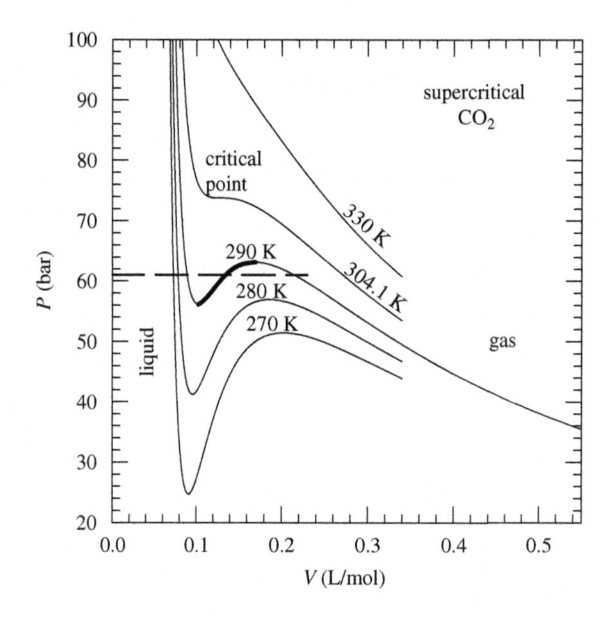

FIGURE 8-22

Van der Waals isotherms for CO_2 in the vicinity of the critical point.

Figure 8-22 shows van der Waals isotherms for CO_2. At and above the critical point, these are similar to the actual isotherms of CO_2 displayed in Figure 8-18. However, below the critical point the van der Waals isotherms are S-shaped (e.g., the 290 K isotherm). The dashed horizontal line intersects the 290 K isotherm three times because the van der Waals equation gives three different values of V_m at the same pressure. This region is the two-phase region inside the coexistence curve (omitted for clarity). The first part of the S-shaped curve (molar volume $V_m \sim 0.214 - 0.170$ L mol^{-1}) represents super saturation of CO_2 gas, which should condense to liquid but does not do so. The last part of the S-shaped curve (V_m ~0.102 – 0.089 L mol^{-1}) represents overexpansion of CO_2 liquid, which should vaporize to gas but does not do so. It is possible to have a supersaturated gas or an overexpanded liquid in some circumstances. However, the center part of the S-shaped curve, shown by the thicker line, has no physical significance. In this region, the pressure decreases as volume decreases; this corresponds to a negative compressibility (and bulk modulus) and is physically unreal.

The location of tie lines across the two-phase region inside the coexistence curve is determined using the *Maxwell equal-area criterion* proposed by James Clerk Maxwell. Inside the coexistence curve, gas and liquid coexist in equilibrium. Thus, the Gibbs free energy change for isothermal vaporization of liquid to gas (or isothermal condensation of gas to liquid) is zero:

$$G_{liq} = G_{gas} \quad (8\text{-}266)$$

The areas enclosed by the van der Waals isotherm above and below the tie line are equal to the Gibbs energies of gas and liquid (integrals of VdP for gas and liquid). The tie lines are drawn such that G_{liq} and G_{gas} are equal, as required by Eq. (8-266).

The van der Waals equation provides a mathematical explanation for the temperature-dependent variations and eventual disappearance of $\Delta_{vap}H$. The enthalpy of vaporization is simply the difference between the enthalpies of the coexisting liquid and vapor:

$$\Delta_{vap}H = H_{gas} - H_{liquid} \tag{8-267}$$

We can rewrite Eq. (8-267) using the definition of enthalpy to obtain

$$H_{gas} - H_{liquid} = E_{gas} - E_{liquid} + P_{gas}(V_{gas} - V_{liquid}) \tag{8-268}$$

We can evaluate Eq. (8-268) using the van der Waals equation. This involves considering the change of internal energy with volume at constant temperature. To do this we first rewrite the thermodynamic equation of state (8-49) using the Maxwell relation in Eq. (8-48) to obtain

$$\left(\frac{\partial E}{\partial V}\right)_T = T\left(\frac{\partial P}{\partial T}\right)_V - P \tag{8-269}$$

Equation (8-269) becomes

$$\left(\frac{\partial E}{\partial V}\right)_T = T\left(\frac{R}{V-b}\right) - P = \frac{a}{V^2} \tag{8-270}$$

for a van der Waals gas. Integrating Eq. (8-270) gives the expression

$$E = -\frac{a}{V} + \text{constant} \tag{8-271}$$

We now substitute Eq. (8-271) back into Eq. (8-268). When we do this the constant term drops out and we are left with

$$\begin{aligned}\Delta_{vap}H &= -\frac{a}{V_{gas}} + \frac{a}{V_{liquid}} + P_{gas}(V_{gas} - V_{liquid}) \\ &= -a\left(\frac{1}{V_{gas}} - \frac{1}{V_{liquid}}\right) + P_{gas}(V_{gas} - V_{liquid})\end{aligned} \tag{8-272}$$

The gas and liquid merge into one phase at the critical point. Consequently, their volumes are identical and Eq. (8-272) is equal to zero.

D. Physical properties at the critical point

As mentioned earlier, the critical point is partially defined by the condition that

$$\left(\frac{\partial P}{\partial V}\right)_T = 0 \quad T = T_C \tag{8-251}$$

The reciprocal of this derivative, that is, (dV/dP), has an infinite anomaly at the critical point. Consequently, other thermodynamic properties that involve either of these two derivative also have anomalies at the critical point. For example, the isothermal compressibility β_T is defined as

$$\beta_T = -\frac{1}{V}\left(\frac{\partial V}{\partial P}\right)_T \tag{2-34}$$

Thus, the isothermal compressibility goes to infinity at the critical point. The isobaric thermal expansion coefficient α_P is defined as

$$\alpha_P = \frac{1}{V}\left(\frac{\partial V}{\partial T}\right)_P \tag{2-33}$$

Using the cyclic rule equation (2-70) we can rewrite Eq. (2-33) as

$$\alpha_P = -\frac{1}{V}\left(\frac{\partial P}{\partial T}\right)_V\left(\frac{\partial V}{\partial P}\right)_T \tag{8-273}$$

Equation (8-273) shows that the thermal expansion coefficient also goes to infinity at the critical point. Using Kirchhoff's equation we can relate $\Delta_{\text{vap}}H$ and C_P for gas and liquid:

$$\left(\frac{\partial \Delta_{vap}H}{\partial T}\right)_P = \Delta C_P = C_P(\text{gas}) - C_P(\text{liq}) = \infty \tag{8-274}$$

Figure 8-20 shows the slope of $\Delta_{\text{vap}}H$ versus temperature becomes infinite as we approach the critical temperature. Thus, ΔC_P and C_P also become infinite at the critical point. The constant-volume heat capacity C_V also goes to infinity at the critical point because of Eq. (4-46):

$$(C_P - C_V)_m = \frac{\alpha_P^2 V_m T}{\beta_T} \tag{4-46}$$

Figure 8-23 shows that C_V approaches infinity at the critical points of CO_2 and SF_6 (Beck et al., 2002). Transport properties such as viscosity and thermal conductivity also show anomalies at the critical point. Levelt Sengers (1975) reviews these anomalies.

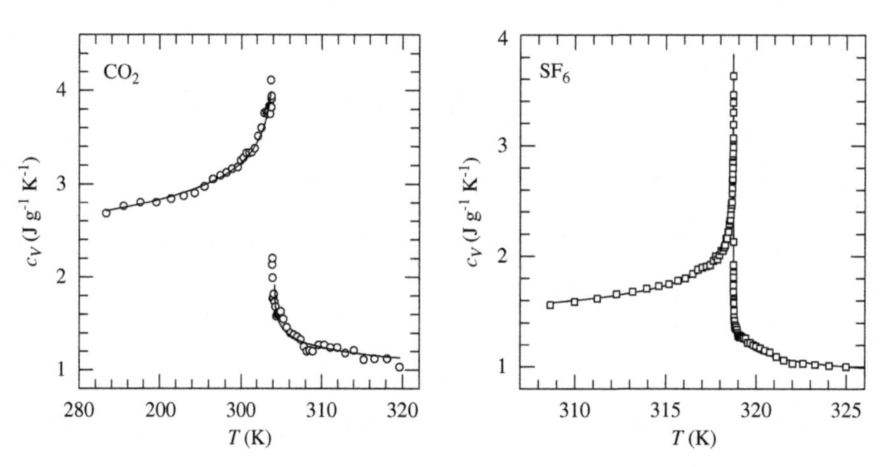

FIGURE 8-23

Specific heat (c_V) near the critical points of CO_2 and SF_6.

Modified from Beck et al. (2002).

We just discussed how $\Delta_{vap}H \rightarrow 0$ for the liquid-vapor phase transition at the critical point as the distinction between the two phases vanishes (see Figure 8-20). As a consequence of this the "wet" and "dry" adiabatic lapse rates in a planetary atmosphere become identical. One expression for the wet adiabatic lapse rate (Γ_w) is

$$\frac{dT}{dz} = \Gamma_w = -\frac{\mu g}{C_{P,m} + \lambda X_i(\lambda/RT^2 - 1/T)} \tag{3-96}$$

where λ is the latent heat of vaporization (or sublimation), X_i is the mole fraction of the condensable species (e.g., water vapor), and μ and $C_{P,m}$ are the mean molecular weight and mean molar heat capacity of the dry gas. As $\lambda = \Delta H_{vap} \rightarrow 0$, Eq. (3-96) reduces to

$$\frac{dT}{dz} = \Gamma_d = -\frac{\mu g}{C_{P,m}} = -\frac{g}{c_P} \tag{3-89}$$

Equation (3-89) is the expression for the dry adiabatic lapse rate. Once this happens, cloud condensation occurs without any change in atmospheric thermal structure. This possibly occurs on Uranus and probably occurs on Neptune because of the large amounts of water in these two planets (Fegley and Prinn, 1986; Lodders and Fegley, 1994).

VI. EQUATIONS OF STATE FOR SOLIDS AND LIQUIDS

A. Some basic definitions

We used the isothermal compressibility β_T in calculations up to this point. However, we can also define a related quantity, the adiabatic compressibility β_S:

$$\beta_S = -\frac{1}{V}\left(\frac{\partial V}{\partial P}\right)_S \tag{8-275}$$

The adiabatic compressibility is the fractional change in volume with pressure under constant entropy (i.e., isentropic or adiabatic) conditions. For example, experiments involving the propagation of sound waves through solids measure the adiabatic compressibility because no heat flow occurs into or out of a small volume of the solid during the time that the pressure pulse of the sound wave compresses the solid. Since $dS = \delta q/T$ there is no entropy change in this experiment and compressibility is measured under constant entropy (adiabatic) conditions.

The isothermal and adiabatic compressibilities are related by the equation

$$\frac{\beta_S}{\beta_T} = \frac{C_V}{C_P} \tag{8-276}$$

and differ by only a few percent because $C_P - C_V$ is only a few percent of C_V for typical solids.

The bulk modulus is the reciprocal of compressibility. The isothermal bulk modulus was defined in Chapter 2. The adiabatic bulk modulus is defined analogously and is

$$K_S = \frac{1}{\beta_S} = -V\left(\frac{\partial P}{\partial V}\right)_S \tag{8-277}$$

The ratio of the two bulk moduli is

$$\frac{K_T}{K_S} = \frac{C_V}{C_P} \tag{8-278}$$

The pressure derivative of the bulk modulus is

$$K' = \left(\frac{\partial K}{\partial P}\right)_T \tag{8-279}$$

A K' value of 4 is typically assumed when experimental data for K' are unavailable. Typical K' values for planetary materials are (Knittle, 1995) 4.0 (forsterite, $K_0 = 135.7$), 5.3 (NH_3 ice, $K_0 = 7.56$), 5.5 (bcc Fe, $K_0 = 162.5$), 6.2 (quartz, $K_0 = 37.1$), and 7.8 (CO_2 ice, $K_0 = 2.93$). Knittle (1995) tabulates the zero-pressure bulk modulus (K_0) and its pressure derivative for many geological materials. Fei (1995) tabulates α_P for many of the same geological materials.

Grüneisen's ratio (or parameter) is a dimensionless parameter given by (Grüneisen, 1926)

$$\gamma_{th} = \frac{\alpha_P K_T}{\rho C_V} = \frac{\alpha_P K_S}{\rho C_P} \tag{8-280}$$

where α_P is the thermal expansion coefficient, C_V is the constant-volume heat capacity, C_P is the constant-pressure heat capacity, K_T is the isothermal bulk modulus, K_S is the adiabatic bulk modulus, the subscript *th* stands for thermal, and ρ is the density. We use γ_{th} for Grüneisen's parameter to avoid confusion with the fugacity coefficient and the adiabatic exponent. Grüneisen's ratio generally has values of 1–2 for solids and 1–1.5 for geological materials in Earth's interior (Stacey, 2005). The Grüneisen ratio is related to the adiabatic gradient via

$$\gamma_{th} = \left(\frac{\partial \ln T}{\partial \ln \rho}\right)_S \tag{8-281}$$

B. Equations of state for solids and liquids

Poirier (2000) and Stacey (2005) review common equations of state used in the geosciences. Two common equations are the Murnaghan and the Birch-Murnaghan equations. The former is

$$P = \frac{K}{K'}\left[\left(\frac{V_T}{V}\right)^{K'} - 1\right] \tag{8-282}$$

The volume at any pressure is V and V_T is volume at ambient temperature and one bar pressure, and the other symbols are defined as before. Holland and Powell (1998) use the Murnaghan equation with $K' = 4$ to compute the contribution of pressure to Gibbs free energy:

$$\int_1^P V dP = \frac{V_T K_T}{3}\left[\left(1 + \frac{4P}{K_T}\right)^{3/4} - 1\right] \tag{8-283}$$

The temperature dependence of the bulk modulus is calculated from

$$K_T = K_{298}\left(1 - 1.5 \times 10^{-4}[T - 298]\right) \tag{8-284}$$

The third-order Birch-Murnaghan equation is

$$P = \frac{3K_T}{2}\left[\left(\frac{V_o}{V}\right)^{7/3} - \left(\frac{V_o}{V}\right)^{5/3}\right]\left\{1 - \frac{3}{4}(4 - K_T')\left[\left(\frac{V_o}{V}\right)^{2/3} - 1\right]\right\} \tag{8-285}$$

In this case, the volume at one bar pressure is V_o and the other symbols are the same as before.

PROBLEMS

1. Consider a continental ice sheet that is one kilometer thick. The temperature at the top surface is 0°C and the temperature at the bottom is –1°C. The enthalpy of melting for ice at the top of the sheet is +6009.5 J mol^{-1}. Calculate the enthalpy of melting at the base of the ice sheet using the following data. *Hints:* Calculate the temperature and pressure variation of the enthalpy of melting.

	Ice	Water
molar volume (cm^3 mol^{-1})	19.654	18.018
heat capacity (J K^{-1} mol^{-1})	37.96	75.99
α (K^{-1})	1.58×10^{-4}	-6.7×10^{-5}

2. The Clapeyron slope for the melting curve of a congruently melting mineral is given by

$$\left(\frac{dT}{dP}\right)_{eq} = \frac{\Delta V}{\Delta S} = \frac{(\Delta V)^o_{m.p.} + \Delta(\alpha_i V_i)(T - T_{m.p.}) - \Delta(\beta_i V_i)(P - 1)}{(\Delta S)^o_{m.p.} + \Delta C_P \ln(T/T_{m.p.}) - \Delta(\alpha_i V_i)(P - 1)} \tag{7-130}$$

Calculate the Clapeyron slope at 2000 K, 2.83 GPa for the diopside fusion curve, which can be treated as congruent melting, using data from Table 6-2 and given here.

$$V_0\ (\text{diopside}) = 69.11\ \text{cm}^3\ \text{mol}^{-1},\ V_0\ (\text{melt}) = 82.64\ \text{cm}^3\ \text{mol}^{-1}$$

$$K_T\ (\text{diopside}) = 90.7\ \text{GPa (isothermal bulk modulus, } 1/\beta),\ K_T\ (\text{melt}) = 21.9\ \text{GPa}$$

$$\alpha\ (\text{diopside}) = 3.2 \times 10^{-5}\ \text{K}^{-1},\ \alpha\ (\text{melt}) = 6.6 \times 10^{-5}\ \text{K}^{-1}$$

$$C_P\ (\text{diopside}) = 221.16 + 2.67 \times 10^{-2}\ T\ \text{J mol}^{-1}\ \text{K}^{-1},\ C_P\ (\text{melt}) = 334.6\ \text{J mol}^{-1}\ \text{K}^{-1}$$

3. Estimate the temperature of the core mantle boundary (CMB) at 2700 km assuming a pure forsterite mantle with an adiabatic temperature gradient below the 670 km discontinuity (T ~2000 K). Use α_P ~5×10^{-5} K^{-1}, c_p ~1.42 J g^{-1} K^{-1}, and g(R) ~10 m s^{-2}.
4. Derive the equation

$$\left(\frac{\partial H}{\partial P}\right)_T = V(1 - \alpha T)$$

used by Kojitani and Akaogi (1995) to determine the enthalpy of melting of basalt in the mantle where the pressure is greater than 1 GPa.

5. Bouhifd et al. (1996) measured $V = V(T)$ for forsterite (Mg_2SiO_4) up to 2160 K. Use a subset of their data (below) to derive an equation for α_P, and use it to compute α_P at each temperature.

T(K)	300	471	696	938	1170	1391	1709	1894	1975	2124
V/V_{ref}	1.000	1.004	1.012	1.021	1.030	1.038	1.051	1.060	1.064	1.070

6. The isothermal variations of C_V with volume and of C_P with pressure are given by

$$\left(\frac{\partial C_V}{\partial V}\right)_T = T\left(\frac{\partial^2 P}{\partial T^2}\right)_V$$

$$\left(\frac{\partial C_P}{\partial P}\right)_T = -T\left(\frac{\partial^2 V}{\partial T^2}\right)_P$$

These two equations are general and apply to all materials. Derive them using the Maxwell relations and the definitions of the constant-volume and constant-pressure heat capacities.

7. Use Bridgman's thermodynamic formulas to derive the general equation for the adiabatic temperature gradient, that is, the variation of T with P at constant entropy:

$$\left(\frac{\partial T}{\partial P}\right)_S = \frac{\alpha_P VT}{C_P}$$

8. The Galileo probe mass spectrometer showed that H_2S is the major S-bearing gas in the atmosphere of Jupiter. However, *PVT* data for H_2S are limited. Use the principle of corresponding states to estimate its compression factor Z at 1120 K and 447 bars.
9. The JANAF tables give Gibbs free energy of formation data for water as an ideal and real gas. Calculate f/P for H_2O at 1000 K and 10 bars pressure from $\Delta_f G^o = -192{,}593$ J mol^{-1} for the ideal gas and $\Delta_f G = -173{,}468$ J mol^{-1} for the real gas at 10 bars pressure.
10. As discussed in Section III-C, another commonly used form of the virial equation is

$$Z = 1 + B'P + C'P^2 + D'P^3 + \cdots$$

Derive Eqs. (8-182) and (8-183), which relate the second and third virial coefficients in the density and pressure power series.

11. Figure 8-4 shows that PH_3 and N_2 have virtually identical reduced vapor pressure curves, which means that their reduced temperatures and pressures are the same. In this case, Eq. (8-78) shows that their reduced volumes must also be identical. Use the law of corresponding states to estimate the critical volume of PH_3.

12. Calculate the reduced vapor pressures for krypton and xenon from the following data and use Eqs. (8-76) and (8-77) to determine if Kr and Xe obey the law of corresponding states.

T(K)	125.6	146.7	167.8	188.4
Kr P(bars)	1.56	5.55	14.3	29.6
T(K)	173.6	203.1	232.0	260.8
Xe P(bars)	1.61	5.85	15.0	31.4

13. The Boyle point is defined by Eq. (8-69). Calculate the Boyle point for the 1000 K isotherm of water using Eqs. (8-136) and (8-137), the polynomials in terms of density and pressure.

14. Calculate V_m, the molar volume of dry air, at –26.5°C and 995.2 millibars (the weather conditions at McMurdo Station in Antarctica during August, the coldest month of their year) using the ideal gas equation and the virial equation and compare your results to those from *PVT* data for air: $V_m = 20{,}586.1$ cm^3 mol^{-1} (Din, 1962).

15. The "hard sphere" collision diameter of an NH_3 molecule is 0.315 nm (Hirschfelder et al., 1954, p. 1200). Use this value to calculate the van der Waals b coefficient for NH_3 and compare your result to the value in Table 8-3.

16. The equation of state of a solid, liquid, or gas can be written in terms of the Helmholtz free energy $A = A(V, T)$ or in terms of the Gibbs free energy $G = G(P, T)$. The former approach is more convenient. Derive and write expressions for pressure P, entropy S, internal energy E, enthalpy H, the Gibbs free energy G, and the constant-volume heat capacity C_V in terms of A and its derivatives with respect to its natural variables V and T.

17. Deming and Shupe (1932) analyzed *PVT* data for hydrogen. Use the subset of their recommended values at 273.15 K to find the second virial coefficient B for H_2 at this temperature.

P (atm)	25	100	200	600	1000	1200
V_m (cm^3/mol)	910.3	238.6	127.1	53.19	38.35	34.57

18. Calculate μ for air at 298 K and one bar pressure using Eq. (8-159). Then compare the result to values from the equation of Joule and Thomson and from Table 8-5.

19. The van der Waals equation and the principle of corresponding states predicts that the Joule-Thomson inversion curve is a parabolic curve given by the equation

$$T_R = 3\left[1 \pm \frac{(9 - P_R)^{1/2}}{6}\right]^2$$

Use this equation to plot the Joule-Thomson inversion curve as a function of T_R and P_R.

20. Evaluate Eq. (8-189) for air at 300 K and $P = 1$, 10, 40, 70, 100 bars to show the pressure range over which Eq. (8-189) is useful.

21. Hydrogen is the major gas in the atmospheres of Jupiter, Saturn, Uranus, and Neptune. Compute f/P for H_2 at 200, 400, 600, 800, 1000, and 1200 bars. The virial coefficients for H_2 at 400 K are $B = 16.9$ cm^3 mol^{-1} and $C = 290$ cm^6 mol^{-2}.

22. Sulfur dioxide and sulfur vapor are the two major constituents of volcanic gases on Io, the innermost Galilean satellite of Jupiter, and the volcanically most active object in the solar system. Use Eq. (8-191) with $B = -125.8$ cm^3 mol^{-1} and $C = 0$ to calculate f/P for SO_2 from 1 to 100 atmospheres at 200°C. The following table gives some PVT data for SO_2 from Kang et al. (1961).

P (atm)	**1**	**10**	**25**	**50**	**75**	**100**
$Z = PV_m/RT$	0.9967	0.9671	0.9157	0.8208	0.7134	0.5891

23. Start with Eq. (8-206) and write down the equation for B_{mix} for a ternary gas mixture.

24. Imagine a Mars-like exoplanet with an atmosphere composed of CO_2 (85.7%), N_2 (7.7%), and Ar (6.6%) and surface $T = 30$°C and $P = 10$ bars. Calculate the mass density of the atmosphere using the virial equation in terms of pressure. Use only the second virial coefficient B_{mix}, which you must calculate, using the values (cm^3 mol^{-1}) listed here.

CO_2 $B_{11} = -119.2$ N_2 $B_{22} = -4.0$ Ar $B_{33} = -14.9$ CO_2-N_2 $B_{12} = -41.4$

CO_2-Ar $B_{13} = -31.8$ N_2-Ar $B_{23} = -9.88$

25. Calculate the Redlich-Kwong coefficients for PH_3.

26. Use the lever rule to calculate the proportions of gaseous and liquid CO_2 at molar volumes of 0.1 and 0.2 L mol^{-1} on the 290 K tie line in Figure 8-18.

27. Calculate $\Delta_{vap}H$ of liquid CO_2 using the Clapeyron equation and the following PVT data from Michels et al. (1937). The densities of liquid and gaseous CO_2 are given in Amagat units.

°C	2.853	10.822	19.874	25.070	25.298	28.052	29.929	30.409	31.013
P (atm)	36.997	45.261	56.268	63.456	63.791	68.017	71.002	71.780	72.797
ρ (liq)	461.8	433.6	393.0	359.8	357.7	331.7	301.0	290.3	259.0
ρ (gas)	54.04	70.21	97.62	122.67	124.47	146.46	172.3	182.22	216.4

28. Calculate $\Delta V = V_{mag} - V_{per}$ at each of the five P-T points studied by Harker and Tuttle (1955) using the following molar volume (V^o_{298}), thermal expansivity (αV), and isothermal compressibility (βV) data for magnesite and periclase (Holland and Powell, 1990).

mineral	V^o_{298} (J bar^{-1})	$10^5 \times \alpha V$ (J bar^{-1} K^{-1})	$10^6 \times \beta V$ (J bar^{-2})
MgO (per)	1.125	4.6	0.7
$MgCO_3$ (mag)	2.803	10.6	3.0

CHAPTER

The Third Law of Thermodynamics

9

If the entropy of each element in some crystalline state be taken as zero at the absolute zero of temperature: every substance has a finite positive entropy, but at the absolute zero of temperature the entropy may become zero, and does so become in the case of perfect crystalline substances.

—Lewis and Randall (1923)

The third law of thermodynamics is the subject of this chapter. The third law is important because it states that entropy is an absolute quantity that can be determined from calorimetric measurements of the heat capacity and enthalpies of transition of a material such as forsterite. We can thus compute the Gibbs free energy of a chemical reaction using only thermochemical data (the absolute entropies and standard enthalpies of formation for reactants and products).

There are four sections in this chapter. Section I describes the historical development of the Nernst heat theorem, which is the forerunner of the third law of thermodynamics. Section II states the third law and examines some of its important consequences. Section III discusses the third law and entropy and describes the different contributions to the total entropy of a material. Section IV discusses calculation of absolute entropies from heat capacity data, introduces the Gibbs function $[(G_T^o - H_{298}^o)/T]$, and describes the use of the third law for computing Gibbs free energies of reaction from calorimetric data.

I. HISTORICAL DEVELOPMENT OF THE NERNST HEAT THEOREM

During the latter half of the 19th century, chemists began to use the first and second laws of thermodynamics in their research. They wanted to predict the direction and driving force of chemical reactions and the extent to which different reactions would occur. Once this information was available for a particular reaction, the optimal pressure and temperature could be used to maximize the amount of a desired product. One example is the synthesis of ammonia, which is essential for production of synthetic fertilizers to increase agricultural production. Initially, chemists thought that the enthalpy of reaction could be used to predict which reactions would occur. Everyday experience showed that many spontaneous reactions, such as coal burning in air, acids dissolving in water, metals dissolving in acids, or explosions, produced heat. Thus, the French chemist Marcellin Berthelot and the Danish chemist Julius Thomsen undertook extensive measurements of the heats of formation for thousands of substances (see their biographical sidebars in Chapter 5). In the 1850s, Thomsen proposed that all spontaneous reactions generate heat. In the 1860s, Berthelot proposed a similar concept, the *principle of maximum work*, which stated that all chemical reactions accomplished without the intervention of external energy move toward the production of the substance or system of substances that release the most heat.

Copyright © 2013 Elsevier Inc. All rights reserved.

However, this principle is incorrect because some spontaneous reactions absorb heat from the surroundings (i.e., they are endothermic), yet they still occur. Endothermic reactions include the solution of ammonium chloride (NH_4Cl) or table salt (NaCl) in water, dissociation of nitrogen tetroxide gas (N_2O_4) to two NO_2 (g) molecules at room temperature, ice melting to liquid water, sublimation of volatile solids like iodine, and limestone ($CaCO_3$) calcination to lime (CaO) plus carbon dioxide during cement manufacture.

At about the same time as Berthelot proposed the principle of maximum work, the German chemist August Friedrich Horstmann (1842–1929) noted that chemical reactions that reach equilibrium before consuming all the reactants are endothermic reactions. In the 1860s, Berthelot and another French chemist, St. Gilles, studied the endothermic reaction of ethyl acetate and liquid water to produce ethyl alcohol (C_2H_5OH) and acetic acid (CH_3COOH):

$$CH_3COOC_2H_5 \text{ (ethyl acetate)} + H_2O = CH_3COOH \text{ (acetic acid)} + C_2H_5OH \text{ (ethanol)} \quad (9\text{-}1)$$

The forward reaction is endothermic ($\Delta_r H^o_{298} = +3.7$ kJ mol^{-1}) and the reverse reaction is exothermic ($\Delta_r H^o_{298} = -3.7$ kJ mol^{-1}). However, Berthelot did not appreciate the important point raised by Horstmann, and he continued to argue for the principle of maximum work.

In the 1870s and 1880s, Willard Gibbs and Hermann von Helmholtz (see biographical sidebars in Chapters 3 and 7, respectively) showed that free energy (and not enthalpy) is the true measure of the thermodynamic favorability of chemical reactions. Their results are expressed by

$$\Delta G = \Delta H - T\Delta S \quad (6\text{-}116)$$

Equation (6-116) shows that both the enthalpy (ΔH) and entropy (ΔS) changes in a chemical reaction contribute to the driving force. We used Eq. (6-116) in Example 6-17 when we computed the free energy of forsterite formation from the oxides. We rewrote Eq. (6-116) in Chapter 7 and used it to compute the position of high-pressure phase boundaries such as the graphite-diamond phase boundary. The rewritten version, Eq. (7-138), gives

$$(\Delta G)^P_T = \Delta H^o_{298} + \int_{298}^{T} \Delta C_P dT - T\Delta S^o_{298} - T\int_{298}^{T} \frac{\Delta C_P}{T} dT + \int_{P=1}^{P=P} \Delta V dP \quad (7\text{-}138)$$

The standard Gibbs free energy change as a function of temperature at one bar pressure is thus

$$\Delta G^o_T = \Delta H^o_{298} + \int_{298}^{T} \Delta C_P dT - T\Delta S^o_{298} - T\int_{298}^{T} \frac{\Delta C_P}{T} dT \quad (9\text{-}2)$$

Equation (9-2) shows that ΔG^o at any temperature is easily calculated from the enthalpy (ΔH^o) and entropy changes (ΔS^o) at 298 K and the difference of the heat capacities of products and reactants (ΔC_P). An equivalent expression is

$$\Delta G^o_T = \Delta G^o_{298} + \int_{298}^{T} \Delta C_P dT - T\int_{298}^{T} \frac{\Delta C_P}{T} dT \quad (9\text{-}3)$$

Equation (9-3) shows that ΔG^o at any temperature is easily calculated from ΔG^o at 298 K and ΔC_P for the reaction. Thus, once we know ΔS^o at 298 K (or some other temperature used as one of the integration limits) we can find ΔG^o at any temperature. Today we typically calculate ΔS^o by looking up the entropies of reactants and products in a table and finding the difference between them (e.g., see Example 6-17). This is only possible because of the third law of thermodynamics, which we take for granted. However, the third law was unheard of in the late 19th and early 20th centuries. The ΔS^o term was an unknown integration constant to chemists of that time. The only way to determine the entropy change for a reaction was to measure both ΔG^o and ΔH^o for the reaction and calculate ΔS^o from

$$\Delta S^o = \frac{\Delta H^o - \Delta G^o}{T} \tag{9-4}$$

Example 6-18 illustrates this type of calculation. The calculated ΔS^o value typically has large uncertainties because the numerator ($\Delta H^o - \Delta G^o$) is a small number comparable in size to the uncertainties in the free energy and enthalpy.

From about 1880 to 1905, a number of great physical chemists, including Henri Le Chatelier, Fritz Haber, J. H. van't Hoff, Theodore W. Richards, and Walther Nernst, considered the question of measuring or calculating the entropy of reaction. Le Chatelier and Haber were unsuccessful, Richards came close to the answer, and Nernst finally solved the problem.

Theodore W. Richards (1868–1928) was the first American to win the Nobel Prize in Chemistry (1914) for his determinations of atomic weights of the elements. In 1902, he considered the free energy ΔG^o and enthalpy ΔH^o for electrochemical reactions such as

$$\text{Fe (metal)} + CuSO_4 \cdot 200\ H_2O\ \text{(aq)} = FeSO_4 \cdot 200\ H_2O\ \text{(aq)} + \text{Cu (metal)} \tag{9-5}$$

Reaction (9-5), and similar reactions involving other pairs of metals and sulfates formed by Zn, Ni, Fe, Cu, and Mg, occurs in galvanic cells known as *Daniell cells.* (A *galvanic cell* is essentially a battery. It uses a chemical reaction to provide electrical energy.) Richards's work was stimulated by the fact that Berthelot's principle is approximately correct for reactions between solids and liquids at room temperature (e.g., in electrochemical cells), and that

$$\Delta G^o - \Delta H^o = -T\Delta S^o \cong 0 \tag{9-6}$$

for these reactions. Richards (1902) calculated the values of ΔG^o and ΔH^o at 291 K and plotted the temperature coefficients for ΔG^o and ΔH^o from 273 K to 323 K. The temperature coefficient of the Gibbs energy of reaction is opposite in sign to the entropy of reaction (see Eq. 7-28):

$$\left(\frac{\partial \Delta G}{\partial T}\right)_P = -\Delta S \tag{9-7}$$

The temperature coefficient of the enthalpy of reaction is the ΔC_P for the reaction (see Eq. 5-25):

$$\left(\frac{\partial \Delta H}{\partial T}\right)_P = \Delta C_P \tag{5-25}$$

Richards found that in many cases the free energy and enthalpy of reaction approached each other as temperature decreased. In other words, the $T\Delta S^o$ term in Eq. (9-6) became smaller with decreasing temperature. Richards also calculated that the temperature coefficients of ΔG^o and ΔH^o have opposite signs. Finally, he noted that the plots of ΔG^o and ΔH^o versus temperature are slightly curved. Unfortunately, he did not fully appreciate the significance of his results and did not take the final step that led to formulating the third law.

This was done shortly thereafter, on December 23, 1905, when the German physical chemist Walther Nernst (see sidebar) publicly read his paper "On the calculation of chemical equilibria from thermal measurements" at the Göttingen Academy. He proposed the *Nernst heat theorem* in this paper. Nernst, like Richards, realized that Berthelot's principle contained a kernel of truth. In Nernst's own words: "We are dealing with a law more or less approximate at ordinary temperature, but true in the neighborhood of absolute zero." His breakthrough was to postulate that the free energy and enthalpy of reaction for chemical reactions between solids and liquids tangentially approach the same value at absolute zero (see Figure 9-1). This means that the slopes of ΔG and ΔH approach zero asymptotically as the temperature approaches 0 K:

$$\text{limit}(\text{as } T \rightarrow 0)\left(\frac{\partial \Delta \text{G}}{\partial \text{T}}\right)_P = \text{limit}(\text{as } T \rightarrow 0)\left(\frac{\partial \Delta \text{H}}{\partial \text{T}}\right)_P = 0 \quad (9\text{-}8)$$

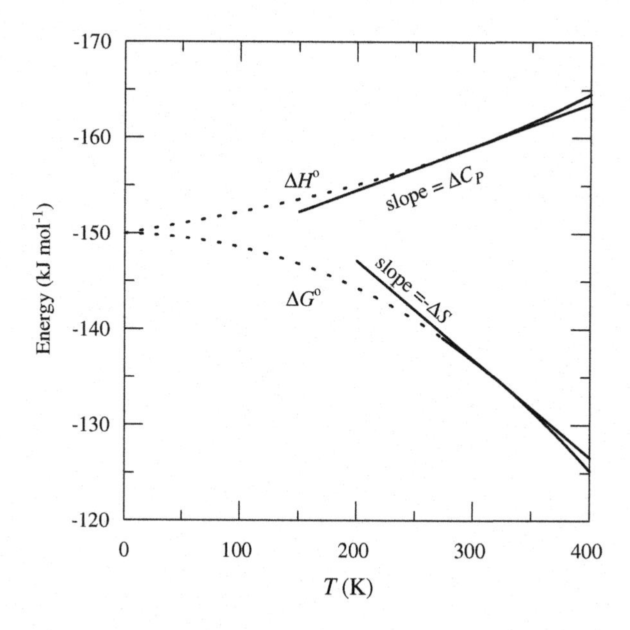

FIGURE 9-1

The asymptotic approach of ΔH^o and ΔG^o to the same value at absolute zero. The temperature coefficients of ΔH^o and ΔG^o are ΔC_P and $-\Delta S^o$, respectively.

HERMANN WALTHER NERNST (1864-1941)

Walther (Walter) Nernst and Fritz Haber were the two preeminent physical chemists in Germany during the first decades of the Twentieth Century. Nernst studied under Helmholtz and other leading scientists and in 1887 became an assistant to Wilhelm Ostwald, who held the world's only professorship in physical chemistry at the University of Leipzig. During this time, Nernst developed the Nernst equation, which relates the emf of a galvanic cell to the equilibrium constant of the chemical reaction taking place inside the cell. In 1889, Nernst invented the concept of the solubility product. In 1891, he developed the Nernst distribution law, which is the basis of the partition coefficients used in geochemistry. Nernst moved to the University of Göttingen in 1890 and became the first professor of physical chemistry there in 1894. He stayed in Göttingen for 10 years. While there, he invented an electric light using a glowing zirconia rod as the light source. Nernst sold the patent rights for one million marks and became a wealthy man. Laboratory infrared spectrometers use the Nernst glower as the IR source. Nernst became professor of physical chemistry at the University of Berlin in 1904. He stayed in Berlin for the rest of his career until he retired in 1933. During his time in Berlin Nernst developed methods for measuring pH with colored indicators (e.g., litmus paper), the degree of ionic hydration in aqueous solutions, and the use of buffer solutions for controlling pH. Just over 100 years ago in 1905, he developed the Nernst Heat Theorem, which is the forerunner of the Third Law of Thermodynamics. Nernst received the 1920 Nobel Prize in Chemistry for the Heat Theorem. However, the Third Law remained controversial until proven by W. F. Giauque's experimental studies, which led to his 1949 Nobel Prize in Chemistry.

Equation (9-8) is a mathematical statement of the Nernst heat theorem. Equations (5-25) and (9-7) give the temperature coefficients of ΔG and ΔH. Thus, the Nernst heat theorem implies

$$\text{limit}(\text{as } T \to 0)\Delta C_p = 0 \tag{9-9}$$

$$\text{limit}(\text{as } T \to 0)\Delta S = 0 \tag{9-10}$$

MAX CARL ERNST LUDWIG PLANCK (1858–1947)

Max Planck was one of the two greatest German physicists of the 20th century (Albert Einstein was the other). Fortunately for physics, Planck chose science over a career in music because he delighted in the search for "something absolute." Planck studied under Helmholtz and Kirchhoff and wrote his doctoral thesis on the second law of thermodynamics (1879). After holding academic positions in Munich and Kiel, Planck became a physics professor in Berlin, where he stayed for the rest of his career.

While in Munich and Kiel, Planck wrote the first edition of his thermodynamics textbook. This book went through nine editions over the next 30 years. During 1897–1901, Planck wrote a series of papers about black-body radiation, which is the radiation emitted from the cavity of a heated object. He explained the energy of the emitted radiation as a function of frequency and temperature by inventing quantum theory. Planck won the Nobel Prize for Physics in 1918 for this work. Ironically, Planck never fully accepted the Heisenberg uncertainty principle, which is a consequence of quantum theory. (The uncertainty principle states that it is impossible to measure simultaneously the position and momentum of subatomic particles because the measurement perturbs the particle's properties.) Planck generalized Nernst's heat theorem by postulating that the entropy of all materials is zero at absolute zero.

Planck remained in Germany after the Nazis came to power in 1933. He witnessed the destruction of the field of physics in Germany as the Nazis drove Einstein and other famous Jewish physicists out of the country. In 1944, Allied bombing destroyed his Berlin home. One of his sons took part in the July 1944 plot to kill Hitler and was murdered by the Nazis in early 1945. Planck survived the Second World War and died in 1947. Today the Max Planck Institute system of scientific research institutes in Germany honors Max Planck's memory.

Equations (9-9) and (9-10) say that the change in heat capacity (ΔC_p) and the entropy change (ΔS) of each reaction approach zero as temperature decreases to absolute zero. This is the key point, which Nernst stated as: "For all reactions involving substances in the condensed state (i.e., liquids, solids), ΔS is zero at the absolute zero of temperature."

The heat theorem allowed Nernst to calculate chemical equilibria at high temperatures from thermal data alone. He had to make several drastic assumptions about heat capacities to do this, because very little data on low-temperature heat capacities of solids were available in 1905. In addition, Nernst assumed that all atomic heat capacities had the value of 1.5 calories (~6.3 J) per mole per degree at 0 K, and that the Neumann-Kopp rule gave the heat capacities of compounds. Nevertheless, his work was a major advance at the time and was of importance for the chemical industry. Over the next decade, the

Nernst heat theorem became what we now know as the third law of thermodynamics. This transformation occurred as Nernst, Einstein, Planck, Debye, G. N. Lewis, and other scientists broadened the heat theorem beyond its original scope. Lewis and Randall (1923), Nernst (1926), Planck (1927), Simon (1956), and Wilks (1961) give accounts of the evolution of the Nernst heat theorem into the third law.

II. THE THIRD LAW AND ITS CONSEQUENCES

A. The third law of thermodynamics

The third law says that entropy is an absolute quantity and that a pure, perfectly ordered material without any internal degrees of freedom has no entropy left at absolute zero:

$$S^o(0\ \text{K}) = 0\ \text{J mol}^{-1}\ \text{K}^{-1} \tag{6-73}$$

Thus, the entropy of a perfect crystalline solid, perfectly ordered liquid helium, or an ideal gas goes to zero at absolute zero (see Simon, 1956; Wilks, 1961; Lewis and Randall, 1923). We can compute the absolute entropy of a material by measuring its heat capacity as a function of temperature and the enthalpies of phase transitions it undergoes. Example 6-10 illustrates this process with an entropy calculation for Fe gas at high temperature. The entropy values can then be combined with calorimetrically measured enthalpy values to calculate ΔG^o values for chemical reactions using Eq. (9-2) or Eq. (9-3).

B. Entropy and heat capacity of solids at low temperatures

When Nernst proposed the heat theorem in 1905, virtually nothing was known about the heat capacities of solids below room temperature. Sir James Dewar and some other workers had measured heat capacities down to the temperature of liquid air (about 90 K), but the data were averaged over rather wide temperature ranges and only a few materials had been studied. Dewar had liquefied hydrogen only seven years earlier (1898). Onnes's great achievement of liquefying helium was three years in the future (1908). No one knew that heat capacity and entropy would go to zero at absolute zero because there were virtually no experimental data available. However, two events changed the situation dramatically.

The first event was Nernst's invention of the vacuum calorimeter. This is the forerunner of the modern adiabatic low-temperature calorimeter illustrated in Figure 4-8. (We described low-temperature adiabatic calorimeters in Section II-E of Chapter 4, and you should go back and reread that discussion if you need to refresh your memory.) Nernst and his colleagues used the vacuum calorimeter for heat capacity measurements of elements and minerals down to liquid hydrogen temperatures (~20 K).

Second, in 1907 Albert Einstein published his model of low-temperature heat capacity of solids. We discussed this in Section III-C of Chapter 4 and showed that it qualitatively explained the dramatic decrease of the heat capacity of diamond with decreasing temperature (see Figure 4-16). However, Einstein's theory also predicts that the constant-volume heat capacity C_V and the entropy S asymptotically approach zero as temperature approaches absolute zero. This is a revolutionary conclusion with important implications in several areas of chemistry and physics.

The lattice heat capacity C_{lat} of an Einstein solid is

$$C_{lat} = 3R\frac{u^2e^u}{(e^u - 1)^2} = 3Rf_E \tag{4-28}$$

In Eq. (4-28) e is an exponential and $u = h\nu/kT = \theta_E/T$, where h is Planck's constant, ν is the characteristic vibrational frequency of the solid, k is Boltzmann's constant, T is the absolute temperature, and θ_E is the *Einstein temperature*. In the case of diamond, the lattice heat capacity is the only contributor to the constant-volume heat capacity C_V. Figure 4-16 illustrates the asymptotic approach of C_V to zero for diamond. An Einstein temperature $\theta_E = 1329$ K was used to compute this figure. The entropy of an Einstein solid can be computed from Eq. (6-57):

$$\left(\frac{\partial S}{\partial T}\right)_V = \frac{C_V}{T} \tag{6-57}$$

Rearranging and integrating from 0 K to a temperature T, we get

$$S_T - S_0 = \int_0^T C_V \frac{dT}{T} \tag{9-11}$$

Substituting Eq. (4-28) into Eq. (9-11) gives the equation for the entropy of an Einstein solid:

$$S_T = 3R\left[\frac{u}{(e^u - 1)} - \ln\left(1 - e^{-u}\right)\right] \tag{9-12}$$

The entropy at zero K (S_0) does not appear in Eq. (9-12) because the exponential terms approach zero as the temperature T approaches zero. The following example, which shows some of the calculations for the Einstein curve in Figure 9-2, illustrates this behavior.

Example 9-1. Figure 9-2 shows the entropy of diamond from 0 K to 1200 K. The Einstein curve was computed using $\theta_E = 1329$ K and the results for some points are tabulated here.

					Entropy (cal mol^{-1} K^{-1})	
T(K)	θ_E/T	$\frac{u}{(e^u-1)}$	$-\ln(1-e^{-u})$	$\frac{S}{3R}$	Einstein	Observed
1200	1.1075	0.54643	0.40105	0.94748	5.648	5.73
600	2.215	0.2714	0.11558	0.38698	2.307	2.42
298	4.4575	0.05228	0.01166	0.06394	0.381	0.57
50	26.58	7.6×10^{-11}	2.9×10^{-12}	7.9×10^{-11}	4.7×10^{-10}	0.0054
15	88.6	3×10^{-37}	~0	3×10^{-37}	2×10^{-36}	0.0002

The results for diamond show that the S_0 term is zero, that is, the entropy of an Einstein solid is zero at absolute zero. However, we also see that Einstein's model qualitatively explains the shape of the entropy curve but underestimates the low-temperature entropy. Figure 4-16 shows similar behavior for Einstein's model of the heat capacity of diamond, and Example 4-8 shows similar behavior for the low-temperature heat capacity of silver metal.

In 1912, Peter Debye modified Einstein's model to include a range of vibrational frequencies and coupled oscillations of the atoms in a solid (see Chapter 4). Figures 4-16 and 9-2 show that the Debye

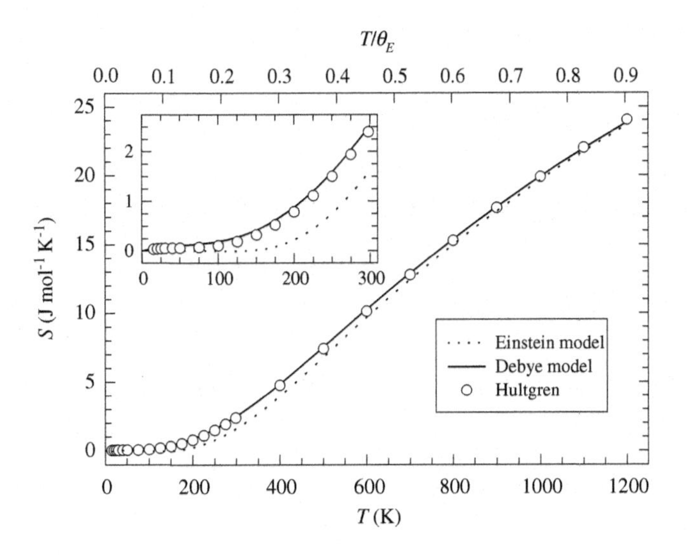

FIGURE 9-2

A comparison of the entropy of diamond from 0 K to 1200 K (Hultgren et al., 1973) with predictions of the Einstein and Debye models. The corresponding heat capacity data are shown in Figure 4-16.

model gives a better fit to the observed heat capacity and entropy of diamond. The Debye model also predicts that the lattice heat capacity and entropy of solids approach zero as temperature approaches absolute zero. At low temperatures, the lattice heat capacity and entropy of a Debye solid are given by the equations

$$C_V = \frac{12\pi^4}{5}R\left(\frac{T}{\theta_D}\right)^3 = 1943.8\left(\frac{T}{\theta_D}\right)^3 = aT^3 \text{ J (g atom)}^{-1}\text{ K}^{-1} \quad (4\text{-}30)$$

$$S = \int_0^T C_V\frac{dT}{T} = \int_0^T aT^3\frac{dT}{T} = \frac{a}{3}T^3 \text{J (g atom)}^{-1}\text{ K}^{-1} \quad (9\text{-}13)$$

The Debye temperature is θ_D and a is a constant equal to $(1943.8/\theta_D^3)$ J (g atom)$^{-1}$ K^{-4}. Equation (4-30) is the *Debye T^3 law*, which is usually valid for $T < \theta_D/50$, which is below 10 K for many materials. If C_V is only due to the lattice heat capacity, the entropy of a solid in the T^3 region is

$$S_T^o = \frac{1}{3}C_V \cong \frac{1}{3}C_P \quad (9\text{-}14)$$

The approximation in Eq. (9-14) arises because the constant-volume and constant-pressure heat capacities are slightly different and are related to one another via

$$(C_P - C_V)_m = \frac{\alpha^2 V_m T}{\beta_T} \quad (4\text{-}46)$$

where α is the isobaric thermal expansion coefficient, β_T is the isothermal compressibility, V_m is the molar volume, and T is the absolute temperature. We usually do not know α and β_T as a function of temperature, so an approximate relationship

$$(C_P - C_V)_m = A(C_P)^2 T \tag{4-47}$$

which is the *Nernst-Lindemann equation*, is often used instead of Eq. (4-46). The empirical constant A is evaluated from Eq. (4-46) at one temperature, usually room temperature, and is remarkably constant with temperature. In the T^3 region, the difference between C_P and C_V is effectively zero, as we illustrate here.

Example 9-2. At 15 K diamond has $C_P = 0.00084$ J mol^{-1} K^{-1} (Hultgren et al., 1973). Show that $C_P - C_V$ is a small fraction of C_P at 298 K and is effectively zero at 15 K, in the T^3 region.

Tables 2-1, 2-12, and 7-1 give the data for calculating $C_P - C_V$ at 298 K from Eq. (4-46):

$$(C_P - C_V)_m = \frac{(0.88 \times 10^{-5}\ \text{K}^{-1})^2 (0.3417\ \text{J mol}^{-1}\ \text{bar}^{-1})(298.15\ \text{K})}{2.2 \times 10^{-7}\ \text{bar}^{-1}} = 0.036\ \text{J mol}^{-1}\ \text{K}^{-1} \tag{9-15}$$

This difference is about 0.6% of $C_P = 6.109$ J mol^{-1} K^{-1} for diamond at 298 K. Note the molar volume is in J mol^{-1} bar^{-1}. The empirical constant A in the Nernst-Lindemann equation is

$$A = \frac{C_P - C_V}{TC_P^2} = \frac{0.036\ \text{J mol}^{-1}\ \text{K}^{-1}}{(298.15\ \text{K})(6.109\ \text{J mol}^{-1}\ \text{K}^{-1})^2} = 3.2 \times 10^{-6}\ \text{mol J}^{-1} \tag{9-16}$$

The calculated $C_P - C_V$ at 15 K is

$$C_P - C_V = A(C_P)^2 T = 3.2 \times 10^{-6}(0.0084)^2(15) \cong 3.4 \times 10^{-11}\ \text{J mol}^{-1}\ \text{K}^{-1} \tag{9-17}$$

The difference between C_P and C_V at 15 K is 4×10^{-8} of the C_P value and is effectively zero.

C. Experimental verification: Calorimetric entropies of gases

The third law was not universally accepted until overwhelming experimental evidence showed that all substances studied followed it. The experimental verification was done by comparing calorimetric entropies of gases with entropies calculated for the same gases from physical properties of their molecules using statistical mechanics. Several books (Mayer and Mayer, 1940; Pitzer and Brewer, 1961; Kelley and King, 1961) explain the general computational methods. Details of the calculations for particular gases are given in the references cited in the following examples. However, the key point is that the statistical entropy values are computed starting from zero entropy at absolute zero. Thus, the calorimetric and statistical entropy values only agree if the third law is correct. The American physical chemist William F. Giauque (1895–1982) won the 1949 Nobel Prize in Chemistry for his verification of the third law (see sidebar). Table 9-1 summarizes the third law entropies for gases at 298 K and shows the good agreement between the measured (calorimetric) values and the calculated (statistical) values.

WILLIAM FRANCIS GIAUQUE (1895–1982)

William F. Giauque was born in Niagara Falls, Canada. He started work at the Hooker Electrochemical Company in 1914 and became interested in chemical engineering. Giauque entered the University of California at Berkeley in 1916 to take their program for chemists and engineers and spent the rest of his career there. As an undergraduate, Giauque discovered that he preferred chemistry to engineering and graduated with a B.S. in chemistry in 1920. He received his Ph.D. two years later. His doctoral research was a study of the low-temperature heat capacity and entropy of crystalline and glassy glycerin. This classic study is the first in a 50-year series of experiments that Giauque did to verify the third law of thermodynamics.

Giauque became an instructor in chemistry in 1922 and a full professor in 1934. In 1927, he proposed the concept of magnetic cooling in his second paper. Magnetic cooling uses a strong magnetic field to orient the magnetic dipoles in a paramagnetic material cooled to liquid helium temperatures. This reduces the magnetic entropy of the material and causes a drop in temperature upon removal of the strong magnetic field. Peter Debye independently proposed the same method at about the same time. The race was then on to be the first to demonstrate magnetic cooling.

Giauque was competing against scientists at low-temperature laboratories in Leiden and Oxford. Over the next six years, Giauque designed, constructed, and installed the heavy equipment needed for producing the necessary low-temperatures and strong magnetic fields. On March 19, 1933, Giauque and D. P. MacDougall performed the first successful magnetic cooling experiment and reached 0.53 K. They had won the race by less than one month. On April 6, 1933, the scientists in Leiden did their first magnetic cooling experiment and reached 0.27 K. Giauque received the 1949 Nobel Prize in Chemistry for his verification of the third law and for the invention of magnetic cooling.

Example 9-3. Argon is the eleventh most abundant element in the solar system and is the most abundant noble gas in the atmospheres of Earth, Venus, Mars, Jupiter, and Titan. Compare the calorimetric entropy of Ar gas at its one atmosphere boiling point (87.3 K) to that calculated for the ideal gas using statistical mechanics (129.2 ± 0.04 J mol^{-1} K^{-1}).

The calorimetric entropy of Ar gas at its one atmosphere boiling point is the sum of the entropy contributions of Ar ice, the entropy of melting, Ar liquid, the entropy of vaporization to the real gas, and the entropy difference between the real and ideal gas. We take all data from Hultgren et al. (1973). The calorimetric data only extend to 2 K, so a Debye T^3 law extrapolation is necessary at lower

Table 9-1 Third Law Entropies of Gases (298.15 K)[a]

	S^o (J mol^{-1} K^{-1})	
Gas	**Calculated**[b]	**Measured**[c]
Ar	154.72 ± 0.04	154.6 ± 0.8
AsH_3	222.7 ± 0.4	222.6 ± 0.4
Br_2	245.4 ± 0.1	245.2 ± 1.7
CH_4	186.2 ± 0.4	186.2 ± 0.8
CO_2	213.7 ± 0.4	213.8 ± 0.4
Cl_2	222.97 ± 0.04	223 ± 0.4
H_2	130.58 ± 0.04	131.1 ± 0.6
HBr	198.7 ± 0.2	199.1 ± 0.6
HCl	186.8 ± 0.2	186.2 ± 0.6
HI	206.4 ± 0.4	206.9 ± 0.6
H_2S	205.6 ± 0.4	205.4 ± 0.4
Kr	163.97 ± 0.04	163.9 ± 0.4
N_2	191.5 ± 0.04	192.2 ± 0.6
NH_3	192.5 ± 0.2	192.3 ± 0.4
Ne	146.23 ± 0.04	146.5 ± 0.4
O_2	205.06 ± 0.04	205.4 ± 0.4
OCS	231.5 ± 0.4	231.2 ± 0.4
PH_3	210.2 ± 0.4	210.7 ± 0.4
SO_2	248.1 ± 0.4	247.9 ± 0.4
Xe	169.58 ± 0.04	170.3 ± 1.3

[a]*All data are from Kelley and King (1961).*
[b]*Statistical value for the ideal gas computed from spectroscopic data.*
[c]*Calorimetric (third law) value for the ideal gas.*

temperatures. Example 4-9 gives $a = 2.8467\times10^{-3}$ J (g atom)$^{-1}$ K^{-4} for Ar. The entropy in the 0–2 K range from the Debye T^3 extrapolation is

$$S_2^o = \frac{a}{3}T^3 = \frac{2.8467}{3} \times 10^{-3}(2^3) = 0.008 \text{ J mol}^{-1} \text{ K}^{-1} \quad (9\text{-}18)$$

The entropy for Ar ice in the 2–83.81 K range (its triple point) is

$$\Delta S^o = S_{83.81}^o - S_2^o = \int_2^{83.81} \frac{C_P(\text{Ar, ice})}{T} dT = 38.854 \text{ J mol}^{-1} \text{ K}^{-1} \quad (9\text{-}19)$$

The entropy of melting at 83.81 K is computed from the enthalpy of melting (1188 J mol^{-1}):

$$\Delta_{fus}S^o = \frac{\Delta_{fus}H^o}{T} = \frac{1188}{83.81} = 14.175 \text{ J mol}^{-1} \text{ K}^{-1} \quad (9\text{-}20)$$

The entropy for liquid Ar from the triple point to the one atmosphere boiling point (87.30 K) is

$$\Delta S^o = S_{87.3} - S_{83.81} = \int_{83.81}^{87.3} \frac{C_P(\text{Ar, liq})}{T} dT = 1.77 \text{ J mol}^{-1} \text{ K}^{-1} \quad (9\text{-}21)$$

The enthalpy of vaporization (6447.5 J mol^{-1}) at 87.3 K gives an entropy of vaporization of

$$\Delta_{vap} S^o = \frac{\Delta_{vap} H^o}{T} = \frac{6447.5}{87.3} = 73.855 \text{ J mol}^{-1} \text{ K}^{-1} \quad (9\text{-}22)$$

Hultgren et al. (1973) give the ΔS between real and ideal Ar gas as 0.544 J mol^{-1} K^{-1} at one atmosphere pressure. The calorimetric entropy of ideal Ar gas at its one atmosphere boiling point is the sum of all these entropy contributions:

$$\begin{aligned} S^o_{87.3}(\text{Ar, ideal gas}) &= 0.008 + 38.854 + 14.175 + 1.77 + 73.855 + 0.544 \\ &= 129.206 \text{ J mol}^{-1} \text{ K}^{-1} \end{aligned} \quad (9\text{-}23)$$

This is identical within error to the entropy of 129.2 ± 0.04 J mol^{-1} K^{-1} calculated for ideal Ar gas using statistical mechanics.

In the preceding example, we used the ΔS between real and ideal gas given by Hultgren et al. (1973). In general, the ΔS between the real and ideal gas is the difference between expansion of the real gas from one atmosphere (or one bar) to zero pressure and compression of the real gas from zero pressure to the same initial pressure. The ΔS is calculated with the Maxwell relation

$$\left(\frac{\partial V}{\partial T}\right)_P = -\left(\frac{\partial S}{\partial P}\right)_T \quad (8\text{-}45)$$

Giauque and his colleagues used the *Berthelot equation* to compute this entropy correction:

$$\Delta S = S_{ideal} - S_{real} = \frac{27 R T_c^3 P}{32 T^3 P_c} \quad (9\text{-}24)$$

The T_c and P_c in Eq. (9-20) are the critical temperature and pressure, respectively. Hultgren et al. (1973) used the virial equation to compute ΔS between real and ideal Ar gas. One of the other equations of state in Chapter 8 could be used, but the Berthelot equation is simpler and suffices in most cases.

Argon is a monatomic gas that has the ideal C_V and C_P values from its boiling point to very high temperatures (see Table 3-4). It is also important to determine if polyatomic gases that have greater deviations from ideality also obey the third law. We give examples for two other gases, CO_2 and Cl_2. Carbon dioxide sublimes at one atmosphere pressure and shows large deviations from ideality, as illustrated in Chapter 8. Molecular chlorine also shows large deviations from ideality and has significant vibrational contributions to its heat capacity at room temperature (ΔE_{vib} ~ 810 K).

Example 9-4. Carbon dioxide is the major gas in the Martian atmosphere and seasonally condenses into and sublimes from the planet's polar caps. Giauque and Egan (1937) measured the heat capacity C_P of solid CO_2 (dry ice) from 15 K to 194.67 K, its one atmosphere sublimation point. Figure 9-3 shows their C_P data plotted versus temperature (top-left panel). The

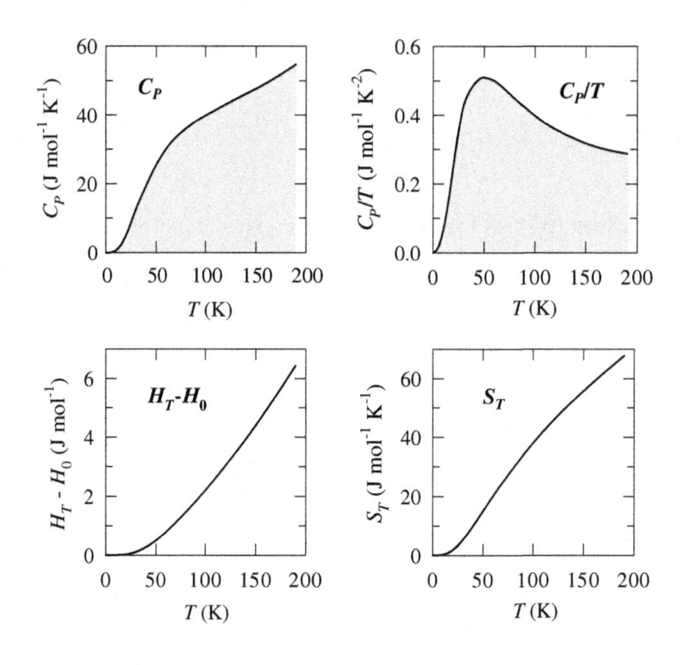

FIGURE 9-3

Heat capacity C_P, enthalpy ($H_T - H_0$), C_p/T, and entropy S for CO_2 from 0 K to 194.67 K from the data of Giauque and Egan (1937). The shaded area under the C_P curve is the enthalpy. The shaded area under the C_P/T curve is the entropy.

shaded area under the C_P curve is the enthalpy ($H_T - H_0$). The top-right panel of Figure 9-3 shows C_P/T plotted versus temperature. The shaded area under this curve is the entropy S. Use their data to compare the calorimetric entropy of CO_2 (g) to the statistical entropy for ideal CO_2 gas (198.95 ± 0.04 J mol^{-1} K^{-1}) at 194.67 K.

Table 9-2 summarizes the calculations for the calorimetric entropy. The heat capacity data only extend to 15 K, so a Debye T^3 extrapolation gives $S = 0.795$ J mol^{-1} K^{-1} from 0 K to 15 K. The

Table 9-2 Entropy of CO_2 Gas at 194.67 K

Entropy Contribution	S (J mol^{-1} K^{-1})
Debye extrapolation, 0–15 K	0.795
area under C_P/T curve, 15–194.67 K	68.325
sublimation, 194.67 K	129.601
correction for nonideality of CO_2 gas	0.367
calorimetric entropy of CO_2 (ideal gas)	199.09 ± 0.42[a]
statistical entropy of CO_2 (ideal gas)	198.95 ± 0.04

[a]*Uncertainty given by Giauque and Egan (1937).*

entropy from 15 K to 194.67 K is 68.325 J mol^{-1} K^{-1} and is the area under the C_P/T curve between these two temperatures (see Figure 9-3). Giauque and Egan (1937) measured the enthalpy of sublimation (6030 ± 5 cal mol^{-1}) at 194.67 K. This gives

$$\Delta_{sub}S = \frac{\Delta_{sub}H}{T_{sub}} = \frac{4.184 \times 6030}{194.67} = 129.601 \text{ J mol}^{-1}\text{ K}^{-1} \quad (9\text{-}25)$$

The ΔS due to nonideality of CO_2 is computed from the Berthelot equation:

$$\Delta S = S_{ideal} - S_{real} = \frac{27RT_c^3P}{32T^3P_c} = \frac{27(8.314)(304.1)^3(1)}{32(72.786)(194.67)^3} = 0.367 \text{ J mol}^{-1}\text{ K}^{-1} \quad (9\text{-}26)$$

The calorimetric entropy of CO_2 (ideal gas) at 194.67 K is 199.09 J mol^{-1} K^{-1}, versus 198.95 ± 0.04 J mol^{-1} K^{-1} for the statistical entropy. The difference of 0.14 J mol^{-1} K^{-1} is smaller than the uncertainty of ±0.42 J mol^{-1} K^{-1} on the calorimetric entropy (Giauque and Egan, 1937). Thus, the calorimetric and statistical entropy values are identical within their uncertainties.

Example 9-5. Chlorine may be present in low-temperature volcanic gases on Io, the most volcanically active body in the solar system. Giauque and Powell (1939) measured the heat capacity C_P of Cl_2 solid, liquid, and gas from 15 K to 239.05 K, its one atmosphere boiling point. Use their data to compare the third law entropy of Cl_2 (g) to the statistical entropy for the ideal gas (215.69 ± 0.04 J mol^{-1} K^{-1}) at 239.05 K.

Table 9-3 summarizes the third law (calorimetric) entropy calculations. The entropy from 0 K to 15 K is calculated using the Debye T^3 law. Chlorine melts at 172.12 K and the enthalpy of melting is 6405.8 J mol^{-1}. Chlorine boils at 239.05 K and the enthalpy of vaporization is 20,409.6 J mol^{-1}. The Berthelot equation gives $\Delta S = 0.491$ for the nonideality correction. The third law and statistical entropies of Cl_2 agree within 0.01 J mol^{-1} K^{-1}, which is smaller than the uncertainties on either value.

Table 9-1 gives a comparison of the third law (i.e., calorimetric or measured) and statistical (or calculated) entropies of common gases at 298 K, including those of Ar, CO_2, and Cl_2. This summary illustrates one of the most important experimental verifications of the third law.

Table 9-3 Entropy of Cl_2 Gas at 239.05 K

Entropy Contribution	S (J mol^{-1} K^{-1})
Debye extrapolation, 0–15 K	1.385
solid Cl_2, 15–172.12 K	69.341
melting, 172.12 K	37.217
liquid Cl_2, 172.12–239.05 K	21.887
vaporization, 239.05 K	85.378
correction for nonideality of Cl_2 gas	0.491
calorimetric entropy of Cl_2 (ideal gas)	215.7 ± 0.42[a]
statistical entropy of Cl_2 (ideal gas)	215.69 ± 0.04

[a] *Uncertainty given by Giauque and Powell (1939).*

D. Experimental verification: Calorimetric entropies of polymorphic solids

The thermodynamic properties of two different crystalline polymorphs (e.g., see Table 7-1) provide another test of the validity of the third law. Assume that we have two different phases of the same element or compound. The stable low-temperature phase α transforms into the stable high-temperature phase β at a transition temperature T_{trans}. We also assume that β can exist as a metastable phase at lower temperatures. At the transition temperature T_{trans}, the entropy of α is

$$S^o_{T_{tr}}(\alpha) = \frac{C_P(\alpha)}{3} + \int_{T_{low}}^{T_{trans}} C_P(\alpha)\frac{dT}{T} \quad (9\text{-}27)$$

The first term is the Debye T^3 extrapolation from 0 to T_{low}, the lowest temperature for calorimetric C_P measurements. This gives the entropy of α at T_{low} (see Eq. 9-14). The second term is the integrated area under the C_P/T curve from T_{low} to T_{trans}. This gives the entropy of α over this temperature range. The sum of the two terms is the entropy of α at T_{trans}. The entropy of the high-temperature phase β at the transition temperature is

$$S^o_{T_{tr}}(\beta) = \frac{C_P(\alpha)}{3} + \int_{T_{low}}^{T_{trans}} C_P(\alpha)\frac{dT}{T} + \Delta_{trans}S(\alpha \rightarrow \beta) \quad (9\text{-}28)$$

The first two terms are the same as in Eq. (9-27). The third term $\Delta_{trans}S$ is

$$\Delta_{trans}S = \frac{\Delta_{trans}H}{T_{trans}} \quad (6\text{-}67)$$

If the third law is valid, the entropies of α and β are both equal to zero at 0 K:

$$S^o_0(\alpha) = S^o_0(\beta) = 0 \quad (9\text{-}29)$$

Then the ΔS between α and β at T_{trans} must be exactly equal to the difference in the calorimetric entropies measured for α and β at T_{trans}. The next example illustrates this point.

Example 9-6. The Pele volcano on Io emits volcanic gas containing several percent S_2 vapor. The S_2 condenses on Io's surface as various forms of sulfur. Orthorhombic (rhombic) sulfur is stable at 298 K but transforms to another crystalline form, monoclinic sulfur, at 368.3 K. However, it is possible to supercool monoclinic sulfur below the 368.3 K transition point and measure its low-temperature heat capacity (e.g., Eastman and McGavock, 1937). Thus, the entropy of monoclinic sulfur at 368.3 K can be calculated from two independent thermochemical cycles. The first cycle involves heating rhombic sulfur from 0 K to 368.3 K and transforming it to monoclinic sulfur at that temperature. This cycle is represented by the changes in state:

$$\text{rhombic S (0 K)} = \text{rhombic S (368.3 K)} \quad (9\text{-}30)$$

$$\text{rhombic S (368.3 K)} = \text{monoclinic S (368.3 K)} \quad (9\text{-}31)$$

In this cycle, we calculate the entropy of monoclinic sulfur at 368.3 K from

$$S^o_{368.3}\ (\text{monoclinic}) = S^o_{12}\ (\text{rhombic}) + \int_{12}^{368.3} C_P\ (\text{rhombic})\frac{dT}{T} + \frac{\Delta_{trans}H}{T_{trans}} \quad (9\text{-}32)$$

$$S^o_{368.3} = 0.222 + 36.793 + \frac{400.4}{368.3} = 38.102\ \text{J mol}^{-1}\ \text{K}^{-1} \quad (9\text{-}33)$$

The second cycle involves heating metastable monoclinic sulfur from 0 K to 368.3 K and measuring its heat capacity as a function of temperature. The entropy from this cycle is

$$\begin{aligned} S^o_{368.3}\ (\text{monoclinic}) &= S^o_{12}\ (\text{monoclinic}) + \int_{12}^{368.3} C_P\ (\text{monoclinic})\frac{dT}{T} = 0.222 + 37.881 \\ &= 38.103\ \text{J mol}^{-1}\ \text{K}^{-1} \end{aligned} \quad (9\text{-}34)$$

The two entropy values agree within 0.001 J mol^{-1} K^{-1}, which is within the experimental uncertainty. Equations (9-32) and (9-33) were evaluated using data from the NIST-JANAF tables (Chase, 1999). The entropies at 12 K are from a Debye T^3 extrapolation to absolute zero.

Two other examples involve the two different crystalline phases of phosphine (PH_3) ice (Stephenson and Giauque, 1937) and cyclohexanol ice (Kelley, 1929). In both cases, the entropy of the high-temperature (β) phase calculated from two different thermodynamic cycles agrees within experimental uncertainties and shows that Eq. (9-29) is valid.

E. Clapeyron slope of melting curves

We derived the Clapeyron equation in Chapter 7. It applies to all first-order phase transitions such as melting and freezing, sublimation, vaporization, and polymorphic transitions:

$$\left(\frac{dP}{dT}\right)_{eq} = \frac{\Delta S}{\Delta V} = \frac{\Delta H}{T\Delta V} \quad (7\text{-}31)$$

(As discussed in Chapter 7, a first-order phase transition has discontinuities in entropy, volume, enthalpy, and heat capacity with $\Delta_{trans}S \neq 0$, $\Delta_{trans}V \neq 0$, $\Delta_{trans}H \neq 0$, $\Delta_{trans}C_P \neq 0$.) The subscript *eq* on the derivative in Eq. (7-31) specifies that the derivative is evaluated along equilibrium lines such as melting, sublimation, solid-state phase transition, and vaporization curves. By definition, $\Delta_{trans}V \neq 0$ for a first-order phase transition. Thus, the third law predicts that the Clapeyron slope goes to zero as temperature decreases to absolute zero where $\Delta S = 0$. The melting curves of ^{3}He and ^{4}He provide tests of this prediction.

F. Liquid helium

We discussed the two stable isotopes of helium, ^{3}He and ^{4}He, in Chapter 7 (Section IV-C). They are the only materials that remain liquid under their saturated vapor pressure down to absolute zero. The maximum densities of liquid ^{3}He and ^{4}He are 0.082 and 0.145 g cm^{-3}, respectively, significantly lower than expected if the He isotopes behaved as normal liquefied gases. The anomalous behavior of ^{3}He and ^{4}He occurs because the zero-point vibrational energy comprises most of their total energy near absolute zero. Quantum mechanical calculations (e.g., Pitzer, 1953) show that the average energy (ε) of an atom oscillating about its position in a solid or liquid is

$$\varepsilon = \frac{1}{2}h\nu_0 + \frac{h\nu_0}{e^{h\nu_0/kT} - 1} \tag{9-35}$$

The h is Planck's constant (6.626×10^{-34} J s) and ν_0 is the oscillation frequency (s^{-1}). At 0 K, ε is

$$\varepsilon = \frac{1}{2}\,h\nu_0 \tag{9-36}$$

Equation (9-36) is the *zero-point vibrational energy.* This energy is a consequence of Heisenberg's uncertainty principle, which states that we cannot simultaneously measure the position and momentum of an atom without changing either by a small amount.

In Section II-B of this chapter, we showed that the heat capacity of solids goes to zero at absolute zero. The presence of the zero point vibrational energy does not change this conclusion. Equation (3-42) defines the constant volume heat capacity C_V as

$$C_V = \left(\frac{\partial E}{\partial T}\right)_V \tag{3-42}$$

However, the zero-point vibrational energy is independent of temperature [$(d\varepsilon/dT) = 0$] and thus does not contribute to the heat capacity.

Figure 7-15 schematically illustrates the melting curve of ^{4}He and Figure 9-4 shows the Clapeyron slope along the melting curve. The Dutch physicist W. H. Keesom (1876–1956) first

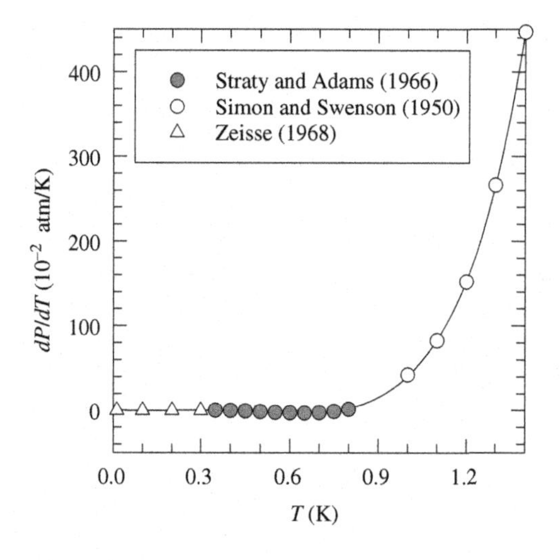

FIGURE 9-4

Clapeyron slope of the ^{4}He melting curve at 0.012–1.4 K.

solidified helium in 1926 and studied the melting curve from 1.2 K to 4.2 K. Subsequent work by Simon and Swenson (1950), Grilly and Mills (1962), Straty and Adams (1966), and Zeisse (1968) determined the shape of the melting curve down to 0.012 K. The melting curve passes through a shallow minimum at 0.775 K and then has a negative Clapeyron slope for about 0.4 K. The enthalpy and entropy of melting of ^{4}He are negative in this region because the $\Delta V$of melting remains positive. Below 0.4 K the Clapeyron slope levels out to zero. Thus, ΔS of melting is also zero in this range. The third law requires that the entropies of solid and liquid ^{4}He also go to zero as temperature approaches absolute zero.

The ^{3}He melting curve is similar to that of ^{4}He, but shifted to much lower temperatures. The minimum in the ^{3}He melting curve is at 0.3 K, and the negative Clapeyron slope extends down to 0.0006 K, the lowest temperature measured (Greywall, 1986). The entropy and enthalpy of melting are negative in this region, and liquid ^{3}He freezes when it is heated. The third law requires that the slope of the melting curve must go to zero at absolute zero, but experimental studies of the melting curve do not yet extend into this very low-temperature region.

G. Thermal expansion coefficient

The third law predicts that all materials have zero entropy at 0 K. Thus, any change in entropy dS must also be zero. Consequently, derivatives such as

$$\left(\frac{\partial S}{\partial P}\right)_T = -\left(\frac{\partial V}{\partial T}\right)_P \tag{9-37}$$

$$\left(\frac{\partial S}{\partial V}\right)_T = \left(\frac{\partial P}{\partial T}\right)_V = -\frac{\alpha_P}{\beta_T} \tag{9-38}$$

must also be zero at absolute zero. The two preceding equations are rewritten versions of two of the Maxwell relations from Chapter 8. Thus, as absolute zero is approached,

$$\lim_{T\to 0} \left(\frac{\partial V}{\partial T}\right)_P = 0 \tag{9-39}$$

$$\lim_{T\to 0} \left(\frac{\partial P}{\partial T}\right)_V = 0 \tag{9-40}$$

These two derivatives occur in the isobaric thermal expansion coefficient α_P (Eq. 2-33) and the thermal pressure coefficient (Eq. 2-35):

$$\alpha_P = \frac{1}{V}\left(\frac{\partial V}{\partial T}\right)_P \tag{2-33}$$

$$\left(\frac{\partial P}{\partial T}\right)_V = \frac{\alpha_P}{\beta_T} \tag{2-35}$$

The isobaric linear thermal expansion coefficient is tabulated for many metals and is defined as

$$\alpha_L = \frac{1}{L}\left(\frac{dL}{dT}\right)_P \tag{9-41}$$

For materials with cubic crystal structure, the volumetric thermal expansion coefficient is equal to three times the linear thermal expansion coefficient:

$$\alpha_P = 3\alpha_L \tag{9-42}$$

Figure 9-5 shows α_L values for copper (Cu) and lead (Pb), which are metals with cubic crystal structure. Their linear thermal expansion coefficients tend to zero as temperature decreases to absolute zero. The two curves in Figure 9-5 look like the low-temperature heat capacity curves plotted in Figures 4-9 (forsterite, Mg_2SiO_4), Figure 4-16 (diamond), and Figure 4-17 (silver, halite NaCl, forsterite, and fluorite CaF_2). This similarity is not coincidental; the heat capacity and thermal expansion coefficient are related to one another via Grüneisen's parameter:

$$\gamma_{th} = \frac{\alpha_P K_T}{\rho C_V} = \frac{\alpha_P K_S}{\rho C_P} \tag{8-280}$$

where K_T is the isothermal bulk modulus, K_S is the adiabatic bulk modulus, *th* stands for thermal, ρ is the density, and the other terms are already defined. Rearranging Eq. (8-280) gives

$$\alpha_P = \frac{\gamma_{th}\rho C_V}{K_T} \tag{9-43}$$

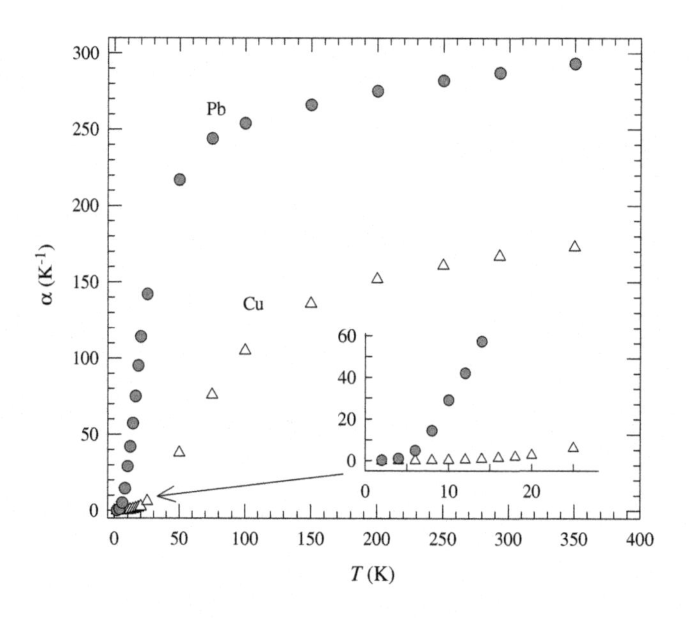

FIGURE 9-5

Linear thermal expansion coefficients of Cu and Pb at 2–293 K from the data of Kirby et al. (1972).

The different shape of the two curves is related to the different Debye temperatures of the two metals (85 K for Pb and 310 K for Cu). Debye temperatures can be computed from the thermal expansion plots in Figure 9-5. For example, Gopal (1966) gives 325 K for Cu from thermal expansion data and 310 K from heat capacity data.

H. The approach to absolute zero with mechanical and magnetic cooling

An important implication of the third law is the impossibility of reaching absolute zero. Walter Nernst emphasized this in his statement of the third law: "*Es ist unmöglich, eine Vorrichtung zu ersinnen, durch die ein Körper völlig der Wärme beraubt, d. h. bis zum absoluten Nullpunkte abgekühlt werden kann*" ("It is impossible to devise an arrangement by which a body may be completely deprived of its heat, i.e. cooled to the absolute zero"). Nevertheless, as of the time of writing, the lowest temperature reached is only 10^{-10} K above absolute zero. This achievement is the result of 125 years of work by many scientists. Mendelssohn (1966) gives an exciting account of the approach to absolute zero. Pobell (1996) reviews work that is more recent.

We briefly described mechanical (*PV*) cooling in Section IV-C of Chapter 6, and illustrated a refrigerator in Figure 6-4. Three steps schematically represent a mechanical cooling cycle:

1. A piston moving inward reversibly compresses gas inside a cylinder. The surroundings absorb heat evolved during compression.
2. The compressed gas is thermally insulated from the surroundings.
3. The compressed gas reversibly expands adiabatically, and the gas cools as it does *PV* work against the piston. This cycle repeats until, in the third step, the gas cools to its boiling point and some liquid forms. The net result of this process is a cold liquefied gas with lower entropy.

Air and other "permanent" gases were liquefied by mechanical (*PV*) cooling in the late 19th century. Working independently and almost simultaneously, the French physicist Louis-Paul Cailletet (1832–1913) and the Swiss physicist Raoul Pictet (1846–1929) liquefied tiny amounts of oxygen, nitrogen, and air in 1877. Shortly thereafter, in 1883, the Polish physicists Karol Olszewski (1846–1915) and Zygmunt Wroblewski (1845–1888) produced larger amounts of liquefied N_2 and O_2. Their work marks the beginning of modern low-temperature research. The Scottish chemist and physicist Sir James Dewar (1842–1923) first liquefied hydrogen in 1898 and reached temperatures of about 20 K. However, helium was much harder to liquefy because of its very low Joule-Thomson inversion temperature. Helium was also difficult to obtain in sufficient quantities prior to its extraction from natural gas wells. Helium was first liquefied (at its normal boiling point, 4.2 K) in 1908 by Kamerlingh Onnes (see sidebar in Chapter 8), who founded a famous cryogenic laboratory at Leiden in The Netherlands. The 360 liters of He gas that Onnes had came from monazite sand, which contains about 0.8–2.4 cm^3 He per 1 gram of sand. By means of pumping, the pressure over liquid He can be reduced and temperatures down to 0.8 K can be produced (e.g., look at the He phase diagram in Figure 7-15, which shows lower pressures along the vaporization curve at lower temperatures).

Magnetic cooling, also known as *adiabatic demagnetization*, proposed almost simultaneously by W. F. Giauque and Peter Debye in the 1920s, was the next major advance in producing low temperatures. Measurements at the Leiden laboratory showed that certain compounds, typically salts of rare earth elements (REE), could easily be magnetized (or demagnetized) by the application (or removal) of a strong external magnetic field. Gadolinium sulfate octahydrate $Gd_2(SO_4)_3 \cdot 8H_2O$ is a rare earth salt that has these properties. Paramagnetic transition metal salts such as chromium potassium alum $CrK(SO_4)_2 \cdot 12H_2O$, and the sulfate minerals bieberite $CoSO_4 \cdot 7H_2O$, melanterite $FeSO_4 \cdot 7H_2O$, chalcanthite $CuSO_4 \cdot 5H_2O$, and retgersite $NiSO_4 \cdot 6H_2O$, also have these properties.

The demagnetized salt → magnetized salt transition is analogous to the vapor → liquid transition because both changes in state go from a higher to a lower entropy state and have a negative ΔS. The magnetized salt is more ordered than the demagnetized salt and the liquefied gas is more ordered than its vapor. Thus, it should be possible to use the magnetic phase transition to produce cooling. Three steps analogous to those in mechanical cooling schematically represent magnetic cooling:

1. A salt such as gadolinium sulfate octahydrate is put into a container with He gas to conduct heat. The container is put in liquid He, cooled down, and then magnetized using a strong electromagnet. The surroundings absorb the heat released during magnetization.
2. The magnetized salt is thermally insulated from the surroundings.
3. The electromagnet is turned off. As the salt demagnetizes it reversibly does magnetic work by inducing an electric current in the windings of the electromagnet. The work is done adiabatically at the expense of internal molecular energy and the salt cools to lower temperature.

Although Giauque proposed magnetic cooling in 1924, it took until 1933 to experimentally demonstrate the process because equipment for liquefaction and storage of liquid air, hydrogen, and helium had to be designed and built. Giauque's first experiment produced a temperature of 0.24 K (Giauque and MacDougall, 1933). Subsequent magnetic cooling experiments over the next few decades produced lower temperatures, with a record low temperature of 1 millikelvin (10^{-3} K). Still lower temperatures of 100 picokelvin (10^{-10} K) were reached in 1999 by applying the magnetic cooling principle to nuclear spins of rhodium atoms. Volumes 14 and 15 of *Handbuch der Physik,* edited by Flugge (1956), provide good technical reviews of low-temperature cooling by mechanical (*PV*) refrigeration and magnetic cooling. Casimir (1961) and Pobell (1996) describe magnetic cooling of paramagnetic salts and atomic nuclei.

III. THE THIRD LAW AND ENTROPY

A. Residual or zero-point entropy of glassy materials

Most people think of window glass when someone mentions glass. Geologists may think of obsidian, which is natural glass (35–80% SiO_2) formed by rapid cooling of viscous lavas. However, *a glass is any amorphous material.* Window glass is a soda lime glass with a typical composition of (in mole percent) of 71.86% SiO_2, 13.13% Na_2O, 9.23% CaO, 5.64% MgO, 0.08% Al_2O_3, 0.04% Fe_2O_3, 0.02% K_2O, and 0.01% TiO_2. In contrast, the glass in fiber optic cables contains no silica, for example, 65% ZrF_4 + 35% BaF_2 in ZB glass. Glassy materials have atomic arrangements with no regularity on a scale a few times larger than the size of their constituents. For example, the average Si-Si distance in silica glass is ~0.36 nanometer (nm), and there is no regularity between the Si atoms on a scale larger than ~1 nm. Glass-forming materials include silica, germania (GeO_2), phosphorus pentoxide (P_2O_5), water, alkali silicates and aluminosilicates, lead diborate (PbB_4O_7), methyl alcohol (CH_3OH), glycerin ($C_3H_8O_3$), other organic compounds, and polymers (Wunderlich, 1960; Doremus, 1994).

Lewis and Randall's statement of the third law heading this chapter says that perfect crystals have zero entropy at absolute zero. However, their statement also implies that imperfect materials, such as glasses, do not have zero entropy at absolute zero. In Chapter 6, we showed that the number of distinguishably different arrangements of particles, such as atoms in a solid material, leads to configurational entropy given by the Boltzmann-Planck equation:

$$S = k \ln \Omega \tag{6-162}$$

The number of distinguishably different arrangements (Ω) in a perfect crystalline lattice is unity and the resulting configurational entropy is zero. However, Ω is larger than unity in glasses, which have disordered atomic arrangements. Thus, glasses posses some finite amount of *residual* or *zero-point entropy* that does not go to zero at absolute zero. This residual entropy is approximately the same as the configurational entropy of the supercooled liquid at the glass transition temperature. Equation (6-162) shows that greater disorder leads to a larger Ω and to greater entropy, whereas less disorder leads to a smaller Ω and to smaller entropy. The existence of residual entropy in glasses is a confirmation of the third law.

In a classic study, Gibson and Giauque (1923) measured the heat capacity and entropy of crystalline and glassy glycerin (glycerol) from 70 K to 300 K. (The antifreeze in cars contains glycerin.) Figure 9-6 illustrates their heat capacity data. Liquid glycerin freezes at 291 K and the enthalpy of fusion is 4370 cal mol^{-1}. However, glycerin, like water, can be supercooled below its freezing point. Figure 9-6 shows that the heat capacity of the supercooled liquid is greater than that of the crystalline solid down to ~183 K, where there is a sudden drop in the heat capacity. Glassy and crystalline glycerins have similar heat capacities down to 70 K, the lowest temperature studied by Gibson and Giauque (1923). Subsequently,

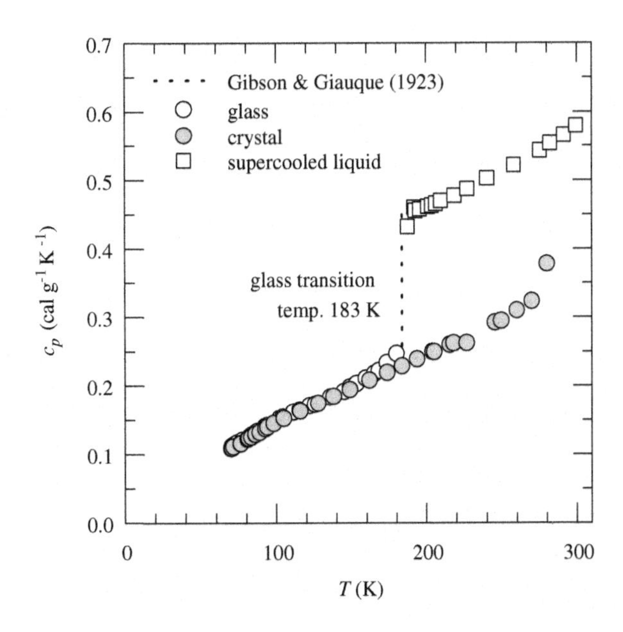

FIGURE 9-6

Specific heat data of Gibson and Giauque (1923) for crystalline and glassy glycerin from 70 K to 300 K. The glass transition temperature of glycerin (~183 K) is shown by the dotted vertical line. Hollow points are glass; shaded points are crystal.

Ahlberg et al. (1937) showed that the two phases have similar heat capacities down to 3 K. The entropy difference between glassy and crystalline glycerin at any temperature T is

$$\Delta S = \frac{\Delta_{fus}H}{T_{m.p.}} - \int_{T}^{T_{m.p.}} \left[C_P\,(\text{glass}) - C_P\,(\text{crystal})\right] \frac{dT}{T} \tag{9-44}$$

Gibson and Giauque (1923) evaluated Eq. (9-44) from their data and found

$$\Delta S = S_0^o\,(\text{glass}) - S_0^o\,(\text{crystal}) = \frac{4370}{291} - 9.39 = 5.63 \pm 0.1 \text{ cal mol}^{-1}\text{ K}^{-1} \tag{9-45}$$

Figure 9-6 shows that the ΔC_P between glassy and crystalline glycerin is approximately constant below the glass transition temperature. Gibson and Giauque (1923) argued it is unlikely that the entropy difference of 5.6 ± 0.1 cal mol^{-1} K^{-1} disappears below 70 K because the heat capacity of the glass would have to be much larger than observed at 70 K. The work of Ahlberg et al. (1937) that we mentioned previously confirms their argument. Gibson and Giauque's result shows that the residual entropy of glassy glycerin at 0 K is ~5.63 cal mol^{-1} K^{-1}. Substituting into Eq. (6-162),

$$\ln \Omega = \frac{S}{R} = \frac{5.63}{1.987} \cong 2.833 \tag{9-46}$$

$$\Omega = \exp(2.833) \cong 17 \tag{9-47}$$

Thus, the glassy glycerin is about 17 times more disordered than the crystal.

Figure 9-6 shows that the supercooled liquid glycerin transforms to glycerin glass at ~183 K. This is the *glass transition temperature* of glycerin. The glass transition temperature (T_g) of a supercooled liquid is the temperature at which physical properties such as the isobaric thermal expansion coefficient, heat capacity, and electrical conductivity show discontinuities (Doremus, 1994). A rule of thumb is that the glass transition temperature is about two-thirds of the melting point ($T_g/T_m \sim 0.67$). Table 9-4 shows that the mean glass transition temperature of 10 common minerals is $T_g/T_m = 0.67 \pm 0.11$, where the error is one standard deviation. The glass transition temperature depends somewhat on the cooling rate and is higher for faster cooling and lower for slower cooling. For example, the glass transition temperature of diopside is 1005 K for the rapid cooling during drop calorimetry and decreases to 950 K for cooling of a few degrees per minute (Richet and Bottinga, 1986).

Typically, the viscosity of a supercooled liquid is about 10^{13} poise (1 poise = 1 P = 1 dyn s cm^{-2} = 1 g cm^{-1} s^{-1} = 0.1 Pascal s) at the glass transition temperature, but the viscosity does not show a discontinuity at T_g. For comparison, at 25°C, water and molasses have viscosities of ~ 0.01 P and ~ 40 P. The viscosity of basalts at their liquidus temperatures is several orders of magnitude larger than these values. For example, MacDonald (1972) lists viscosities of $(0.2 - 2) \times 10^4$ P at 1100–940°C for basalts erupted at Mauna Loa and Kilauea in Hawaii. Thus, the viscosity of a supercooled liquid is very high at its glass transition temperature.

Even so, changes in the structure of the supercooled liquid at T_g occur on timescales of minutes to hours. Oblad and Newton (1937) observed that glassy glycerin had a relaxation time of about one week (168 hours) at 175 K and that the resulting material behaved like the supercooled liquid. Ahlberg et al. (1937) found that supercooled liquid glycerin cooled at different rates had slightly different heat

Table 9-4 Silicates and Glasses: Entropies, Glass, and Melting Temperatures[a]

Mineral[b]	Formula	$S_{crystal}$	S_{glass}	ΔS(mol)[c]	ΔS(g atom)[c]	T_g	T_m	T_g/T_m[d]
albite	$NaAlSi_3O_8$	207.40	251.90	44.50	3.42	1096	1393	0.79
anorthite	$CaAlSi_2O_8$	199.30	237.30	38.00	3.17	1160	1830	0.63
diopside	$CaMgSi_2O_6$	142.70	166.0	23.30	2.33	1005	1670	0.60
enstatite	$MgSiO_3$	66.30	74.10	7.80	1.56	1020	1834	0.56
jadeite	$NaAlSi_2O_6$	133.50	170.50	37.00	3.70	1130	–	–
microcline	$KAlSi_3O_8$	214.20	261.60	47.40	3.65	1221	1473	0.83
nepheline	$NaAlSiO_4$	124.35	134.50	10.15	1.45	1033	1750	0.57
pyrope	$Mg_3Al_2Si_3O_{12}$	266.27	346.30	80.03	4.00	1020	1570	0.65
quartz	SiO_2	41.46	48.50	7.04	2.35	1607	1999	0.80
wollastonite	$CaSiO_3$	81.69	94.80	13.11	2.62	1065	1821	0.58

[a]*Data from Robie and Hemingway (1995) and Richet and Bottinga (1986).*
[b]*Stable phase at 298.15 K.*
[c]*Entropy (J/K) at 298.15 K per mole or gram atom, for example, one mole SiO_2 = three gram atoms.*
[d]*Mean $T_g/T_m = 0.67 \pm 0.11$ (one sigma).*

capacities. The glassy glycerin still had residual entropy of 4.64 cal mol^{-1} K^{-1} at 0 K, corresponding to about 10 times more disorder than the crystal.

In Chapter 4 we mentioned that anorthite, alkali feldspars, nepheline, wollastonite, diopside, and a number of other silicates form glasses upon cooling. Glass formation complicates drop calorimetric measurements of their enthalpies of fusion because the quenched melts do not have the same thermodynamic properties as the crystalline silicates. Figure 9-7 illustrates the different heat capacities and entropies of crystalline diopside $CaMgSi_2O_6$ and diopside glass (Krupka et al., 1985; Richet et al., 1986). At 0 K, the ΔS between glassy and crystalline diopside is 23.3 J mol^{-1} K^{-1}. Substituting the ΔS into Eq. (6-162) we find that glassy diopside is about 16.5 times more disordered than crystalline diopside. Table 9-4 lists entropies at 298.15 K for several crystalline and glassy silicates. The mean ΔS (glass-crystal) is ~ 2.8 ± 0.9 J (g atom)$^{-1}$ K^{-1}, but there is a trend between ΔS and the type of melt. Two review papers by Richet and Bottinga (1986) and Stebbins et al. (1984) discuss relationships between the heat capacities and entropies of silicate glasses and liquids, and their composition, structure, and viscosity.

B. Entropy of mixing in solutions at absolute zero

Gaseous, liquid, and solid solutions possess residual or zero-point entropy at absolute zero because of their configurational entropy, or entropy of mixing. No gaseous solutions exist at 0 K because no gases exist at 0 K. The ^3He-^4He liquid solution is the only liquid solution that exists at 0 K, and its behavior follows the third law. Thus, in practical terms we are interested in the configurational entropy of solid solutions as temperature decreases to absolute zero. The entropy of mixing for a binary solution is

$$S_{mix} = -R[X_1 \ln X_1 + X_2 \ln X_2] \quad (6\text{-}178)$$

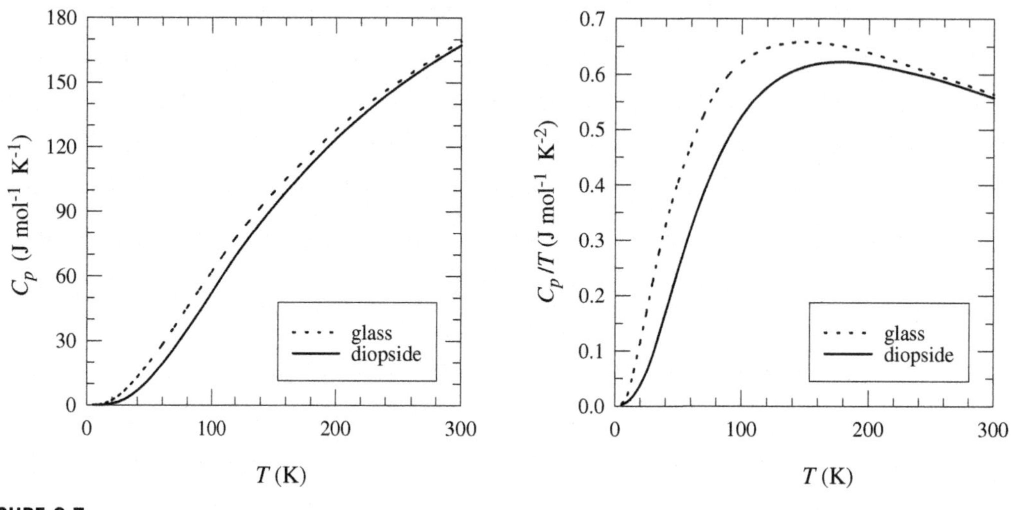

FIGURE 9-7

A comparison of the heat capacity and entropy of crystalline diopside ($CaMgSi_2O_6$) and diopside glass from the data of Krupka et al. (1985) and Richet et al. (1986).

The mole fractions of the two components in the solution are X_1 and X_2. In general, the entropy of mixing for any number n of components in a solution is

$$S_{mix} = -R\sum_{i=1}^{n} X_i \ln X_i \tag{6-179}$$

Equations (6-178) and (6-179) are general equations that are valid for any ideal solution of gases, liquids, and solids where mixing is completely random and only one substitution is possible.

Example 9-7. Eastman and Milner (1933) studied the entropy of a solid solution of silver chloride (AgCl) and silver bromide (AgBr) from 15 K to 298 K and found that it retained the configurational entropy predicted by Eq. (6-178) at 0 K. They synthesized solid solutions containing 72.8 mole % AgBr and 27.2 mole % AgCl. Prior experimental studies cited by them indicate that the Br^- and Cl^- ions randomly substitute for one another in the crystalline lattice of the AgBr-AgCl solid solution. Eastman and Milner (1933) used solution calorimetry to measure $\Delta H^o = 81 \pm 10$ cal mol^{-1} and a galvanic cell to measure $\Delta G^o = -254 \pm 20$ cal mol^{-1} for formation of the solid solution from the pure silver halides at 298 K:

$$0.272 \text{ AgCl (c)} + 0.728 \text{ AgBr (c)} = \text{AgCl}_{0.272}\text{Br}_{0.728} \text{ (c)} \tag{9-48}$$

They calculated $\Delta S^o = 1.12 \pm 0.1$ cal mol^{-1} K^{-1} for Eq. (9-48) at 298 K using Eq. (9-4):

$$\Delta S^o_{298} = \frac{\Delta H^o_{298} - \Delta G^o_{298}}{T} = \frac{81 + 254}{298.15} = 1.12 \text{ cal mol}^{-1}\text{ K}^{-1} \tag{9-49}$$

Within error, this ΔS value is identical to that from Eq. (6-178) for the $AgCl_{0.272}Br_{0.728}$ solution:

$$S_{mix} = -R[X_1 \ln X_1 + X_2 \ln X_2] = -R[0.728 \ln (0.728) + 0.272 \ln (0.272)] \\ = 0.585R = 1.16 \text{ cal mol}^{-1} \text{ K}^{-1} \quad (9\text{-}50)$$

The agreement between the two entropy values indicates that the solid solution has ideal entropy of mixing, that is, mixing is completely random and only one substitution is possible. As we discuss later in Chapter 11, the nonzero ΔH and ΔG of solid solution formation show that the AgBr-AgCl solid solution is a regular solution. A *regular solution* has ideal ΔS of mixing but nonzero ΔH and ΔG of solution formation.

Eastman and Milner measured the heat capacities of AgBr, AgCl, and the $AgCl_{0.272}Br_{0.728}$ solid solution from 15 K to 298 K by low-temperature adiabatic calorimetry. Within experimental error, they found $\Delta C_P = 0$ for reaction (9-48) over the entire temperature range, that is, solid solution formation via Eq. (9-48) obeyed the Neumann-Kopp rule (Chapter 4, Section III-A). This means that the ΔS of mixing is constant and does not change with temperature:

$$\Delta S_T^o = \Delta S_{298}^o + \int_{298}^{T} \Delta C_P \frac{dT}{T} = \Delta S_{298}^o \quad (9\text{-}51)$$

Thus, their heat capacity measurements show that the solid solution retained the configurational entropy of 1.12 cal mol^{-1} K^{-1} at 15 K (the lowest temperature they could reach) and presumably at 0 K as well. Solutions that remain mixed retain the residual or zero-point entropy given by Eq. (6-179) at 0 K. In contrast, the ^{3}He-^{4}He liquid solution reduces its entropy by unmixing into ordered liquids as T approaches 0 K.

C. Configurational entropy of minerals

Configurational entropy is the entropy from mixing metal atoms in alloys or mixing ions in minerals and mineral solid solutions. As described in Chapter 6, configurational entropy arises from mixing of Fe, Ni, and Co in meteoritic metal, of Al and Si in high albite $NaAlSi_3O_8$, of Mg and Fe in olivine $(Mg,Fe)_2SiO_4$, and of F, Cl, and OH in apatite $Ca_5(PO_4)_3(OH,F,Cl)$. Equations (6-178) and (6-179) have to be modified to apply them to minerals with mixing on more than the crystallographic site. In these cases the configurational entropy is

$$S_{config} = -R\sum_j \sum_i X_{i,j} \ln X_{i,j} \quad (6\text{-}186)$$

The first summation is for all sites j in a mineral and the second summation is for all atoms i with mole fractions X_i on each site in the mineral. We illustrated the use of this equation in Section IX-E of Chapter 6 when we calculated the configurational entropy of gehlenite ($Ca_2Al_2SiO_7$). Waldbaum (1972) and Ulbrich and Waldbaum (1976) discuss configurational entropy contributions to minerals in more detail and compute G for a mineral as a function of S_{config} by using an ordering parameter.

Many spinel minerals have configurational entropy. Normal 2–3 spinels have the general formula $A^{2+}B^{3+}{}_2O_4$ with all divalent ions in the tetrahedral A site and all trivalent ions in the octahedral B site. Spinel ($MgAl_2O_4$), hercynite ($FeAl_2O_4$), gahnite ($ZnAl_2O_4$), and galaxite ($MnAl_2O_4$) are normal 2–3 spinels. Inverse 2–3 spinels have the general formula $A^{3+}B^{2+}B^{3+}O_4$ with all divalent ions on the

B site, half of the trivalent ions on the A site, and half of the trivalent ions on the B site. Magnetite (Fe_3O_4), magnesioferrite ($MgFe_2O_4$), and trevorite ($NiFe_2O_4$) are examples of inverse 2–3 spinels. Normal 4–2 spinels have the general formula $A^{4+}B^{2+}{}_2O_4$ with all quadrivalent ions in the tetrahedral A site and all divalent ions in the octahedral B site. Ringwoodite (γ-Mg_2SiO_4), which is a high-pressure form of Mg_2SiO_4 in Earth's lower mantle, is a normal 4–2 spinel. Ülvospinel (Fe_2TiO_4) is an inverse 4–2 spinel. Magnetite-ülvospinel solid solutions are common in igneous rocks. These definitions and formulas are for the end-member normal and inverted spinels.

However, natural spinels are generally not end-members and their cation distribution depends on temperature. The general formula for a 2–3 spinel with a variable cation distribution is $^{tet}(A_{1\text{-}x}B_x)$ $^{oct}(A_yB_{1\text{-}y})_2O_4$ where the superscripts *tet* and *oct* denote the tetrahedral A and octahedral B sites, and $y = x/2$. The inversion parameter x is the mole fraction of B^{3+} ions in the tetrahedral A site. End-member normal spinels have $x = 0$, end-member inverse spinels have $x = 1$, and $x = 2/3$ in the random, maximum entropy cation distribution. Both normal and inverse spinels move toward $x = 2/3$ with increasing temperature.

There is no configurational entropy for a normal 2–3 spinel because there is no cation mixing on the two sites: all divalent cations are on the tetrahedral A site, and all trivalent cations are on the B site. Inverse 2–3 spinels have configurational entropy because of the cation mixing on the B site. Intermediate 2–3 spinels that are only partially normal have cations mixing on both sites. The general equation for S_{config} of spinels derived from Eq. (6-186) is

$$S_{config} = -R\left[\left(\sum x_i \ln x_i\right)_A + 2\left(\sum y_i \ln y_i\right)_B\right] \tag{9-52}$$

The x_i and y_i terms in Eq. (9-52) are the mole fractions of the divalent and trivalent ions on each of the sites and are the same as the x and y terms in the general formula $^{tet}(A_{1-x}B_x)^{oct}(A_yB_{1-y})_2O_4$. The sum of mole fractions for each site is unity.

Example 9-8. Calculate S_{config} for $MgAl_2O_4$ spinel that is 15% inverted, that is, $x = 0.15$. This amount of inversion is typical for synthetic spinels or for natural spinels heated above ~1300 K.

The formula for this spinel is $^{tet}(Mg_{0.85}Al_{0.15})^{oct}(Mg_{0.075}Al_{0.925})_2O_4$. Rewriting Eq. (9-52),

$$S_{config} = -R\left[\left(\sum x_i \ln x_i\right)_A + 2\left(\sum y_i \ln y_i\right)_B\right] \tag{9-52}$$

$$S_{config} = -R[(x \ln x + (1 - x) \ln (1 - x)) + 2(y \ln y + (1 - y) \ln (1 - y))] \tag{9-53}$$

Substituting the numerical values from the spinel formula into Eq. (9-53) and solving gives

$$\begin{aligned} S_{config} &= -R[(0.15 \ln 0.15 + 0.85 \ln 0.85) + 2(0.075 \ln 0.075 + 0.925 \ln 0.925)] \\ &= -R[(-0.2846 - 0.1381) + 2(-0.1943 - 0.0721)] \\ S_{config} &= 0.9555R = 7.944 \text{ J mol}^{-1} \text{ K}^{-1} \end{aligned} \tag{9-54}$$

The total entropy of an inverse spinel, or another mineral with configurational entropy, is the sum of the thermal and configurational entropy terms. For example, at 298 K,

$$S^o_{298} = S^o_{thermal} + S^o_{config} = \int_0^{298.15} C_P \frac{dT}{T} + S^o_{config} = \left(S^o_{298} - S^o_0\right) + S^o_{config} \tag{9-55}$$

Configurational entropy is a residual or zero-point entropy retained by the inverse spinel at absolute zero. This occurs because cation rearrangement into the normal spinel structure requires more energy than the thermal energy available at 298 K and lower temperatures. Thus, the inverse spinel cation distribution is frozen in at higher temperatures, and cation rearrangement from the inverse to normal spinel structure does not occur in the 0–298 K temperature range.

Magnetite is an important example of an inverse spinel without zero-point entropy. Reordering of the Fe^{2+} and Fe^{3+} cations occurs at ~115 K via electron transfer, and the ions do not move. As a result, the inverse spinel structure is not frozen in at absolute zero (Waldbaum, 1972). The entropy for magnetite at 298 K is simply the calorimetric entropy ($S^o_{298} = 146.15$ J mol^{-1} K^{-1}; Westrum and Grønvold, 1969). High-temperature chemical equilibria verify this calorimetric value (Rau, 1972; O'Neill, 1987, 1988) and show the absence of zero-point entropy in magnetite.

Example 9-9. Copper ferrite ($CuFe_2O_4$) is an inverse spinel. Low-temperature calorimetry gives $S^o_{298} - S^o_0 = 135.19$ J mol^{-1} K^{-1}, but this does not take into account the configurational entropy, which the spinel retains at absolute zero. Compute S^o_{298} including the configurational entropy.

Assuming a perfect inverse spinel structure, the formula of copper ferrite is rewritten as $Fe^{3+}(Cu^{2+}{}_{0.5}Fe^{3+}{}_{0.5})_2O_4$, showing that the Cu^{2+} ion and one Fe^{3+} ion are on the octahedral B site, and the other Fe^{3+} ion is on the tetrahedral A site. The mole fractions of the cations on the two sites are $x_{3+} = 1$ on the A site, and $y_{2+} = y_{3+} = 0.5$ on the B site. Equation (9-52) gives

$$\begin{aligned} S_{config} &= -R\left[\left(\sum x_i \ln x_i\right)_A + 2\left(\sum y_i \ln y_i\right)_B\right] = -R[(1\cdot \ln 1) + 2(0.5\cdot \ln 0.5 + 0.5\cdot \ln 0.5)] \\ &= -R[0 + 2\ln 0.5] = 2R\cdot \ln 2 = 11.53 \text{ J mol}^{-1}\text{ K}^{-1} \end{aligned} \quad (9\text{-}56)$$

The absolute entropy at 298 K is thus

$$S^o_{298} = \left(S^o_{298} - S^o_0\right) + S_{config} = 135.19 + 11.53 = 146.72 \text{ J mol}^{-1}\text{K}^{-1} \quad (9\text{-}57)$$

D. Configurational entropy of imperfect crystals

Configurational entropy also arises from the disorder in an imperfect crystal. In many cases, molecules can have more than one possible configuration in a crystal. If the time required for the molecules to orient themselves is longer than the cooling time, some disorder is "frozen into" the crystal (Parsonage and Staveley, 1978). For example, there are two possible orientations for a CO molecule in crystalline CO at low temperatures:

OC OC OC OC OC OC OC

CO CO CO CO CO CO CO

These two orientations lead to configurational entropy S_{config} of

$$S_{config} = \frac{R}{N_A}\ln \Omega = \frac{R}{N_A}\ln 2^{N_A} = R\cdot \ln 2 = 5.76 \text{ J mol}^{-1}\text{ K}^{-1} \quad (9\text{-}58)$$

In this equation, N_A is Avogadro's number and Ω is the possible number of orientations for each CO molecule. Clayton and Giauque (1932) measured the heat capacity and enthalpies of phase transitions for CO from 13 K to 81.61 K, which is its normal boiling point. They calculated the entropy of ideal CO gas from their experimental data and from the molecular structure and spectrum of the CO molecule via statistical mechanics. However, in contrast to the examples listed in Table 9-1, the two entropy values do not agree. The calculated (statistical) entropy of ideal CO gas is 160.32 J mol^{-1} K^{-1} at 81.61 K, whereas the experimental (calorimetric) value from their measurements is 155.64 J mol^{-1} K^{-1}. The calorimetric entropy is 4.68 J mol^{-1} K^{-1} smaller than the statistical entropy because CO ice retains this much residual or zero-point entropy at absolute zero. The residual entropy is equal to $R \cdot \ln(1.76)$, which is slightly smaller than the configurational entropy of $R \cdot \ln(2)$ calculated for a completely random orientation, indicating some ordering of the molecules in solid CO. Subsequent work by Gill and Morrison (1966) verified the conclusions of Clayton and Giauque (1932).

Nitrous oxide (N_2O) ice and nitric oxide (NO) ice also retain residual or zero-point entropy at absolute zero. Nitrous oxide is a linear molecule with two possible orientations in N_2O ice (NNO or ONN). The configurational entropy for completely random molecular orientation is again $R \cdot \ln(2)$. Blue and Giauque (1935) observed residual entropy of ~ $R \cdot \ln(1.775)$, indicating some ordering of the N_2O molecules. Nitric oxide is also a linear molecule, but the molecules in NO ice exist as N_2O_2 dimers with two possible orientations (Parsonage and Staveley, 1978):

$$\begin{matrix} N=O & O=N \\ O=N & N=O \end{matrix}$$

Johnston and Giauque (1929) studied NO and found residual entropy of $0.52 \cdot R \cdot \ln(2)$ versus the predicted value of $0.50 \cdot R \cdot \ln(2)$. The difference between the two values is within the experimental error and indicates a completely random orientation of the N_2O_2 dimers in NO ice.

Perchloryl fluoride (ClO_3F) also retains residual entropy at absolute zero (Koehler and Giauque, 1958). This molecule has one fluorine atom and three oxygen atoms at the vertices of a distorted tetrahedron centered on the chlorine atom. The four atoms bound to the central chlorine atom are distinct from one another because of the distorted tetrahedral molecular structure. Random orientation of ClO_3F molecules in the ice gives $S_{config} = R \cdot \ln(4) \sim 11.53$ J mol^{-1} K^{-1}. Koehler and Giauque (1958) observed residual entropy of ~10.12 J mol^{-1} K^{-1} ~ $R \cdot \ln(3.38)$. Thus, some of the ClO_3F molecules are ordered in the ice.

Ethane ice also has residual entropy, but for a different reason. The calorimetric entropy of ideal ethane (C_2H_6) gas at its boiling point (184.1 K) is 207.3 ± 0.6 J mol^{-1} K^{-1} (Witt and Kemp, 1937). However, the calculated (statistical) entropy is 213.8 J mol^{-1} K^{-1} (Kemp and Pitzer, 1937). Ethane is a symmetrical linear molecule, and the large difference between the calorimetric and statistical entropy values is not due to the same type of configurational disorder as observed for CO, N_2O, and ClO_3F. The methyl groups in ethane (H_3C-CH_3) cannot rotate freely in the ice at low temperatures because the energy required is larger than the thermal energy. Thus, different ethane molecules have different orientations of the two methyl groups with respect to each other (e.g., 1 H atom up versus 1 H atom down). The hindered rotation of methyl (CH_3) groups in ethane leads to residual entropy of ~6.5 J mol^{-1} K^{-1}. Kemp and Egan (1938) studied propane (H_3C-CH_2-CH_3) and observed residual entropy of ~14.2 J mol^{-1} K^{-1}. In both cases, the calorimetric and statistical entropy values agree by considering a small energy barrier for rotation of the methyl groups in the hydrocarbons. Chapter 27 of Pitzer and Brewer (1961) describes the internal rotation of methyl groups in hydrocarbons and analogous effects in more detail.

E. Isotopic entropy

Many elements have two or more stable isotopes. Thus, many natural materials are isotopic mixtures and have isotopic mixing entropy. For example, carbon has two stable isotopes ^{12}C (98.89%) and ^{13}C (1.11%). Atmospheric CO_2 with these average terrestrial abundances of ^{12}C and ^{13}C (and with pure ^{16}O in this example) has isotopic mixing entropy of

$$S_{iso} = -R[X_1 \ln X_1 + X_2 \ln X_2] \tag{6-178}$$

$$\begin{aligned} S_{iso} &= -R[(0.9889) \ln (0.9889) + (0.0111) \ln (0.0111)] \\ &= -R[-0.01104 - 0.04996] = 0.061R = 0.507 \text{ J mol}^{-1} \text{ K}^{-1} \end{aligned} \tag{9-59}$$

In comparison, the entropy of CO_2 is 213.785 J mol^{-1} K^{-1} at 298 K and one bar pressure. Thus, the entropy of isotopic mixing is an insignificant fraction of the total entropy of CO_2 gas. However, isotopic mixing entropy is a much larger contribution to the total entropy of elements with several stable isotopes (e.g., S, Ca, Ti, Cr, Ni, Se, Mo, Ru, Pd, Cd, Sn, Te, Ba, several REE and Pt metals) and to the minerals formed by these elements. The isotopic mixing entropy of a mineral composed of two or more elements is the sum of the isotopic mixing entropy of each element times the stoichiometric coefficient of the element in the mineral's chemical formula.

Example 9-10. Large deposits of native platinum occur in the Ural Mountains of Russia. Calculate the isotopic entropy S_{iso} of platinum and the entropy contribution from each isotope.

The average isotopic composition of terrestrial platinum is listed in the following table.

isotope	190	192	194	195	196	198
% abundance	0.01	0.79	32.9	33.8	25.3	7.2
$S_{iso}/R = -X_i \ln X_i$	9×10^{-4}	0.038	0.366	0.367	0.348	0.189

The isotopic entropy of Pt is the sum of the entropy contributions from each isotope:

$$\begin{aligned} S_{iso} &= R\left[9 \times 10^{-4} + 0.038 + 0.366 + 0.367 + 0.348 + 0.189\right] \\ &= 1.309R = 10.88 \text{ J mol}^{-1} \text{ K}^{-1} \end{aligned} \tag{9-60}$$

In general, the isotopic mixing entropy of an element or compound cancels out in a chemical reaction. Geochemists usually disregard isotopic mixing entropy. However, the isotopic mixing entropy of the reactants and products can be very different for reactions that involve large isotopic enrichments or depletions. The decay of radioactive isotopes also changes the isotopic mixing entropy of the parent and daughter nuclides, for example, see the time dependence of S_{iso} for lead and uranium illustrated by Ulbrich and Waldbaum (1976).

F. Magnetic entropy

Magnetic entropy in minerals arise from ordering and disordering of unpaired *d* and/or *f* electron spins in the outermost electron shell of a transition metal, lanthanide, or actinide atom or ion. Each spinning electron has an associated magnetic moment that can be ordered or disordered with the magnetic moments of the other unpaired electrons. Thermal energy destroys the magnetic orientation present at lower temperatures as the temperature increases to some critical value that is the magnetic transition temperature of a mineral. For example, the ferromagnetism of Fe metal disappears at the Curie temperature of 1043 K, where the ambient thermal energy disorients the electron spin (and magnetic moment) alignment. Iron metal is paramagnetic with disordered adjacent electron spins above 1043 K (see the C_P curve in Figure 4-22). The ΔS for the ferromagnetic to paramagnetic transition in Fe metal is the magnetic entropy. Ulbrich and Waldbaum (1976) list the predicted magnetic entropy values for many transition metal–bearing minerals. Spinel minerals often have magnetic entropy, and inverse spinels also have configurational entropy (see Table 9-5).

The magnetic transitions of some minerals such as iron metal (Figure 4-22) and FeS (Figure 4-10) are at or above room temperature. Other minerals such as Fe_2SiO_4, Ni_2SiO_4, Co_2SiO_4, and Mn_2SiO_4 have low-temperature magnetic transitions (Figure 4-21). As described in Chapter 4, much of the older heat capacity data for minerals only extended to ~50 K and missed magnetic transitions at lower temperatures. For example, the entropies of chromite ($FeCr_2O_4$) and magnesiochromite ($MgCr_2O_4$) were too low until Klemme et al. (2000) extended heat capacity measurements to liquid helium temperatures and measured the magnetic phase transitions and their associated entropy (see Figure 9-8).

The magnetic entropy of an atom (or ion) depends on the number of distinguishably different arrangements of the magnetic moments via the Boltzmann-Planck equation (6-162):

$$S = k \ln \Omega \qquad (6\text{-}162)$$

The quantity $(2S + 1)$ where S is the spin angular momentum quantum number of the atom (or ion) gives the value of Ω per gram atom of transition metal atom (or ion) in a mineral:

$$S_{mag} = R \ln (2S + 1) \text{J mol}^{-1}\ \text{K}^{-1} \qquad (9\text{-}61)$$

Table 9-5 Entropy Contributions for Some Spinel Minerals (J mol^{-1} K^{-1})

Spinel	Formula	Type[a]	S^o_{298}	S^o_{lat}	S^o_{config}	S^o_{mag}	Ref.
chromite	$FeCr_2O_4$	N	152.2	146	0	6.2	1, 2
hercynite	$FeAl_2O_4$	N	113.9	106.3	0	7.6	3, 4
magnesiochromite	$MgCr_2O_4$	N	118.3	105.9	0	12.4	1, 2
magnetite	Fe_3O_4	I	146.1	140.3	5.8	0	5
nickel chromite	$NiCr_2O_4$	N	140	124[b]	0	~16	6
spinel	$MgAl_2O_4$	N[a]	88.7	80.6	8.1	0	7
zinc chromite	$ZnCr_2O_4$	N	128.6	116.4[b]	0	12.2	8
zinc ferrite	$ZnFe_2O_4$	N	150.7	134.7	0	16.0	3, 9

References: 1. Shomate, 1944, 2. Klemme et al., 2000, 3. King, 1956, 4. Klemme and van Miltenburg, 2003, 5. Westrum and Gronvold, 1969, 6. Klemme and van Miltenburg, 2002, 7.NIST-JANAF tables (Chase, 1999), 8. Klemme and van Miltenburg, 2004, 9. Westrum and Grimes, 1957.

[a]N – normal, I – inverse, inversion parameter x ~ 0.05 for the $MgAl_2O_4$ sample.

[b]Estimated by Kubaschewski et al. (1993).

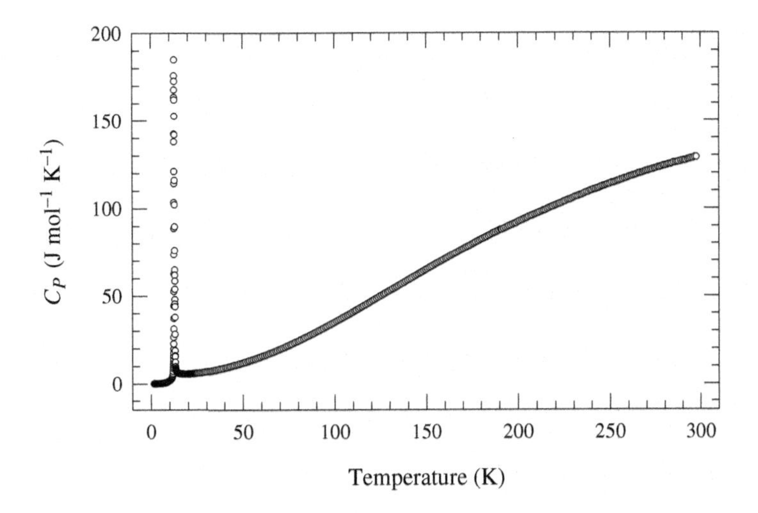

FIGURE 9-8

The low-temperature heat capacity data of Klemme et al. (2000) for magnesiochromite ($MgCr_2O_4$) from 1.5 K to 340 K. The antiferromagnetic transition at 12.55 K contributes magnetic entropy of ~$2R\ln(2S+1) = 2R\ln(4)$.

The magnetic entropy of lanthanide or actinide compounds depends on the total angular momentum quantum number J of an atom (or ion) in a similar manner:

$$S_{mag} = R\ln(2J+1)\text{J mol}^{-1}\text{ K}^{-1} \tag{9-62}$$

Equations (9-61) and (9-62) give the maximum magnetic entropy due to complete disordering of the magnetic moments of an atom or ion. Magnetic entropy is sometimes called *electronic entropy* (e.g., Burns, 1993). Sherman (1988) correctly noted that the equation $R \cdot N \cdot \ln(N)$ for electronic entropy given in the mineralogical literature (e.g., Wood, 1981), where N is the number of unpaired electrons, is incorrect. Equations (9-61) or (9-62) must be used instead.

The values of S and J depend on two factors: (1) the electron configuration of the atom or ion, and (2) the perturbing effects of external electrostatic fields generated by negatively charged anions surrounding a metal cation in its site in a mineral (the *crystal field*). The electron configuration and the crystal field are also important for the Schottky and nuclear spin entropy of materials, and we review them in the next two sections.

Electron configurations and quantum numbers

Four different *quantum numbers* describe the properties of electrons in an atom. These are the principal quantum number n, the azimuthal quantum number l, the magnetic quantum number m_l, and the spin quantum number m_s. The principal quantum number is related to the average distance and energy level of an electron from the nucleus of an atom. It has integral values of 1, 2, 3, 4, 5, 6, and so on. These values correspond to the *K, L, M, N, O, P*, ... electron shells in an atom and to the different rows in the periodic table. The elements H and He in the first row of the periodic table have $n = 1$, the elements Li, Be, B,...,F, Ne in the second row have $n = 2$, the elements Na, Mg,...,Cl, Ar in the third row have $n = 3$, the elements K, Ca, Sc,...,Br, Kr in the fourth row have $n = 4$, and so on. Geochemists sometimes refer to the different electron shells as the helium shell ($n = 1$, K shell), neon shell ($n = 2$, L), Ar shell ($n = 3$, M), Kr shell ($n = 4$, N), and so on. There are two electrons in the He shell ($n = 1$), eight electrons in the Ne shell ($n = 2$), 18 electrons in the Ar shell ($n = 3$), 32 electrons in the Kr shell ($n = 4$), and $2n^2$ electrons in any shell with principal quantum number n.

The azimuthal quantum number l describes the orbital angular momentum of an electron. It has integral values ranging from 0 to $(n - 1)$, that is, integral values of $0 \leq l \leq (n - 1)$ are allowed. The different l values correspond to different types of orbitals occupied by electrons:

$$\begin{aligned} l \text{ values} &= 0, 1, 2, 3, 4, 5\ldots \\ \text{orbitals} &: s, p, d, f, g, h\ldots \end{aligned}$$

The elements H and He only have s orbitals because $l = n - 1 = 0$ is the only possible l value for $n = 1$. Elements in the second row ($n = 2$) have s and p orbitals because $0 \leq l \leq (n - 1)$ gives two allowed values of 0 (s orbitals) and 1 (p orbitals). Third row elements ($n = 3$) have s, p, and d orbitals because $l = 0, 1, 2$ for $n = 3$. Fourth row elements ($n = 4$) have s, p, d, and f orbitals ($l = 0, 1, 2, 3$), fifth row elements ($n = 5$) have s, p, d, f, and g orbitals ($l = 0, 1, 2, 3, 4$), and so on.

The magnetic quantum number m_l describes the energy of an electron in a magnetic field and relates to the spatial orientation of orbitals. It has $(2l + 1)$ integral values ranging from $-l$ to $+l$, which correspond to the different numbers of orbitals:

$$\begin{aligned} l \text{ values} &= 0, 1, 2, 3, 4, 5\ldots \\ \text{orbitals} &: s, p, d, f, g, h\ldots \\ m_l &= \pm l : 0, 3, 5, 7, 9, 11\ldots \end{aligned}$$

Atoms can have one s ($m_l = 0$), three p ($m_l = -1, 0, +1$), five d ($m_l = -2, -1, 0, +1, +2$), seven f ($m_l = -3, -2, -1, 0, +1, +2, +3$), nine g ($m_l = -5, -3, -2, -1, 0, +1, +2, +3, +5$) orbitals, and so on per electron shell. In the absence of strong electric or magnetic fields, the energies of the p, d, f, g ... orbitals with the same principal and azimuthal quantum numbers are the same. Thus, the p orbitals have three-fold degeneracy (or multiplicity), the d orbitals have five-fold degeneracy, the f orbitals have seven-fold degeneracy, the g orbitals have nine-fold degeneracy, and so on.

The spin quantum number m_s relates to the spin angular momentum of an electron. When filled, each orbital denoted by the three quantum numbers n, l, m_l holds two electrons with opposite spins of $+½$ and $-½$. The *Pauli exclusion principle* requires that spins of two electrons in the same orbital cancel out because no two electrons can have the same four quantum numbers. The fine structure of atomic spectra shows that electrons spin about their axes. For example, the yellow double lines of Na at 5889.95 Å and 5895.92 Å, which are seen in sodium vapor streetlights, are due to the different spins of the two electrons that emit light during their transition from the $3p$ to the $3s$ orbitals of Na atoms.

The magnetic and spin quantum numbers of the unpaired electrons in the outermost electron shell are important for the magnetic properties of an atom or ion because the m_l and m_s values for paired electrons in the innermost and outermost electron shells cancel out. The quantum number S represents the resultant spin angular momentum of the atom or ion. It is the absolute value of the sum of the m_s values of the unpaired electrons:

$$S = \left|\sum m_s\right| \tag{9-63}$$

The quantum number L represents the resultant orbital angular momentum of the atom or ion. It is the absolute value of the sum of the m_l values of the unpaired electrons:

$$L = \left|\sum m_l\right| \quad (9\text{-}64)$$

L values of 0, 1, 2, 3, 4... have the symbols *S, P, D, F, G,* and so on. The quantum number J represents the total angular momentum of the atom or ion. It is equal to $|L+S|$ or $|L-S|$ depending on whether the outermost electron shell is more or less than half filled, respectively. The S, L, and J values for an atom or ion are written in the format $^{2S+1}L_J$ known as *Russell-Saunders terms, L-S terms*, or *spectroscopic terms*. The superscript gives the multiplicity (degeneracy) for the quantum number S. The L quantum number is written as the corresponding letter (*S, P, D, F,* and so on). The subscript gives the J quantum number. *Hund's rules* state that the lowest energy state has the largest L and smallest S values, that electrons go into unfilled orbitals before filling half-occupied orbitals, and that half-filled sets of orbitals have maximum stability. Several texts explain the Pauli exclusion principle, Hund's rules, and Russell-Saunders terms in more detail than possible here (Herzberg, 1944; Pauling, 1960).

Example 9-11. Calcium, aluminum-rich inclusions (CAIs) found in the Allende carbonaceous chondrite contain trivalent titanium (Ti^{3+}) in fassaitic pyroxenes [Ca(Mg,Al,Ti)$(Si,Al)_2O_6$]. Neutral Ti has a [Ar]$4s^2 3d^2$ electron configuration with $n = 4$ and $l = 2$. The [Ar] is the electron configuration of the completely filled Ar (*M*) shell. The two 4*s* electrons and one 3*d* electron are removed in trivalent Ti (Ti^{3+}), which has one unpaired 3*d* electron. Write down the four quantum numbers for the outermost electron, the S, L, and J values, and the magnetic entropy for Ti^{3+}.

The electron configuration of Ti^{3+} is [Ar]$3d^1$. There are five possible values for m_l and two possible values for m_s. According to Hund's rules, the electron goes into a *d* orbital with $m_l = 2$ and $m_s = -½$. This gives $S = |\pm ½| = ½$, $2S + 1 = 2$, $L = |\pm 2| = 2$. The five 3*d* orbitals are less than half filled, so $J = |L - S| = 3/2$. The Russell-Saunders term symbol for the Ti^{3+} ion is thus $^2D_{3/2}$. The magnetic entropy of Ti^{3+} is

$$S_{mag} = R \ln(2S + 1) = R \ln\left(2 \cdot \frac{1}{2} + 1\right) = R \ln 2 = 5.76 \text{ J mol}^{-1} \text{ K}^{-1} \quad (9\text{-}65)$$

We find the quantum numbers, S, L, and J values, and the magnetic entropy for ions of the other first-row transition metals in a similar manner. Electron arrow diagrams that represent occupancy of the five *d* orbitals simplify this process.

Example 9-12. Chromium (III) causes the red color of ruby $(Al,Cr)_2O_3$. The electron configuration of Cr^{3+} is [Ar]$3d^3$. Find the quantum numbers, S, L, and J values, and the magnetic entropy for Cr^{3+}. Hund's rules state that the three *d* electrons will partially fill three of the 3*d* orbitals $\underline{\uparrow}\ \underline{\uparrow}\ \underline{\uparrow}$ with parallel spins and will maximize L. The values of S, L, and J are

$$S = \left|\sum m_s\right| = 3\left(\frac{1}{2}\right) = \frac{3}{2} \quad (9\text{-}66)$$

$$L = \left|\sum m_l\right| = |-2 - 1 + 0| = 3 \quad (9\text{-}67)$$

We get the same results if we use $m_s = -½$ and $m_l = 2, 1, 0$ instead of the preceding values. The total angular momentum $J = |L - S| = 3/2$ and the multiplicity $(2S + 1) = 4$. The Russell-Saunders term symbol for Cr^{3+} is thus $^4F_{3/2}$ and the magnetic entropy is

$$S_{mag} = R \ln (2S + 1) = R \ln\left(2 \cdot \frac{3}{2} + 1\right) = R \ln 4 = 11.53 \text{ J mol}^{-1} \text{ K}^{-1} \quad (9\text{-}68)$$

Crystal field and mineral site symmetry

The mineral site symmetry produces an external electrical field (the crystal field or ligand field) that alters the energy levels of partially filled *d* and *f* orbitals and the number of unpaired electrons in the outermost electron shell of an atom or ion in the mineral from those of the free atom or ion. The Nobel Prize–winning physicist Hans Bethe (1906–2005) developed crystal field theory in a classic paper (Bethe, 1929). He showed that lower-symmetry fields arising from distorted sites of irregular shape produced more orbital splitting than higher-symmetry fields arising from sites with regular polyhedral shapes. Splitting of the five *d* orbitals about their energetic center of gravity is different in different types of crystal fields. In a spherical crystal field, the five *d* orbitals are energetically equivalent and no energy-level splitting occurs. However, the five *d* orbitals divide into two groups because of their angular shape (i.e., their symmetry) in tetrahedral (four-fold), octahedral (six-fold), and body-centered cubic (eight-fold) crystal fields. The d_{xy}, d_{yz}, and d_{xz} orbitals form one group of orbitals denoted t_{2g} or t_2 orbitals and the $d_{x^2-y^2}$, d_{z^2} orbitals form another group denoted as e or e_g. The t_{2g} denotes three orbitals of the same energy (i.e., three-fold degeneracy) that are symmetric upon inversion and rotation. The e_g means there are two orbitals of the same energy (i.e., two-fold degeneracy) that are symmetric upon inversion. Tetrahedral crystal fields have no center of symmetry and orbitals split into t_2 and e groups.

Electrons in the two e_g orbitals are repelled by an octahedral crystal field more than the electrons in the three t_{2g} orbitals, because in octahedral symmetry the e_g orbitals and the anions surrounding the transition metal ion are both along the *x, y,* and *z* axes. However, the situation is reversed in cubic (and tetrahedral) crystal fields where the electrons in the two e_g (or e) orbitals are repelled less than the electrons in the three t_{2g} (or t_2) orbitals.

The energy-level splitting is the *crystal field splitting*, and it is denoted by Δ with subscripts *c, t, o* for cubic, tetrahedral, and octahedral crystal fields, respectively. In an octahedral crystal field, the t_{2g} orbitals are $0.4\Delta_o$ lower and the e_g orbitals are $0.6\Delta_o$ higher than the energetic center of gravity in a spherical crystal field. Conversely, in cubic (or tetrahedral) crystal fields, the e_g (or e) orbitals are $0.6\Delta_c$ (or $0.6\Delta_t$) lower and the t_{2g} (or t_2) orbitals are $0.4\Delta_c$ (or $0.4\Delta_t$) higher than the energetic center of gravity in a spherical crystal field. The Δ values are different in tetrahedral, octahedral, and cubic fields with ratios $\Delta_t : \Delta_o : \Delta_c = -4/9 : 1 : -8/9$. The minus signs indicate the e (or e_g) orbitals have lower energy than the t_2 (or t_{2g}) orbitals. Absorption spectroscopy (visible, IR) measures the crystal field–splitting energy.

According to Hund's rules, electrons singly fill empty *d* orbitals with parallel spins up or down (↑ ↑). This is the normal situation and is a *high-spin configuration* because it gives the maximum value of the spin quantum number *S*. The opposite case has two electrons completely filling one orbital with antiparallel spins (↑↓). This is a *low-spin configuration* because it gives the minimum value of the spin quantum number *S*. High-spin configurations occur when the crystal field–splitting energy is less than the repulsive energy of two electrons in one orbital (the weak field situation). Conversely, low-spin configurations occur when the crystal field–splitting energy is greater than the repulsive energy of two electrons in one orbital (the strong field situation). Depending on the mineral site symmetry, low-spin states occur for ions with different numbers of *d* or *f* electrons. For example, transition metal ions with d^4 – d^7 electron configurations can be either high spin or low spin in octahedral fields. Ferrous (Fe^{2+}) and ferric (Fe^{3+}) minerals typically have high-spin configurations at ambient pressure. An interesting question is whether Fe-bearing minerals undergo spin state conversion in Earth's lower mantle. Such a conversion would have important consequences for the physical properties and thermal structure of the lower mantle (Burns, 1993). At the time of writing, this question remains unresolved. Interestingly, conversion from high- to low-spin Fe^{2+} occurs during oxygenation of hemoglobin, the oxygen-containing protein in red blood cells. The low-spin Fe^{2+} converts back to the high-spin state during deoxygenation of hemoglobin.

Example 9-13. Cobalt spinel ($Co^{2+}Co^{3+}{}_2O_4$) is a normal 2–3 spinel with Co^{2+} in the tetrahedral site and Co^{3+} in the octahedral sites. It is noteworthy because of the low-spin state of Co^{3+}. Give the term symbol and magnetic entropy for Co^{3+}, which has a [Ar]$3d^6$ electron configuration.

The low-spin of Co^{3+} state has six electrons in the three t_{2g} lower energy orbitals. Thus, $S = 0$, $(2S + 1) = 1$, $L = |-2-2-1-1+0+0| = 6$, $J = |L + S| = 6$, the term symbol is $^1I_6$, and the magnetic entropy is $S_{mag} = R \cdot \ln(2S + 1) = 0$.

G. Schottky entropy

Schottky (1922) proposed heat capacity (C_{Sch}) and entropy (S_{Sch}) contributions arising from temperature-dependent excitation of low-lying energetic states of an atom or molecule. The easiest example is an atom or molecule with two energy levels separated by ΔE. The lower level (E_0) is the ground state and the upper level (E_1) is a higher-energy state. The energy-level separation $\Delta E = E_1 - E_0$ corresponds to T/k, where k is Boltzmann's constant. The multiplicity (or degeneracy) of the ground state energy level is g_0 and that of the upper level is g_1. Only the lower energy level is filled at low temperatures where $kT << \Delta E$. Both energy levels are equally filled at high temperatures where $kT >> \Delta E$. The two energy levels are partially filled at intermediate temperatures where $kT \sim \Delta E$. The Schottky contributions to heat capacity and entropy are zero at low temperatures, rise to a maximum at intermediate temperatures, and decrease back to zero at high temperatures.

Equation (6-162) shows why this behavior occurs. At low temperatures ($kT << \Delta E$), the number of distinguishably different configurations (Ω) is unity because only the lower energy level is filled. Likewise, at high temperatures ($kT >> \Delta E$), Ω is unity because the two energy levels are equally filled. At intermediate temperatures ($kT \sim \Delta E$), the value of Ω initially increases with temperature as the upper energy level becomes occupied and then decreases as the filling of the two energy levels equalizes. The Schottky heat capacity contribution for an atom or molecule with two energy levels is

$$C_{Sch} = R\left[\left(\frac{g_1}{g_0}\right)\left(\frac{\theta}{T}\right)^2 \frac{\exp(-\theta/T)}{(1 + (g_1/g_0)\exp(-\theta/T))^2}\right] \tag{9-69}$$

The θ in Eq. (9-69) is a characteristic temperature at which $\Delta E = k\theta$ (in other words $\theta = \Delta E/k$). The θ/T is a dimensionless parameter analogous to θ_E/T or θ_D/T where θ_E and θ_D are the Einstein and Debye temperatures (e.g., see Figures 4-16, 4-17, and 9-2). The total Schottky entropy contribution for this two energy–level system is

$$S^o_{Sch} = R \ln \left(1 + \frac{g_1}{g_0}\right) \tag{9-70}$$

For example, the total Schottky entropy for $g_0 = g_1 = 1$ is $R \cdot \ln(2) = 0.693R$. Figure 9-9 shows the Schottky heat capacity and entropy plotted versus T/θ for a two energy level system with $g_0 = g_1 =$ unity. The Schottky heat capacity peaks at $T/\theta = 0.42$ where $C_{Sch} = 0.44 \cdot R$.

The most common Schottky heat capacity and entropy effects are from crystal field splitting of the magnetic ground states of rare earth and transition metal ions. Westrum (1983) and Gopal (1966) discuss Schottky effects for lanthanide and transition metal compounds, respectively. The Schottky entropy contribution can be as large as $R \cdot \ln(2J + 1)$ where J is the total angular momentum quantum number of the magnetic ground state. Several minerals, including bronzite [$(Mg_{0.8}Fe_{0.2})SiO_3$],

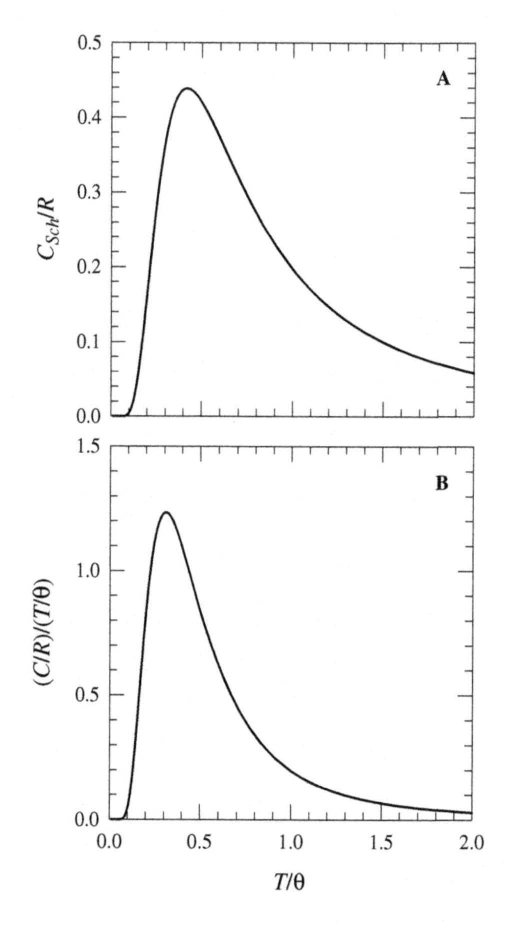

FIGURE 9-9

Schottky heat capacity and entropy for a two energy–level system.

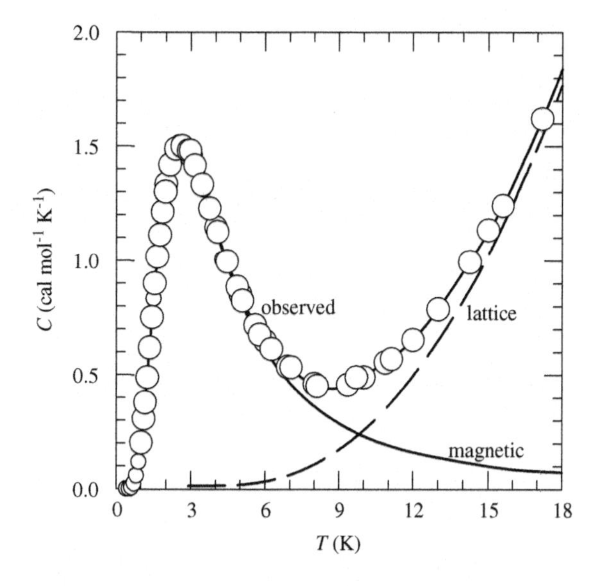

FIGURE 9-10

Low-temperature heat capacity data for retgersite (α-$NiSO_4 \cdot 6\,H_2O$).

(Stout and Hadley, 1964; Stout et al., 1966; Fisher et al., 1968)

staurolite $[(Fe^{2+},Mg)_2(Al,Fe^{3+})_9O_6[SiO_4]_4(O,OH)_2]$, retgersite ($\alpha$-$NiSO_4 \cdot 6\,H_2O$), and monazite ($MPO_4$ where M is Ce – Gd), have Schottky heat capacity and entropy contributions (Krupka et al., 1985; Hemingway and Robie, 1984; Stout and Hadley, 1964; Thiriet et al., 2005). Figure 9-10 illustrates the total heat capacity, the lattice heat capacity, and the Schottky heat capacity for retgersite (α-$NiSO_4 \cdot 6\,H_2O$). In this case, the crystal field split the five *d* orbitals of the Ni^{2+} ion into three energy levels (E_0, E_1, E_2) with $(E_1 - E_0)/k = \theta_1 = 6.44$ K, $(E_2 - E_0)/k = \theta_2 = 7.26$ K, and equal degeneracies $g_0 = g_1 = g_2 = 1$ (Stout and Hadley, 1964).

H. Nuclear spin entropy

The spectral "lines" of some elements are actually split into groups of several individual lines that are visible at very high resolution. For example, the 412.2 nm "line" of atomic Bi is actually a group of four lines falling within a width of only 0.044 nm. This hyperfine structure in atomic spectra was the first evidence for nuclear spin (Herzberg, 1944). *Nuclear spin* (*I*) is the total angular momentum of the nucleus of an atom. Nuclei with even atomic numbers ($Z =$ the number of protons) and even numbers of neutrons (*N*) have no nuclear spin. Nuclei with odd mass numbers ($A = Z + N$) have half-integral spins (1/2, 3/2, 5/2, 7/2...), and nuclei with odd numbers of protons and neutrons have integral spins (1, 2, 3...). These guidelines do not specify the exact *I* values for nuclei with half-integral or integral spins, and these have to be measured empirically.

According to quantum mechanics there are $(2I + 1)$ possible orientations for a nucleus with nuclear spin *I*. Each orientation has the same energy in the absence of a magnetic field, but a different energy in the presence of a magnetic field such as an externally applied field or the magnetic field generated by the interaction of the nucleus with unpaired electrons in *d* or *f* orbitals. Nuclear spin entropy arises from the number of distinguishably different spin orientations for atomic nuclei with nonzero spin. The separation between the energies of the different nuclear spin orientations is small with respect to kT, and the nuclear spin entropy becomes the largest entropy term at very low temperatures close to absolute zero. It is probably most important for the rare earth elements (REE) and their compounds. The nuclear spin entropy (S_N) of monoisotopic elements such as ^{127}I ($I = 5/2$), ^{141}Pr ($I = 5/2$), ^{159}Tb ($I = 3/2$), ^{165}Ho ($I = 7/2$), ^{169}Tm ($I = ½$), and ^{209}Bi ($I = ½$) is

$$S_N = R \ln (2I + 1) \tag{9-71}$$

Thus, Eq. (9-71) predicts $S_N = R \cdot \ln(8) = 17.29$ J mol^{-1} K^{-1} for ^{165}Ho. Krusius et al. (1969) observed $S_N = 17.22$ J mol^{-1} K^{-1} for ^{165}Ho metal, which is identical to the predicted value within their 1% experimental error. The nuclear spin entropy of multi-isotopic elements such as La, Nd, Sm, Eu, Gd, Dy, Yb, and Lu is

$$S_N = R \sum_j n_j \left[\sum_k X_{j,k} \ln (2I_{j,k} + 1) \right] \tag{9-72}$$

The *n* and *j* in Eq. (9-72) stand for the number of atoms of element *j* in a compound. The $X_{j,k}$ is the atom fraction of isotope *k* of element *j*.

Example 9-14. Terrestrial neodymium has seven isotopes and two have nonzero nuclear spins: ^{143}Nd (12.19%, $I = 5/2$) and ^{145}Nd (8.30%, $I = 5/2$). Calculate the nuclear spin entropy of Nd.

The nuclear spin entropy of Nd metal is

$$S_N = R[0.1219 \ln 6 + 0.0830 \ln 6] = 0.367R = 3.05 \text{ J mol}^{-1} \text{ K}^{-1} \tag{9-73}$$

Some thermodynamic data compilations combine the isotopic mixing entropy and nuclear spin entropy into one value called the *nuclear entropy*. This is simply an addition of Eqs. (6-178) and (9-72). Nuclear spin entropy of the REE is important at relatively high temperatures of a few Kelvins and it must be separated from lattice, magnetic, and Schottky contributions to obtain the correct practical entropy values for REE elements and compounds.

I. Ortho- and parahydrogen

Ortho- and para-H_2 are the two different nuclear spin modifications of H_2. *Orthohydrogen* (o-H_2) has parallel nuclear spins (↑↑), odd rotational quantum numbers ($J = 1, 3, 5\ldots$), and is a triplet state. *Parahydrogen* (p-H_2) has antiparallel nuclear spins (↑↓), even rotational quantum numbers ($J = 0, 2, 4\ldots$), and is a singlet state. Heisenberg postulated ortho- and para-H_2 in 1926 and Bonhoeffer and Harteck experimentally confirmed their existence in 1929 (Farkas, 1935).

In principle, other homonuclear diatomic gases such as N_2 and O_2 also have two different nuclear spin modifications. However, the rotational energy-level separation in the other homonuclear diatomic gases such as N_2 and O_2 corresponds to temperatures where the "gases" are solid. For example, $\theta_{rot} = 2.86$ K for N_2, which is well below its triple point (63.14 K). Likewise, $\theta_{rot} = 2.07$ K for O_2, which is below its triple point (54.36 K). Molecular H_2, D_2, and T_2 (tritium, radioactive hydrogen) exist as ortho- and para-isomers in the gas phase because their rotational energy-level separation corresponds to temperatures above their triple points $T_{t.p.}$ ($\theta_{rot} = 85.4$ K versus $T_{t.p.} = 13.80$ K for H_2,

$\theta_{rot} = 43.0$ K versus $T_{t.p.} = 18.72$ K for D_2, and $\theta_{rot} = 29.1$ K versus $T_{t.p.} = 20.6$ K for T_2). Symmetrical polyatomic molecules such as H_2O, H_2CO, and CH_4 also exist as nuclear spin isomers (Farkas, 1935; Pitzer and Brewer, 1961; Chapovsky and Hermans, 1999).

Orthohydrogen converts into parahydrogen with decreasing temperature because the lowest-energy state of p-H_2 has $J = 0$. As temperatures decrease, the proportion of p-H_2 increases until it reaches 100% at absolute zero. The equilibrium between ortho- and parahydrogen is

$$H_2\,(\text{ortho}) = H_2\,(\text{para}) \tag{9-74}$$

Reaction (9-74) does not depend on the total pressure. The standard enthalpy, entropy, and Gibbs free energy of reaction for Eq. (9-74) are

$$\Delta_r H_T^o = \left(H_T^o - H_0^o\right)_{para} - \left(H_T^o - H_0^o\right)_{ortho} \tag{9-75}$$

$$\Delta_r S_T^o = S_T^o(\text{para}) - S_T^o(\text{ortho}) \tag{9-76}$$

$$\Delta_r G_T^o = \Delta_r H_T^o - T\Delta_r S_T^o \tag{9-77}$$

The equilibrium abundances of ortho- and para-H_2 are related to the Gibbs energy of reaction via

$$\Delta_r G_T^o = -RT\ln\left(\frac{P_{para}}{P_{ortho}}\right) = -RT\ln\left(\frac{X_{para}}{X_{ortho}}\right) \tag{9-78}$$

The partial pressures of orthohydrogen (P_{ortho}) and parahydrogen (P_{para}) in Eq. (9-78) are equal to their mole fractions (X_{ortho} or X_{para}) times the total pressure, which cancels out of the equation. Figure 9-11

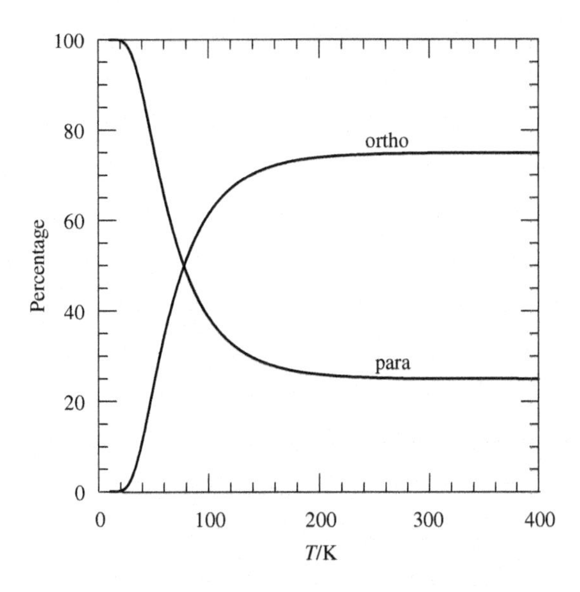

FIGURE 9-11

The temperature-dependent ortho-para ratio in equilibrium hydrogen (e-H_2).

shows the equilibrium abundances of ortho- and para-H_2 from 10 K to 350 K. These data show the ortho-para molar ratio approaches a value of 3:1 (i.e., 75% o-H_2 and 25% p-H_2) with increasing temperature. *Equilibrium hydrogen* (e-H_2) is hydrogen with an equilibrated ortho-para ratio at the ambient temperature. However, the equilibration of ortho- and parahydrogen is slow in the absence of paramagnetic catalysts. When H_2 gas cools, it retains the high-temperature 3:1 ortho-para ratio because the conversion reaction is so slow. Hydrogen with the 3:1 ortho-para ratio is *normal hydrogen* (n-H_2).

The physical properties of ortho- and para-H_2, including their vapor pressure, boiling points, triple points, thermal conductivity, heat capacity (C_p), entropy (S), and spectra, are measurably different at low temperatures because their rotational energy-level separation is greater than kT (where k is the Boltzmann constant) due to the low moment of inertia of the H_2 molecule. This difference is illustrated in Figure 9-12, which shows the constant-volume heat capacity (as C_V/R) for normal, ortho-, and para-H_2 from 0 K to 350 K. The three curves are computations and the dots are experimental values measured by different scientists.

The enthalpy change for reaction (9-74) affects the thermal structure of the upper atmospheres of Jupiter, Saturn, Uranus, and Neptune, which are at low temperatures where the equilibrium ortho–para ratio is significantly different from the 3:1 high-temperature ratio. The largest effects occur in the upper atmospheres of Uranus (minimum T ~ 53 K) and Neptune (minimum T ~ 52 K). Convective mixing upward of a gas parcel containing ortho and para H_2 moves H_2 with the high temperature ortho–para ratio (i.e., normal H_2) into a lower temperature region of an atmosphere. As the n-H_2

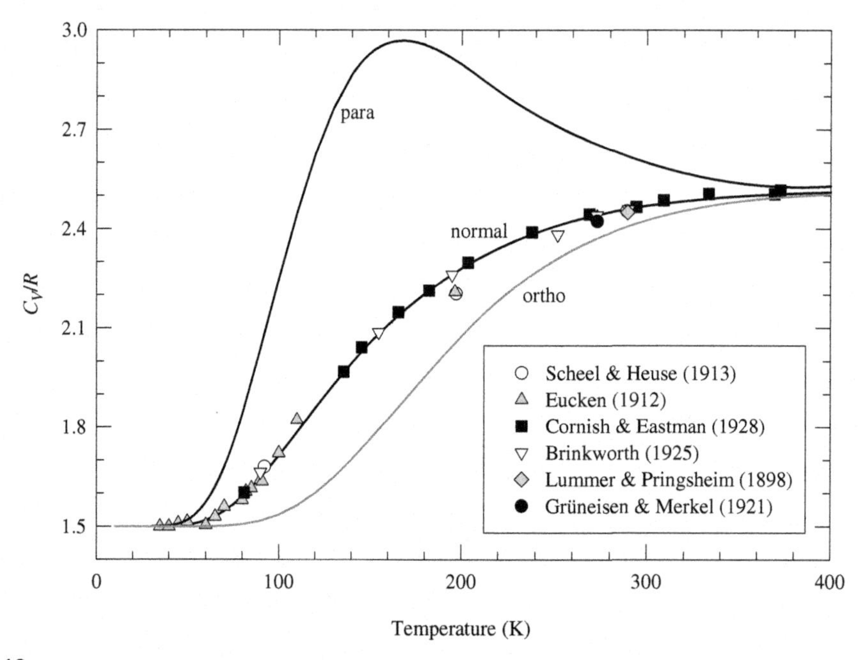

FIGURE 9-12

Constant-volume heat capacity (given as C_V/R) for ortho-, para-, and normal H_2. The C_V/R data of Brinkworth (1925), Cornish and Eastman (1928), Eucken (1912), Grüneisen and Merkel (1921), Lummer and Pringsheim (1898), and Scheel and Heuse (1913) are also plotted. The latter four data sets are from Partington and Shilling (1925).

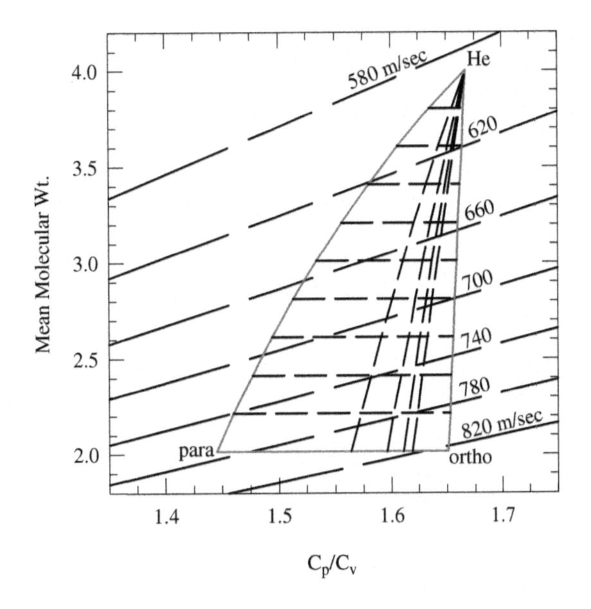

FIGURE 9-13

Sound speed triangle for a ternary mixture of ortho-H_2, para-H_2, and He.

approaches the low temperature ortho-para equilibrium ratio (i.e., e-H_2), the latent heat released drives the atmosphere toward stability.

As discussed in Chapter 4, the sound speed (w, in m s^{-1}) in an ideal gas or gas mixture is related to γ, the ratio of C_P/C_V, by the equation $w^2 = RT\gamma/\mu$ where μ is the molecular weight of the gas or mean molecular weight of the gas mixture. The relationship between the sound speed w and the composition of a three-component gas mixture is conveniently expressed as a triangle on a plot of mean molecular weight μ versus γ.

Figure 9-13 shows such a triangular plot for o-H_2, p-H_2, and He. The dashed curves labeled from 580 m/sec to 820 m/sec are contours of constant sound velocity. The vertical dashed curves within the triangle are contours of constant ortho-para ratio, and constant He/H_2 contours are the horizontal dashed lines within the triangle. The three sides of the triangle are slightly curved. In principle, measurement of the sound speed and either the mean molecular weight or γ yields the mole fractions of o-H_2, p-H_2, and He in the gas. In practice, the mean molecular weight would be used because it is obtained from He abundance measurements by a mass spectrometer or refractive index experiment, for example, on the *Galileo* entry probe into Jupiter.

IV. ABSOLUTE ENTROPY VALUES AND THE THIRD LAW

A. Practical entropy values

As discussed in Section II, the third law states that entropy is an absolute quantity and that

$$S^o(0\,K) = 0\,\text{J mol}^{-1}\,\text{K}^{-1} \tag{6-73}$$

for a pure, perfectly ordered crystalline solid. Thus, we compute absolute entropies of solids, liquids, and gases from integration of heat capacity data using equations such as

$$S_T^o = \int_0^T \frac{C_P}{T} dT \tag{6-74}$$

Equation (6-74) is valid for materials without any phase transitions. If we consider entropy values relative to 298.15 K and use a modified Maier-Kelley equation, Eq. (6-74) becomes

$$S_T^o = S_{298}^o + \int_{298}^T C_P \frac{dT}{T} \tag{9-79}$$

$$S_T^o = S_{298}^o + a \ln\left(\frac{T}{298.15}\right) + b(T - 298.15) - \frac{1}{2}c\left(\frac{1}{T^2} - \frac{1}{298.15^2}\right) + \frac{1}{2}d(T^2 - 298.15^2) \tag{9-80}$$

If the material undergoes one or more phase transitions, the ΔS terms due to these transitions also are included (see Example 6-10 in Chapter 6).

Pitzer and Brewer (1961) call absolute entropy values calculated in this manner practical entropy values. *Practical entropy values* are absolute entropy values that ignore the isotopic mixing and nuclear spin entropy contributions. The absolute entropy values calculated on this practical basis are the entropy values used every day in chemical thermodynamics.

B. Calculations with practical entropy values

We can calculate the ΔH of a chemical reaction using ΔS values from the third law, that is, from calorimetric data, if we know the ΔG of reaction. Conversely, we can compute the ΔG of a chemical reaction using ΔS values from the third law if we know the ΔH of reaction. We illustrate these calculations using Gerke's (1922) results for AgCl (cerargyrite) formation.

Example 9-15. Cerargyrite is a supergene (or secondary) silver ore mineral deposited by hydrothermal fluids. Until deposits ran out, it was widely mined in the western United States, including Leadville, Colorado, and Tombstone, Arizona. Gerke (1922) electrochemically measured $\Delta_f G_{298}^o$, $\Delta_f H_{298}^o$, and $\Delta_f S_{298}^o$ for AgCl formation from its constituent elements:

$$\text{Ag (metal)} + \frac{1}{2}\text{Cl}_2\text{ (g)} = \text{AgCl (cerargyrite)} \tag{9-81}$$

(We describe electrochemical measurements in Chapter 10.) Gerke's results (with one standard deviation uncertainties) are $\Delta_f G_{298}^o = -109{,}627 \pm 27$ J mol^{-1}, $\Delta_f H_{298}^o = -126{,}743 \pm 260$ J mol^{-1}, and $\Delta_f S_{298}^o = -57.41 \pm 0.87$ J mol^{-1}K^{-1}. The standard Gibbs free energy of formation is the most accurate of these three values and the standard entropy change is the least accurate.

Appendix 1 lists the third law entropies for Ag metal, Cl_2 gas, and AgCl at 298 K. Using these values we compute $\Delta_f S_{298}^o = -57.97$ J mol^{-1}K^{-1}. This agrees with Gerke's result within the uncertainties. We use this $\Delta_f S^o_{298}$ value to compute either $\Delta_f H^o_{298}$ or $\Delta_f G^o_{298}$ for Eq. (9-81):

$$\begin{aligned}\Delta_f H^o_{298} &= \Delta_f G^o_{298} + T\Delta_f S^o_{298} = -109,627 + (298.15)(-57.97)\\ &= -126,911 \text{ J mol}^{-1}\text{ K}^{-1}\end{aligned} \quad (9\text{-}82)$$

$$\begin{aligned}\Delta_f G^o_{298} &= \Delta_f H^o_{298} - T\Delta_f S^o_{298} = -126,743 - (298.15)(-57.97)\\ &= -109,459 \text{ J mol}^{-1}\text{ K}^{-1}\end{aligned} \quad (9\text{-}83)$$

The ΔH and ΔG values calculated using the third law are within 170 J of Gerke's values.

C. Gibbs free energy function

Because of the third law, Eq. (9-2) can be used to calculate Gibbs free energies using only calorimetric data for heat capacities, entropy, and the enthalpy of reaction at one temperature:

$$\Delta G^o_T = \Delta H^o_{298} + \int_{298}^{T} \Delta C_P dT - T\Delta S^o_{298} - T\int_{298}^{T} \frac{\Delta C_P}{T} dT \quad (9\text{-}2)$$

However, Eq. (9-2) is cumbersome because two different heat capacity integrals are involved. The Gibbs free energy function simplifies these calculations without using heat capacity integrals. The Gibbs free energy function (J mol^{-1} K^{-1}) defined relative to 298.15 K is

$$\frac{G^o_T - H^o_{298}}{T} = \frac{H^o_T - H^o_{298}}{T} - S^o_T = \frac{1}{T}\int_{298}^{T} C_P dT - \int_{0}^{T} C_P \frac{dT}{T} \quad (9\text{-}84)$$

Free energy functions for all reactants and products in a chemical reaction are computed from their heat capacities. At 298 K, the free energy function is equal to $-S^o_{298}$:

$$\frac{G^o_{298} - H^o_{298}}{298.15} = -S^o_{298} \quad (9\text{-}85)$$

For low-temperature equilibria, the Gibbs free energy function is defined relative to 0 K:

$$\frac{G^o_T - H^o_0}{T} = \frac{H^o_T - H^o_0}{T} - S^o_T \quad (9\text{-}86)$$

The Gibbs free energy functions relative to 0 K and 298 K are related by the equation

$$\frac{G^o_T - H^o_{298}}{T} = \frac{G^o_T - H^o_0}{T} - \frac{H^o_{298} - H^o_0}{T} \quad (9\text{-}87)$$

Example 9-16. Use the C_P coefficients in Table 3-3 and $S^o_{298} = 94.11$ J mol^{-1} K^{-1} to calculate the heat capacity, enthalpy $(H^o_T - H^o_{298})$, entropy S^o_T, and free energy function $[(G^o_T - H^o_{298})/T]$ for forsterite (Mg_2SiO_4) at 1000 K. The modified Maier-Kelley equation for C_P (forsterite) is

$$C_P = 143.967 + 39.362 \times 10^{-3}T - 32.525 \times 10^5/T^2 - 5.672 \times 10^{-6}T^2 \quad (9\text{-}88)$$

We use Eq. (4-3) to compute the enthalpy $(H_T^o - H_{298}^o)$

$$H_T^o - H_{298}^o = \int_{298.15}^{T} C_P dT$$

$$= a(T - 298.15) + \frac{b}{2}\left(T^2 - 298.15^2\right) - c\left(\frac{1}{T} - \frac{1}{298.15}\right) + \frac{d}{3}\left(T^3 - 298.15^3\right) \tag{9-89}$$

Equation (9-80) gives the entropy and Eq. (9-84) gives the free energy function. Our results agree almost exactly with the values at 1000 K from Table 6-4, which follows.

Property	Calculated	Table 6-4
C_P (J mol^{-1} K^{-1})	174.4	174.64
$(H_T^o - H_{298}^o)$ kJ mol^{-1}	109.48	109.51
S_T^o (J mol^{-1} K^{-1})	276.71	276.53
$[(G_T^o - H_{298}^o)/T]$ J mol^{-1} K^{-1}	−167.23	−167.02

Example 9-17. Anhydrous $NiCl_2$ has an antiferromagnetic to paramagnetic transition at 52.3 K. The Ni^{2+} ion has two unpaired *d* electrons and S_{mag} associated with this transition is

$$S_{mag} = R \ln(2S + 1) = R \ln(2 \cdot 1 + 1) = R \ln 3 = 9.13 \text{ J mol}^{-1} \text{ K}^{-1} \tag{9-90}$$

The magnetic entropy is an integral part of the total entropy of $NiCl_2$ and must be included to get the correct thermodynamic properties of this compound.

Busey and Giauque (1953) studied the chemical equilibrium:

$$NiCl_2 \text{ (c, hexagonal)} + H_2 \text{ (g)} = \text{Ni (metal)} + 2 \text{ HCl (g)} \tag{9-91}$$

Use a subset of their equilibrium data for Eq. (9-91) and auxiliary thermodynamic data to compute $\Delta_f H_{298}^o$ of $NiCl_2$. Compare the calculated value to $\Delta_f H_{298}^o = -304.78$ kJ mol^{-1} measured by Lavut et al. (1984) for direct reaction of Cl_2 (g) with Ni metal.

We compute $\Delta_r H_{298}^o$ for Eq. (9-91) and then solve for $\Delta_f H_{298}^o$ of $NiCl_2$ (*c*, hexagonal). The enthalpy of reaction at 298 K is calculated from the equilibrium data for reaction (9-91) via

$$\Delta_r G_T^o = \Delta_r H_{298}^o + T\Delta\left[\frac{G_T^o - H_{298}^o}{T}\right] \tag{9-92}$$

The $\Delta_f H_T^o$ values for Ni metal and H_2 (g) are zero by definition. Thus, $\Delta_r H_{298}^o$ for Eq. (9-91) is

$$\Delta_r H_{298}^o = 2\Delta_f H_{298}^o(\text{HCl, g}) - \Delta_f H_{298}^o(NiCl_2) \tag{9-93}$$

The difference of the Gibbs energy functions for products and reactants is

$$\Delta\left[\frac{G_T^o - H_{298}^o}{T}\right] = 2\left[\frac{G_T^o - H_{298}^o}{T}\right]\text{HCl} + \left[\frac{G_T^o - H_{298}^o}{T}\right]\text{Ni} - \left[\frac{G_T^o - H_{298}^o}{T}\right]\text{H}_2 - \left[\frac{G_T^o - H_{298}^o}{T}\right]\text{NiCl}_2 \tag{9-94}$$

We interpolated values of the Gibbs free energy functions for HCl, Ni, H_2, and $NiCl_2$ from the NIST-JANAF tables. The calculations for three data points spanning the 630–738 K range studied by Busey and Giauque (1953) are given here (the units are J mol^{-1} and J mol^{-1} K^{-1}).

T (K)	$\Delta_r G^o$	$\Delta[(G_T^o - H_{298}^o)/T]$	$\Delta_r H_{298}^o$
645.33	10,222	−170.6589	120,353
692.14	2728.6	−169.9793	120,378
738.02	−4654.2	−169.3213	120,308
			Mean ±1σ = 120,346±40 J mol^{-1}

We combine the mean $\Delta_r H_{298}^o$ value (120,346 ± 40 J mol^{-1}) with $\Delta_f H_{298}^o$ for HCl gas to compute

$$\Delta_f H_{298}^o(\text{NiCl}_2) = 2(-92{,}312) - 120{,}346 = -304{,}970\ \text{J mol}^{-1} \tag{9-95}$$

This is only 190 J mol^{-1} more negative than the calorimetric value (– 304,780 J mol^{-1}) measured by Lavut et al. (1984). The small difference is well within the uncertainties of the two values and demonstrates that the magnetic entropy of $NiCl_2$ must be included in its total entropy. This result also demonstrates that there is no residual entropy in Ni metal or $NiCl_2$ at absolute zero.

PROBLEMS

1. Hydrogen sulfide, H_2, and S_2 occur in terrestrial volcanic gases, for example, 1.065% H_2, 1.89% S_2, and 3.21% H_2S in the January 15, 1983 E. Rift eruption at Kilauea (Table 6.13 in Lodders and Fegley, 1998). The vent temperature of this gas was 1208 K, which is close to T = 1216 K in two experiments of Preuner and Schupp (1909). Use Eq. (9-2) to compute ΔG^o at 1216 K for

$$2\text{H}_2 + \text{S}_2 = 2\text{H}_2\text{S}$$
$$\Delta H_{298}^o = -169{,}762\ \text{J mol}^{-1} \text{ and } \Delta S_{298}^o = -77.917\ \text{J mol}^{-1}\ \text{K}^{-1}$$

Heat capacity equations for H_2 and H_2S are in Table 3-4. The C_P equation for S_2 (Kelley, 1960) is

$$C_P = 36.484 + 0.669 \times 10^{-3}T - 3.766 \times 10^5 T^{-2}\ \text{J mol}^{-1}\ \text{K}^{-1}$$

Compare your result to the experimental value of −60,590 ± 671 J mol^{-1} obtained by Preuner and Schupp (1909), and to the tabular value of −61,079 J mol^{-1}.

2. Jamieson (1953) measured the calcite-aragonite phase boundary as a function of pressure and temperature. At 302.1 K he found the phase boundary pressure is 3980 kg cm^{-2} (3903 bars). Bäckström (1925) measured $\Delta H^o_{298} = -30 \pm 20$ cal mol^{-1} for the conversion of calcite to aragonite at one bar pressure. Staveley and Linford (1969) measured $\Delta S^o_{298} = -0.89 \pm 0.05$ cal mol^{-1} K^{-1} and $\Delta C_P = -0.285$ cal mol^{-1} K^{-1} for the same reaction at one bar pressure. (a) Use Eq. (7-138) and Table 7-1 to compute the phase boundary at 302.1 K from their data. (b) Is the result consistent, within the ± 25 cal mol^{-1} uncertainty, with Jamieson's phase boundary pressure?
3. Low-temperature C_P data for quartz (Arkansas novaculite) are given below (Hemingway et al., 1991). (a) Compute the Debye temperature of quartz. (b) Use the Debye T^3 law to compute the heat capacity, entropy, and enthalpy from 0 K to 10 K. (c) Use the Nernst-Lindemann equation to compute $C_P - C_V$ at each temperature. Necessary auxiliary data are in Tables 2-1, 2-12, and 7-1.

T (K)	10	15	20	25	30
C_P (J mol^{-1} K^{-1})	0.028	0.216	0.655	1.289	2.077

4. Gray tin transforms to white tin at 286.2 K with $\Delta_{trans}H = 538$ cal mol^{-1}. Naumov et al. (1979) calorimetrically measured the heat capacity and entropy of white tin from 1.8 K to 311 K and found its entropy $S_{286.2} = 11.967$ cal mol^{-1} K^{-1}. Compute $S^o_{286.2}$ for gray tin (J mol^{-1} K^{-1}).
5. Derive Eq. (9-42) starting from Eqs. (2-33) and (9-41).
6. The *Voyager 2* spacecraft took pictures showing active plumes erupting on Triton, the largest satellite of Neptune. Cryogenic volcanism of liquefied gases is a possible explanation for the observed plumes. Use the following data (Gladun, 1966) to compute C_P, adiabatic compressibility, the thermal pressure coefficient, and the adiabatic Joule-Thomson coefficient for liquid neon. For reference, Triton's average surface temperature is 38 K, and N_2 ice is present on its surface. Neon ice cannot be detected spectroscopically, but the solar elemental abundance of Ne (3.29×10^6 atoms) is slightly higher than that of N (2.12×10^6 atoms).

T/K	31	35	39	43
V_m (cm^3 mol^{-1})	17.794	19.342	21.692	27.174
α_P (K^{-1})	0.0184	0.0252	0.0424	0.18
β_T (atm^{-1})	7.9×10^{-4}	1.4×10^{-3}	3.4×10^{-3}	2.6×10^{-2}
C_V (J mol^{-1} K^{-1})	17.11	16.65	16.53	18.94

7. In Chapter 4, we stated that the heat capacity of volatile liquids along their saturation curve is C_σ, which is different from either C_P or C_V. Use Bridgman's equations to show that the difference between C_σ and C_P is

$$C_P - C_\sigma = \alpha_P VT \left(\frac{dP}{dT}\right)_{eq} \tag{9-100}$$

Note that derivatives taken along equilibrium (*eq*) or saturation (σ) curves are equivalent to derivatives at constant Gibbs free energy *G*.

8. Stout and Robie (1963) measured the heat capacity of a natural dolomite sample from 11 K to 500 K. Use their tabulated value of $C_P = 0.028$ cal mol^{-1} K^{-1} at 10 K, to calculate (a) the Debye temperature for dolomite, (b) the extrapolated portion of the entropy ($S^o_{10} - S^o_0$), and (c) the enthalpy ($H^o_{10} - H^o_0$). Assume $C_V \sim C_P$ and pure dolomite [$CaMg(CO_3)_2$] in your calculations.

9. The cation distribution of a spinel is temperature dependent, and different cation distributions exist in samples quenched at different temperatures. Calculate and plot S_{config}/R in a 2–3 spinel as a function of the inversion parameter (x) from 0.05 to 0.95.

10. Rosen and Muan (1966) derived the Gibbs free energy of $MgAl_2O_4$ spinel from high-temperature chemical equilibria involving $MgAl_2O_4$-$CoAl_2O_4$ solid solutions. Their results give $\Delta G^o = -35.15$ kJ mol^{-1} and $\Delta H^o = -16.74$ kJ mol^{-1} at 1673 K for spinel formation from the oxides:

$$MgO\ (\text{periclase}) + Al_2O_3\ (\text{corundum}) = MgAl_2O_4\ (\text{spinel})$$

The calculated $\Delta G^o = -39.5$ kJ mol^{-1} at the same temperature for this reaction taking all data from Robie and Hemingway (1995). Assuming that the difference in ΔG^o is solely due to the entropy term, use your results from Problem 9-9 to compute the inversion parameter for the spinel studied by Rosen and Muan (1966). Robie and Hemingway (1995) adopted the NIST-JANAF S^o_{298} value for $MgAl_2O_4$ spinel with $x = 0.15$.

11. Nitrogen is the sixth most abundant element in the galaxy (H, He, O, C, Ne, N) and is the most abundant gas in Earth's atmosphere. Its atomic number is 7 and its two stable isotopes have abundances of 99.63% (^{14}N) and 0.37% (^{15}N). Give the nuclear spins of the two isotopes (zero, half integral, or integral) and state why.

12. The saturated vapor pressure of liquid ^3He (see below) is larger than that over liquid ^4He at the same temperature. A temperature of 0.25 K was reached using vacuum pumps to reduce the vapor pressure over liquid ^3He. Compute the saturated vapor pressure of ^3He at this temperature.

T (K)	0.6	0.7	0.8	0.9	1
P (kPa)	0.071	0.18	0.378	0.695	1.16

13. Trivalent vanadium causes the amethystine color of tanzanite, which is gem-quality zoisite $Ca_2(Al,V)_3Si_3O_{12}(OH)$ (Burns, 1993). Give the values of *S*, *L* and *J*, the magnetic entropy, and the Russell-Saunders term symbol for V^{3+}, which has the electronic configuration [Ar] $3d^2$.

14. Geologically important ions (and electron configurations) of several first-row transition metals are Cr^{6+} [Ar]$3d^0$, Cu^{2+} [Ar]$3d^9$, Ni^{2+} [Ar]$3d^8$, Co^{2+} [Ar]$3d^7$, Fe^{2+} [Ar]$3d^6$, Fe^{3+} [Ar]$3d^5$, and Mn^{3+} [Ar]$5d^4$. Give electron arrow diagrams showing *d* orbital occupancy, *S*, *L* and *J* values, and Russell-Saunders term symbols for these ions. Use high-spin states where relevant.

15. High- and low-spin state transitions of Fe^{2+} and Fe^{3+} may be important in Earth's mantle. (a) Write down the Russell-Saunders term symbols for the high- and low-spin configurations for both ions. (b) Use Eq. (9-61) to compute the magnetic entropy change between the high- and low-spin configurations of each ion. Assume octahedral coordination.

16. The French geochemist and mineralogist G. A. Daubree (1814–1896) discovered lawrencite ($FeCl_2$) in the Tazewell iron meteorite from Tennessee and named it after the American chemist J. Lawrence Smith (1818–1883). The origin of lawrencite is a matter of debate. Many geologists think it forms by corrosion of meteoritic iron metal.

(a) Calculate its standard enthalpy of formation from a subset of equilibrium data (Kangro and Petersen, 1950) for the reaction:

$$FeCl_2\ (\text{lawrencite}) + H_2\ (\text{g}) = Fe\ (\text{metal}) + 2\ HCl\ (\text{g})$$

The $\Delta_r G_T^o$ values (in J mol^{-1}) and the $\Delta[(G_T^o - H_{298}^o)/T]$ values (in J mol^{-1} K^{-1}) for this are shown here.

T (K)	$\Delta_r G_T^o$	$\Delta[(G_T^o - H_{298}^o)/T]$
757	50,893	−145.167
833	39,516	−143.898
883	32,655	−143.119
931	28,015	−142.412

(b) The electron configuration of Fe^{2+} is [Ar]3d^6. Compute the magnetic entropy of $FeCl_2$.

17. Gypsum ($CaSO_4 \cdot 2H_2O$) is the most abundant sulfate mineral on Earth and is commonly associated with anhydrite ($CaSO_4$) in evaporite deposits. Posnjak (1938) measured the solubility of gypsum and anhydrite in pure water and found the one-bar transition temperature is 42 ± 1°C (315.2 ± 1 K). This is the temperature at which $\Delta_r G^o$ is zero for the dehydration reaction

$$CaSO_4 \cdot 2H_2O\ (\text{gypsum}) = CaSO_4\ (\text{anhydrite}) + 2H_2O\ (\text{liquid})$$

Compute the transition point for this reaction and determine whether Posnjak (1938) is correct or not. Enthalpy ($H_{298}^o - H_0^o$) values and Gibbs free energy functions $-[(G_T^o - H_0^o)/T]$ for gypsum, anhydrite, and water (in J mol^{-1} and J mol^{-1} K^{-1}, respectively) are given here (Robie et al., 1989; CODATA). The heat of dehydration at 298 K is 4030 ± 20 cal mol^{-1} (Kelley et al., 1941).

T (K)	273.15	280	290	300	320	$(H_{298}^o - H_0^o)$
water	21.656	22.697	24.214	25.709	28.657	13,272.5
anhydrite	44.51	45.86	47.83	49.78	53.63	17,295.7
gypsum	80.6	83.03	86.57	90.07	97.02	31,126.9

18. Calculate the nuclear spin entropy of ^{159}Tb ($I = 3/2$).

19. Konings et al. (2005) measured the magnetic entropy of Gd_2O_3 as 4.14·*R* (34.42 J mol^{-1} K^{-1}). Their experimental error is about 1%. Does their result agree with the predicted magnetic entropy for Gd^{3+} (seven unpaired 4*f* electrons)?

20. Determine the different electron configurations for which transition metal ions (d electrons) have low-spin states in tetrahedral sites, such as the *A* site in spinels.

21. The isothermal bulk modulus K_T is the inverse of the isothermal compressibility β_T. Use the third law and fundamental equation for dE to show that at absolute zero K_T is

$$K_T = V\left(\frac{\partial^2 E}{\partial V^2}\right)_T \tag{9-101}$$

22. Measurements by the *Voyager* 2 spacecraft show the upper atmosphere of Uranus is 82.5% H_2, 15.2% He, and 2.3% CH_4. Calculate the dry adiabatic lapse rate (dT/dZ) at the 100 K level in the convective region of Uranus's atmosphere for (a) the equilibrium ortho-para ratio and (b) the 3:1 high-temperature ortho-para ratio. The average gravitational acceleration on Uranus is 8.85 m s^{-2}. Heat capacities (J mol^{-1} K^{-1}) at 100 K are listed here.

gas	o-H_2	p-H_2	He	CH_4
C_P	21.082	26.997	20.786	33.258

23. Magnetite (Fe_3O_4) occurs in carbonaceous and unequilibrated ordinary chondrites, which are meteorites formed in the early solar system. Magnetite may have formed by the reaction

$$0.75\ \text{Fe (metal)} + H_2O\ (g) = 0.25\ Fe_3O_4 + H_2\ (g) \tag{9-103}$$

at low temperatures on the meteorite parent bodies (Hong and Fegley, 1998). Rau (1972) measured the equilibrium constant (K_P) for reaction (9-103) from 583 K to 810 K:

$$K_P = \frac{P_{H_2}}{P_{H_2O}} = \exp\left(-\Delta_r G_T^o/RT\right) \tag{9-104}$$

Use a subset of his data (shown here) and Eq. (9-92) to calculate the average value of $\Delta_r H_{298}^o$ for reaction (9-103). The change in Gibbs energy functions for reaction (9-103) is

$$\Delta\left[\frac{G_T^o - H_{298}^o}{T}\right] = 47.4473 - 0.013556T - 1.675 \times 10^{-6}T^2 \tag{9-105}$$

Give the individual and mean values of $\Delta_r H_{298}^o$ (kJ mol^{-1}) to two decimal places.

T (K)	583	641	730.5	810
K_P	0.0539	0.0902	0.1858	0.2884

24. Calculate the standard enthalpy of formation for Fe_3O_4 (magnetite) using your result from Problem 9-23 and the $\Delta_f H_{298}^o$ value for H_2O (g) from Appendix 1.

CHAPTER

10

Chemical Equilibria

Tout systéme en équilibre chimique éprouve, du fait de la variation d'un seul des facteurs de l'équilibre, une transformation dans uns sens tel que, si elle produisait seul, elle amènerait uns variation de signe contraire du facteur considéré.
(Any system in chemical equilibrium that is subjected to a stress adjusts to a new equilibrium position to relieve the stress.)
—Henri Le Chatelier (1887)

This chapter discusses chemical equilibria involving gases, liquids, and solids and has three sections. Section I deals with chemical equilibria of gases. Section II discusses heterogeneous chemical equilibria between gases and condensed phases. Section III discusses electrochemical equilibria.

I. CHEMICAL EQUILIBRIA OF GASES

It is worth reviewing some facts about the Gibbs free energy before using it in our discussion of chemical equilibria. The Gibbs free energy G was introduced in Chapter 6. It is an *extensive variable* defined by the equations

$$G = E + PV - TS \qquad (6\text{-}113)$$

$$G = H - TS \qquad (6\text{-}114)$$

The fundamental equation for the Gibbs free energy G is

$$dG = VdP - SdT \qquad (7\text{-}28)$$

Equation (7-28) gives the change in G as a function of P and T once we substitute for VdP and/or SdT and integrate between initial (T_1, P_1) and final (T_2, P_2) states. An isothermal change ($dT = 0$) in the Gibbs free energy of an ideal gas between any two pressures (P_1, P_2) is thus

$$G_2 - G_1 = nRT \ln \frac{P_2}{P_1} \qquad (8\text{-}163)$$

If we now take the pressure P_1 as the standard-state pressure of one bar, G_1 is equal to G^o, the standard Gibbs free energy and Eq. (8-163) can be rearranged to give

$$G = G^o + nRT \ln P \qquad (8\text{-}164)$$

Copyright © 2013 Elsevier Inc. All rights reserved.

Equation (8-164) gives the free energy of a pure, ideal gas as a function of temperature, pressure, and the number of moles n. The logarithm in Eq. (8-164) is dimensionless because it is the logarithm of the ratio of pressure P to the standard-state pressure of one bar.

A. Derivation of the equilibrium constant (K_P)

We use the water-gas reaction to illustrate derivation of the equilibrium constant K_P for a gas phase reaction involving ideal gases. The *water-gas reaction* occurs in volcanic gases and many other naturally occurring high-temperature gases and fluids containing water and CO_2 (e.g., steam atmospheres formed during and/or after the accretion of Earth-like planets):

$$H_2 + CO_2 = H_2O + CO \tag{10-1}$$

Reaction (10-1) also occurs in exhaust gases of car and truck engines and industrial furnaces where it affects the amount of CO released into the air. The assumption of ideality is reasonable for the water-gas reaction, because in almost all cases of interest to us, the gases behave ideally to a very good first approximation.

The total Gibbs free energy G for the water-gas system is the sum of the Gibbs energies of the products and reactants in the gas mixture:

$$G = n_{H_2O}\overline{G}_{H_2O} + n_{CO}\overline{G}_{CO} + n_{CO_2}\overline{G}_{CO_2} + n_{H_2}\overline{G}_{H_2} \tag{10-2}$$

The $\overline{G}_i$ terms in Eq. (10-2) are the partial molar Gibbs energies of each gas. The *partial molar Gibbs energy* is defined as the change in the total Gibbs energy of the gas mixture at constant P and T when the number of moles n_i of gas i is varied holding the number of moles of all other gases constant (i.e., at otherwise constant composition):

$$\overline{G}_i = \left(\frac{\partial G}{\partial n_i}\right)_{P,T,n \neq n_i} \tag{10-3}$$

The definition of the partial molar Gibbs energy and Eq. (10-3) are analogous to the definition of the partial molar volume and Eq. (8-194). Because we are dealing with a mixture of ideal gases, the partial molar Gibbs energy of each gas in the mixture is the same as the molar Gibbs energy ($G_{m,i}$) of the pure gas:

$$\frac{G_i}{n_i} = G_{m,i} = \overline{G}_i = \left(\frac{\partial G}{\partial n_i}\right)_{P,T,n \neq n_i} \tag{10-4}$$

Thus, using Eq. (8-164) and Eq. (10-4), we can write the partial molar Gibbs free energies of the four gases involved in the water-gas reaction as

$$\overline{G}_{H_2} = G^o_{m,H_2} + RT \ln P_{H_2} \tag{10-5}$$

$$\overline{G}_{CO_2} = G^o_{m,CO_2} + RT \ln P_{CO_2} \tag{10-6}$$

$$\overline{G}_{CO} = G^o_{m,CO} + RT \ln P_{CO} \tag{10-7}$$

$$\overline{G}_{H_2O} = G^o_{m,H_2O} + RT \ln P_{H_2O} \tag{10-8}$$

As discussed in Chapters 6 and 7, $\Delta G = 0$ at equilibrium. In the case of the water-gas system, $\Delta G = 0$ when the Gibbs energy of the products is the same as that of the reactants:

$$\Delta G = G_{products} - G_{reactants} = 0 \tag{10-9}$$

$$\Delta G = n_{H_2O}\overline{G}_{H_2O} + n_{CO}\overline{G}_{CO} - n_{CO_2}\overline{G}_{CO_2} - n_{H_2}\overline{G}_{H_2} \tag{10-10}$$

Substituting Eqs. (10-5) to (10-8) into Eq. (10-10) we obtain

$$\begin{aligned}\Delta G = {} & n_{H_2O}(G^o_{m,H_2O} + RT \ln P_{H_2O}) + n_{CO}(G^o_{m,CO} + RT \ln P_{CO}) \\ & - n_{CO_2}(G^o_{m,CO_2} + RT \ln P_{CO_2}) - n_{H_2}(G^o_{m,H_2} + RT \ln P_{H_2})\end{aligned} \tag{10-11}$$

All of the n_i terms in Eqs. (10-10) and (10-11) are unity for the water-gas reaction. However, in general the number of moles of each gas in a chemical reaction can be different for each gas and different from unity. (Or in other words, the stoichiometric coefficients for the reactants and products can be different from unity and from each other.) Rewriting Eq. (10-11) to group together like terms we obtain

$$\Delta G = (G^o_{H_2O} + G^o_{CO} - G^o_{CO_2} - G^o_{H_2}) + RT(\ln P_{H_2O} + \ln P_{CO} - \ln P_{CO_2} - \ln P_{H_2}) \tag{10-12}$$

Equation (10-12) shows that the ΔG for the water-gas reaction is the sum of two terms.

The first term is the standard Gibbs free energy change ($\Delta_r G^o$) with all gases behaving ideally at the standard-state pressure of one bar. The $\Delta_r G^o$ for the water-gas reaction is

$$\begin{aligned}\Delta_r G^o &= G^o_{H_2O} + G^o_{CO} - G^o_{CO_2} - G^o_{H_2} \\ &= n_{H_2O}G^o_{m,H_2O} + n_{CO}G^o_{m,CO} - n_{CO_2}G^o_{m,CO_2} - n_{H_2}G^o_{m,H_2}\end{aligned} \tag{10-13}$$

As discussed in Chapter 6, we can only measure differences in the Gibbs free energy but not its absolute value. Thus, we measure the standard Gibbs free energy (G^o) values of all compounds relative to the standard Gibbs free energy of formation of the constituent elements in their reference states. The latter are defined as zero at all temperatures. Thus, for example

$$G^o_{m,CO} = \Delta_f G^o_{m,CO} \tag{10-14}$$

$$G^o_{m,H_2} = \Delta_f G^o_{m,H_2} = 0 \tag{10-15}$$

The standard molar Gibbs energies of formation ($\Delta_f G^o_m$) at the reaction temperature are tabulated in thermodynamic compilations or can be calculated as illustrated in earlier chapters.

The second term in Eq. (10-12) is the ΔG due to expansion (or compression) of gases from the standard-state pressure of one bar. It can be written as a quotient:

$$RT(\ln P_{H_2O} + \ln P_{CO} - \ln P_{CO_2} - \ln P_{H_2}) = RT \ln\left(\frac{P_{H_2O}P_{CO}}{P_{H_2}P_{CO_2}}\right) \tag{10-16}$$

An equivalent and more commonly used form of Eq. (10-12) is

$$\Delta G = \Delta_r G^o + RT \ln\left(\frac{P_{H_2O}P_{CO}}{P_{H_2}P_{CO_2}}\right) \tag{10-17}$$

Equation (10-17) is the *reaction isotherm* and is a very important equation that we will use frequently. It gives the Gibbs free energy change ΔG for transformation of reactants at some arbitrary partial pressures into products at other arbitrary partial pressures. The term in parenthesis is the *reaction quotient* Q:

$$Q = \left(\frac{P_{H_2O}P_{CO}}{P_{H_2}P_{CO_2}}\right) \tag{10-18}$$

$$\Delta G = \Delta_r G^o + RT \ln Q \tag{10-19}$$

Equation (10-19) is another version of the *reaction isotherm.* The value of the reaction quotient Q varies throughout the reaction because the partial pressures of the gases change as H_2 and CO_2 are consumed to form H_2O and CO.

At equilibrium, $\Delta G = 0$ and Eq. (10-17) reduces to

$$\Delta_r G^o = -RT \ln\left[\frac{P_{H_2O}P_{CO}}{P_{CO_2}P_{H_2}}\right]_{eq} = -RT \ln K_P \tag{10-20}$$

In Eq. (10-20) the value of the reaction quotient at equilibrium is the equilibrium constant K_P:

$$K_P = \left[\frac{P_{H_2O}P_{CO}}{P_{CO_2}P_{H_2}}\right]_{eq} \tag{10-21}$$

The subscript *eq* in Eqs. (10-20) and (10-21) indicates that the reaction quotient involves the equilibrium partial pressures of each gas and is equal to the equilibrium constant for the water-gas reaction. The equilibrium constant for the water-gas reaction and for other gas-phase reactions is written K_P because the reaction quotient is the ratio of gas partial pressures. It can be calculated from the standard Gibbs free energy change ($\Delta_r G^o$) for the water-gas reaction using

$$K_P = \exp\left(-\frac{\Delta_r G^o}{RT}\right) \tag{10-22}$$

Equation (10-23) is another important equation that we will use frequently.

We now substitute Eqs. (10-20) and (10-21) into Eq. (10-19) for the reaction isotherm:

$$\Delta G = -RT \ln K_P + RT \ln Q \tag{10-23}$$

Equation (10-23) is a third version of the *reaction isotherm.* It shows that at equilibrium ΔG is zero because the gas partial pressures in the reaction quotient Q correspond to equilibrium partial pressures, so Q becomes equal to K_P, and they cancel out.

Example 10-1. Hahn (1903) studied the water-gas reaction at one atmosphere total pressure from 686°C to 1405°C. Gas mixtures of $H_2 + CO_2$ (forward reaction) or of $CO + H_2O$ (reverse reaction) flowed through a heated tube containing a platinum catalyst. The German physical chemist Wilhelm

Ostwald (see sidebar) did pioneering research on catalysis for which he won the Nobel Prize in Chemistry in 1909. Ostwald defined a *catalyst* as "a substance that changes the velocity of a reaction without itself being changed by the process." A catalyst speeds up the forward and reverse reactions and does not change the equilibrium constant of a reaction. The metals iridium, osmium, palladium, platinum, rhodium, and ruthenium are good catalysts for chemical reactions.

FRIEDERICH WILHELM OSTWALD (1853–1932)

Wilhelm Ostwald, Svante Arrhenius, and J. H. van't Hoff are the three founders of modern physical chemistry. Ostwald was born in 1853 to German parents living in Riga, the capital of modern-day Latvia, but part of Czarist Russia at that time. He entered the University of Dorpat (now the University of Tartu, Estonia) as a chemistry student in 1872. However, Ostwald was not conscientious and skipped lectures to pursue his interests in landscape painting, music, and other activities. Eventually his father told him to shape up or ship out!

Ostwald took this advice to heart. He taught himself what he missed in the lectures he'd skipped, did an original research project about the law of mass action, and passed his final examinations in 1875. Ostwald continued his graduate studies at Dorpat. He received his doctorate in 1878 and was a high school teacher for three years before becoming a professor at the Polytechnic Institute in Riga. During this period (1881–1887) Ostwald did some of his most important research (on catalysis) using homemade equipment because of the acute lack of facilities and funds. He also began his lifelong friendship with Svante Arrhenius and cofounded (with J. H. van't Hoff) the *Zeitschrift für Physikalische Chemie* (*Journal of Physical Chemistry*), which Ostwald edited for many years.

In 1887, Ostwald moved to the University of Leipzig, Germany. During 1887–1905 Ostwald's laboratory in Leipzig was a Mecca for many physical chemistry students from Germany, England, and the United States. Ostwald published numerous textbooks and hundreds of scientific papers throughout his life, translated Gibbs's papers into German (1892), and in 1889 founded the series *Klassiker der exakten Wissenschaften* (*Classics of Exact Science*). Number 1 in this series is "Über die Erhaltung der Kraft," by Helmholtz. Ostwald received the 1909 Nobel Prize in Chemistry for his work on catalysis. He resigned his professorship in 1905 and devoted the rest of his life to color theory, the history and the philosophy of science, and pacifism.

Table 10-1 Water Gas Reaction at 786°C (Hahn, 1903)

No.	Equil. Conc. (mole %)			K_P
	CO_2	CO (= H_2O)	H_2	
1	42.34	22.10	13.46	0.857
2	42.25	22.14	13.47	0.861
3	42.42	22.10	13.38	0.861
4	31.86	24.03	20.08	0.903
5	31.50	24.06	20.38	0.902

Mean $K_P \pm 1\sigma = 0.877 \pm 0.024$.

Hahn analyzed the gas mixture after it flowed through the furnace to measure the equilibrium concentrations (mole %) of H_2, CO_2, CO, and H_2O. Table 10-1 lists his results at 786°C (1059 K). (a) Calculate K_P for the first experiment, and (b) compute $\Delta_r G^o$ from the mean K_P value.

(a) We first calculate equilibrium partial pressures from the equilibrium concentrations:

$$P_{CO_2} = \frac{42.34\%}{100\%} \times 1 = 0.4234 \text{ atm} \tag{10-24}$$

$$P_{H_2} = \frac{13.46\%}{100\%} \times 1 = 0.1346 \text{ atm} \tag{10-25}$$

$$P_{CO} = P_{H_2O} = \frac{22.10\%}{100\%} \times 1 = 0.2210 \text{ atm} \tag{10-26}$$

The equilibrium partial pressures of CO and H_2O are identical and only one value is listed in Table 10-1. Substituting numerical values into Eq. (10-21) K_P for the first experiment is

$$K_P = \left[\frac{P_{H_2O}P_{CO}}{P_{H_2}P_{CO_2}}\right]_{eq} = \frac{(0.2210)\cdot(0.2210)}{(0.4234)\cdot(0.1346)} = 0.857 \tag{10-27}$$

In general, the equilibrium partial pressure of a gas is its mole fraction (X_i) times the total pressure (P_T). The mole fraction of a gas is the number of moles of the gas divided by the total number of moles of all gases and is the mole percentage of a gas divided by 100.

(b) Hahn's five experiments give a mean K_P value ($\pm 1\ \sigma$) of 0.877 ± 0.024. The Gibbs free energy of reaction at 1059 K is

$$\Delta_r G^o = -RT \ln K_P = -8.314 \cdot 1059.15 \cdot \ln 0.877 = 1156 \text{ J mol}^{-1} \tag{10-28}$$

The ± 0.024 uncertainty on K_P translates into an uncertainty of about ± 240 J mol^{-1} in the Gibbs free energy of reaction ($\Delta_r G^o$). Data from the NIST-JANAF tables (Chase, 1999) give $K_P = 0.881$ and $\Delta_r G^o = 1118$ J mol^{-1} at 1059 K for the water-gas reaction. These agree with Hahn's average K_P and $\Delta_r G^o$ values within his errors.

B. Illustration that K_P is a constant

The equilibrium constant K_P is a constant at a given temperature. We illustrate this point by calculating K_P for experiments 2–5 in Table 10-1. (We computed $K_P(1) = 0.857$ in the last section.) The K_P values for experiments 2–5 are

$$K_P(2) = \left[\frac{P_{H_2O}P_{CO}}{P_{H_2}P_{CO_2}}\right]_{eq} = \frac{(0.2214)\cdot(0.2214)}{(0.1347)\cdot(0.4225)} = 0.861 \quad (10\text{-}29)$$

$$K_P(3) = \left[\frac{P_{H_2O}P_{CO}}{P_{H_2}P_{CO_2}}\right]_{eq} = \frac{(0.2210)\cdot(0.2210)}{(0.1338)\cdot(0.4242)} = 0.861 \quad (10\text{-}30)$$

$$K_P(4) = \left[\frac{P_{H_2O}P_{CO}}{P_{H_2}P_{CO_2}}\right]_{eq} = \frac{(0.2403)^2}{(0.2008)\cdot(0.3186)} = 0.903 \quad (10\text{-}31)$$

$$K_P(5) = \left[\frac{P_{H_2O}P_{CO}}{P_{H_2}P_{CO_2}}\right]_{eq} = \frac{(0.2406)^2}{(0.2038)\cdot(0.3150)} = 0.902 \quad (10\text{-}32)$$

The average of the five K_P values ($\pm 1\sigma$) is $K_p = 0.877 \pm 0.024$. Table 10-1 shows that different percentages of CO_2, H_2, H_2O, and CO are present at equilibrium in each of the five experiments, but the K_P values for the five experiments are (essentially) the same. This demonstrates the constancy of K_P at a given temperature.

C. Initial and equilibrium concentrations

In the last two sections we calculated K_P values using equilibrium compositions for the water-gas reaction. Hahn (1903) studied this reaction by passing either $H_2 + CO_2$ or $H_2O + CO$ gas mixtures through heated tubes. Thus, the initial and equilibrium compositions of the gas were different. His experimental procedure is the same as that used in many other studies. In general, we want to know how to use mass balance to relate the initial composition of the gas mixture to the equilibrium composition. Then we can use the initial gas composition and analytical data to calculate K_p. Or, if we have the K_P value already (e.g., calculated from thermal data via the third law), then we can calculate the composition of the equilibrated gas mixture.

In the case of the water-gas reaction, heating a $H_2 + CO_2$ mixture produces a mixture of the four gases H_2, CO_2, H_2O, and CO. The mass balance (i.e., the stoichiometry) of reaction (10-1) shows that some of the H_2 in the initial gas mixture reacts to produce H_2O and some of the CO_2 in the initial gas mixture reacts to give CO. Denoting the initial concentrations by a superscript I and equilibrium concentrations by a superscript *eq,* we then have

$$P_{H_2}^{eq} = P_{H_2}^{I} - P_{H_2O}^{eq} \quad (10\text{-}33)$$

$$P_{CO_2}^{eq} = P_{CO_2}^{I} - P_{CO}^{eq} \quad (10\text{-}34)$$

$$P_{H_2O}^{eq} = P_{CO}^{eq} = x \quad (10\text{-}35)$$

For simplicity, we also define the following variables:

$$P^I_{CO_2} = I_{CO_2} \tag{10-36}$$

$$P^I_{H_2} = I_{H_2} \tag{10-37}$$

Then we can write the equilibrium constant expression for the water-gas reaction as

$$K_P = \left[\frac{P_{H_2O}P_{CO}}{P_{H_2}P_{CO_2}}\right]_{eq} = \frac{x^2}{(I_{H_2} - x)\cdot(I_{CO_2} - x)} \tag{10-38}$$

Rearranging, we get

$$K_P(I_{H_2} - x)(I_{CO_2} - x) - x^2 = 0 \tag{10-39}$$

$$K_P(I_{CO_2}I_{H_2} - I_{CO_2}x - I_{H_2}x + x^2) - x^2 = 0 \tag{10-40}$$

$$(K_P - 1)x^2 - K_P(I_{CO_2} + I_{H_2})x + K_P I_{H_2} I_{CO_2} = 0 \tag{10-41}$$

Equation (10-41) is the quadratic equation with a, b, and c given by

$$ax^2 + bx + c = 0 \tag{10-42}$$

$$a = (K_P - 1) \tag{10-43}$$

$$b = -K_P(I_{CO_2} + I_{H_2}) \tag{10-44}$$

$$c = K_P I_{H_2} I_{CO_2} \tag{10-45}$$

The general solution to the quadratic equation is

$$x = \frac{-b \pm \sqrt{b^2 - 4ac}}{2a} \tag{10-46}$$

and x must be positive in this case because it is the equilibrium partial pressure of H_2O (or CO) that is made in reaction (10-1). We now work an example using data from Table 10-2.

Example 10-2. Hahn (1903) used five different $H_2 + CO_2$ mixtures containing 10–70 mole % CO_2 to measure K_P for the water-gas reaction at 986°C (1259 K). His average $K_P = 1.601$ for two experiments done using 10.1% CO_2 and 89.9% H_2 at one atmosphere total pressure. Use this K_P value to calculate the equilibrium partial pressures of H_2, CO_2, H_2O, and CO.

We need to solve for x in Eq. (10-41). The initial H_2 and CO_2 partial pressures are

$$P^I_{CO_2} = \frac{10.1}{100} \times 1 = 0.101 \text{ atm} = I_{CO_2} \tag{10-47}$$

Table 10-2 WaterGas Reaction at 986 °C (Hahn, 1903)

Initial Conc. (mole %)		Equil. Conc. (mole %)			
CO_2	H_2	CO_2	H_2	CO (= H_2O)	K_p
10.1	89.9	0.70	80.38	9.46	1.591
		0.67	80.67	9.33	1.611
30.1	69.9	7.18	46.82	23.00	1.574
		7.12	47.04	22.92	1.568
49.1	51.9	21.52	22.82	27.83	1.577
		20.78	23.14	28.04	1.635
60.9	39.1	34.67	12.77	26.28	1.560
		34.20	12.58	26.61	1.646
70.3	29.7	47.66	6.76	22.79	1.612
		47.35	6.95	22.85	1.587

Mean $K_P \pm 1\sigma = 1.596 \pm 0.029$.

$$P^I_{H_2} = \frac{89.9}{100} \times 1 = 0.899 \text{ atm} = I_{H_2} \quad (10\text{-}48)$$

The a, b, and c terms in the quadratic equation are

$$a = (K_P - 1) = 1.601 - 1 = 0.601 \quad (10\text{-}49)$$

$$b = -K_P(I_{CO_2} + I_{H_2}) = -1.601(0.101 + 0.899) = -1.601 \quad (10\text{-}50)$$

$$c = K_P I_{H_2} I_{CO_2} = 1.601 \times 0.101 \times 0.899 = 0.1454 \quad (10\text{-}51)$$

Substituting these values into Eq. (10-46), the general solution for a quadratic equation gives:

$$x = \frac{-b \pm \sqrt{b^2 - 4ac}}{2a} = \frac{1.601 \pm \sqrt{(-1.601)^2 - 4(0.601)(0.1454)}}{2(0.601)}$$
$$= \frac{1.601 \pm \sqrt{2.2137}}{1.202} = \frac{1.601 \pm 1.4878}{1.202} = \frac{1.601 - 1.4878}{1.202} = 0.0941 \quad (10\text{-}52)$$

This is the only physically realistic solution. The alternative solution gives $x = 2.570$ and is physically unrealistic because x must be less than the total pressure of one atmosphere. Thus, the equilibrium partial pressures are

$$P^{eq}_{H_2O} = P^{eq}_{CO} = x = 0.0941 \text{ atm} \quad (10\text{-}53)$$

$$P^{eq}_{H_2} = P^I_{H_2} - P^{eq}_{H_2O} = 0.899 - 0.0941 = 0.8049 \text{ atm} \quad (10\text{-}54)$$

$$P_{CO_2}^{eq} = P_{CO_2}^{I} - P_{CO}^{eq} = 0.101 - 0.0941 = 0.0069 \text{ atm} \quad (10\text{-}55)$$

The corresponding equilibrium concentrations are 9.41% each CO and H_2O, 80.49% H_2, and 0.69% CO. These results are virtually the same as the average concentrations from the first two rows of Table 10-2, which are 9.40% each CO and H_2O, 80.52% H_2, and 0.68% CO.

This example also illustrates *Le Chatelier's principle.* The French physical chemist Henri Le Chatelier (see sidebar) stated that a system under stress acts to relieve the stress and, if possible, to return to equilibrium. In this case, the system is the $H_2 + CO_2$ gas mixture in the heated tube. The initial gas mixture contains significantly more H_2 and CO_2 than their equilibrium concentrations. Thus, upon heating, the water-gas reaction proceeds to the right:

$$CO_2 + H_2 \rightarrow H_2O + CO \quad (10\text{-}56)$$

to relieve the stress (excess H_2 and CO_2) and return the system to equilibrium.

The equilibrium gas composition—but not the equilibrium constant K_p—is significantly different in the five sets of experiments. The equilibrium gas composition is different because the five different gas mixtures have five different C-H-O elemental ratios. Thus, the final distribution of carbon between CO_2 and CO, the final distribution of oxygen between CO_2, CO, and H_2O, and the final distribution of hydrogen between H_2 and H_2O, must be different in the five sets of experiments. However, K_p is the same because it is given by the ratio of these gases:

$$K_P = \left[\frac{P_{H_2O}P_{CO}}{P_{H_2}P_{CO_2}}\right]_{eq} = \left(\frac{P_{H_2O}}{P_{H_2}}\right)_{eq}\left(\frac{P_{CO}}{P_{CO_2}}\right)_{eq} \quad (10\text{-}57)$$

HENRI-LOUIS LE CHATELIER (1850–1936)

Henri-Louis Le Chatelier came from a long line of scientists and engineers and grew up surrounded by opportunities to pursue scientific endeavors. Le Chatelier was always a promising student, but he had to interrupt his studies to serve as an officer during the Franco-Prussian War in 1870.

Le Chatelier graduated from the École des Mines in 1875. Two years later G. A. Daubrée, the director of the École des Mines, appointed him as professor of general chemistry, a position he held until 1887, when he became professor of industrial chemistry. Le Chatelier held this position until his retirement in 1919. At various times during this period, he also held other professorships at the College de France, the Sorbonne, and the École des Mines. Le Chatelier served on numerous national committees, including the Commission on Weights and Measures, and in 1904 founded the *Revue de Métallurgie* (*Review of Metallurgy*).

Much of Le Chatelier's research was in materials science and metallurgy. He studied the hydration and phase relations of the important constituents of cement. This work involved high-temperature measurements, which led to his invention of the platinum-rhodium thermocouple and improvements in optical pyrometry. His book *High Temperature Measurements* went through two editions and was translated into English. As a result of his work on cement hydration, Le Chatelier became interested in polymorphic transformations in solids. He used an improved dilatometer to discover the low-high (α-β) transition in quartz. After he became professor of metallurgy at the École des Mines he invented a new type of metallurgical microscope that improved photography of metals and alloys. Le Chatelier's principle resulted from his work on reversible equilibria in materials science and metallurgy and was described in a 126-page memoir published in the *Annales des Mines* in 1888.

D. Temperature dependence of equilibrium constants

Tables 10-1 and 10-2 show that K_p values are temperature dependent. The effect of temperature on the equilibrium constant is described by the *van't Hoff equation*, which was originally derived by the Dutch physical chemist J. H. van't Hoff in 1885 (see sidebar).

JACOBUS HENRICUS VAN'T HOFF (1852–1911)

J. H. van't Hoff, Wilhelm Ostwald, and Svante Arrhenius are the three founders of modern physical chemistry. Van't Hoff was born in Rotterdam in 1852 and studied at universities in Bonn, Delft, Leiden, Paris, and Utrecht, where he received his doctorate in 1874.

Around this time, he published his famous paper about stereochemistry of carbon compounds. Earlier, Louis Pasteur had discovered that tartaric acid exists in two forms that are chemically indistinguishable but that, when in solution, will rotate a beam of polarized light passing through to the right or the left. Van't Hoff showed that this optical activity (or isomerism) arose from the tetrahedral arrangement in space of four atoms around a carbon atom.

There are two possible ways, which are mirror images of one another, to position four different atoms in a tetrahedron around carbon.

In 1878 van't Hoff became professor of chemistry, mineralogy, and geology at the newly founded University of Amsterdam. During his 18 years in Amsterdam, he wrote the first book on chemical kinetics, developed the thermodynamic theory of solutions that explains osmotic pressure, freezing point depression, and boiling point elevation, and derived the van't Hoff equation for the temperature dependence of equilibrium constants.

Along with Wilhelm Ostwald, van't Hoff founded the *Zeitschrift für Physikalische Chemie* (*Journal of Physical Chemistry*) in 1887. Van't Hoff moved to the University of Berlin in 1896; there he used phase equilibria to study formation of the Stassfurt salt deposits. This work founded modern experimental petrology and was the impetus for starting the Geophysical Laboratory at the Carnegie Institution. Van't Hoff won the first Nobel Prize for Chemistry in 1901. This was awarded for his work on chemical kinetics, chemical equilibria, and thermodynamics of solutions. He died at age 59 in 1911 from tuberculosis.

We first discuss the temperature dependence of ΔG and start with Eq. (6-116), which is one form of the *Gibbs-Helmholtz equation*:

$$\Delta G = \Delta H - T\Delta S \tag{6-116}$$

Equation (6-116) describes the Gibbs free energy change for a reversible isothermal change in state. Dividing through by temperature gives

$$\frac{\Delta G}{T} = \frac{\Delta H}{T} - \Delta S \tag{10-58}$$

Differentiating Eq. (10-58) with respect to temperature at constant pressure gives

$$\left[\frac{\partial(\Delta G/T)}{\partial T}\right]_P = -\frac{\Delta H}{T^2} \tag{10-59}$$

Equation (10-59) is another version of the *Gibbs-Helmholtz equation*. The intermediate steps in its derivation are the subject of Problem 10-18 at the end of this chapter. Equation (10-59) is often derived by assuming that ΔH and ΔS are temperature independent, which is incorrect and also unnecessary. We can rewrite Eq. (10-59) by using the identity

$$d\frac{1}{T} = -\frac{1}{T^2}dT \tag{7-50}$$

$$\left[\frac{\partial(\Delta G/T)}{\partial(1/T)}\right]_P = \Delta H \tag{10-60}$$

Example 10-3. Use the Gibbs-Helmholtz equation (10-60) to calculate the average ΔH for the water-gas reaction from Hahn's K_P data over the 1059−1359 K range (see Table 10-3).

We compute the $\Delta_r G^o$ values using Eq. (10-20). The corresponding $\Delta_r G^o/T$ values are graphed versus inverse temperature in Figure 10-1. A linear least-squares fit to these data is

$$\frac{\Delta_r G^o}{T} = -29.5(\pm 0.1) + \frac{32,439(\pm 700)}{T} \tag{10-61}$$

Table 10-3 K_P for the Water-Gas Reaction (Hahn, 1903)

T (K)	K_P	$\Delta_r G^o$ (J mol^{-1})	$\Delta_r G^o/T$ (J mol^{-1} K^{-1})
1059	0.872	1205.9	1.1387
1159	1.208	−1820.9	−1.5711
1259	1.596	−4893.5	−3.8868
1359	1.956	−7580.3	−5.5778

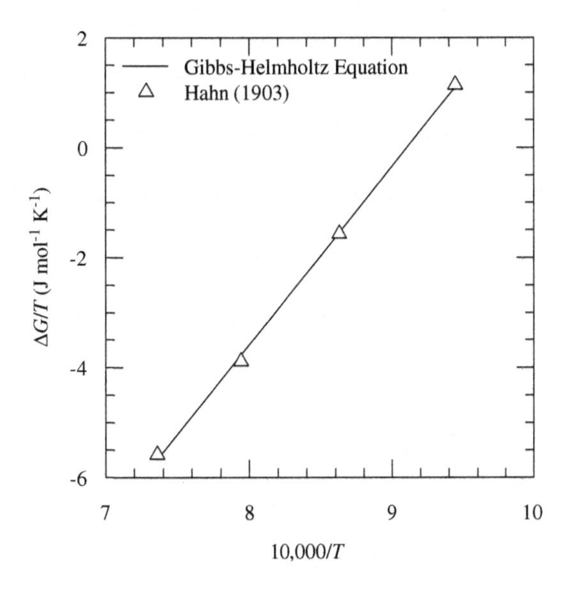

FIGURE 10-1

An illustration of the Gibbs-Helmholtz equation (10-53). A plot of $\Delta_r G/T$ for the water-gas reaction versus inverse temperature. The line slope gives the average reaction enthalpy (32,429 J mol^{-1}) over the 1059–1359 K temperature range. The equilibrium constant data of Hahn (1903) were used to calculate the Gibbs free energy of reaction (see Table 10-3).

The average $\Delta_r H^o$ for the water-gas reaction is equal to the slope of $32,439 \pm 700$ J mol^{-1}. Within error, the $\Delta_r H^o$ calculated from the data of Hahn (1903) is identical to the average $\Delta H = 32,912$ J mol^{-1} calculated from the NIST-JANAF tables over the same temperature range (Chase, 1999).

We now consider the temperature dependence of K_P and start by rearranging Eq. (10-20):

$$-R \ln K_P = \frac{\Delta_r G^o}{T} \tag{10-62}$$

Differentiating Eq. (10-62) with respect to temperature at constant pressure gives

$$\left[\frac{\partial(-R\ln K_P)}{\partial T}\right]_P = \left[\frac{\partial(\Delta_r G^o/T)}{\partial T}\right]_P = -\frac{\Delta_r H^o}{T^2} \quad (10\text{-}63)$$

Equation (10-63) is a form of the *van't Hoff isobar*, which is so named because differentiation is done at constant pressure. Equivalent versions of the van't Hoff isobar are

$$\left(\frac{\partial \ln K_P}{\partial T}\right)_P = \frac{\Delta_r H^o}{RT^2} \quad (10\text{-}64)$$

$$\left[\frac{\partial \ln K_P}{\partial(1/T)}\right]_P = \frac{-\Delta_r H^o}{R} \quad (10\text{-}65)$$

Equations (10-63)–(10-65) apply only to equilibrium constants defined in terms of partial pressures (or fugacities). When an equilibrium constant is defined in terms of concentrations (as discussed in a few pages), differentiation is done at constant volume and the slope of ln K versus $1/T$ is the internal energy of reaction ($\Delta_r E$).

The van't Hoff equation predicts three qualitatively different types of behavior for chemical equilibria depending on the value of $\Delta_r H^o$. The equilibrium constant will be a weak function of temperature for reactions that are essentially thermoneutral, that is, $\Delta_r H^o \sim 0$. The hydrolysis of ethyl acetate that was studied by Berthelot and St. Gilles is the classic example of this type of reaction:

$$CH_3COOC_2H_5 \text{ (ethyl acetate)} + H_2O = CH_3COOH \text{ (acetic acid)} + C_2H_5OH \text{ (ethanol)} \quad (9\text{-}1)$$

Reaction (9-1) has $\Delta_r H^o_{298} = -3.7$ kJ mol^{-1}, and its K_{eq} (which is given by molar concentrations):

$$K_{eq} = \frac{[C_2H_5OH]\cdot[CH_3COOH]}{[H_2O]\cdot[CH_3COOC_2H_5]} \approx 4 \quad (10\text{-}66)$$

is essentially temperature independent (Parks and Huffman, 1932). The water-gas reaction (10-1) has a larger average $\Delta_r H^o$, but it behaves as a thermoneutral reaction from 1000 K to 1900 K because the slope of ln K_P versus $1/T$ is shallow. Thus, K_P for the water-gas reaction only varies from 0.7 to 4.2 over this temperature range. Isotopic exchange equilibria (Urey, 1947) are also examples of reactions that are essentially thermoneutral.

In contrast, the equilibrium constant K_P increases with increasing temperature for endothermic reactions (positive $\Delta_r H^o$ values). Nitric oxide formation from heated air is the classic example of this type of reaction:

$$N_2 + O_2 = 2NO \quad (10\text{-}67)$$

Finally, K_P decreases with increasing temperature for exothermic reactions (negative $\Delta_r H^o$ values). Sulfur trioxide formation from SO_2 and O_2 exemplifies this behavior:

$$SO_2 + \frac{1}{2}O_2 = SO_3 \quad (10\text{-}68)$$

Table 10-4 Nernst's Data for the NO Equilibrium

T (K)	% N_2	% O_2	% NO
1811	78.92	20.72	0.37
1877	78.89	20.69	0.42
2033	78.78	20.58	0.64
2195	78.61	20.42	0.97
2580	78.08	19.88	2.05
2675	77.98	19.78	2.23

Example 10-4. Nitric oxide (NO) is formed by high-temperature combustion of air. This occurs in car engines, furnaces, lightning bolts, atomic bomb explosions, and large impacts on the earth at Tunguska, Siberia, in 1908 and at the end of the Cretaceous period, 65 million years ago (Prinn and Fegley, 1987). Nernst and colleagues measured K_P for NO formation from 1811 K to 2675 K via reaction (10-67). Their data are given in Table 10-4 (Nernst, 1907). Use the van't Hoff isobar equation (10-65) to compute the average $\Delta_r H^o$ over the temperature range studied.

The equilibrium constant K_P for reaction (10-67) is

$$K_P = \left[\frac{P_{NO}^2}{P_{N_2}P_{O_2}}\right]_{eq} = \left[\frac{X_{NO}^2}{X_{N_2}X_{O_2}}\right]_{eq} \tag{10-69}$$

The number of molecules on both sides of Eq. (10-67) is the same. Consequently, the total pressure cancels out of Eq. (10-69), which can be written in terms of the mole fractions of NO, N_2, and O_2 ($P_i = X_i P_T$). From data in Table 10-4 we compute $K_P = 2.53 \times 10^{-4}$ at 2033 K. The other K_P values are calculated the same way and are plotted as ln K_P versus inverse temperature in Figure 10-2. The data in this figure and in Table 10-4 show that K_P increases with increasing temperature. Thus, NO formation is endothermic. The equation of the line in Figure 10-2 is

$$\ln K_P = 2.24(\pm 0.08) - \frac{21{,}239(\pm 500)}{T} \tag{10-70}$$

The K_P value is for the formation of two moles of NO. Thus, the slope of $-21{,}239(\pm 500)$ K corresponds to an average $\Delta_r H^o = 88{,}290(\pm 2080)$ J mol^{-1} of NO formed. To first approximation this is the $\Delta_r H^o$ for NO formation at 2243 K, the midpoint of the temperature range studied by Nernst and colleagues. For comparison, the NIST-JANAF tables (Chase, 1999) give $\Delta_r H^o = 90{,}420$ J mol^{-1} of NO formed at the same temperature.

As mentioned above, the formation of SO_3 from SO_2 and O_2 is an exothermic reaction:

$$SO_2 + \frac{1}{2}O_2 = SO_3 \tag{10-68}$$

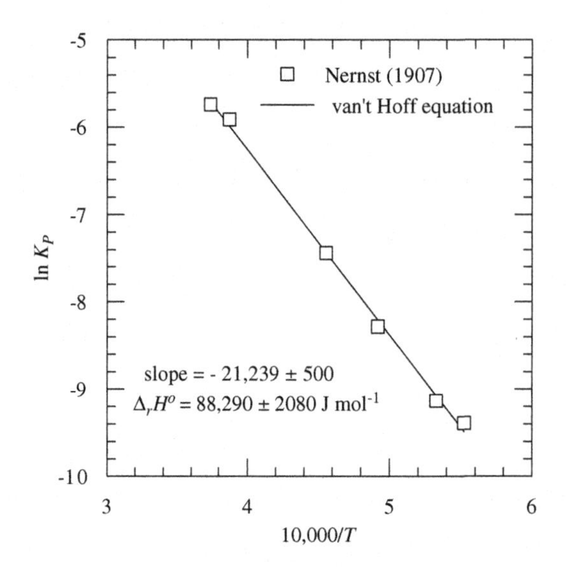

FIGURE 10-2

An illustration of the van't Hoff equation (10-58). A plot of ln K_P for NO formation versus inverse temperature. The line slope gives $\Delta_r H^o = 88{,}290 \pm 2080$ J mol^{-1} of NO formed.

Reaction (10-68) controls the equilibrium abundance of SO_3 in volcanic gases on Venus, Earth, Io, and Mars (when it was volcanically active). The equilibrium constant for reaction (10-68) is

$$K_P = \left[\frac{P_{SO_3}}{P_{SO_2}P_{O_2}^{1/2}}\right]_{eq} \tag{10-71}$$

Three different measurements of K_P (Bodenstein and Pohl, 1905; Bödlander and Köppen, 1903; Knietsch, 1901) are plotted in Figure 10-3, which shows that K_P for this exothermic reaction decreases with increasing temperature, in accord with the van't Hoff equation.The line is the best fit to all three data sets and has the equation

$$\log K_P = -4.60(\pm 0.12) + \frac{4950(\pm 185)}{T} \tag{10-72}$$

Base 10 logarithms are used in Figure 10-3 and Eq. (10-65) is rewritten as

$$\left[\frac{\partial \log K_p}{\partial(1/T)}\right]_P = \frac{-\Delta_r H^o}{(\ln 10)R} \sim \frac{-\Delta_r H^o}{2.30258 \cdot R} \tag{10-73}$$

The approximation arises because the natural logarithm of 10 is a transcendental number that is approximately equal to 2.30258, but cannot be exactly specified. The enthalpy for Eq. (10-68) is

$$\Delta_r H^o = -(\ln 10) \cdot 8.314 \cdot 4950(\pm 185) = -94{,}760(\pm 3540)\text{J mol}^{-1}\text{SO}_3 \tag{10-74}$$

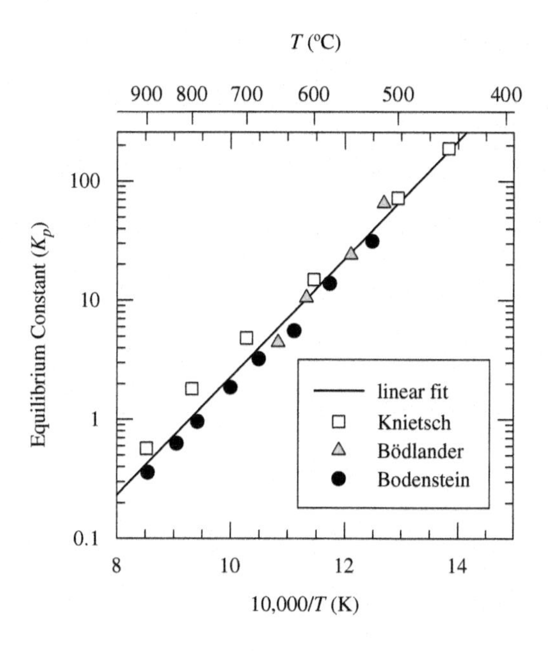

FIGURE 10-3

A plot of log K_P versus inverse temperature for the reaction $2\ SO_2 + O_2 = 2\ SO_3$ using the data of Knietsch (1901), Bödlander and Köppen (1903), and Bodenstein and Pohl (1905). The line slope gives $\Delta_r H^o = 94{,}760 \pm 3{,}560$ J mol^{-1} of SO_3 formed.

The midpoint temperature for the K_P values is 950 K where $\Delta_r H^o = -97{,}940$ J mol^{-1} according to the NIST-JANAF tables. The two $\Delta_r H^o$ values agree within the uncertainty of the data.

Example 10-5. Bodenstein and Pohl (1905) studied the thermal decomposition of SO_3:

$$2\ SO_3 = 2\ SO_2 + O_2 \tag{10-75}$$

from 800 K to 1170 K at a total pressure of about one atmosphere. Reaction (10-75) is the reverse of Eq. (10-68) and involves twice as many moles of all gases. Bodenstein and Pohl (1905) give the concentrations of SO_3, SO_2, and O_2 in moles per liter and calculated the equilibrium constant:

$$K'_C = \left[\frac{c^2_{SO_2} c_{O_2}}{c^2_{SO_3}}\right]_{eq} \tag{10-76}$$

The prime denotes the reverse reaction. Recompute their data in terms of K_P for reaction (10-68).

The K'_C values as a function of temperature from Bodenstein and Pohl (1905) are given in the first two columns of Table 10-5. All three gases SO_3, SO_2, and O_2 behave ideally at the P-T conditions studied and their concentrations (mol L^{-1}) are computed with the ideal gas law:

$$c_i = \frac{n_i}{V_i} = \frac{P_i}{RT} \tag{10-77}$$

Table 10-5 SO_3 Equilibrium Data (Bodenstein and Pohl, 1905)

T (K)	K'_C	(RT)	K'_P	$(1/K'_P)^{1/2}$
801	1.55×10^{-5}	65.78	1.02×10^{-3}	31.3
852	7.55×10^{-5}	69.97	5.28×10^{-3}	13.8
900	3.16×10^{-4}	73.91	2.34×10^{-2}	5.54
953	1.12×10^{-3}	78.27	8.77×10^{-2}	3.24
1000	3.54×10^{-3}	82.13	0.291	1.86
1062	1.26×10^{-2}	87.22	1.099	0.956
1105	2.80×10^{-2}	90.75	2.54	0.627
1170	8.16×10^{-2}	96.09	7.84	0.358

Rewriting Eq. (10-76) using Eq. (10-77) gives a simple relationship between K'_C and K'_P:

$$K'_C = \left[\frac{c_{SO_2}^2 c_{O_2}}{c_{SO_3}^2}\right]_{eq} = \left[\left(\frac{P_{SO_2}^2}{RT}\right)\left(\frac{P_{O_2}}{RT}\right)\Big/\left(\frac{P_{SO_3}^2}{RT}\right)\right]_{eq} \quad (10\text{-}78)$$

$$K'_C = \left[(RT)\left(\frac{P_{SO_2}^2}{RT}\right)\left(\frac{P_{O_2}}{RT}\right)\Big/P_{SO_3}^2\right]_{eq} \quad (10\text{-}79)$$

$$K'_C = \frac{1}{(RT)}\left[\frac{P_{SO_2}^2 P_{O_2}}{P_{SO_3}^2}\right]_{eq} = \frac{1}{(RT)}K'_P \quad (10\text{-}80)$$

The third and fourth columns in Table 10-5 give (RT), and the calculated K'_P value. The fifth column gives K_P for reaction (10-68). This is equal to $1/(K'_P)^{1/2}$ because reaction (10-75) is for two moles of each gas and is the reverse of reaction (10-68). The K_P values in the fifth column are plotted in Figure 10-3 as the points labelled *Bodenstein*. The general equation for converting between equilibrium constants for forward (K_f) and reverse reactions (K_r) is

$$K_f = \frac{1}{K_r} \quad (10\text{-}81)$$

The general equation for converting between equilibrium constants in terms of concentrations (K_C) and in terms of partial pressures (K_P) is

$$K_P = K_C(RT)^{\Delta\nu} \quad (10\text{-}82)$$

where $\Delta\nu$ is the total moles of products minus the total moles of reactants.

As mentioned earlier, when equilibrium constants are defined in terms of concentrations, the van't Hoff equation is differentiated at constant volume. The resulting equation is analogous to Eq. (10-65) and shows that the slope of ln K_C versus $1/T$ is the internal energy of reaction ($\Delta_r E$):

$$\left[\frac{\partial \ln K_C}{\partial(1/T)}\right]_V = \frac{-\Delta_r E^o}{R} \quad (10\text{-}83)$$

E. Integration of the van't Hoff equation

The van't Hoff equation (10-64) is similar to the Clausius-Clapeyron equation (7-45) and is integrated in a similar manner. The simplest assumption of a constant $\Delta_r H^o$ implies $\Delta C_P = 0$ for the reaction being studied. Even if this assumption is incorrect, it is useful as long as $\Delta_r H^o$ is much larger than the ($\Delta T \cdot \Delta C_P$) product. Integration of Eq. (10-64) yields

$$\ln K_P = -\frac{\Delta_r H^o}{RT} + c \quad (10\text{-}84)$$

This derivation shows why Eq. (10-84), or its analog using $\log_{10} K_P$ were used to fit the K_P data plotted in Figures 10-2 and 10-3. If the enthalpy of reaction is known (e.g., from calorimetry), the integration constant c can be evaluated from a single K_P value. Integrating Eq. (10-64) between two temperatures T_1 and T_2 gives

$$\int_{T_1}^{T_2} d \ln K_P = \int_{T_1}^{T_2} \frac{\Delta_r H^o}{RT^2} dT = \frac{\Delta_r H^o}{R} \int_{T_1}^{T_2} \frac{1}{T^2} dT \quad (10\text{-}85)$$

$$\ln\left(\frac{K_{P,2}}{K_{P,1}}\right) = -\frac{\Delta_r H^o}{R}\left(\frac{1}{T_2} - \frac{1}{T_1}\right) = \frac{\Delta_r H^o}{R}\left(\frac{1}{T_1} - \frac{1}{T_2}\right) = \frac{\Delta_r H^o}{R}\left(\frac{T_2 - T_1}{T_1 T_2}\right) \quad (10\text{-}86)$$

The T_1 and T_2 are generally the low (T_1) and high (T_2) temperatures of the experimental study.

Example 10-6. Essentially all chlorine in terrestrial volcanic gases is emitted as HCl (Symonds et al., 1994). The distribution of Cl between HCl and Cl_2 is controlled by the Deacon reaction:

$$HCl + \frac{1}{4} O_2 = \frac{1}{2} H_2O + \frac{1}{2} Cl_2 \quad (10\text{-}87)$$

The equilibrium constant for this reaction is

$$K_P = \left[\frac{P_{Cl_2}^{1/2} P_{H_2O}^{1/2}}{P_{HCl} P_{O_2}^{1/4}}\right]_{eq} \quad (10\text{-}88)$$

The American physical chemist G. N. Lewis (see sidebar) led the development of chemical thermodynamics in the United States. In an early paper he measured equilibrium constants for the Deacon reaction (Lewis, 1906). Compute K_P for reaction (10-87) at 1400 K, which is a typical volcanic vent temperature, using Lewis's (1906) data.

GILBERT NEWTON LEWIS (1875–1946)

G. N. Lewis was a precocious child who learned to read at the age of three and was home-schooled until the age of nine. He received his bachelor's, master's, and doctorate degrees from Harvard and studied with Ostwald in Leipzig and Nernst in Gottingen. In 1905, Lewis joined the Research Laboratory for Physical Chemistry at MIT. During his seven years at MIT, Lewis developed the key concepts of activity and fugacity and developed the field of modern chemical thermodynamics. In 1912, Lewis moved to Berkeley as dean of the College of Chemistry and chair of the Chemistry Department. He remained there until his death in 1946. Harold Urey (Nobel Prize in Chemistry, 1934), Joseph Mayer, and Glenn Seaborg (Nobel Prize in Chemistry, 1951) were students or postdoctoral researchers with him.

In 1923, Lewis published two books that are among the most important chemistry books ever written. *Thermodynamics and the Free Energy of Chemical Substances*, by Lewis and Randall, summarized 25 years of work by Lewis and colleagues and became the standard text in chemical thermodynamics for decades. The second book, *Valence and the Structure of Atoms and Molecules*, describes Lewis's theories of chemical bonding. Out of this theory came the familiar Lewis dot structures, which are drawings of the atoms and bonds in a molecule showing only the valence shell electrons.

The concept of Lewis acids and bases is related to his valence bond theory. An *acid* is a substance that accepts electrons and a *base* is a substance that donates electrons. Lewis's definition of acids and bases has broad applicability and is widely used today.

Table 10-6 lists the five K_P values measured by Lewis (1906) from 625–692 K. He also gives $\Delta_r H^o = -6{,}900 \text{ cal mol}^{-1}$, which is based on calorimetric measurements by Julius Thomsen (see sidebar in Chapter 5). The integration constant c was evaluated for each point using Eq. (10-84) giving an average value of $c = -1.808$. The temperature dependence of K_P is

$$\log K_P = \frac{1508}{T} - 1.808 \quad (10\text{-}89)$$

Equation (10-89) gives $K_P = 0.19$ at 1400 K. This is close to $K_P = 0.21$ at the same temperature calculated from the thermochemical data in the NIST-JANAF tables. Problem 10-19 at the end of this

Table 10-6 Lewis's Data for the Deacon Reaction

T (K)	K_P	c
625	3.95	−1.816
625	4.15	−1.795
659	2.94	−1.820
659	3.01	−1.810
692	2.40	−1.799

chapter uses our derived K_P equation to calculate the Cl_2 abundance in volcanic gas emitted by the Erta' Ale volcano in Ethiopia.

More sophisticated integration of the van't Hoff equation is done using Kirchhoff's equation (5-26) and a ΔC_P value that is assumed to be constant with temperature. Then the temperature-dependent enthalpy of reaction is given by

$$\Delta_r H_T^o = \Delta_r H_{ref}^o + \int_{T_{ref}}^{T} \Delta C_P dT = (\Delta_r H_{ref}^o - T_{ref}\Delta C_P) + T\Delta C_P \tag{10-90}$$

In Eq. (10-90) T_{ref} is a reference temperature such as 298.15 K and $\Delta_r H_{ref}^o$ is the standard enthalpy of reaction at this temperature. Substituting Eq. (10-90) into Eq. (10-64) gives

$$\left(\frac{\partial \ln K_p}{\partial T}\right)_P = \frac{\Delta_r H_T^o}{RT^2} = \frac{(\Delta_r H_{ref}^o - T_{ref}\Delta C_P) + T\Delta C_P}{RT^2} \tag{10-91}$$

Rearranging and integrating Eq. (10-91) gives a three-term equation for $\ln K_P$:

$$\ln K_P = A + \frac{B}{T} + C \ln T \tag{10-92}$$

Equation (10-92) is analogous to Eq. (7-64), the three-term vapor pressure equation. The intercept *A* is an integration constant while

$$B = \frac{T_{ref}\Delta C_P - \Delta_r H_{ref}^o}{R} \tag{10-93}$$

$$C = \frac{\Delta C_P}{R} \tag{10-94}$$

Example 10-7. The average $\Delta C_P = -0.23$ cal mol^{-1} K^{-1} for the Deacon reaction over the 600−1400 K range. Lewis (1906) gives $\Delta_r H^o = -6900$ cal mol^{-1}. Assume this applies to $T_{ref} = 600$ K. Use these data to derive an equation like Eq. (10-92) for the temperature dependence of K_P for the Deacon reaction from 600 K to 1400 K. The *B* and *C* terms are calculated first:

$$B = \frac{T_{ref}\Delta C_P - \Delta_r H_{ref}^o}{R} = \frac{(600)(-0.23) + 6900}{1.987} = 3403 \text{ K} \tag{10-95}$$

$$C = \frac{\Delta C_P}{R} = \frac{-0.23}{1.987} = -0.12 \tag{10-96}$$

Then an average A value is computed from the five A values derived from Lewis's five K_P values in Table 10-6. The derived equation for K_P as a function of temperature is

$$\ln K_P = -3.28 + \frac{3403}{T} - 0.12 \ln T \tag{10-97}$$

Equation (10-97) gives $K_P = 0.18$ at 1400 K, which is very close to the value calculated with Eq. (10-87). The small ΔC_P for the Deacon reaction justifies the assumption of a constant enthalpy of reaction for the derivation and use of Eq. (10-89). However, ΔC_P is larger for other reactions and in these cases a three-term equation such as Eq. (10-92) is necessary.

Finally, the van't Hoff equation can be integrated using Kirchhoff's equation and a ΔC_P value that varies with temperature. Depending on the accuracy of the experimental data, a linear equation or more complicated polynomial is used for ΔC_P. For example, if a modified Maier-Kelly heat capacity polynomial with four terms is used,

$$\Delta C_P = \Delta a + \Delta b \times 10^{-3} T + \Delta c \times 10^5 T^{-2} + \Delta d \times 10^{-6} T^2 \tag{5-28}$$

$$\Delta_r H^o = \int \Delta C_P dT = \int \Delta a + \Delta b \times 10^{-3} T + \Delta c \times 10^5 T^{-2} + \Delta d \times 10^{-6} T^2 dT \tag{10-98}$$

$$\Delta_r H^o = \Delta_r H_0^o + \Delta a T + \frac{1}{2} \Delta b \times 10^{-3} T^2 - \Delta c \times 10^5 T^{-1} + \frac{1}{3} \Delta d \times 10^{-6} T^3 \tag{10-99}$$

The $\Delta_r H_0^o$ is an integration constant, but it is not the actual enthalpy of reaction at absolute zero. Substituting Eq. (10-99) into the van't Hoff isobar (10-64),

$$\left(\frac{\partial \ln K_p}{\partial T}\right)_P = \frac{\Delta_r H_0^o}{RT^2} + \frac{\Delta a}{RT} + \frac{\Delta b \times 10^{-3}}{2R} - \frac{\Delta c \times 10^5}{RT^3} + \frac{\Delta d \times 10^{-6} T}{3R} \tag{10-100}$$

Integrating gives an expression for K_P as a function of temperature:

$$\ln K_P = -\frac{\Delta_r H_0^o}{RT} + \frac{\Delta a}{R} \ln T + \frac{\Delta b \times 10^{-3}}{2R} T + \frac{\Delta c \times 10^5}{2R} T^{-2} + \frac{\Delta d \times 10^{-6}}{6R} T^2 + I \tag{10-101}$$

The I in Eq. (10-101) is the second integration constant. Equation (10-101) gives K_P as a function of temperature and using the third law it can be computed using only calorimetric data for heat capacities, the enthalpy of reaction at one temperature, and entropies. If we rearrange Eq. (10-101) and multiply through by the gas constant R, we get

$$-R \ln K_P + \Delta a \ln T + \frac{1}{2} \Delta b \times 10^{-3} T + \frac{1}{2} \Delta c \times 10^5 T^{-2} + \frac{1}{6} \Delta d \times 10^{-6} T^2 = \frac{\Delta_r H_0^o}{T} - IR \tag{10-102}$$

The left side of Eq. (10-102) is called the Σ *function*:

$$\Sigma = \frac{\Delta_r H_0^o}{T} - IR \quad (10\text{-}103)$$

The two integration constants in the Σ function are evaluated by plotting Σ versus $1/T$. This gives a straight line with a slope equal to $\Delta_r H_0^o$ and an intercept equal to IR.

Example 10-8. Ultraviolet sunlight continually converts CH_4 in Jupiter's upper atmosphere into C_2H_6 and smaller amounts of other hydrocarbons. In fact, this photolysis would destroy all CH_4 in Jupiter's upper atmosphere within about 10^5 years unless it were replenished by CH_4-bearing gas mixed upward from Jupiter's deep atmosphere. Atmospheric mixing also transports most of the photochemically produced C_2H_6 downward to hotter, higher-pressure regions where it is reconverted into methane by a series of reactions starting with the net thermochemical reaction:

$$C_2H_6 \text{ (ethane)} = C_2H_4 \text{ (ethylene)} + H_2 \quad (10\text{-}104)$$

As Problem 10-27 illustrates, conversion of ethane back to methane becomes rapid around the 1000 K, 454 bar level in Jupiter's deep atmosphere. Kistiakowsky et al. (1935) calorimetrically measured the enthalpy of reaction (10-104) at 355 K and obtained $\Delta_r H^o = 32{,}824(\pm 50)$ cal mol^{-1}. Subsequently, Kistiakowsky and Nickle (1951) measured the ethane-ethylene equilibrium and obtained $K_P = 4.04\ (\pm 0.17) \times 10^{-5}$ at 653.2 K and $K_P = 5.13\ (\pm 0.13) \times 10^{-4}$ at 723.2 K (relative to a standard state of one atmosphere). Using heat capacity equations from Table 3-4, the ΔC_P (298–1000 K) for reaction (10-104) is

$$\Delta C_P = 33.685 - 45.678 \times 10^{-3}T - 1.870 \times 10^5 T^{-2} + 14.121 \times 10^{-6} T^2 \text{ J mol}^{-1}\text{ K}^{-1} \quad (10\text{-}105)$$

Use these data to derive an equation for K_P as a function of temperature.

We first use the measured enthalpy of reaction and Eq. (10-99) to find the value of $\Delta_r H^o{}_0$:

$$\Delta_r H_0^o = \Delta_r H_{355}^o - \Delta a T - \frac{1}{2}\Delta b \times 10^{-3} T^2 + \Delta c \times 10^5 T^{-1} - \frac{1}{3}\Delta d \times 10^{-6} T^3 \quad (10\text{-}106)$$

$$\Delta_r H_0^o = 4.184(32{,}824) - 11958.2 + 2878.3 - 526.8 - 210.6 = 127{,}518 \text{ J mol}^{-1} \quad (10\text{-}107)$$

Next we use Eq. (10-102) to calculate the Σ function at 653.2 K (Σ = 288.319 J mol^{-1} K^{-1}) and 723.2 K (Σ = 269.287 J mol^{-1} K^{-1}). We then use Eq. (10-103) to calculate the corresponding IR values at 653.2 K(−93.099 J mol^{-1} K^{-1}) and 723.2 K (−92.962 J mol^{-1} K^{-1}). The mean of the two values gives $I = -11.190$ J mol^{-1} K^{-1}. Finally, we substitute our numerical results back into Eq. (10-101) and obtain the desired equation for K_P as a function of temperature from 298 K to 1000 K:

$$\ln K_P = -\frac{15{,}338}{T} + 4.052 \ln T - 2.75 \times 10^{-3} T - \frac{11{,}200}{T^2} + 2.83 \times 10^{-7} T^2 - 11.190 \quad (10\text{-}108)$$

We deliberately neglected the small difference between the one atmosphere standard state used by Kistiakowsky and Nickle (1951) and the one bar standard state used today because this is well within the uncertainty of their K_P values. A comparison with tabular data (Gurvich et al., 1989, 1991) shows that Eq. (10-108) almost exactly reproduces K_P for the ethane-ethylene reaction over the 298 K to

1000 K range. At 1000 K Eq. (10-108) gives $K_P = 0.36$ versus 0.37 from tabular data, and at 300 K Eq. (10-108) gives $K_P = 3.73 \times 10^{-18}$ versus 3.56×10^{-18} from the tables.

F. Pressure effects on equilibrium constants

The equilibrium constant K_P for reactions of ideal gases is constant at a given temperature and is unaffected by the total pressure. However, the equilibrium partial pressures of reactants and products are affected by the total pressure. These effects are particularly important for molecular equilibria of P, As, Sb, Bi, S, and Se elemental vapors (e.g., $As_4 = 2\ As_2$), dissociation of molecules into their constituent atoms (e.g., $H_2 = 2\ H$), thermal ionization of atoms, radicals, molecules to positively charged ions and electrons (e.g., $Cs = Cs^+ + e^-$), and recombination of monatomic gases into diatomic molecules (e.g., $2\ Na = Na_2$).

We begin by considering thermal dissociation of diatomic iodine vapor I_2 into iodine atoms:

$$I_2\ (g) = 2\,I\ (g) \tag{10-109}$$

Iodine is a trace element with an average abundance of about 1.5 micrograms per gram in Earth's crust and oceans. About 80% of Earth's iodine is concentrated in marine organisms and sediments and it is emitted from the oceans as methyl iodide (CH_3I) and from volcanoes as HI, I, and I_2 (Lodders and Fegley, 1997, 1998). The atmospheric chemistry of iodine and its compounds is important in Earth's troposphere and stratosphere. For example, reactions involving I, IO, and HOI gases destroy stratospheric ozone more efficiently than reactions involving the other halogens (F, Cl, Br). Perlman and Rollefson (1941) studied the thermal dissociation of I_2 gas and their results are in excellent agreement with other thermodynamic data for the same reaction.

The equilibrium constant K_P for reaction (10-109) is

$$K_p = \left[\frac{P_I^2}{P_{I_2}}\right]_{eq} = \left[\frac{X_I^2}{X_{I_2}}\right]_{eq} P_T \tag{10-110}$$

The P_T is the total pressure of a gas mixture containing iodine molecules and atoms:

$$P_T = P_I + P_{I_2} \tag{10-111}$$

Perlman and Rollefson (1941) measured P_T as a function of temperature from 872 K to 1274 K and used several different total pressures at each temperature. However, they did not directly measure the abundances of I_2 molecules and I atoms in the gas. Instead they computed K_P from mass balance. Their approach is a general method that applies to many dissociation reactions.

Depending on pressure and temperature, some fraction of the I_2 molecules decomposes to form monatomic iodine. We call this fraction α and it ranges from $\alpha = 0$ for no decomposition to $\alpha = 1$ for complete decomposition of I_2 to I atoms. Alpha is measurable and is equal to the difference in pressure between the observed total pressure P_T and the theoretical pressure (from the ideal gas law) if all iodine remains as I_2 and does not dissociate. However, in most problems of interest to us we will compute α from thermodynamic data. At equilibrium there are $(1 - \alpha)$ moles of I_2 molecules and 2α moles of I atoms, and the total number of moles is

$$n_T = n_I + n_{I_2} = 2\alpha + (1 - \alpha) = (1 + \alpha) \tag{10-112}$$

Table 10-7 K_P for I_2 Dissociation

T (K)	K_P
872	1.47×10^{-4}
973	1.705×10^{-3}
1073	0.0107
1173	0.0477
1274	0.1674

The mole fractions of I atoms and I_2 molecules are thus

$$X_I = \frac{\text{moles I}}{(\text{moles } I_2 + \text{moles I})} = \frac{n_I}{n_T} = \frac{2\alpha}{1+\alpha} \tag{10-113}$$

$$X_{I_2} = \frac{\text{moles } I_2}{(\text{moles } I_2 + \text{moles I})} = \frac{n_{I_2}}{n_T} = \frac{1-\alpha}{1+\alpha} \tag{10-114}$$

$$X_I + X_{I_2} = 1 \tag{10-115}$$

Rewriting the equilibrium constant expression (10-110) using these equations gives

$$K_P = \frac{X_I^2}{X_{I_2}} P_T = \frac{(2\alpha/1+\alpha)^2}{(1-\alpha/1+\alpha)} P_T = \frac{4\alpha^2}{(1+\alpha)^2} \cdot \frac{(1+\alpha)}{(1-\alpha)} P_T = \frac{4\alpha^2}{(1-\alpha^2)} P_T \tag{10-116}$$

Perlman and Rollefson (1941) used Eq. (10-116) to calculate K_P from their measurements of P_T. Table 10-7 gives the average K_P values that we calculated from their data. Conversely, if K_P is known, Eq. (10-116) can be used to compute α (also listed in Table 10-7):

$$\alpha = \left[\frac{K_P}{K_P + 4P_T}\right]^{1/2} \tag{10-117}$$

The relationship between K_P and the mole fractions of I and I_2 (see Eq. 10-116) is graphically illustrated in Figure 10-4. This shows the measurements of Perlman and Rollefson (1941) at 1173 K (the triangles) compared to the mole fractions of I and I_2 calculated from thermodynamic data (Gurvich et al., 1989) for reaction (10-109). If we substitute values from Figure 10-4 back into Eq. (10-116) we get K_P at 1173 K (see Problem 10-29 at the end of this chapter).

G. Reversibility of equilibria

So far, we have been thinking about gas-phase equilibria in terms of reactants being converted to products. However, chemical equilibrium is a dynamic situation in which the reactants are being converted to products and products are also being converted back into reactants. The net effect is to establish equilibrium concentrations of both reactants and products.

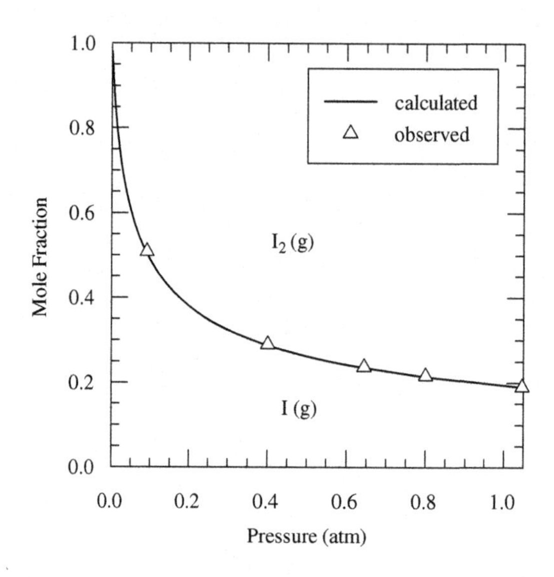

FIGURE 10-4

The thermal dissociation of I_2 molecules to I atoms at 1173 K as a function of the total pressure. The triangular points are observations by Perlman and Rollifson (1941) and the curve is calculated using thermodynamic data for I_2 and I gases from Gurvich et al. (1989).

Thus, we would get the same equilibrium constant if we started with a mixture of the product gases instead of starting with the reactants. This is illustrated by the work of Terres and Wesemann (1932) on the reaction

$$H_2S\ (g) + CO_2\ (g) = H_2O\ (g) + OCS\ (g) \qquad (10\text{-}118)$$

at 350–900°C in a flow system. Mixtures composed of either $H_2S + CO_2$ (forward reaction) or $H_2O + OCS$ (reverse reaction) were passed through a hot tube filled with a catalyst and the outlet gas was then analyzed. Reaction (10-118) converts H_2S to OCS (and vice versa) in terrestrial volcanic gases and its equilibrium constant is

$$K_p = \left[\frac{P_{H_2O}P_{OCS}}{P_{H_2S}P_{CO_2}}\right]_{eq} = \left[\frac{X_{H_2O}X_{OCS}}{X_{H_2S}X_{CO_2}}\right]_{eq} \qquad (10\text{-}119)$$

The total pressure cancels out of the equilibrium constant expression because the number of molecules on both sides of reaction (10-118) is the same.

The K_P values calculated using Eq. (10-119) from reacting mixtures of $H_2S + CO_2$ or from reacting mixtures of $H_2O + OCS$ are given in Tables 10-8 and 10-9, respectively. These two tables also give the compositions in mole % (= volume %) of the inlet (unreacted) and outlet (equilibrated) gas mixtures. Table 10-8 shows that the initial $H_2S + CO_2$ gas mixture is close to equilibrium and that only ~10% of each reactant is consumed to reach equilibrium. Conversely, Table 10-9 shows that the initial

Table 10-8 Forward Reaction $H_2S + CO_2 = H_2O + OCS$[a]

Temp.	Inlet Gas (%)		Outlet Gas (%)				K_P ($\times10^{-3}$)	
K	H_2S	CO_2	H_2S	CO_2	H_2O	OCS		Mean ± 1σ
623	50.3	49.7	47.5	47.0	2.80	2.72	3.41	3.35 ± 0.18
	51.3	48.7	48.6	46.4	2.66	2.67	3.17	
	50.3	49.7	47.5	46.9	2.82	2.74	3.47	
673	50.3	49.7	46.6	46.2	3.70	3.43	5.89	5.61 ± 0.47
	50.0	50.0	46.5	46.8	3.53	3.34	5.42	
	50.5	49.5	46.5	46.3	3.92	3.32	6.04	
	51.5	48.5	47.8	45.4	3.69	2.98	5.07	
723	50.9	49.1	46.5	44.7	4.40	3.93	8.32	8.50 ± 0.56
	51.3	48.7	46.6	44.0	4.75	3.5	8.11	
	51.2	48.8	46.5	44.0	4.66	3.98	9.06	
773	51.3	48.7	45.5	43.0	5.78	4.04	11.9	12.0 ± 0.4
	50.9	49.1	45.3	43.5	5.60	4.14	11.8	
	50.7	49.3	45.2	43.5	5.64	4.32	12.4	
823	52.0	48.0	45.9	41.9	6.12	4.93	15.7	15.8 ± 0.6
	50.9	48.9	44.7	43.5	6.24	5.1	16.4	
	51.2	48.8	45.3	43.2	5.95	5.06	15.4	
873	50.2	49.8	42.8	42.4	7.38	5.55	22.6	23.4 ± 1.2
	50.9	49.1	43.0	41.3	7.89	5.54	24.6	
	50.8	49.2	42.1	40.7	8.67	4.52	22.9	

[a] *Terres and Wesemann (1932).*

Table 10-9 Reverse Reaction H_2O (g) + OCS (g) = H_2S (g) + CO_2 (g)[a]

Temp.	Inlet gas (%)		Outlet gas (%)				K_P ($\times10^{-3}$)	
K	H_2O	OCS	H_2S	CO_2	H_2O	OCS		Mean ± 1σ
723	48.7	51.3	45.2	45.2	3.62	4.97	8.81	9.2 ± 0.6
	48.6	51.4	45.1	45.1	3.51	5.52	9.53	
773	50.1	49.9	44.9	44.9	5.23	4.70	12.2	12.6 ± 0.8
	51.3	48.7	44.1	44.2	7.23	3.52	13.1	
823	47.9	52.1	43.4	43.4	4.49	6.81	16.2	16.6 ± 0.7
	50.3	49.7	44.1	44.1	6.31	5.24	17.0	
873	46.6	53.4	41.7	41.7	4.91	9.05	25.6	25.1 ± 0.9
	50.4	49.6	42.4	42.4	8.02	5.52	24.6	

[a] *Terres and Wesemann (1932).*

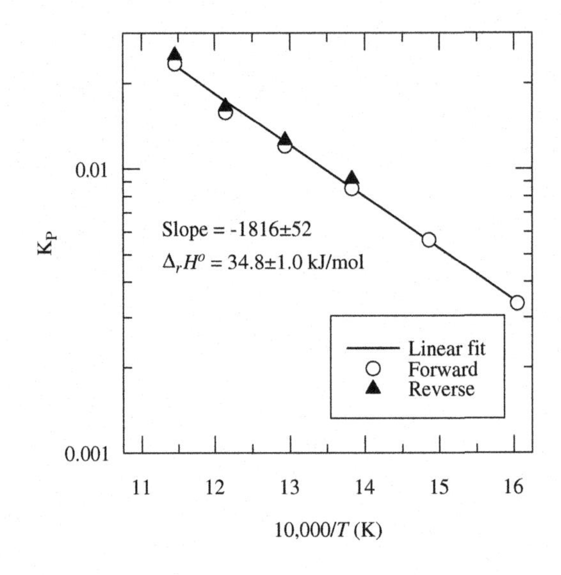

FIGURE 10-5

A plot of log K_P versus inverse temperature for the reaction $H_2S + CO_2 = OCS + H_2O$ using the data of Terres and Wesemann (1932). The line slope gives $\Delta_r H^o = 34.8 \pm 1.0$ kJ mol^{-1}.

$H_2O + OCS$ mixture is far from equilibrium and about 90% of each gas has to be consumed to reach equilibrium.

Figure 10-5 compares the average K_P values calculated at each temperature from the forward and reverse reactions. The uncertainties are about the size of the data points and are not shown. There is good agreement between the two data sets at each temperature. These data demonstrate that the same K_P value is obtained starting from a mixture of reactants and from a mixture of products. This *reversibility* demonstrates that equilibrium is established. Figure 10-5 also shows the application of the van't Hoff equation (10-64) to the K_P data. The linear least-squares line is

$$\log K_P = 0.44 \pm 0.02 - \frac{1816 \pm 52}{T} \tag{10-120}$$

Equation (10-95) gives a mean $\Delta_r H^o = 34{,}800(\pm 1000)$ J mole^{-1} at the midpoint temperature of 748 K, identical to the value calculated from the NIST-JANAF tables (Chase, 1999).

We emphasize that *reversibility is the only proof that equilibrium has been attained.* If the same equilibrium constant is not obtained starting from the products and from the reactants, then equilibrium has not been reached.

H. Chemical reaction rates and rate constants

Many experimental measurements show that forward and reverse reactions proceed at different rates and that it takes longer to approach equilibrium from one side than from the other side (e.g., Haber and

Le Rossignol, 1907, for NH_3; Taylor and Lenher, 1931, for SO_3). This qualitative observation shows that there is a relationship between chemical reaction rates and the concentrations of reactants (forward rate) or the concentrations of products (reverse rate).

As discussed briefly in Chapter 1, the reactions that actually take place between atoms, radicals (e.g., OH), and molecules are *elementary reactions*. A *net thermochemical reaction* is the result of a sequence of elementary reactions (the *reaction mechanism*). For example, the elementary reactions that occur when a $H_2 + CO_2$ gas mixture reacts to form $H_2O + CO$ are

$$H_2 + M \rightarrow H + H + M \quad (10\text{-}121)$$

$$H + CO_2 \rightarrow CO + OH \quad (10\text{-}122)$$

$$OH + H_2 \rightarrow H_2O + H \quad (10\text{-}123)$$

$$H + H + M \rightarrow H_2 + M \quad (10\text{-}124)$$

The M in Eqs. (10-121) and (10-124) is a third body, which can be any other gas (e.g., H, H_2, CO, CO_2). The arrows show that the reactions proceed in the specified direction. The sum of reactions (10-121) to (10-124) is the water-gas reaction in the forward direction:

$$H_2 + CO_2 \rightarrow H_2O + CO \quad \text{Net Reaction} \quad (10\text{-}56)$$

The elementary reactions that occur when a mixture of $H_2O + CO$ reacts to form $H_2 + CO_2$ (the water-gas reaction in the reverse direction) are

$$H_2O + M \rightarrow H + OH + M \quad (10\text{-}125)$$

$$OH + CO \rightarrow CO_2 + H \quad (10\text{-}126)$$

$$H + H_2O \rightarrow H_2 + OH \quad (10\text{-}127)$$

$$H + OH + M \rightarrow H_2O + M \quad (10\text{-}128)$$

As you can verify for yourself, the net result of reactions (10-125) to (10-128) is the water-gas reaction in the reverse direction:

$$H_2O + CO \rightarrow H_2 + CO_2 \quad \text{Net Reaction} \quad (10\text{-}129)$$

The *rate* of an elementary reaction is immediately evident from its stoichiometry and is the rate at which reactants are consumed and at which products are formed. Thus, the rate of reaction (10-127) can be expressed in terms of four equivalent differential equations:

$$R_{127} = -\frac{d[H]}{dt} = -\frac{d[H_2O]}{dt} = +\frac{d[H_2]}{dt} = +\frac{d[OH]}{dt} \quad (10\text{-}130)$$

The $d[i]/dt$ terms in Eq. (10-130) are the rate of change with time of the concentrations of H, H_2O, H_2, and OH. The minus signs show that the concentrations of H and H_2O decrease with time as they are consumed. The plus signs show that the concentrations of H_2 and OH increase with time as they are formed. In general, only the minus signs are written and the plus signs are implied (they were used above to emphasize a point). Likewise, the rate of reaction (10-126) is

$$R_{126} = -\frac{d[\mathrm{OH}]}{dt} = -\frac{d[\mathrm{CO}]}{dt} = \frac{d[\mathrm{CO_2}]}{dt} = \frac{d[\mathrm{H}]}{dt} \tag{10-131}$$

However, the rates of reactions (10-121) and (10-124) are

$$R_{121} = -\frac{d[\mathrm{H_2}]}{dt} = \frac{1}{2}\frac{d[\mathrm{H}]}{dt} \tag{10-132}$$

$$R_{124} = -\frac{1}{2}\frac{d[\mathrm{H}]}{dt} = \frac{d[\mathrm{H_2}]}{dt} \tag{10-133}$$

The concentration of the third body M is constant and it does not enter into the differential form of the rate equation. The square brackets in Eqs. (10-130) to (10-133) denote *number densities*, which are the concentrations of atoms, radicals, or molecules per cm^3. The number density of gas i is defined as

$$[i] = \frac{P_i N_A}{RT} \tag{10-134}$$

where N_A in Eq. (10-134) is Avogadro's number and P_i is the partial pressure of gas i.

The rate of an elementary reaction is proportional to the product of the concentrations of the reactants. The stoichiometric coefficient for each species becomes the power to which its concentration is raised in the rate law. The proportionality constant in the equation is the *rate constant* (k), which is written in lowercase to distinguish it from the equilibrium constant K. Thus, the rates of reactions (10-126) and (10-127) can also be written as

$$R_{126} = k_{126}[\mathrm{OH}][\mathrm{CO}] \tag{10-135}$$

$$R_{127} = k_{127}[\mathrm{H}][\mathrm{H_2O}] \tag{10-136}$$

Elementary reactions involving two reactants are *bimolecular reactions*. Atmospheric chemists generally express reaction rates in units of molecules cm^{-3} s^{-1}, so bimolecular rate constants must have units of cm^3 $molecule^{-1}$ s^{-1}.

Elementary reactions involving three reactants or two reactants and a third body M are *termolecular reactions*. Reaction (10-124) is a termolecular reaction. Its reaction rate has the same units (molecules cm^{-3} s^{-1}) and its rate equation is

$$R_{124} = -\frac{1}{2}\frac{d[\mathrm{H}]}{dt} = \frac{d[\mathrm{H_2}]}{dt} = k_{124}[\mathrm{H}]^2[\mathrm{M}] \tag{10-137}$$

The $[\mathrm{H}]^2$ term in Eq. (10-137) occurs because two H atoms are reacting. Reaction (10-128) is also a termolecular reaction and its rate equation is

$$R_{128} = -\frac{d[\mathrm{H}]}{dt} = -\frac{d[\mathrm{OH}]}{dt} = k_{128}[\mathrm{H}][\mathrm{OH}][\mathrm{M}] \quad (10\text{-}138)$$

Termolecular rate constants must have units of cm^6 $molecule^{-2}$ s^{-1} for the reaction rates to have units of molecules cm^{-3} s^{-1}. There are no elementary reactions involving four reactants. However, *unimolecular reactions* are elementary reactions with only one reactant. Unimolecular rate constants have units of s^{-1}. The dissociation of N_2O_5 is a unimolecular reaction,

$$\mathrm{N_2O_5} \rightarrow \mathrm{NO_2} + \mathrm{NO_3} \quad (10\text{-}139)$$

The rate of reaction (10-139) is

$$R_{139} = -\frac{d[\mathrm{N_2O_5}]}{dt} = \frac{d[\mathrm{NO_2}]}{dt} = \frac{d[\mathrm{NO_3}]}{dt} = k_{139}[\mathrm{N_2O_5}] \quad (10\text{-}140)$$

Radioactive decay is another example of a unimolecular reaction. In this case the reaction rate constant is the *decay constant* (λ) of the radioactive isotope.

The *chemical lifetime* (t_{chem}) of a species is its concentration divided by its rate of change. For example, the chemical lifetime of H atoms in reaction (10-127) is

$$t_{chem}(\mathrm{H}) = \frac{[\mathrm{H}]}{R_{127}} = \frac{[\mathrm{H}]}{k_{127}[\mathrm{H}][\mathrm{H_2O}]} = \frac{1}{k_{127}[\mathrm{H_2O}]} \quad (10\text{-}141)$$

Likewise, the chemical lifetime of H_2O molecules in reaction (10-127) is

$$t_{chem}(\mathrm{H_2O}) = \frac{[\mathrm{H_2O}]}{R_{127}} = \frac{[\mathrm{H_2O}]}{k_{127}[\mathrm{H}][\mathrm{H_2O}]} = \frac{1}{k_{127}[\mathrm{H}]} \quad (10\text{-}142)$$

In general, the chemical lifetimes of reactive atoms and radicals such as H and OH are much shorter than the chemical lifetimes of less reactive molecules such as H_2O.

I. The equilibrium constant and reaction rate constants

Motivated by Berthelot's and St. Gilles's study of ethyl acetate hydrolysis (see Chapter 9), two Norwegian professors (and brothers-in-law!), Cato Maximilian Guldberg (1836–1902) and Peter Waage (1833–1900), realized that the overall or net reaction rate (R_{net}) is the difference of the forward (R_f) and reverse (R_r) reaction rates:

$$R_{net} = R_f - R_r \quad (10\text{-}143)$$

However, once chemical equilibrium is attained, the net reaction rate is zero because the forward and reverse reaction rates are equal to one another:

$$R_{net}^{eq} = R_f^{eq} - R_r^{eq} = 0 \quad (10\text{-}144)$$

$$R_f^{eq} = R_r^{eq} \quad (10\text{-}145)$$

Equations (10-144) and (10-145) are two versions of the *law of mass action*, which was stated by Guldberg and Waage in 1867.

For example, we can consider Eq. (10-122), the forward reaction denoted by subscript f, and Eq. (10-126), its reverse reaction denoted by subscript r. The rate equations for these two elementary reactions are

$$R_f = k_f[H]\cdot[CO_2] \tag{10-146}$$

$$R_r = k_r[CO]\cdot[OH] \tag{10-147}$$

When we substitute Eqs. (10-146) and (10-147) into the mass action law Eq. (10-145), we find

$$R_f^{eq} = k_f([H]\cdot[CO_2])_{eq} = R_r^{eq} = k_r([CO]\cdot[OH])_{eq} \tag{10-148}$$

The eq subscripts denote the equilibrium concentrations of reactants and products. The two sides of Eq. (10-148) are equal, so we can rearrange it to get

$$\frac{R_f^{eq}}{R_r^{eq}} = 1 = \frac{k_f}{k_r}\cdot\left(\frac{[H]\cdot[CO_2]}{[CO]\cdot[OH]}\right)_{eq} \tag{10-149}$$

The quotient in Eq. (10-148) is the inverse of the quotient in the expression for the equilibrium constant K_C of reaction (10-122):

$$K_C = \left(\frac{[CO]\cdot[OH]}{[H]\cdot[CO_2]}\right)_{eq} \tag{10-150}$$

Thus, upon combining Eqs. (10-149) and (10-150) we obtain the important conclusion that the equilibrium constant in terms of concentrations K_C is the ratio of the forward k_f and reverse k_r rate constants in the same concentration units:

$$K_C = \frac{k_f}{k_r} \tag{10-151}$$

Equation (10-151) is an important equation relating equilibrium constants and reaction rate constants. It is frequently used to calculate either k_f or k_r from K_C and the other rate constant. The uncertainties in rate constants are typically larger than the uncertainties in equilibrium constants. Thus, small disagreements between calculated and measured rate constant values are not unusual.

Example 10-9. As discussed, reactions (10-122) and (10-126) are a pair of forward and reverse reactions. Calculate the rate constant for Eq. (10-126), the reverse reaction, using the rate constant for Eq. (10-122), the forward reaction, and the equilibrium constant K_P. Baulch et al. (1976) review experimental measurements of the rate constants for these two reactions. Their recommended equation for the rate constant k_f from 1000 K to 3000 K (with $\pm$ 20% uncertainty) is

$$k_f = 2.50 \times 10^{-10} \exp(-13{,}300/T) \text{ cm}^3 \text{ molecule}^{-1} \text{ s}^{-1} \tag{10-152}$$

Reaction (10-122) has two molecules on each side of the equation. Equation (10-82) shows that $K_C = K_P$ in this case. We calculated K_P using data from Gurvich et al. (1989) and the updated dissociation energy for the OH radical (Ruscic et al., 2002). Table 10-10 summarizes our calculations. The reverse rate constant is for the reaction

$$CO + OH \rightarrow H + CO_2 \tag{10-126}$$

Table 10-10 Rate and Equilibrium Constants for $CO_2 + H \rightarrow CO + OH$

T(K)	k_f	K_P	$k_r = k_f/K_P$	k_r (Baulch)
1000	4.19×10^{-16}	1.98×10^{-3}	2.12×10^{-13}	2.78×10^{-13}
1500	3.52×10^{-14}	8.70×10^{-2}	4.05×10^{-13}	2.12×10^{-13}
2000	3.24×10^{-13}	5.16×10^{-1}	6.28×10^{-13}	6.89×10^{-13}
2500	1.22×10^{-12}	1.41×10^{0}	8.65×10^{-13}	1.08×10^{-12}

The recommended equation for this rate constant from 250 K to 2500 K (with ± 50% uncertainty) from Baulch et al. (1976) is

$$\log_{10} k_r = 3.94 \times 10^{-4}T - 12.95 \text{ cm}^3 \text{ molecule}^{-1} \text{ s}^{-1} \quad (10\text{-}153)$$

The k_r value calculated from the forward rate constant and the equilibrium constant agrees with their recommended value within 24% over the 1000–2500 K range. This agreement is within the 50% uncertainty of Eq. (10-153).

What if there are different numbers of molecules on each side of the reaction? In this case, there is a difference between the equilibrium constants written in terms of concentration K_C and partial pressure K_P. Furthermore, the concentration units (mol L^{-1}, molecules cm^{-3}, etc.) must cancel out in going from K_C to K_P or else calculations done with Eq. (10-151) will be incorrect.

Example 10-10. Nitrogen pentoxide N_2O_5 is a pollutant formed in urban areas. Calculate the equilibrium constant K_P at 288 K (Earth's global mean surface temperature) for decomposition of N_2O_5 from the rate constants given by Atkinson et al. (1997). These are for 200–400 K and have uncertainties of a factor of three. The reactions and their recommended rate constants are

$$N_2O_5 \rightarrow NO_2 + NO_3 \quad (10\text{-}139)$$

$$k_{139} = k_f = 5.49 \times 10^{14} T^{0.10} \exp(-11{,}080/T) \text{ s}^{-1} \quad (10\text{-}154)$$

$$NO_2 + NO_3 \rightarrow N_2O_5 \quad (10\text{-}155)$$

$$k_{155} = k_r = 6.40 \times 10^{-13} T^{0.20} \text{ cm}^3 \text{ molecule}^{-1} \text{s}^{-1} \quad (10\text{-}156)$$

Evaluating the rate constant equations at 288 K and substituting into Eq. (10-151) gives

$$K_C = \frac{k_f}{k_r} = \frac{0.0189 \text{ s}^{-1}}{1.99 \times 10^{-12} \text{ cm}^3 \text{ s}^{-1}} = 9.5 \times 10^9 \text{ molecule cm}^{-3} \quad (10\text{-}157)$$

This K_C value is in terms of number density, which must cancel out to compute K_P correctly. We do this by dividing K_C by the molecular number density of an ideal gas at 288 K, 1 bar pressure:

$$K_P = \frac{RT}{PN_A} K_C = \frac{83.145 \text{ cm}^3 \text{ bar mol}^{-1} \text{ K}^{-1} \times 288 \text{ K}}{1 \text{ bar} \cdot 6.022 \times 10^{23} \text{ mol}^{-1}} 9.50 \times 10^9 \text{ cm}^{-3} = 3.78 \times 10^{-10} \quad (10\text{-}158)$$

The calculated K_P value agrees within 27% with $K_P = 2.98 \times 10^{-10}$ from Gurvich et al. (1989).

J. Gas mixing and oxygen fugacity control

An important application of gas-phase chemical equilibria in experimental geochemistry and petrology is the control of oxygen fugacity (i.e., the oxygen partial pressure) in high-temperature furnaces. This is done to study minor and trace element partitioning, crystallization of minerals from melts, and subsolidus phase equilibria under conditions relevant to igneous processes on Earth, Moon, Mars, and parent bodies of differentiated meteorites.

The general principle of gas mixing for oxygen fugacity control is quite simple. A two-component gas mixture, such as CO and CO_2 (or H_2 and H_2O), flows through a high-temperature furnace, which has an experimental sample inside the isothermal hot zone. The gases react and produce a constant equilibrium partial pressure of O_2 that is controlled by their equilibrium ratio. For example, in the case of CO and CO_2, the net thermochemical reaction involved is

$$\mathrm{CO} + \frac{1}{2}\mathrm{O_2} = \mathrm{CO_2} \tag{10-159}$$

The equilibrium constant K_P for reaction (10-159) is

$$K_P = \left[\frac{P_{CO_2}}{P_{CO}P_{O_2}^{1/2}}\right]_{eq} = \left[\frac{X_{CO_2}}{X_{CO}}\right]_{eq} \cdot \left(\frac{1}{P_{O_2}^{1/2}}\right)_{eq} \tag{10-160}$$

Rearranging this, and solving for the equilibrium partial pressure of O_2 gives

$$P_{O2} = f_{O_2} = \left[\frac{X_{CO_2}}{X_{CO}}\right]_{eq}^2 \cdot \left(\frac{1}{K_P}\right)^2 \tag{10-161}$$

Thus, the oxygen fugacity (f_{O_2}) is proportional to the CO_2/CO ratio squared.

Example 10-11. Calculate the CO/CO_2 ratio and the volume (mole) percentages of CO and CO_2 in the inlet gas mixture that are required to give an O_2 equilibrium partial pressure of 10^{-10} bars from reaction (10-159) at 1227°C (1500 K) using the Gibbs free energy data in Table 10-11.

The standard Gibbs free energy change for reaction (10-159) at 1500 K is

$$\begin{aligned}\Delta_r G^o_{1500} &= \Delta_f G^o_{1500}(\mathrm{CO_2}) - \Delta_f G^o_{1500}(\mathrm{CO}) \\ &= -396,265 - (-243,596) = -152,669 \text{ J mol}^{-1}\end{aligned} \tag{10-162}$$

Remember that O_2 is the reference state for oxygen, so by definition its standard Gibbs free energy of formation from itself is zero. The K_P for reaction (10-159) is thus

$$K_P = \exp(-\Delta_r G/RT) = \exp(152{,}193/8.314 \cdot 1500) = 2.07 \times 10^5 \tag{10-163}$$

Substituting numerical values into the equilibrium constant expression (10-160) and rearranging it to solve for the CO/CO_2 ratio, we find that

$$\left[\frac{X_{CO_2}}{X_{CO}}\right]_{eq} = (f_{O_2})^{1/2} K_P = 10^{-5} \cdot 2.07 \times 10^5 = 2.07 \tag{10-164}$$

Table 10-11 Standard Gibbs Energies for Some Gas Reactions $\Delta G^o = A + BT\log T + CT (\text{J mol}^{-1})$[a]

Reaction	A	B	C
2 C (graphite) + O_2 = 2 CO	−214,104	25.2183	−262.1545
C (graphite) + O_2 = CO_2	−392,647	4.5855	−16.9762
C (graphite) + 2 H_2 = CH_4	−77,437	22.5098	29.5967
2 C (graphite) + H_2 = C_2H_2	229,047	10.7501	−90.8942
2 C (graphite) + 2 H_2 = C_2H_4	50,070	21.8889	3.5519
2 C (graphite) + 3 H_2 = C_2H_6	−89,966	25.4738	124.5782
C (graphite) + H_2 + N_2 = 2 HCN	265,591	11.5634	−103.9072
C (graphite) + S_2 (g) = CS_2	−13,617	0.0912	−6.9480
2 CO + S_2 (g) = 2 OCS	−195,781	−20.3931	225.7796
2 H_2 + O_2 = 2 H_2O	−483,095	25.3687	21.9563
3 O_2 = 2 O_3	277,958	−20.3935	204.7286
2 H_2 + S_2 (g) = 2 H_2S	−172,187	13.5394	49.4010
S_2 (g) + 2 O_2 = 2 SO_2	−724,016	−1.0153	149.7695
S_2 (g) + 3 O_2 = 2 SO_3	−926,651	−18.8280	396.0757
2 S_2 (g) + O_2 = 2 S_2O	−370,023	0.8024	125.8593
S_2 (g) + O_2 = 2 SO	−119,036	0.6689	−12.1517
2 NH_3 = N_2 + 3 H_2	96,712	−23.1989	−151.1944
N_2 + O_2 = 2 NO	182,578	−0.5669	−23.3503
N_2 + 2O_2 = 2 NO_2	64,932	−6.5412	147.7654
N_2 + 2 O_2 = N_2O_4	3833	−33.7650	403.9426
2 N_2 + O_2 = 2 N_2O	157,344	−21.9819	220.5689
H_2 + Cl_2 = 2 HCl	−184,269	8.9922	−44.2428
H_2 + F_2 = 2 HF	−544,349	14.3326	−56.9916
H_2 + Br_2 = 2 HBr	−100,868	12.4981	−57.8975
H_2 + I_2 = 2 HI	−9809	6.3716	−37.1395

[a]*From 298 K to 2500 K.*

The gas inside the furnace is a binary mixture of CO and CO_2 to a very good first approximation because the O_2 partial pressure is so small. Thus, the mole fractions of CO and CO_2 are given by

$$X_{CO_2} + X_{CO} = 2.07\, X_{CO} + X_{CO} = 1 \quad (10\text{-}165)$$

$$X_{CO} = \frac{1}{3.07} = 0.326 \quad (10\text{-}166)$$

$$X_{CO_2} = 1 - X_{CO} = 0.674 \quad (10\text{-}167)$$

Our calculations show that an inlet gas mixture containing about 32.6% CO and 67.4% CO_2 (by volume) will give the desired O_2 equilibrium partial pressure at 1500 K. We do not have to worry about adjusting the initial CO and CO_2 concentrations for the amount of O_2 produced because the equilibrium O_2 partial pressure is so small. However, this is not always the case.

Other gas mixtures are used to control the fugacities (i.e., partial pressures) of other volatile elements. For example, H_2-H_2S mixtures regulate the S_2 fugacity, H_2-NH_3 mixtures control the N_2 fugacity, and H_2-HCl mixtures control the Cl_2 fugacity. Ternary gas mixtures are used to simultaneously control the fugacities of two or more elements, for example, the CO-CO_2-SO_2 mixture controls O_2 and S_2 fugacities and the H_2-H_2O-H_2S mixture controls H_2, O_2, and S_2 fugacities.

Nafziger et al. (1971), Kubaschewski and Alcock (1979), and Huebner (1987) review fugacity control by gas mixtures, give numerous references to experimental work, and discuss the practical problems that limit the utility of gas mixing in some cases. These problems include graphite precipitation from CO-CO_2 gas mixtures at low-oxygen fugacities, sulfur preciptation from CO-CO_2-SO_2 mixtures at high-sulfur fugacities and low temperatures, and water condensation from H_2O-rich H_2-H_2O gas mixtures. These problems limit the range of oxygen (or sulfur) fugacities that can be obtained from gas mixing. As discussed later, solid-state oxygen fugacity buffers can provide O_2 fugacities lower (or higher) than ambient-pressure gas mixtures.

K. Simultaneous equilibria

Hydrogen-CO_2 gas mixtures are often used to control the O_2 fugacity in high-temperature gas mixing furnaces in geochemistry laboratories. The method is the same as that described earlier for CO-CO_2 gas mixtures. The water-gas reaction (10-1) does not directly involve oxygen. However, O_2 is involved in equilibria between CO-CO_2 and H_2-H_2O that also occur:

$$CO + \frac{1}{2}O_2 = CO_2 \quad (10\text{-}159)$$

$$2\,H_2 + O_2 = 2\,H_2O\,(g) \quad (10\text{-}168)$$

In fact, all possible reactions among C-O-H gases occur simultaneously in the hot H_2-CO_2 mixture. For example, reactions forming methane from CO_2 and CO also take place:

$$CO_2 + 4\,H_2 = CH_4 + 2\,H_2O\,(g) \quad (10\text{-}169)$$

$$CO + 3\,H_2 = CH_4 + H_2O\,(g) \quad (10\text{-}170)$$

The water-gas reaction and these additional four reactions are usually the most important ones, but the complete set of possible reactions is much larger and includes formation of H atoms, hydroxyl (OH) radicals, methyl (CH_3) radicals, and other gases, for example, formaldehyde (H_2CO), formic acid (HCOOH), and methanol (CH_3OH):

$$H_2O\,(g) = H + OH \quad (10\text{-}171)$$

$$CH_4 = H + CH_3 \quad (10\text{-}172)$$

$$CO + H_2 = H_2CO\,(g) \quad (10\text{-}173)$$

$$CO_2 + H_2 = HCOOH\ (g) \tag{10-174}$$

$$CO + 2\,H_2 = CH_3OH\ (g) \tag{10-175}$$

We generally do not need to consider most of these reactions, because the equilibrium partial pressures of their products are very small. However, they all take place, and different reactions become important at different pressure and temperature conditions. How do we know which reactions must be considered to correctly calculate the equilibrium partial pressure of O_2, CH_4, or other gases of interest? The answer is that any reactions involving the gases of interest can be used because all gases are in equilibrium with one another.

Example 10-12. Use the H_2-H_2O and CO-CO_2 reactions to calculate the O_2 equilibrium partial pressure in an H_2-CO_2 gas mixture heated at 886°C and one bar pressure. At equilibrium the gas contains 11.54% CO_2, 40.62% H_2, 23.92% CO, and 23.92% H_2O. The necessary data for standard Gibbs free energies of reaction are given in Table 10-11.

We first calculate the O_2 equilibrium partial pressure for the H_2-H_2O reaction (10-168). The $\Delta_r G^o$ for this reaction at 1159 K is −367,556 J and the equilibrium constant K_P is

$$K_P = \left[\left(\frac{P_{H_2O}}{P_{H_2}}\right)^2 \frac{1}{P_{O_2}}\right]_{eq} = \exp\left(-\frac{\Delta_r G^o}{RT}\right) = \exp\left(\frac{367,556}{8.314 \times 1159}\right) = 3.68 \times 10^{16} \tag{10-176}$$

Substituting the equilibrium percentages of H_2 and H_2O and rearranging gives

$$P_{O_2} = \left[\frac{P_{H_2O}}{P_{H_2}}\right]^2 \frac{1}{K_P} = \left[\frac{23.92}{40.62}\right]^2 \cdot 2.72 \times 10^{-17} = 9.42 \times 10^{-18}\ \text{bars} \tag{10-177}$$

We now compute the O_2 equilibrium partial pressure for the CO-CO_2 reaction (10-159). The $\Delta_r G^o$ for this reaction at 1159 K is −181,846 J and the equilibrium constant K_P is

$$K_P = \left[\left(\frac{P_{CO_2}}{P_{CO}}\right) \frac{1}{P_{O_2}^{1/2}}\right]_{eq} = \exp\left(-\frac{\Delta_r G^o}{RT}\right) = \exp\left(\frac{181,846}{8.314 \times 1159}\right) = 1.572 \times 10^{8} \tag{10-178}$$

Substituting the equilibrium percentages of CO and CO_2 and rearranging gives

$$P_{O_2} = \left[\frac{P_{CO_2}}{P_{CO}}\right]^2 \cdot \frac{1}{K_P^2} = \left[\frac{11.54}{23.92}\right]^2 \cdot 4.047 \times 10^{-17} = 9.42 \times 10^{-18}\ \text{bars} \tag{10-179}$$

Both equilibria give the same O_2 equilibrium partial pressure. Likewise, either reaction (10-169) or (10-170) will give the same CH_4 equilibrium partial pressure.

L. Ammonia synthesis

Ammonia is ubiquitous throughout the solar system, and NH_3 formation from its constituent elements is one of the most important chemical reactions in the solar system. Recombination of N_2 and H_2 to NH_3 in the hot, high-pressure deep atmospheres of Jupiter, Saturn, Uranus, and Neptune replaces the NH_3 destroyed by UV sunlight in the cold, low-pressure upper atmospheres of these planets. Ammonia ice is present in some comets and on the surface of Pluto's satellite Charon.

Ammonia-bearing fluids and NH_3 ice are probably present inside Saturn's satellite Titan and inside Jupiter's satellites Europa, Ganymede, and Callisto. On Earth, NH_3 formation is the basis for production of artificial fertilizers that sustain modern agriculture. Ammonia synthesis is so important that two Nobel Prizes were awarded for its development: the 1918 Nobel Prize in Chemistry to Fritz Haber (see sidebar) and half of the 1931 Nobel Prize in Chemistry to Carl Bosch (1874–1940). Topham (1985) describes the history of ammonia synthesis and the stupendous problems that had to be overcome by Bosch and his colleagues at the BASF chemical company to industrialize the process.

The formation of NH_3 from the elements occurs via the reaction

$$\frac{1}{2}N_2 + \frac{3}{2}H_2 = NH_3 \tag{10-180}$$

The K_P for reaction (10-180) is deceptively simple:

$$K_P = \left[\frac{P_{NH_3}}{P_{N_2}^{1/2}P_{H_2}^{3/2}}\right]_{eq} = \left[\frac{X_{NH_3}}{X_{N_2}^{1/2}X_{H_2}^{3/2}}\right]_{eq} \times \frac{1}{P_T} \tag{10-181}$$

FRITZ HABER (1868–1934)

Fritz Haber was a German physical chemist, best known for his development of catalyzed ammonia synthesis from the elements. He did this work while he was a professor at Karlsruhe, Germany, during the happiest and most brilliant period of his career. During this time he also studied glass electrodes (used for pH measurements), high-temperature fuel cells, and flame chemistry, and wrote a pioneering text in chemical thermodynamics.

Haber moved to Berlin and became the director of the newly formed Kaiser Wilhelm Institut für Physikalische Chemie and Electrochemie in 1911. The First World War started three years later. Haber's sense of patriotism and national pride led him to develop chemical weapons and gas masks for the German army during World War I. This work killed Otto Sackur, a very talented and young physical chemist, and led Haber's first wife to commit suicide.

The Allied nations (France, Britain, and the United States, which entered the war in 1917) responded in kind with their own chemical weapons and later branded Haber a war criminal.

Despite this history, Haber was awarded the Nobel Prize for Chemistry in 1919 for his ammonia synthesis work. After the war, Haber attempted to extract gold from the oceans to pay Germany's war reparations, which amounted to 50,000 tons of gold. This attempt failed because the amount of gold in seawater is much lower than Haber's method could extract. In principle his idea was sound because there is about 0.03–4.9 nanograms gold per kilogram seawater, corresponding to 50,000 to 8,154,000 metric tons of gold in the oceans.

Sadly, after the rise of the Nazi regime, Haber, who was Jewish, was forced to resign his post and go into exile (first in England, then Switzerland, where he died), leaving the country he had previously worked to protect.

Measurements of the correct K_P values for NH_3 formation are very difficult because of the small amount of NH_3 produced at temperatures where reaction (10-180) proceeds at an appreciable rate. The French chemist Claude Louis Berthollet (1748–1822) established the formula of ammonia in 1785. However, it took until the early 20th century for Haber and coworkers to finally measure K_P accurately. The results of Haber and LeRossignol (1907) are given in Table 10-12. They measured the equilibrium percentage of NH_3 in a 3:1 mixture of H_2 and N_2 at one atmosphere total pressure. Subsequently, Haber and Maschke (1915), Haber et al. (1915), Larson and Dodge (1923), Larson (1924), Schulz and Schaefer (1966), and Winchester and Dodge (1956) extended measurements to lower temperatures and higher pressures. The large amount of experimental data for ammonia synthesis make reaction (10-180) a good example for us to study.

An *approximate* K_P for NH_3 formation is easy to calculate. The total pressure for reaction (10-180) is the sum of the partial pressures of N_2, H_2, and NH_3:

$$P_{N_2} + P_{H_2} + P_{NH_3} = P_T \tag{10-182}$$

The experimental data in Table 10-12 show that at high temperatures the percentage of NH_3 is very much smaller than the percentages of N_2 (~25%) or of H_2 (~75%). To first approximation we can neglect the NH_3 partial pressure in Eq. (10-182) and write

$$P_{N_2} + P_{H_2} \cong P_T \tag{10-183}$$

If the H_2 to N_2 molar ratio is exactly 3:1, mass balance gives

$$P_{N_2} + P_{H_2} = P_{N_2} + 3P_{N_2} = 4P_{N_2} = P_T \tag{10-184}$$

Table 10-12 NH_3 Equilibrium Abundances[a]

T (°C)	T(K)	% NH_3
700	973	0.0221
750	1023	0.0152
850	1123	0.0091
930	1203	0.0065
1000	1273	0.0048

[a]*Data of Haber and LeRossignol (1907) for a H_2/N_2 molar ratio of 3:1, 1 atm total pressure.*

Table 10-13 NH_3 K_P values

T (K)	Haber[a]	Accepted[b]
973	6.81×10^{-4}	6.86×10^{-4}
1023	4.68×10^{-4}	4.97×10^{-4}
1123	2.80×10^{-4}	2.83×10^{-4}
1203	2.00×10^{-4}	1.92×10^{-4}
1273	1.48×10^{-4}	1.43×10^{-4}

[a] *Haber and LeRossignol (1907).*
[b] *Calculated from the ΔG^o data in Table 10-11.*

Substituting back into Eq. (10-181) and rearranging gives an *approximate* expression for K_P in terms of the NH_3 equilibrium mole fraction and the total pressure:

$$K_P = \left[\frac{P_{NH_3}}{P_{N_2}^{1/2}P_{H_2}^{3/2}}\right]_{eq} = \left[\frac{X_{NH_3}}{X_{N_2}^{1/2}X_{H_2}^{3/2}}\right]_{eq} \times \frac{1}{P_T} \cong \frac{X_{NH_3}}{(0.25)^{1/2}(0.75)^{3/2}} \times \frac{1}{P_T} \tag{10-185}$$

$$K_P \cong 3.07920\,\frac{X_{NH_3}}{P_T} \tag{10-186}$$

We used Eq. (10-186) to compute K_P from the NH_3 abundances (Table 10-12) measured by Haber and LeRossignol (1907) and give the results in Table 10-13. This table illustrates two important points. First, there is fairly good agreement between the currently accepted values (from the ΔG^o equation in Table 10-11) and the experimental results of Haber and LeRossignol (1907). Second, K_P increases with decreasing temperature. The van't Hoff equation (10-65) shows that this happens because NH_3 synthesis is exothermic (negative ΔH^o of reaction).

The *exact* K_P for ammonia formation is more complex (Gillespie and Beattie, 1930), but it is an excellent mass balance example. We want to express the equilibrium constant K_P in terms of the initial mole fractions of H_2 and N_2, the equilibrium mole fraction of NH_3, and the total pressure. To do this we need mass balance equations for the molar abundances and the mole fractions of all three gases. We begin by defining the number of moles of each gas as

$$A = \text{moles } H_2 \tag{10-187}$$

$$B = \text{moles } N_2 \tag{10-188}$$

$$C = \text{moles } NH_3 \tag{10-189}$$

We represent the mole fractions of H_2, N_2, and NH_3 by x, y, z, respectively. Initially (denoted by subscript 0), the mole fractions of H_2, N_2, and NH_3 are

$$x_0 = \frac{A_0}{A_0 + B_0} = \text{initial } H_2 \text{ mole fraction} \tag{10-190}$$

$$y_0 = \frac{B_0}{A_0 + B_0} = \text{initial } N_2 \text{ mole fraction} \tag{10-191}$$

$$z_0 = 0 = \text{initial } NH_3 \text{ mole fraction} \tag{10-192}$$

At equilibrium the mole fractions of the three gases are

$$x = \frac{A}{A + B + C} = X_{H_2} \tag{10-193}$$

$$y = \frac{B}{A + B + C} = X_{N_2} \tag{10-194}$$

$$z = \frac{C}{A + B + C} = X_{NH_3} \tag{10-195}$$

Finally, the initial and equilibrium mole fractions sum to unity as

$$x_0 + y_0 = 1 \tag{10-196}$$

$$x + y + z = 1 \tag{10-197}$$

The initial abundances of H_2 (A_0 moles) and N_2 (B_0 moles) are reduced by the formation of C moles of NH_3. The stoichiometry of the NH_3 synthesis reaction (10-180) shows that

$$A = A_0 - \frac{3}{2}C \tag{10-198}$$

$$B = B_0 - \frac{1}{2}C \tag{10-199}$$

By combining these two expressions and rearranging we get

$$A + B = \left(A_0 - \frac{3}{2}C\right) + \left(B_0 - \frac{1}{2}C\right) = A_0 + B_0 - 2C \tag{10-200}$$

Substituting Eq. (10-200) back into Eq. (10-195) and rearranging gives

$$C = z(A + B + C) = z(A_0 + B_0 - 2C + C) = z(A_0 + B_0 - C) \tag{10-201}$$

$$C(1 + z) = z(A_0 + B_0) \tag{10-202}$$

Dividing by $(1+z)$ gives an equation for C, the number of moles of NH_3, in terms of its mole fraction (z) and the initial molar abundances of H_2 and N_2:

$$C = \frac{z(A_0 + B_0)}{(1+z)} \tag{10-203}$$

The next step is to replace C by Eq. (10-203) in our two earlier equations (10-198) and (10-199):

$$A = A_0 - \frac{3}{2}C = A_0 - \frac{3}{2}\frac{z(A_0 + B_0)}{(1+z)} \tag{10-204}$$

$$B = B_0 - \frac{1}{2}C = B_0 - \frac{1}{2}\frac{z(A_0 + B_0)}{(1+z)} \tag{10-205}$$

The A_0 in Eq. (10-204) is now replaced by Eq. (10-190), and the B_0 in Eq. (10-205) is replaced by Eq. (10-191), to give two expressions that relate z to A_0 plus x_0, or to B_0 plus y_0, respectively:

$$A = x_0(A_0 + B_0) - \frac{3}{2}\frac{z(A_0 + B_0)}{(1+z)} = (A_0 + B_0)\left[x_0 - \frac{3}{2}\frac{z}{(1+z)}\right] \tag{10-206}$$

$$B = y_0(A_0 + B_0) - \frac{1}{2}\frac{z(A_0 + B_0)}{(1+z)} = (A_0 + B_0)\left[y_0 - \frac{1}{2}\frac{z}{(1+z)}\right] \tag{10-207}$$

At this point we have equations for the number of moles of each gas (A, B, and C) in terms of the initial mole fractions of H_2 and N_2 (A_0 and B_0) and the mole fraction of NH_3 (z). However, the equilibrium constant expression (10-181) is written in terms of the equilibrium mole fractions (x, y, z) and the total pressure. Thus, we have to evaluate Eqs. (10-193) to (10-195) using Eq. (10-205) for A, Eq. (10-207) for B, and Eq. (10-203) for C. The first part of this is done here:

$$\begin{aligned}
\Sigma = A + B + C &= (A_0 + B_0)\left[x_0 - \frac{3}{2}\frac{z}{1+z} + y_0 - \frac{1}{2}\frac{z}{1+z} + \frac{z}{1+z}\right] \\
&= (A_0 + B_0)\left[x_0 + y_0 + \frac{z}{1+z}\left(1 - \frac{3}{2} - \frac{1}{2}\right)\right] \\
&= (A_0 + B_0)\left[1 + \frac{z}{1+z}(-1)\right] \\
\Sigma &= (A_0 + B_0)\left[\frac{1}{1+z}\right]
\end{aligned} \tag{10-208}$$

Equation (10-208) uses Eq. (10-196), which defines $x_0 + y_0 = 1$. We can rewrite the equilibrium constant expression Eq. (10-181) in terms of the desired variables:

$$K_P = \left[\frac{X_{NH_3}}{X_{N_2}^{1/2}X_{H_2}^{3/2}}\right]_{eq} \times \frac{1}{P_T} = \frac{z}{x^{3/2}y^{1/2}} \cdot \frac{1}{P_T}$$
$$= \frac{(C/\Sigma)}{(A/\Sigma)^{3/2}(B/\Sigma)^{1/2}} \cdot \frac{1}{P_T} = \frac{C}{A^{3/2}B^{1/2}} \cdot \frac{\Sigma^{3/2}\Sigma^{1/2}}{\Sigma P_T} = \frac{C}{A^{3/2}B^{1/2} \cdot \frac{\Sigma}{P_T}} \tag{10-209}$$

Replacing A, B, and C by Eqs. (10-206), (10-207), and (10-203), respectively, and rearranging:

$$K_P = \frac{z/(1+z)}{(x_0 - (\frac{3}{2})z/(1+z))^{3/2}(y_0 - (\frac{1}{2})z/(1+z))^{1/2}} \cdot \frac{\Sigma}{P_T} \cdot \frac{(A_0+B_0)}{(A_0+B_0)^2} \tag{10-210}$$

Replacing Σ by Eq. (10-208) and canceling out terms:

$$K_P = \frac{z/(1+z)}{(x_0 - (\frac{3}{2})z/(1+z))^{3/2}(y_0 - (\frac{1}{2})z/(1+z))^{1/2}} \cdot \frac{1}{P_T} \cdot \frac{(A_0+B_0)}{1+z} \cdot \frac{(A_0+B_0)}{(A_0+B_0)^2}$$
$$= \frac{z/(1+z)}{(x_0 - (\frac{3}{2})z/(1+z))^{3/2}(y_0 - (\frac{1}{2})z/(1+z))^{1/2}} \cdot \frac{1}{P_T} \cdot \frac{1}{1+z} \tag{10-211}$$

Some more rearranging yields an expression for the equilibrium constant K_P in terms of the initial mole fractions of H_2 (x_0) and N_2 (y_0), the equilibrium mole fraction of NH_3 (z), and the total pressure (P_T):

$$K_P = \frac{z/(1+z)^2}{(x_0 - (\frac{3}{2})z/(1+z))^{3/2}(y_0 - (\frac{1}{2})z/(1+z))^{1/2}} \cdot \frac{1}{P_T} \tag{10-212}$$

Problem 9 at the end of this chapter illustrates the use of Eq. (10-212).

M. Temperature and pressure dependence of K_P for ammonia synthesis

We now use the data of Haber and LeRossignol (1907) to calculate equilibrium constants for NH_3 synthesis from 298 K to 2000 K. The equations and methods used apply to all chemical equilibria. Each of the five K_P values measured by Haber and LeRossignol (1907) gives the standard Gibbs free energy of reaction at the same temperature:

$$-R \ln K_P = \frac{\Delta_r G_T^o}{T} \tag{10-213}$$

Table 10-14 Third Law Analysis of Haber and LeRossignol (1907) Data

T (°C)	T (K)	$-R \ln K_P$	Δ(fef)	$\Delta_r H^o_{298}/T$	$\Delta_r H^o_{298}$
700	973	60.628	107.416	−46.788	−45,525
750	1023	63.740	107.897	−44.157	−45,173
850	1123	68.006	108.789	−40.783	−45,799
930	1203	70.804	109.435	−38.631	−46,473
1000	1273	73.327	109.951	−36.624	−46,622
					Mean $\Delta_r H^o_{298} \pm 1\sigma = -45{,}918 \pm 618$ J mol^{-1}

Furthermore, Eq. (10-213) is equal to the difference in Gibbs free energy functions (fef) of products and reactants plus the standard enthalpy of reaction at 298 K:

$$-R \ln K_P = \frac{\Delta_r G^o_T}{T} = \Delta\left[\frac{G^o_T - H^o_{298}}{T}\right] + \frac{\Delta_r H^o_{298}}{T} \tag{10-214}$$

$$\Delta\left[\frac{G^o_T - H^o_{298}}{T}\right] = \left[\frac{G^o_T - H^o_{298}}{T}\right]_{NH_3} - \frac{3}{2}\left[\frac{G^o_T - H^o_{298}}{T}\right]_{H_2} - \frac{1}{2}\left[\frac{G^o_T - H^o_{298}}{T}\right]_{N_2} \tag{10-215}$$

The Gibbs free energy functions are calculated from C_P data (e.g., see Table 3-4) via Eq. (9-84). The numerical values for the Δ(fef) from 298 K to 2000 K are given by

$$\Delta\left[\frac{G^o_T - H^o_{298}}{T}\right] = 93.4149 + 1.893 \times 10^{-2} T - 4.666 \times 10^{-6} T^2 J\,\text{mol}^{-1} K^{-1} \tag{10-216}$$

We rearrange Eq. (10-214) and solve for the standard heat of reaction at 298 K ($\Delta_r H^o_{298}$):

$$\Delta_r H^o_{298} = T\left(-R \ln K_P - \Delta\left[\frac{G^o_T - H^o_{298}}{T}\right]\right) \tag{10-217}$$

This calculation uses the K_P values from Haber and LeRossignol (1907) and the Δ(fef) values from Eqs. (10-215) and (10-216). Table 10-14 shows the calculations and results. The mean value $\pm 1\sigma$ error for $\Delta_r H^o_{298} = -45{,}918 \pm 618$ J mol^{-1}, which is almost identical to the accepted value of $-45{,}940 \pm 350$ J mol^{-1} (Gurvich et al., 1989). Substituting Eq. (10-216) and the calculated $\Delta_r H^o_{298}$ value into Eq. (10-214) and rearranging gives this equation for log K_P from 298 K to 2000 K:

$$\log K_P = \frac{2{,}399}{T} - 4.880 - 9.89 \times 10^{-4} T + 2.44 \times 10^{-7} T^2 \tag{10-218}$$

Figure 10-6 shows the temperature dependence for ammonia synthesis at one atmosphere total pressure. The equilibrium percentage of NH_3 calculated from Eqs. (10-218) and (10-212) is plotted versus temperature (°C). The experimental measurements of Haber and LeRossignol (1907), Haber and Maschke (1915), and Schulz and Schaefer (1966) are shown for comparison. The results in Figure 10-6 show that ammonia formation is favored by lower temperatures, as predicted by the van't

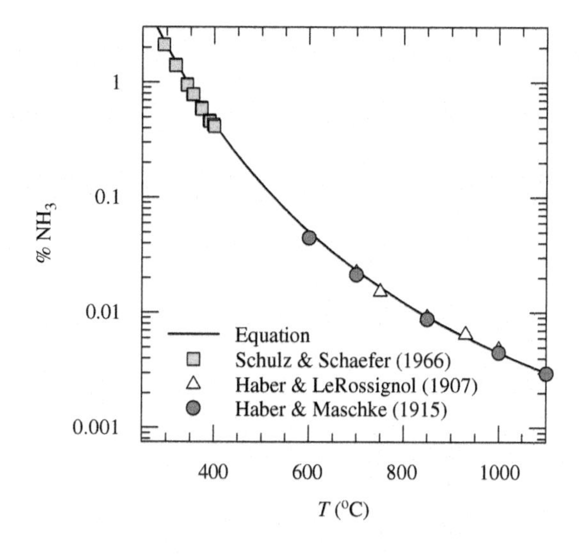

FIGURE 10-6

A plot of the equilibrium percentage of NH_3 versus temperature for ammonia synthesis from a 3:1 mixture of H_2 and N_2 at one atmosphere pressure. The data of Haber and LeRossignol (1907), Haber and Maschke (1915), and Schulz and Schaefer (1966) are compared to the curve calculated using Eq. (10-218).

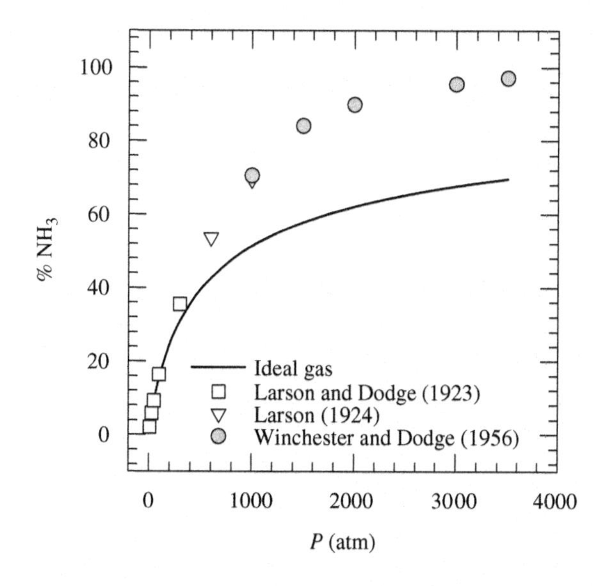

FIGURE 10-7

A plot of the equilibrium percentage of NH_3 versus pressure for ammonia synthesis from a 3:1 mixture of H_2 and N_2 at 450°C. The data of Larson and Dodge (1923), Larson (1924), and Winchester and Dodge (1956) are compared to the curve for an ideal gas mixture.

Table 10-15 Effect of Pressure on Ammonia Synthesis[a]

P (atm)	Eq. % NH_3	$K_P \times 10^3$	Empirical K_Y	Theoretical K_Y^c
1	0.215	6.66[b]	1.00	0.999
10	2.08	6.69	0.996	0.995
30	5.80	6.76	0.985	0.983
50	9.17	6.90	0.965	0.970
100	16.35	7.27	0.916	0.934
300	35.5	8.92	0.747	0.764
600	53.6	13.2	0.505	0.524
1000	70.5	25.0	0.266	0.302
1500	84.1	68.0	0.098	0.170
2000	89.8	134	0.050	0.116
3500	97.2	1075	0.0062	0.055

[a]*The % NH_3 values at 10–3500 atm pressure are from Larson (1924), Larson and Dodge (1923), and Winchester and Dodge (1956). The one atm % NH_3 value is calculated using Eq. (10-212).*
[b]*The one atm value is from Stephenson and McMahon (1939). A K_P value of 6.66 is 6.66×10^{-3}, and so on.*
[c]*Calculated with the Redlich-Kwong equation in Chapter 8.*

Hoff equation (10-65). However, the equilibrium percentage of NH_3 remains small at low temperatures and is only about 2% at 300°C.

The pressure dependence of ammonia synthesis is illustrated in Figure 10-7 and Table 10-15. The equilibrium constant expression (10-181) shows that the NH_3 equilibrium mole fraction is linearly proportional to the total pressure if K_P and the equilibrium mole fractions of H_2 and N_2 remain constant. Thus, the ratio of the NH_3 equilibrium mole fractions (X_1, X_2) at two different total pressures (P_1, P_2) is simply

$$\frac{X_1}{X_2} = \frac{P_1}{P_2} \tag{10-219}$$

The results of Haber and Le Rossignol (1907, 1908) at 700°C illustrate the use of Eq. (10-219). They found 0.0221% NH_3 at one atmosphere ($X_1 = 0.000221$ at $P_1 = 1$) and 0.65% NH_3 at 30 atmospheres ($X_2 = 0.0065$ at $P_2 = 30$), which is an increase in ammonia of 29.4 times for a 30 times increase in total pressure. The amount of NH_3 formed in their experiments is so small that the equilibrium mole fractions of H_2 and N_2 remain constant in both experiments.

However, as the amount of ammonia formed increases the equilibrium mole fractions of H_2 and N_2 decrease with total pressure and Eq. (10-219) is no longer valid. This breakdown is illustrated by the ideal gas curve in Figure 10-7. At low pressures, the curve is almost linear, but it begins to curve noticeably above 50 atmospheres pressure.

N. Nonideality and ammonia synthesis

Figure 10-7 also shows that deviations from ideality increase with increasing pressure for the NH_3 synthesis reaction. The K_P for NH_3 synthesis would be independent of total pressure at constant temperature if NH_3, H_2, and N_2 were ideal. However, the results in Table 10-15 show that this is not the case. Instead, K_P increases by about a factor of 161 as the total pressure is increased from 1 to 500 atmospheres. The change in K_P occurs because the pressure and the fugacity of H_2, N_2, and NH_3 diverge more and more as pressure is increased. As discussed in Chapter 8, fugacity and pressure are related via the fugacity coefficient γ:

$$\gamma = \frac{f}{P} \tag{8-166}$$

We can rewrite the equilibrium constant expression (10-181) for NH_3 synthesis in terms of the fugacities of H_2, N_2, and NH_3:

$$K_f = \left[\frac{f_{NH_3}}{f_{N_2}^{1/2} f_{H_2}^{3/2}}\right]_{eq} \tag{10-220}$$

where K_f stands for an equilibrium constant in terms of fugacities. Using Eq. (8-166), we can then rewrite K_f as the product of the equilibrium constants K_P and K_γ:

$$K_f = \left[\frac{f_{NH_3}}{f_{N_2}^{1/2} f_{H_2}^{3/2}}\right]_{eq} = \left[\frac{(\gamma P)_{NH_3}}{(\gamma P)_{N_2}^{1/2} (\gamma P)_{H_2}^{3/2}}\right]_{eq} \tag{10-221}$$

$$K_f = \left[\frac{\gamma_{NH_3}}{\gamma_{N_2}^{1/2} \gamma_{H_2}^{3/2}}\right]_{eq} \left[\frac{P_{NH_3}}{P_{N_2}^{1/2} P_{H_2}^{3/2}}\right]_{eq} = K_\gamma K_P \tag{10-222}$$

Fugacity is defined such that K_f is independent of pressure. As a result, the increasing K_P values with increasing pressure are counterbalanced by decreasing K_γ values (see Table 10-15).

In general, the K_f value is calculated from standard thermodynamic data:

$$\Delta_r G_T^o = -RT \ln K_f = -(\ln 10) RT \log K_f \tag{10-223}$$

For example, the Gibbs energy equation in Table 10-11 gives $K_f = 6.49 \times 10^{-3}$ for NH_3 synthesis from its constituent elements via reaction (10-180). However, for consistency we use a slightly different K_f value of 6.66×10^{-3} in Table 10-15, which Stephenson and McMahon (1939) derived from analysis of the data of Larson and Dodge (1923).

Table 10-15 gives two different K_γ values. The empirical K_γ value is calculated from

$$K_\gamma = \frac{K_f}{K_P} = \frac{6.66 \times 10^{-3}}{K_P} \tag{10-224}$$

Ammonia synthesis is one of the few reactions for which empirical K_γ values can be calculated. We calculated the theoretical K_γ value with the Redlich-Kwong equation (8-235). For example, at 300 atmospheres pressure the Redlich-Kwong equation gives fugacity coefficients of 0.957 for NH_3, 1.126 for H_2, and 1.101 for N_2. The corresponding K_γ value is

$$K_{\gamma} = \left[\frac{\gamma_{NH_3}}{\gamma_{N_2}^{1/2} \gamma_{H_2}^{3/2}} \right]_{eq} = \frac{0.957}{(1.101)^{1/2}(1.126)^{3/2}} = 0.764 \quad (10\text{-}225)$$

If another equation of state were used different theoretical K_{γ} values would result (e.g., see Winchester and Dodge, 1956). The comparison in Table 10-15 shows agreement within 3.8% between the empirical and theoretical K_{γ} values at 1−600 atmospheres pressure, and increasing disagreement with increasing pressure. The empirical K_{γ} values for ammonia synthesis are more accurate and preferable at pressures above 600 atmospheres.

However, in most cases no empirical K_{γ} data exist for the gas mixtures or pure gases of interest. Then we are limited to using K_{γ} values calculated from an equation of state or using reduced variables and generalized charts such as Figure 8-16. The Lewis and Randall fugacity rule must also be used in many cases because fugacity coefficient data for gas mixtures of interest do not exist (e.g., the important natural gas mixtures listed in Table 8-6).

II. GAS-CONDENSED PHASE EQUILIBRIA

A. Gas-solid reactions

We use the reaction of H_2S gas with Fe metal to illustrate equilibrium calculations for reactions involving gases and solids. This reaction is

$$H_2S\ (g) + Fe\ (metal) = FeS\ (troilite) + H_2\ (g) \quad (10\text{-}226)$$

Reaction (10-226) is important for several reasons. It involves two (Fe, S) of the four most abundant rock-forming elements (Mg, Si, Fe, S) and is the reaction that incorporated sulfur into Earth and other terrestrial planets (Fegley, 2000). The presence of sulfur is important for the differentiation and evolution of rocky planets because the lowest melting point (the eutectic point) in a Fe-FeS mixture is 1260 K. This is 550 degrees below the melting point of pure iron (1810 K), and the separation of a Fe-FeS eutectic melt is the beginning of core formation on rocky planets and asteroids. Troilite is stoichiometric iron monosulfide and is the iron-rich endmember of the pyrrhotite solid solution series (Hansen and Anderko, 1958).

At equilibrium,

$$\Delta G = G_{products} - G_{reactants} = 0 \quad (10\text{-}9)$$

$$\Delta G = n_{H_2}\overline{G}_{H_2} + n_{FeS}\overline{G}_{FeS} - n_{Fe}\overline{G}_{Fe} - n_{H_2S}\overline{G}_{H_2S} \quad (10\text{-}227)$$

The partial molar Gibbs energies of the gases in reaction (10-226) are

$$\overline{G}_{H_2} = G^{o}_{m,H_2} + RT \ln f_{H_2} = G^{o}_{m,H_2} + RT(\ln P_{H_2} + \ln\gamma_{H_2}) \quad (10\text{-}228)$$

$$\overline{G}_{H_2S} = G^{o}_{m,H_2S} + RT \ln f_{H_2S} = G^{o}_{m,H_2S} + RT(\ln P_{H_2S} + \ln\gamma_{H_2S}) \quad (10\text{-}229)$$

These equations are analogous to Eqs. (10-5) to (10-8) but are written using fugacities instead of partial pressures (see Eq. 8-167). In many cases of interest to us such as volcanic gases, planetary atmospheres, metamorphic reactions at shallow depth in the Earth's crust, the solar nebula, and protoplanetary accretion disks, most gases behave ideally and fugacities can be replaced by partial pressures with little or no error.

The partial molar Gibbs energies of Fe metal and solid FeS in reaction (10-226) are evaluated using Eq. (8-160) and are given by the equations

$$\overline{G}_{FeS} = G^o_{m,FeS} + \int_1^P V_T^P(FeS)dP \tag{10-230}$$

$$\overline{G}_{Fe} = G^o_{m,Fe} + \int_1^P V_T^P(Fe)dP \tag{10-231}$$

The integrals in Eqs. (10-230) and (10-231) can be evaluated using the molar volume at 298 K, isobaric thermal expansion coefficient α_P, and isothermal compressibility β_T for each phase, as illustrated in Section V of Chapter 7. Substituting Eqs. (10-228) to (10-231) into Eq. (10-227),

$$\begin{aligned}\Delta G = {} & n_{H_2}(G^o_{m,H_2} + RT\ln f_{H_2}) + n_{FeS}\left(G^o_{m,FeS} + \int_1^P V_T^P(FeS)dP\right) \\ & - n_{H_2S}(G^o_{m,H_2S} + RT\ln f_{H_2S}) - n_{Fe}\left(G^o_{m,Fe} + \int_1^P V_T^P(Fe)dP\right)\end{aligned} \tag{10-232}$$

Reaction (10-226) involves one mole of each substance. Rewriting Eq. (10-232) gives

$$\Delta G = \Delta_r G^o + RT\left(\ln\frac{P_{H_2}}{P_{H_2S}} + \ln\frac{\gamma_{H_2}}{\gamma_{H_2S}}\right) + \int_1^P \Delta V_T^P dP \tag{10-233}$$

The standard Gibbs energy of reaction term in Eq. (10-233) is

$$\Delta_r G^o_m = G^o_{m,H_2} + G^o_{m,FeS} - G^o_{m,H_2S} - G^o_{m,Fe} \tag{10-234}$$

The integral term in Eq. (10-233) is

$$\int_1^P \Delta V_T^P dP = \int_1^P (V_{FeS} - V_{Fe})dP \tag{10-235}$$

The reaction quotient term in Eq. (10-233) is the ratio of the H_2 and H_2S fugacities:

$$Q = \frac{f_{H_2}}{f_{H_2S}} = \left(\frac{P_{H_2}}{P_{H_2S}} \times \frac{\gamma_{H_2}}{\gamma_{H_2S}}\right) \tag{10-236}$$

Substituting Eq. (10-236) into Eq. (10-233) gives the reaction isotherm:

$$\Delta G = \Delta_r G^o + RT \ln Q + \int_1^P \Delta V_T^P dP \tag{10-237}$$

If reaction (10-226) occurs at the standard state pressure of one bar, Eq. (10-237) simplifies to

$$\Delta G = \Delta_r G^o + RT \ln Q \tag{10-238}$$

The next example illustrates the use of Eq. (10-233).

Example 10-13. Rosenqvist (1954) studied formation of FeS via reaction (10-226). At 800°C iron metal and troilite coexisted in equilibrium with gas having a H_2S/H_2 molar ratio $= 0.00183$. Use these data and the 298 K molar volumes of Fe (0.709 J bar^{-1}) and FeS (1.82 J bar^{-1}) to compute the standard Gibbs energy of reaction at this temperature. You should assume that the H_2S/H_2 gas mixture is at one atmosphere pressure.

Reaction (10-226) is in equilibrium under the conditions given ($\Delta G = 0$), so we can rearrange Eq. (10-233) to solve for the standard Gibbs energy of reaction:

$$\Delta_r G^o = -RT\left(\ln\frac{P_{H_2}}{P_{H_2S}} + \ln\frac{\gamma_{H_2}}{\gamma_{H_2S}}\right) - \int_1^P \Delta V_T^P dP \tag{10-239}$$

The three terms contributing to $\Delta_r G^o$ are the partial pressures and fugacity coefficients of the gases and the integral term for the ΔV of the solids. The partial pressure quotient in Eq. (10-239) is the single most important contribution to the standard Gibbs energy of reaction and is

$$-RT \ln\frac{P_{H_2}}{P_{H_2S}} = RT \ln\frac{P_{H_2S}}{P_{H_2}} = RT \ln(1.83 \times 10^{-3}) = -6.30\,RT = -56{,}232.49 \text{ J mol}^{-1} \tag{10-240}$$

The Redlich-Kwong equation (8-235) for gas mixtures gives fugacity coefficients of $\gamma = 1.0002$ for H_2 and $\gamma = 1.00025$ for H_2S at 1073 K and one atm total pressure. The fugacity coefficient quotient in Eq. (10-239) is the second most important contribution to $\Delta_r G^o$, but is much smaller:

$$-RT \ln\frac{\gamma_{H_2}}{\gamma_{H_2S}} = RT \ln\frac{1.00025}{1.0002} = 0.00005RT = 0.45 \text{ J mol}^{-1} \tag{10-241}$$

In this case the integral term is the third most important contribution to $\Delta_r G^o$, but it is very small:

$$\int_1^P \Delta V_T^P dP \cong \Delta V_{298}^o (P-1) = (-1.11 \text{ J bar}^{-1})(1.01325 - 1 \text{ bars}) = -0.015 \text{ J mol}^{-1} \quad (10\text{-}242)$$

Summing up these three contributions, we get

$$\Delta_r G^o = -56{,}232.49 + 0.45 - 0.015 \text{ J mol}^{-1} = -56{,}232.06 \text{ J mol}^{-1} \quad (10\text{-}243)$$

This result is only 265 J mol^{-1} more negative than that (– 55,967 J mol^{-1}) calculated from the calorimetric data of Grønvold and Stølen (1992), which we used in Chapter 4. The experimental uncertainties in temperature and gas partial pressures in Rosenqvist (1954) probably correspond to uncertainties of about the same size (a few hundred J mol^{-1}) in $\Delta_r G^o$.

In general, the partial pressure quotient dominates $\Delta_r G^o$ for gas-solid reactions at low pressure (e.g., up to tens or perhaps hundreds of bars pressure) and the fugacity coefficient and integral terms can be neglected because they are much smaller than the $RT \ln P$ term.

In contrast, the fugacity coefficient and integral terms are more important for gas-solid reactions taking place at high pressures like those encountered inside Earth and other terrestrial planets, the Moon and other large satellites, Ceres and other large asteroids, and icy bodies such as Pluto and Kuiper Belt objects. The pressures at which the fugacity coefficient and integral terms become significant relative to the standard Gibbs free energy change vary with the particular reaction. Holland and Powell (1998) tabulate $RT \ln f$ for H_2O and CO_2, and their tables can be used to show the P-T conditions where nonideality of these two gases is important. The importance of the integral term depends on the volume change between the solid reactants and products. For example, a ΔV of 10 cm^3 mol^{-1} (1 J bar^{-1}) corresponds to 1000 J mol^{-1} at one kilobar (0.1 GPa) pressure and to 10,000 J mol^{-1} at ten kilobars (1 GPa) pressure. Several problems at the end of this chapter deal with high-pressure reactions and illustrate the importance of the fugacity coefficient and ΔV_{solid} integral terms.

B. Definition of thermodynamic activity

The general version of Eq. (10-233) applies to all gas-solid reactions, but it is often convenient to rewrite Eq. (10-233) and its analogs in terms of the activities of the solid phases. Thus we now introduce the concept of thermodynamic activity that was proposed by G. N. Lewis (1907).

The *activity* of a material is the ratio between its fugacity f at some temperature, pressure, and concentration, and its fugacity f^* in a standard state, which is defined as the pure material at its saturation vapor pressure p^* and the same temperature T:

$$a = \frac{f}{f^*} \cong \frac{p}{p^*} \quad (10\text{-}244)$$

The p and p^* are the vapor pressure of a material at some P, T and concentration, and the saturation vapor pressure of the pure material at the same temperature, respectively. For example, the saturation vapor pressure p^* is 23.756 mm Hg for pure water at 25°C. In contrast, the vapor pressure p of pure water sitting in a cup or glass on your desktop at one bar total pressure is 24.636 mm Hg (see Example 7-9 in Chapter 7). Water vapor is nearly ideal at these pressures and temperatures, so the fugacity $f \sim p$ and $f^* \sim p^*$. (You can verify this for yourself using one of the equations of state or the reduced variable chart in Chapter 8.)

As discussed in Chapter 8, fugacity is an effective pressure. The concept of fugacity applies to all materials because all materials have a finite vapor pressure, even though it may be vanishingly small in some cases. For example, at 25°C the calculated vapor pressure of iron metal is about 1.6×10^{-65} bars. In contrast, at the same temperature naphthalene, which is the active ingredient in mothballs, has a distinctive odor and a measured vapor pressure of 0.08 mm Hg. In general the fugacity of a solid or liquid is equal to the vapor pressure of the solid or liquid if the vapor is ideal and is equal to the vapor pressure times a fugacity coefficient if the vapor is nonideal.

Example 10-14. Zellars and colleagues measured the vapor pressure over pure liquid Fe (Morris et al., 1957) and over Fe–Ni melts (Zellars et al., 1959). At 1600°C the vapor pressure over pure liquid Fe is 0.0572 mm Hg and the Fe vapor pressure over a 90% Fe–10% Ni melt ($Fe_{0.90}Ni_{0.10}$) is 0.05124 mm Hg. Calculate the activity of pure liquid Fe and Fe in the melt from these data.

Iron vapor is probably ideal at these low pressures and high temperatures so fugacities are taken equal to the vapor pressures.The activity of pure liquid Fe is unity:

$$a_{Fe}(\text{liquid}) = \frac{f}{f^*} = \frac{p}{p^*} = \frac{0.0572}{0.0572} = 1.0 \tag{10-245}$$

The activity of Fe dissolved in the $Fe_{0.90}Ni_{0.10}$ melt at 1600°C is less than unity and is proportional to its concentration in the molten alloy:

$$a_{Fe}(\text{melt}) = \frac{f}{f^*} = \frac{p}{p^*} = \frac{0.05142}{0.0572} = 0.899 \tag{10-246}$$

The activity of a pure solid, pure liquid, or pure melt in equilibrium with its vapor at a fugacity equal to the fugacity of the saturated vapor is unity. Pure solids, liquids, and melts are thermodynamically unstable under conditions in which their activities are less than unity.

C. Effect of total pressure on thermodynamic activity

Thermodynamic activity depends on the total pressure because fugacity depends on pressure. Equation (8-168) describes the pressure dependence of fugacity:

$$VdP = nRT(d\ln f) \tag{8-168}$$

Rewriting this in terms of the molar volume ($V_m = V/n$), separating variables, and integrating between the saturation vapor pressure p^* and an arbitrary pressure P gives

$$\int_{f^*}^{f} d\ln f = \ln\frac{f}{f^*} = \int_{p^*}^{P} \frac{V_m}{RT} dP \tag{10-247}$$

Substituting into Eq. (10-247) using the definition of activity Eq. (10-244) gives

$$\ln a = \ln\frac{f}{f^*} = \int_{p^*}^{P} \frac{V_m}{RT} dP \quad (10\text{-}248)$$

Rearranging yields

$$RT \ln a = \int_{p^*}^{P} V_m dP \quad (10\text{-}249)$$

If the pressure P is not too large and/or the material of interest has a small compressibility (large bulk modulus), Eq. (10-249) is approximated by

$$RT \ln a \cong V_m(P - p^*) \quad (10\text{-}250)$$

Equations (10-249) and (10-250) give the effect of total pressure on activity. The activity a in Eq. (10-250) is the activity at an arbitrary pressure P. If $P = p^*$, the activity of the pure material is unity. However, if $P > p^*$, the activity is greater than unity.

Example 10-15. The *S-L-V* triple point of cristobalite (SiO_2) is 1726°C and 2.6 Pa. The density of liquid silica as a function of temperature from 1935°C to 2165°C (Bacon et al., 1960) is

$$\rho(\text{g cm}^{-3}) = 2.508 - 2.13 \times 10^{-4} T(°\text{C}) \quad (10\text{-}251)$$

Calculate the activity of liquid silica at the triple-point temperature and one bar total pressure (e.g., molten silica in a crucible inside a furnace in air).

We calculate the molar volume of silica from its density (2.140 g cm^{-3}) and molecular weight (60.084 g mol^{-1}), take $R = 83.145$ cm^3 bar mol^{-1} K^{-1}, and substitute into Eq. (10-250):

$$RT \ln a \cong 28.08 \text{ cm}^3 \text{ mol}^{-1}(1 - 2.6 \times 10^{-5}) \text{ bars} \quad (10\text{-}252)$$

$$a = \exp\left(\frac{28.08 \times 0.999974}{83.145 \times 1999}\right) = 1.00017 \quad (10\text{-}253)$$

To a good first approximation, there is no difference between the activity of liquid silica under its saturation vapor pressure (e.g., in a crucible inside a vacuum furnace) and at one bar pressure.

Likewise, there is no significant difference between the activity of other geological materials under their saturation vapor pressure and at one bar pressure, as long as their saturation vapor pressure is less than one bar ($p^* < 1$). However, if p^* is greater than one bar pressure, the activity of the material is less than unity at one bar total pressure, and it is thermodynamically unstable.

Example 10-16. Subterranean reservoirs of hot, high-pressure liquid sulfur may exist on Io, the volcanically active satellite of Jupiter. At 1000°C, the vapor pressure and density of liquid sulfur are 144.5 atmospheres and 0.263 g cm^{-3}, respectively (Rau et al., 1973a, b). The molecular weight of sulfur is 32.066 g mol^{-1}. Calculate the activity of liquid sulfur at one bar pressure.

Substituting into Eq. (10-250), rearranging, and solving, we get

$$\ln a = \frac{(121.92 \text{ cm}^3 \text{ mol}^{-1})}{(83.145 \text{ cm}^3 \text{ bar mol}^{-1} \text{ K}^{-1})(1273 \text{ K})}(1 \text{ bar} - 144.5 \text{ atm} \times 1.01325 \text{ bar atm}^{-1}) \quad (10\text{-}254)$$

$$a = \exp(-0.1675) = 0.846 \quad (10\text{-}255)$$

Thus, the activity of liquid sulfur at one bar pressure and 1000°C is 0.846 relative to an activity of unity at the saturation vapor pressure of 144.5 atm (146.4 bars). In other words, liquid sulfur is unstable at 1000°C unless the pressure of sulfur vapor equals the saturation vapor pressure.

D. Relationship between activity and Gibbs energy

The variation of activity with pressure is also related to the variation of G with pressure and thus to the difference $(G - G^o)$ between the Gibbs energy G at an arbitrary pressure P and the standard Gibbs energy G^o at the standard-state pressure of one bar for a material. According to Eq. (8-167),

$$dG = nRT(d \ln f) \quad (8\text{-}167)$$

Combining Eqs. (8-167) and (8-168) and rearranging, we get

$$\left(\frac{\partial \ln f}{\partial P}\right)_T = \frac{1}{nRT}\left(\frac{\partial G}{\partial P}\right)_T = \frac{V}{nRT} \quad (10\text{-}256)$$

Separating variables and integrating Eq. (10-256) between the standard-state pressure $P^o = 1$ bar and an arbitrary pressure P, we get

$$\int_{G^o}^{G} dG = \int_{P^o}^{P} \frac{V}{nRT} dP \quad (10\text{-}257)$$

$$G - G^o = \frac{1}{nRT}\int_{P^o}^{P} VdP \quad (10\text{-}258)$$

Integration of Eq. (8-168) gives an analogous result, which we can combine with Eq. (10-258),

$$G - G^o = nRT \ln\frac{f}{f^o} = \int_{P^o}^{P} VdP \quad (10\text{-}259)$$

Substituting into this using Eq. (10-244) gives an expression relating activity and the pressure dependence of the Gibbs energy:

$$G - G^o = nRT \ln a = \int_{P^o}^{P} VdP \quad (10\text{-}260)$$

The analogous expression in terms of molar properties is

$$\overline{G} - G_m^o = RT \ln a = \int_{P^o}^{P} V_m dP \quad (10\text{-}261)$$

Equation (10-261) is similar to Eq. (10-249), but their lower integration limits are different. Equation (10-249) is integrated from the saturation vapor pressure p^* of the pure material, whereas Eq. (10-261) is integrated from the standard-state pressure $P^o = 1$ bar. However, as illustrated in Example 10-15 there is essentially no difference between the activity at p^* and $P^o = 1$ bar. The corresponding difference in the Gibbs energy is also small, for example, for pure molten SiO_2:

$$G_m^o - G_m^* = RT \ln a = (8.314)(1999) \ln 1.00017 \sim 3 \text{ J mol}^{-1} \quad (10\text{-}262)$$

For comparison, the standard Gibbs energy of formation $\Delta_f G^o$ of pure molten silica from its constituent elements is about $-$ 550,595 J mol^{-1} at the same temperature. The uncertainty on this value is about 1000 J mol^{-1}. The 3 J mol^{-1} difference between the saturation vapor pressure of pure SiO_2 (liq) and one bar pressure is insignificant and can be neglected without any error. In general, we can safely neglect the difference between Eqs. (10-249) and (10-261) and use the latter equation for our calculations.

E. Oxygen fugacity buffers

Background information

Gas-solid equilibria that control oxygen fugacity (fO_2) in experimental petrology are called *buffer reactions* or *solid-state buffers*. Many buffer reactions involve transition metal–oxide pairs such as nickel and nickel oxide (bunsenite):

$$\text{Ni (metal)} + \frac{1}{2} O_2 \text{ (g)} = \text{NiO (bunsenite)} \quad (10\text{-}263)$$

Other metal-oxide pairs used as fO_2 buffers include Co-CoO, Cu-Cu_2O (cuprite), Cr-Cr_2O_3 (eskolaite), and Fe-$Fe_{0.95}O$ (wüstite). Wüstite is stable, but stoichiometric FeO is unstable at ambient pressure. Iron-rich wüstite has a temperature-dependent Fe/O atomic ratio ranging from ~ 0.950 at 1371°C, the eutectic temperature where γ-Fe metal + Fe-rich wüstite are in equilibrium with a melt, to ~0.932 at 574°C, where wüstite decomposes to Fe metal plus magnetite (the wüstite eutectoid temperature).

Oxide-oxide pairs such as Cu_2O (cuprite)–CuO (tenorite), $Fe_{0.95}O$ (wüstite)–Fe_3O_4 (magnetite), Fe_3O_4 (magnetite)–Fe_2O_3(hematite), and MnO (manganosite)–Mn_3O_4 (hausmannite) are also used as fO_2 buffers. Many of these oxides become nonstoichiometric at high temperatures and high oxygen fugacities. For example, at 1400°C magnetite in equilibrium with wüstite is stoichiometric, while O-rich magnetite ($Fe_3O_{4.116}$) is in equilibrium with hematite. This example illustrates that oxygen fugacities for metal-oxide and oxide-oxide buffers involving nonstoichiometric phases are slightly different than calculated from thermochemical data. Also, the extent of nonstoichiometry at a given T and fO_2 is different for each oxide and has to be determined experimentally. Sykor and Mason (1987) compare NiO and CoO at 1200°C and fO_2 ~1 bar. The NiO is essentially stoichiometric (Ni/O = 0.99977) while CoO is noticeably nonstoichiometric (Co/O = 0.99). As the fO_2 increases at constant temperature, the non-stoichiometry of each oxide increases but remains greater for CoO than for NiO.

In principle, solid-state buffers give fO_2 values that are much lower or much higher than possible from gas mixtures such as CO-CO_2 or H_2-H_2O. For example, at 1000°C solid-state buffers span fO_2 values over 26 orders of magnitude from ~$10^{1.04}$ bars (Pd-PdO) to $10^{-25.0}$ bars (Nb-NbO). In practice, some metal-oxide pairs and oxide-oxide pairs are not useful as solid-state buffers. The alkali metals are very reactive, hard to handle, and have low melting points. Mercury (Hg) and gallium (Ga) also have low melting points (−39°C and 29°C, respectively) and Hg is too volatile. Lead, some other metals, and some metal oxides are also too volatile for use as oxygen buffers at high temperatures.

A number of other oxygen fugacity buffers are also important in petrology, such as the QFM buffer involving quartz (SiO_2, Q), fayalite (Fe_2SiO_4, F), and magnetite (Fe_3O_4, M):

$$3\,Fe_2SiO_4\,(\text{fayalite}) + O_2\,(g) = 2\,Fe_3O_4\,(\text{magnetite}) + 3\,SiO_2\,(\text{quartz}) \tag{10-264}$$

and the quartz, fayalite, iron metal (QFI) buffer:

$$2\,Fe\,(\text{metal}) + SiO_2\,(\text{quartz}) + O_2\,(g) = Fe_2SiO_4\,(\text{fayalite}) \tag{10-265}$$

Iron-titanium oxides are used as fO_2 geobarometers and geothermometers in petrology (O'Neill et al., 1988; Buddington and Lindsley, 1962). The common Fe-Ti oxides in terrestrial rocks are ilmenite ($FeTiO_3$), which can be dissolved with hematite, and ulvöspinel (Fe_2TiO_4), which is commonly dissolved in magnetite. Ilmenite is common in lunar rocks, constitutes up to 18% (by volume) of TiO_2-rich lunar basalts, and is a potential source of lunar oxygen and water. The two important fO_2 equilibria involving Fe-Ti oxides are the iron-rutile-ilmenite (IRI) buffer and the iron-ilmenite-ulvöspinel (IIU) buffer:

$$2\,Fe\,(\text{metal}) + 2\,TiO_2\,(\text{rutile}) + O_2\,(g) = 2\,FeTiO_3\,(\text{ilmenite}) \tag{10-266}$$

$$2\,Fe\,(\text{metal}) + 2\,FeTiO_3 + O_2\,(g) = 2\,Fe_2TiO_4\,(\text{ulvöspinel}) \tag{10-267}$$

Several different methods can be used to measure the temperature-dependent oxygen fugacities of solid-state buffers. For example, Smyth and Roberts (1920) manometrically measured the O_2 pressure for the Cu_2O (cuprite)—CuO (tenorite) buffer from 1173 K to 1333 K. O'Neill (1988) electrochemically measured the standard Gibbs energy change from 800 K to 1300 K for the same buffer. The Co-CoO buffer is another example. Emmett and Schultz (1929, 1930) measured the equilibrium H_2/H_2O (608−843 K) and CO/CO_2 (723−843 K) ratios for Co-CoO, whereas O'Neill (1987b) electrochemically measured the standard Gibbs energy change from 800 K to 1397 K. Finally, the fO_2 for solid-state buffers involving stoichiometric phases can be calculated from calorimetric data (S^o_{298}, $\Delta_f H^o_{298}$, C_P) using the third law. Fritsch and Navrotsky (1995) did this for the MnO_2 (pyrolusite)—Mn_2O_3 (bixbyite) buffer and their results agree with the fO_2 measured manometrically by Otto (1965).

Application of the Gibbs phase rule to oxygen fugacity buffers

The Gibbs phase rule (Eq. 7-70) helps us to understand the operation of an oxygen fugacity buffer. The Ni-NiO reaction (10-263) and other equilibria involving metal-oxide pairs or oxide-oxide pairs have three phases: O_2 gas, and two solid phases (metal + oxide or two oxides). Each of these reactions also

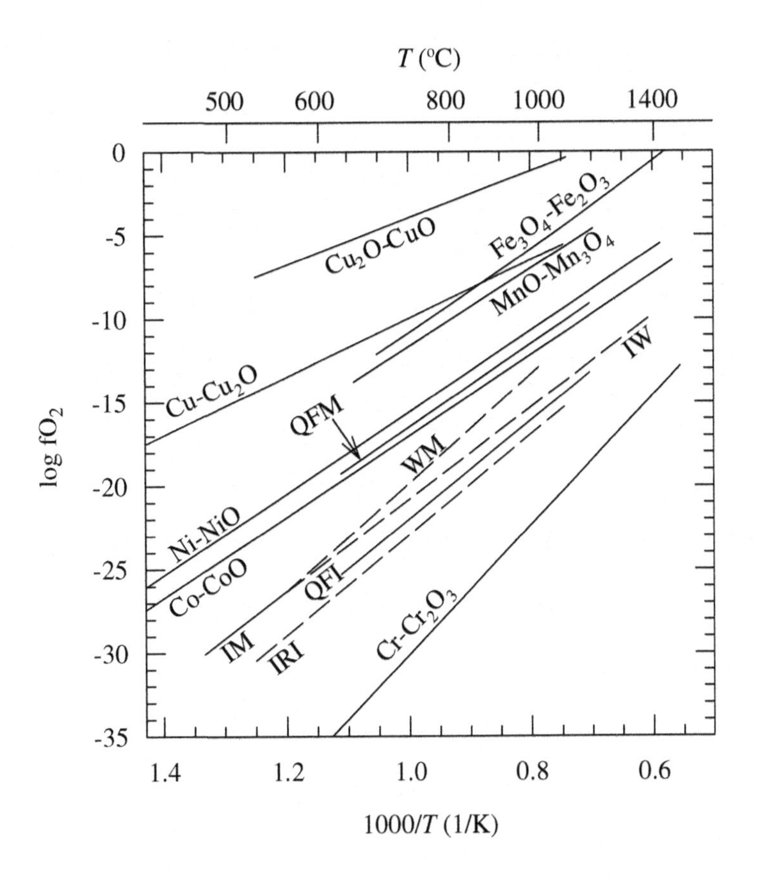

FIGURE 10-8

A plot of log fO_2 versus inverse temperature for solid-state oxygen fugacity buffers.

has two components: the metallic element and oxygen. Using the Gibbs phase rule ($F = C - P + 2$) we find that each reaction has one degree of freedom. Thus, the O_2 fugacity is uniquely fixed at constant temperature by the coexistence of the two solid phases. The QFM, QFI, and Fe-Ti oxide buffers have three components (Fe, O, Si or Fe, O, Ti, respectively), four phases (gas and three solids), and one degree of freedom. The O_2 fugacity is uniquely fixed at constant temperature by the coexistence of the three solid phases.

Because oxygen fugacity buffer reactions are univariant (i.e., only one degree of freedom), it is convenient to plot $\log_{10} f_{O_2}$ versus $1/T$ as on Figure 10-8. This graph is similar to Figure 7-11 and has a similar interpretation, which we explain using the Ni-NiO line as an example.

First we consider oxygen fugacities below the Ni-NiO line. Nickel metal coexists with O_2 at the very small O_2 fugacities in this P-T region. Bunsenite is thermodynamically unstable in this region and decomposes to Ni + O_2 gas. However, NiO coexists with O_2 at oxygen fugacities above the Ni-NiO line. Nickel metal is thermodynamically unstable in this region and reacts with O_2 to form NiO. Along the Ni-NiO line, Ni metal, NiO, and O_2 gas coexist and the oxygen fugacity at constant temperature is uniquely fixed by the coexistence of the two solid phases.

Removal of O_2 from a chemically unreactive gas such as Ar or N_2 is a practical application of the univariant curves shown in Figure 10-8. Compressed gases typically contain 0.1 − 1 parts per million by volume (ppmv) O_2, equivalent to O_2 fugacities of 10^{-7} to 10^{-6} bars at one bar total pressure. The O_2 fugacity is reduced to much lower levels by passing the gas through heated copper metal turnings that react with the gas forming cuprite (Cu_2O). The gas can be passed through several different heated metals in series (e.g., Cu, Ni, Cr) to reduce the O_2 fugacity to extremely low levels. Conversely, the discoloration of pure metals or metal alloys (e.g., Cu − Pt, or Cu − Au) due to their oxidation can also indicate the O_2 fugacity of a gas.

Thermodynamic activity and buffer reactions

We reach the same conclusion about a unique oxygen fugacity at a given temperature using the concept of thermodynamic activity. The Gibbs energy change for the Ni-NiO buffer is

$$\Delta G = n_{NiO}\overline{G}_{NiO} - n_{Ni}\overline{G}_{Ni} - n_{O_2}\overline{G}_{O_2} \tag{10-268}$$

The stoichiometry of reaction (10-263) shows the number of moles $n = 1$ for Ni and NiO and $n = ½$ for O_2 gas. The partial molar Gibbs energies of O_2, Ni, and NiO are given by the equations

$$\overline{G}_{O_2} = G^o_{m,O_2} + RT \ln f_{O_2} \tag{10-269}$$

$$\overline{G}_{Ni} = G^o_{m,Ni} + \int_{P^o}^{P} V_{Ni} dP \tag{10-270}$$

$$\overline{G}_{NiO} = G^o_{m,NiO} + \int_{P^o}^{P} V_{NiO} dP \tag{10-271}$$

The standard-state pressure $P^o = 1$ bar in the preceding equations. We use Eq. (10-261) to rewrite the equations for the partial molar Gibbs energies of Ni and NiO:

$$\overline{G}_{Ni} = G^o_{m,Ni} + RT \ln a_{Ni} \tag{10-272}$$

$$\overline{G}_{NiO} = G^o_{m,NiO} + RT \ln a_{NiO} \tag{10-273}$$

We can use Eqs. (10-269), (10-272), and (10-273) to rewrite Eq. (10-268) as

$$\Delta G = \Delta_r G^o + RT \ln a_{Ni} - RT \ln a_{NiO} - \frac{1}{2} RT \ln f_{O_2} \tag{10-274}$$

The standard Gibbs energy of reaction term in Eq. (10-274) is

$$\Delta_r G^o = G^o_{m,NiO} - G^o_{m,Ni} - \frac{1}{2} G^o_{m,O_2} = G^o_{m,NiO} \tag{10-275}$$

The $\Delta_r G^o$ in Eq. (10-275) reduces to the standard molar Gibbs energy of formation for NiO because the Gibbs energies of Ni and O_2 in their reference states are defined as zero. Rearranging Eq. (10-274) gives the reaction isotherm for the Ni-NiO buffer reaction:

$$\Delta G = \Delta_r G^o + RT \ln\left(\frac{a_{NiO}}{a_{Ni} f_{O_2}^{1/2}}\right) \quad (10\text{-}276)$$

Along the univariant curve where Ni metal and NiO (bunsenite) coexist, $\Delta G = 0$. Both Ni metal and NiO are pure solids because they do not dissolve in one another (i.e., the solubility of oxygen in Ni metal and of Ni metal in NiO are both insignificant). Thus, both solids have unit activity ($a_{Ni} = a_{NiO} = 1$). The reaction isotherm becomes

$$\Delta G = 0 = \Delta_r G^o + RT \ln\left(\frac{1}{f_{O_2}^{1/2}}\right) \quad (10\text{-}277)$$

The term in parenthesis is the reaction quotient Q. The equilibrium constant K_f is the value of the reaction quotient Q at equilibrium, which is the value of the equilibrium oxygen fugacity:

$$K_f = \left(\frac{1}{f_{O_2}^{1/2}}\right)_{eq} = \exp\left(\frac{-\Delta_r G^o}{RT}\right) \quad (10\text{-}278)$$

Pressure dependence of oxygen fugacity buffers

We use the Ni-NiO buffer as an example to illustrate the pressure dependence of oxygen fugacity buffers. Substituting Eqs. (10-269) to (10-271) for the partial molar Gibbs energies of O_2, Ni, and NiO into Eq. (10-268) and rearranging to solve for the equilibrium O_2 fugacity gives

$$\log f_{O_2} = \frac{2}{(\ln 10)RT}\left(\Delta_r G^o + \int_{P^o}^{P} \Delta V_T^P dP\right) = \frac{0.1045}{T}\left(\Delta_r G^o + \int_{P^o}^{P} \Delta V_T^P dP\right) \quad (10\text{-}279)$$

To first approximation the integral term can be replaced by ΔV^o at 298 K:

$$\int_{P^o}^{P} \Delta V_T^P dP \cong \Delta V_{298}^o (P - 1) \quad (10\text{-}280)$$

$$\Delta V_{298}^o = (V_{NiO}^o - V_{Ni}^o) = (10.97 - 6.589)\ \text{cm}^3\ \text{mol}^{-1} = 0.438\ \text{J bar}^{-1} \quad (10\text{-}281)$$

The $\Delta_r G^o$ from 700 K to 1728 K (O'Neill and Pownceby, 1993a) is

$$\Delta_r G^o = -239{,}483.5 - 11.2782T \log T + 124.257T\ \text{J mol}^{-1} \quad (10\text{-}282)$$

Table 10-16 Oxygen Buffer Reactions[a]

$$\mathrm{Log}_{10} f_{O_2}\ (\mathrm{bars}) = A - \frac{B}{T} + C\log_{10}T + \frac{D(P-1)}{T}$$

Buffer[b]	*A*	*B*	*C*	*D*	*T* Range (K)[c]	Source
Co-CoO	6.7107	24,214	0.1620	0.0519	700 – 1768	1
Cr-Cr_2O_3	10.9243	39,084	−0.6383	0.0764	600 – 1800	2
Cu-Cu_2O	12.8529	18,160	−1.5520	0.0962	700 – 1338	1
Cu_2O-CuO	19.6938	15,266	−2.7902	0.0105	800 – 1348	1
Fe-$Fe_{0.95}O$	5.2614	27,341	0.4291	0.0556	833 – 1650	1
Fe-Fe_3O_4	16.2052	29,493	−2.4174	0.0607	298 – 833	3
$Fe_{0.95}O$-Fe_3O_4	−3.4270	30,392	4.6589	0.0878	833 – 1270	4
Fe_3O_4-Fe_2O_3	−25.7139	19,375	11.4191	0.0185	950 – 1870	5–7
MnO-Mn_3O_4	5.5783	22,424	1.7357	0.0761	917 – 1433	8
Ni-NiO	12.9850	25,026	−1.1786	0.0458	700 – 1728	1
Fe-SiO_2-Fe_2SiO_4	4.5468	29,194	0.8868	0.0497	900 – 1400	9
Fe_3O_4-SiO_2-Fe_2SiO_4	5.5976	24,505	0.8099	0.0937	900 – 1400	9
Fe-TiO_2-$FeTiO_3$	5.0032	29,826	0.6360	0.0604	800 – 1340	10
Fe-$FeTiO_3$-Fe_2TiO_4	10.2320	27,745	−1.2749	0.0840	750 – 1500	10

[a]*The calculated log fO_2 values correspond to the univariant curves graphed in Figure 10-8.*
[b]*All reactions are denoted in a similar fashion. Thus, Co-CoO denotes 2 Co + O_2 = 2 CoO, Fe_3O_4-Fe_2O_3 denotes 4 Fe_3O_4 + O_2 = 6 Fe_2O_3, Fe-SiO_2-Fe_2SiO_4 denotes the QFM buffer 2 Fe + SiO_2 + O_2 = Fe_2SiO_4, and so on.*
[c]*The temperature ranges are generally those studied experimentally. In several cases the T range is limited by melting or phase equilibria as follows: Co m.p. 1768 K, Cu-Cu_2O eutectic 1338 K, Cu_2O-CuO eutectic 1348 K, Ni m.p. 1728 K, wüstite m.p. 1650 K, and the wüstite eutectoid 833 K, which is the lower T limit for the iron-wüstite buffer and the upper temperature limit for the iron-magnetite buffer.*

Sources: 1. O'Neill and Pownceby (1993a), 2. Holzheid and O'Neill (1995), 3. O'Neill (1988), 4. Robie et al. (1978), 5. Crouch et al. (1971), 6. Darken and Gurry (1946), 7. Jacobsson (1985), 8. O'Neill and Pownceby (1993b), 9. O'Neill (1987a), 10. O'Neill et al. (1988).

The O_2 fugacity as a function of T and P for the Ni-NiO buffer is then

$$\log f_{O_2} = 12.985 - \frac{25,026}{T} - 1.1786\log T + \frac{0.0458(P-1)}{T}\ \text{bars} \tag{10-283}$$

Table 10-16 gives equations similar to Eq. (10-283) for several oxygen fugacity buffers. At sufficiently large O_2 pressures (e.g., for the Ag-Ag_2O and MnO_2-Mn_2O_3 buffers), another correction for the difference between the fugacity and pressure of O_2 is necessary.

F. Sulfur fugacity buffers

Control of sulfur fugacity is important for studies of sulfide ore mineralogy, sulfide formation in meteorites, and for partitioning experiments between metal and sulfide or silicate and sulfide. In principle, solid and liquid sulfur in equilibrium with its saturated vapor is a sulfur fugacity buffer.

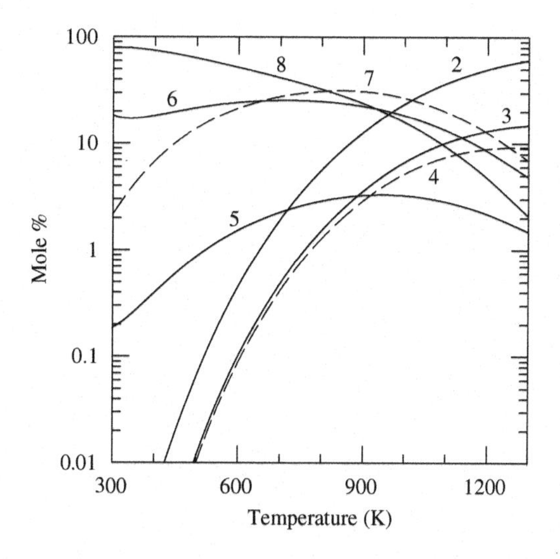

FIGURE 10-9

The composition of saturated sulfur vapor from 300 K to 1300 K calculated from the data of Rau et al. (1973a, b). The abundance of S gas is less than 0.01% and it is not shown.

However, in practice, solid or liquid sulfur are not useful buffers because the composition of saturated sulfur vapor is complex and depends on temperature. Figure 10-9 shows the composition of saturated sulfur vapor from 300 K to 1300 K. At 300 K the vapor is mainly S_8 molecules plus smaller amounts of S_6, S_7, and S_5 molecules. As temperature increases the smaller allotropes become more important, and at 1300 K the vapor is 60% S_2 and 40% S_3 to S_8 molecules. (The abundance of S gas is less than 0.01% and it is not shown in Figure 10-9.) At any temperature below the sulfur critical point (1313 K), S_2 becomes more important as pressure is decreased, but the total pressure is no longer buffered by the presence of coexisting solid or liquid sulfur once we move off the sublimation or vaporization curves. In contrast, the vapor in equilibrium with metal + sulfide or sulfide + sulfide pairs is almost 100% S_2 gas. The iron-troilite buffer provides a good example.

The reaction between iron metal and troilite is

$$\text{Fe (metal)} + \frac{1}{2}\,\text{S}_2\text{ (g)} = \text{FeS (troilite)} \tag{10-284}$$

The Gibbs energy change for reaction (10-284) is

$$\Delta G = n_{\text{FeS}}\overline{G}_{\text{FeS}} - n_{\text{Fe}}\overline{G}_{\text{Fe}} - n_{\text{S}_2}\overline{G}_{\text{S}_2} = \overline{G}_{\text{FeS}} - \overline{G}_{\text{Fe}} - \frac{1}{2}\,\overline{G}_{\text{S}_2} \tag{10-285}$$

The partial molar Gibbs energies of S_2, Fe, and FeS are

$$\overline{G}_{\text{S}_2} = G^o_{m,\text{S}_2} + RT\ln f_{\text{S}_2} \tag{10-286}$$

$$\overline{G}_{\text{Fe}} = G^{o}_{m,\text{Fe}} + \int_{P^o}^{P} V_{\text{Fe}} dP = G^{o}_{m,\text{Fe}} + RT \ln a_{\text{Fe}} \tag{10-287}$$

$$\overline{G}_{\text{FeS}} = G^{o}_{m,\text{FeS}} + \int_{P^o}^{P} V_{\text{FeS}} dP = G^{o}_{m,\text{FeS}} + RT \ln a_{\text{FeS}} \tag{10-288}$$

The standard-state pressure $P^o = 1$ bar in the preceding equations, which are analogous to Eqs. (10-269) to (10-273) for O_2, Ni, and NiO. Proceeding as done for the Ni-NiO reaction we can write the standard Gibbs free energy and the reaction isotherm for the Fe-FeS buffer as

$$\Delta_r G^o = G^o_{m,\text{FeS}} - G^o_{m,\text{Fe}} - \frac{1}{2} G^o_{m,\text{S}_2} = G^o_{m,\text{FeS}} - \frac{1}{2} G^o_{m,\text{S}_2} \tag{10-289}$$

$$\Delta_r G^o = -165,671 - 21.0648T \log T + 131.51T \text{ J mol}^{-1} \tag{10-290}$$

$$\Delta G = \Delta_r G^o + RT \ln \left(\frac{a_{\text{FeS}}}{a_{\text{Fe}} f_{\text{S}_2}^{1/2}} \right) \tag{10-291}$$

In contrast to Eq. (10-275) for the Ni-NiO buffer, we must include the molar Gibbs energy of S_2 (g) in the standard Gibbs energy for the Fe-FeS buffer. The reason is that O_2 is the reference state for oxygen at all temperatures, but S_2 is not always the reference state for sulfur. The four different reference states for sulfur are orthorhombic sulfur (0–368.54 K), monoclinic sulfur (368.54–388.36 K), liquid sulfur (388.36–882.117 K), and S_2 gas ($T \geq 882.117$ K).

Equation (10-290) is valid from 298 K to 1463 K, the melting point of FeS. At 1200 K the standard Gibbs energy of reaction and the sulfur fugacity over coexisting Fe + FeS are

$$\Delta_r G^o = -85,694 \text{ J mol}^{-1} \tag{10-292}$$

$$f_{\text{S2}} = \exp\left(\frac{2\Delta_r G^o}{RT}\right) = \exp(-17.18) = 3.46 \times 10^{-8} \text{ bars} \tag{10-293}$$

The equilibrium partial pressure of S_3 over coexisting Fe + FeS is controlled by the reaction

$$3\ \text{S}_2 = 2\ \text{S}_3 \tag{10-294}$$

The standard Gibbs energy of reaction (298–2000 K) and the equilibrium constant for this are

$$\Delta_r G^o = -96,384 - 3.5378T \log T + 141.5674T \text{ J mol}^{-1} \tag{10-295}$$

$$K_P = \left[\frac{f_{\text{S}_3}^2}{f_{\text{S}_2}^3}\right]_{eq} \cong \left[\frac{P_{\text{S}_3}^2}{P_{\text{S}_2}^3}\right]_{eq} \tag{10-296}$$

Table 10-17 Metal-Sulfide Reactions and Sulfur Saturation Curve

$$\log_{10} f_{S_2} \text{ (bars)} = A - \frac{B}{T} + C \log_{10} T$$

Reaction	A	B	C	T Range
Ag-Ag_2S	13.3917	10,330	−2.5654	298−451
	12.9375	9855	−2.7844	451−1109
Co-CoS_2	15.4694	14,883	−1.9975	298−1299
Cu-CuS	14.8058	12,589	−2.1960	298−780
Cu-Cu_2S	22.8309	15,999	−5.7784	298−1358
Fe-FeS_2	12.7698	15,772	−0.7545	298−1015
Mn-MnS	4.2646	28,799	0.7213	298−1803
Mo-MoS_2	14.3151	21,117	−1.4121	298−2023
Ni-Ni_3S_2	11.7995	18,108	−0.7599	298−829
	28.2989	17,245	−6.7717	829−1062
Ni-NiS	22.4486	16,933	−4.2809	298−1066
Pb-PbS	6.9199	16,943	0.6788	298−1386.5
Zn-ZnS	4.0312	27,628	1.9245	298−1180
S(s,liq)-S_2	20.8089	7320	−4.2631	298−1313

The upper temperature limits for the equations are set by phase transitions or melting: α (acanthite) – β (argentite) 451 K, m.p. 1109 K; CoS_2 (cattierite) incongruent melting to $Co_{1-x}S$ + melt 1299 K; CuS (covellite) incongruent melting to $Cu_{1.732}S$ (high digenite) + melt 780 K; Cu m.p. 1358 K; pyrite incongruent melting to pyrrhotite + melt at 10 bars sulfur vapor pressure; MnS (alabandite) m.p. 1803 K; MoS_2 (molybdenite) incongruent melting 2023 K; Ni_3S_2 (heazlewoodite) hexagonal-cubic transition 829 K, incongruent melting 1062 K; NiS (millerite) m.p. 1066 K; PbS (galena) m.p. 1386.5 K; Zn b.p. 1180 K; sulfur critical point 1313 K.

At 1200 K, $\Delta_r G^o = 60{,}425$ J mol^{-1}, $K_P = 2.34 \times 10^{-3}$, and the S_3 partial pressure is 3.1×10^{-13} bars.

Other common sulfur fugacity buffers involve metal-sulfide pairs and coexisting copper and/or iron sulfides (see Table 10-17). The pyrite-pyrrhotite buffer is probably the most important example. Pyrite is stoichiometric FeS_2. Pyrrhotite in equilibrium with it has a temperature-dependent Fe/S ratio of Fe_7S_8 to first approximation. The pyrite-pyrrhotite buffer can be written as

$$7\ FeS_2 \text{ (pyrite)} = Fe_7S_8 \text{ (pyrrhotite)} + 3\ S_2 \text{ (g)} \tag{10-297}$$

The S_2 fugacity in equilibrium with coexisting pyrite and pyrrhotite from 567 K to 1016 K is

$$\log_{10} f_{S_2} \text{ (bars)} = 19.41 - \frac{21{,}004}{T} + \frac{2{,}061{,}770}{T^2} \tag{10-298}$$

Equation (10-298) is based on measurements by Toulmin and Barton (1964), Schneeberg (1973), and Lusk and Bray (2002). Toulmin and Barton (1964) used the tarnishing of electrum (gold-silver) alloys by sulfur vapor to indicate the S_2 fugacity. Schneeberg (1973) and Lusk and Bray (2002) used electrochemical sulfur fugacity sensors. Barton and Skinner (1979) review thermodynamic data for many sulfur fugacity buffer reactions. Many sulfides are nonstoichiometric and some are like wüstite $Fe_{1-x}O$ with different metal/sulfur ratios in equilibrium with metal or in equilibrium with a higher sulfide. Copper sulfides exemplify this behavior with anilite ($Cu_{1.75}S$), digenite ($Cu_{1.765}S$), and djurleite ($Cu_{1.934}S$) being nonstoichiometric and chalcocite (Cu_2S) existing over a range of Cu/S ratios.

G. Mineral buffer reactions

In the previous sections we discussed gas-solid reactions that buffer oxygen and sulfur fugacity. However, gas-solid reactions are also important in metamorphic petrology on Earth and for atmosphere-surface chemistry on Venus. We briefly discuss these two topics now and start with the mineral buffer reactions that are important on Venus.

Atmosphere-surface reactions on Venus

The global average temperature and total pressure on Venus are 740 K and 95.6 bars. These values are for the modal radius (equal to 6051.4 km), which is the reference point for measuring elevations on Venus. The modal radius is the most common radius (i.e., the average defined by the mode) and is used as a reference point because there is no sea level on Venus. Temperature and pressure decrease with elevation on Venus, and at the summit of Maxwell Montes (10.4 km above the modal radius) the temperature is 660 K and the pressure is 48 bars.

An important consequence of Venus's high surface temperature is that several elements found in rocks on Earth are present in Venus's atmosphere. Carbon dioxide (96.5% of the atmosphere), SO_2 (150 ppmv = 0.015%), HCl (0.5 ppmv), and HF (5 ppbv) are the prime examples; other rock-forming elements such as Br, I, As, Hg, and Sb may also be present at low concentrations in Venus's near-surface atmosphere. The levels of CO_2, SO_2, HCl, and HF in Venus's atmosphere are much higher than their concentrations in Earth's atmosphere. Furthermore, in some cases, almost *all* gaseous compounds of an element (e.g., fluorine) in the terrestrial atmosphere are humanmade (anthropogenic) and not natural products.

The high temperatures, high pressures, and the corrosive gases present in Venus's atmosphere lead to chemical reactions between the atmosphere and surface. Some of the most important reactions involve CO_2, H_2O, HCl, and HF. The similarity between *P-T* conditions on Venus's surface and in low-pressure metamorphism on Earth were first noticed in the 1960s and led to the idea that chemical equilibria between surface minerals buffer Venus's atmospheric composition.

Carbonate-silicate buffers for CO_2

Univariant reactions between oxides and carbonates (e.g., CaO-$CaCO_3$, MgO-$MgCO_3$) buffer (or regulate) the CO_2 fugacity as a function of temperature (e.g., go back to our discussion of the CaO-CO_2 system in Section III-H of Chapter 7). Reactions between silicates and carbonates also buffer the CO_2 fugacity and are important geologically for lithospheric carbon speciation and storage, high-temperature metamorphic reactions, and the global carbon cycles on Earth and Venus. The Norwegian geochemist Viktor Goldschmidt (1888–1947) first recognized the importance of the reaction among calcite, quartz, wollastonite, and CO_2 for buffering the CO_2 pressure during contact metamorphism, but the first measurements of the univariant T-P_{CO2} curve for this reaction were made by Harker and Tuttle (1956):

$$CaCO_3\ (\text{calcite}) + SiO_2\ (\text{quartz}) = CaSiO_3\ (\text{wollastonite}) + CO_2\ (g) \tag{10-299}$$

The quartz-calcite-wollastonite buffer involves three components (CaO, CO_2, SiO_2), and four phases (three solids plus a gas). The Gibbs phase rule ($F = C - P + 2$) shows that coexistence of calcite, quartz, and wollastonite uniquely fixes the equilibrium CO_2 fugacity at a given temperature. The Gibbs energy change for reaction (10-299) is

$$\Delta G = n_{CO_2}\overline{G}_{CO_2} + n_{wo}\overline{G}_{wo} - n_{qtz}\overline{G}_{qtz} - n_{cc}\overline{G}_{cc} \tag{10-300}$$

Equation (10-300) uses the abbreviations *wo*, *qtz*, and *cc* for wollastonite, quartz, and calcite, respectively. The partial molar Gibbs energies of CO_2, wollastonite, quartz, and calcite are

$$\overline{G}_{CO_2} = G^o_{m,CO_2} + \int_{P^o}^{P} V_{CO_2}dP = G^o_{m,CO_2} + RT\ \ln f_{CO_2} \tag{10-301}$$

$$\overline{G}_{wo} = G^o_{m,wo} + \int_{P^o}^{P} V_{wo}dP = G^o_{m,wo} + RT\ \ln a_{wo} \tag{10-302}$$

$$\overline{G}_{qtz} = G^o_{m,qtz} + \int_{P^o}^{P} V_{qtz}dP = G^o_{m,qtz} + RT\ \ln a_{qtz} \tag{10-303}$$

$$\overline{G}_{cc} = G^o_{m,cc} + \int_{P^o}^{P} V_{cc}dP = G^o_{m,cc} + RT\ \ln a_{cc} \tag{10-304}$$

Substituting these four expressions back into Eq. (10-300) and doing some rearranging yields

$$\Delta G = \Delta_r G^o + RT\ \ln f_{CO_2} + RT\ \ln\left[\frac{a_{wo}}{a_{qtz}a_{cc}}\right] \tag{10-305}$$

The standard Gibbs energy change in Eq. (10-305) is

$$\Delta_r G^o = G^o_{m,CO_2} + G^o_{m,wo} - G^o_{m,qtz} - G^o_{m,cc} \tag{10-306}$$

The $G^o{}_m$ values in Eq. (10-306) are the temperature-dependent standard molar Gibbs energies of formation, which can be calculated from the standard entropy and enthalpy of formation values at 298.15 K and heat capacity equations as described earlier. These calculations give

$$\Delta_r G^o = 85{,}304 - 152.58T\ \text{J mol}^{-1} \tag{10-307}$$

We now solve for the CO_2 equilibrium pressure. Equation (10-305) reduces to

$$\Delta G = \Delta_r G^o + RT\ \ln f_{CO_2} \tag{10-308}$$

because calcite, quartz, and wollastonite are pure crystalline phases with unit activity. As shown in Chapter 8, CO_2 is ideal at the P-T conditions of Venus's surface. So, we can replace the fugacity of CO_2

by its pressure in Eq. (10-308). Substituting Eq. (10-307) for the standard Gibbs energy change into Eq. (10-308), rearranging, and solving at equilibrium ($\Delta G = 0$) gives

$$\log P_{CO2} = 7.97 - \frac{4456}{T} \tag{10-309}$$

The global average temperature on Venus is 740 K and Eq. (10-209) shows the CO_2 equilibrium pressure at this temperature is 89 bars. The observed total pressure (P_T) and CO_2 concentration in Venus's atmosphere are 95.6 bars and 96.5% (by volume), respectively. Thus, the CO_2 partial pressure at Venus's surface is 92 bars, which (within error) is identical to the CO_2 equilibrium pressure from the calcite-quartz-wollastonite buffer at Venus's average surface temperature. This agreement and independent evidence for the presence of rocks containing calcite, quartz, and wollastonite on Venus suggests that the CO_2 pressure in Venus's atmosphere is controlled by the surface mineralogy (see Fegley, 2004, for a review). This prediction can be tested by *in situ* X-ray diffraction measurements of Venus's surface mineralogy or by Venus sample return.

Other important CO_2 buffers include the magnesite-quartz-enstatite, magnesite-enstatite-forsterite, and siderite-hematite-magnetite buffers (Aranovich and Newton, 1999; Koziol, 2004; Koziol and Newton, 1995, 1998):

$$MgCO_3\ (\text{magnesite}) + SiO_2\ (\text{quartz}) = MgSiO_3\ (\text{enstatite}) + CO_2\ (g) \tag{10-310}$$

$$MgCO_3\ (\text{magnesite}) + MgSiO_3\ (\text{enstatile}) = Mg_2SiO_4\ (\text{forsterite}) + CO_2\ (g) \tag{10-311}$$

$$FeCO_3\ (\text{siderite}) + Fe_2O_3\ (\text{hematite}) = Fe_3O_4\ (\text{magnetite}) + CO_2\ (g) \tag{10-312}$$

These reactions are important for terrestrial metamorphic reactions, CO_2 storage and carbonate stability in Earth's mantle, the formation of siderite in the Martian meteorite ALH84001, and atmosphere-surface reactions on Venus-like planets in other planetary systems.

Mineral buffer reactions for HCl on Venus

The two most abundant halogen elements are chlorine and fluorine. These are highly reactive and occur naturally as HCl and HF in terrestrial volcanic gases (Symonds et al., 1994). The HCl abundance in Venus's atmosphere is 0.5 ppmv ($5 \times 10^{-5}\%$). This is ~140 times larger than the total concentration of all chlorine gases in Earth's atmosphere, which is 3.6 parts per billion by volume ($3.6 \times 10^{-7}\%$). Furthermore, most of the Cl gases in Earth's atmosphere are humanmade (e.g., CF_2Cl_2, $CFCl_3$, CCl_4) and would not be present otherwise. The disparity is even greater if we compare the total number of Cl atoms in the two atmospheres. There are about 10,000 times more Cl atoms per cm^2 area in Venus's atmosphere than in Earth's atmosphere.

In the late 1960s, shortly after HCl was discovered in the atmosphere of Venus, scientists suggested that the HCl partial pressure in Venus's atmosphere was buffered by minerals on Venus's surface. There are several possible mineral assemblages that can do this because their equilibrium HCl fugacities (identical to the HCl equilibrium pressure under Venus *P-T* conditions) overlap the observed HCl pressure. Lewis (1970) suggested the reaction

$$\begin{aligned} 2\ HCl\ (g) + 8\ NaAlSiO_4\ (\text{nepheline}) &= 2\ Na_4[AlSiO_4]_3\ Cl\ (\text{sodalite}) \\ &+ Al_2SiO_5\ (\text{andalusite}) + SiO_2\ (\text{quartz}) + H_2O\ (g) \end{aligned} \tag{10-313}$$

Reaction (10-313) is divariant and has two degrees of freedom at constant pressure and temperature. The two degrees of freedom are the equilibrium fugacities of HCl and H_2O. Thus, the fugacity of HCl depends on that of H_2O, and vice versa. Proceeding as we have done for other gas-solid reactions, assuming that all minerals are pure phases with unit activity and ideal gas behavior, we derive the equilibrium constant expression

$$K_P = \frac{P_{H_2O}}{P^2_{HCl}} = \frac{X_{H_2O}}{X^2_{HCl}} \cdot \frac{1}{P_T} \quad (10\text{-}314)$$

$$\log_{10} K_P = \frac{15,718}{T} - 14.31 \quad (10\text{-}315)$$

Earth-based observations and spacecraft measurements show that the H_2O concentration in Venus's subcloud atmosphere is about 30 ppmv (0.003%). We already mentioned that Venus's average surface temperature is 740 K and that the average surface pressure is 95.6 bars. Thus, we have all the data we need to calculate the equilibrium HCl pressure from Lewis's proposed buffer reaction. Rearranging equation (10-314), we obtain

$$P_{HCl} = \left(\frac{P_{H_2O}}{K_P}\right)^{1/2} \quad (10\text{-}316)$$

We substitute into Eq. (10-316) using the observed H_2O concentration of 30 ppmv and the K_P value at 740 K from Eq. (10-315). This gives the HCl equilibrium pressure for reaction (10-313):

$$P_{HCl} = \left[\frac{(30 \times 10^{-6})(95.6)}{10^{6.93}}\right]^{1/2} = 1.84 \times 10^{-5} \text{ bars} \quad (10\text{-}317)$$

This corresponds to an HCl mole fraction of $1.92 \times 10^{-7} \sim 0.2$ ppmv. Our calculated HCl abundance matches the observed HCl abundance within a factor of 2.5. This is within the mutual errors of the thermodynamic calculations and the astronomical observations of HCl.

Mineral buffer reactions for HF on Venus

The situation is similar to that described for HCl. Although the observed HF abundance on Venus is only 5 ppbv, it is much greater than the HF abundance in Earth's atmosphere. In fact, the total abundance of all fluorine gases in Earth's atmosphere is equivalent to about 2.1 ppbv total F and is dominated by the humanmade chlorofluorocarbon gases. The amount of HF in Venus's atmosphere (expressed as HF molecules per cm^2 area) is about 200 times larger than the total number of F atoms per cm^2 in the terrestrial atmosphere.

Shortly after HF was discovered in Venus's atmosphere in the late 1960s, scientists also proposed mineral buffers for HF. Their reasoning is simple: HF is a highly corrosive gas that etches glass and thus must be reacting with silicates on Venus's surface. One possible HF buffer suggested by Lewis (1970) is the reaction

$$2\ HF\ (g) + KAlSi_2O_6\ (\text{leucite}) + 2\ Mg_2SiO_4\ (\text{forsterite})$$
$$= MgSiO_3\ (\text{enstatite}) + KMg_3AlSi_3O_{10}F_2\ (\text{fluorphlogopite}) + H_2O\ (g) \quad (10\text{-}318)$$

Reaction (10-318) is divariant and has two degrees of freedom at constant pressure and temperature. The two degrees of freedom are the equilibrium fugacities of HF and H_2O. Thus, the fugacity of HF depends on that of H_2O, and vice versa. The equilibrium expression for Eq. (10-318), assuming pure minerals and ideal gas behavior, is

$$K_P = \frac{P_{H2O}}{P_{HF}^2} \tag{10-319}$$

$$\log_{10} K_P = \frac{13,948}{T} - 8.71 \tag{10-320}$$

Rearranging Eq. (10-319) to solve for the HF equilibrium pressure and substituting using the data for H_2O, temperature, and total pressure discussed earlier, we get

$$P_{HF} = \left(\frac{P_{H_2O}}{K_P}\right)^{1/2} = \left(\frac{2.87 \times 10^{-3}}{10^{10.14}}\right)^{1/2} = 4.56 \times 10^{-7} \text{ bars} \tag{10-321}$$

The corresponding HF mole fraction is 4.77×10^{-9} (4.77 ppbv), in almost exact agreement with the observed HF abundance (5 ppbv).

Metamorphic reactions

An important distinction between atmosphere-surface reactions on Venus and metamorphic reactions on Earth is that the generally higher pressures involved in terrestrial metamorphic reactions lead to greater nonideality. The fugacity coefficients for metamorphic fluids (H_2O, CO_2, their mixtures, NaCl-bearing fluids, etc.) need to be known as a function of temperature, pressure, and fluid composition to calculate phase boundaries accurately. For example, the phase boundary between quartz + calcite (low T, high P) and wollastonite + CO_2 (high T, low P) varies with the CO_2 concentration in H_2O-CO_2 mixtures. We refer the reader to the books by Philpotts (1990) and Turner and Verhoogen (1960) for introductory discussions and to the collection of papers edited by Kerrick (1991) for more advanced descriptions of metamorphic reactions.

III. GIBBS FREE ENERGY AND ELECTROCHEMISTRY

A. Electron transfers during chemical reactions

All chemical reactions involve electron transfers. For example, consider the oxidation of sulfite (SO_3^{2-}) ions dissolved in cloud water droplets by ozone:

$$SO_3^{2-}\text{ (aq)} + O_3\text{ (g)} = SO_4^{2-}\text{ (aq)} + O_2\text{ (g)} \tag{10-322}$$

This is an important reaction forming acid rain. The sulfur atom in SO_3^{2-} has a valence of +4; the sulfur atom in SO_4^{2-} has a valence of +6. The sulfur lost two electrons when it was oxidized in reaction (10-322). Another example is the reduction of nitrate (NO_3^-) ions to N_2 gas by denitrifying bacteria in soils

$$2\,NO_3^-\text{ (aq)} + 10e^-\text{ (aq)} + 12\,H^+\text{ (aq)} = N_2\text{ (g)} + 6\,H_2O\text{ (liq)} \tag{10-323}$$

This reaction is important in the nitrogen biogeochemical cycle on Earth. The nitrogen atom in NO_3^- has a valence of +5 while the nitrogen atoms in N_2 have a valence of 0. Each nitrogen atom gained five electrons when it was reduced.

A third example is the bacterially mediated oxidation and reduction of sulfur in wet soils:

$$4\,S\,(s) + O_2\,(g) + 4\,H_2O\,(liq) = 2\,HSO_3^-(aq) + 2\,H_2S\,(g) + 2\,H^+\,(aq) \tag{10-324}$$

Reaction (10-324) is important for the sulfur biogeochemical cycle. The elemental sulfur has a valence of zero while the sulfur in H_2S has a valence of −2 and the sulfur in HSO_3^- has a valence of +4. In this case both oxidation and reduction occur. In general, the element, ion, or compound being oxidized loses electrons and the species being reduced gains electrons. The three preceding equations are examples of reduction-oxidation chemistry, which is often called *redox chemistry*.

We can write down the standard Gibbs free energy change for each of these reactions. For example, the $\Delta_r G^o$ at 298.15 K for nitrate reduction to N_2 gas is

$$\begin{aligned}\Delta_r G^o_{298} &= 6\Delta_f G^o_{298}(H_2O, liq) - 2\Delta_f G^o_{298}(NO_3^-, aq) \\ &= 6(-237.14) - 2(-110.79) = -1201.26\text{ kJ mol}^{-1}\end{aligned} \tag{10-325}$$

Equation (10-325) does not include $\Delta_f G^o$ values for H^+ ions, electrons, or N_2 because these are zero by definition. In this example we used data from Appendix 1 to compute the standard Gibbs energy change. In general, we can relate the ΔG^o of a reaction to the electron transfers that occur, and measure the standard Gibbs energy change of a chemical reaction from the electrical work done. Electrochemical measurements of Gibbs free energy are done in aqueous solutions, high-temperature melts, and for high-temperature gas-solid reactions.

B. A review of some electrical concepts and units

We briefly discussed electrical work in Chapter 4 when we described calorimetric measurements of heat capacity (see Section II-A in Chapter 4). Equation (4-8) defines electrical energy (in joules) as the product of voltage, amperage, and time. The same definition applies to electrical work. One joule of electrical work is equivalent to

$$1\text{ joule} = (1\text{ volt}) \times (1\text{ ampere}) \times (1\text{ second}) \tag{10-326}$$

The volt (*V*) and ampere (*A*) are the SI units for electric potential and current. The watt (*W*) is the SI unit for electric power. One watt is one joule per one second ($J\ s^{-1}$). Your electric bill lists the number of kilowatt hours of electricity used. The kilowatt hour (KWH) is another unit for measuring electrical energy. Table 3-2 shows that one kilowatt hour is equivalent to 3.60×10^6 J.

Example 10-17. A furnace in a geochemistry laboratory is connected to a 20 ampere, 240 volt line and runs continuously for three days during an experiment to study metal/sulfide partitioning of Ni. How much electrical energy is consumed during this time? The energy consumed is

$$(240\text{ V})(20\text{ A})(3 \cdot 24 \cdot 60 \cdot 60\text{ s}) = 1.24416 \times 10^9\text{ J} = 345.6\text{ KWH} \tag{10-327}$$

The product of current and time is defined as the *coulomb* (C), which is the SI unit for electric charge. One coulomb is equal to a current of one ampere flowing for one second:

$$1 \text{ coulomb} = (1 \text{ ampere}) \times (1 \text{ second}) \tag{10-328}$$

Thus, Eq. (10-326) for electrical work can be rewritten as

$$1 \text{ joule} = (1 \text{ volt}) \times (1 \text{ coulomb}) \tag{10-329}$$

The coulomb is named after the French physicist Charles-Augustin de Coulomb (1736–1806), who discovered that the electrostatic force between two point charges is directly proportional to the size of each charge and inversely proportional to the square of the distance between the two charges (Coulomb's law).

Because chemical reactions involve electron transfers, the amount of charge transferred in a chemical reaction is directly related to the number of electrons (or ions) involved. One mole of electrons carries 96,485 coulombs of electric charge. One mole of univalent ions (e.g., Ag^+, Na^+, Cl^-, NO_3^-) also carries 96,485 coulombs of electric charge. The quantity of charge carried by one mole of electrons or univalent ions is called *Faraday's constant* $\mathcal{F}$ in honor of the English scientist Michael Faraday (1791–1867), who did pioneering research in electrochemistry:

$$1\,\mathcal{F} = 96{,}485 \text{ C mol}^{-1} = 96{,}485 \text{ J mol}^{-1} \text{ volt}^{-1} \tag{10-330}$$

The amount of charge carried by a mole of ions depends on the valence of the ions. One mole of divalent ions (e.g., Mg^{2+}, SO_4^{2-}) carries 192,970 C = 2 $\mathcal{F}$ of charge, one mole of trivalent ions (e.g., Al^{3+}, PO_4^{3-}) carries 289,455 C = 3 $\mathcal{F}$, and so on. In all cases an equivalent weight of ions (the atomic or molecular weight divided by the valence) carries 96,485 C = 1 $\mathcal{F}$ of charge.

Faraday's laws of electrolysis describe the amount of reaction and electric charge transferred in electrochemical reactions. The first law states that the amount of material produced is proportional to the amount of electric charge transferred. The second law states that the amounts of different materials produced are proportional to their equivalent weights. Faraday's laws of electrolysis are summarized by the equation

$$m = \frac{I \cdot t \cdot \mu}{|z|\mathcal{F}} = \frac{C \cdot \mu}{|z|\mathcal{F}} \tag{10-331}$$

Equation (10-331) shows the mass (m) of material formed is proportional to electric current I (in amperes), time t (in seconds), and the molecular weight μ (in g mol^{-1}) divided by Faraday's constant $\mathcal{F}$ and the absolute value of the valence (z).

Example 10-18. A *coulometer* measures the amount of electric charge transferred in a chemical reaction and is essentially an electric meter like the one on your house or apartment building. The silver coulometer consists of two platinum electrodes immersed in aqueous silver nitrate ($AgNO_3$) and the reaction taking place is

$$Ag^+ \text{ (aq)} + e^- \text{ (aq)} \rightarrow Ag \text{ (metal)} \tag{10-332}$$

(a) How many grams of silver metal are deposited on the cathode (positive electrode) when a 20 ampere current passes through the coulometer for one hour? Substituting into Eq. (10-331) gives

$$m_{Ag} = \frac{I \cdot t \cdot \mu}{z\mathcal{F}} = \frac{20 \text{ A} \cdot 3600 \text{ s} \cdot 107.8682 \text{ g mol}^{-1}}{1 \cdot 96{,}485 \text{ C mol}^{-1}} = 80.49 \text{ g silver} \tag{10-333}$$

(b) How many grams of silver are deposited per coulomb? Using the definition of a coulomb in Eq. (10-328) we find that the total charge passed through the coulometer in one hour is

$$20\ \text{A} \cdot 3600\ \text{s} = 72{,}000\ \text{C} \tag{10-334}$$

Thus, the amount of silver deposited per coulomb is about 1.118×10^{-3} grams.

C. Electrical work and the first law

As discussed in Chapter 3, the differential amount of work done is the product of an intensive variable and the differential of an extensive variable (see Section V and Table 3-6 in Chapter 3). Electrical work is the product of electric potential in volts and the differential electric charge in coulombs (Eq. 10-328). We denote the electric potential difference in volts by $\Delta\phi$. The differential charge transferred in coulombs is the product of the ionic (or electronic) valence z, the Faraday constant $\mathcal{F}$, and the number of moles (dn). The differential amount of electrical work done (δw) is thus

$$\delta w = (z\mathcal{F}\,dn)\Delta\phi \tag{10-335}$$

Electrochemical reactions occur reversibly when essentially no current is supplied by the reaction. In this case the electric potential difference $\Delta\phi$ is replaced by $\mathcal{E}$, the *electromotive force* (emf), and Eq. (10-335) is rewritten as

$$\delta w = z\mathcal{F}\mathcal{E}\,dn \tag{10-336}$$

In practice, electrochemical reactions are conducted reversibly by measuring the electromotive force using a very high-resistance voltmeter. *Ohm's law* relates current, voltage, and resistance:

$$1\ \text{volt} = (1\ \text{ampere}) \times (1\ \text{ohm}) \tag{10-337}$$

If we use a 10^{15} ohm voltmeter to measure the emf $\mathcal{E} = 1$ volt of an electrochemical reaction, the current in the measuring circuit is only 10^{-15} amperes. This is an insignificant value compared to the typical current of ~1 ampere supplied by the battery in a laptop computer.

We now want to relate electrical work to the Gibbs energy. We do this by starting with the combined first and second law Eq. (6-47) and include terms for PV and electrical work in it:

$$dE = TdS - \delta w \tag{6-47}$$

$$dE = TdS - (PdV + z\mathcal{F}\mathcal{E}\,dn) \tag{10-338}$$

Recall that the Gibbs energy G is defined as

$$G = E + PV - TS \tag{6-113}$$

Differentiating Eq. (6-113), substituting for dE with Eq. (10-338), and simplifying gives

$$dG = VdP - SdT - z\mathcal{F}\mathcal{E}\,dn \tag{10-339}$$

Equation (10-339) is analogous to the second fundamental equation (7-28), but it also includes an electrical work term. If an electrochemical reaction is conducted reversibly at constant P and T, for example, Eq. (10-339) reduces to

$$dG = -z\mathscr{F}\mathscr{E}dn \tag{10-340}$$

The ΔG change for a reaction involving one mole of ions is thus

$$\Delta G = -z\mathscr{F}\mathscr{E} \tag{10-341}$$

Equation (10-341) is a very important equation that relates the Gibbs energy change of a reaction to the electromotive force produced by the reaction.

Example 10-19. Benz and Wagner (1961) measured $\mathscr{E} = 461 \pm 4$ millivolts at 700°C for

$$CaO\ (lime) + SiO_2\ (silica) = CaSiO_3\ (wollastonite) \tag{10-342}$$

Calculate the ΔG for wollastonite formation from its constituent oxides.

We substitute into Eq. (10-341) to find ΔG. Reaction (10-342) involves one mole of CaO containing one mole of divalent Ca^{2+} ions, so $z = 2$:

$$\Delta G = -z\mathscr{F}\mathscr{E} = -2(96,485)(0.461) = -89.0\ \text{kJ mol}^{-1} \tag{10-343}$$

The ± 4 millivolt uncertainty on the emf corresponds to an uncertainty of ± 0.8 kJ mol^{-1} in ΔG. For reference, thermochemical data from Robie and Hemingway (1995) give $\Delta G^o = -89.4$ kJ mol^{-1} at 973 K, which agrees with the value from the emf data within error.

If all reactants and products are pure phases in their standard states, the electromotive force is the *standard emf* (also called the *standard potential*) $\mathscr{E}^o$ and the derived ΔG value is the standard Gibbs energy ΔG^o for wollastonite formation from its constituent oxides. In this case,

$$\Delta G^o = -zFE^o \tag{10-344}$$

The standard states for CaO, SiO_2, and $CaSiO_3$ at 973 K are their stable phases, which are lime, cristobalite, and wollastonite. Benz and Wagner (1961) used lime and wollastonite, but they did not specify if the silica was quartz or cristobalite. The ΔG difference between the two phases of silica is ~740 J mol^{-1} at 973 K, comparable to the uncertainty in the emf measurements.

D. Electrochemical measurements of $\Delta_f G^o$ from the elements

Electrochemical measurements are often used to measure standard Gibbs energies of formation of compounds from their constituent elements. For example, Wachter and Hildebrand (1930) electrochemically measured the $\Delta_f G^o$ of pure molten $PbCl_2$ from its constituent elements:

$$Pb\ (liquid) + Cl_2\ (gas) = PbCl_2\ (liquid) \tag{10-345}$$

Their data are important for the thermodynamic properties of $PbCl_2$ (cotunnite), which condenses at Mount Vesuvius and around other terrestrial volcanic vents. The electrochemical cell that Wachter and Hildebrand (1930) used is schematically illustrated in Figure 10-10. The negative electrode

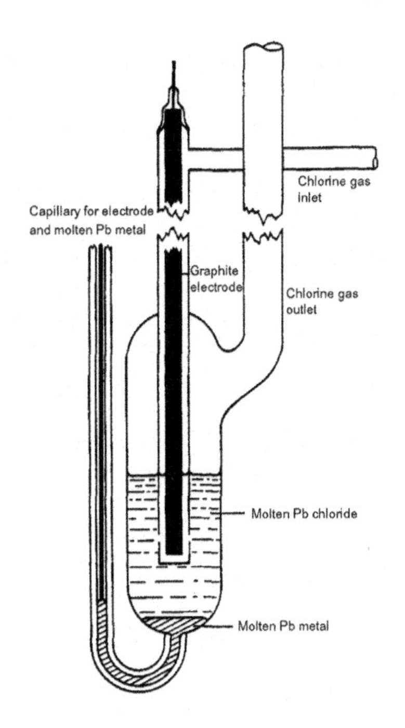

FIGURE 10-10

A schematic diagram of the electrochemical cell used by Wachter and Hildebrand (1930) to measure $\Delta_f G^o$ for the formation of molten $ZnCl_2$ from Cl_2 gas and molten zinc.

(anode) is a platinum wire inserted into the molten lead at the bottom of the glass container. The reaction at this electrode is

$$\text{Pb (liquid)} = \text{Pb}^{2+} + 2\bar{e} \tag{10-346}$$

The electrons produced at the Pb electrode travel through the electrical circuit and high-resistance voltmeter (not shown) to the positive electrode (cathode). The cathode is Cl_2 gas at one bar pressure bubbling over a graphite rod immersed in the molten $PbCl_2$. The electrons react with the Cl_2 gas forming chloride ions in the molten $PbCl_2$ electrolyte:

$$\text{Cl}_2\ (\text{g}) + 2\text{e}^- = 2\ \text{Cl}^- \tag{10-347}$$

Ionic conduction of the Cl^- ions through the molten $PbCl_2$ completes the electrical circuit. The electrochemical cell shown in Figure 10-10 is represented by this notation:

$$\text{Pb (liquid)} \mid \text{PbCl}_2\ \text{(liquid)} \mid \text{Cl}_2\ \text{(gas)}$$

Wachter and Hildebrand (1930) measured the emf from 500°C to 580°C. These temperatures are at or above the melting points of Pb metal (328°C) and $PbCl_2$ (501°C), which are insoluble in each other. Consequently, both Pb metal and $PbCl_2$ are in their standard states (the stable form of the pure element or compound at the ambient temperature). As defined in Chapter 5, Cl_2 gas at one bar pressure behaving ideally is the standard state for chlorine at all

Table 10-18 Emf Measurements for Liquid $PbCl_2$

			$\Delta_f G^o$ (kJ mol^{-1})	
T(°C)	T(K)	emf (volts)	Measured	Third Law
499.1	772.2	1.2731	−245.54	−245.34
517.5	790.6	1.2631	−243.33	−243.25
525.3	798.4	1.2563	−242.39	−242.37
536.8	810.0	1.2507	−241.00	−241.08
541.8	815.0	1.2469	−240.40	−240.52
549.5	822.6	1.2417	−239.49	−239.66
566.0	839.2	1.2310	−237.50	−237.82
578.5	851.6	1.2249	−236.01	−236.45

temperatures. Using methods described in Chapter 8 we can show that the deviation of Cl_2 gas from ideality at one bar and 500−580°C is negligible. Thus, the emf data of Wachter and Hildebrand (1930) give $\mathscr{E}^o$ and $\Delta_f G^o$ for formation of molten $PbCl_2$ from its constituent elements in their standard states.

Table 10-18 gives a subset of the emf measurements of Wachter and Hildebrand (1930). A linear least-squares fit to these emf values gives the equation

$$\mathscr{E}^o = 1.7526 - 6.20 \times 10^{-4} T \text{ volts} \tag{10-348}$$

The $\Delta_f G^o$ for formation of molten $PbCl_2$ is calculated using Eq. (10-344) with z = 2 because Pb^{2+} is divalent. The resulting equation is

$$\Delta_f G^o(\text{PbCl}_2) = 120T - 338{,}200 \text{ J mol}^{-1} \tag{10-349}$$

The $\Delta_f G^o$ values corresponding to the emf data are listed in Table 10-18. They agree with $\Delta_f G^o$ values for molten $PbCl_2$ computed from thermal data using the third law of thermodynamics.

E. Entropy and enthalpy of reaction from emf measurements

We now show that the temperature coefficient of the emf ($d\mathscr{E}/dT$) gives the ΔS and ΔH of reaction. Thus, in principle emf measurements provide all the necessary thermodynamic data for a chemical reaction. However, in practice, very accurate emf data are needed because errors in the emf measurements are magnified in the first and second derivatives. This situation is analogous to that described in Chapter 4 (Section II-D) for derivation of heat capacity data from enthalpy measurements (e.g., 0.1% errors in enthalpy data give 1% errors in the derived C_P data).

Equation (8-26) gives the isobaric temperature coefficient of the Gibbs energy:

$$\left(\frac{\partial G}{\partial T}\right)_P = -S \tag{8-26}$$

It is easily shown that the isobaric temperature coefficient of ΔG for any reaction is given by

$$\left(\frac{\partial \Delta G}{\partial T}\right)_P = -\Delta S \tag{10-350}$$

Substituting for ΔG using Eq. (10-341) we get

$$\left(\frac{\partial \Delta G}{\partial T}\right)_P = \frac{\partial}{\partial T}[-z\mathscr{F}\mathscr{E}] = -z\mathscr{F}\left(\frac{\partial \mathscr{E}}{\partial T}\right)_P = -\Delta S \tag{10-351}$$

Neither the ionic valence z nor Faraday's constant $\mathscr{F}$ depend on temperature. Rearranging Eq. (10-351) to solve for the temperature coefficient of the emf gives

$$\left(\frac{\partial \mathscr{E}}{\partial T}\right)_P = \frac{\Delta S}{z\mathscr{F}} \tag{10-352}$$

Equation (10-352) shows that the variation of the electromotive force $\mathscr{E}$ with temperature gives information about ΔS for the reaction being studied. If $\mathscr{E}$ decreases with temperature the ΔS is negative, if $\mathscr{E}$ increases with temperature the ΔS is positive, and if $\mathscr{E}$ is constant the ΔS is zero. If all reactants and products are in their standard states, as is the case for the $PbCl_2$ measurements previously discussed, the standard emf $\mathscr{E}^o$ is measured and the temperature coefficient gives the standard entropy of reaction.

The Gibbs-Helmholtz equation (6-116) is the starting point for deriving the relationship between the ΔH of reaction and the emf $\mathscr{E}$. We rearrange Eq. (6-116) to solve for ΔH, and substitute for ΔS using Eq. (10-350):

$$\Delta G = \Delta H - T\Delta S \tag{6-116}$$

$$\Delta H = \Delta G + T\Delta S \tag{10-353}$$

$$\Delta H = \Delta G - T\left(\frac{\partial \Delta G}{\partial T}\right)_P \tag{10-354}$$

Now we substitute for ΔG using Eq. (10-341):

$$\Delta H = -z\mathscr{F}\mathscr{E} + Tz\mathscr{F}\left(\frac{\partial \mathscr{E}}{\partial T}\right)_P \tag{10-355}$$

$$\Delta H = z\mathscr{F}\left[T\left(\frac{\partial \mathscr{E}}{\partial T}\right)_P - \mathscr{E}\right] \tag{10-356}$$

Equation (10-356) shows that exothermic reactions have negative temperature coefficients, endothermic reactions have positive temperature coefficients, and thermoneutral reactions have constant emf values independent of temperature.

Example 10-20. Calculate the $\Delta_f H^o$ and $\Delta_f S^o$ for the formation of molten $PbCl_2$ from the emf data for reaction (10-345) and compare the results to values from thermal data.

Equation (10-348) is a convenient starting point. The temperature coefficient of $\mathscr{E}^o$ is

$$\left(\frac{\partial \mathscr{E}^o}{\partial T}\right)_P = \frac{\partial}{\partial T}(1.7526 - 6.2 \times 10^{-4}T) = -6.2 \times 10^{-4} \text{ volts K}^{-1} \quad (10\text{-}357)$$

Rearranging Eq. (10-352) gives the standard entropy of formation of molten $PbCl_2$:

$$\Delta_f S^o(\text{PbCl}_2, \text{ liq}) = z\mathscr{F}\left(\frac{\partial \mathscr{E}^o}{\partial T}\right)_P = -120 \text{ J mol}^{-1} \text{ K}^{-1} \quad (10\text{-}358)$$

For comparison, thermal data for reaction (10-345) give an average $\Delta_f S^o = -112 \text{ J mol}^{-1} \text{ K}^{-1}$ over the temperature range studied by Wachter and Hildebrand (1930). The difference between the two values is about 6.7% of the standard entropy of formation.

The standard enthalpy of formation of molten $PbCl_2$ is calculated from Eq. (10-356) and is constant if $\mathscr{E}^o$ is a linear function of temperature. At 813 K, the midpoint of the temperature range studied by Wachter and Hildebrand (1930), the $\Delta_f H^o$ is given by

$$\Delta_f H^o = z\mathscr{F}\left[T\left(\frac{\partial \mathscr{E}^o}{\partial T}\right)_P - \mathscr{E}^o\right] = 2(96{,}485)[813(-6.2 \times 10^{-4}) - 1.2485] \quad (10\text{-}359)$$

$$\cong -338{,}190 \text{ J mol}^{-1}$$

For comparison, thermal data give an average $\Delta_f H^o = -331{,}750 \text{ J mol}^{-1}$ over the 500–580°C temperature range. The difference of 6440 J mol^{-1} is about 1.9% of the average $\Delta_f H^o$ value. The uncertainties in the emf and thermal data are a few percent, which is about the same size as the disagreement between the entropy and enthalpy values from the emf and thermal data.

F. Effect of concentration on electromotive force

The discussion above used formation of pure liquid $PbCl_2$ from its constituent elements in their standard states to illustrate derivation of Gibbs energy, entropy, and enthalpy data from emf measurements. However, if the reactants or products of a reaction are not in their standard states we need to use the activities (or fugacities) of the reactants and products to convert the measured emf values into standard $\mathscr{E}^o$ values. This is done using the *Nernst equation.*

For example, Salstrom and Hildebrand (1930) electrochemically measured the $\Delta_f G^o$ from 711–829 for the formation of pure molten $PbBr_2$ from pure liquid Pb metal and Br_2 gas:

$$\text{Pb (liquid)} + \text{Br}_2 \text{ (gas)} = \text{PbBr}_2 \text{ (liquid)} \quad (10\text{-}360)$$

The electrochemical cell for this reaction is written as Pb (liquid) | $PbBr_2$ (liquid) | Br_2 (gas). Their data for reaction (10-360) are represented by the equation

$$\mathscr{E}^o = 1.5012 - 6.06 \times 10^{-4}T \text{ volts} \quad (10\text{-}361)$$

Some $\mathscr{E}^o$ values calculated using this equation are listed in Table 10-19. The corresponding equation for the $\Delta_f G^o$ of pure molten $PbBr_2$ is

$$\Delta_f G^o = -2(96{,}485)\,\mathscr{E}^o = -289{,}687 + 117T \text{ J mol}^{-1} \quad (10\text{-}362)$$

Table 10-19 Nernst Equation and emf Measurements for Liquid $PbBr_2$

P (cm/Hg)	74.96	60.8	60.04	55.21	43.80
P (bars)	0.9994	0.8106	0.8005	0.7361	0.5840
$\mathscr{E}$ (obs volts)	1.0671	1.0609	1.0634	1.0576	1.0511
$\mathscr{E}$ (calcd volts)	1.0671	1.0615	1.0630	1.0576	1.0511
T(K)	716.4	714.8	711.8	716.4	715.4
$\mathscr{E}^o$ (volts)	1.0671	1.0680	1.0698	1.0671	1.0677
$(RT/zF)\ln P_{Br_2}$	−0.0000	−0.0065	−0.0068	−0.0095	−0.0166

The results of Salstrom and Hildebrand (1930) give $\Delta_f G^o$ values that agree very well with tabular data for molten $PbBr_2$ in the temperature range they studied.

Salstrom and Hildebrand (1930) also measured emf values as a function of the Br_2 pressure at (approximately) constant temperature (see the observed $\mathscr{E}$ values listed in Table 10-19). The standard state for Br_2 gas is the ideal gas at one bar pressure. However, the deviation of Br_2 gas from ideality at their experimental conditions is insignificant. (You can show this for yourself by computing the reduced T and P from the critical constants of 588 K, 103.4 bars and estimating the fugacity coefficient using Figure 8-16.) Thus, the fugacity of Br_2 is given by its pressure.

The reaction isotherm for formation of pure molten $PbBr_2$ from its constituent elements is

$$\Delta G = \Delta_r G^o + RT \ln\left(\frac{a_{PbBr_2}}{a_{Pb} f_{Br_2}}\right) \quad (10\text{-}363)$$

This equation can be derived using the methods described earlier for the Ni-NiO buffer reaction. The activities of the molten Pb metal and molten $PbBr_2$ are equal to one and thus the ΔG for $PbBr_2$ formation depends only on the Br_2 fugacity, which is equal to its pressure:

$$\Delta G = \Delta_r G^o + RT \ln\left(\frac{1}{f_{Br_2}}\right) = \Delta_r G^o - RT \ln P_{Br_2} \quad (10\text{-}364)$$

We can use Eq. (10-364) to calculate ΔG as a function of the Br_2 gas pressure in Table 10-19. However, we now use this equation to calculate the emf values for $PbBr_2$ formation as a function of the Br_2 gas pressure. The calculated $\mathscr{E}$ values can be compared to those observed by Salstrom and Hildebrand (1930).

Equations (10-341) and (10-344) relate the Gibbs energy and emf values to one another. Substituting into the reaction isotherm (10-364) we obtain

$$\mathscr{E} = \mathscr{E}^o + \frac{RT}{z\mathscr{F}} \ln P_{Br_2} \text{ volts} \quad (10\text{-}365)$$

Equation (10-365) is one form of the *Nernst equation*, which gives the effect of concentration (or gas pressure) on emf values. We used Eq. (10-365) to calculate the $\mathscr{E}$ values listed in the fourth row of Table 10-19. The agreement between our calculated emf values and those measured by Salstrom and Hildebrand (1930) is very good.

The general form of the Nernst equation is

$$E = E^o - \frac{RT}{zF}\ln Q \text{ volts} \quad (10\text{-}366)$$

where Q is the reaction quotient.

G. Oxygen fugacity (partial-pressure) sensors

Oxygen fugacity sensors are electrochemical sensors that are used to measure the oxygen fugacity of gas mixtures such as volcanic gases and gas mixtures in a laboratory furnace. These sensors measure the Gibbs energy change for O_2 due to an isothermal change in its fugacity. To a good first approximation, the fugacity and partial pressure of O_2 are identical in volcanic gases and gas mixtures in laboratory furnaces. Thus, we substitute partial pressure for fugacity in the equations that follow. The ΔG measured by an oxygen fugacity sensor is for the change in state:

$$O_2 \text{ (g, } T\text{, pressure } P_1) = O_2 \text{ (g, } T\text{, pressure } P_2) \quad (10\text{-}367)$$

From Chapter 8, the Gibbs free energy change for n moles of an ideal gas between two different pressures P_1 and P_2 at constant temperature T is

$$G_2 - G_1 = nRT \ln\frac{P_2}{P_1} \quad (8\text{-}163)$$

If we now consider one mole of gas and take the pressure P_1 as the standard-state pressure of one bar, G_1 is equal to G^o, the standard Gibbs free energy, and Eq. (8-163) can be rewritten as

$$G - G^o = RT \ln\frac{P}{1} \quad (10\text{-}368)$$

Equation (10-368) can be rewritten as the Nernst equation, as done for Br_2 (g) above, giving

$$\mathscr{E} - \mathscr{E}^o = -\frac{RT}{z\mathscr{F}}\ln\left[\frac{f_{O2\text{ (at }P)}}{f_{O2\text{ (at 1 bar)}}}\right] \quad (10\text{-}369)$$

The $\mathscr{E}^o$ in Eq. (10-369) is the standard emf for the reaction

$$O_2 \text{ (g, } T\text{, } P_1 = 1 \text{ bar}) = O_2\text{(g, } T\text{, } P_1 = 1 \text{ bar}) \quad (10\text{-}370)$$

Both $\Delta G^o = 0$ and $\mathscr{E}^o = 0$ for this reaction. The $\mathscr{E}$ in Eq. (10-369) is the emf for the reaction

$$O_2\text{(g, } T\text{, } P_1 = 1 \text{ bar }) = O_2 \text{ (g, } T\text{, } P_2 = P) \quad (10\text{-}371)$$

The $\mathscr{E}^o$ term drops out of Eq. (10-369), which reduces to

$$\mathscr{E} = -\frac{RT}{z\mathscr{F}}\ln f_{O2} \cong -\frac{RT}{z\mathscr{F}}\ln P_{O2} \text{ volts} \quad (10\text{-}372)$$

Thus, the emf $\mathscr{E}$ measured by an oxygen fugacity sensor is proportional to the oxygen fugacity in a gas sample. The O_2 fugacity and partial pressure are identical if the fugacity coefficient $\gamma = 1$, which is true in most applications of interest to us.

Oxygen fugacity sensors operate at high temperatures (typically above 700°C) and use ceramic oxygen ion (O^{2-}) conductors such as solid solutions of 8 mole % yttria (Y_2O_3), 16 mole % lime (CaO), or 16 mole % magnesia (MgO) in zirconia (ZrO_2). These solid solutions are *solid electrolytes*. Kiukkola and Wagner (1957) pioneered the use of zirconia solid electrolytes for high-temperature electrochemical measurements. Pure zirconia is monoclinic at room temperature but transforms to a denser tetragonal form at ~1000°C. This phase transition involves a large volume change and cracks the ceramic. The addition of yttria, lime, or magnesia to zirconia increases its electrical conductivity. Also, the solid solutions have cubic structure and melt at very high temperatures (~2500°C) without any solid-state phase transitions. A typical sensor is a gas-tight zirconia tube closed at one end. The outside of the tube is in the gas being studied and a stream of dry O_2 flows through the inside of the sensor. The two electrodes are platinum wires connected to the outside and inside of the tube.

The positive electrode is the platinum wire connected to the outside of the zirconia tube. This is the sample gas electrode. The negative electrode is the platinum wire connected to the inside of the zirconia tube. This is the O_2 reference electrode. The reactions at the two electrodes are

$$O_2\ (\text{gas}) + 4e^- = 2O^{2-}\ (\text{at the } O_2 \text{ reference electrode}) \quad (10\text{-}373)$$

$$2O^{2-} = O_2\ (\text{gas}) + 4e^-\ (\text{at the sample gas electrode}) \quad (10\text{-}374)$$

The electrons produced at the sample gas electrode travel through the electrical circuit and high-resistance voltmeter to the O_2 gas reference electrode. The two O^{2-} ions produced at the O_2 reference electrode travel through the zirconia solid solution to the sample gas electrode. This electrochemical cell is written as: sample gas | solid electrolyte | O_2.

The two O^{2-} ions transfer 4 coulombs of charge, so $z = 4$ in the Nernst equation. Rewriting the Nernst equation (10-372) in terms of base 10 logarithms, inserting numerical values for the gas constant and Faraday's constant, and solving for the oxygen fugacity gives

$$\mathscr{E} = -\frac{RT}{z\mathscr{F}}\ln f_{O_2} = -\frac{(\ln 10)8.314T}{(4)96{,}485}\log f_{O_2}\ \text{volts} \quad (10\text{-}375)$$

$$\log f_{O_2}(\text{bars}) = \frac{-20{,}160\,\mathscr{E}\ (\text{volts})}{T} \quad (10\text{-}376)$$

Example 10-21. A gas mixture containing 2.28% CO and 97.72% CO_2 is used in a gas-mixing furnace at 1200°C (1473 K). A yttria-zirconia sensor is used to measure the fO_2 of the gas mixture, relative to pure O_2 at one bar pressure. The sensor records an emf $\mathscr{E} = 0.5655$ volts. What is the oxygen fugacity of the gas mixture inside the furnace?

Substituting into Eq. (10-376) and solving we find

$$\log f_{O2}(\text{bars}) = \frac{-20{,}160\,\mathscr{E}\ (\text{volts})}{T} = \frac{-20{,}160(0.5655)}{1473} = -7.74 \quad (10\text{-}377)$$

The oxygen fugacity measured by the sensor is virtually identical to the value of log $fO_2 = -7.72$ calculated from thermodynamic data for reaction (10-159) between CO and CO_2.

For safety and convenience, dry air at one bar is used as the reference gas instead of pure O_2. Dry air contains 20.946% O_2 (see Table 2-11) so the oxygen fugacity of dry air at one bar pressure is 0.20946 bars. Substituting the fO_2 of dry air into the denominator of Eq. (10-369), and solving for the fO_2 of the sample gas gives a modified version of the Nernst equation:

$$\log f_{O2}(\text{bars}) = \frac{-20{,}160\,\mathscr{E}\,(\text{volts})}{T} - 0.679 \tag{10-378}$$

The new term in Eq. (10-378) is the logarithm of the oxygen fugacity in dry air because the emf being measured by the sensor is now relative to 0.20946 bar O_2 instead of 1 bar O_2 pressure.

Example 10-22. The experiment in Example 10-21 is repeated using dry air instead of pure O_2 for the reference electrode. What is the emf value read by the sensor?

Substituting into Eq. (10-378) and rearranging, we find

$$\mathscr{E} = \frac{T(\log f_{O2} + 0.679)}{-20{,}160} = \frac{1473(-7.74 + 0.679)}{-20{,}160} = 0.5159 \text{ volts} \tag{10-379}$$

In principle, a solid-state buffer such as a metal-oxide pair can be used as the reference electrode in an oxygen fugacity sensor. In this case the sensor measures an emf relative to the fO_2 of the solid-state buffer and a term equal to log fO_2 of the buffer has to be added into the Nernst equation (10-376). In general, the Nernst equation for an oxygen sensor is written as

$$\log f_{O_2}(\text{sample}) = \frac{-20{,}160\,\mathscr{E}\,(\text{volts})}{T} + \log f_{O_2}\,(\text{reference}) \tag{10-380}$$

The oxygen fugacities of the reference and sample are both in the same units, generally bars.

Example 10-23. Rosen et al. (1993) measured the oxygen fugacity of fumarolic gases from the Kudrjaviy volcano in the Kuril islands. They used a sensor with the reference electrode at the oxygen fugacity of the Cu-Cu_2O buffer (see Table 10-16). They measured emf values of 0.260 volts at 720°C and 0.280 volts at 890°C. What are the corresponding oxygen fugacities?

We use Table 10-16 to compute the fO_2 values for the Cu-Cu_2O buffer at the two temperatures, substitute the sensor readings and reference fO_2 values into Eq. (10-380), and solve for the oxygen fugacities of the fumarolic gases. At 720°C (993 K) the Cu-Cu_2O buffer and the fumarolic gas have log fO_2 values (in bars) of −10.09 and −15.36, respectively. At 890°C (1163 K) the buffer and fumarolic gas have log fO_2 values (in bars) of −7.52 and −12.37, respectively.

PROBLEMS

1. The sun is mainly hydrogen (92.6% H) with some helium (7.3% He), and 0.1% of all other elements combined (atomic percentages; Lodders, 2003). Calculate the fractional dissociation (α) of H_2, and the mole fractions of monatomic H and H_2 from 1000 K to 4000 K at 10^{-4} bars total pressure using the dissociation energy of H_2 ($D_0^o = 432.070$ kJ mol^{-1}) and the Gibbs free energy functions given below (Gurvich et al., 1989). These *P-T* conditions are representative of temperature and pressure in the photosphere of the sun.

T(K)	1000	2000	3000	4000
H $[(G^o_T - H^o_0)/RT]$	−14.3233	−16.0563	−17.0700	−17.7893
H_2 $[(G^o_T - H^o_0)/RT]$	−16.4862	−18.9687	−20.5058	−21.6512

2. Thermal ionization of atomic H occurs in the photospheres of the sun and other stars:

$$H(g) = H^+ (g) + e^- (g)$$

Calculate the fractional ionization of H gas and the mole fractions of H, H^+, and e^- at temperatures of 4000, 6000, 8000, and 10,000 K using equilibrium constants given by the equation $\log_{10} K_P = 4.114 - 75{,}011/T$. The initial pressure of the monatomic H gas is 10^{-4} bar.

3. The equilibrium constant for the water-gas reaction

$$H_2 + CO_2 = H_2O + CO \quad (10\text{-}1)$$

is 1.596 at 1259 K. The average heat of reaction is $\Delta_r H^o = 32{,}439$ J mol^{-1}. (a) Calculate K_P at 1473 K, a typical temperature for a gas-mixing furnace in a geochemistry laboratory. (b) Calculate the equilibrium composition of gas inside the furnace at 1473 K if the initial gas mixture (at one bar pressure) contains 85% H_2 and 15% CO_2.

4. Approximately 1.6 parts per billion of CO is observed in Jupiter's atmosphere. This is very much higher that the equilibrium abundance in the observed region of Jupiter's atmosphere and results from rapid vertical mixing from a much deeper level where CO is much more abundant. Calculate the equilibrium abundance of CO resulting from the reaction

$$CH_4 + H_2O = CO + 3\ H_2$$

at the 1100 K, 530 bar level of Jupiter's atmosphere. Express your result as the CO mole fraction $X_{CO} = P_{CO}/P_{total}$ and use the observational data that follow. The abundances of H_2, CH_4, and H_2O in Jupiter's atmosphere are 86.2% H_2, 0.181% CH_4, and 0.052% H_2O.

5. Sulfur monoxide gas (SO) is produced by high-temperature volcanism on Io, one of the four Galilean satellites of Jupiter. Calculate the equilibrium percentage of SO outgassed by the Pele volcano at 1430 K and $10^{-4.6}$ bar pressure from the reaction

$$0.5\ SO_2 + 0.25\ S_2 = SO$$

and from the observed SO_2/S_2 ratio of 3:12. Use both the low and high values of the SO_2/S_2 ratio to calculate two values for the equilibrium percentage of SO. *Hint:* You can assume $P_{total} = SO_2 + S_2$ and that the observed ratio is the equilibrium value.

6. A geochemist is interested in experimentally studying trace element partitioning as a function of oxygen fugacity and temperature to understand petrogenesis of the alkaline basalts from the October 1964 eruptions of Surtsey, a volcano in the North Atlantic near Iceland. He plans to do experiments in a gas-mixing furnace at one bar total pressure using a H_2-CO_2 mixture to control the oxygen fugacity. This eruption of Surtsey had a temperature of 1125°C and log $f_{O_2} = -9.80$ (log bar units). (a) What are the equilibrium partial pressures of H_2 and CO_2 inside the furnace? (b) What is the composition of the gas mixture that he has to use, taking into account formation of H_2O and CO via the water-gas reaction?

7. Oldhamite (ideally CaS) occurs in the highly reduced enstatite chondrites and achondrites (aubrites). Larimer (1968) measured the oxygen fugacity from 800°C to 1000°C for the reaction

$$\text{CaS (oldhamite)} + 2\text{O}_2\text{ (g)} = \text{CaSO}_4\text{ (anhydrite)}$$

His data give the upper limit on f_{O_2} for CaS stability. Use his results (given below) to (a) derive a linear equation for K_P from 800°C to 1000°C, (b) calculate the average $\Delta_r H^o$ over this T range, and (c) calculate $\Delta_r H^o$ at 298 K using C_P data from Chapter 3 and Appendix 2. (You can assume that $\Delta_r H^o$ from part b is the reaction enthalpy at 900°C.)

T(°C)	800	850	900	950	1000
$-\log f_{O_2}$	15.068	14.012	13.053	11.972	11.483

8. Nitrogen oxides (NO_x) are emitted from cars because atmospheric N_2 is combusted at high temperatures (~2000 K) inside car engines via the reaction

$$N_2 + O_2 = 2\text{ NO} \quad (10\text{-}67)$$

at one bar total pressure. Assuming that dry air is 79% N_2 and 21% O_2 by volume and that the average heat of reaction is +90.5 kJ per mole NO, predict (a) the effect of increased temperature, (b) the effect of increased pressure, and (c) the effect of a lower oxygen abundance (< 21%) on the equilibrium NO mole fraction. You can answer these questions using only the data given.

9. Show that the amount of NH_3 produced via the reaction

$$N_2 + 3\text{ H}_2 = 2\text{ NH}_3$$

is maximized when the starting materials are a 3:1 ratio of H_2 to N_2 (i.e., $X_{H_2} = 3/4$). You can assume that the amount of NH_3 produced is sufficiently small that we can approximate X_{H2} (final) = X_{H_2} (initial) and X_{N_2} (final) = X_{N_2} (initial).

10. At equilibrium, H_2S is the major sulfur-bearing gas over a wide P-T range in protoplanetary accretion disks (e.g., the solar nebula from which our solar system formed). The equilibrium between H_2S and H_2, the dominant gas, controls the S_2 fugacity and also determines which sulfides are stable. Preuner and Schupp (1909) studied the thermal decomposition of H_2S:

$$2\text{H}_2\text{S} = 2\text{H}_2 + \text{S}_2\text{ (g)}$$

Calculate the sulfur fugacity at 700 K in the solar nebula ($H_2S/H_2 = 3.66 \times 10^{-5}$) from their data.

T(K)	1023	1103	1218	1338	1405
$K_P \times 10^{-4}$	0.89	3.8	24.5	118	260

A value of 3.8 is 3.8 × 10^{-4}, and so on.

11. Calculate the ΔG for $PbBr_2$ formation from its constituent elements using the data in Table 10-19 and Eq. (10-364).

12. The geochemist from Problem 10-6 decides to check the f_{O2} inside his furnace during his experiment with a zirconia oxygen fugacity sensor. His sensor has an air reference electrode and gives a reading of 632.6 millivolts. Is the f_{O2} inside the furnace the same as the calculated value?

13. If 0.25 ampere of electric current is passed through an aqueous solution of copper sulfate for 36 hours, how many grams of Cu metal are deposited at the cathode from the Cu^{2+} in solution?

14. Zirconia oxygen sensors can also be used as oxygen pumps to deliver O_2 for a reaction or to control f_{O_2} in a gas mixture. This is done by applying a voltage to the sensor, which causes O_2 to be released into the sample gas. Calculate the necessary voltage to produce an f_{O2} of 10^{-6} bar in a gas mixture heated at 800 K. Assume that the other side of the sensor is in air.

15. The temperature coefficient of emf is

$$\left(\frac{\partial \mathscr{E}}{\partial T}\right)_P = \frac{\Delta S}{z\mathscr{F}} \tag{10-352}$$

What is the second derivative equal to?

16. What would be the CO_2 pressure on a Venus-like exoplanet in another planetary system if it were buffered by the reaction

$$MgCO_3 \text{ (magnesite)} + SiO_2 \text{ (quartz)} = MgSiO_3 \text{ (clinoenstatite)} + CO_2 \text{ (gas)}$$

Assume a surface temperature of 740 K. Use the second law method (i.e., the van't Hoff equation) and data in Appendix 1 to answer this question.

17. The water vapor pressure at Venus's surface is about 2.90×10^{-3} bar. Use the second law method (i.e., the van't Hoff equation) and data in Appendix 1 to determine if talc is stable against dehydration on Venus's surface at 740 K. The relevant reaction is

$$Mg_3Si_4O_{10}(OH)_2 \text{ (talc)} = SiO_2 \text{ (quartz)} + 3\,MgSiO_3 \text{ (enstatite)} + H_2O \text{ (g)}$$

18. Show the intermediate steps in the derivation of the Gibbs-Helmholtz equation (10-59).

19. The Erta' Ale volcano in Ethiopia has a vent temperature of 1403 K. Use Eq. (10-79) for the equilibrium constant K_P of the Deacon reaction

$$HCl + \frac{1}{4}O_2 = \frac{1}{2}H_2O + \frac{1}{2}Cl_2 \tag{10-77}$$

to calculate the Cl_2 equilibrium partial pressure in volcanic gas emitted by Erta' Ale in January 1974. Chemical analysis of this gas gave 77.24% H_2O, 0.42% HCl, and $\log f_{O_2} = -9.16$. Assume one bar total pressure in your calculations.

20. Calculate the average $\Delta_r H^o_{298}$ for I_2 dissociation to monatomic I using the equilibrium constants in Table 10-7, Eq. (9-92), and this quadratic fit for the Δ(fef) of Eq. (10-82):

$$\Delta\left[\frac{G^o_T - H^o_{298}}{T}\right] = \Delta(\text{fef}) = -99.2464 - 4.8876 \times 10^{-3}T + 1.0243 \times 10^{-6}T^2$$

21. Astronomers use I_2 absorption cells as wavelength standards for measuring the "wobbles" in stellar motions caused by planets orbiting other stars (Marcy and Butler, 1992). Typically the

I_2 absorption cells operate at 50°C and have a total pressure of 2.13 mmHg. What is the fractional dissocation of I_2 to I in an iodine absorption cell heated to 500°C by mistake?

22. Many scientists believe Earth had a steam atmosphere early in its history. The Nobel Prize–winning chemist Irving Langmuir measured the thermal dissociation (α) of water vapor at high temperatures in his doctoral thesis (Langmuir, 1906). This proceeds via the reaction

$$H_2O = H_2 + \frac{1}{2} O_2$$

Calculate the equilibrium constant K_P for this reaction and the oxygen fugacity f_{O_2} from a subset of his data listed here, which are for one atmosphere total pressure.

T (K)	1000	1200	1400
α	2.8×10^{-7}	7.45×10^{-6}	7.87×10^{-5}

23. Gaseous nitric acid HNO_3 is a pollutant in urban areas and forms by the elementary reaction

$$OH + NO_2 \rightarrow HNO_3$$

Atkinson et al. (1997) give the rate constant for this reaction from 200 K to 400 K as

$$k = 2.29 \times 10^{-9} T^{-0.60} \text{ cm}^3 \text{ mol}^{-1} \text{ s}^{-1}$$

Calculate the rate constant (units of s^{-1}) of the reverse reaction at 298 K from $K_P = 6.05\times10^{27}$.

24. The NH_3 and H_2S in Jupiter's atmosphere react to form solid ammonium hydrosulfide (NH_4SH) clouds. This occurs at low temperatures where the partial pressures of the gases exceed their vapor pressure over solid NH_4SH. Vapor pressure data for NH_4SH (Walker and Lumsden, 1897) are tabulated here. Use their data to calculate $\Delta_r G^o = A + B\times T$ for NH_4SH sublimation:

$$NH_4SH \text{ (solid)} = NH_3 \text{ (gas)} + H_2S\text{(gas)}$$

T (°C)	7.7	11.2	12.2	15.2	17.6	22.4	24.7	27.6
P (torr)	153	198	217	263	311	417	479	572

25. The reaction $H_2S + CO_2 = H_2O + OCS$ forms carbonyl sulfide (OCS) in terrestrial volcanic gases. Calculate the OCS abundance in volcanic gases emitted during the December 1971 eruption of the Erta' Ale volcano in Ethiopia using Eq. (10-95) and observed gas abundances in volume percent from Symonds et al. (1994): 69.41% H_2O, 17.16% CO_2, and 1.02% H_2S. The vent temperature is 1075°C. Assume a vent pressure of one bar. Erta'Ale is a divergent plate volcano erupting tholeiitic lava.

26. Terres and Wesemann (1932) studied the reaction $H_2S + OCS = CS_2 + H_2O$, which occurs in volcanic gases. Use their data given here to calculate an equation of the form $\ln K_P = A + B/T$. Then use your equation to estimate the CS_2 mole fraction in the 1959 eruption of the Nyiragongo volcano for which $T = 970°C$, $P = 1$ bar, 1.72% H_2S, 0.09% OCS, and 43.50% H_2O by volume.

(°C)	700	750	800	850	900
$10^3 \times K_P$	3.36[a]	3.69	5.53	7.69	10.8

[a] *$K_P = 3.36 \times 10^{-3}$, and so on.*

27. Ethane C_2H_6 is produced by photochemistry in Jupiter's upper atmosphere and is destroyed by thermochemistry in Jupiter's deep atmosphere. The first step is the elementary reaction

$$H + C_2H_6\ (\text{ethane}) \rightarrow H_2 + C_2H_5\ (\text{ethyl radical})$$

The reaction rate R is the number of C_2H_6 molecules per cm^3 destroyed per second and is

$$R = -\frac{d[C_2H_6]}{dt} = k[H][C_2H_6]$$

Tsang and Hampson (1986) give the bimolecular reaction rate constant from 300 K to 2500 K as

$$k = 9.19 \times 10^{-22} T^{3.5} \exp(-2600/T) \text{cm}^3\ \text{mol}^{-1}\ \text{s}^{-1}$$

Calculate the chemical lifetime (seconds) for C_2H_6 loss via this reaction in Jupiter's deep atmosphere using the H atom number densities here from chemical equilibrium calculations. What is the temperature at which the C_2H_6 lifetime is only one hour?

T (K)	600	700	800	900	1000
[H] cm^{-3}	5.0×10^3	3.0×10^6	3.6×10^8	1.5×10^{10}	3.2×10^{11}

28. Assemblages of Fe metal, cohenite (Fe_3C), and graphite are found in some meteorites. Calculate the equilibrium temperature of these assemblages using ΔG^o equations given by Kelley (1935) for methane decomposition and iron carburization:

$$CH_4 = 2H_2 + C\ (\text{graphite})$$

$$\Delta G^o = 15{,}560 - 21.14T \log T + 2.98 \times 10^{-3} T^2 + \frac{2600}{T} + 40.39T \text{ cal mol}^{-1}$$

$$CH_4 + 3Fe\ (\alpha, \text{metal}) = Fe_3C\ (\beta, \text{cohenite}) + 2H_2$$

$$\Delta G^o = 19{,}060 - 48.66T \log T + 13.13 \times 10^{-3} T^2 - \frac{55{,}900}{T} + 109.70T \text{ cal mol}^{-1}$$

29. Calculate the average K_P value at 1173 K for the thermal dissociation of I_2 molecules to I atoms using data from Figure 10-4 tabulated here.

P_T (atm)	0.09181	0.40087	0.64446	0.80042	1.04680
X_I	0.5091	0.2901	0.2374	0.2164	0.1906
X_{I_2}	0.4909	0.7099	0.7626	0.7836	0.8094

30. The elementary reaction (10-122) is important for atmospheric chemistry on Earth and Mars:

$$H + CO_2 \rightarrow CO + OH \tag{10.122}$$

(a) Write down four equivalent differential equations for its rate. **(b)** Write down expressions for the chemical lifetimes of H and CO_2.

31. Manganese is the 12th most abundant element in Earth's continental crust after O, Si, Al, Fe, Ca, Na, Mg, K, Ti, C, and P (Wedepohl, 1995) with an abundance (716 μg/g) similar to those of phosphorus (757 μg/g) and sulfur (697 μg/g). Rhodochrosite ($MnCO_3$) is an important Mn-bearing mineral that occurs in hydrothermal, metasomatic, and sedimentary ore deposits and also in solid solution with calcite, magnesite, and siderite. Goldsmith and Graf (1957) measured the thermal decomposition of rhodochrosite via the reaction

$$MnCO_3\ (\text{rhodochrosite}) = MnO\ (\text{manganosite}) + CO_2(g)$$

as a function of pressure and temperature from 376°C to 750°C. Calculate $\Delta_r G^o$ as a function of temperature from their *P-T* data for the univariant curve tabulated here. The ΔV is for the solids and the CO_2 fugacity coefficient was calculated using the Kerrick-Jacobs equation.

T (°C)	376	400	450	500	550	600	650	700	750
P (bar)	1.01	2.10	8.28	26.6	75.9	190	428	855	1552
ΔV (J bar^{-1})	−1.793	−1.793	−1.794	−1.795	−1.796	−1.797	−1.798	−1.798	−1.797
$\gamma(CO_2)$	1.00	1.00	1.00	1	1.00	1.02	1.08	1.24	1.57

32. Emmett and Schultz (1930) measured K_P from 450°C to 570°C for the reaction

$$CoO\ (c) + CO\ (g) = Co\ (\text{metal}) + CO_2\ (g)$$

Use their results to calculate the oxygen fugacity of the Co-CoO buffer at these temperatures.

T (°C)	450	515	570
K_P	489.6	245.9	148.4

33. The Fe (iron)—FeS (troilite) buffer controls the transport and vapor pressure of sulfur during meteorite metamorphism on their parent bodies. (a) Derive an equation for $\log_{10} f_{S_2}$ similar to those in Table 10-16 for f_{O_2} buffers. (b) Use this equation to calculate the S_2 fugacity during meteorite metamorphism at 850°C and 150 bars total pressure (typical conditions for grade 6 ordinary chondrites). The 298 K molar volumes of Fe and FeS are 7.092 and 18.200 cm^3 mol^{-1}, respectively. You can neglect the thermal expansion and compressibility terms.

34. Reaction (10-313) is divariant at a given temperature and pressure. Write down the components and phases for this reaction. Use the Gibbs phase rule to show this reaction is divariant.

CHAPTER

Solutions

11

The alkahest, or universal solvent imagined by the alchemists.
—Sir Humphry Davy (1812)

There are noble Arcanas in Nature preparable by the great Dissolvent, the liquor Alchahest.
—G. Starkey, *Helmont's Vind.* 294 (1657), cited by *The Oxford English Dictionary*

This chapter discusses the thermodynamic properties of solutions. Section I reviews basic concepts including the different methods and units for specifying the composition of solutions. Section II describes partial molal properties and the chemical potential. Section III covers physical and thermodynamic properties of ideal and dilute solutions. Section IV describes thermodynamic properties of real nonelectrolyte solutions. Aqueous solutions are discussed in Section V.

I. BASIC CONCEPTS

A. Types of solutions

Gases, liquids, and solids all form solutions. We discussed equations of state for gaseous solutions in Chapter 8. Dry air exemplifies solutions of gases in gases. Terrestrial volcanic gases and the atmospheres of the other planets are other examples of gaseous solutions (see Table 8-6). Moist air exemplifies the solution of a liquid in gas. The sublimation of naphthalene (the active ingredient in mothballs) exemplifies the solution of a solid in gas. Examples of liquid solutions are provided by air-saturated water (gas in liquid), aqueous ethanol (liquid in liquid), and seawater (solid in liquid). Examples of solid solutions are H_2 gas dissolved in Pd metal (gas in solid), mercury amalgams (liquid in solid), and Fe-Ni-Co alloy in meteorites (solid in solid).

B. Electrolyte and nonelectrolyte solutions

Solutions can be divided into electrolyte and nonelectrolyte solutions. Electrolyte solutions conduct electricity. Water is the most important (but not the only) ionizing solvent, and most electrolyte solutions of interest in the earth and planetary sciences are aqueous solutions. Electrolytes are divided into strong and weak electrolytes. Strong electrolytes are completely or almost completely ionized in dilute and concentrated solutions. Solutions of strong electrolytes have high molar electrical conductivities in dilute and concentrated solutions because the fraction ionized is always unity (or close to it). Hydrochloric, nitric, and sulfuric acids and table salt (NaCl) are examples of strong electrolytes. Weak electrolytes are only partially ionized, and the fraction ionized varies

Copyright © 2013 Elsevier Inc. All rights reserved.

inversely with the concentration of the electrolyte. The molar electrical conductivities of weak electrolyte solutions increase dramatically as their concentrations decrease and their fractional ionization increases. Acetic acid (the acid in vinegar), citric acid (the acid in citrus fruits), and other organic acids are weak electrolytes. However, the division into strong and weak electrolytes depends on the solvent because a strong electrolyte in water can be a weak electrolyte in another solvent, and vice versa.

Electrolytes are also classified by the charges on the ions they form when dissolved in water. For example, HCl, KCl, NaCl, HNO_3, and acetic acid CH_3COOH are 1:1 electrolytes because they ionize to singly charged ions when dissolved in water. Calcium carbonate $CaCO_3$ and $CuSO_4$ (chalcocyanite) are 2:2 electrolytes because they ionize to divalent ions. A 3:3 electrolyte ionizes to trivalent ions when dissolved in water. Berlinite ($AlPO_4$) and lanthanide phosphates are 3:3 electrolytes, but they have very small solubilities in water at room temperature. The 1:1, 2:2, and 3:3 electrolytes are symmetrical electrolytes because their aqueous ions have the same valences. In contrast, unsymmetrical electrolytes ionize to cations and anions with different valences. Examples of 2:1 electrolytes include $MgCl_2$ (chlormagnesite) and $CaCl_2$ (hydrophilite). Other examples of unsymmetrical electrolytes are Na_2SO_4 (thenardite) and K_2SO_4 (arcanite) (1:2), $La(NO_3)_3$ lanthanum nitrate (3:1), $La_2(SO_4)_3$ lanthanum sulfate (3:2), and $Th(NO_3)_4$ thorium nitrate (4:1). In each case, the first number gives the cation charge and the second number gives the anion charge.

Examples of electrolytic solutions include seawater, molten metals (such as Earth's outer core), molten oxides and silicates (such as basaltic magma), anhydrous nitric and sulfuric acids, liquified polar gases (NH_3, SO_2, H_2S, HCN, HF), and some organic solvents (alcohols, benzene, dimethyl sulfoxide, formamide, and pyridine). Liquid SO_2 may be an important solvent on Io, the innermost Galilean satellite of Jupiter. Observations show SO_2 ice and gas and photogeological interpretations of *Galileo* and *Voyager* spacecraft images suggest the presence of liquid SO_2. Compositions of aqueous electrolytic solutions are generally given in terms of molality (m), formality (f), molarity (M), or molar concentration (c).

Nonelectrolyte solutions do not conduct electricity. Examples include solutions of nonpolar gases (H_2, noble gases, CH_4, gaseous hydrocarbons, SF_6, air), nonpolar organic compounds (liquid and solid hydrocarbons), nonpolar liquified gases, and mineral solid solutions (olivine, pyroxene, feldspar). Photogeological interpretations of *Cassini-Huygens* radar images suggest that cryogenic hydrocarbon solutions are present on Titan, Saturn's largest satellite. The compositions of nonaqueous solutions and nonelectrolyte solutions are usually given in terms of mole fractions.

C. Definitions and conversion formulae for solution compositions

Salt (NaCl) is the major constituent of seawater, human blood plasma, and brines in the crust and mantle. Thus, we use salt dissolved in water to illustrate the different composition units for solutions. At 25°C, the density (ρ) of pure water is 0.99707 g mL^{-1}, the solubility of NaCl in water is 26.43% (by mass), and the density of the saturated solution is 1.1979 g mL^{-1} (Linke, 1965). Solubility in mass % is defined as

$$\text{mass }\% = \frac{100\cdot\text{grams NaCl}}{(\text{grams NaCl} + \text{grams H}_2\text{O})} = \frac{100\cdot\text{grams solute}}{\text{grams solution}} \tag{11-1}$$

The salt/water mass ratio in the saturated solution is then

$$\frac{\text{grams NaCl}}{\text{grams H}_2\text{O}} = \frac{\text{mass \% NaCl}}{(100 - \text{mass \% NaCl})} = \frac{26.43}{73.57} = 0.3592 \tag{11-2}$$

The concentration in grams NaCl per milliliter H_2O is

$$\frac{\text{grams NaCl}}{\text{mL H}_2\text{O}} = \frac{\text{grams NaCl}}{\text{grams H}_2\text{O}}\rho_{\text{H}_2\text{O}} = \frac{26.43}{73.57}0.99707 = 0.3582 \text{ g mL}^{-1} \tag{11-3}$$

In other words, one gram of NaCl dissolves in about 2.79 mL of water. The mole fractions of NaCl and H_2O in the saturated solution are given by

$$\begin{aligned} X_{\text{NaCl}} &= \frac{\text{moles NaCl}}{\text{moles NaCl} + \text{moles H}_2\text{O}} \\ &= \frac{(\text{grams NaCl}/\text{mol. wt. NaCl})}{(\text{grams NaCl}/\text{mol. wt. NaCl}) + (\text{grams H}_2\text{O}/\text{mol. wt. H}_2\text{O})} \\ &= \frac{(26.43/58.443)}{(26.43/58.443) + (73.57/18.015)} = \frac{0.4522}{0.4522 + 4.084} = 0.0997 \end{aligned} \tag{11-4}$$

$$X_{\text{H}_2\text{O}} = (1 - X_{\text{NaCl}}) = 0.9003 \tag{11-5}$$

Molality (m) is defined as moles of solute per kilogram of solvent. Thus, the molal concentration of a solution is independent of temperature. The molality of the saturated NaCl solution in water at 25°C is

$$\begin{aligned} \text{molality } (m) &= \frac{\text{moles NaCl}}{1000 \text{ g H}_2\text{O}} = \frac{\text{grams NaCl} \times 1000}{\text{mol. wt. NaCl} \times \text{grams H}_2\text{O}} \\ &= \frac{26.43 \times 1000}{58.443 \times 73.57} = 6.147\ m \end{aligned} \tag{11-6}$$

Formality (f) is defined as moles of solute per kilogram of solution. Thus, the saturated NaCl solution has a formality of

$$\begin{aligned} \text{formality } (f) &= \frac{\text{moles NaCl}}{1000 \text{ g solution}} = \frac{\text{grams NaCl} \times 1000}{\text{mol. wt. NaCl} \times \text{grams solution}} \\ &= \frac{26.43 \times 1000}{58.443 \times 100} = 4.522\ f \end{aligned} \tag{11-7}$$

Molarity (M) or molar concentration (c) is defined as moles of solute per liter of solution. Typically, molarity is used in older literature while molar concentration is used in more recent literature. In contrast to molality, the molarity (or molar concentration) of a solution depends on temperature because the density of the solution depends on temperature. The molarity of the saturated NaCl solution is

$$\text{molarity } (M) = \frac{\text{moles NaCl}}{1000 \text{ mL solution}} = \frac{\text{grams NaCl} \times 1000}{\text{mol. wt. NaCl} \times V_{soln.}}$$
$$= \frac{26.43 \times 1000}{58.443 \times (100 \text{ g}/1.1979 \text{ g mL}^{-1})} = 5.417\, M \qquad (11\text{-}8)$$

The molality and formality of a solution are related via the formula

$$\frac{\text{molality}}{\text{formality}} = \frac{\text{solution wt.}}{\text{water wt.}} = \frac{\text{solution wt.}}{(\text{solution wt.} - \text{solute wt.})} \qquad (11\text{-}9)$$

The ratio of the molality (6.147 *m*) and formality (4.522 *f*) for the saturated NaCl aqueous solution is 1.359, identical to the value given by the formula

$$\frac{m\,(\text{NaCl})}{f\,(\text{NaCl})} = \frac{6.147\, m}{4.522\, f} = 1.359 = \frac{100 \text{ grams}}{(100 - 26.43) \text{ grams}} \qquad (11\text{-}10)$$

The ratio of the molality (6.147 *m*) and molarity (5.417 *M*) for the saturated NaCl aqueous solution is 1.135, identical to the value given by the conversion formula

$$\frac{\text{molality}}{\text{molarity}} = \frac{\text{solution wt.}}{(\text{solution wt.} - \text{solute wt.})} \frac{1}{\rho_{soln.}}$$
$$= \frac{100 \text{ grams}}{(100 - 26.43) \text{ grams}} \frac{1}{1.1979} = 1.135 \qquad (11\text{-}11)$$

Example 11-1. The major salts in seawater are NaCl (77.76%), $MgCl_2$ (10.88%), $MgSO_4$ (4.74%), $CaSO_4$(3.60%), K_2SO_4 (2.46%), $MgBr_2$ (0.22%), and $CaCO_3$ (0.34%). Earth's global mean surface temperature is 15°C and at this temperature 35.1 grams of $MgCl_2$ (chlormagnesite) dissolves per 100 grams of saturated solution. Calculate the molality of the saturated solution.

Equation (11-6) gives the desired answer. Substituting $MgCl_2$ in place of NaCl and using its molecular weight of 95.21 g mol^{-1} we find

$$\text{molality } (m) = \frac{\text{moles MgCl}_2}{1000 \text{ g H}_2\text{O}} = \frac{\text{grams MgCl}_2 \times 1000}{\text{mol. wt. MgCl}_2 \times \text{grams H}_2\text{O}}$$
$$= \frac{35.1 \times 1000}{95.21 \times 64.9} = 5.68\, m \qquad (11\text{-}12)$$

Example 11-2. At 15°C and one atmosphere pressure, 1.5 *m* NaCl aqueous solution has a density of 1.05795 g cm^{-3} (Chen et al., 1980). Calculate the molar concentration of this solution.

Either Eq. (11-8) or (11-11) gives the answer to this question. Rearranging and substituting into the latter equation, we find

$$\begin{aligned} \text{molarity} &= \frac{\rho_{soln.}(\text{wt. solution} - \text{wt. solutes})}{\text{wt. solution}}\text{molality} \\ &= \frac{1.05795(1000 - 1.5 \times 58.443)}{1000}1.5 = 1.448\ M \end{aligned} \qquad (11\text{-}13)$$

II. PARTIAL MOLAL PROPERTIES

A. Definition and illustration using partial molal volume

Partial molal (molar) quantities express the dependence of extensive thermodynamic properties such as volume V, internal energy E, enthalpy H, entropy S, Gibbs free energy G, and Helmholtz free energy A on the composition of a solution. We illustrate the physical meaning of partial molal quantities, and their calculation using volume as an example.

Intuitively, it is easy to understand that the volume of a mixture is an additive function of the volumes of its components. For example, the volume of a physical mixture of salt (NaCl, $V_m = 27.015\ \text{cm}^3\ \text{mol}^{-1}$) and sulfur (S, $V_m = 15.511\ \text{cm}^3\ \text{mol}^{-1}$) is the sum of the molar volumes times the number of moles (n_i) of each component:

$$V_{mix} = n_{\text{NaCl}}V_{m,\text{NaCl}} + n_S V_{m,S} \qquad (11\text{-}14)$$

An equation analogous to Eq. (11-14) gives the volume of any mixture. In general, the volume of a mixture of n components is the weighted sum of their molar volumes:

$$V_{mix} = \sum_{i=1}^{n} n_i V_{m,i} \qquad (11\text{-}15)$$

The volume of a solution also depends on the volumes of its constituents. The volume of an ideal solution is identical to the volume of the physical mixture. In this case, the partial molal volumes are equal to the molar volumes of the pure components. However, the volume of a real (nonideal) solution is larger or smaller than the weighted sum of the molar volumes of its constituents.

Figure 11-1 shows the volume of an aqueous solution of NaCl at 25°C and one atmosphere pressure (Chen et al., 1980). The volume of the solution in Figure 11-1 is the weighted sum of the partial molal volumes of NaCl and water (W):

$$V = n_W \overline{V}_W + n_{\text{NaCl}}\overline{V}_{\text{NaCl}} \qquad (11\text{-}16)$$

The molar volumes of pure water and pure salt are not used in Eq. (11-16) or in the analogous equations for the volumes of other solutions. The partial molal volumes of NaCl and H_2O in the aqueous solution are different from the molar volumes of the pure compounds. The n_i in Eq. (11-16)

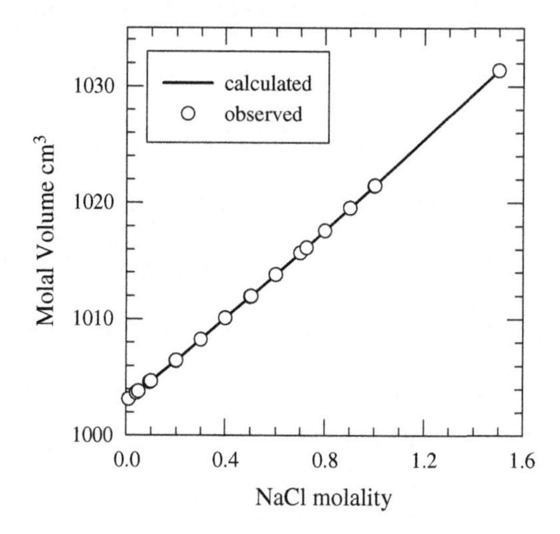

FIGURE 11-1

Molal volume of an aqueous solution of NaCl at 25°C and one atmosphere pressure based on the data of Chen et al. (1980).

are the number of moles of NaCl and water. The number of moles of NaCl in the solution is the same as the molality (m) of the solution, so the volume can be expressed by

$$V = 1002.961 + 16.6084m + 1.9291m^{3/2} - 0.1219m^2 + 0.0849m^{5/2}\ \text{cm}^3 \tag{11-17}$$

Equation (11-17) gives the volume of the solution per one kilogram of water because molal concentration is used. The constant term is the molar volume of pure water at 25°C. Equation (11-17) is derived from careful measurements of density as a function of the NaCl molality of the solution. In general, densities of solutions are measured and are converted into volumes.

We defined the *partial molal volume* $\overline{V}_i$ in Chapter 8 with equation (8-194):

$$\overline{V}_i = \left(\frac{\partial V}{\partial n_i}\right)_{P,T,n \neq n_i} \tag{8-194}$$

Thus, the partial molal volume of NaCl in water at 25°C is

$$\overline{V}_{\text{NaCl}} = 16.6084 + 2.8937m^{1/2} - 0.2438m + 0.2123m^{3/2}\ \text{cm}^3\ \text{mol}^{-1} \tag{11-18}$$

The partial molal volume is an intensive variable. The partial molal volume of NaCl is the change in volume of the solution per mole of NaCl added to (or removed from) an infinitely large amount of the aqueous salt solution *at constant temperature and pressure*. The same interpretation applies to the partial molal volume of water. In the case of a finite amount of solution, the partial molal volume of NaCl is the change in volume of the solution as an infinitesimally small amount of NaCl is added to (or removed from) the solution.

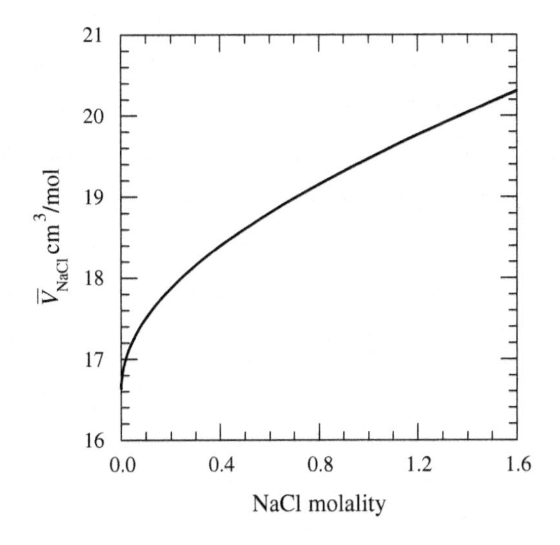

FIGURE 11-2

Partial molal volume of NaCl in aqueous salt solution at 25°C and one atmosphere pressure.

The partial molal volume of NaCl in the aqueous salt solution is not the same as the molar volume of pure NaCl, which is 27.015 cm^3 mol^{-1}. At infinite dilution, that is, when the molality of NaCl is zero, Eq. (11-18) shows that the partial molal volume of the dissolved NaCl is ~16.61 cm^3 mol^{-1}. As shown in Figure 11-2, as the concentration of NaCl in the solution increases, its partial molal volume also increases. The values for the partial molal volume of NaCl are more uncertain than the values for the volume of the aqueous solution because errors are enhanced during differentiation. We faced this same problem in calculating heat capacities from enthalpy data (Chapter 4), calculating Joule-Thomson coefficients from *PVT* data (Chapter 8), and in calculating the entropy and enthalpy of reaction from emf data (Chapter 10).

Figure 11-3 shows the molar volume of an ethanol (C_2H_5OH) aqueous solution at 25°C and ambient pressure (~1 atm). In this figure, we plotted the molar volume as a function of the ethanol mole fraction X_E in solution. The curve in Figure 11-3 is

$$V_m = X_W\overline{V}_W + X_E\overline{V}_E \tag{11-19}$$

$$V_m = 18.064 + 34.855X_E + 8.4871X_E^2 - 2.771X_E^3 \tag{11-20}$$

In this case, calculation of the partial molal volumes of water and ethanol is more involved because the mole fractions of two components in a binary solution are related to one another via

$$dX_1 = -dX_2 \tag{11-21}$$

The partial molal volumes of water and ethanol are thus given by the equations

$$\overline{V}_E = V_m + X_W\left(\frac{dV_m}{dX_E}\right) \tag{11-22}$$

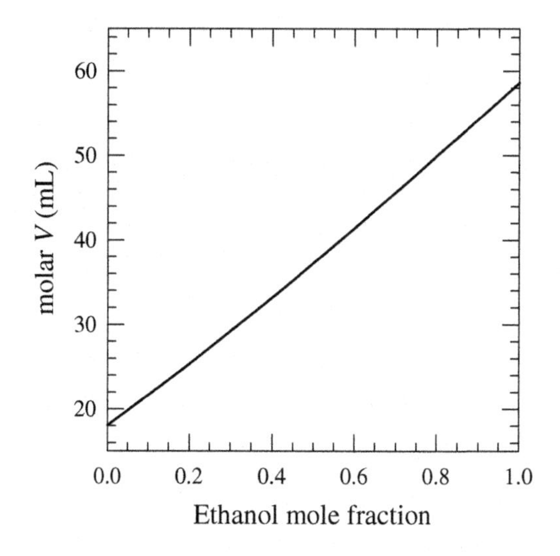

FIGURE 11-3

Molal volume of an aqueous solution of ethanol (C_2H_5OH) at 25°C and one atmosphere pressure. The curve is calculated using Eq. (11-20).

$$\overline{V}_W = V_m + X_E\left(\frac{dV_m}{dX_W}\right) \tag{11-23}$$

The general forms of these equations that apply to the partial molal volumes of the two components in any binary solution are

$$\overline{V}_1 = V_m + X_2\left(\frac{\partial V_m}{\partial X_1}\right)_{P,T} \tag{11-24}$$

$$\overline{V}_2 = V_m + X_1\left(\frac{\partial V_m}{\partial X_2}\right)_{P,T} \tag{11-25}$$

B. Partial molal Gibbs free energy and other partial molal properties

The complete differential (*dG*) of the Gibbs free energy with respect to temperature (*T*), pressure (*P*), and the number of moles (n_1, n_2, ..., n_i) of the different components in an open system is

$$dG = dT\left(\frac{\partial G}{\partial T}\right)_{P,n_1,\ldots} + dP\left(\frac{\partial G}{\partial P}\right)_{T,n_1,\ldots} + dn_1\left(\frac{\partial G}{\partial n_1}\right)_{T,P,n_i\neq n_1} + dn_2\left(\frac{\partial G}{\partial n_2}\right)_{T,P,n_i\neq n_2} + \cdots \tag{11-26}$$

The subscripts $n_i \neq n_1$ and so on in Eq. (11-26) mean that the partial derivative is evaluated with respect to a change in the number of moles of a particular component (e.g., n_1 moles of the first component, n_2 moles of the second component, etc .), whereas the number of moles of all other components are held constant. In other words, each partial derivative with respect to composition is the differential change of *G* due only to a differential change in the number of moles of that particular component of the system. The ellipsis (...) in Eq. (11-26) indicates that partial differentiation with respect to composition continues until all components of the system have been included. A binary system has only the two terms shown (dn_1 and dn_2), a ternary system would also have a third term for dn_3, and so on. Equation (11-26) is analogous to Eq. (2-47), which gives the dependence of volume on temperature, pressure, and composition. (At this point you may find it useful to review Sections IX and XII of Chapter 2 to remind yourself about complete differentials and their interpretation.)

Equation (11-26) shows that the total differential change in Gibbs free energy for an open system is the sum of the separate differential changes in G due to differential changes in the temperature, pressure, and molar amounts of all the components of the system. The partial derivatives of G with respect to T and P in Eq. (11-26) are the entropy and volume, respectively:

$$S = -\left(\frac{\partial G}{\partial T}\right)_{P,n_1,\ldots} \tag{11-27}$$

$$V = \left(\frac{\partial G}{\partial P}\right)_{T,n_1,\ldots} \tag{11-28}$$

Each of the partial derivatives of G with respect to composition is the *partial molal Gibbs free energy* with respect to one of the components in an open system:

$$\overline{G}_1 = \left(\frac{\partial G}{\partial n_1}\right)_{T,P,n_i \neq n_1} \tag{11-29}$$

$$\overline{G}_2 = \left(\frac{\partial G}{\partial n_2}\right)_{T,P,n_i \neq n_2} \tag{11-30}$$

$$\overline{G}_3 = \left(\frac{\partial G}{\partial n_3}\right)_{T,P,n_i \neq n_3} \tag{11-31}$$

The partial molal Gibbs free energy is interpreted in the same way as the partial molal volume. It is also an intensive variable. We can now rewrite the second fundamental equation (7-28) as

$$dG = VdP - SdT + \overline{G}_1 dn_1 + \overline{G}_2 dn_2 + \cdots \tag{11-32}$$

Equation (11-32) is valid for an open system, and is a more general version of the second fundamental equation (7-28).

The partial molal internal energy, enthalpy, Helmholtz free energy, and entropy are defined similarly to the partial molal volume and partial molal Gibbs free energy, namely:

$$\overline{E}_i = \left(\frac{\partial E}{\partial n_i}\right)_{T,P,n \neq n_i} \tag{11-33}$$

$$\overline{H}_i = \left(\frac{\partial H}{\partial n_i}\right)_{T,P,n \neq n_i} \tag{11-34}$$

$$\overline{A}_i = \left(\frac{\partial A}{\partial n_i}\right)_{T,P,n\neq n_i} \tag{11-35}$$

$$\overline{S}_i = \left(\frac{\partial S}{\partial n_i}\right)_{T,P,n\neq n_i} \tag{11-36}$$

In each case, the partial molal property is defined at *constant temperature and pressure*. All of the partial molal properties are intensive variables.

C. Gibbs-Duhem equation

As discussed earlier, at constant temperature and pressure, the total volume of a solution is the weighted sum of the partial molal volume of its components. Likewise, at constant temperature and pressure, the total Gibbs free energy of a solution is the weighted sum of the partial molal Gibbs free energies of its components. Thus, G for a ternary solution of components 1, 2, 3 is

$$G = n_1\overline{G}_1 + n_2\overline{G}_2 + n_3\overline{G}_3 \tag{11-37}$$

The differential of Eq. (11-37) is

$$dG = n_1 d\overline{G}_1 + dn_1\overline{G}_1 + n_2 d\overline{G}_2 + dn_2\overline{G}_2 + n_3 d\overline{G}_3 + dn_3\overline{G}_3 \tag{11-38}$$

Equation (11-38) must be equal to Eq. (11-32) at constant temperature and pressure,

$$\overline{G}_1 dn_1 + \overline{G}_2 dn_2 + \overline{G}_3 dn_3 = n_1 d\overline{G}_1 + dn_1\overline{G}_1 + n_2 d\overline{G}_2 + dn_2\overline{G}_2 + n_3 d\overline{G}_3 + dn_3\overline{G}_3 \tag{11-39}$$

However, for the two equations to be equal the following terms must sum to zero:

$$n_1 d\overline{G}_1 + n_2 d\overline{G}_2 + n_3 d\overline{G}_3 = 0 \tag{11-40}$$

Equation (11-40) is one form of the *Gibbs-Duhem equation*. Using Y to denote any one of the extensive variables (V, E, H, S, G, A) the general form of the Gibbs-Duhem equation is

$$\sum_i n_i d\overline{Y}_i = 0 \tag{11-41}$$

Equation (11-41) is valid for any extensive property that is a continuous single-valued function of composition. Gibbs derived this equation in his 1876 paper on heterogeneous phase equilibria. The French physicist Pierre Duhem (1861–1916) independently derived it in 1886. Later, we discuss the Duhem-Margules equation, which is a special case of the Gibbs-Duhem equation.

D. Chemical potential

We now examine the compositional dependence of the internal energy E when its complete differential is written in terms of its natural variables, entropy and volume. We have already done this for the Gibbs free energy because its natural variables are temperature and pressure.

The complete differential for the internal energy with respect to entropy, volume, and the number of moles of the different components in an open system is

$$dE = dS\left(\frac{\partial E}{\partial S}\right)_{V,n_1,\ldots} + dV\left(\frac{\partial E}{\partial V}\right)_{S,n_1,\ldots} + dn_1\left(\frac{\partial E}{\partial n_1}\right)_{S,V,n_i\neq n_1} + dn_2\left(\frac{\partial E}{\partial n_2}\right)_{S,V,n_i\neq n_2} + \cdots \quad (11\text{-}42)$$

Equation (11-42) is analogous to Eq. (11-26), which gives the dependence of the Gibbs free energy on its natural variables (temperature and pressure) and composition.

A comparison of Eq. (11-42) with the fundamental equation (6-48) shows that the partial derivatives of E with respect to S and V are temperature and pressure, respectively:

$$T = \left(\frac{\partial E}{\partial S}\right)_{V,n_1,\ldots} \quad (11\text{-}43)$$

$$P = -\left(\frac{\partial E}{\partial V}\right)_{S,n_1,\ldots} \quad (11\text{-}44)$$

The partial derivatives of E with respect to composition do not have counterparts in the fundamental equation, which applies to a closed system. Furthermore, as you can see by comparing Eq. (11-33) and Eq. (11-42), these partial derivatives are *not* partial molal internal energy terms. The partial molal internal energy is defined at constant temperature and pressure, while the compositional terms in Eq. (11-42) are at constant entropy and volume. The partial derivatives of E with respect to composition at *constant entropy and volume* define the chemical potentials (μ) of the different components:

$$\mu_1 = \left(\frac{\partial E}{\partial n_1}\right)_{S,V,n_i\neq n_1} \quad (11\text{-}45)$$

$$\mu_2 = \left(\frac{\partial E}{\partial n_2}\right)_{S,V,n_i\neq n_2} \quad (11\text{-}46)$$

In general, the *chemical potential* (μ_i) of any component i is defined by

$$\mu_i = \left(\frac{\partial E}{\partial n_i}\right)_{S,V,n\neq n_i} \quad (11\text{-}47)$$

This definition of the chemical potential is similar to that of Gibbs, who defined μ_i per unit mass, instead of per mole as we have done. His definition of the chemical potential is: "If to any homogeneous mass we suppose an infinitesimal quantity of any substance to be added, the mass remaining homogeneous and its entropy and volume remaining unchanged, the increase of the energy of the mass divided by the quantity of the substance added is the *potential* for that substance in the mass considered" (Gibbs, 1928, p. 93). This definition is a description of the partial derivative in Eq. (11-47) and makes clear that chemical potential is an intensive quantity. We can use Eqs. (11-43) to (11-47) to rewrite Eq. (11-42) as

$$dE = TdS - PdV + \mu_1 dn_1 + \mu_2 dn_2 + \cdots \quad (11\text{-}48)$$

which is valid for open systems with any number of components. Equation (11-48) is useful for relating the chemical potential as defined by Gibbs to other definitions involving H, G, and A.

We now consider enthalpy H. In Chapter 8, we showed that

$$dH = TdS + VdP \tag{8-3}$$

Equation (8-3) is the *third fundamental equation* and it expresses the dependence of H on its natural variables S and P for a closed system. The complete differential for enthalpy with respect to entropy, pressure, and composition in an open system is given by

$$dH = dS\left(\frac{\partial H}{\partial S}\right)_{P,n_1,\ldots} + dP\left(\frac{\partial H}{\partial P}\right)_{S,n_1,\ldots} + dn_1\left(\frac{\partial H}{\partial n_1}\right)_{S,P,n_i\neq n_1} + dn_2\left(\frac{\partial H}{\partial n_2}\right)_{S,P,n_i\neq n_2} + \cdots \tag{11-49}$$

Equation (11-49) is analogous to Eq. (11-26) for the Gibbs free energy and to Eq. (11-42) for the internal energy. The partial derivatives of H with respect to composition are *not* partial molal enthalpy terms, because they are not evaluated at constant temperature and pressure. Instead, the partial derivatives in Eq. (11-49) are evaluated at constant entropy and pressure. We can show that they are equivalent to the chemical potentials of the different components by substituting Eq. (11-48) into Eq. (8-1):

$$dH = (TdS - PdV + \mu_1 dn_1 + \mu_2 dn_2 + \cdots) + PdV + VdP \tag{11-50}$$

At constant entropy ($dS = 0$) and constant pressure ($dP = 0$), Eq. (11-50) reduces to

$$dH = (\mu_1 dn_1 + \mu_2 dn_2 + \cdots) \tag{11-51}$$

Likewise, at constant entropy and constant pressure, Eq. (11-49) becomes

$$dH = dn_1\left(\frac{\partial H}{\partial n_1}\right)_{S,P,n_i\neq n_1} + dn_2\left(\frac{\partial H}{\partial n_2}\right)_{S,P,n_i\neq n_2} + \cdots \tag{11-52}$$

Equations (11-51) and (11-52) must be equal to each other. Thus, the partial derivatives of H with respect to composition are identical to the chemical potentials defined earlier, for example

$$\mu_1 = \left(\frac{\partial H}{\partial n_1}\right)_{S,P,n_i\neq n_1} \tag{11-53}$$

Equation (11-49) can be rewritten as

$$dH = TdS + VdP + \mu_1 dn_1 + \mu_2 dn_2 + \cdots \tag{11-54}$$

which applies to an open system. Likewise, we can show that the partial molal Gibbs free energy is equivalent to the chemical potential:

$$\overline{G}_1 = \mu_1 = \left(\frac{\partial G}{\partial n_1}\right)_{T,P,n_i\neq n_1} \tag{11-55}$$

Equation (11-55) is very important, and we will use it more than the other definitions of μ.

As stated earlier, the chemical potential is an intensive quantity. Consequently, the chemical potential is the same for every part of a phase, depends only on the composition of the phase, and does not depend on the amount of the phase. For example, the chemical potential of H_2O in a glass of pure water ($\mu_{H_2O}^{water}$) kept at constant *P-T* conditions is the same everywhere in the water and does not depend on the amount of water in the glass. In this example we have a pure phase made of one component. With the exception of very small droplets mentioned earlier (see Chapter 7, Section I-F), the chemical potential of a pure one-component phase is also independent of the size and shape of the phase and its state of aggregation. In general, for any system consisting of a single pure phase made of one component, the chemical potential (and the partial molal Gibbs free energy) is identical to the molar Gibbs free energy (G_m) of the phase.

Second, the difference in chemical potential (or the chemical potential gradient) is the driving force for phase equilibrium and chemical equilibrium. Third, once phase or chemical equilibrium is reached, the chemical potential of a component is the same in all the phases in which it occurs. This is true at constant temperature and pressure (Eq. 11-55), but also at constant entropy and volume (Eq. 11-47), at constant entropy and pressure (Eq. 11-53), and at constant temperature and volume. We can reiterate the same point by giving this equation for the chemical potential in terms of *E*, *H*, *G*, and *A*:

$$\mu_i = \left(\frac{\partial E}{\partial n_i}\right)_{S,V,n\neq n_i} = \left(\frac{\partial H}{\partial n_i}\right)_{S,P,n\neq n_i} = \left(\frac{\partial G}{\partial n_i}\right)_{T,P,n\neq n_i} = \left(\frac{\partial A}{\partial n_i}\right)_{T,V,n\neq n_i} \quad (11\text{-}56)$$

The key point is that the chemical potential is defined in terms of the compositional dependence of *E, H, A,* or *G when their natural variables are held constant.* The definition of μ_i in terms of the Helmholtz free energy *A* is a problem at the end of this chapter.

E. Some properties of the chemical potential

Equation (11-55) shows that the chemical potential is the same as the partial molal Gibbs free energy. Thus, the variation with temperature of the chemical potential (μ_i) of component *i* is

$$\left(\frac{\partial \mu_i}{\partial T}\right)_{P,n\neq n_i} = \left(\frac{\partial \overline{G}_i}{\partial T}\right)_{P,n\neq n_i} \quad (11\text{-}57)$$

As stated earlier, the Gibbs free energy is a state function. Thus, the order of differentiation is irrelevant to the final result. (You should go back and review Section XII-A in Chapter 2 if you do not understand this point.) The derivative for μ_i with temperature in Eq. (11-57) is therefore

$$\left(\frac{\partial \mu_i}{\partial T}\right)_{P,n\neq n_i} = \left(\frac{\partial \overline{G}_i}{\partial T}\right)_{P,n\neq n_i} = \frac{\partial}{\partial T}\left(\frac{\partial G}{\partial n_i}\right)_{P,T} = \frac{\partial}{\partial n_i}\left(\frac{\partial G}{\partial T}\right)_{P,n_i} = -\left(\frac{\partial S}{\partial n_i}\right)_{P,T} = -\overline{S}_i \quad (11\text{-}58)$$

Equation (11-58) shows that the variation of chemical potential with temperature at constant pressure and composition is equal to the *partial molal entropy*, but with opposite sign:

$$\overline{S}_i = -\left(\frac{\partial \mu_i}{\partial T}\right)_{P,n\neq n_i} \quad (11\text{-}59)$$

Using a similar chain of equations and logic, we can show that the variation of the chemical potential with pressure at constant temperature and constant composition is given by

$$\overline{V}_i = \left(\frac{\partial \mu_i}{\partial P}\right)_{T,n\neq n_i} \quad (11\text{-}60)$$

The variation of chemical potential with pressure at constant temperature and composition is equal to the *partial molal volume* of the constituent. Equations (11-59) and (11-60) are general equations that apply to a solid, liquid, or gas.

However, if we assume ideal gas behavior, we can relate the chemical potential of a solid or liquid component to its vapor pressure. The partial molal volume of an ideal gas i is

$$\overline{V}_i = \left[\frac{\partial(n_i RT/P_i)}{\partial n_i}\right]_{P,T,n\neq n_i} = \frac{RT}{P_i} \quad (11\text{-}61)$$

Using Eq. (11-61) to substitute into Eq. (11-60) we find that

$$\left(\frac{\partial \mu_i}{\partial P}\right)_{T,n\neq n_i} = \overline{V}_i = \frac{RT}{P_i} \quad (11\text{-}62)$$

If the pressure change dP is due solely to the change in partial pressure dP_i of gas i, we have

$$d\mu_i = RT\frac{dP_i}{P_i} \quad (11\text{-}63)$$

Integrating Eq. (11-63) between the standard-state pressure of one bar and pressure P_i (which can be less than or greater than one bar) then gives a very useful relationship:

$$\mu_i = \mu_i^o + RT\ln(P_i/1) = \mu_i^o + RT\ln P_i \quad (11\text{-}64)$$

The logarithmic term in Eq. (11-64) is dimensionless because the pressure P_i is divided by the standard state pressure of one bar. The μ_i^o term in Eq. (11-64) is the chemical potential of species i under standard-state conditions and is identical to the standard molal Gibbs energy of the material. Equation (11-64) applies to solids and liquids in equilibrium with vapor behaving ideally. A large number of substances fall into this category, including many planetary materials (e.g., ices of H_2O, CH_4, NH_3, N_2, CO, H_2S; Fe metal, Mg_2SiO_4, $MgSiO_3$). The pressure P_i in Eq. (11-64) is replaced by the fugacity f_i for nonideal vapors.

III. THERMODYNAMIC PROPERTIES OF IDEAL AND DILUTE SOLUTIONS

Traditionally, ideal solutions are defined as solutions that follow the ideal solution laws (i.e., Raoult's and Henry's laws) at all concentrations (see Sections A and B that follow). Dilute solutions are real solutions that follow the ideal solution laws at sufficiently low solute concentrations. Solutions of gases in liquids (water, liquid metals, magmas) and trace element solutions in minerals are good examples of dilute solutions.

A. Raoult's law

In an *ideal solution*, the partial vapor pressure of a component in the solution is proportional to its mole fraction in the solution. This relationship is *Raoult's law:*

$$P_i = X_i P_i^o \quad (11\text{-}65)$$

RAOULT, FRANÇOIS MARIE (1830-1901)

Raoult obtained his doctorate from the University of Paris in 1863. He later taught at Grenoble for many years.

Raoult's research was largely concerned with vapor pressures over solutions and their freezing point depressions. He found that the reduction in vapor pressure over a liquid when a small amount of salt is dissolved in it is nearly proportional to its freezing point depression. He also found a method for determining the molecular weights of organic compounds by dissolving them in water and measuring the freezing point depression. He found that this principle could also be applied to salts, once the valence of the ions was taken into account. He formulated Raoult's law, which describes the vapor pressure over an ideal solution. The rule states that a given molar amount of any non-saline solute in a given molar amount of any volatile solvent will produce the same decrease in vapor pressure.

Raoult took an experimental approach to his work. He even built most of his own apparatus, to ensure that it was made properly. At the same time, he was pleased that the theoretical work of van't Hoff confirmed his own experimental results.

The French chemist Francois Marie Raoult (see sidebar) discovered this relationship in 1887. The mole fraction of species i is X_i, P_i^o is the vapor pressure of pure i at some temperature, and P_i is the partial vapor pressure of i in the solution at the same temperature. The partial vapor pressures of components 1 and 2 and the total vapor pressure (P_T) for a binary ideal solution are

$$P_1 = X_1 P_1^o \quad (11\text{-}66)$$

$$P_2 = X_2 P_2^o \quad (11\text{-}67)$$

$$P_T = P_1 + P_2 = X_1 P_1^o + X_2 P_2^o \quad (11\text{-}68)$$

Figure 11-4 is a pressure-composition (P-X) diagram illustrating the partial and total vapor pressures over the ethylene bromide ($C_2H_4Br_2$)–propylene bromide ($C_3H_6Br_2$) solution at 85.1°C

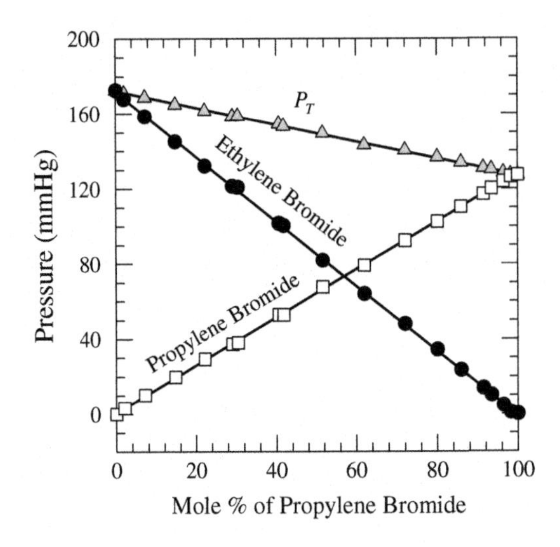

FIGURE 11-4

A pressure-composition (P-X) diagram showing the partial and total vapor pressures over a solution of ethylene bromide ($C_2H_4Br_2$) and propylene bromide ($C_3H_6Br_2$) at 85.1°C. Based on the data of Zawidzki (1900). This solution is ideal or nearly so.

(Zawidzki, 1900). Ethylene bromide is a fumigant for grain and soil, and propylene bromide is an industrial solvent. Table 11-1 gives a subset of Zawidzki's data for this system, which is a classic example of an ideal solution. The data in Table 11-1 and Figure 11-4 obey Raoult's law as you can verify for yourself using Eqs. (11-65) to (11-68).

Equation (11-65) implicitly assumes ideal gases. In many cases, this is a good assumption because the nonideality of the gas phase is negligible or is small relative to experimental errors in measuring the partial vapor pressures of the different components. However, Raoult's law is not restricted to ideal gases, and its most general form is

$$f_i = X_i f_i^o \tag{11-69}$$

The mole fraction of species i is X_i, f_i^o is the fugacity of pure i at some temperature, and f_i is the fugacity of i in a solution at the same temperature. The rearranged version of Eq. (11-69) is

$$X_i = \frac{f_i}{f_i^o} \tag{11-70}$$

This is identical to the definition of thermodynamic activity given by Eq. (10-244). Thus, Raoult's law tells us that the thermodynamic activity of a component in an ideal solution is equal to its mole fraction:

$$a_i = X_i = \frac{f_i}{f_i^o} \cong \frac{P_i}{P_i^o} \tag{11-71}$$

Equations (11-65), (11-69), (11-70), and (11-71) are equivalent statements of Raoult's law. As shown later, the thermodynamic properties of ideal solutions can be derived from Raoult's law.

Table 11-1 Ethylene Bromide–Propylene Bromide Solution at 85.1°C[a]

Mole % Propylene Bromide		Pressure (mmHg)		
Solution	**Vapor**	**Total**	$C_3H_6Br_2$	$C_2H_4Br_2$
0	0	172.6	0	172.6
2.02	1.85	171	3.2	167.8
7.18	6.06	168.8	10.2	158.6
14.75	12.09	165	19.9	145.1
22.21	18.22	161.6	29.4	132.2
29.16	23.5	158.7	37.3	121.4
30.48	23.96	158.9	38.1	120.8
40.62	34.25	154.6	52.9	101.7
41.8	34.51	153.4	52.9	100.5
51.63	45.28	149.6	67.7	81.9
62.03	55.35	143.3	79.3	64
72.03	65.86	140.5	92.5	48
80.05	74.94	136.8	102.5	34.3
85.96	82.45	133.9	110.4	23.5
91.48	86.5	130.9	117.1	13.8
93.46	92.31	130.2	120.1	10.1
96.41	96.41	128.4	123.8	4.6
98.24	99.39	127.3	126.5	0.8
100	100	127.2	127.2	0

[a]*Zawidzki (1900), ethylene bromide ($C_2H_4Br_2$,)–propylene bromide ($C_3H_6Br_2$.)*

The *Duhem-Margules equation* shows that if Raoult's law applies to one component of a binary solution, it also applies to the other. This equation, which is valid for all solutions, is

$$\left(\frac{\partial \ln P_1}{\partial \ln X_1}\right)_{P,T} = \left(\frac{\partial \ln P_2}{\partial \ln X_2}\right)_{P,T} \tag{11-72}$$

The French physicist Pierre Duhem (see sidebar) and the Austrian meteorologist Max Margules (1856–1920) independently derived Eq. (11-72) in 1886 and 1895, respectively. If the vapors are nonideal, Eq. (11-72) is rewritten in terms of the fugacities of the two components.

DUHEM, PIERRE-MAURICE-MARIE (1861-1916)

Duhem was a French scientist who took a theoretical approach. He wrote on the history and philosophy of science, and he sought a general scientific theory that would explain electricity and magnetism as well as classical mechanics. He never achieved this, and suffered set backs from the very beginning. His thesis on the potential of electrochemical cells, which he presented at the École Normale in Paris in 1884, was rejected due to the influence of Berthelot. Duhem angered Berthelot by stating, correctly, in the thesis that the enthalpy of a reaction does not predict its spontaneity, which contradicted Berthelot's principle of maximum work. Duhem published the thesis as a book, at the risk of damaging his career, and wrote a second dissertation, this one on magnetism, which passed in 1888. Duhem and Berthelot remained at odds, which might be the reason that Duhem was never offered a professorship in Paris.

Example 11-3. Determine if Zawidzki's data for the ethylene bromide–propylene bromide system obey the Duhem-Margules equation.

We rewrite Eq. (11-72) using the mathematical identity $d(\ln u) = du/u$ and rearrange it:

$$\frac{X_1}{P_1}\left(\frac{\partial P_1}{\partial X_1}\right)_{P,T} = \frac{X_2}{P_2}\left(\frac{\partial P_2}{\partial X_2}\right)_{P,T} \tag{11-73}$$

$$\frac{(\partial P_1/\partial X_1)_{P,T}}{(\partial P_2/\partial X_2)_{P,T}} = \frac{P_1/X_1}{P_2/X_2} = \frac{P_1^o}{P_2^o} \tag{11-74}$$

Equation (11-74) says the ratio of the line slopes is equal to the ratio of the vapor pressures of the pure liquids. The slopes are easily determined using linear least-squares fits and the vapor pressures given in Table 11-1. Arbitrarily calling ethylene bromide component 1 and propylene bromide component 2, we find that the ratio of the two slopes is

$$\frac{(\partial P_1/\partial X_1)_{P,T}}{(\partial P_2/\partial X_2)_{P,T}} = \frac{172.6}{127.6} = 1.353 \tag{11-75}$$

The ratio of the vapor pressures of the pure liquids is almost identical:

$$\frac{P_1^o}{P_2^o} = \frac{172.6}{127.2} = 1.357 \tag{11-76}$$

Zawidzki's data obey the Duhem-Margules equation within 0.3%. This is excellent agreement.

B. Henry's law

The English chemist William Henry (1775–1836) studied the solubility of gases in liquids and proposed *Henry's law* in 1803. This law states that the solubility of a gas in a liquid is proportional to the partial pressure of the gas:

$$P_i = k_i' X_i \tag{11-77}$$

The partial pressure and mole fraction of the gas are P_i and X_i, respectively. The proportionality constant k' is Henry's law constant. This is determined experimentally. Although originally proposed for gas solubility in liquids, Henry's law applies to very dilute solutions of all species and is important for trace element partitioning between coexisting phases. An important question is the concentration range over which Henry's law is valid for the trace element(s) of interest.

Figure 11-5 illustrates Henry's law for O_2 solubility in water (23°C). Gas solubility in liquids is important for the alkalinity of fresh water and seawater, biochemical reactions in living organisms, anesthesia during surgical operations, artificial blood substitutes, biological oxygen demand in natural water and in waste water, and breathing mixtures for deep-sea divers.

Gas solubilities in liquids are expressed in several different units, named after different scientists who studied gas solubility. It is important to understand the different units because they are used in

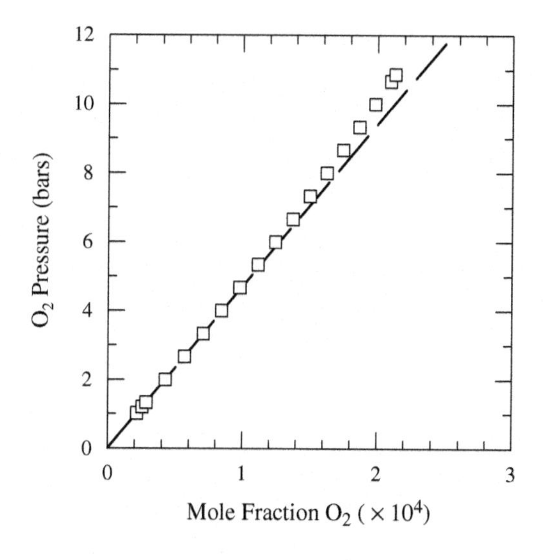

FIGURE 11-5

Solubility of O_2 in water at 23°C plotted to illustrate Henry's law and the deviations from it as pressure increases. The dashed line is Henry's law constant defined by Eq. (11-77). Based on data from Loomis (1928).

reference books and scientific papers. Table 11-2 gives the pressure-dependent solubility of O_2 in water at 23°C (Loomis, 1928) in some of these units. In the mid-1800s, the German chemist Robert Bunsen (1811–1899) developed methods for chemical analyses of gases, analyzed volcanic gases, and studied gas solubility in water. The *Bunsen coefficient* (α) is the gas volume at one atmosphere gas

Table 11-2 O_2 Solubility in Water at 23°C[a] (STP cm^3 L^{-1})

P (mmHg)	*P*(bars)	cm^3 O_2	$10^3 n_{O_2}$ [b]	$10^5 X_{O_2}$ [c]	$10^{-4} k'$ [d]
1000	1.33	27.1	1.575	2.837	4.69
1500	2	27.09	2.361	4.254	4.7
2000	2.67	27.06	3.145	5.665	4.71
2500	3.33	27	3.922	7.066	4.71
3000	4	26.89	4.688	8.444	4.74
4000	5.33	26.57	6.176	11.12	4.79
6000	8	25.86	9.016	16.24	4.93
8000	10.66	25.04	11.6	20.97	5.08

[a] cm^3 O_2 (computed at 0°C, 1 atm) per one liter of water.
[b] Millimoles, for example, 1.575 is 1.575×10^{-3} moles.
[c] Mole fraction, for example, 2.837 is a mole fraction of 2.837×10^{-5}
[d] Henry's law constant, for example, 4.69 is 46,900 and so on.

pressure and 0°C (standard temperature and pressure, denoted as STP) dissolved per volume of liquid at the same gas pressure and temperature T:

$$\alpha = \frac{V_{gas}}{V_{liquid}} \cdot \frac{273.15}{T} \quad (11\text{-}78)$$

The third column from the left in Table 11-2 gives the Bunsen coefficient for O_2. The Dutch physicist Johannes Petrus Kuenen (1866–1922) originated the *Kuenen coefficient* (α'). This is the gas volume at one atmosphere gas pressure and 0°C dissolved per unit mass of liquid at the same gas pressure and temperature T:

$$\alpha' = \frac{V_{gas}}{m_{liquid}} \cdot \frac{273.15}{T} \quad (11\text{-}79)$$

Kuenen was probably Kammerlingh Onnes's best student and did pioneering studies of critical phenomena in gas mixtures. He discovered *retrograde condensation*, which is the formation of liquid hydrocarbons in a gas reservoir as the pressure decreases below the critical point of the hydrocarbon gas mixture. This is important in oilfields on Earth and may occur on Titan, the largest satellite of Saturn.

The German physical chemist Wilhelm Ostwald (see sidebar in Chapter 10) studied gas solubility in the late 1880s. The *Ostwald coefficient* (L) is the volume of gas dissolved per volume of liquid at one atmosphere gas pressure and ambient temperature:

$$L = \frac{V_{gas}}{V_{liq}} \quad (11\text{-}80)$$

The gas volume in the Ostwald coefficient is not corrected to STP, standard temperature (0°C) and pressure (one atmosphere gas pressure), as done in the Bunsen and Kuenen coefficients.

The Ostwald coefficient (L_i) and mole fraction (X_i) of the dissolved gas are related via

$$X_i = \left[\frac{RT}{L_i P_i V^o_{liq}} + 1\right]^{-1} \quad (11\text{-}81)$$

The P_i in Eq. (11-81) is the partial pressure of gas i and V^o_{liq} is the molar volume of the pure liquid at ambient pressure and temperature. Equation (11-81) assumes ideal gas behavior, which is generally valid at low-gas partial pressures.

Example 11-4. The first row of Table 11-2 gives the solubility of O_2 in water equilibrated with O_2 gas at a pressure of 1000 mmHg and 23°C. The solubility is equivalent to 1.575×10^{-3} moles O_2 per one liter of water. Derive the Ostwald coefficient, Bunsen coefficient, mole fraction, and Henry's law constant for O_2 from these data.

Assuming ideality, the amount of O_2 dissolved in the water corresponds to a volume of

$$V = \frac{nRT}{P} = \frac{(1.575 \times 10^{-3}\ \text{mol})(83.145\ \text{cm}^3\ \text{bar}\ \text{K}^{-1}\ \text{mol}^{-1})(296.15\ \text{K})}{(1000\ \text{mm}/750.062\ \text{mm}\ \text{bar}^{-1})} = 29.09\ \text{cm}^3 \quad (11\text{-}82)$$

This is the O_2 solubility per 1000 cm^3 of water. The Ostwald coefficient L is calculated using Eq. (11-80) and is 0.02909. The Bunsen coefficient (α) is calculated from Eq. (11-78) and is

$$\alpha = \frac{V_{gas}}{V_{liquid}} \cdot \frac{273.15}{T} = 0.02909 \cdot \frac{273.15}{296.15} = 0.027 \quad (11\text{-}83)$$

The mole fraction of O_2 is the number of moles of O_2 divided by the number of moles of solution:

$$X_{O_2} = \frac{n_{O_2}}{n_{O_2} + n_{H_2O}} = \frac{0.001575}{0.001575 + 55.509} = 2.837 \times 10^{-5} \quad (11\text{-}84)$$

The Henry's law constant is calculated from Eq. (11-77) and is

$$k'_{O_2} = \frac{P_{O_2}}{X_{O_2}} = \frac{1.33}{2.837 \times 10^{-5}} = 4.69 \times 10^{4} \quad (11\text{-}85)$$

This is dimensionless because the O_2 partial pressure is a ratio relative to the standard-state pressure of one bar.

Example 11-5. Use the solubilities of N_2, O_2, and Ar in water at 0°C to calculate the mole fraction, Ostwald coefficient, and Henry's law constant of air in water saturated with air at the same temperature and one atmosphere air pressure. Wilhelm et al. (1977) give these solubilities: $X_{N_2} = 1.914 \times 10^{-5}$, $X_{O_2} = 3.941 \times 10^{-5}$, and $X_{Ar} = 4.309 \times 10^{-5}$ at $P = 1$ atm for each gas. The composition of dry air is noted in Table 2-11.

Dry air is 78.084% N_2, 20.946% O_2, and 0.946% Ar. The mole fraction of air (X_{air}) in air-saturated water at 0°C is the weighted sum of the mole fractions of each gas:

$$X_{air} = 0.78084(1.914 \times 10^{-5}) + 0.20946(3.941 \times 10^{-5}) + 9.46 \times 10^{-3}(4.309 \times 10^{-5}) \quad (11\text{-}86)$$

Equation (11-86) gives $X_{air} = 2.36 \times 10^{-5}$ for air-saturated water at 0°C. Henry's law constant is the reciprocal of this (42,373). The Ostwald coefficient is $L_{air} = 29.4 \times 10^{-3}$, that is, 29.4 milliliters of air dissolve per liter of water. This was computed using Eq. (11-81).

C. Thermodynamics of gas solubility in aqueous solutions

We use the solubility of CH_4 in pure water to illustrate the thermodynamics of gas solubility in aqueous solutions. Methane is an important greenhouse gas in Earth's atmosphere with a mole fraction of about 1.7×10^{-6} (1.7 ppmv = parts per million by volume). Several anthropogenic, biological, and natural sources produce methane. Reactions with OH radicals in Earth's troposphere and methane utilizing bacteria (methanotrophs) such as *Methylocossus*, *Methylomonas*, and *Methylobacter* destroy CH_4. Methane dissolves in water via the reaction

$$CH_4 \text{ (gas)} = CH_4 \text{ (aqueous)} \quad (11\text{-}87)$$

The molar Gibbs energy change for reaction (11-87) is

$$\Delta G_m = \overline{G}_{CH_4(aq)} - \overline{G}_{CH_4(g)} \quad (11\text{-}88)$$

The partial molar Gibbs energies of CH_4 gas and CH_4 dissolved in aqueous solution are

$$\overline{G}_{CH_4(g)} = G^o_{m,CH_4(g)} + RT \ln f_{CH_4(g)} \tag{11-89}$$

$$\overline{G}_{CH_4(aq)} = G^o_{m,CH_4(aq)} + RT \ln a_{CH_4(aq)} \tag{11-90}$$

The reaction isotherm for reaction (11-87) is

$$\Delta G_m = \left(G^o_{m,CH_4(aq)} - G^o_{m,CH_4(g)}\right) + RT \ln\left(\frac{a_{CH_4(aq)}}{f_{CH_4(g)}}\right) \tag{11-91}$$

To a good first approximation, methane behaves ideally at the low CH_4 partial pressures in Earth's atmosphere. Thus, the methane fugacity in Eqs. (11-89) and (11-91) can be replaced by its partial pressure. In addition, the activity of CH_4 dissolved in aqueous solution can be replaced by its molal concentration because the concentrations of dissolved CH_4 are so small. At equilibrium, the ΔG change for reaction (11-87) is zero and the reaction isotherm becomes

$$\Delta G_m = 0 = \Delta_r G^o_m + RT \ln K_{eq} \tag{11-92}$$

The standard Gibbs energy change in Eq. (11-92) is

$$\Delta_r G^o_m = \left(G^o_{m,CH_4(aq)} - G^o_{m,CH_4(g)}\right) = \Delta_f G^o_{CH_4(aq)} - \Delta_f G^o_{CH_4(g)} \tag{11-93}$$

Equation (11-93) shows that the standard molar Gibbs energies of CH_4 dissolved in aqueous solution and of CH_4 (g) are identical to the standard Gibbs energies of formation of these two species. The equilibrium constant K_{eq} in Eq. (11-92) is

$$K_{eq} = \frac{a_{CH_4(aq)}}{f_{CH_4(g)}} \cong \frac{[CH_4]}{P_{CH_4}} \tag{11-94}$$

The square brackets denote molal concentration (m). The CH_4 partial pressure is a dimensionless ratio relative to the hypothetical ideal gas at one bar pressure. Likewise, the CH_4 concentration is a dimensionless ratio relative to a standard-state concentration of CH_4 dissolved in water. The standard state for CH_4 and other solutes dissolved in water is a hypothetical one-molal solution that behaves ideally ($a_i = m_i$). This is true at infinite dilution and is extrapolated linearly to a one-molal solution of the solute, CH_4 in this case. Wilhelm et al. (1977) give the CH_4 mole fraction in CH_4-saturated water at 25°C and one atmosphere CH_4 pressure as 2.507×10^{-5}. This gives

$$m_{CH_4} = [CH_4] = 55.509 \cdot 1.01325 \cdot X_{CH_4} = 1.41 \times 10^{-3} \tag{11-95}$$

The factor of 55.509 is the number of moles of water in 1000 grams and converts between the CH_4 mole fraction and molality. The factor of 1.01325 converts between the one atmosphere standard state used by Wilhelm et al. (1977) and the onebar standard state used here. We now substitute back into Eq. (11-92) and solve for the standard Gibbs energy change:

$$\Delta_r G^o = -RT \ln\frac{[CH_4]}{P_{CH_4}} = -(8.314)(298.15) \ln\frac{1.41 \times 10^{-3}}{1} \cong 16{,}270 \text{ J mol}^{-1} \tag{11-96}$$

Using this result and the standard Gibbs energy of formation for CH_4 (g) at 298.15 K from Appendix 1 we find that the $\Delta_f G^o$ for CH_4 dissolved in aqueous solution is

$$\Delta_f G^o_{298}(\mathrm{CH_4}, aq) = 16{,}270 - 50{,}760 = -34{,}490 \text{ J mol}^{-1} \quad (11\text{-}97)$$

This value is the standard Gibbs energy of formation for a one-molal solution of CH_4 in water that behaves ideally. It is listed in thermodynamic data compilations and is used in thermodynamic calculations for aqueous CH_4 solutions. The tabulated values for the standard Gibbs energies of formation for aqueous solutions of other gases have a similar meaning.

D. Effect of high pressure on gas solubility in liquids

The high-pressure solubility of gases in water is important in planetary science. For example, aqueous solution clouds form deep in the atmospheres of Uranus and Neptune and dissolve appreciable amounts of NH_3, CH_4, H_2, and other gases (Fegley and Prinn, 1986). Figure 11-5 illustrates that Henry's law (11-77) is a good approximation at relatively low pressures but fails at moderately higher pressures. The range over which Henry's law is valid depends on the gas-liquid pair, but large deviations typically occur at gas pressures of 10–100 bars and dissolved gas mole fractions of 0.01–0.1. However, a modified, and more exact, form of Henry's law predicts gas solubilities to high pressures in the 100–10,000 bar range. This equation is the *Krichevsky-Kasarnovsky equation*, named after the two Russian scientists who derived it in the 1930s (Krichevsky and Kasarnovsky, 1935):

$$\ln\left(\frac{f_i}{X_i}\right) = \ln k' + \frac{\overline{V}_i P_i}{RT} \quad (11\text{-}98)$$

The $\overline{V}_i$ term is the partial molal volume of the dissolved gas, which is assumed to be constant with pressure to a first approximation. Table 11-3 and Figure 11-6 give data for H_2 solubility in water at 25°C from 25 to 1000 atmospheres pressure (Wiebe et al., 1932) and illustrate the use of the Krichevsky-Kasarnovsky equation.

Table 11-3 H_2 Solubility in Water at High Pressure (298 K)[a]

P(atm)	$cm^3\ g^{-1}$	obs. $X_{H_2}(\times 10^3)$	calcd. $X_{H_2}(\times 10^3)$
25	0.436	0.35	0.35
50	0.867	0.696	0.696
100	1.728	1.387	1.386
200	3.39	2.72	2.72
400	6.57	5.25	5.25
600	9.58	7.64	7.64
800	12.46	9.91	9.91
1000	15.2	12.07	12.06

[a]*Wiebe et al. (1930); Krichevsky and Kasarnovsky (1935).*

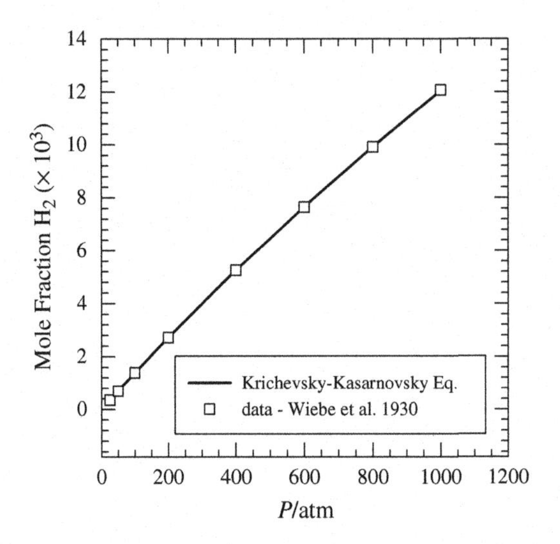

FIGURE 11-6

Solubility of H_2 in water from 25 to 1000 atmospheres pressure at 298 K. The data points of Wiebe et al. (1930) are compared to a curve calculated using the Krichevsky-Kasarnovsky equation (11-98).

E. Boiling-point elevation of dilute solutions

The Dutch physical chemist J. H. van't Hoff (see sidebar in Chapter 10) developed the solution laws in the late 1880s. These describe the effect of solutes on the colligative properties (vapor pressure, freezing-point depression, boiling-point elevation, and osmotic pressure) of dilute solutions. Colligative properties depend on the concentration of a solute in a solution but not on the type of solute. At about the same time, F. M. Raoult was developing the cryoscopic method (i.e., freezing-point method) for molecular weight determination. A few years later Ernst Beckmann (1853–1923), the first director of the Kaiser Wilhelm Institut für Chemie in Germany, developed the ebullioscopic method (i.e., boiling-point method) for molecular weight determination.

Raoult's law and Henry's law predict that the addition of a solute lowers the vapor pressure of the solvent at constant temperature. Thus, the boiling point of the solvent is elevated to a higher temperature. The *normal boiling point* (T_b) of a liquid is the temperature on the vaporization curve at which its vapor pressure is one atmosphere. The corresponding change in state for aqueous solutions is

$$H_2O\ (\text{solution}) = H_2O\ (\text{vapor}) \tag{11-99}$$

The equilibrium constant for reaction (11-99) is

$$K_{eq} = \frac{P_w}{a_w} \tag{11-100}$$

The subscript w denotes water as liquid or vapor. The van't Hoff equation gives the temperature dependence of the equilibrium constant. We rewrite this as

$$\left(\frac{\partial \ln K_{eq}}{\partial T}\right)_P = \frac{\Delta_{vap}H^o}{RT_b^2} \tag{11-101}$$

The enthalpy of reaction is the standard enthalpy of vaporization ($\Delta_{vap}H^o$) in this case. We are interested in the change of the normal boiling point T_b, where the pressure (P_w) remains constant at one atmosphere. Thus, we can rewrite K_{eq} and Eq. (11-101) as follows:

$$K_{eq} = \frac{P_w}{a_w} = \frac{1}{a_w} \tag{11-102}$$

$$\left(\frac{\partial \ln K_{eq}}{\partial T}\right)_P = \left[\frac{\partial(-\ln a_w)}{\partial T}\right]_P = \frac{\Delta_{vap}H^o}{RT_b^2} \tag{11-103}$$

To first approximation $\Delta_{vap}H^o$ is constant over a small ΔT. Then integrating Eq. (11-103) gives

$$-\ln a_w = \frac{\Delta_{vap}H^o}{R}\int_{T_b}^{T}\frac{dT}{T^2} \tag{11-104}$$

$$-\ln a_w = \frac{\Delta_{vap}H^o}{R}\left[\frac{1}{T}-\frac{1}{T_b}\right] = \frac{\Delta_{vap}H^o}{RTT_b}(T_b - T) \tag{11-105}$$

The solvent follows Raoult's law to a very good first approximation in dilute solutions. Thus, we can replace the activity of water by its mole fraction (X_w). Furthermore, because X_w is much larger than the mole fraction of the solute (X_2), we can use the following mathematical identity:

$$-\ln X_w = -\ln(1 - X_2) \cong X_2 \tag{11-106}$$

Also, for $T_b >> \Delta T$, the denominator in Eq. (11-105) is $\sim T_b^2$. Rewriting Eq. (11-105) gives

$$X_2 \cong \frac{\Delta_{vap}H^o}{RT_b^2}(T - T_b) \tag{11-107}$$

$$\Delta T_b = (T - T_b) = X_2\left[\frac{RT_b^2}{\Delta_{vap}H^o}\right] \tag{11-108}$$

Equation (11-107) gives the mole fraction of solute from the observed boiling-point elevation, and Eq. (11-108) gives the boiling-point elevation from the mole fraction of solute. The term in square brackets in Eq. (11-108) is the *ebullioscopic constant*, which is characteristic of a solvent. The molecular weights of crude oil fractions, fulvic acids in soils, kerogen, natural polymers, and synthetic polymers can be determined by boiling-point elevation measurements.

Example 11-6. Vellut et al. (1998) determined the average molecular weights of crude oil fractions by measuring the boiling-point elevation of their solutions in toluene. In their calibration experiments, Vellut et al. (1998) observed that 50.25 grams of *n*-tetradecane ($C_{14}H_{30}$) dissolved in 1000 grams of toluene gave a boiling-point elevation $\Delta T_b = 0.861$ K. The normal boiling point of toluene is 383.77 K, its $\Delta_{vap}H^o = 33{,}192$ J mol^{-1}, and its molecular weight is 92.14 g mol^{-1}. Calculate the average molecular weight of *n*-tetradecane. (The *n* stands for *normal* and denotes a straight chain hydrocarbon.)

Substituting the data into Eq. (11-107) gives the mole fraction of n-tetradecane (X_2):

$$X_2 \cong \frac{\Delta_{vap}H^o}{RT_b^2}\Delta T_b = \frac{33,192}{(8.314)(383.77)^2}0.861 = 0.0233 \tag{11-109}$$

We now solve for M_2, the molecular weight of n-tetradecane:

$$X_2 = 0.0233 = \frac{n_2}{n_1+n_2} = \frac{50.25/M_2}{(10^3/92.14)+(10/M_2)} \cong \frac{50.25/M_2}{10.853} \tag{11-110}$$

$$M_2 \cong \frac{50.25}{10.853 \times 0.0233} \cong 198.7 \text{ g mol}^{-1} \tag{11-111}$$

The molecular weight of n-tetradecane ($C_{14}H_{30}$) is 198.4 g mol^{-1}, 0.2% lower than our result.

In practice, scientists measure the boiling-point elevation for several different weights of the solute (per 1000 grams solvent) as shown for n-tetradecane in Figure 11-7. A linear least-squares fit to the line in Figure 11-7 gives the equation

$$\Delta T_b = 4.319 \times 10^{-3} + 0.01706 w_2 \tag{11-112}$$

The w_2 in Eq. (11-112) is the number of grams solute per 1000 grams solvent. The intercept in Eq. (11-112) should be zero if there were no errors in the data because there is no change in the boiling point of the solvent if nothing is dissolved in it. Equation (11-112) is simply the combination of Eqs. (11-108) and (11-110) and has the general form

$$\Delta T_b = \left(\frac{M_1}{M_2 w_1}\left[\frac{RT_b^2}{\Delta_{vap}H^o}\right]\right) \tag{11-113}$$

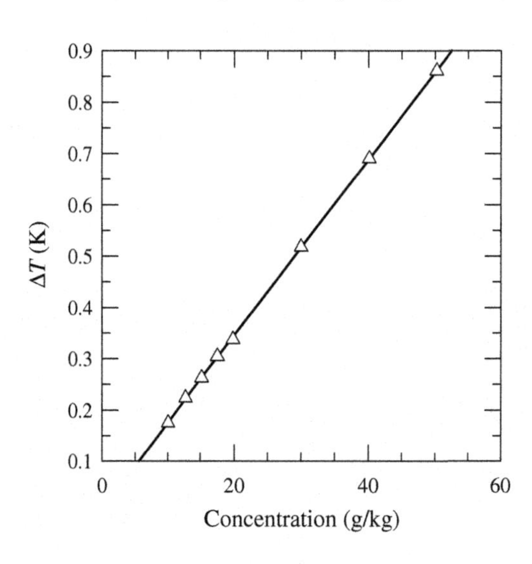

FIGURE 11-7

Boiling-point elevation of toluene as a function of the n-tetradecane concentration in the solution. The line slope gives the molecular weight of the solute (n-tetradecane).

The line slope is the term in parenthesis in Eq. (11-113). For n-tetradecane dissolved in toluene,

$$M_2 = \frac{3.399}{\text{slope}} = \frac{3.399}{0.01706} = 199.2 \text{ g mol}^{-1} \tag{11-114}$$

This value is the average of the molecular weights determined from each of the eight measurements done by Vellut et al. (1998) and is 0.4% higher than the actual molecular weight.

F. Freezing-point depression of dilute solutions

As mentioned earlier, Raoult's law and Henry's law predict that the addition of a solute lowers the vapor pressure of the solvent at constant temperature. The liquid and solid phases of a material have the same vapor pressure at the freezing point. Thus, solutes depress the freezing point of the solvent. We illustrate this point using aqueous solutions as an example. Aqueous solutions freeze to pure ice:

$$H_2O\ (\text{solution}) = H_2O\ (\text{ice}) \tag{11-115}$$

The equilibrium constant for reaction (11-115) is

$$K_{eq} = \frac{a_{ice}}{a_w} = \frac{1}{a_w} \tag{11-116}$$

Once again, the van't Hoff equation (10-64) gives the temperature dependence of K_{eq}:

$$\left(\frac{\partial \ln K_{eq}}{\partial T}\right)_P = \frac{\Delta_m H^o}{RT_m^2} \tag{11-117}$$

Making the same assumptions as before and integrating, we obtain

$$-\ln a_w = \frac{\Delta_m H^o}{R}\int_{T_m}^{T}\frac{dT}{T^2} = \frac{\Delta_m H^o}{RT_m T}(T_m - T) \tag{11-118}$$

The freezing-point depression of a dilute aqueous solution with a solute mole fraction X_2 is thus

$$\Delta T_m = (T_m - T) = -X_2\left[\frac{RT_m^2}{\Delta_m H^o}\right] \tag{11-119}$$

The term in square brackets is the *cryoscopic constant*, which is characteristic of a solvent.

Example 11-7. As stated in Chapter 7, pure, air-free water freezes at 0.0099°C. However, air-saturated water freezes 0.0024°C lower. Calculate the mole fraction of air dissolved in water. The enthalpy of fusion of water is 6009.5 J mol^{-1} and is effectively constant over the small ΔT.

We arrange Eq. (11-119) to solve for X_2 and substitute the appropriate numerical values:

$$X_2 = -\frac{\Delta_m H^o}{RT_m^2}\Delta T = -\frac{6009.5}{8.314(273.16)^2}(-0.0024) = 2.32 \times 10^{-5} \tag{11-120}$$

This is nearly identical to our value of 2.36×10^{-5} calculated earlier.

G. Gibbs energy

The Gibbs energy and other thermodynamic properties of an ideal solution come from Raoult's law. In Chapter 10 we showed that the activity is related to the difference between the molar Gibbs energy of a pure material ($G^o_{m,i}$) and its partial molal Gibbs energy ($\overline{G}_i$) in a solution:

$$\overline{G}_i - G^o_{m,i} = RT \ln a_i \tag{10-261}$$

Combining Eq. (10-261) with Eq. (11-71), the definition of an ideal solution, we find that

$$a_i = X_i \tag{11-71}$$

$$\overline{G}_i - G^o_{m,i} = RT \ln X_i \tag{11-121}$$

Equation (11-121) is valid for every component in an ideal solution.

For simplicity, we now consider the general case of a binary solution formed by n_1 moles of component 1 and n_2 moles of component 2. (The number of moles n_1 and n_2 can have any positive value.) At constant temperature and pressure, the total Gibbs energy of a physical mixture of two pure components is the weighted sum of their Gibbs energy values:

$$G_{mix} = n_1 G^o_{m,1} + n_2 G^o_{m,2} \tag{11-122}$$

Graphically, G_{mix} is a straight line between the Gibbs energies of the two pure components. The total Gibbs energy of the binary solution of the same number of moles of the two components is

$$G_{soln} = n_1 \overline{G}_1 + n_2 \overline{G}_2 \tag{11-123}$$

The difference between the Gibbs energy of the binary solution (G_{soln}) and that of the physical mixture (G_{mix}) is the *integral Gibbs energy of mixing (or solution)* ΔG^M and is

$$\Delta G^M = G_{soln} - G_{mix} = n_1 \left(\overline{G}_1 - G^o_{m,1}\right) + n_2 \left(\overline{G}_2 - G^o_{m,2}\right) \tag{11-124}$$

Henceforth, we call ΔG^M the Gibbs energy of mixing. Equation (11-124) applies to all ideal and nonideal binary solutions. Combining it with Eq. (11-121) gives the Gibbs energy change for the formation of $(n_1 + n_2)$ moles of a binary ideal solution:

$$\Delta G^M = RT[n_1 \ln X_1 + n_2 \ln X_2] \tag{11-125}$$

Dividing through by the total number of moles, the ΔG^M per mole of solution is

$$\Delta G^M = RT[X_1 \ln X_1 + X_2 \ln X_2] \tag{11-126}$$

Figure 11-8 illustrates ΔG^M for an ideal binary solution and plots the quantity $\Delta G^M/T$ versus mole fraction. The points on the graph show the actual $\Delta G^M/T$ values for several different solutions that are ideal, or nearly so.

Figure 11-8 shows that the Gibbs energy of mixing is negative. This is true for all solutions that are thermodynamically stable. The ΔG^M is positive for some nonideal solutions, which are thermodynamically unstable relative to the physical mixture of their components. In these cases, the solution unmixes into its components. It is possible for ΔG^M to change sign from negative to positive to

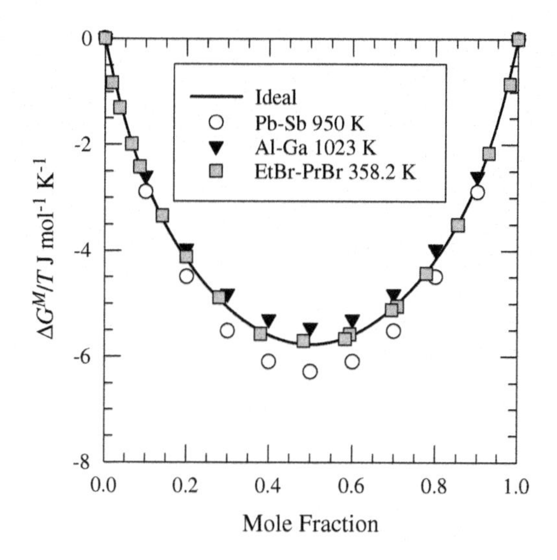

FIGURE 11-8

The Gibbs energy of mixing curve ($\Delta G^M/T$) for an ideal binary solution is compared to values for several solutions that are ideal or nearly so. The solutions are lead-antimony (Pb-Sb), aluminum-gallium (Al-Ga), and ethylene bromide–propylene bromide (EtBr-PrBr). The curve is calculated from Eq. (11-126).

negative again as a function of composition at constant temperature and pressure. The compositions with negative ΔG^M values form stable solutions, while the compositions with positive ΔG^{M} values are more stable as physical mixtures. As illustrated in Chapter 12, the shape of the temperature-composition (T-X) phase diagram for these solutions resembles the critical curve for a pure gas (see Figure 8-21). This resemblance is not coincidental because the unmixing of a solution into its constituents occurs below the *critical unmixing temperature*.

Equation (10-261) also defines the partial molal Gibbs energy of mixing (or solution),

$$\Delta\overline{G}_i^M = \overline{G}_i - G^o_{m,i} = RT \ln a_i \tag{11-127}$$

Equation (11-127) applies to all solutions. In the case of an ideal solution it becomes

$$\Delta\overline{G}_i^M = \overline{G}_i - G^o_{m,i} = RT \ln X_i \tag{11-128}$$

By substituting Eq. (11-127) into Eq. (11-124) we obtain an alternative equation for the Gibbs energy of mixing that applies to all ideal and nonideal binary solutions:

$$\Delta G^M = G_{soln} - G_{mix} = n_1\Delta\overline{G}_1^M + n_2\Delta\overline{G}_2^M \tag{11-129}$$

All these derivations are easily extended to solutions containing three, four, or more components. For example, the ΔG^M for an ideal ternary solution of n_1 moles of component 1, n_2 moles of component 2, and n_3 moles of component 3 is

$$\Delta G^M = RT[n_1 \ln X_1 + n_2 \ln X_2 + n_3 \ln X_3] \tag{11-130}$$

Dividing through by the total number of moles, the ΔG^M per mole of an ideal ternary solution is

$$\Delta G^M = RT[X_1 \ln X_1 + X_2 \ln X_2 + X_3 \ln X_3] \tag{11-131}$$

In general, the ΔG^M for a (ideal or real) solution formed by mixing n_i moles of i components is

$$\Delta G^M = \sum_i n_i\left(\overline{G}_i - G^o_{m,i}\right) = \sum_i n_i \Delta \overline{G}_i^M \tag{11-132}$$

In the case of ideal solutions this equation becomes

$$\Delta G^M = \sum_i n_i\left(\overline{G}_i - G^o_{m,i}\right) = \sum_i n_i \Delta \overline{G}_i^M = RT\sum_i n_i \ln X_i \tag{11-133}$$

Per mole of solution, this becomes

$$\Delta G^M = \sum_i X_i\left(\overline{G}_i - G^o_{m,i}\right) = \sum_i X_i \Delta \overline{G}_i^M = RT\sum_i X_i \ln X_i \tag{11-134}$$

We now rewrite these equations using the chemical potential. The partial molal Gibbs energy of mixing for component i is equivalent to a chemical potential difference:

$$\mu_i - \mu_i^o = \overline{G}_i - G^o_{m,i} = \Delta \overline{G}_i^M = RT \ln a_i \tag{11-135}$$

Substituting Eq. (11-135) into Eq. (11-132) gives a general equation that applies to all solutions:

$$\Delta G^M = \sum_i n_i(\mu_i - \mu_i^o) = \sum_i n_i \Delta \overline{G}_i^M \tag{11-136}$$

For ideal solutions, Eq. (11-136) becomes

$$\Delta G^M = \sum_i n_i(\mu_i - \mu_i^o) = RT\sum_i n_i \ln X_i \tag{11-137}$$

The corresponding equation per mole of solution is

$$\Delta G^M = \sum_i X_i(\mu_i - \mu_i^o) = RT\sum_i X_i \ln X_i \tag{11-138}$$

As illustrated next, all the other thermodynamic properties of solutions can be obtained by differentiation of the equations for ΔG^M, the Gibbs energy of mixing. However, in practice it is often better to measure the heat capacity, entropy, and enthalpy directly because differentiation magnifies errors and gives values less accurate than those measured experimentally.

Table 11-4 illustrates the application of these concepts to the ethylene bromide–propylene bromide solution (at 85.1°C = 358.25 K) studied by Zawidski (1900). This solution is nearly ideal, and earlier we used it as an example of a solution obeying Raoult's law. In Table 11-4 we take component 1 as

Table 11-4 Some Properties of Ethylene Bromide–Propylene Bromide Solution at 85.1°C

1−Propylene Bromide $C_3H_6Br_2$				2−Ethylene Bromide $C_2H_4Br_2$				Solution
X_1	a_1	γ_1	$\Delta\overline{G}_1^M$	X_2	a_2	γ_2	$\Delta\overline{G}_2^M$	ΔG^M
0	0	–	–	1	1	1	0	0
0.0202	0.025	1.245	−10,969	0.9798	0.972	0.992	−84	−304
0.0718	0.08	1.117	−7516	0.9282	0.919	0.99	−252	−774
0.1475	0.156	1.061	−5525	0.8525	0.841	0.986	−517	−1256
0.2221	0.231	1.041	−4363	0.7779	0.766	0.985	−794	−1587
0.2916	0.293	1.006	−3654	0.7084	0.703	0.993	−1048	−1808
0.3048	0.3	0.983	−3591	0.6952	0.7	1.007	−1063	−1833
0.4062	0.416	1.024	−2613	0.5938	0.589	0.992	−1576	−1997
0.418	0.416	0.995	−2613	0.582	0.582	1	−1611	−2030
0.5163	0.532	1.031	−1879	0.4837	0.475	0.981	−2220	−2044
0.6203	0.623	1.005	−1407	0.3797	0.371	0.977	−2955	−1995
0.7203	0.727	1.01	−949	0.2797	0.278	0.994	−3812	−1750
0.8005	0.806	1.007	−643	0.1995	0.199	0.996	−4813	−1475
0.8596	0.868	1.01	−422	0.1404	0.136	0.9	−5939	−1197
0.9148	0.921	1.006	−246	0.0852	0.08	0.938	−7525	−867
0.9346	0.944	1.01	−171	0.0654	0.059	0.895	−8455	−713
0.9641	0.973	1.01	−81	0.0359	0.027	0.742	−10,797	−465
0.9824	0.994	1.012	−16	0.0176	0.005	0.263	−16,007	−298
1	1	1	0	0	0	–	–	0

Calculated using the partial pressure data in Table 11-1.

propylene bromide and component 2 as ethylene bromide. Their mole fractions are denoted by X_1 and X_2, respectively. The activity of each component (a_i) is computed using Eq. (11-71) and is the ratio of its partial pressure above the solution to the vapor pressure of the pure material at the same temperature. The agreement between activity and mole fraction is better at large mole fractions of each component because the major component (i.e., solvent) approaches ideality and obeys Raoult's law. Conversely, the agreement between activity and mole fraction is worse at small mole fractions of each component because the dilute component (i.e., solute) approaches Henry's law behavior. The generally good agreement—and close approach to ideality for this solution—is also shown in Figure 11-4. The activity coefficients ($\gamma_i = a_i/X_i$) give the deviation of activity from mole fraction for each component. Table 11-4 shows that the activity coefficients are generally close to unity, except at small mole fractions of either component. Equation (11-127) is used to calculate the partial molal Gibbs energy of mixing for each component. You should do this for yourself. The partial molal Gibbs energy of the solute decreases dramatically as the solute mole fraction approaches zero and formally goes to minus infinity in an infinitely dilute ($X_{solute} = 0$) solution. The integral Gibbs energy of mixing for the solution is calculated with Eq. (11-134) and the partial molal Gibbs energies. You should do this for yourself, and you should also convince yourself that Eq. (11-134) is only one of the several equivalent equations that can be used to compute ΔG^M.

H. Entropy

The integral and partial molal entropies of mixing are given by the temperature coefficients of the integral and partial molal Gibbs energies of mixing at constant pressure and constant composition, that is, by applying Eq. (8-26) at constant composition. For example, the *integral entropy of mixing (or solution)* ΔS^M per mole of an ideal binary solution is

$$\Delta S^M = -\left(\frac{\partial \Delta G^M}{\partial T}\right)_{P,X} = -R\frac{\partial}{\partial T}(T[X_1 \ln X_1 + X_2 \ln X_2])_{P,X} \tag{11-139}$$

The mole fractions of the two components do not depend on temperature and their temperature derivatives are zero. Thus, Eq. (11-139) simplifies to

$$\Delta S^M = -\left(\frac{\partial \Delta G^M}{\partial T}\right)_{P,X} = -R[X_1 \ln X_1 + X_2 \ln X_2] \tag{11-140}$$

Equation (11-140) probably looks familiar because it is identical to Eq. (6-178) derived (from a completely different argument) in Chapter 6 for the entropy of mixing of a binary mixture. Figure 11-9 illustrates ΔS^M for an ideal binary solution. The points on the graph show the actual ΔS^M values for several different solutions that are ideal, or nearly so. A comparison of Figures 11-8 and 11-9 shows that the ideal ΔG^M and ΔS^M curves are mirror images of each other. This happens because the ideal ΔH^M is zero as we show in the next section.

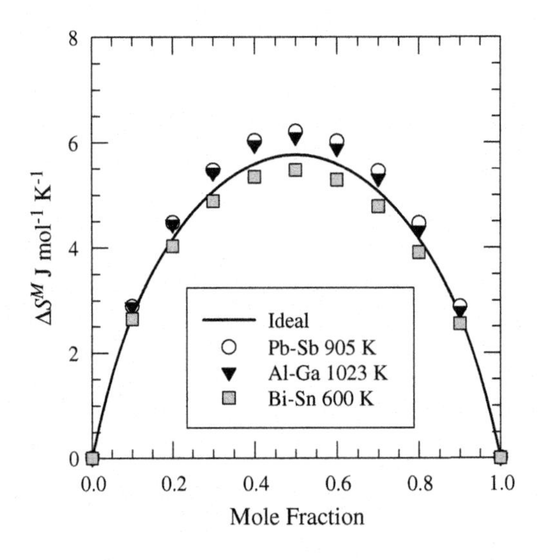

FIGURE 11-9

The entropy of mixing curve (ΔS^M) for an ideal binary solution is compared to ΔS^M values for several solutions that are ideal or nearly so. The curve is calculated from Eq. (11-140).

The *partial molal entropy of mixing (or solution)* for each component in a solution is the temperature coefficient of the corresponding partial molal Gibbs energy at constant pressure, that is,

$$\Delta \overline{S}_i^M = -\left(\frac{\partial \Delta \overline{G}_i^M}{\partial T}\right)_{P,X} = -\left[\frac{\partial(\mu_i - \mu_i^o)}{\partial T}\right]_{P,X} = -R \ln X_i \quad (11\text{-}141)$$

Substituting Eq. (11-141) into Eq. (11-140) and rearranging, we see that ΔS^M is given by

$$\Delta S^M = -R\left(X_1 \Delta \overline{S}_1^M + X_2 \Delta \overline{S}_2^M\right) \quad (11\text{-}142)$$

Again, these equations can be generalized to solutions of three, four, or more components. For example, the ΔS^M per mole of an ideal ternary solution of components 1, 2, and 3 is

$$\Delta S^M = -R[X_1 \ln X_1 + X_2 \ln X_2 + X_3 \ln X_3] \quad (11\text{-}143)$$

This is rewritten in terms of the partial molal entropies as

$$\Delta S^M = -R\left(X_1 \Delta \overline{S}_1^M + X_2 \Delta \overline{S}_2^M + X_3 \Delta \overline{S}_3^M\right) \quad (11\text{-}144)$$

In general, the ΔS^M per mole of an ideal solution formed by mixing i components is

$$\Delta S^M = -R \sum_i X_i \ln X_i \quad (11\text{-}145)$$

I. Enthalpy

The *integral enthalpy of mixing (or solution)* ΔH^M is derived from the integral Gibbs energy and entropy of mixing by rearranging and substituting into the Gibbs-Helmholtz equation:

$$\Delta G^M = \Delta H^M - T \Delta S^M \quad (11\text{-}146)$$

$$\Delta H^M = \Delta G^M + T \Delta S^M \quad (11\text{-}147)$$

Thus, ΔH^M is zero because the ΔG^M and ΔS^M terms cancel each other. This is easily seen in the case of a binary ideal solution,

$$\Delta H^M = RT[X_1 \ln X_1 + X_2 \ln X_2] + T[-R(X_1 \ln X_1 + X_2 \ln X_2)] = 0 \quad (11\text{-}148)$$

Figure 11-10 shows ΔH^M for an ideal binary solution and a comparison to the actual ΔH^M values for several different solutions that are close to ideality.

The *partial molal enthalpy of mixing (or solution)* is derived in an analogous manner and it is also zero because the partial molal Gibbs energy and entropy cancel each other. In general, the integral enthalpy of mixing per mole of an ideal (or real) binary solution is

$$\Delta H^M = X_1 \Delta \overline{H}_1^M + X_2 \Delta \overline{H}_2^M \quad (11\text{-}149)$$

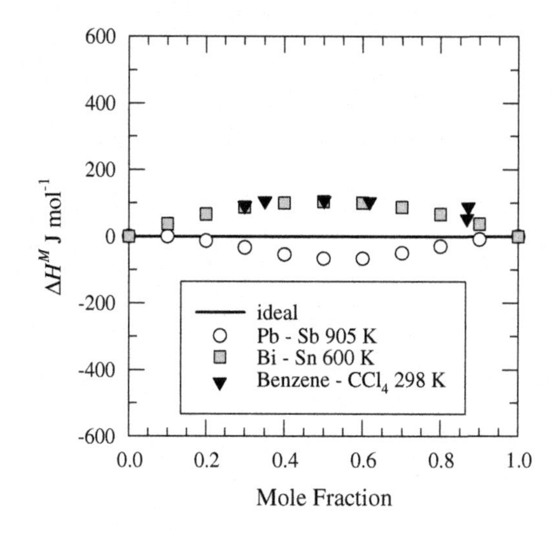

FIGURE 11-10

The enthalpy of mixing for an ideal binary solution is compared to ΔH^M values for several solutions that are nearly ideal.

In the case of an ideal binary solution:

$$\Delta H^M = \Delta \overline{H}_i^M = 0 \tag{11-150}$$

This requires that

$$\frac{H_i}{n_i} = \overline{H}_i = \left(\frac{\partial H_i}{\partial n_i}\right)_{P,T,n \neq n_i} \tag{11-151}$$

In other words, the partial molal enthalpy of a component of an ideal solution is identical to the molar enthalpy of the pure material. In contrast, the integral and partial molal enthalpies of mixing are generally nonzero for nonideal (i.e., real) solutions and can be positive or negative.

J. Heat capacity

The heat capacity difference between a solution and a physical mixture of the same components with the same bulk composition as the solution is the *integral* ΔC_P^M *of mixing (or solution)*. This is the temperature coefficient of ΔH^M at constant pressure and composition:

$$\Delta C_P^M = \left(\frac{\partial \Delta H^M}{\partial T}\right)_{P,X} \tag{11-152}$$

The ΔC_P^M per mole of an ideal (or real) binary solution of components 1 and 2 is the mole fraction weighted sum of the partial molal $\Delta \overline{C}_P^M$ values:

$$\Delta C_P^M = X_1 \Delta \overline{C}_{P,1}^M + X_2 \Delta \overline{C}_{P,2}^M \tag{11-153}$$

The integral enthalpy of mixing ΔH^M is zero for ideal solutions. Thus, the integral and partial molal ΔC_P^M values are also zero for ideal solutions:

$$\Delta C_P^M = \Delta \overline{C}_{P,i}^M = 0 \tag{11-154}$$

This requires that the partial molal $\overline{C}_P$ is identical to the molal C_P of the pure material:

$$\frac{C_{P,i}}{n_i} = \overline{C}_{P,i} = \left(\frac{\partial C_P}{\partial n_i}\right)_{P,T,n \neq n_i} \tag{11-155}$$

The integral and partial molal ΔC_P^M values are generally not equal to zero for nonideal (real) solutions and can be positive or negative. Figure 11-11 illustrates this point. It shows the observed (Dachs et al., 2007) and calculated C_P values for forsterite (Mg_2SiO_4)–fayalite (Fe_2SiO_4) olivine solid solutions. If the olivine solid solution were ideal, their data points would be on the straight line calculated from Eqs. (11-153) and (11-155). Although this is not the case, the difference between the observed and ideal heat capacities is fairly small.

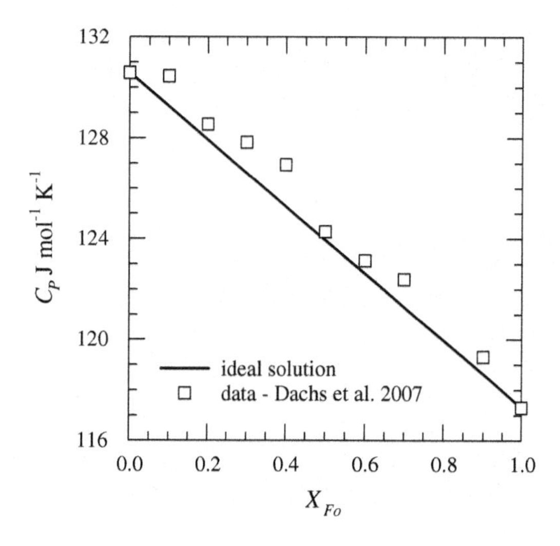

FIGURE 11-11

The heat capacity of forsterite (Mg_2SiO_4)–fayalite (Fe_2SiO_4) solutions from Dachs et al. (2007) compared to the ideal solution line calculated from a linear combination of the heat capacities of forsterite and fayalite, that is, using Eqs. (11-153) and (11-155).

K. Volume

The *integral volume of mixing (or solution)* ΔV^M is the pressure coefficient of the integral Gibbs energy of mixing at constant temperature and constant composition, that is, using Eq. (8-25) at constant composition. The ΔV^M for an ideal binary solution is

$$\Delta V^M = \left(\frac{\partial \Delta G^M}{\partial P}\right)_{T,X} = R\frac{\partial}{\partial P}(T[X_1 \ln X_1 + X_2 \ln X_2]) = R(0) = 0 \quad (11\text{-}156)$$

The same conclusion holds for an ideal solution of any number of components because none of the terms in ΔG^M depends on the pressure. Thus, the ΔV^M for an ideal solution is zero, and the solution has the same total volume as a physical mixture of the pure components. Molten Na_2CO_3-K_2CO_3 (Liu and Lange, 2003), which is a model system for carbonatite magmas, is an example of a binary solution with ideal volume of mixing. Equivalently, as discussed in Chapter 8 for ideal gas mixtures, the partial molal volume of any component i in an ideal solution is identical to the molal volume of the pure material:

$$\frac{V_i}{n_i} = \overline{V}_i = \left(\frac{\partial V}{\partial n_i}\right)_{P,T,n \neq n_i} \quad (8\text{-}201)$$

IV. THERMODYNAMIC PROPERTIES OF NONIDEAL (REAL) SOLUTIONS

A. Deviations from Raoult's law

The thermodynamic activity a_i of a component i in a nonideal solution is proportional to its mole fraction X_i with the proportionality constant being the activity coefficient γ_i:

$$\gamma_i = \frac{a_i}{X_i} = \frac{f_i}{f_i^o}\frac{1}{X_i} \cong \frac{P_i}{X_i P_i^o} \quad (11\text{-}157)$$

The activity coefficient γ_i can be greater than one or less than one and is a function of composition, temperature, and pressure. Equation (11-157) shows that activity coefficients less than unity are due to negative deviations from Raoult's law, that is, the activity of a component is smaller than its mole fraction. This is the case for a solution of acetone and chloroform.

Figure 11-12 is a pressure-composition diagram for the acetone-chloroform solution at constant temperature (35.2°C). The data are those of Zawidzki (1900). Acetone $(CH_3)_2CO$ is used to remove nailpolish, paint, and varnish. Chloroform $CHCl_3$ is used to make CFC-22, is in some fire extinguishers, and was formerly used as an anesthetic.

Figure 11-12 shows that the partial pressures of acetone and chloroform are generally both less than expected from ideality (the diagonal dashed lines) and that the total vapor pressure over the solution is less than expected from ideality (the dashed line at the top). However, the partial pressures of acetone and chloroform approach ideality as their concentrations approach 100%. In other words, each compound approaches Raoult's law as it becomes the major part of the solution. Likewise, the partial pressures of acetone and chloroform approach zero with a constant slope—as predicted by Henry's law—as their concentrations approach zero.

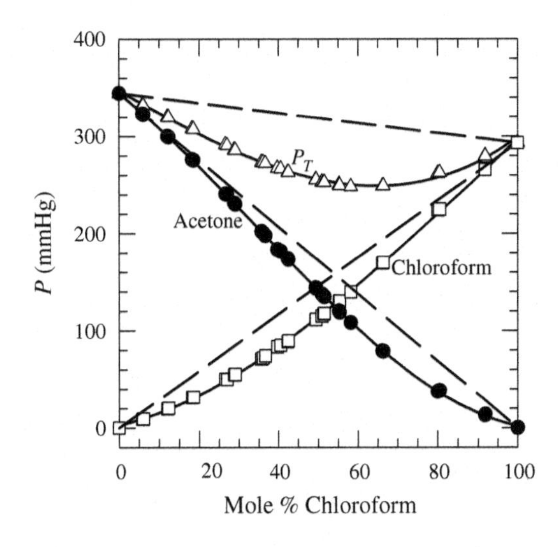

FIGURE 11-12

A pressure-composition (*P-X*) diagram showing the partial and total vapor pressures over a solution of acetone $(CH_3)_2CO$ and chloroform $CHCl_3$ at 35.2°C. Based on the data of Zawidzki (1900). This solution is nonideal and shows negative deviations from Raoult's law. The dashed lines show the behavior for an ideal solution.

Figure 11-13 is the activity-composition diagram corresponding to Figure 11-12. It shows that the activities of acetone and chloroform are both less than expected from ideality (the diagonal dashed lines). The activities of acetone and chloroform approach ideality as their concentrations approach 100% as both compounds approach Raoult's law. Both compounds also approach Henry's law as their concentrations approach zero.

Conversely, activity coefficients greater than unity give positive deviations from Raoult's law. This is the case for a solution of ethyl acetate and ethyl iodide. Ethyl acetate $CH_3COOC_2H_5$ is a fruity-smelling solvent used in perfumes and ethyl iodide CH_3CH_2I is a heavy liquid with a density of 1.95 g cm^{-3}. Figure 11-14, which is also based on the data of Zawidzki (1900), is a pressure-composition diagram at constant temperature (50°C). It shows that the partial pressures of ethyl acetate and ethyl iodide are both greater than expected from ideality (the diagonal dashed lines), and the total vapor pressure over their solution is thus greater than the ideal value. Figure 11-15 is the activity-composition diagram corresponding to Figure 11-14. It shows that the activities of ethyl acetate and ethyl iodide are both greater than expected from ideality (the diagonal dashed lines). Again, Raoult's and Henry's laws are approached by each component as its concentration approaches unity or zero, respectively.

Excess functions are the easiest way to represent deviations from ideality and the activity coefficient γ_i as a function of temperature, pressure, and composition. The concept of excess functions was developed by the American physical chemist George Scatchard (1892–1973) in the early 1930s and is widely used today. Different models used to describe thermodynamics of real solutions can be thought of as ways to represent the excess functions of the solutions.

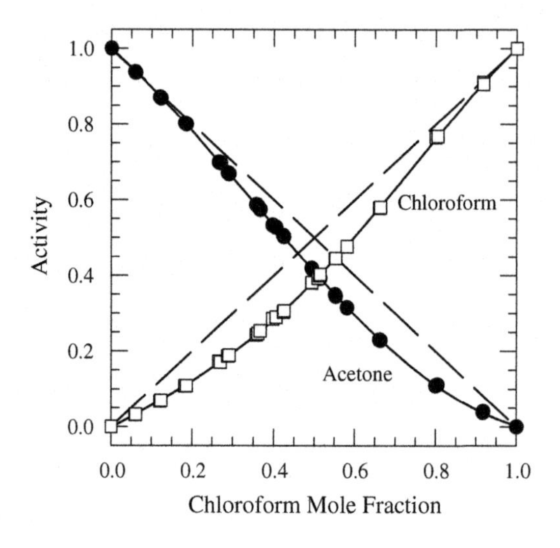

FIGURE 11-13

The activity-composition (a-X) diagram corresponding to Figure 11-12. The dashed lines show the behavior for an ideal solution.

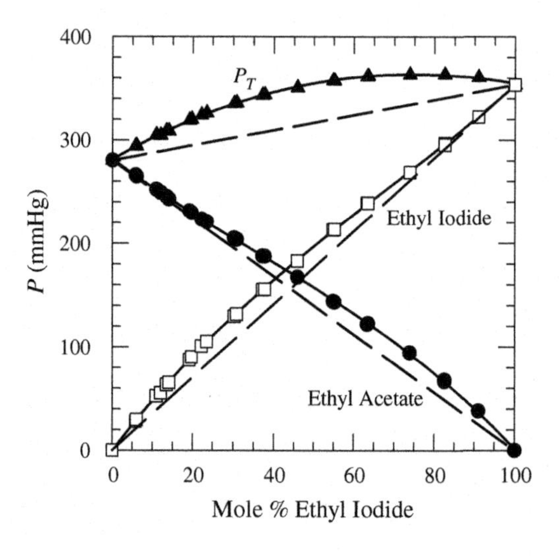

FIGURE 11-14

A pressure-composition (P-X) diagram showing the partial and total vapor pressures over a solution of ethyl acetate $CH_3CH_2CO_2$ and ethyl iodide CH_3CH_2I at 50°C. Based on the data of Zawidzki (1900). This solution is nonideal and shows positive deviations from Raoult's law. The dashed lines show the behavior for an ideal solution.

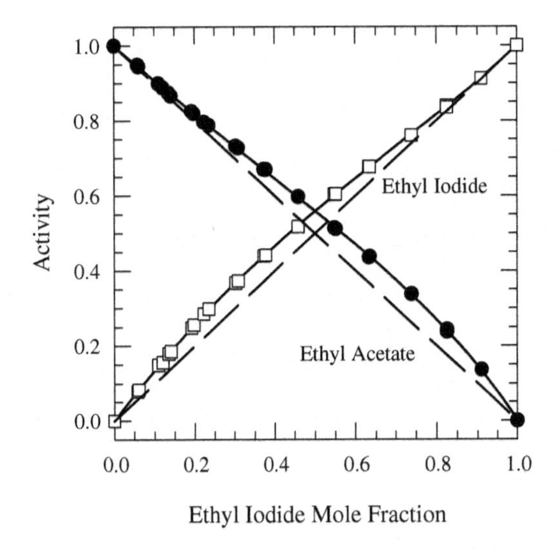

FIGURE 11-15

The activity-composition (*a*-*X*) diagram corresponding to Figure 11-14. The dashed lines show the behavior for an ideal solution.

B. Excess functions

An *excess function* is the difference between the real and ideal mixing properties of a solution at the same temperature, pressure, and composition. Excess functions are used to describe the properties of all types of solutions, for example, liquified gases, hydrocarbons, other organic compounds, aqueous solutions, magmas, metallic alloys and melts, and mineral solid solutions. For example, the *excess volume* V^E is the difference between the integral volume of mixing ΔV^M of the real solution and the ideal volume of mixing ΔV^I for a solution:

$$V^E = \Delta V^M - \Delta V^I \quad (11\text{-}158)$$

As shown earlier, the ideal volume of mixing ΔV^I is zero, so the entire volume change on mixing (ΔV^M) is V^E, the excess volume. The excess volume is written V^E even though it is the difference between the observed and ideal volumes of mixing. Likewise, the *excess enthalpy* is written H^E, the *excess Gibbs energy* as G^E, the *excess entropy* as S^E, and so on, even though there are also differences. This simplifies our equations somewhat.

The excess enthalpy and the excess heat capacity are equal to the respective mixing properties because the enthalpy and heat capacity of ideal mixing are zero as shown earlier:

$$H^E = \Delta H^M - \Delta H^I = \Delta H^M \quad (11\text{-}159)$$

$$C_P^E = \Delta C_P^M - \Delta C_P^I = \Delta C_P^M \quad (11\text{-}160)$$

The excess heat capacity is the temperature coefficient of the excess enthalpy at constant pressure and constant composition (denoted by n, the number of moles):

$$C_P^E = \left(\frac{\partial H^E}{\partial T}\right)_{P,n} \tag{11-161}$$

In general, all the thermodynamic relationships described earlier apply to the excess functions with the additional constraint of differentiation at constant composition.

In contrast to the ideal enthalpy and ideal heat capacity of mixing, the ideal Gibbs energy and ideal entropy of mixing are not zero but are functions of the number of moles and mole fractions of the components in the solution. The excess Gibbs energy and excess entropy are defined as

$$G^E = \Delta G^M - \Delta G^I \tag{11-162}$$

$$S^E = \Delta S^M - \Delta S^I \tag{11-163}$$

We discuss the Gibbs energy, enthalpy, and entropy of mixing and the activity coefficients of nonideal solutions in the next sections of this chapter.

C. Gibbs energy

Equation (11-136) is the general equation for the Gibbs energy of mixing of a solution. Combining this with Eq. (11-135) gives an equivalent equation in terms of the number of moles n_i and thermodynamic activities a_i of the different components in a solution:

$$\Delta G^M = RT\left[n_1 \ln a_1 + n_2 \ln a_2 + n_3 \ln a_3 + \cdots\right] \tag{11-164}$$

We could also derive Eq. (11-164) by combining Eqs. (10-261), (11-71), and (11-132). Other routes also lead to the same result. Per mole of solution Eq. (11-164) becomes

$$\Delta G^M = RT[X_1 \ln a_1 + X_2 \ln a_2 + X_3 \ln a_3 + \cdots] \tag{11-165}$$

The ΔG^M of solution is often represented by polynomial equations such as

$$\Delta G^M = aX_1 + bX_1^2 + cX_1^3 + dX_1^4 + \cdots \tag{11-166}$$

The X_1 denotes the mole fraction of component one and the coefficients in Eq. (11-166) are empirical and are derived from a least-squares fit to the data points. Similar equations are used to represent the excess Gibbs energy of mixing and activity coefficient of each component.

As discussed earlier, the Gibbs energy of mixing is the sum of the ideal and excess parts. We divide Eq. (11-164) into two parts by rewriting the activity of each component i in terms of its mole fraction X_i and activity coefficient γ_i. The two parts correspond to the ideal Gibbs energy of mixing ΔG^I and the excess Gibbs energy G^E:

$$\Delta G^M = G^E + \Delta G^I \tag{11-167}$$

$$\Delta G^M = RT\left[n_1 \ln(\gamma_1 X_1) + n_2 \ln(\gamma_2 X_2) + n_3 \ln(\gamma_3 X_3) + \cdots\right] \tag{11-168}$$

$$\Delta G^M = RT\left[n_1 \ln \gamma_1 + n_2 \ln \gamma_2 + n_3 \ln \gamma_3 + \cdots\right] + RT\left[n_1 \ln X_1 + n_2 \ln X_2 + n_3 \ln X_3 + \cdots\right] \tag{11-169}$$

$$G^E = RT\left[n_1 \ln \gamma_1 + n_2 \ln \gamma_2 + n_3 \ln \gamma_3 + \cdots\right] \tag{11-170}$$

$$\Delta G^I = RT\left[n_1 \ln X_1 + n_2 \ln X_2 + n_3 \ln X_3 + \cdots\right] \tag{11-171}$$

As discussed earlier, the integral Gibbs energy of mixing is equal to the weighted sum of the partial molal Gibbs energies of its individual components. The ΔG^M for $(n_1 + n_2 + n_3)$ moles of a ternary solution of components 1, 2, and 3 is

$$\Delta G^M = n_1 \Delta \overline{G}_1^M + n_2 \Delta \overline{G}_2^M + n_3 \Delta \overline{G}_3^M \tag{11-172}$$

Likewise, the ideal and excess parts of ΔG^M are equal to the weighted sums of the partial molal functions of the individual components:

$$\Delta G^I = n_1 \Delta \overline{G}_1^I + n_2 \Delta \overline{G}_2^I + n_3 \Delta \overline{G}_3^I \tag{11-173}$$

$$G^E = n_1 \overline{G}_1^E + n_2 \overline{G}_2^E + n_3 \overline{G}_3^E \tag{11-174}$$

The preceding equations are written for an arbitary total number of moles of solution. Dividing through by the total number of moles gives equations per one mole of solution, for example,

$$\Delta G^M = RT\left[X_1 \ln \gamma_1 + X_2 \ln \gamma_2 + X_3 \ln \gamma_3 + \cdots\right] + RT\left[X_1 \ln X_1 + X_2 \ln X_2 + X_3 \ln X_3 + \cdots\right] \tag{11-175}$$

$$G^E = RT\left[X_1 \ln \gamma_1 + X_2 \ln \gamma_2 + X_3 \ln \gamma_3 + \cdots\right] \tag{11-176}$$

$$\Delta G^I = RT\left[X_1 \ln X_1 + X_2 \ln X_2 + X_3 \ln X_3 + \cdots\right] \tag{11-177}$$

We immediately see that the activity coefficients are related only to the excess Gibbs energy. The ideal Gibbs energy of mixing and the activity coefficients of the different components are independent of each other. The activity coefficient of each component in a solution is given by its *partial molal excess Gibbs energy*. For example,

$$\overline{G}_1^E = RT \ln \gamma_1 \tag{11-178}$$

for component 1 in a solution. An analogous equation exists for each component in the solution.

In many cases the activity coefficients of the two components in a binary solution can be represented by polynomial equations such as

$$RT \ln \gamma_1 = \overline{G}_1^E = a_1 X_2 + b_1 X_2^2 + c_1 X_2^3 + d_1 X_2^4 + \cdots \tag{11-179}$$

$$RT \ln \gamma_2 = \overline{G}_2^E = a_2 X_1 + b_2 X_1^2 + c_2 X_1^3 + d_2 X_1^4 + \cdots \tag{11-180}$$

The coefficients in the polynomial equations are empirical and are chosen to fit the data. In general, the coefficients are different for each component of the solution (e.g., $a_1 \neq a_2$) and are functions of temperature and pressure. As we discuss later, simplified expressions can be used for certain types of solutions that have either no or very little excess entropy of mixing.

D. Entropy

Differentiation of the partial molal excess Gibbs energy with respect to temperature at constant pressure and composition (denoted by n) gives the *partial molal excess entropy* $\overline{S}_i^E$:

$$\overline{S}_i^E = -\left(\frac{\partial \overline{G}_i^E}{\partial T}\right)_{P,n} \tag{11-181}$$

Equation (11-181) gives the temperature coefficient for the partial molal excess Gibbs energy of a component i in a solution at constant composition and pressure. It is analogous to Eq. (8-26) and a plot of $\overline{G}_i^E$ versus T has a slope equal to the negative value of the excess entropy $(-\overline{S}_i^E)$.

E. Enthalpy

Differentiation of the partial molal excess Gibbs energy with respect to inverse temperature at constant pressure and composition gives the *partial molal excess enthalpy* $\overline{H}_i^E$:

$$\overline{H}_i^E = -\left[\frac{\partial(\overline{G}_i^E/T)}{\partial(1/T)}\right]_{P,n} \tag{11-182}$$

In turn, differentiation of the partial molal excess enthalpy with respect to temperature at constant pressure and composition gives the *partial molal excess heat capacity* $\overline{C}_{P,i}^E$:

$$\overline{C}_{P,i}^E = \left(\frac{\partial \overline{H}_i^E}{\partial T}\right)_{P,n} \tag{11-183}$$

Equation (11-182) is analogous to the Gibbs-Helmholtz equation (10-60). We get another very useful expression by combining Eqs. (11-178) and (11-182) to obtain

$$\left[\frac{\partial \ln \gamma_i}{\partial(1/T)}\right]_{P,n} = \frac{\overline{H}_i^E}{R} \tag{11-184}$$

Equation (11-184) shows that a plot of $\ln \gamma_i$ versus $1/T$ has a slope given by $-\overline{H}_i^E/R$. To first approximation this plot is linear. Nonlinearity arises from differences in the heat capacities of the pure material and the same material dissolved in a solution. Equation (11-184) also predicts three qualitatively different types of behavior for the activity coefficient of a component in a solution depending on the values of its partial molal excess enthalpy:

If $\gamma_i < 1$ then $\overline{H}_i^E$ is negative, that is, negative deviations from Raoult's law (11-185)

$$\text{If } \gamma_i = 1 \text{ then } \overline{H}_i^E \text{ is zero, that is, ideal mixing} \tag{11-186}$$

$$\text{If } \gamma_i > 1 \text{ then } \overline{H}_i^E \text{ is positive, that is, positive deviations from Raoult's law} \tag{11-187}$$

Finally, the analogies between the preceding equations and those in Chapter 10 show that measurements of Gibbs energies of reaction and of chemical equilibria involving solutions give values for the excess partial molal enthalpies and activity coefficients of components in the solution. We return to this point later.

F. Volume

As discussed earlier, the ideal volume of mixing is zero. Thus, the observed volume change upon mixing is entirely due to the excess volume of mixing. By analogy with Eq. (7-80) the excess volume of mixing is the pressure coefficient of the Gibbs energy of mixing:

$$V^E = \Delta V^M = \left(\frac{\partial \Delta G^M}{\partial P}\right)_{T,n} \tag{11-188}$$

Likewise, the *partial molal excess volume* $\overline{V}_i^E$ is the pressure coefficient for the partial molal excess Gibbs energy of component i in the solution

$$\overline{V}_i^E = \left(\frac{\partial \overline{G}_i^E}{\partial P}\right)_{T,n} = RT\left(\frac{\partial \ln \gamma_i}{\partial P}\right)_{T,n} \tag{11-189}$$

In most cases the volume of mixing is measured at or near ambient conditions and is used to calculate the pressure dependence of the Gibbs energy terms.

G. An example of a real solution

Bronze is a Cu-Sn alloy and is one of the most important alloys known to man. Today, bronze alloys are widely used for making many electrical and mechanical components. Historically, bronze alloys date to about 4000 B.C. and gave the Bronze Age its name. We use molten bronze as an example to illustrate experimental measurements and their use in deriving thermodynamic properties of real solutions.

Hager et al. (1970) used Knudsen effusion mass spectrometry to measure the activities of Cu and Sn in molten bronze at 1593 K. Their results are listed in columns 2 and 6 of Table 11-5. The activity coefficients for Cu and Sn are calculated from the measured activity values and the molten alloy compositions using Eq. (11-157). For example, the activity coefficients for Sn and Cu in a molten alloy with a tin mole fraction (X_{Sn}) of 0.4 are

$$\gamma_{Sn} = \frac{a_{Sn}}{X_{Sn}} = \frac{0.333}{0.4} = 0.8325 \tag{11-190}$$

$$\gamma_{Cu} = \frac{a_{Cu}}{X_{Cu}} = \frac{0.362}{0.6} = 0.603 \tag{11-191}$$

Columns 3 and 7 of Table 11-5 give calculated values for the activity coefficients of Sn and Cu.

Table 11-5 Some Thermodynamic Properties of Molten Bronze at 1593 K[a]

X_{Sn}	a_{Sn}	γ_{Sn}	$-\Delta\overline{G}_{Sn}^{M}$	$-\overline{G}_{Sn}^{E}$	a_{Cu}	γ_{Cu}	$-\Delta\overline{G}_{Cu}^{M}$	$-\overline{G}_{Cu}^{E}$	$-\Delta G^{M}$	$-\Delta G^{I}$	$-G^{E}$
0	0	0.055	∞	38,415	1	1	0	0	0	0	0
0.1	0.015	0.15	55,622	25,126	0.853	0.948	2106	707	7458	4309	3149
0.2	0.077	0.385	33,959	12,642	0.644	0.805	5828	2873	11,454	6627	4827
0.3	0.201	0.67	21,250	5304	0.472	0.674	9944	5220	13,336	8091	5245
0.4	0.333	0.8325	14,564	2428	0.362	0.603	13,458	6692	13,900	8913	4987
0.5	0.457	0.914	10,372	1191	0.279	0.558	16,907	7726	13,640	9180	4460
0.6	0.577	0.962	7283	517	0.21	0.525	20,670	8534	12,638	8913	3725
0.7	0.692	0.989	4876	152	0.15	0.5	25,127	9181	10,951	8091	2860
0.8	0.799	0.999	2972	17	0.097	0.485	30,900	9583	8558	6627	1931
0.9	0.9	1	1395	0	0.048	0.48	40,218	9721	5277	4306	971
1	1	1	0	0	0	0.479	∞	9749	0	0	0

The values in the table are rounded to the nearest J mol^{-1} and are computed from the following equations:

$\Delta\overline{G}_i^{M} = RT \ln a_i$

$\overline{G}_i^{E} = RT \ln \gamma_i$

$\Delta G^{M} = \sum X_i \Delta\overline{G}_i^{M}$

$G^{E} = \sum X_i \overline{G}_i^{E}$

$\Delta G^{I} = \Delta G^{M} - G^{E}$

[a]*Based on activity measurements of Hager et al., (1970). Gibbs energies in J mol^{-1}.*

We now use Eq. (11-127) and the activities of Sn and Cu to calculate their partial molal Gibbs energies of mixing. Once again using the molten alloy with $X_{Sn} = 0.4$ as an example,

$$\Delta\overline{G}^M_{Sn} = RT \ln a_{Sn} = (8.314)(1593) \ln 0.333 = -14{,}563 \text{ J mol}^{-1} \quad (11\text{-}192)$$

$$\Delta\overline{G}^M_{Cu} = RT \ln a_{Cu} = (8.314)(1593) \ln 0.362 = -13{,}458 \text{ J mol}^{-1} \quad (11\text{-}193)$$

The partial molal Gibbs energies of mixing for Sn and Cu are displayed in Figure 11-16 and listed in columns 4 and 8 of Table 11-5, which gives the $-\Delta\overline{G}^M_i$ values in J mol^{-1}.

The formation of one mole of molten bronze alloy from pure molten Sn and Cu occurs via

$$X_1\text{Sn (liq)} + (1 - X_1)\text{Cu (liq)} = \text{Cu}_{1-X_1}\text{Sn}_{X_1} \text{ (liq)} \quad (11\text{-}194)$$

The integral Gibbs energy of mixing ΔG^M for molten bronze is the ΔG for this reaction and it is given by a rewritten version of Eq. (11-129):

$$\Delta G^M = X_{Sn}\Delta\overline{G}^M_{Sn} + X_{Cu}\Delta\overline{G}^M_{Cu} \quad (11\text{-}195)$$

Once again considering the molten alloy with a tin mole fraction of 0.4, we calculate

$$\Delta G^M = (0.4)(-14,564) + (0.6)(-13,458) = -13,900 \text{ J mol}^{-1} \quad (11\text{-}196)$$

Column 10 of Table 11-5 lists the integral Gibbs energy of mixing for molten bronze; these values are negative (or zero) and are listed as $-\Delta G^M$ in J mol^{-1}. Figure 11-16 shows ΔG^M, which is not symmetric about the midpoint composition.

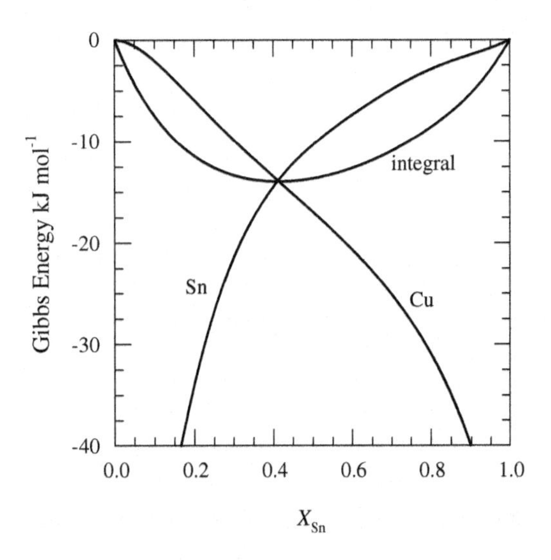

FIGURE 11-16

The integral and partial Gibbs energies of mixing for molten bronze (Cu-Sn) at 1400 K.

As discussed earlier, the ΔG^M is the sum of ideal and excess terms. Columns 11 and 12 of Table 11-5 give the ideal ΔG^I and excess G^E terms for the integral Gibbs energy of mixing. The ideal Gibbs energy of mixing is calculated from a rewritten version of Eq. (11-134):

$$\Delta G^I = RT(X_{\mathrm{Sn}} \ln X_{\mathrm{Sn}} + X_{\mathrm{Cu}} \ln X_{\mathrm{Cu}}) \tag{11-197}$$

At a tin mole fraction of 0.4:

$$\Delta G^I = (8.314)(1593)[0.4 \ln 0.4 + 0.6 \ln 0.6] = -8914 \text{ J mol}^{-1} \tag{11-198}$$

The excess Gibbs energy is computed using Eq. (11-162) and is –4986 J mol^{-1} at $X_{\mathrm{Sn}} = 0.4$.

H. Regular solutions and Margules parameters

In the 1920s, the American physical chemist Joel Hildebrand (1881–1983) observed that simplified versions of Eqs. (11-179) and (11-180) describe the activity coefficients for many binary solutions. In these cases the activity coefficients are given by the equations:

$$RT \ln \gamma_1 = \overline{G}_1^E = bX_2^2 = b(1 - X_1)^2 \tag{11-199}$$

$$RT \ln \gamma_2 = \overline{G}_2^E = bX_1^2 = b(1 - X_2)^2 \tag{11-200}$$

The empirical b coefficient in Eqs. (11-199) and (11-200) is the same for both components of the solution ($b_1 = b_2 = b$) and is independent of temperature. Hildebrand named solutions obeying these equations regular solutions. A *regular solution* has these properties:

$$\Delta H^M = H^E \tag{11-201}$$

$$S^E = 0 \tag{11-202}$$

Equation (11-202) is equivalent to saying that the entropy of mixing ΔS^M for regular solutions is equal to the ideal entropy of mixing ΔS^I:

$$\Delta S^M = \Delta S^I \tag{11-203}$$

Many solutions formed by inorganic compounds, liquid metals, minerals, molten oxides, and organic compounds are regular solutions or can be approximated as regular solutions.

The activity coefficients and excess functions can also be written using *Margules equations*. These are popular in geological sciences and are named after Margules of Duhem-Margules equation fame. For example, the Margules equation for the excess Gibbs energy G^E of a binary regular solution formed by components denoted as 1 and 2 is

$$G^E = W_G X_1 X_2 \tag{11-204}$$

The W_G in Eq. (11-204) is the *Margules parameter* for Gibbs energy. The Margules equations for the other excess functions and activity coefficients of binary regular solutions are

$$H^E = W_H X_1 X_2 = \Delta H^M \tag{11-205}$$

$$S^E = W_S X_1 X_2 \tag{11-206}$$

$$V^E = W_V X_1 X_2 \tag{11-207}$$

$$RT \ln \gamma_1 = W_G X_2^2 = W_G (1 - X_1)^2 \tag{11-208}$$

$$RT \ln \gamma_2 = W_G X_1^2 = W_G (1 - X_2)^2 \tag{11-209}$$

By comparing Eqs. (11-199) and (11-200) with Eqs. (11-208) and (11-209), we see that Hildebrand's temperature-independent constant b is the same as the W_G Margules parameter. In general, the Margules parameters (W_G, W_H, W_S, W_V, etc.) behave like their respective thermodynamic functions (G, H, S, V, etc.) and obey the same formulas derived earlier.

The preceding Margules equations show that the excess functions of binary regular solutions are symmetric about a mole fraction $X_1 = X_2 = 0.5$ where they have maxima. Thus, the excess functions of regular solutions are often tabulated at a mole fraction of 0.5, and in principle the Margules parameters can be calculated simply from measurements made at the midpoint.

According to Hildebrand's definition, the excess entropy of a regular solution is zero and the entropy of mixing is equal to the ideal entropy of mixing. In this case the Margules parameter W_S in Eq. (11-206) would also be equal to zero. The corollary of this is that the Margules parameters W_H and W_G are equal to one another because they obey the Gibbs-Helmholtz equation:

$$W_G = W_H - TW_S \tag{11-210}$$

However, many solutions that closely approach regular solution behavior have nonzero excess entropies and nonzero W_S parameters. This observation led scientists to divide regular solutions into two categories: symmetric (or strictly regular) solutions, which have no excess entropy of mixing ($S^E = 0$), and regular solutions, which have small S^E values. We use this nomenclature throughout the rest of this chapter.

Example 11-8. Titan, the largest satellite of Saturn, has an extensive atmosphere containing N_2, Ar, CH_4, C_2H_6, and other organic compounds. Its surface is partially covered with hydrocarbon lakes that are modeled as CH_4-C_2H_6 liquid solutions at 94 K in this example. Experimental measurements give $G^E = 116$ J mol^{-1} and $H^E = 86.6$ J mol^{-1} at a mole fraction of 0.5 for this system. Use the strictly regular, or symmetric, solution model to compute the activity coefficients for CH_4 and C_2H_6 and plot the results as a function of composition.

The excess entropy $S^E = 0$ for a strictly regular, or symmetric, solution. Actually the excess entropy is nonzero because G^E and H^E are different and the Gibbs-Helmholtz equation shows that $S^E = -0.31$ J mol^{-1} K^{-1} in this case. However, we will ignore it and model the CH_4-C_2H_6 solution as a strictly regular solution. We find the Margules W_G parameter from

$$W_G = \frac{G^E}{X_{CH_4} X_{C_2H_6}} = \frac{116}{0.5^2} = 464 \text{ J mol}^{-1} \tag{11-211}$$

The activity coefficients for CH_4 and C_2H_6 are then calculated using Eqs. (11-208) and (11-209):

$$RT \ln \gamma_{CH_4} = W_G X_{C_2H_6}^2 = W_G (1 - X_{CH_4})^2 \tag{11-212}$$

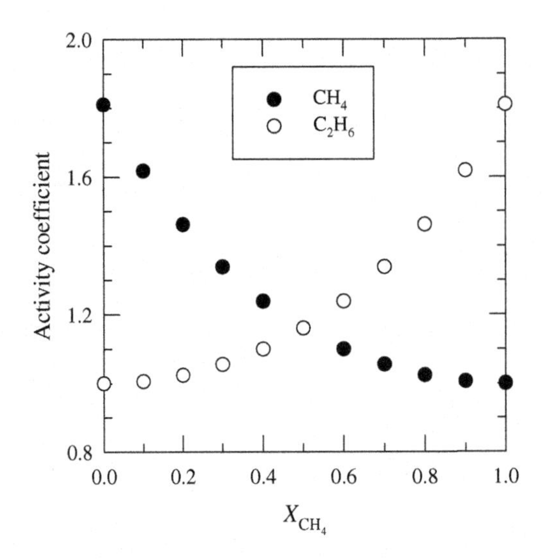

FIGURE 11-17

The activity coefficients for CH_4 and C_2H_6 in a binary solution at 94 K for a regular solution with $W_G = 464$ J mol^{-1}.

$$RT \ln \gamma_{C_2H_6} = W_G X_{CH_4}^2 = W_G (1 - X_{C_2H_6})^2 \tag{11-213}$$

The CH_4 and C_2H_6 activity coefficients are plotted in Figure 11-17, which shows the symmetric form of the activity coefficients. The activity coefficients for a regular solution can also be plotted as ln (γ_1/γ_2) versus composition (either X_1 or X_2). This type of plot gives a straight line passing through zero at $X_1 = X_2 = 0.5$.

I. Subregular or asymmetric solutions

If a solution is strongly asymmetric, for example, like the phlogopite-eastonite mica system in Figure 11-18, it is called an *asymmetric solution* or a *subregular solution*. A variety of equations exist for describing the thermodynamic properties of asymmetric solutions. The Redlich-Kister equation (Redlich and Kister, 1948a, b) for the excess Gibbs energy of mixing of a binary solution is

$$\frac{G^E}{RT} = X_1 X_2 \left[A + B(X_1 - X_2) + C(X_1 - X_2)^2 + D(X_1 - X_2)^3 + \cdots\right] \tag{11-214}$$

The A, B, C, D, etc. are empirically determined coefficients. All coefficients are zero for an ideal solution, which has no excess Gibbs energy of mixing. A strictly regular, or symmetric, solution has $B = C = D = \ldots = 0$ but A is nonzero. An asymmetric or subregular solution has at least two nonzero coefficients (A and B) with the higher-order coefficients equal to zero. For now we consider binary asymmetric solutions where the excess Gibbs energy is

$$\frac{G^E}{RT} = X_1 X_2 [A + B(X_1 - X_2)] \tag{11-215}$$

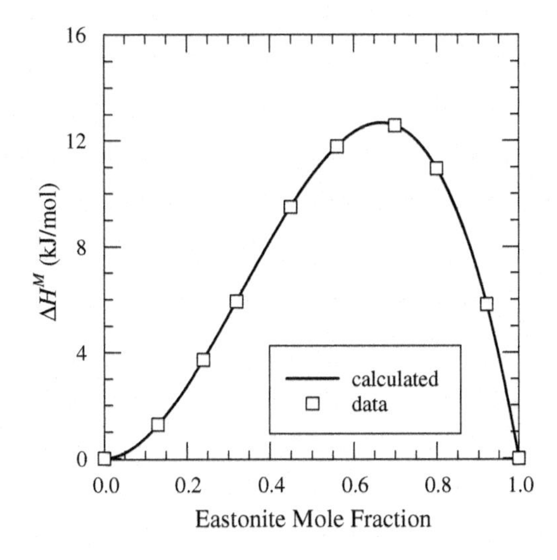

FIGURE 11-18

The integral enthalpy of mixing ΔH^M at 977 K for eastonite-phlogopite micas derived from the data of Circone and Navrotsky (1992) is compared to the ΔH^M calculated using a subregular solution model.

The activity coefficients for this type of solution are given by the equations

$$\ln \gamma_1 = (A + 3B)X_2^2 - 4BX_2^3 \tag{11-216}$$

$$\ln \gamma_2 = (A - 3B)X_1^2 + 4BX_1^3 \tag{11-217}$$

Derivation of Eqs. (11-216) and (11-217) is one of the problems at the end of this chapter. The Redlich-Kister equation can be extended to ternary and higher-order systems, but the equations quickly become more complicated (Redlich and Kister, 1948b).

Margules equations for the excess functions of subregular binary solutions are

$$G^E = X_1X_2(X_1W_{G2} + X_2W_{G1}) \tag{11-218}$$

$$H^E = X_1X_2(X_1W_{H2} + X_2W_{H1}) \tag{11-219}$$

$$S^E = X_1X_2(X_1W_{S2} + X_2W_{S1}) \tag{11-220}$$

$$V^E = X_1X_2(X_1W_{V2} + X_2W_{V1}) \tag{11-221}$$

The Margules parameters can be calculated from experimental measurements of excess enthalpy and other excess properties using equations exemplified by

$$\frac{H^E}{X_1X_2} = W_{H1} + (W_{H2} - W_{H1})X_1 \tag{11-222}$$

Table 11-6 Enthalpy of Mixing (kJ mol^{-1}) for Eastonite-Phlogopite Micas[a]

X_E	0.13	0.24	0.32	0.45	0.56	0.7	0.8	0.92
ΔH^M	1.28	3.73	5.93	9.49	11.78	12.57	10.95	5.8

[a]*The enthalpy of mixing calculated from the data of Circone and Navrotsky (1992) as a function of the eastonite mole fraction.*

The Margules parameter W_{H1} is the intercept and $(W_{H2} - W_{H1})$ is the slope of the straight line derived by linear least-squares fitting of the data points.

Example 11-9. Eastonite $K(Mg_2Al)(Al_2Si_2)O_{10}(OH)_2$ and phlogopite $KMg_3(AlSi_3)O_{10}(OH)_2$ are trioctahedral micas that occur in a variey of igneous and metamorphic rocks. Circone and Navrotsky (1992) measured the enthalpy of mixing for solid solutions of synthetic eastonite-phlogopite micas. This is a subregular solution and H^E is given by Eq. (11-219). Their results are listed in Table 11-6 and graphed in Figure 11-18. Derive the two Margules parameters W_{H1} and W_{H2} from their enthalpy measurements.

As discussed earlier, the enthalpy of mixing is equal to the excess enthalpy of mixing. We used Eq. (11-222) to analyze the data and calculated $W_{H1} = 85.535$ kJ mol^{-1} for phlogopite and $W_{H2} = -0.0349$ kJ mol^{-1} for eastonite. The resulting equation for the excess enthalpy of mixing is

$$H^E = X_1X_2(-0.0349\,X_1 + 85.535\,X_2)\ \text{kJ mol}^{-1} \tag{11-223}$$

Our calculated curve for H^E is also plotted in Figure 11-18 and agrees well with the data points.

The Margules equations for ternary and higher-order solutions are more complicated. The simplest case is a symmetric ternary solution, which is formed by mixing three symmetric binary solutions. The excess volume for a symmetric ternary solution is

$$V^E = X_2X_3W_{V1} + X_1X_3W_{V2} + X_1X_2W_{V3} \tag{11-224}$$

Analogous equations represent the excess enthalpy, entropy, Gibbs energy, and other excess functions. An asymmetric ternary solution does not have bilaterial symmetry about its component binary solutions. In this case the excess volume is

$$V^E = X_1X_2(X_1W_{V12} + X_2W_{V21}) + X_1X_3(X_1W_{V13} + X_3W_{V31}) + X_2X_3(X_2W_{V23} + X_3W_{V32}) \tag{11-225}$$

V. AQUEOUS SOLUTIONS

A. Ionic strength

Lewis and Randall (1921) developed the concept of *ionic strength*, which measures the average electrostatic interactions between all ions in an electrolytic solution. The ionic strength I depends on the molality (m_i) and valence (z_i) of all ions in an electrolytic solution:

$$I = \frac{1}{2}\sum_i m_i z_i^2 \tag{11-226}$$

The ionic strength can also be defined in terms of molarity by using the conversion formula given earlier, and the difference becomes larger as molality and molarity increasingly deviate from one another. Activity coefficients of ions, the solubilities of salts, and rates of ionic reactions are functions of the ionic strength of solutions.

Example 11-10. Calculate the ionic strength of the following aqueous solutions: 1.5 m NaCl, 5.68 m saturated $MgCl_2$, saturated $CuSO_4$ (22 g per 100 g water at 25°C), saturated $LaCl_3$ (95.7 g per 100 g water at 25°C), and saturated $La_2(SO_4)_3 \cdot 9\ H_2O$ (2.7 g per 100 g water at 20°C).

The ionic strength of the 1.5 m NaCl solution is

$$I = \frac{1}{2}\sum_i m_i z_i^2 = \frac{1}{2}\left[2 \times 1.5 \times 1^2\right] = 1.5 = m_i \tag{11-227}$$

The ionic strength of the saturated $MgCl_2$ solution is

$$I = \frac{1}{2}\sum_i m_i z_i^2 = \frac{1}{2}\left[5.68(2)^2 + 2 \times 5.68(1)^2\right] = 17.04 = 3m_i \tag{11-228}$$

The formula weight of $CuSO_4$ is 159.61 g mol^{-1}. The saturated solution is 1.378 molal and has an ionic strength of

$$I = \frac{1}{2}\sum_i m_i z_i^2 = \frac{1}{2}\left[1.378(2)^2 + 1.378(2)^2\right] = 5.512 = 4m_i \tag{11-229}$$

The formula weight of $LaCl_3$ is 245.264 g mol^{-1} and the saturated solution is 3.902 molal with an ionic strength of

$$I = \frac{1}{2}\sum_i m_i z_i^2 = \frac{1}{2}\left[3.902(3)^2 + 3 \times 3.902(1)^2\right] = 23.412 = 6m_i \tag{11-230}$$

The formula weight of $La_2(SO_4)_3 \cdot 9\ H_2O$ is 728.139 g mol^{-1}. The saturated solution is thus a 0.0371 m solution with an ionic strength of

$$I = \frac{1}{2}\sum_i m_i z_i^2 = \frac{1}{2}\left[2 \times 0.0371(3)^2 + 3 \times 0.0371(2)^2\right] = 0.5565 = 15m_i \tag{11-231}$$

In general, the ionic strength of 1:1 electrolytes such as NaCl is equal to the molality, the ionic strength of 2:1 electrolytes such as $MgCl_2$ is $3m$, the ionic strength of 2:2 electrolytes such as $CuSO_4$ is $4m$, the ionic strength of a 3:1 electrolyte is $6m$, and the ionic strength of a 3:2 electrolyte such as $La_2(SO_4)_3$ is $15m$.

B. Solubility and solubility product

The solubility of a material is the amount that dissolves in a solvent to form a saturated solution coexisting in equilibrium with the material. Solubility data are important for geochemical models of many natural processes, including the types of solutes in terrestrial clouds, the alkalinity of Earth's oceans, the origin of salt deposits, mineral dissolution in fresh and salt water, the precipitation of solids from fresh and salt water, and rock-water chemical interactions in the hydrologic cycle. Figure 11-19

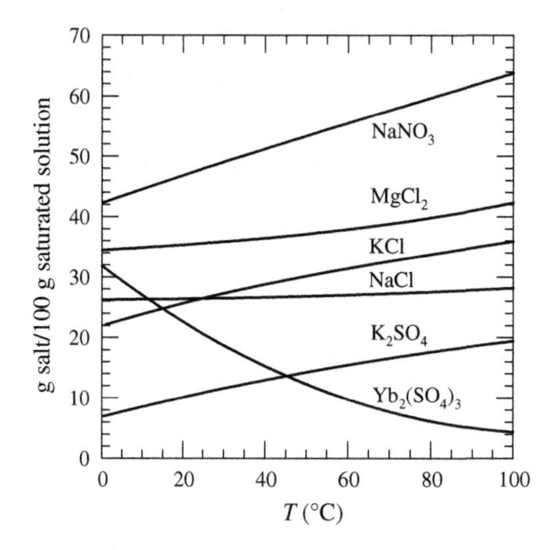

FIGURE 11-19

Solubility curves for several natural salts.

shows solubility curves for several natural salts that occur in seawater, caliche (the sodium nitrate–rich deposits found in the Atacama desert in Chile), and the Stassfurt salt deposits in Germany. Ytterbium sulfate $Yb_2(SO_4)_3$ is not a natural salt, but it exemplifies *retrograde solubility*: decreasing solubility with increasing temperature. Thenardite Na_2SO_4, lithium sulfate Li_2SO_4, lanthanide sulfates, calcium acetate $Ca(CH_3COO)_2$, and some other salts also have retrograde solubility curves.

In principle, each solubility curve is a univariant curve along which a saturated solution with a unique composition (e.g., x grams anhydrous salt per 100 grams water), a pure solid, and water vapor coexist in equilibrium. This can be seen from the Gibbs phase rule because there is one degree of freedom when three phases coexist in a two-component system. A solubility curve thus gives an equilibrium constant denoted K_s at each temperature. A discontinuity in a solubility curve arises from a phase change (e.g., the 305.4 K transition in NH_4NO_3) or a chemical change (e.g., dehydration of a hydrated salt). The slopes of solubility curves depend on the enthalpy of solution via the van't Hoff equation:

$$\left[\frac{\partial \log K_s}{\partial T}\right]_P = \frac{\Delta_{sol}H^o}{(\ln 10)RT^2} \tag{11-232}$$

The dissolution of most salts in water is endothermic ($\Delta_{sol}H^o$ is positive) and thus solubility increases with increasing temperature. Conversely, salts with exothermic enthalpies of solution ($\Delta_{sol}H^o$ is negative) have retrograde solubility curves.

In practice, the solubility curves are often determined at a constant total pressure of one atmosphere (or close to it) in air. Thus, the solutions are saturated with air and are not two-component systems, strictly speaking. Also, the water vapor pressure above the saturated solution is very slightly different from that in the absence of the external pressure due to air because of the Poynting effect (go back to Eq. (7-90) in Chapter 7).

In the case of halite (NaCl), the solubility curve is the reaction

$$\text{NaCl (halite)} = \text{Na}^+ \text{ (satd. aq. soln.)} + \text{Cl}^- \text{ (satd. aq. soln.)} \quad (11\text{-}233)$$

The abbreviation *sat. aq. soln.* denotes saturated aqueous solution. The curve labelled NaCl in Figure 11-19 and Eq. (11-233) indicates that pure halite is in equilibrium with a saturated aqueous NaCl solution containing a specific amount of dissolved NaCl that is dissociated into Na^+ and Cl^- ions at each temperature along the curve. Likewise, the solubility curves for $NaNO_3$ (soda niter), KCl (sylvite), and K_2SO_4 (arcanite) represent the chemical equilibria:

$$\text{NaNO}_3 \text{ (soda niter)} = \text{Na}^+ \text{ (satd. aq. soln.)} + \text{NO}_3^- \text{ (satd. aq. soln.)} \quad (11\text{-}234)$$

$$\text{KCl (sylvite)} = \text{K}^+ \text{ (satd. aq. soln.)} + \text{Cl}^- \text{ (satd. aq. soln.)} \quad (11\text{-}235)$$

$$\text{K}_2\text{SO}_4 \text{ (arcanite)} = 2\text{K}^+ \text{ (satd. aq. soln.)} + \text{SO}_4^{2-} \text{ (satd. aq. soln.)} \quad (11\text{-}236)$$

However, the solubility curve for $MgCl_2$ (chlormagnesite) represents chemical equilibrium between $MgCl_2$ hexahydrate and a saturated solution of $MgCl_2$ dissociated into Mg^{2+} and Cl^- ions:

$$\text{MgCl}_2 \cdot 6\,\text{H}_2\text{O (c)} = \text{Mg}^{2+} \text{ (satd. aq. soln.)} + 2\,\text{Cl}^- \text{ (satd. aq. soln.)} + 6\,\text{H}_2\text{O (satd. aq. soln.)} \quad (11\text{-}237)$$

Many other salts behave like $MgCl_2$ and their solubility curves are chemical equilibria between saturated aqueous solutions and hydrated salts with varying numbers of water molecules.

The intersection of the solubility curves of two hydrated salts or of an anhydrous and a hydrated salt gives the transition temperature between the two solid phases. Anhydrous sodium sulfate (Na_2SO_4, thenardite) and sodium sulfate decahydrate ($Na_2SO_4 \cdot 10\ H_2O$, mirabilite) are a classic example. These minerals occur in deposits from saline lakes, playas, and springs, for example, at Mono Lake, Searles Lake, and Soda Lake in California. Figure 11-20 shows their solubility curves. The solubility of mirabilite in water increases with increasing temperature while that of thenardite decreases with increasing temperature (retrograde solubility). The two solubility curves intersect at 32.38°C (305.53 K), which is the transition temperature at one bar pressure between the two salts. This transition temperature is close, but not identical, to the quadruple-point temperature of 32.6°C in the Na_2SO_4-H_2O binary system where thenardite, mirabilite, the saturated aqueous solution, and water vapor coexist ($F = C - P + 2 = 0$) at a total pressure equal to the water vapor pressure. The solubility curve for mirabilite represents the equilibrium

$$\text{Na}_2\text{SO}_4 \cdot 10\,\text{H}_2\text{O (mirabilite, crystal)} = 2\,\text{Na}^+ \text{(satd. aq. soln.)} + \text{SO}_4^{2-} \text{ (satd. aq. soln.)} + 10\,\text{H}_2\text{O (satd. aq. soln.)} \quad (11\text{-}238)$$

The solubility curve for thenardite represents the equilibrium

$$\text{Na}_2\text{SO}_4 \text{ (thenardite, crystal)} = 2\,\text{Na}^+ \text{(satd. aq. soln.)} + \text{SO}_4^{2-} \text{(satd. aq. soln.)} \quad (11\text{-}239)$$

At the intersection point in Figure 11-19, the two salts have the same solubility and the two solubility reactions are in equilibrium. A summation of the two reactions gives the net reaction

$$\text{Na}_2\text{SO}_4 \cdot 10\,\text{H}_2\text{O (mirabilite, c)} = \text{Na}_2\text{SO}_4 \text{ (thenardite, c)} + 10\,\text{H}_2\text{O (satd. aq. soln.)} \quad (11\text{-}240)$$

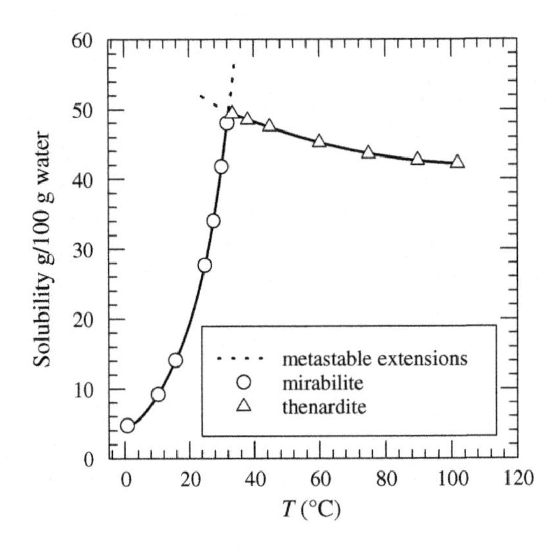

FIGURE 11-20

Solubility curves for thenardite (Na_2SO_4) and mirabilite ($Na_2SO_4 \cdot 10\ H_2O$).

The Gibbs energy change for Eq. (11-240) is

$$\Delta G = 10\overline{G}_w + \overline{G}_{then} - \overline{G}_{mir} \tag{11-241}$$

The partial molal Gibbs energies of liquid water (w), thenardite (then), and mirabilite (mir) are

$$\overline{G}_w = G^o_{m,w} + RT \ln a_w \tag{11-242}$$

$$\overline{G}_{then} = G^o_{m,then} + RT \ln a_{then} \tag{11-243}$$

$$\overline{G}_{mir} = G^o_{m,mir} + RT \ln a_{mir} \tag{11-244}$$

The reaction isotherm and standard Gibbs energy change for Eq. (11-240) are

$$\Delta G = \Delta G^o + RT \ln \left(\frac{a_{then} a_w^{10}}{a_{mir}} \right)_{eq} \tag{11-245}$$

$$\Delta G^o = 10G^o_{m,w} + G^o_{m,then} - G^o_{m,mir} \tag{11-246}$$

The activities of thenardite and mirabilite are unity because both salts are pure phases. However, the activity of water in the saturated solution is less than unity and according to Eq. (10-247) is given by the ratio of the water vapor pressure over the saturated solution (30.875 ± 0.055 mmHg) and over pure water (36.477 mmHg) at 32.38°C:

$$a_w = \frac{f_w}{f_w^*} \cong \frac{p_w}{p_w^*} = \frac{30.875}{36.477} = 0.84642 \tag{11-247}$$

At the transition point $\Delta G = 0$ and we can solve for ΔG^o by substituting numerical values into the reaction isotherm Eq. (11-245):

$$\Delta G^o = -RT \ln a_w^{10} = -8.314 \cdot 305.53 \cdot \ln(0.18874) = 4235 \pm 45 \text{ J mol}^{-1} \tag{11-248}$$

We can use the ΔG^o of reaction and $\Delta_f G^o$ values for liquid water and thenardite to calculate the standard Gibbs energy of formation of mirabilite at this temperature. In addition, the metastable extensions of the two solubility curves can be used to calculate the Gibbs energy difference between the stable and metastable salt in the vicinity of their transition point. We used similar arguments in Chapter 7 when we used Brønsted's solubility data to calculate the transition temperature between orthorhombic and monoclinic sulfur (see Example 7-2). Furthermore, we can use the equilibrium water vapor pressures over coexisting thenardite + mirabilite to compute the ΔG^o of reaction and $\Delta_f G^o$ of mirabilite at lower temperatures. In principle, the ΔG^o values obtained from the solubility data and water vapor pressure measurements should agree.

The *solubility product* K_s is the equilibrium constant for the dissolution of a solid in water or another solvent. For example, $BaSO_4$ (barite) dissolution in water is represented by

$$BaSO_4 \text{ (barite)} = Ba^{2+} \text{ (satd. aq. soln.)} + SO_4^{2-} \text{ (satd. aq. soln.)} \tag{11-249}$$

The equilibrium constant for reaction (11-249) is

$$K_s = \left[\frac{a_{Ba^{2+}} a_{SO_4^{2-}}}{a_{BaSO_4}}\right]_{eq} = a_{Ba^{2+}} a_{SO_4^{2-}} = (m_{Ba^{2+}} m_{SO_4^{2-}})(\gamma_{Ba^{2+}} \gamma_{SO_4^{2-}}) \cong m_{Ba^{2+}} m_{SO_4^{2-}} \tag{11-250}$$

The a_i terms in Eq. (11-250) are the activities of pure barite and the aqueous ions. The activity of pure barite is unity and those of the aqueous ions are proportional to their concentration in the solution. The proportionality constants are the activity coefficients of the ions. We describe the measurement and calculation of ionic activity coefficients later in this chapter. For now we simply state that at sufficiently low concentrations (such as those for barite dissolved in water at or near room temperature), the activity coefficients can be taken as unity to first approximation. Then the solubility product K_s is the product of the ionic (molal) concentrations in solution. This product is dimensionless because the standard state for aqueous solutions is chosen as the hypothetical one-molal solution obeying the laws of the perfect (i.e., infinitely) dilute solution (the activity coefficient $\gamma_i = 1$ for each solute, such as the individual ions).

Electrical conductivity measurements show that a saturated solution of barite contains 9.56 micromoles $BaSO_4$ per liter of saturated solution. From the stoichiometry of reaction (11-249) we see that each mole of dissolved $BaSO_4$ gives one mole Ba^{2+} ions and one mole sulfate ions. Rewriting Eq. (11-250) for the solubility product we obtain

$$K_s \cong m_{Ba^{2=}} m_{SO_4^{2-}} = \left(9.56 \times 10^{-6}\right)^2 = 9.14 \times 10^{-11} \tag{11-251}$$

Solubility products are often reported as negative logarithmic values:

$$pK_s = -\log_{10} K_s \tag{11-252}$$

Our result from Eq. (11-251) corresponds to $pK_s = 10.04$ for $BaSO_4$. The accepted value for the solubility product of $BaSO_4$ at 25°C is $pK_{s0} = 9.96 \pm 0.03$ (Butler, 1964), where the subscript zero denotes extrapolation to an infinitely dilute solution. Our calculated value agrees within 20%.

The van't Hoff equation (10-65) also applies to the temperature dependence of the solubility product K_s, and can be used to estimate the standard enthalpy of solution:

$$\left[\frac{\partial \log K_s}{\partial T}\right]_P = \frac{\Delta_{sol}H^o}{(\ln 10)RT^2} \tag{11-232}$$

$$\left[\frac{\partial \log K_s}{\partial(1/T)}\right]_P = \frac{-\Delta_{sol}H^o}{(\ln 10)R} \tag{11-253}$$

The $\Delta_{sol}H^o$ values determined from solubility data are for solutions with finite concentrations of dissolved solutes. In contrast, the enthalpy of solution values calculated from calorimetric data, such as those in Appendix 1, are for infinitely dilute solutions. The agreement between the two $\Delta_{sol}H^o$ values is better for relatively insoluble minerals such as the silver halides and is worse for halite and other soluble salts.

Example 11-11. Owen (1938) measured the solubility product of AgCl (cerargyrite) from 5°C to 45°C at one atmosphere total pressure. Use his data to calculate the $\Delta_{sol}H^o$ for AgCl in water:

$$\text{AgCl (cerargyrite)} = Ag^+ \text{ (aq)} + Cl^- \text{ (aq)} \tag{11-254}$$

Table 11-7 and Figure 11-21 give Owen's values. As a first approximation the log K_s values fall on a straight line and are fit with a constant enthalpy of solution ($\Delta C_P = 0$). However, better agreement is obtained with a three-term equation that takes the temperature dependence of the enthalpy of solution into account:

$$\log K_s = 9.21289 - \frac{4540.05}{T} - 0.01253T \tag{11-255}$$

The derivative of this equation is

$$\left[\frac{\partial \log K_s}{\partial T}\right]_P = \frac{4540.05}{T^2} - 0.01253 \tag{11-256}$$

The enthalpy of solution is calculated from the van't Hoff equation (11-232):

$$\Delta_{sol}H^o = RT^2(\ln 10)\left[\frac{\partial \log K_s}{\partial T}\right]_P = 86913 - 0.23987T^2 \tag{11-257}$$

Table 11-7 AgCl Solubility Product K_{sp} (Owen, 1938)

T (°C)	5	15	25	35	45
$-\log K_{sp}$	10.595	10.1536	9.7508	9.3818	9.044

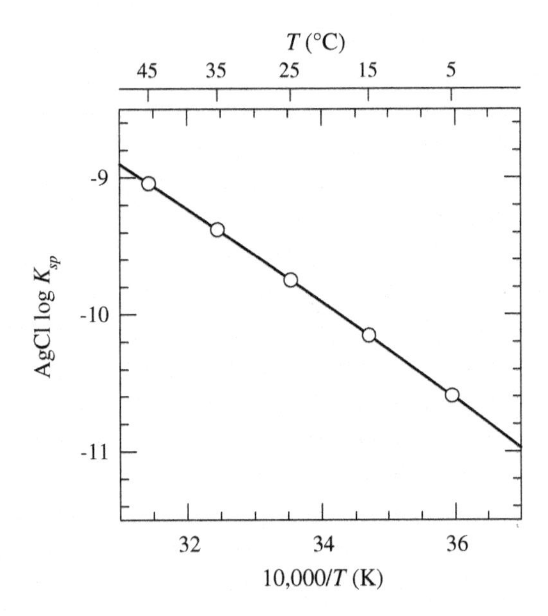

FIGURE 11-21

Solubility product of AgCl in water as a function of temperature.

At 25°C (298.15 K), this equation gives $\Delta_{sol}H^o = 65{,}590$ J mol^{-1}, which is only 134 J smaller than the calorimetric value of 65,724 J mol^{-1} (Wagman and Kilday, 1973). The small difference is probably within the combined uncertainty of the two values.

The calculation of the solubility product K_s for $Mg(OH)_2$ (brucite) illustrates the general case in which different numbers of positively charged and negatively charged ions are produced when a compound or mineral dissolves in water. Dissolution of brucite in water occurs via

$$Mg(OH)_2 \text{ (brucite)} = Mg^{2+} \text{ (satd. aq. soln.)} + 2\,OH^- \text{(satd. aq. soln.)} \tag{11-258}$$

At 25°C, a saturated solution of $Mg(OH)_2$ contains 1.11×10^{-4} moles of $Mg(OH)_2$ per liter of solution. The stoichiometry of reaction (11-258) shows that each mole of dissolved $Mg(OH)_2$ gives one mole of Mg^{2+} cations and two moles of OH^- anions. The solubility product is

$$K_s = \left[\frac{a_{Mg^{2+}} a_{OH^-}^2}{a_{Mg(OH)_2}}\right]_{eq} = a_{Mg^{2+}} a_{OH^-}^2 \cong m_{Mg^{2+}} m_{OH^-}^2 \tag{11-259}$$

The activity of pure brucite is unity and it drops out of the equilibrium constant expression. As a first approximation, the activities of the Mg^{2+} and OH^- aqueous ions are taken equal to their molal concentrations. We also neglect the small difference between the molar and molal concentration of $Mg(OH)_2$ in its saturated solution. Substituting the molar solubility S of $Mg(OH)_2$ into Eq. (11-259) we find

$$K_s \cong S(2S)^2 = \left(1.11 \times 10^{-4}\right)\left(2.22 \times 10^{-4}\right)^2 = 5.47 \times 10^{-12} \tag{11-260}$$

This corresponds to $pK_s = 11.26$, which agrees with the value of 11.15 ± 0.2 in Butler (1964).

In general, the dissolution of a compound C_xA_y (C = cation, A = anion) is represented by

$$C_xA_y \text{ (solid)} = xC^+(\text{aq}) + yA^-(\text{aq}) \quad (11\text{-}261)$$

The solubility product K_s is given by

$$K_s = a^x_{C^+}a^y_{A^-} \cong m^x_{C^+}m^y_{A^-} = (xS)^x(yS)^y \quad (11\text{-}262)$$

This discussion considers solubility of an ionic compound in pure water. However, under otherwise constant conditions, the solubility of the same compound is generally lower in an aqueous solution containing one of the ions in the compound than in pure water. This is the *common ion effect*, and it is a consequence of the law of mass action introduced in Chapter 10.

Example 11-12. Nikitin and Tolmatscheff (1933) measured the solubility of radium sulfate $RaSO_4$ as 2.1×10^{-4} g per 100 mL water at 20°C. Calculate the solubility of $RaSO_4$ in Na_2SO_4 aqueous solution (total SO_4^{2-} concentration = 5×10^{-4} molal) at the same temperature. Assume that the ionic activity coefficients are unity in this approximate calculation.

We first write down the solubility reaction and use it to calculate K_s from the solubility data:

$$RaSO_4 \text{ (s)} = Ra^{2+} \text{ (aq)} + SO_4^{2-} \text{ (aq)} \quad (11\text{-}263)$$

The molecular weight of $RaSO_4$ is 322.09 g mol^{-1} and the density of water at 20°C is 0.998206 g mL^{-1}. The solubility corresponds to a 6.53×10^{-6} molal solution of $RaSO_4$ in water. The K_s is

$$K_s \cong m_{Ra^{2+}}m_{SO_4^{2-}} = \left(6.52 \times 10^{-6}\right)^2 = 4.25 \times 10^{-11} = 10^{-10.37} \quad (11\text{-}264)$$

We now calculate the solubility of $RaSO_4$ in aqueous sodium sulfate. In this case the sulfate ion concentration is fixed at 5×10^{-4} molal because Na_2SO_4 is much more soluble in water than $RaSO_4$. We substitute this concentration into Eq. (11-264) and rearrange it to solve for the molal concentration of Ra^{2+}, which gives the solubility of $RaSO_4$:

$$m_{Ra^{2+}} = \frac{K_s}{m_{SO_4^{2-}}} = \frac{4.25 \times 10^{-11}}{5 \times 10^{-4}} = 8.5 \times 10^{-8} \quad (11\text{-}265)$$

A molality of 8.5×10^{-8} is equivalent to about 2.7×10^{-6} g $RaSO_4$ per 100 mL water.

C. Ionization equilibrium of water and pH

Liquid water undergoes a small amount of self-ionization via the reaction

$$H_2O \text{ (liquid)} = H^+ \text{ (aq)} + OH^- \text{ (aq)} \quad (11\text{-}266)$$

The self-ionization (or self-dissociation) constant for water is denoted K_w and is given by

$$K_w = \frac{a_{H^+}a_{OH^-}}{a_{H_2O}} = a_{H^+}a_{OH^-} \quad (11\text{-}267)$$

The activity of pure liquid water is unity and it drops out of the equilibrium constant expression. At 25°C and a pressure equal to the saturated vapor pressure over pure liquid water (23.75 mmHg), $K_w = 1.0116 \times 10^{-14}$ ($pK_w = -\log K_w = 13.995$). If activities of the H^+ and OH^- ions are equal,

$$a_{H^+} = a_{OH^-} = (K_w)^{1/2} = 1.0058 \times 10^{-7} \quad (11\text{-}268)$$

Table 11-8 Some Physical Properties of Liquid Water

T (°C)	ρ g cm^{-3}	c_p J g^{-1}	P_{vap} (mm)	$\Delta_{vap}H$ J mol^{-1}	pK_w
0	0.999843	4.2174	4.58	45,054	14.96
10	0.999702	4.1919	9.2	44,629	14.53
15	0.999102	4.1867	12.78	44,416	14.36
20	0.998206	4.1816	17.53	44,204	14.17
25	0.997048	4.1799	23.75	43,990	13.997
30	0.995651	4.1782	31.82	43,778	13.83
40	0.99222	4.1783	55.34	43,350	13.53
50	0.98803	4.1804	92.56	42,918	13.26
60	0.9832	4.1841	149.47	42,482	13.02
70	0.97778	4.1893	233.81	42,037	12.8
80	0.97182	4.1961	355.31	41,585	12.61
90	0.96535	4.2048	525.92	41,127	12.43
100	0.9584	4.2156	760	40,657	12.27

Sources: Density and enthalpy of vaporization, 83rd ed. Handbook of Chemistry and Physics; specific heat, vapor pressure, Millero (2001) Physical Chemistry of Natural Waters; *pK_w, Butler (1998)* Ionic Equilibrium.

This is the neutral pH point at 298.15 K, where pH is defined as

$$pH = -\log a_{H^+} \tag{11-269}$$

As shown in Table 11-8, the self-dissociation constant of water varies with temperature. Thus, the neutral pH point also varies with temperature.

D. Ionization equilibria of electrolytes

As mentioned at the start of this chapter, weak electrolytes are only partially ionized and their fractional ionization varies inversely with the concentration of the electrolyte. Monoprotic weak electrolytes are acids or bases that release or accept only one proton in an ionization reaction. Acetic acid CH_3COOH (the acid in vinegar) is a typical weak electrolyte and ionizes via

$$CH_3COOH\ (aq) = CH_3COO^-\ (aq) + H^+\ (aq) \tag{11-270}$$

The equilibrium constant K_a for this reaction is

$$K_a = \frac{a_{CH_3COO^-} a_{H^+}}{a_{CH_3COOH}} \tag{11-271}$$

At 0°C, $K_a = 1.657 \times 10^{-5}$ ($pK_a = -\log K_a = 4.78$). In sufficiently dilute solution the activities become equal to the molalities and K_a becomes

$$K_a = \frac{m_{CH_3COO^-} m_{H^+}}{m_{CH_3COOH}} \tag{11-272}$$

By analogy with our treatment of thermal dissociation of molecular I_2 in Chapter 10, we denote the fractional ionization by α, which varies from 0 for no ionization to 1 for complete ionization. The value of α also equals the relative concentration of the acetate ion and the hydrogen ion. We substitute α into Eq. (11-272) and rewrite it to get

$$K_a = \frac{(\alpha m_{HAc})^2}{(1-\alpha)m_{HAc}} \tag{11-273}$$

The *HAc* in Eq. (11-273) stands for unionized acetic acid. Equation (11-273) is very useful because it relates the fractional ionization to the equilibrium constant K_a and the (known) molality of acetic acid. Freezing-point depression measurments on aqueous acetic acid and other weak electrolytes give the α values in Eq. (11-273). This can be seen by rewriting Eq. (11-119) in terms of the solute molality m and the cryoscopic constant c for water:

$$\frac{\Delta T_m}{m(1+\alpha)} = c = 1.858 \text{ K kg mol}^{-1} \tag{11-274}$$

The ionization equilibria of bases are handled analogously. Aqueous ammonia is a weak base and it ionizes via the reaction

$$NH_3 \text{ (aq)} + H_2O \text{ (liq)} = NH_4^+ \text{ (aq)} + OH^- \text{ (aq)} \tag{11-275}$$

The ionization constant K_b for this reaction is

$$K_b = \frac{a_{NH_4^+}a_{OH^-}}{a_{NH_3(aq)}a_{H_2O}} = \frac{a_{NH_4^+}a_{OH^-}}{a_{NH_3(aq)}} \cong \frac{m_{NH_4^+}m_{OH^-}}{m_{NH_3(aq)}} \tag{11-276}$$

At 25°C the ionization constant $K_b = 1.74 \times 10^{-5}$ ($pK_b = -\log K_b = 4.76$) for ammonia.

Aqueous ammonia NH_3 (a base) and the aqueous ammonium ion NH_4^+ (an acid) are an example of a *conjugate acid-base pair.* Likewise, acetic acid and the acetate ion (a base) are a conjugate acid-base pair. According to the Brønsted-Lowry definition of acids and bases, an acid is a proton donor and a base is a proton acceptor. The ionization of a base produces its conjugate acid and the ionization of an acid produces its conjugate base. The ionization constants for conjugate acids K_a and bases K_b are related via the equation

$$pK_a + pK_b = pK_w \tag{11-277}$$

The addition of a common ion affects the ionization equilibrium of a weak acid or base and moves the pH of the solution toward neutrality (less acidic or less basic as the case may be). For example, the addition of sodium acetate CH_3COONa to a solution of acetic acid CH_3COOH increases the acetate ion concentration. This drives the ionization equilibrium back toward the unionized acid and gives a less acidic solution with a higher pH value.

In contrast to acetic acid and other weak acids, strong acids such as hydrochloric acid HCl, nitric acid HNO_3, and perchloric acid $HClO_4$ are completely ionized in solution, and their ionization constants are larger than unity. Thus, their pK_a values are negative. For example, at 20°C the ionization constant for nitric acid is $K_a \sim 25.1$ and $pK_a = -\log K_a \sim -1.4$. Likewise, in contrast to ammonia and other weak bases, strong bases such as sodium hydroxide NaOH have ionization constants greater than unity and negative pK_b values. Tables 11-9 and 11-10 list the ionization constants (pK_a and pK_b values) for some aqueous acids and bases at 25°C.

Polyprotic electrolytes are acids or bases that release or accept more than one proton in ionization equilibria. Carbonic acid H_2CO_3 and boric acid H_3BO_3 are important polyprotic acids that control the alkalinity of seawater. Carbonic acid also plays an important role in the chemistry of groundwater and fresh water. Sulfuric acid H_2SO_4, which occurs in mine wastewaters, phosphoric acid H_3PO_4, which is used to acidify sodas, and citric acid $CH_3C(OH)(COOH)_3$, which occurs in citrus fruit, are examples of other polyprotic acids.

Table 11-9 Ionization Constants of Some Aqueous Acids at 25°C and Zero Ionic Strength[a]

Acid	Formula	pK_a	Notes
acetic acid	CH_3COOH	4.756	in soils, excreted by Acetobacter bacteria
ammonium	NH_4^+	9.245	conjugate acid for NH_3
arsenic acid	H_3AsO_4	2.31	pK_1 for arsenic acid, toxic pollutant in groundwater
	$H_2AsO_4^-$	7.05	pK_2 for arsenic acid, toxic pollutant in groundwater
	$HAsO_4^{2-}$	11.9	pK_3 for arsenic acid, toxic pollutant in groundwater
bisulfate ion	HSO_4^-	1.987	from H_2SO_4, in acid mine waters
boric acid	H_3BO_3	9.237	helps regulate alkalinity of oceans
carbonic acid	H_2CO_3	3.76[b]	pK_1 for carbonic acid, regulates alkalinity of oceans
	HCO_3^-	10.329	pK_2 for carbonic acid, bicarbonate
formic acid	HCOOH	3.745	excreted by ants
hydrofluoric acid	HF	3.17	emitted in volcanic gases
hypobromous acid	HOBr	8.63	in cloud droplets, strong oxidant
hypochlorous acid	HOCl	7.53	in cloud droplets, strong oxidant
lactic acid	$CH_3CH(OH)COOH$	3.86	in sour milk
nitrous acid	HNO_2	3.25	in cloud droplets
oxalic acid	HOOC-COOH	1.27	pK_1 for oxalic acid
	$HOOC\text{-}COO^-$	4.266	pK_2 for oxalic acid
phosphoric acid	H_3PO_4	2.148	pK_1 for phosphoric acid
	$H_2PO_4^-$	7.198	pK_2 for phosphoric acid
	HPO_4^{2-}	12.35	pK_3 for phosphoric acid
sulfurous acid	H_2SO_3	1.857	pK_1 for sulfurous acid, in cloud droplets
	HSO_3^-	7.172	pK_2 for sulfurous acid, in cloud droplets

[a] *Data from Butler (1998), Goldberg et al. (2002), Lide (2002).*
[b] *The pK_1 for ionization of aqueous CO_2 is 6.35; see text.*

Table 11-10 Ionization Constants of Some Aqueous Bases at 25°C and Zero Ionic Strength[a]

Name	Formula	pK_b	Notes
ammonia	NH_3	4.752	in aqueous solutions inside icy satellites
arsenic trioxide	As_2O_3	3.96	toxic pollutant in groundwater
ethyl amine	$CH_3CH_2NH_2$	3.36	in comet Wild 2
hydroxyl amine	NH_2OH	7.97[b]	in cloud droplets
methyl amine	CH_3NH_2	3.352	in comet Wild 2
portlandite	$Ca(OH)_2$	2.43	in cement, mortar, plaster

[a]Data from Butler (1998), Goldberg et al. (2002), Lide (2002).
[b]At 20°C.

E. Carbonic acid–calcium carbonate system

We consider ionization and solubility equilibria of carbon dioxide, carbonic acid, and calcium carbonate in some detail because of their importance for geochemistry of natural waters. The dissolution of CO_2 gas in water occurs via the reaction

$$CO_2\ (g) = CO_2\ (aq) \tag{11-278}$$

$$K_H = \frac{a_{CO_2(aq)}}{f_{CO_2(g)}} \cong \frac{m_{CO_2(aq)}}{P_{CO_2(g)}} \tag{11-279}$$

$$\log K_H = 108.3865 + 1.985076 \times 10^{-2} T - \frac{6919.53}{T} - 40.45154 \log T + \frac{669{,}365}{T^2} \tag{11-280}$$

At 25°C, $K_H = 10^{-1.47}$ (relative to one molal and one bar standard states for aqueous and gaseous CO_2). Equation (11-280) from Plummer and Busenberg (1982) is valid from 0°C to 100°C at one bar pressure. Table 11-11 lists values of pK_H (= − log K_H) calculated from this equation (and the other equilibrium constants defined below) as a function of temperature from 5°C to 45°C.

Table 11-11 Equilibrium Constants for H_2CO_3-$CaCO_3$ System[a]

T (°C)	T (K)	pK_H	pK_{CO2}	pK_1	pK_2	pK_s
5	278.15	1.19	2.71	3.81	10.56	8.39
15	288.15	1.34	2.67	3.75	10.43	8.43
25	298.15	1.47	2.59	3.76	10.33	8.48
35	308.15	1.58	2.53	3.78	10.25	8.54
38	311.15	1.61	2.5	3.8	10.23	8.56
45	318.15	1.67	2.49	3.8	10.2	8.62

[a]Harned and Owen (1958), Harned and Scholes (1941), Plummer and Busenberg (1982).

A small fraction of the CO_2 dissolved in the water is hydrated to form carbonic acid:

$$CO_2\ (aq) + H_2O\ (liq) = H_2CO_3\ (aq) \tag{11-281}$$

The equilibrium constant for hydration of aqueous CO_2 is

$$K_{CO_2} = \frac{a_{H_2CO_3}}{a_{CO_2(aq)}a_{H_2O}} = \frac{a_{H_2CO_3}}{a_{CO_2(aq)}} \cong \frac{m_{H_2CO_3}}{m_{CO_2}} \tag{11-282}$$

The activity of water is unity and it drops out of the equilibrium constant expression but the $\Delta_f G^o$ of liquid water is needed to compute the standard Gibbs energy change for this reaction. At 25°C, $K_{CO_2} = 10^{-2.59}$ and the molar CO_2/H_2CO_3 ratio is about 389. Furthermore, hydration of CO_2 proceeds slowly. Thus, most of the "carbonic acid" in solution is actually aqueous CO_2.

Reactions (11-278) and (11-281) are often combined to give the net reaction

$$CO_2\ (g) + H_2O\ (liq) = H_2CO_3\ (aq) \tag{11-283}$$

The equilibrium constant for reaction (11-283) is

$$K'_{CO_2} = \frac{a_{H_2CO_3}}{f_{CO_2}a_{H_2O}} = \frac{a_{H_2CO_3}}{f_{CO_2}} \cong \frac{m_{H_2CO_3}}{P_{CO_2}} \tag{11-284}$$

The equilibrium constant K'_{CO_2} is the product of K_H and K_{CO_2}:

$$K'_{CO_2} = K_H K_{CO_2} = \frac{a_{CO_2(aq)}}{f_{CO_2(g)}} \cdot \frac{a_{H_2CO_3}}{a_{CO_2(aq)}} = \frac{a_{H_2CO_3}}{f_{CO_2}} \cong \frac{m_{H_2CO_3}}{P_{CO_2}} \tag{11-285}$$

Thus, at 25°C, $K'_{CO_2} = 10^{-4.06}$. Neutral carbonic acid ionizes via the reaction

$$H_2CO_3\ (aq) = HCO_3^-\ (aq) + H^+\ (aq) \tag{11-286}$$

The equilibrium constant for this reaction, which is the first ionization constant (K_1) for carbonic acid, and its value at 298.15 K are

$$K_{H_2CO_3} = K_1 = \frac{a_{HCO_3^-(aq)}a_{H^+(aq)}}{a_{H_2CO_3(aq)}} \cong \frac{m_{HCO_3^-}m_{H^+}}{m_{H_2CO_3}} = 10^{-3.76} \tag{11-287}$$

The sum of reactions (11-281) and (11-286) gives the ionization of aqueous CO_2 in water:

$$CO_2\ (aq) + H_2O\ (liq) = HCO_3^-\ (aq) + H^+\ (aq) \tag{11-288}$$

$$K'_1 = \frac{a_{HCO_3^-(aq)}a_{H^+(aq)}}{a_{CO_2(aq)}} \cong \frac{m_{HCO_3^-}m_{H^+}}{m_{CO_2}} = 10^{-6.35} \tag{11-289}$$

The equilibrium constant K'_1 is often quoted as the first ionization constant of carbonic acid, but it is the ionization constant for aqueous CO_2 dissolved in water.

Ionization of the aqueous bicarbonate ion occurs via

$$HCO_3^-\ (aq) = CO_3^{2-}\ (aq) + H^+\ (aq) \tag{11-290}$$

The ionization constant for the bicarbonate ion, which is the second ionization constant for carbonic acid, and its value at 298.15 K are

$$K_{HCO_3^-} = K_2 = \frac{a_{CO_3^{2-}} a_{H^+}}{a_{HCO_3^-}} \cong \frac{m_{CO_3^{2-}} m_{H^+}}{m_{HCO_3^-}} = 10^{-10.33} \quad (11\text{-}291)$$

The dissolution of $CaCO_3$ (calcite), its solubility product, and its value at 298.15 K are

$$CaCO_3 \text{ (calcite)} = Ca^{2+} \text{ (aq)} + CO_3^{2-} \text{ (aq)} \quad (11\text{-}292)$$

$$K_s = a_{Ca^{2+}} a_{CO_3^{2-}} \cong m_{Ca^{2+}} m_{CO_3^{2-}} = 10^{-8.48} \quad (11\text{-}293)$$

Plummer and Busenberg (1982) give the following equation for the temperature dependence of the solubility product for calcite between 0°C and 90°C:

$$\log K_s = -171.9065 - 7.7993 \times 10^{-2} T + \frac{2839.319}{T} + 71.595 \log T \quad (11\text{-}294)$$

These reactions and equilibrium constants allow us to consider a number of interesting problems applicable to the geochemistry of natural waters. One example is given here, and others are given in problems at the end of this chapter.

Example 11-13. Calculate the pH of rainwater equilibrated with atmospheric CO_2 at the global mean temperature and pressure of 15°C and one atmosphere (~1.013 bar). As of August 2008, Earth's atmosphere contained about 385 parts per million by volume (385 ppmv) CO_2. We assume that dissolved carbonic acid is the only source of acidity in the rainwater and that activities and fugacities can be approximated by molalities and partial pressures.

We first calculate the CO_2 partial pressure in Earth's atmosphere. This is

$$P_{CO_2} = X_{CO_2} P_T = (385 \times 10^{-6})(1.013 \text{ bar}) = 3.9 \times 10^{-4} \text{ bar} \quad (11\text{-}295)$$

Equilibration of rainwater with this CO_2 partial pressure gives a carbonic acid molality of

$$m_{H_2CO_3} \cong (K'_{CO_2})(P_{CO_2}) = (10^{-4.01})(3.9 \times 10^{-4}) = 10^{-7.42} = 3.81 \times 10^{-8} \quad (11\text{-}296)$$

The carbonic acid molality in rainwater remains constant because any H_2CO_3 that ionizes is replenished by CO_2 from air. Ionization of carbonic acid produces hydrogen ions and bicarbonate ions on a one-to-one basis. Hence, the hydrogen ion molality is given by

$$m_{H^+} = [(K_1)(m_{H_2CO_3})]^{1/2} = (10^{-3.75} \cdot 10^{-7.42})^{1/2} = 10^{-5.59} \quad (11\text{-}297)$$

The pH of the rainwater is 5.59 due only to atmospheric carbon dioxide. You should convince yourself that you get the same answer by calculating the molality of aqueous CO_2 and its subsequent ionization.

VI. ACTIVITY COEFFICIENTS OF AQUEOUS ELECTROLYTES

We cannot measure activities and activity coefficients for individual ions in solution because aqueous solutions are electrically neutral. It is impossible to add an excess of either positive or negative ions. Even if it were possible, the ionic properties in a charged solution would be different than those in a neutral solution. However, it is possible to define mean activities and mean activity coefficients for symmetrical electrolytes such as NaCl (halite), KCl (sylvite), $NaNO_3$ (soda niter), and unsymmetrical electrolytes such as $MgCl_2$ (chlormagnesite) and K_2SO_4 (arcanite). We illustrate the basic concepts, starting with symmetrical electrolytes.

The following equation represents ionization of KCl (sylvite) dissolved in water:

$$\text{KCl (aq)} = \text{K}^+\text{(aq)} + \text{Cl}^-\text{(aq)} \tag{11-298}$$

We define the mean ionic molality of KCl dissolved in water as the geometric mean molality

$$m_\pm = (m_+ m_-)^{1/2} \tag{11-299}$$

The m_+ and m_- terms in Eq. (11-299) are the molalities of the aqueous K^+ and Cl^- ions, respectively. From stoichiometry,

$$m_{\text{KCl}} = m_{\text{K}^+} = m_{\text{Cl}^-} \tag{11-300}$$

Thus for KCl, the mean ionic molality is simply

$$m_\pm = m_{\text{KCl}} \tag{11-301}$$

Equations analogous to Eq. (11-301) are true for any symmetrical electrolyte (e.g., NaCl, $NaNO_3$, $CuSO_4$, and so on).

Likewise, the mean ionic activity and mean ionic activity coefficient of KCl are also defined as the geometric means:

$$a_\pm = (a_+ a_-)^{1/2} = (a_{\text{KCl}})^{1/2} \tag{11-302}$$

$$\gamma_\pm = (\gamma_+ \gamma_-)^{1/2} \tag{11-303}$$

$$\gamma_\pm = \left(\frac{a_+}{m_+} \cdot \frac{a_-}{m_-}\right)^{1/2} \tag{11-304}$$

$$\gamma_\pm = \frac{a_\pm}{m_\pm} = \frac{a_{\text{KCl}}^{1/2}}{m_{\text{KCl}}} \tag{11-305}$$

Analogous equations apply to other symmetrical electrolytes. The definitions of the geometric mean molality, activity, and activity coefficient are equivalent to defining the standard Gibbs energy change $\Delta G^o = 0$ for the ionization of KCl dissolved in water in a hypothetical one-molal solution obeying the laws of the perfect dilute solution (the activity coefficient $\gamma_i = 1$ for each solute, such as the individual ions) as the standard state for aqueous solutions. Hence, $K_{eq} = 1$ for reaction (11-298) from Eq. (10-22):

$$K_{eq} = \exp\left(-\frac{\Delta_r G^o}{RT}\right) \tag{10-22}$$

The equilibrium constant expression for KCl ionization in water,

$$K_{eq} = \frac{a_+ a_-}{a_{KCl}} = 1 \tag{11-306}$$

then leads immediately to Eq. (11-302), which in turn leads to the definitions of mean ionic molality and mean ionic activity coefficients given previously.

The mean ionic molality, ionic activity, and ionic activity coefficient for unsymmetrical electrolytes are defined using $CaCl_2$ (hydrophilite) as an example. Ionization of $CaCl_2$ in water proceeds via the reaction

$$CaCl_2\ (aq) = Ca^{2+}\ (aq) + 2\ Cl^-\ (aq) \tag{11-307}$$

The stoichiometry of reaction (11-306) gives the relationship

$$m_{CaCl_2} = m_{Ca^{2+}} = \frac{1}{2} m_{Cl^-} \tag{11-308}$$

The mean ionic molality is then

$$m_\pm = m_{CaCl_2}[(\nu_+)^{\nu_+}(\nu_-)^{\nu_-}]^{1/\nu} = 4^{1/3} m_{CaCl_2} \tag{11-309}$$

where ν_+ is the number of cations (1 Ca^{2+}), ν_- is the number of anions (2 Cl^-), and ν is sum of $\nu_+ + \nu_-$ $(1 + 2 = 3)$. The mean ionic activity of aqueous $CaCl_2$ solutions is

$$a_\pm = [(a_+)^{\nu_+}(a_-)^{\nu_-}]^{1/\nu} = (a_{CaCl_2})^{1/3} \tag{11-310}$$

$$a_{CaCl_2} = \left(a_{Ca^{2+}} a_{Cl^-}^2\right) \tag{11-311}$$

The mean ionic activity coefficient is

$$\gamma_\pm = \frac{a_\pm}{m_\pm} = [(\gamma_+)^{\nu_+}(\gamma_-)^{\nu_-}]^{1/\nu} \tag{11-312}$$

Hydrophilite is a 2:1 electrolyte, but analogous equations apply to other unsymmetrical electrolytes such as 3:1 electrolytes (e.g., $LaCl_3$), 3:2 electrolytes (e.g., $La_2(SO_4)_3$), and so on. Equations (11-309), (11-310), and (11-312) give general definitions of the mean ionic molality, activity, and activity coefficient. These equations are derived for a hypothetical one-molal solution obeying the laws of the perfect dilute solution, but are very useful and are defined to be valid at all molalities in real solutions.

Example 11-14. Write down the expressions for the mean molality and activity coefficient of the 3:2 unsymmetrical electrolyte $La_2(SO_4)_3$. The mean molality is

$$m_{\pm} = m[(\nu_+)^{\nu_+}(\nu_-)^{\nu_-}]^{1/\nu} = m[2^2 \cdot 3^3]^{1/5} = m \cdot 108^{1/5} \quad (11\text{-}313)$$

The m in Eq. (11-313) is the molality of $La_2(SO_4)_3$. The mean activity coefficient is

$$\gamma_{\pm} = \frac{a_{\pm}}{m_{\pm}} = 0.392\frac{a_{\pm}}{m} \quad (11\text{-}314)$$

Different methods are used to measure activities and activity coefficients of electrolytes in aqueous solutions. These include measuring the partial pressures of volatile electrolytes such as HCl and HNO_3, emf measurements (aqueous acids), freezing-point depressions, partitioning between water and another solvent, solubility measurements ($a_{\pm}$ and $\gamma_{\pm}$ for the saturated solution), and measuring the water vapor pressure as a function of molality. Several texts give detailed examples of these measurements (e.g., Harned and Owen, 1958; Pitzer and Brewer, 1961; Robinson and Stokes, 1970). Table 11-12 and Figure 11-22 give activity coefficients for some electrolytes and for sucrose ($C_{12}H_{22}O_{11}$, cane sugar) as a function of molality at 25°C. In most cases the partial pressure of the dissolved electrolyte is too small and cannot be measured directly (as is the case for NaCl, $CaCl_2$, $CuSO_4$, and many other salts dissolved in water). The partial pressure of water vapor over the aqueous solution is measured in these cases. The ratio of the water vapor partial pressure over the solution to the vapor pressure of pure liquid water at the same temperature gives the activity of water in the aqueous solution:

$$a_w = \frac{f_w}{f_w^*} \cong \frac{P_w}{P_w^*} \quad (11\text{-}315)$$

In fact, as shown by Eq. (11-315), the H_2O partial pressure and vapor pressure should be corrected for the nonideality of water vapor and it is the fugacity ratio that gives the activity of water in the aqueous solution.

Then the activity coefficient of the electrolyte is calculated using a version of the Gibbs-Duhem equation. For example, the partial molal Gibbs energies, chemical potentials, activities, and activity coefficients of aqueous NaCl and water (w) are related via the equations

$$n_w d\overline{G}_w + n_{NaCl} d\overline{G}_{NaCl} = 0 \quad (11\text{-}316)$$

$$n_w \mu_w + n_{NaCl} \mu_{Nacl} = 0 \quad (11\text{-}317)$$

$$n_w d\ln a_w + n_{NaCl} d\ln a_{NaCl} = 0 \quad (11\text{-}318)$$

$$n_w d\ln \gamma_w + n_{NaCl} d\ln \gamma_{NaCl} = 0 \quad (11\text{-}319)$$

The activity coefficient of water in aqueous solutions is often expressed in terms of the *osmotic coefficient* (ϕ). This is related to the activity of water via

$$\phi = -55.51\frac{\ln a_w}{\nu m} \quad (11\text{-}320)$$

Table 11-12 Mean Molal Activity Coefficients at 298.15 K in Aqueous Solution

Molality	**0.001**	**0.002**	**0.005**	**0.01**	**0.02**	**0.05**	**0.1**	**0.2**	**0.5**	**1**	**2**	**5**
HCl	0.966	0.952	0.929	0.905	0.876	0.83	0.796	0.767	0.757	0.809	1.009	2.38
HNO_3	0.965	0.951	0.927	0.902	0.871	0.823	0.785	0.748	0.715	0.72	0.783	0.982[a]
H_2SO_4	0.83	0.757	0.639	0.544	0.453	0.34	0.265	0.209	0.154	0.13	0.124	0.212
LiCl	0.963	0.948	0.921	0.89	0.86	0.82	0.79	0.757	0.739	0.774	0.921	2.02
NaCl	0.965	0.952	0.927	0.902	0.871	0.819	0.778	0.735	0.681	0.657	0.668	0.874
KCl	0.965	0.952	0.927	0.901	0.866	0.815	0.77	0.718	0.649	0.604	0.573	0.588[b]
KBr	0.965	0.952	0.927	0.903	0.872	0.822	0.772	0.722	0.657	0.617	0.593	0.626
KI	0.965	0.951	0.927	0.905	0.88	0.84	0.778	0.733	0.676	0.645	0.637	0.683[c]
$ZnSO_4$	0.7	0.608	0.477	0.387	0.298	0.202	0.15	0.104	0.063	0.044	0.035	0.048[d]
$LaCl_3$	0.790	0.729	0.636	0.56	0.483	0.388	0.314	0.274	0.266	0.342	0.825	

[a] *4 m.*
[b] *4.8 m.*
[c] *4.5 m.*
[d] *3.5 m.*

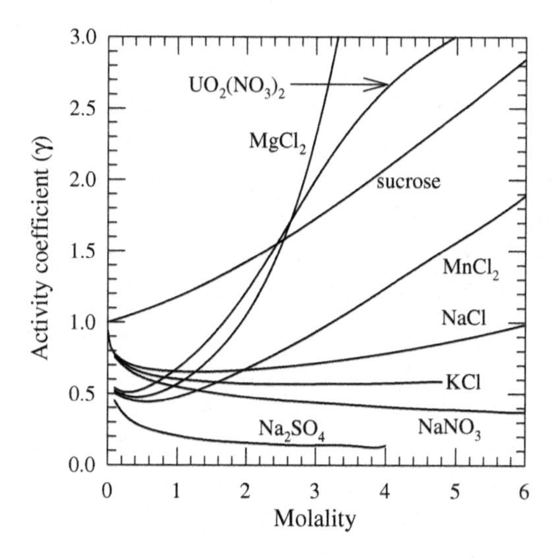

FIGURE 11-22

Activity coefficients for some aqueous solutions at 25°C.

For example, $a_w = 0.996668$ and $\phi = 0.9263$ in 0.1 m KCl solution (25°C). Rewriting Eq. (11-319) in terms of the osmotic coefficient gives

$$(\phi - 1)\frac{dm}{m} + d\phi = d\ln\gamma_{\text{NaCl}} \tag{11-321}$$

As mentioned earlier, it is impossible to measure activities or activity coefficients of individual ions. However, activities and activity coefficients can be measured relative to those of KCl in a solution of the same ionic strength by assuming that $\gamma_{K^+} = \gamma_{Cl^-}$ in KCl aqueous solutions. The K^+ and Cl^- aqueous ions have the same charge and similar sizes, and presumably they have similar activity coefficients. This assumption is supported by data in Table 11-12, which shows that LiCl, NaCl, KCl, KBr, and KI have similar mean activity coefficients in dilute solutions (≤ 0.02 m). This procedure is known as the *mean salt method*. The mean activity coefficients for monovalent chlorides (MCl) are then

$$\gamma_{\pm\text{MCl}} = (\gamma_+\gamma_-)^{1/2} = (\gamma_+\gamma_{\pm\text{KCl}})^{1/2} \tag{11-322}$$

Rearranging Eq. (11-322) gives the activity coefficient for the M^+ aqueous ion:

$$\gamma_+ = \frac{\gamma_{\pm\text{MCl}}^2}{\gamma_{\pm\text{KCl}}} \tag{11-323}$$

A similar procedure is used for divalent chlorides MCl_2. In this case we have

$$\gamma_{\pm\text{MCl}_2} = \left(\gamma_+\gamma_-^2\right)^{1/3} = \left(\gamma_+\gamma_{\pm\text{KCl}}^2\right)^{1/3} \tag{11-324}$$

$$\gamma_+ = \frac{\gamma_{\pm\text{MCl}_2}^3}{\gamma_{\pm\text{KCl}}^2} \tag{11-325}$$

The procedure for arcanite K_2SO_4 and other K-bearing salts is this:

$$\gamma_{\pm K_2SO_4} = \left(\gamma_+^2\gamma_-\right)^{1/3} = \left(\gamma_{\pm KCl}^2\gamma_-\right)^{1/3} \tag{11-326}$$

$$\gamma_- = \frac{\gamma_{\pm K_2SO_4}^3}{\gamma_{\pm KCl}^2} \tag{11-327}$$

A more involved procedure is needed for other salts such as $CuSO_4$. In this case Eq. (11-327) substitutes for the γ_- term and γ_+ is then

$$\gamma_+ = \frac{\gamma_{\pm CuSO_4}^2\gamma_{\pm KCl}^2}{\gamma_{\pm K_2SO_4}^3} \tag{11-328}$$

Example 11-15. Use the mean salt method to estimate γ_+ for aqueous Mg^{2+} in 0.1 m $MgCl_2$, which has $\gamma_\pm = 0.528$ and ionic strength $I = 0.3$. The mean activity coefficient $\gamma_\pm$ of KCl in 0.3 m solution is 0.688. The estimated value of γ_+ for aqueous Mg^{2+} is

$$\gamma_{Mg^{2+}} = \frac{\gamma_{\pm MgCl_2}^3}{\gamma_{\pm KCl}^2} = \frac{0.528^3}{0.688^2} = 0.31 \tag{11-329}$$

For comparison, the extended Debye-Hückel equation and the Davies equation (both described below) give γ_+ values of 0.45 and 0.25, respectively, for Mg^{2+} in the same solution.

In 1923 Peter Debye (see sidebar in Chapter 4) and the German physical chemist Erich Hückel (1896–1980) developed a model for mean activity coefficients of electrolytes and for activity coefficients of individual ions. Their model is based on the electrostatic interactions of electrolytes with the solvent and assumes that ions are point charges in a continuous medium with a dielectric constant equal to that of the solvent. The Debye-Hückel limiting law for the mean activity coefficient of an electrolyte is

$$-\log\gamma_\pm = Az_+z_-\sqrt{I} \tag{11-330}$$

The activity coefficient for an individual ion (either positively or negatively charged) is given by

$$-\log\gamma_i = Az_i^2\sqrt{I} \tag{11-331}$$

The limiting law equations are valid in dilute solutions ($I \leq 0.001$).

The extended Debye-Hückel equation, valid in more concentrated solutions ($I \leq 0.1$), is

$$-\log\gamma_\pm = \frac{Az_+z_-\sqrt{I}}{I + \left(\frac{\mathbf{a}_+ + \mathbf{a}_-}{2}\right)B\sqrt{I}} \tag{11-332}$$

$$-\log\gamma_i = \frac{Az_i^2\sqrt{I}}{I + \mathbf{a}_iB\sqrt{I}} \tag{11-333}$$

The terms in the Debye-Hückel limiting law and extended equation are the ionic strength (I), the valence z_i of ion i, the size of hydrated ions in Angstroms ($\mathbf{a}_i$), and two constants (A and B) that depend on the temperature, density (ρ), and the dielectric constant (ε) of the solvent:

$$A = 1.825 \times 10^6 \rho^{1/2} (\varepsilon T)^{-3/2} \tag{11-334}$$

$$B = 50.29 \rho^{1/2} (\varepsilon T)^{-1/2} \tag{11-335}$$

For aqueous solutions at 298.15 K, $A = 0.509$ and $B = 0.328$. Table 11-13, based on Kielland (1937), gives values of the $\mathbf{a}_i$ parameter for many common ions.

Two related equations are the Guntelberg equation, using a constant value of $\mathbf{a} = 3$:

$$-\log\gamma_{\pm} = A z_+ z_- \frac{\sqrt{I}}{1 + \sqrt{I}} \tag{11-336}$$

and the Davies equation developed for 1:1 and 1:2 electrolytes:

$$-\log\gamma_{\pm} = A z_+ z_- \left(\frac{\sqrt{I}}{1 + \sqrt{I}} - 0.2I \right) \tag{11-337}$$

$$-\log\gamma_i = A z_i^2 \left(\frac{\sqrt{I}}{1 + \sqrt{I}} - 0.2I \right) \tag{11-338}$$

The Guntelberg and Davies equations are useful for solutions with $I \leq 0.1$ and $I \leq 0.5$, respectively. Models of electrolyte activities in more concentrated aqueous solutions and their comparison to experimental data are described in Pitzer (1991).

Example 11-16. Calculate the mean activity coefficient for a saturated solution of $BaSO_4$ using the Debye-Hückel limiting law. As mentioned earlier, the saturated solution contains 9.56 micromoles $BaSO_4$ per liter solution. Neglecting the difference between molarity and molality, and using Eq. (11-226), $I = 3.824 \times 10^{-5}$ for the saturated solution. Substituting values for $A = 0.509$, $z_+ = z_- = 2$, and I into the Debye-Hückel limiting law gives

$$\log\gamma_{\pm} = -A z_+ z_- \sqrt{I} = -(0.509 \cdot 2 \cdot 2)\left(3.824 \times 10^{-5}\right)^{1/2} = -0.0126 \tag{11-339}$$

Table 11-13 Effective Diameters (**a**) for Some Common Aqueous Ions

Effective Diameter (a)	Ions
3	K^+, Cl^-, Br^-, I^-, CN^-, NO_2^-, NO_3^-, $HCOO^-$, Rb^+, Cs^+, NH_4^+, Tl^+, Ag^+, OH^-, F^-, NCS^-, NCO^-, HS^-, ClO_3^-, ClO_4^-, BrO_3^-, IO_4^-, MnO_4^-
4	Hg_2^{2+}, SO_4^{2-}, $S_2O_3^{2-}$, $S_2O_6^{2-}$, SeO_4^{2-}, CrO_4^{2-}, PO_4^{3-}, HPO_4^{2-}, CH_3COO^-, Na^+, $CdCl^+$, ClO_2^-, IO_3^-, HCO_3^-, $H_2PO_4^-$, HSO_3^-, $H_2AsO_4^-$
5	Sr^{2+}, Ba^{2+}, Ra^{2+}, Cd^{2+}, Hg^{2+}, S^{2-}, $S_2O_4^{2-}$, WO_4^{2-}, Pb^{2+}, CO_3^{2-}, SO_3^{2-}, MoO_4^{2-}
6	Li^+, Ca^{2+}, Cu^{2+}, Zn^{2+}, Sn^{2+}, Mn^{2+}, Fe^{2+}, Ni^{2+}, Co^{2+}
8	Mg^{2+}, Be^{2+}
9	H^+, Al^{3+}, Fe^{3+}, Cr^{3+}, Sc^{3+}, Y^{3+}, La^{3+}, In^{3+}, Ce^{3+}, Pr^{3+}, Nd^{3+}, Sm^{3+}
11	Th^{4+}, Zr^{4+}, Ce^{4+}, Sn^{4+}

The $\gamma_{\pm} = 0.971$ for $BaSO_4$ and is sufficiently close to unity that we were justified in neglecting it in our earlier calculations of the solubility product.

Example 11-17. Calculate the mean activity coefficient of 0.1 m KNO_3 using the limiting law and the extended Debye-Hückel equation. The limiting law gives $\gamma_{\pm} = 0.69$, the extended Debye-Hückel equation gives $\gamma_{\pm} = 0.75$, and the experimental value is $\gamma_{\pm} = 0.739$.

PROBLEMS*

1. Calculate the Henry's law constant at 25°C with units of mol kg^{-1} atm^{-1} from solubility data for N_2 (P = 1 atm.) in water summarized by Dorsey (1940). Also use the temperature-dependent solubility to compute ΔH^o_{298} for N_2 solubility in water: N_2 (g) = N_2 (aN).

T (°C)	0	10	20	30	40	50
N_2 (mg N_2/kg water)	28.945	23.08	19.01	16.895	15.105	13.835

2. Brown and Morris (1888) measured the freezing-point depression (ΔT) of D-glucose aqueous solutions. D-glucose, or dextrose, is also known as blood sugar, corn sugar, or grape sugar and is a major energy source for living organisms. Use a subset of their data given here to calculate the gram formula weight (molecular weight) of D-glucose.

ΔT (°C)	−1.45	−0.9417	−0.4433
grams dextrose	12.616	8.3704	4.114
grams water	92.25	94.86	97.47

3. Iron meteorites, the Earth's core, and, presumably, the cores of Mercury and Venus, are predominantly Fe-Ni alloy. Calculate and plot as a function of composition the activities of Fe and Ni in the molten alloy at 1600°C using the data here. (a) Is the molten alloy ideal? (b) If not, is it a regular solution? (c) If not, why?

X_{Ni}	ΔH^M (J mol^{-1})	ΔH^M_{Ni} (J mol^{-1})	ΔS^E (J mol^{-1} K^{-1})	ΔS^E_{Ni} (J mol^{-1} K^{-1})
0	0	−10,041.6	0	−2.13
0.1	−1025.1	−9,761.3	−0.19	−2.13
0.3	−2928.8	−9,723.6	−0.59	−2.35
0.5	−4602.4	−7,213.2	−0.94	−1.76
0.7	−4853.4	−3,832.5	−1	−1.16
0.9	−2426.7	−205.0	−0.6	−0.01

*pK_a and pK_b values are given in Tables 11-9 to 11-11.

4. Several groups have measured the enthalpy of mixing for olivine (Mg_2SiO_4-Fe_2SiO_4). The data of Kojitani and Akaogi (1994) are considered here. Calculate W_H from their data and plot their data points and your calculated $\Delta H^M = \Delta H^E$ values as a function of composition.

olivine	ΔH^M (kJ mol^{-1})
Fa_{100}	0 ± 0.86
Fa_{80}	1.658 ± 1.83
Fa_{60}	2.606 ± 1.720
Fa_{40}	2.294 ± 1.130
Fa_{20}	1.982 ± 1.51
Fa_0 (Fo_{100})	0 ± 0.62

5. Use the average composition of seawater (milligrams per kilogram of seawater) to compute its ionic strength. The density of seawater is 1.025 g cm^{-3} at 18°C. We summarize the data in the following table.

species	Cl^-	Na^+	SO_4^{2-}	Mg^{2+}	Ca^{2+}	K^+	HCO_3^-	Br^-
mg kg^{-1}	19,353	10,781	2690	1280	415	399	142	67

6. At 298.15 K, the ionization constant (K_W) for the reaction

$$H_2O\ (l) = H^+\ (aq) + OH^-\ (aq)$$

in pure water is $K_W = 1.008 \times 10^{-14}$. Estimate K_W and neutral pH for pure water at 0°C from the standard enthalpies of formation here. Assume $\Delta C_P = 0$ in your calculations.

$$\Delta_f H^o_{298}(OH^-,\ aq) = -230.015\ \text{kJ mol}^{-1}$$
$$\Delta_f H^o_{298}(H_2O,\ l) = -285.83\ \text{kJ mol}^{-1}$$

7. Calculate the freezing-point depression and degree of ionization of aqueous acetic acid solutions. (a) Calculate α to four decimal places at $m_{acid} = 0.001$. (b) Find m_{acid} at which $\alpha = 0.99$.

8. Calculate and plot the equilibrium abundances of H_2CO_3, HCO_3^-, and CO_3^{2-} in aqueous solution with total dissolved carbonate concentration of 0.001 m (i.e., $m_{H2CO3} + m_{CO3} + m_{CO_3^{2-}} = 0.001$ m) as a function of pH from 0 to14 at 25°C. The pH is an independent variable. You can assume $a_i = m_i$ in your work.

9. The average composition of North American river water is tabulated here. Use these data for the following computations. (a) Calculate the molality of each species. (b) Calculate the ionic strength. (c) Estimate individual ionic activity coefficients using the extended Debye-Hückel equation. Necessary auxiliary data are also listed.

species	Na^+	Mg^{2+}	K^+	Ca^{2+}	HCO_3^-	NO_3^-	SO_4^{2-}	Cl^-	SiO_2	Fe
mg/liter	9	5	1.4	21	68	1	20	8	9	0.16
ion size parameter (Å)	4	8	3	6	4	3	4	3		

10. Oxalic acid (HOOC-COOH) is an organic acid produced by fungi and is one of many organic acids that convert minerals to soil. Calculate and plot the concentrations of neutral, singly, and doubly ionized oxalic acid in soil water at a total concentration (neutral $+ OX^- + OX^{2-}$) of 10^{-6} m from pH 0-8.

11. Lead levels in groundwater around abandoned mines may be controlled by solubility of cerussite ($PbCO_3$), anglesite ($PbSO_4$), and other secondary minerals formed by oxidation of galena (PbS). Calculate the molality of Pb^{2+} in groundwater equilibrated with (a) anglesite and (b) cerussite at 25°C. Assume ideality and groundwater concentrations of 50 mg L^{-1} SO_4^{2-} and 118 mg L^{-1} HCO_3^- at neutral pH. At 25°C, $K_{sp} = 1.382 \times 10^{-8}$ for $PbSO_4$ and $K_{sp} = 7.4 \times 10^{-14}$ for $PbCO_3$.

12. Calculate the pH at which acetic acid (CH_3COOH) and acetate (CH_3COO^-) have equal concentrations in aqueous solution, in which these species behave ideally.

13. Calculate the solubility of fluorite (CaF_2) in pure water at 18°C, the average surface temperature on Earth. The solubility product K_{sp} (273–373 K) is

$$\log K_{sp} = 66.348 - \frac{4298.2}{T} - 25.271 \log T$$

14. What is the mole fraction (X_2) and mole ratio (X_2/X_1) of a solute in a 1.5 molal aqueous binary solution? The solvent is component 1, the solute component 2.

15. Vellut et al. (1998) obtained the results below for a synthetic mixture of C_{11}-C_{30} hydrocarbons dissolved in toluene. Calculate the molecular weight of this material.

g solute/kg toluene	10.47	19.74	39.03	52.21
ΔT_b (K)	0.1119	0.2594	0.5599	0.7907

16. Derive the conversion formula between the molarity and formality of an aqueous solution.

17. Calculate the mole fraction, molality, molarity, and formality for 30% NH_3 aqueous solution, which has a density of 0.904 g mL^{-1} at 0°C. These conditions correspond to those predicted for aqueous ammonia clouds in the atmospheres of gas-giant planets such as Jupiter.

18. Use the fourth fundamental equation $dA = -SdT - PdV$ to show that the chemical potential can also be defined as $\mu_i = (\partial A/\partial n_i)_{T,V,n \neq n_i}$ (i.e., demonstrate the equality of this partial derivative with Gibbs' definition of the chemical potential in terms of internal energy).

19. Cristobalite melts at 1713°C. Calculate its average enthalpy of melting using all four data points for freezing-point depression in alkali metal silicates (Kracek, 1930) given here.

X_{silica}	0.9766	0.974	0.9665	0.965
$T_{liq.}$ (°C)	1617	1607	1575	1569

20. The melting point of pure silver (1235 K) is reduced to 1224 K in air (0.21 bar O_2) because the molten Ag contains O_2 at a mole fraction of 0.00981. (a) Calculate the enthalpy of fusion of Ag from these data. (b) Assuming ideal gas behavior, recompute the O_2 solubility in terms of cm^3 O_2 per cm^3 of liquid silver, which has $V_m \sim 11.54$ cm^3 mol^{-1} at the melting point.

21. The following table shows the acetone partial pressure (mmHg) over an acetone-chloroform solution at 35.2°C as a function of the acetone mole fraction in the solution. Assuming this is a regular solution, calculate and graph the activity coefficients for acetone and chloroform as a function of composition.

$X_{acetone}$	1	0.9405	0.8783	0.8165	0.7103	0.575	0.3378	0.1978	0.0823
$P_{acetone}$	344.5	322.9	299.7	275.8	230.7	173.7	79.1	38	13.4

22. The NaCl concentration in human blood plasma is ~ 0.1 mol L^{-1}. Calculate its mean activity coefficient ($\gamma_{\pm}$) using the extended Debye-Hückel, Guntelberg, and Davies equations.

23. At 298.15 K $\Delta G^o = -10.05$ kJ mol^{-1} for NH_3 dissolution in water:

$$NH_3\ (gas) = NH_3\ (aqueous)$$

(a) Calculate the solubility of NH_3 in aqueous water clouds on Saturn at the 298.15 K, 10 bar level. The NH_3 mole fraction in Saturn's atmosphere is 10^{-4}. **(b)** Calculate the pH of the aqueous NH_3 solution from part (a).

24. Calculate the pH of pre-Cambrian rainwater assuming $P_{CO2} = 0.2$ bar and $T = 288$ K.

25. Hager et al. (1970) give the following values (cal mol^{-1}) for the partial molal excess enthalpies of solution of Cu and Sn in molten bronze at 1320°C (1593 K). (a) Calculate the integral enthalpy of mixing ΔH^M (in J mol^{-1}) as a function of the Sn mole fraction. (b) Use your results and data in Table 11-5 to calculate the integral ΔS^M and excess S^E entropy of mixing as a function of the Sn mole fraction. (c) Calculate the Sn and Cu activity coefficients at 1400 K at a Sn mole fraction of 0.4.

X_{Sn}	0	0.1	0.2	0.3	0.4	0.5	0.6	0.7	0.8	0.9	1
H^E_{Sn}	−8380	−4850	−1700	690	1200	690	660	370	170	40	0
H^E_{Cu}	0	−190	−740	−1530	−1800	−1600	−1230	−690	−60	640	1440

26. Show that Eqs. (11-216) and (11-217) for the activity coefficients of an asymmetric solution are consistent with Eq. (11-215) for the excess Gibbs energy of this type of solution.

27. Konings et al. (2008) measured the excess properties for the $LaPO_4$-$CaTh(PO_4)_2$ solid solution because of its potential importance for radioactive waste disposal. Use their data given in the following table to calculate the Margules parameter W_H assuming regular solution behavior.

X_{LaPO4}	0	0.1	0.2	0.3	0.4	0.5	0.6	0.7	0.8	1
H^E(J mol^{-1})	0	3043	5141	6835	6684	7383	6633	5811	4629	0

CHAPTER

Phase Equilibria of Binary Systems 12

The concept of the eutectic was early seized upon by petrologists and has been one of great utility in petrogenic theory. It accounted for the low melting temperatures of mixtures of minerals that are individually highly refractory. It threw light on some of the factors governing the separation of minerals from their mutual solution. But most of all, it stimulated the tendency to think of magmas in the light of the laws of solutions, or, better, of phase equilibrium, and encouraged experimental research whose expected result was the location of the composition of the eutectics for chosen mineral mixtures.

—Bowen (1922), *The Reaction Principle in Petrogenesis*

This chapter discusses phase diagrams and phase equilibria of binary systems and explains how to interpret phase diagrams, derive thermodynamic data from them, and use thermodynamic data to draw phase diagrams. It extends our prior discussion of phase equilibria of pure materials in Chapter 7 and complements our discussion of solutions in Chapter 11. This chapter describes binary phase diagrams and begins with some general information about conversion between molar and weight percentages and the use of isobaric sections for plotting temperature-composition (*T-X*) phase diagrams. The bulk of Section I is a review of several common types of phase diagrams. This review is not exhaustive but focuses on simpler diagrams containing features (e.g., eutectics, peritectics, and binodal curves) that are often found in more complex phase diagrams. For example, we describe eutectics (simple and with partial solid solubility), monotectics, binary water-salt phase diagrams, systems with intermediate compounds (stoichiometric and nonstoichiometric), and mutual liquid and solid solutions with continuous liquidus and solidus curves such as albite-anorthite or with minimum melting points such as åkermanite-gehlenite. We also describe systems with unmixing of liquid and/or solid solutions. We give introductory descriptions of thermodynamic modeling and extraction from phase diagrams, including the calculation of Margules parameters. Several of the topics discussed in this chapter are closely related to solution thermodynamics (Chapter 11), chemical equilibrium (Chapter 10), critical behavior of fluids (Chapter 8), and the Gibbs phase rule (Chapter 7). It might be useful to review the material in these chapters if you are unsure about these topics before proceeding with this one.

I. BINARY PHASE DIAGRAMS

A. General discussion

As illustrated in Chapter 7, two variables, such as temperature and pressure, are sufficient to plot phase diagrams of pure materials such as water, CO_2, and sulfur (e.g., see Figures 7-1, 7-2, and 7-10).

Copyright © 2013 Elsevier Inc. All rights reserved.

However, three variables (temperature, pressure, and composition) are needed to plot phase diagrams of binary systems. In many cases the phase diagrams of geologically interesting materials involve solids and liquids with low vapor pressures at their triple-point temperatures. For example, the vapor pressures of iron and SiO_2 are about 3×10^{-5} bars and 2×10^{-5} bars, respectively, at their triple-point temperatures of 1811 K and 1999 K where solid, liquid, and vapor coexist. Other molten metals and silicates have similarly low vapor pressures at their triple points.

Hence, the phase diagrams of geologic materials are generally plotted as functions of temperature (*T*) and composition (*X*) on isobaric sections at one bar pressure of the *P-T-X* three-dimensional surfaces. Temperature-composition phase diagrams (using either mole fractions as done here or mass fractions to denote composition) are of interest for understanding the petrology of the earth and other rocky bodies. In contrast, pressure-composition (*P-X*) phase diagrams are often used for liquid-vapor equilibria such as those of interest on Titan and other volatile-rich satellites of the outer planets. However, *P-X* diagrams are also used to show phase relations of mineral assemblages at high temperatures encountered in the lower crust and mantle (e.g., phase diagrams in Presnall (1995) for the $MgSiO_3$-Al_2O_3, Mg_2SiO_4-Fe_2SiO_4, $MgSiO_3$-$CaSiO_3$ joins).

B. Interconversion of molar and weight percentages

Phase diagrams are plotted in terms of molar percentages or weight percentages and it is important to convert between the two ways of expressing compositions. Phase diagrams in the geological literature are often plotted in terms of weight percentages, whereas phase diagrams in the chemical and metallurgical literature are often graphed in terms of molar percentages. Hansen and Anderko (1958) give conversion formulas, which are

$$W = \frac{100M}{M + \frac{B}{A}(100 - M)} \tag{12-1}$$

$$M = \frac{100W}{W + \frac{A}{B}(100 - W)} \tag{12-2}$$

where *W* is the weight percentage of component A, *M* is the molar percentage of component A, *A* denotes the molecular weight of component A, and *B* denotes the molecular weight of component B. We illustrate the use of these formulas using the anorthite-diopside eutectic listed in Table 12-1.

Example 12-1. (a) The eutectic point in the anorthite-diopside binary is at 1274°C and 42 weight percent anorthite. Calculate the molar percentages of anorthite and diopside at the eutectic composition. We call anorthite component A because it melts at a higher temperature than diopside. We use the same system in Table 12-1 and subsequent tables, that is, component A is always the one with the higher melting point. The molecular weights of the two silicates are 278.211 g mol^{-1} (anorthite) and 216.553 g mol^{-1} (diopside). We use Eq. (12-2) to compute the molar percentage of anorthite:

$$M = \frac{100W}{W + \frac{A}{B}(100 - W)} = \frac{100 \cdot 42}{42 + \frac{278.211}{216.553}(100 - 42)} = 36 \tag{12-2}$$

Table 12-1 Examples of Simple Eutectic Systems

Components				Eutectic Point	
A	**m.p. (°C)**	**B**	**m.p. (°C)**	**Mass (molar) %**	**Eutectic *T* (°C)**
corundum	2054	$KAlSiO_4$	>1750	A_8B_{92} ($A_{12}B_{88}$)	1680
$KAlSiO_4$	>1750	leucite	1686	$A_{41.1}B_{58.9}$ ($A_{49.0}B_{51.0}$)	1615
corundum	2054	leucite	1686	$A_{7.8}B_{92.2}$ ($A_{15.3}B_{84.7}$)	1588
$MgAl_2O_4$	2135	leucite	1686	$A_{88.5}B_{11.5}$ ($A_{92.2}B_{7.8}$)	1553
$MgAl_2O_4$	2135	gehlenite	1593	$A_{16.5}B_{83.5}$ ($A_{27.6}B_{72.4}$)	1527
$CaTiO_3$	1915	α-$CaSiO_3$	1548	$A_{76}B_{24}$ ($A_{73}B_{27}$)	1425
leucite	1686	anorthite	1557	$A_{55}B_{45}$ ($A_{61}B_{39}$)	1413
α-$CaSiO_3$	1548	akermanite	1454	$A_{43}B_{57}$ ($A_{64}B_{36}$)	1400
SiO_2	1726	anorthite	1557	$A_{49.5}B_{50.5}$ ($A_{81.9}B_{18.1}$)	1368
gehlenite	1593	α-$CaSiO_3$	1548	$A_{36.7}B_{63.3}$ ($A_{19.7}B_{80.3}$)	1318
anorthite	1557	α-$CaSiO_3$	1548	$A_{48}B_{52}$ ($A_{28}B_{72}$)	1307
anorthite	1557	titanite	1397	$A_{37}B_{73}$ ($A_{26}B_{74}$)	1301
leucite	1686	diopside	1397	$A_{38.5}B_{61.5}$ ($A_{38.3}B_{61.7}$)	1300
anorthite	1557	diopside	1397	$A_{42}B_{58}$ ($A_{36}B_{64}$)	1274
lime	2572	$CaCO_3$	1320	A_7B_{93} ($A_{12}B_{88}$)	1248
corundum	2054	albite	1120	$A_{1.5}B_{98.5}$ ($A_{96.2}B_{3.8}$)	1108
SiO_2	1726	albite	1120	$A_{31.5}B_{68.5}$ ($A_{66.7}B_{33.3}$)	1062
fayalite	1217	albite	1120	$A_{16}B_{84}$ ($A_{20}B_{80}$)	1050
β-spodumene	1423	albite	1120	$A_{15}B_{85}$ ($A_{20}B_{80}$)	1045
$KAlSiO_4$	>1750	$K_2Si_2O_5$	1045	$A_{23.9}B_{76.1}$ ($A_{29.9}B_{70.1}$)	923
leucite	1686	$K_2Si_2O_5$	1045	$A_{31.7}B_{68.3}$ ($A_{31.3}B_{68.7}$)	918
NaF	996	CsF	703	A_8B_{92} ($A_{24}B_{76}$)	810
albite	1120	$Na_2Si_2O_5$	874	$A_{38}B_{62}$ ($A_{30}B_{70}$)	767
NaF	996	RbF	793	$A_{16}B_{84}$ ($A_{33}B_{67}$)	667
Si	1412	Al	660	$A_{12.7}B_{87.3}$ ($A_{12.3}B_{87.7}$)	577.2
KF	857	LiF	848	$A_{68}B_{32}$ ($A_{49}B_{51}$)	492
KCl	770	AgCl	457	$A_{17.9}B_{82.1}$ ($A_{29.5}B_{70.5}$)	318.6
Al	660	Sn	231.9	$A_{0.5}B_{99.5}$ ($A_{2.2}B_{97.8}$)	228.3
Cd[a]	321.0	Bi[a]	271.4	$A_{40}B_{60}$ ($A_{55}B_{45}$)	145.5
caffeine	236.3	paracetamol[b]	169.3	$A_{44}B_{56}$ ($A_{38}B_{62}$)	141.0
KNO_3	337	$LiNO_3$	255	$A_{66.8}B_{33.2}$ ($A_{57.8}B_{42.2}$)	125
phenobarbital	175.0	aspirin	135.0	$A_{38.5}B_{61.5}$ ($A_{33}B_{67}$)	121.5
Zn	419.5	Ga	29.8	$A_{3.5}B_{96.5}$ ($A_{3.7}B_{96.3}$)	25.25
Na	97.8	Rb	39.5	$A_{5.5}B_{94.5}$ ($A_{17.9}B_{82.1}$)	−4.5
KCl	770	Ice	0	$A_{19.5}B_{80.5}$ ($A_{5.5}B_{94.5}$)	−10.7
$NaNO_3$	310	Ice	0	$A_{38.1}B_{61.9}$ ($A_{11.5}B_{88.5}$)	−18.1
$(NH_4)_2SO_4$	280[c]	Ice	0	$A_{38.4}B_{61.6}$ ($A_{7.8}B_{92.2}$)	−19.05

[a] *Up to 2.75 mole % Bi dissolves in Cd (max at eutectic T).*
[b] *Also known as acetaminophen, an over-the-counter remedy for headaches.*
[c] *Decomposition temperature.*

Sources: Bowen and Schairer, 1929, 1935, 1936); Cook and McMurdie, 1989; DeVries et al., 1955; Hultgren et al., 1973; Levin et al., 1964; Linke, 1965; Nurse and Stutterheim, 1950; Prince, 1943; Sangster, 1999; Schairer, 1955; Schairer and Bowen, 1942, 1955, 1956.

The eutectic is at 36 mole % anorthite and 64 mole % diopside. Table 12-1 and subsequent tables give compositions in terms of weight and molar percentages with the molar percentages in parenthesis after the weight percentages.

(b) The eutectic point for the bismuth (Bi)–cadmium (Cd) system is at 145.5°C and 55 mole % cadmium (Hultgren et al., 1973). Compute the weight percentage of Cd at the eutectic point. We use the atomic weights of Bi (208.98 g mol^{-1}), Cd (112.41 g mol^{-1}), and Eq. (12-1) to do this:

$$W = \frac{100M}{M + \frac{B}{A}(100 - M)} = \frac{100 \cdot 55}{55 + \frac{208.98}{112.41}(100 - 55)} = 39.7 \tag{12-1}$$

The eutectic is at 40 weight % Cd and 60 weight % Bi.

C. Simple eutectic diagrams

The anorthite ($CaAl_2Si_2O_8$)–diopside ($CaMgSi_2O_6$) phase diagram in Figure 12-1 exemplifies the phase diagrams of materials that form simple eutectic systems. As in the case of anorthite and diopside, the two components are insoluble in one another as solids, do not form intermediate compounds, and are completely soluble in one another as liquids. Many elements and inorganic and organic compounds form simple eutectics with one another. Table 12-1 lists several examples of simple eutectics arranged in order of decreasing eutectic temperature from 1680°C to −19°C. The examples in Table 12-1

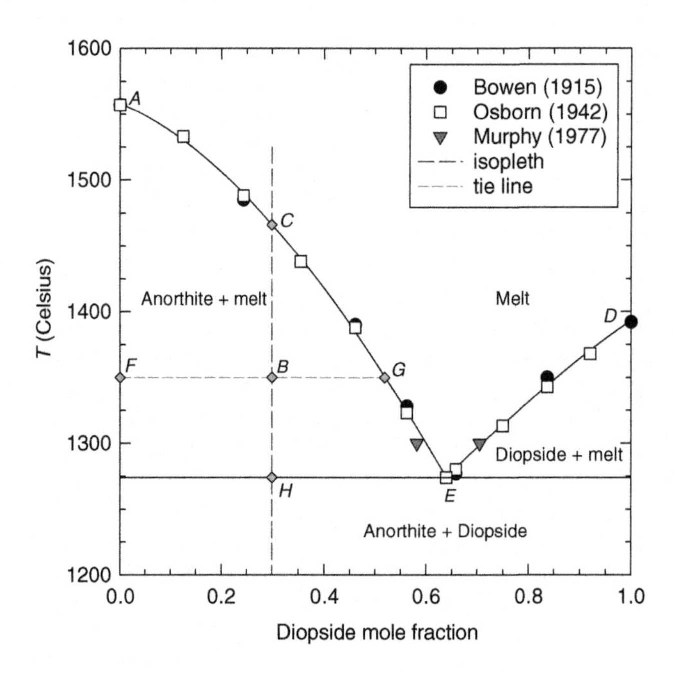

FIGURE 12-1

Anorthite ($CaAl_2Si_2O_8$)–diopside ($CaMgSi_2O_6$) phase diagram.

deliberately include a wide variety of materials. The eutectic composition is given as mass and molar percentages and the higher melting point component is always component A. More examples of simple eutectics relevant to geology and materials science are in the compilations of Hansen and Anderko (1958), Hultgren et al. (1973), and Levin et al. (1964).

A *eutectic point*, which is the minimum melting point for the solids (or equivalently a minimum freezing point for the melt), is the characteristic feature of these phase diagrams. The word *eutectic* is derived from the Greek word *eutēktos*, which means easily melted. In Figure 12-1 the eutectic point (*E*) is at 1274°C (1547 K) and a diopside mole fraction of ~0.64, where the two freezing-point curves intersect one another. Applying the Gibbs phase rule, Eq. (7-70), we see that the eutectic point is an invariant point (degrees of freedom $F = 0$) where four phases coexist in equilibrium in a two-component system. In this case the two components are anorthite and diopside and the four phases are melt, anorthite, diopside, and vapor. The anorthite-diopside diagram in Figure 12-1 is drawn at a constant pressure of one bar. It would not be greatly different if it were drawn at the equilibrium vapor pressure of the eutectic melt, which is much less than one bar. However, because Figure 12-1 is an isobaric section through the *P-T-X* surface for the anorthite-diopside system, one degree of freedom has already been used. The reduced phase rule ($F = C - P + 1$) is typically applied to *T-X* phase diagrams drawn at constant pressure, such as the anorthite-diopside diagram in Figure 12-1. Thus, it is equally correct to regard the eutectic point as an invariant point because three phases (melt and two solids) are at equilibrium in a two-component system at a specified constant pressure.

The curves *AE* and *DE* in Figure 12-1 are the *freezing-point curves* of anorthite (An) and diopside (Di), respectively. The freezing point of anorthite (1557°C, 1830 K) is depressed by the addition of diopside and moves along curve *AE*. Likewise, the freezing point of diopside (1397°C, 1670 K) is depressed by the addition of anorthite and moves along curve *DE*. (The triple-point temperatures of anorthite and diopside, where the solid, liquid, and vapor coexist in equilibrium, are close, but not identical, to their freezing points at one bar pressure. The small difference can be computed using the Clapeyron equation as discussed in Chapter 7.) The anorthite and diopside freezing-point curves are also known as *liquidus curves* because they show the temperatures at which melting is complete (i.e., the last infinitesimal amount of solid anorthite or solid diopside is melted). The *AE* and *DE* curves are also the *solubility curves* for anorthite dissolving in melt (curve *AE*) and for diopside dissolving in melt (curve *DE*). We use these terms interchangeably in the rest of this chapter.

As mentioned earlier, the eutectic point *E* is the intersection of the anorthite liquidus curve *AE* and the diopside liquidus curve *DE*. Thus, the eutectic point can be located by calculating and plotting the two liquidus curves. For example, Morey (1952) did this to locate the eutectic point (at 725°C, 46 wt % $Ca[PO_3]_2$) for the sodium metaphosphate ($NaPO_3$)–calcium metaphosphate ($Ca[PO_3]_2$) system. The eutectic point can also be located by experimental methods such as quenching of heated samples and determination of the phases present or from the behavior of heating (or cooling) curves. In practice, the quenching method is preferable for silicates because of their tendency for undercooling.

The horizontal line at the eutectic temperature 1274 K (line *HE*) is the *solidus line*. This is the temperature at which the first infinitesimal amount of solid melts upon heating or the last infinitesimal amount of melt freezes upon cooling. At lower temperatures the region labeled anorthite + diopside is a two-phase field. The regions labeled anorthite + melt and diopside + melt are also two-phase fields in which the pure solids coexist with melt. In contrast, the melt region above the liquidus curve *AED* is a single-phase region.

We now discuss cooling and equilibrium crystallization at constant composition ($X_{Di} = 0.3$) along the vertical dashed line *CBH*, which is an *isopleth*. The first infinitesimal amount of solid appears at point C (1466°C). It has the composition of pure anorthite and coexists in equilibrium with melt having the composition at point *C* ($An_{70}Di_{30}$). The appearance of pure anorthite also causes a change in the slope of the heating (or cooling) curve (*dT/dt*) with units of degrees per second) because crystallization of anorthite from the melt releases heat. The amount of heat released (ΔH_T^o) at a temperature *T* depends on the moles of anorthite crystallized (N_{An}) and the difference between the enthalpy of fusion at the melting point (T_m) of anorthite and at temperature *T*:

$$\Delta H_T^o = N_{An}\left[\Delta_{fus}H_{T_m}^o - \int_T^{T_m} \Delta C_P dT\right] \quad (12\text{-}3)$$

The ΔC_P in Eq. (12-3) is the heat capacity difference between molten and crystalline anorthite. Average values of ΔC_P for anorthite and diopside from their melting points to the eutectic temperature are 68.6 and 73.8 J mol^{-1} K^{-1}, respectively.

With continued cooling along the isopleth *CBH* the amount of anorthite increases, the amount of melt decreases, and the composition of the melt moves along the liquidus curve *AE*. At 1350°C, melt of composition *G* is in equilibrium with pure anorthite and the equilibrium phases are connected by the *tie-line* (or *conode*) *FBG*. The amounts of melt and solid are given by the *lever rule*, which we introduced in Chapter 8:

$$\frac{melt}{solid} = \frac{|BF|}{|BG|} = \left|\frac{0.30 - 0}{0.519 - 0.30}\right| = \frac{0.30}{0.219} = 1.370 \quad (12\text{-}4)$$

From mass balance the mole fractions and melt and solid are related via the equation

$$X_{melt} + X_{solid} = 1 = 1.370X_{solid} + X_{solid} \quad (12\text{-}5)$$

Thus, $X_{solid} = 0.422$ and $X_{melt} = 0.578$ at point *B*, that is, point *B* is a mixture of 57.8 mole % melt plus 42.2 mole % anorthite. This agrees with our intuition that point *B* is richer in melt because it is closer to the melting curve than to pure anorthite. We could also have said that the bulk composition at point *B* is given by

$$1 \text{ mole } B = \left(\frac{BG}{FG}\right) \text{moles An} + \left(\frac{FB}{FG}\right) \text{moles melt} \quad (12\text{-}6)$$

The lever rule applies in all the two-phase regions in Figure 12-1 and can be used to calculate the amounts of diopside and melt in the two-phase region between curve *DE* and the solidus line, and of anorthite and diopside below the solidus line.

The amount of solid anorthite or diopside crystallized per degree cooling and how this varies with temperature at constant composition is also of interest. The Dutch physical chemist Bakhuis Roozeboom (see the biographical sidebar in Chapter 7) demonstrated that the amount of solid crystallized per degree cooling is largest when crystallization commences and decreases with continued cooling. This is true for all concave liquidus curves, like those of anorthite and diopside. (In other words, the derivative *dN/dT* where *N* is the number of moles of solid crystallized and *T* is temperature decreases with decreasing temperature.) Only very strongly convex liquidus curves display the opposite behavior, that is, the amount of solid crystallized increases with continued cooling.

At point *H* pure anorthite is in equilibrium with eutectic melt of composition E. As you can verify for yourself using the lever rule, the composition at point *H* consists of 52.8 mole % anorthite and 47.2 mole % melt. The eutectic melt is *congruent* with respect to anorthite and diopside because its composition is intermediate between them. Equivalently, the composition of the eutectic melt is the linear combination of positive quantities of anorthite and diopside.

Further cooling below the eutectic temperature, for example, to 1225°C, leads to a physical mixture of anorthite (52.8 mole %) and solidified eutectic melt (47.2 mole %). The solidified eutectic melt is called the *eutectic mixture*. The eutectic mixture is not a compound but is a physical mixture of anorthite and diopside in the eutectic proportions ($An_{36}Di_{64}$). The overall proportions of anorthite and diopside in the cooled sample correspond to the bulk composition ($An_{70}Di_{30}$), that is, 70 mole % anorthite and 30 mole % diopside. All the diopside occurs in the eutectic mixture, while anorthite occurs both in the eutectic mixture and as separate anorthite crystals. This is easy to show using mass balance: The eutectic mixture is 47.2% of the sample and contains 64% diopside, giving 30 mole % diopside.

Cooling along an isopleth to the right of the eutectic point, that is, at diopside mole fractions greater than 0.64, is interpreted analogously. The major difference is that a cooled sample in the two-phase region (below 1274 K) consists of diopside crystals plus the eutectic mixture. All anorthite occurs in the eutectic mixture.

Cooling along an isopleth at the eutectic composition has a different result. No solids form until the eutectic temperature of 1274°C is reached. Then three phases—melt (with eutectic composition E), anorthite, and diopside—form in equilibrium with one another. All of the anorthite and diopside occur in the eutectic mixture. Initially, the melt comprises almost all the mass of the system, and only infinitesimal amounts of the two solids are present. As we try to continue cooling, the temperature remains constant due to the heat released by freezing of the eutectic melt. At the same time the amount of melt decreases and the amount of the eutectic mixture ($An_{36}Di_{64}$) increases. Finally, after all melt is consumed, cooling continues and only the eutectic mixture of anorthite plus diopside remains. The cooling (or heating) curve at the eutectic point has a sharp break and becomes flat ($dT/dt = 0$) until all melt has crystallized (or all eutectic mixture has melted). This is analogous to cooling curve behavior at the melting points of pure anorthite and pure diopside, but is different from that along isopleths in the An + melt and Di + melt two-phase regions discussed earlier.

The preceding discussion relates to equilibrium crystallization. However, anorthite and diopside are denser than the melt. If their crystals settle due to gravity and sit at the bottom of a magma chamber instead of remaining in contact with the melt, then crystallization may proceed differently. For example, consider the *FBG* tie-line in Figure 12-1. If the anorthite crystals are removed from contact with the melt (e.g., by gravitational sedimentation), the bulk composition of the system becomes richer in diopside and shifts from point *B* to point *G* on the liquidus curve. If this process continues with cooling, the bulk composition is continually shifted over to the liquidus curve and ends up at the eutectic composition (point *E*).

Finally, we briefly discuss *eutectoid points*, which are the solid-state analog to eutectic points. A *eutectoid* is an invariant point involving only solid phases. For example, wüstite ($Fe_{0.945}O$) is unstable below ~570°C (843 K) where it decomposes to Fe metal and magnetite (Fe_3O_4). Another example is the decomposition of C-bearing austenite (γ-Fe) to C-bearing ferrite (α-Fe) plus cementite (Fe_3C) at 723°C (996 K). Hansen and Anderko (1958) show the phase diagrams for the Fe-O and Fe-C systems.

D. Thermodynamics of simple eutectic diagrams

We now use the LiF-LiOH phase diagram in Figure 12-2 (modified from Cook and McMurdie, 1989) to illustrate calculation of enthalpies of fusion and liquidus curves for simple eutectic systems that behave ideally or nearly so. Lithium fluoride crystals are transparent to ultraviolet (UV) light and are used as UV optics. Lithium fluoride is also used in radiation detectors. Lithium hydroxide is used for removing CO_2 from the air on manned spacecraft (e.g., as on Apollo 13) and on submarines.

Along its liquidus line, crystalline LiF is in equilibrium with LiF in the melt

$$\text{LiF (crystal)} = \text{LiF (in melt)} \tag{12-7}$$

The equilibrium constant for reaction (12-7) is

$$K_{eq} = \frac{a_{\text{LiF}(melt)}}{a_{\text{LiF}(c)}} = a_{\text{LiF}(melt)} \cong X_{\text{LiF}(melt)} \tag{12-8}$$

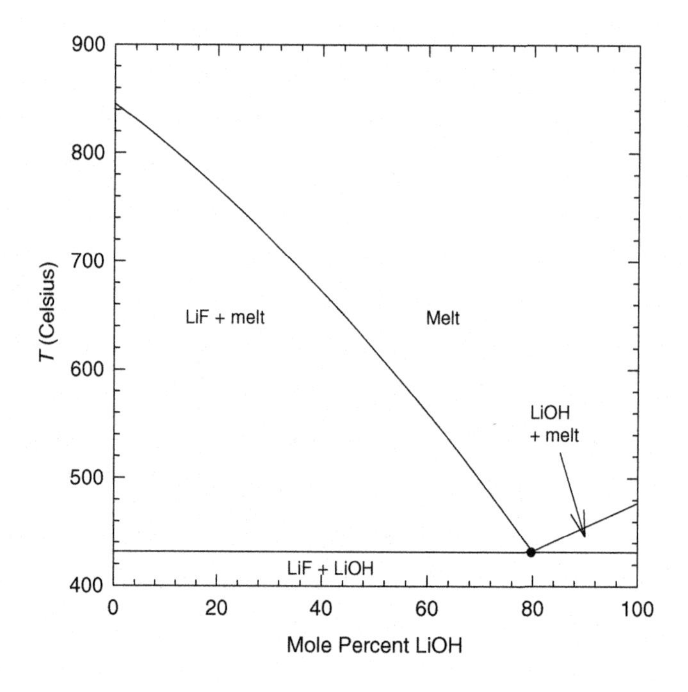

FIGURE 12-2

Lithium fluoride (LiF)–lithium hydroxide (LiOH) phase diagram.

The activity of pure crystalline LiF is unity and the equilibrium constant is equal to the activity of LiF in the melt. Thermodynamic modeling of the LiF-LiOH phase diagram shows that the melt is nearly ideal, so the activity of LiF in the melt can be replaced by its mole fraction in the melt. According to the van't Hoff equation (10-64) and (10-65), the temperature dependence of K_{eq} is given by

$$\left[\frac{\partial \ln K_{eq}}{\partial T}\right]_P = \left[\frac{\partial \ln X_{\mathrm{LiF}}}{\partial T}\right]_P = \frac{\Delta_{fus}H^o}{RT^2} \tag{12-9}$$

$$\left[\frac{\partial \ln K_{eq}}{\partial (1/T)}\right]_P = \left[\frac{\partial \ln X_{\mathrm{LiF}}}{\partial (1/T)}\right]_P = -\frac{\Delta_{fus}H^o}{R} \tag{12-10}$$

Thus, a plot of ln X_{LiF} versus T is curved and a plot of ln X_{LiF} versus $1/T$ (in kelvins) is a line with a slope equal to $-\Delta_{fus}H^o/R$. Equation (12-9) shows that the curvature of liquidus lines occurs in ideal systems and is not only due to nonideality effects.

Example 12-2. (a) Use the LiF liquidus line in the LiF-LiOH phase diagram to calculate the enthalpy of fusion of LiF. The following table summarizes the calculations.

T (°C)	848	824	807	788	765
X_{LiF}	1	0.9382	0.8993	0.8521	0.8021
T (K)	1121	1097	1080	1061	1038
$\frac{10,000}{T}$	8.9206	9.1158	9.2593	9.4251	9.6339
ln X_{LiF}	0	−0.0638	−0.1061	−0.1601	−0.2205

A linear least-squares fit to the data gives the equation

$$\ln X_{\mathrm{LiF}} = 2.758 - \frac{3094.1}{T} \tag{12-11}$$

The calculated $\Delta_{fus}H^o$ is ~ 25,720 J mol^{-1}, which is about 5% smaller than the actual value of ~ 27,090 J mol^{-1}. The small discrepancy is due to two factors. First, the LiF-LiOH melt is slightly nonideal with an excess enthalpy of mixing H^E ~ 100 J mol^{-1}. Second, the $\Delta_{fus}H^o$ varies with temperature according to Kirchhoff's equation (5-25) because crystalline and liquid LiF have different heat capacities.

(b) Use the melting point and enthalpy of fusion of LiOH to derive an equation for its liquidus line. Lithium hydroxide melts at 750 K and its $\Delta_{fus}H^o$ is 20,900 J mol^{-1} (Gurvich et al., 1996; Cook and McMurdie, 1989). The slope of the liquidus line is given by

$$-\frac{\Delta_{fus}H^o}{R} = -\frac{20,900}{8.314} = -2513.8 \tag{12-12}$$

The requirement that the LiOH mole fraction is unity at its melting point gives the intercept of 3.352. The equation for the LiOH liquidus line is thus

$$\ln X_{\mathrm{LiOH}} = 3.352 - \frac{2513.8}{T} \tag{12-13}$$

Equation (12-13) gives a good fit to the liquidus line, for example, the calculated eutectic point is 704 K (431°C) and 80.4 mole % LiOH (versus the observed value of 80.0 mole %).

If there is a reason for higher accuracy in a liquidus line calculation, the temperature dependence of the enthalpy of fusion can be included using the heat capacity difference between melt and solid. In general, a constant ΔC_P value is sufficient and an equation of the following form is used for the enthalpy of fusion:

$$\Delta_{fus}H_T^o = \Delta_{fus}H_{T_m}^o - \int_T^{T_m} \Delta C_P dT = \Delta_{fus}H_{T_m}^o - \Delta C_P(T_m - T) \tag{12-14}$$

The T_m in Eq. (12-14) is the melting point, $\Delta_{fus}H_{T_m}^o$ is the enthalpy of fusion at the melting point, and ΔC_P is an average over the temperature range considered.

The LiF-LiOH system is representative of the simple eutectic phase diagrams for many inorganic compounds, metals, and organic compounds for which ideal solution modeling gives good estimates of enthalpies of fusion. In contrast, the anorthite-diopside phase diagram is typical of silicate phase diagrams, which, with a few exceptions, cannot be modeled as ideal solutions.

We show this by considering the anorthite liquidus curve in Figure 12-1. Along curve *AE* pure anorthite is in equilibrium with melt:

$$\text{Anorthite (crystal)} = \text{Anorthite (melt)} \tag{12-15}$$

The standard Gibbs energy for reaction (12-15) is the standard Gibbs energy of fusion,

$$\Delta_{fus}G^o = -RT \ln K_{eq} = -RT \ln a_{\text{An}(melt)} \tag{12-16}$$

The anorthite activity in the melt is the product of the anorthite mole fraction and activity coefficient. Rewriting Eq. (12-16), we get

$$\Delta_{fus}G^o = -RT \ln a_{\text{An}(melt)} = -RT \ln X_{\text{An}(melt)} - RT \ln \gamma_{\text{An}(melt)} \tag{12-17}$$

If the melt were ideal, $\gamma_{\text{An}(melt)}$ would be unity and the activity coefficient term would be zero. However, this is not the case and we need to consider both the activity coefficient and mole fraction terms. The temperature dependence of $\Delta_{fus}G^o$ is given by

$$\Delta_{fus}G^o = \Delta_{fus}H_{T_m}^o - \int_T^{T_m} \Delta C_P dT - T\Delta_{fus}S_{T_m}^o + T\int_T^{T_m} \frac{\Delta C_P}{T} dT \tag{12-18}$$

The standard enthalpy and entropy of fusion for anorthite are 133.0 kJ mol^{-1} and 72.68 J mol^{-1} K^{-1}, respectively. The integration limits are the anorthite freezing point ($T_m = 1830$ K) and any temperature T along the liquidus curve. The mean ΔC_P (melt-solid) is ~68.6 J mol^{-1} K^{-1} over the temperature range spanned by the liquidus curve.

The activity coefficient term is evaluated assuming that the anorthite-diopside melt is a regular solution (see Chapter 11):

$$RT \ln \gamma_{\text{An}(melt)} = W_G(1 - X_{\text{An}})^2 \tag{12-19}$$

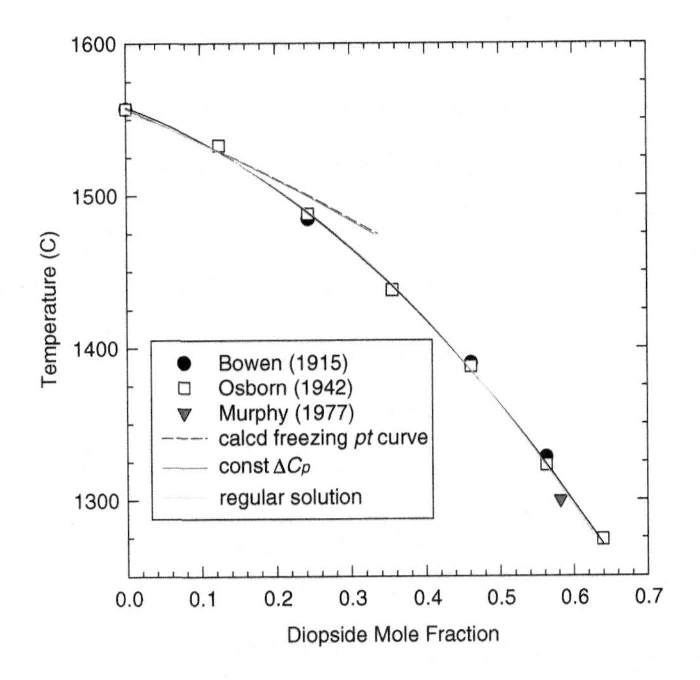

FIGURE 12-3

A comparison of the observed and calculated liquidus curves for anorthite in the anorthite-diopside phase diagram.

This assumption, and hence Eq. (12-19), are not strictly correct because the excess enthalpy of solution for the An-Di melt is asymmetric (e.g., Sugawara et al., 2009), but they nevertheless provide a good fit to the anorthite liquidus curve.

Equations (12-16) and (12-18) are used to compute anorthite activity in the melt at the temperatures of the data points measured by scientists studying the phase diagram (Bowen, 1915; Osborn, 1942; Murphy 1977). The activity coefficients are then computed from the calculated anorthite activities and observed mole fractions. The Margules W_G parameter is then calculated at each temperature using Eq. (12-19). The average W_G value from the 10 data points is $-15{,}730\ \text{J mol}^{-1}$. Equations (12-17)–(12-19) are then used to solve for X_{An} along the liquidus curve. This is done iteratively starting with the $a_{An(melt)}$ values because X_{An}, which is unknown, enters into the calculations. The initial and calculated X_{An} values converge within a few iterations. Figure 12-3 shows the result and compares it to the observed liquidus curve and calculations, assuming ideal solution with $\Delta C_P = 0$ (Eq. 11-119) or 68.6 J $\text{mol}^{-1}\ \text{K}^{-1}$ (Eqs. 12-17 and 12-18, considering only the mole fraction term).

E. Monotectics

Some simple eutectic systems have eutectic points very close to the lower melting-point component in the phase diagram. For example, the gallium (Ga)–zinc (Zn) and sodium (Na)–rubidium (Rb) systems (see Table 12-1) show this behavior. A *monotectic* can be regarded as the extreme case where the eutectic point moves completely to one side of the phase diagram and the liquidus curve rises continuously from the melting point of the lower melting component to that of the higher melting component. The solidus curve is a horizontal line at the melting point of the lower melting component. Three examples of monotectic systems are the silicon (Si)–tin (Sn), bismuth (Bi)–copper (Cu), and aluminum (Al)–tin (Sn) phase diagrams.

F. Systems with intermediate compounds

Many binary phase diagrams (including those for metals, organics, oxides, salts, and silicates) include intermediate compounds in addition to eutectics. The intermediate compounds can decompose at a temperature below the eutectic temperature, decompose above the eutectic temperature, or melt congruently. The decomposition of Ni_2SiO_4 at 1545°C to NiO + cristobalite in the NiO-SiO_2 system is an example of the first case. Enstatite and forsterite exemplify the second and third cases, respectively. The intermediate compounds can be stoichiometric (i.e., with a fixed composition such as forsterite Mg_2SiO_4 and enstatite $MgSiO_3$ in the MgO-SiO_2 system) or nonstoichiometric (often with a range of compositions that varies with temperature, such as $MgAl_2O_4$ spinel in the MgO-Al_2O_3 system, wüstite $Fe_{1-x}O$ in the Fe-O system, and pyrrhotite $Fe_{1-x}S$ in the Fe-S system). Many borides (e.g., hafnium, titanium, and zirconium) and carbides (hafnium, silicon, titanium, and zirconium) are also nonstoichiometric with a range of compositions that varies with temperature.

Figure 12-4 shows the MgO-SiO_2 phase diagram. This was originally determined by Bowen and Andersen (1914) and modified by Greig (1927), who showed the existence of liquid immiscibility at the silica-rich end of the diagram. The MgO-SiO_2 system exemplifies phase diagrams with

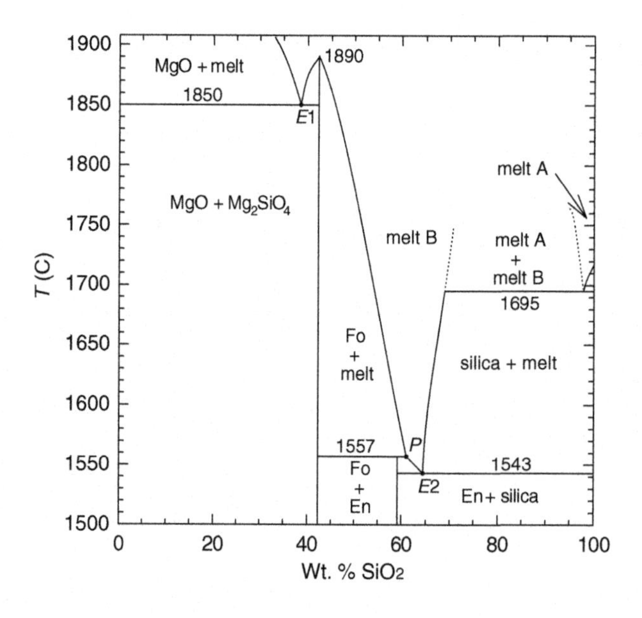

FIGURE 12-4

Periclase (MgO)–silica (SiO_2) phase diagram.

intermediate compounds that have stoichiometric composition, called *line compounds* because they appear as a straight vertical line on the phase diagram. The heights of the lines correspond to the melting points of the compounds. The four compounds in Figure 12-4 are periclase (MgO), forsterite (Mg_2SiO_4, represented by the vertical line at 42.71 wt. % SiO_2), enstatite ($MgSiO_3$, represented by the vertical line at 59.85 wt. % SiO_2), and silica (SiO_2). The melting points of these compounds are 2825°C (periclase), 1890°C (forsterite), 1557°C (enstatite), and 1726°C (silica). The periclase melting point (off scale) is depressed by the addition of forsterite. The lower part of the MgO liquidus curve is shown and it extends to the periclase-forsterite eutectic point $E1$ (1850°C, 38.1 wt. % SiO_2). A two-phase field containing MgO + melt exists in the region between pure MgO and its liquidus curve. Likewise, a two-phase field containing forsterite + melt exits in the region between pure Mg_2SiO_4 and its liquidus curve. This is the small region to the left of the Mg_2SiO_4 line because the eutectic is at the forsterite-rich end of the periclase-forsterite system. The MgO + melt and the Mg_2SiO_4 + melt fields both extend down to 1850°C, where the eutectic melt crystallizes. A two-phase field containing periclase + forsterite exists between MgO and Mg_2SiO_4 at lower temperatures because the two minerals are insoluble in one another. There is a second two-phase field containing forsterite + melt immediately to the right of the forsterite line. This field is labeled Fo + melt and extends down to 1557°C. At lower temperatures forsterite + enstatite (Fo + En) coexist in the region between the Mg_2SiO_4 and $MgSiO_3$ lines.

Periclase, forsterite, and silica each melt to a liquid of the same composition as the solid. This is called *congruent melting* and hence their melting points are congruent melting points. In contrast, enstatite does not melt to a liquid of the same composition. Instead it melts to a solid and a liquid, both having different compositions. This is called *incongruent melting* and hence enstatite has an *incongruent melting point*. The incongruent melting point of enstatite is 1557°C, where it melts to a small amount of forsterite and a larger amount of silica-rich liquid. The composition of the silica-rich liquid (at the point labeled P) is 60.9 wt. % SiO_2 + 39.1 wt. % MgO. This liquid coexists with forsterite, enstatite, and vapor, and thus point P is an invariant point.

Point P is called a *peritectic point*, which is an invariant point where the composition of the liquid cannot be expressed in terms of positive quantities of the solid phases involved in the equilibrium reaction. You can see this from Figure 12-4 because the peritectic point P lies to the silica-rich side of both enstatite and forsterite. This is qualitatively different than a eutectic point, which lies between the two solid phases involved in the reaction (e.g., the eutectic point in the anorthite-diopside phase system).

In general, a peritectic point, like a eutectic point, is the intersection of two different liquidus (solubility, freezing-point) curves. In this case point P is the intersection of the forsterite liquidus at higher temperatures and the enstatite liquidus at lower temperatures. Thermodynamic calculations show that the metastable extension of the enstatite liquidus extends to 1561°C, which is the fictive congruent melting point of enstatite (Swamy et al., 1994).

At a eutectic point the melt reacts to form two solids, for example,

$$\text{Eutectic melt} = \text{Anorthite} + \text{Diopside} \quad (12\text{-}20)$$

In contrast, at a peritectic point one solid reacts to form melt and a second, different solid or one solid reacts with melt to form a second, different solid, for example,

$$\text{Enstatite} = \text{Forsterite} + \text{Melt} \quad (12\text{-}21)$$

Reaction (12-21) proceeds to the right upon heating (enstatite melts incongruently) and to the left upon cooling (forsterite reacts with melt giving enstatite).

Peritectic reactions are common and occur in a large number of petrologically important systems. Some examples are the incongruent melting of hibonite ($CaAl_{12}O_{19}$) at 1883°C to corundum + CaO-rich liquid, of $Ca_3Al_2O_6$ to lime (CaO) + Al_2O_3-rich liquid, of mullite ($Al_6Si_2O_{13}$) at 1810°C to corundum (Al_2O_3) + silica-rich liquid, of $Ca_3Ti_2O_7$ at 1750°C to perovskite ($CaTiO_3$) + CaO-rich liquid, of monticellite ($CaMgSiO_4$) at 1503°C to periclase (MgO) + liquid, of rhodonite ($MnSiO_3$) at 1291°C to silica (cristobalite) + MnO-rich liquid, of orthoclase ($KAlSi_3O_8$) at 1150°C to leucite ($KAlSi_2O_6$) + silica-rich liquid, and of acmite ($NaFeSi_2O_6$) at 988°C to hematite (Fe_2O_3) + Na_2O-rich and SiO_2-rich liquid. Some systems such as CaO-Al_2O_3 have more than one compound with incongruent melting points. A *peritectoid* reaction is the solid-state analog to a peritectic reaction. In the case of a peritectoid, one solid reacts to form two different solids.

The small four-sided wedge below point *P* is a two-phase region in which enstatite + melt coexist. The curve between *P* and *E2* is the liquidus curve of enstatite. It ends at the enstatite-silica eutectic point *E2* at 1543°C and a composition of 65 wt. % silica. Enstatite and silica coexist in the region below eutectic point *E2*. The stable phases of enstatite and silica that coexist with one another depend on temperature, but these details are not shown, to simplify the phase diagram.

The silica liquidus curve ascends from the eutectic point at 1543°C to 1695°C. Silica + melt coexist in the region to the right of the liquidus curve. The silica liquidus curve ends at 1695°C, where unmixing of the melt into two separate liquids occurs (Greig, 1927). The dashed lines that begin at 1695°C (at 68.9 wt % and 97.9 wt % SiO_2) are the two sides of the coexistence (or *binodal*) curve. The coexistence curve for MgO-SiO_2 liquids has a similar appearance to the coexistence curve for CO_2 (Figures 8-1 and 8-18 in Chapter 8). Two different *conjugate liquids*—one being nearly pure silica and the other one containing more MgO—coexist inside the dome. Tie-lines, which are parallel to isotherms, connect the compositions of the conjugate liquids, which approach each other with increasing temperature. The distinction between the two liquids vanishes at the top of the binodal at ~1970°C and 89.0 wt % SiO_2 (Hageman and Oonk, 1986). This temperature is variously known as the *upper consulate temperature* or *upper critical solution temperature*. Finally, the silica liquidus, barely visible on Figure 12-4, extends from the silica melting point (1726°C) to 1695°C (at ~97.9% SiO_2).

Unmixing and separation into two different immiscible liquids commonly occurs at the high silica end of metal oxide–silica binary systems (e.g., BaO, CaO, CoO, FeO, K_2O, La_2O_3, Li_2O, MgO, MnO, Na_2O, Sm_2O_3, SrO, ThO_2, Y_2O_3, Yb_2O_3, ZrO_2). One liquid is silica-rich and the other is richer in the metal oxide. The unmixing occurs because the Gibbs energy of the silicate melt is positive with respect to the Gibbs energy of the two separate, immiscible liquids. The binodal curve is the locus of (*T, X*) points where the first derivative of the integral Gibbs energy of mixing (ΔG^M) zero equals zero:

$$\left(\frac{\partial \Delta G_M}{\partial X_{SiO_2}}\right)_{P,T} = 0 \tag{12-22}$$

The activities of silica and of the metal oxide are equal in the SiO_2-rich (A) and MgO-rich (B) melts on either side of the binodal curve, for example, for the MgO-SiO_2 system:

$$a_{SiO_2}\ (melt\ A) = a_{SiO_2}(melt\ B) \tag{12-23}$$

$$a_{MgO}\ (melt\ A) = a_{MgO}\ (melt\ B) \tag{12-24}$$

The *spinodal* curve is inside the binodal curve. At a given temperature the sides of the spinodal curve are the compositions where the second derivative of the integral Gibbs energy of mixing (ΔG^M) equals zero, that is,

$$\left(\frac{\partial^2 \Delta G_M}{\partial X^2_{SiO_2}}\right)_{P,T} = 0 \quad (12\text{-}25)$$

It is possible that unmixing does not occur in the region between the binodal and spinodal curves and a one-phase metastable melt may persist. This is because the ΔG^M curve is still concave downward in the region between the binodal and spinodal. The ΔG^M curve is convex upward throughout the entire region inside the spinodal boundaries, and unmixing should occur unless prevented because of slow kinetics. The spinodal and binodal curves coincide at the upper critical solution temperature where the second and third derivatives of the integral Gibbs energy of mixing are simultaneously equal to zero:

$$\left(\frac{\partial^2 \Delta G^M}{\partial^2 X^2_{SiO_2}}\right)_{P,T} = \left(\frac{\partial^3 \Delta G^M}{\partial^3 X^3_{SiO_2}}\right)_{P,T} = 0 \quad (12\text{-}26)$$

These equations are not specific to metal oxide–silica melts and apply to all other immiscible liquids and to immiscible solid solutions.

Bowen and Andersen (1914) first discussed crystallization behavior in the vicinity of the enstatite peritectic point. Forsterite is the first phase to crystallize out of melts with compositions lying between forsterite and enstatite. Further cooling leads to larger amounts of forsterite until the incongruent melting point of enstatite at 1557°C is reached. Then forsterite begins to dissolve back into the melt (with composition P). As we continue cooling below 1557°C a mixture of forsterite and enstatite forms. If a melt of enstatite composition is cooled, forsterite is still the first phase formed. At 1557°C the forsterite begins to redissolve in the melt and it disappears with further cooling, giving only enstatite. Forsterite is also the first precipitate from melts that lie between enstatite and the peritectic point (~ 60.9 wt % SiO_2) in composition. However, at 1557°C it reacts with the melt quantitatively and completely disappears. As cooling continues enstatite crystallizes and the melt composition moves along the enstatite liquidus until it reaches the eutectic with silica.

G. Thermodynamics of systems with intermediate compounds

We can also describe the MgO-SiO_2 phase diagram (and other phase diagrams with intermediate compounds) in terms of chemical equilibria and the activities of the two end-members. This approach is important because measurements of thermodynamic activities as a function of composition along different isotherms can be used to help determine phase diagrams. For example, Zaitsev et al. (2006) measured MgO and SiO_2 activities by Knudsen effusion mass spectroscopy (KEMS) and used their data to calculate the positions and temperatures of phase boundaries in the MgO-SiO_2 phase diagram. Other methods that have been used to measure activities (and examples of their application) include electrochemical measurements of the CaO activity in the CaO-TiO_2 system, Zn vapor pressure measurements in Cu-Zn alloys, and equilibration of CO/CO_2 gas mixtures with iron + wüstite in the Fe-O system. Kubaschewski and Alcock (1979) review experimental methods applied to the determination of phase diagrams.

We consider chemical reactions along the 1527°C (1800 K) isotherm. This spans three different two-phase fields (periclase + forsterite, forsterite + enstatite, and enstatite + silica) and four different compounds (MgO, Mg_2SiO_4, $MgSiO_3$, and SiO_2). We start with the MgO + Mg_2SiO_4 two-phase field at the MgO-rich end of the phase diagram. The coexistence of periclase + forsterite uniquely fixes the activities of MgO and SiO_2 at any given temperature. Neither periclase nor forsterite dissolves in the other phase, and the activities of MgO and Mg_2SiO_4 are unity. The amounts of periclase and forsterite vary according to the lever rule as we move from pure MgO to pure Mg_2SiO_4 across the 1527°C isotherm, but their thermodynamic activities remain constant. Silica is not present in the periclase + forsterite two-phase field. However, its activity can be calculated from the equilibrium constant for the reaction

$$2\,\text{MgO (periclase)} + \text{SiO}_2\text{ (silica)} = \text{Mg}_2\text{SiO}_4\text{ (forsterite)} \tag{12-27}$$

$$K_{\text{Fo}} = \frac{a_{\text{Mg}_2\text{SiO}_4}}{a^2_{\text{MgO}}a_{\text{SiO}_2}} = \frac{1}{a_{\text{SiO}_2}} \tag{12-28}$$

$$a_{\text{SiO}_2} = \frac{1}{K_{\text{Fo}}} = \exp\left(\frac{\Delta_r G^o_T}{RT}\right) \tag{12-29}$$

The equilibrium constant for reaction (12-27) can be calculated from the standard Gibbs energy change using Eq. (10-22) and tabulated data. This was done using data for forsterite, periclase, and β-cristobalite, which is the stable silica polymorph (up to its melting point of 1726°C) at the temperatures shown in the phase diagram. The results are

$$\Delta_r G^o_T = -67{,}606 + 13.8273T - 3.335 \times 10^{-3}T^2\ \text{J mol}^{-1} \tag{12-30}$$

$$\log_{10}K_{\text{Fo}} = \frac{-\Delta G^o_T}{19.1437T} = \frac{3531.5}{T} - 0.7223 + 1.742 \times 10^{-4}T \tag{12-31}$$

These equations are valid from 1700 K to 1999 K (1427°C to 1726°C). At 1800 K,

$$a_{\text{SiO}_2} = \frac{1}{35.75} = \exp\left(\frac{-53{,}522}{RT}\right) = 0.0280 \tag{12-32}$$

As stated previously, the silica activity is constant (at a given temperature) throughout the MgO + Mg_2SiO_4 two-phase field. However, the activity coefficient (γ) of silica decreases across the two-phase field because the mole fraction of silica increases. For example, at a silica mole fraction of 0.2 its activity coefficient is

$$\gamma_{\text{SiO}_2} = \frac{a_{\text{SiO}_2}}{X_{\text{SiO}_2}} = \frac{0.0280}{0.20} \sim 0.14 \tag{12-33}$$

This decreases to 0.0840 just to the right of the forsterite line ($X_{silica} \sim 0.333$).

The forsterite + enstatite two-phase field is next along the 1527°C isotherm. The two silicates are insoluble in one another and they each have unit activity. The activities of MgO and SiO_2 are less than unity, but are constant across the two-phase field. The amounts (i.e., mole fractions) of each do vary across the two-phase field according to the lever rule and thus their activity coefficients covary to maintain constant MgO and silica activities along the isotherm. The activity of MgO is calculated from the equilibrium constant for the reaction

$$MgO\ (periclase) + MgSiO_3\ (enstatite) = Mg_2SiO_4\ (forsterite) \qquad (12\text{-}34)$$

$$K_{En} = \frac{a_{Mg_2SiO_4}}{a_{MgSiO_3}} \cdot \frac{1}{a_{MgO}} \qquad (12\text{-}35)$$

After rearranging, we have

$$a_{MgO} = \frac{1}{K_{En}} = \exp\left(\frac{\Delta_r G_T^o}{RT}\right) \qquad (12\text{-}36)$$

The values for the standard Gibbs energy and equilibrium constant (1700–1830 K) are

$$\Delta_r G_T^o = -33{,}380 + 9.898T - 2.64 \times 10^{-3} T^2\ \mathrm{J\ mol^{-1}} \qquad (12\text{-}37)$$

$$\log_{10} K_{En} = \frac{-\Delta G_T^o}{19.1437T} = \frac{1743.7}{T} - 0.5170 + 1.379 \times 10^{-4} T \qquad (12\text{-}38)$$

The MgO activity at 1800 K in the forsterite + enstatite two phase field is

$$a_{MgO} = \frac{1}{5.01} = \exp\left(\frac{-24{,}117}{RT}\right) = 0.200 \qquad (12\text{-}39)$$

The silica activity is computed using Eq. (12-28), the MgO activity above, and $K_{Fo} = 35.75$. The result is

$$a_{SiO_2} = \frac{a_{Mg_2SiO_4}}{a_{MgO}^2 K_{Fo}} = \frac{1}{(0.200)^2 35.75} = 0.140 \qquad (12\text{-}40)$$

This is larger than the silica activity inside the periclase + forsterite field. This is expected because the enstatite + forsterite field occurs at larger silica contents. We could also have computed the silica activity from the reaction forming enstatite from the oxides.

Finally, we consider the enstatite + silica two-phase field. In this case the activities of enstatite and silica are unity and their coexistence fixes the MgO activity at a unique value at each temperature. The MgO activity is computed from the equilibrium constant for the reaction forming enstatite from its constituent oxides:

$$MgO\ (periclase) + SiO_2\ (silica) = MgSiO_3\ (enstatite) \qquad (12\text{-}41)$$

$$K_{silica} = \frac{a_{MgSiO_3}}{a_{MgO} a_{SiO_2}} = \frac{1}{a_{MgO}} \qquad (12\text{-}42)$$

$$a_{MgO} = \frac{1}{K_{silica}} = \exp\left(\frac{\Delta_r G_T^o}{RT}\right) \qquad (12\text{-}43)$$

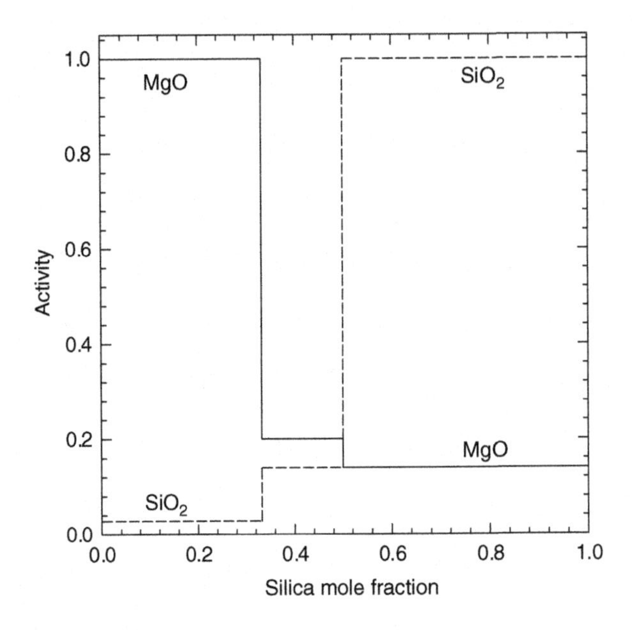

FIGURE 12-5

MgO and SiO_2 activities along the 1527°C (1800 K) isotherm in the MgO-SiO_2 phase diagram.

Equations valid from 1700 K to 1816 K for the standard Gibbs energy and K_{silica} are

$$\Delta_r G_T^o = -32{,}097 + 1.496T \text{ J mol}^{-1} \tag{12-44}$$

$$\log_{10} K_{silica} = \frac{-\Delta G_T^o}{19.1437T} = \frac{1676.6}{T} - 0.078 \tag{12-45}$$

The MgO activity at 1800 K in the enstatite + silica two-phase field is

$$a_{\text{MgO}} = \frac{1}{7.13} = \exp\left(\frac{-29{,}404}{RT}\right) = 0.14 \tag{12-46}$$

Figure 12-5 shows the variation of MgO and SiO_2 activities across the 1800 K (1527°C) isotherm. The activities are fixed within the two-phase fields and show discontinuities at silica mole fractions corresponding to the four compounds (periclase, forsterite, enstatite, and silica) on the phase diagram. In contrast, activities of MgO and SiO_2 in the melts vary with both composition and temperature and must be measured as a function of both.

H. Binary phase diagrams of water and salts

Phase diagrams of water and salts, such as bromides, carbonates, chlorides, fluorides, iodides, nitrates, and sulfates, are important for low-temperature geochemistry (e.g., the origin of evaporite minerals on

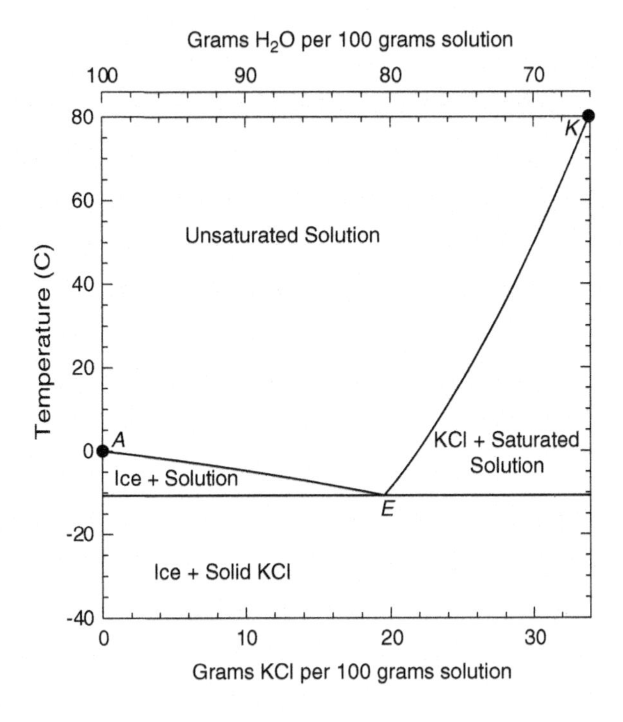

FIGURE 12-6

Sylvite (KCl)-water ice phase diagram.

Earth, Mars, and Europa). Some of these phase diagrams, such as sylvite (KCl)–water ice and mascagnite $(NH_4)_2SO_4$–water ice, which are listed in Table 12-1, are simple eutectics. Other systems such as halite (NaCl)–water ice involve the formation of intermediate hydrated salts, (e.g., $NaCl \cdot 2\ H_2O$). We discuss the interpretation of these phase diagrams using the sylvite (KCl)–water ice diagram in Figure 12-6 as an example. The general appearance of the sylvite-ice phase diagram is the same as that of the anorthite-diopside phase diagram in Figure 12-1. Curves *AE* and *KE* are solubility curves for KCl in water. The solubility data are listed in Table 12-2 and are in order of increasing solubility (grams KCl dissolved in 100 grams of solution). Curve *AE* extends from −0.34°C to −10.70°C, which is the eutectic point, where 19.54 grams of KCl dissolves in 100 grams of solution. Curve *AE* is also the solubility curve for ice dissolving in the solution (see the top scale in Figure 12-6). At the eutectic temperature of −10.70°C, 80.46 grams of ice dissolves in the solution. Thus, the solution at the eutectic point *E* is saturated with respect to both ice and KCl.

The two-phase region labeled ice + solution shows the region where pure water ice coexists in equilibrium with solution. Tie-lines are parallel to isotherms in this region and the lever rule can be used to calculate the amounts of ice and solution in equilibrium with one another at any temperature. The composition of the solution in equilibrium with ice varies with temperature and moves along curve *AE*. This solution is unsaturated with respect to KCl but is saturated with respect to ice.

Table 12-2 KCl Solubility in Water

T (°C)	Solubility g/100 g Solution	Coexisting Phases
−0.34	0.7456	Ice + solution
−0.858	1.864	Ice + solution
−1.681	3.728	Ice + solution
−2.24	4.95	Ice + solution
−3.07	7.09	Ice + solution
−4.60	9.48	Ice + solution
−6.88	13.70	Ice + solution
−10.31	19.02	Ice + solution
−10.70	19.54	Ice + KCl + solution
0	21.92	KCl + solution
5	22.90	KCl + solution
10	23.80	KCl + solution
15	24.70	KCl + solution
20	25.50	KCl + solution
25	26.40	KCl + solution
30	27.10	KCl + solution
40	28.60	KCl + solution
50	30.0	KCl + solution
60	31.40	KCl + solution
70	32.70	KCl + solution
80	33.90	KCl + solution

Source: Linke (1965).

Curve *KE* shows the solubility of KCl in water as a function of temperature and extends from the eutectic point at −10.70°C upward to the boiling point (108.6°C) of KCl-saturated solution (446 grams per liter). In fact, curve *KE* extends up to the melting point of KCl at 770°C and also gives the solubility of KCl in steam. However, it is terminated at 80°C in Figure 12-6 and Table 12-2.

The two-phase region labeled KCl + saturated solution shows where pure KCl coexists in equilibrium with solution and the two-phase region labeled ice + solid KCl shows where KCl and ice coexist in equilibrium below the eutectic temperature.

In many cases salts form one or more hydrates and the resulting phase diagrams have intermediate compounds that form eutectics and/or may also incongruently decompose (to another hydrate + solution or to anhydrous salt + solution) at low temperatures. For example, NaCl, NaBr, and NaI form the hydrates $NaCl \cdot 2\ H_2O$, $NaBr \cdot 2\ H_2O$, $NaI \cdot 5\ H_2O$ and the ice-hydrate eutectics are at −21.2°C (23.3 wt. % NaCl), −28°C (40.3 wt. % NaBr), and −31.5°C (47.1 wt. % NaI), respectively. An example of peritectic decomposition is $NaCl \cdot 2\ H_2O$, which incongruently decomposes at 0.15°C to halite (NaCl) + aqueous solution. Examples of peritectic reactions involving hydrated salts that occur at higher temperatures are the incongruent dehydration of $Na_2SO_4 \cdot 10\ H_2O$ (mirabilite) to Na_2SO_4 (thenardite) + solution at

32.38°C (also discussed in Chapter 11), of $Na_2CO_3 \cdot 10\,H_2O$ (natron) to $Na_2CO_3 \cdot 7\,H_2O$ + solution, and of $Na_2CO_3 \cdot 7\,H_2O$ to $Na_2CO_3 \cdot H_2O$ (thermonatrite) + solution at ~32°C.

I. Mutual liquid and solid solutions

There are three types of phase diagrams displayed by materials that are completely soluble in the solid and liquid states (*isomorphous* materials). The simplest, and possibly most common, type has a continuous variation of the solidus and liquidus curves from the melting point of the lower melting component to that of the higher melting component. The albite ($NaAlSi_3O_8$)–anorthite ($CaAl_2Si_2O_8$) phase diagram in Figure 12-7 exemplifies binary phase diagrams of this type. Table 12-3 lists some other examples of systems displaying similar behavior.

The second, less common type of isomorphous phase diagram has a minimum melting point at some intermediate composition. The åkermanite ($Ca_2MgSi_2O_7$)–gehlenite ($Ca_2Al_2SiO_7$) phase diagram in Figure 12-8 exemplifies this behavior. Other examples are listed in Table 12-4. Finally, the third, least common type has a maximum melting point at some intermediate composition. The wulfenite ($PbMoO_4$)-$Pr_2(MoO_4)_3$ phase diagram is apparently of this type (Zambonini, 1923) but should be reinvestigated. The classic example is the phase diagram of the enantiomers (i.e., optical isomers denoted *R* (for rectus, or right) and *S* (for sinister, or left) of carvoxime ($C_{10}H_{14}NOH$), which melt at 72°C while the racemic mixture (1:1 *R*:*S*) melts at 91°C. Carvoxime is a derivative of carvone ($C_{10}H_{14}O$), which smells like spearmint (as the *R* optical isomer) or caraway (as the *S* optical isomer). We discuss the first two types here. Calvet and Oonk (1995) discuss the carvoxime phase diagram.

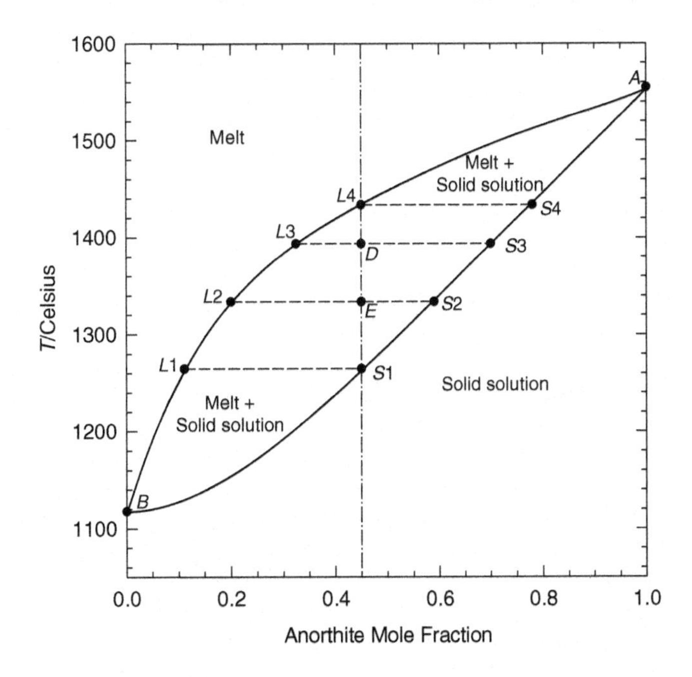

FIGURE 12-7

Albite ($NaAlSi_3O_8$)–anorthite ($CaAl_2Si_2O_8$) phase diagram.

Table 12-3 Some Isomorphous Solid Solutions (Cigar Phase Diagram)

A	m.p. (°C)	B	m.p. (°C)
CaO	2572	MnO	1810
eskolaite	2239	karelianite	1976
eskolaite	2239	corundum	2054
forsterite	1890	fayalite	1217
forsterite	1890	tephroite	1251
Si	1685	Ge	1210
anorthite	1557	albite	1120
Pd	1552	Ag	961
tephroite	1347	fayalite	1217
K_2SO_4	1069	K_2CO_3	901
Au	1063	Ag	961
KF	857	KOH	404

Sources: Bowen, 1913; Bowen and Schairer, 1935; Cook and McMurdie, 1989; Dessurault et al., 1990; Glasser and Osborn, 1960; Hultgren et al., 1973; Levin et al., 1964.

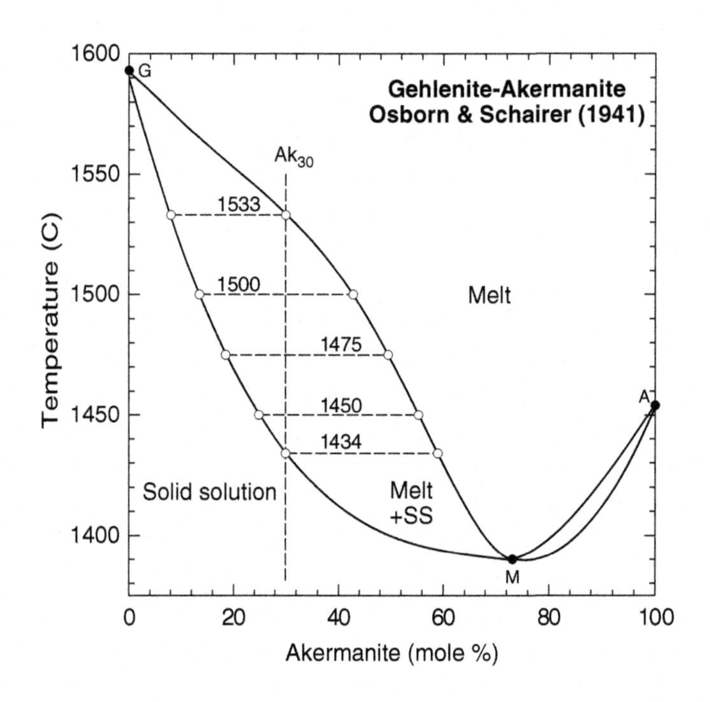

FIGURE 12-8

The åkermanite ($Ca_2MgSi_2O_7$)–gehlenite ($Ca_2Al_2SiO_7$) phase diagram.

Table 12-4 Examples of Solid Solutions with a Minimum Melting Point

Components				Minimum Melting Point	
A	m.p. (°C)	B	m.p. (°C)	Mass (molar) Comp.	*T* (°C)
$SrSiO_3$	1578	α-$CaSiO_3$	1548	$A_{56}B_{44}$ ($A_{47}B_{53}$)	1474
gehlenite	1593	akermanite	1454	$A_{27}B_{73}$ ($A_{27}B_{73}$)	1390
Pd	1552	Ni	1453	$A_{60}B_{40}$ ($A_{45}B_{55}$)	1237
sanidine	1200	albite	1120	$A_{35}B_{65}$ ($A_{34}B_{66}$)	1063
K_2SO_4	1069	Na_2SO_4	884	$A_{29}B_{71}$ ($A_{25}B_{75}$)	832
Na_2SO_4	884	Na_2CO_3	858	$A_{43.8}B_{56.2}$ ($A_{36.8}B_{63.2}$)	828
K_2CO_3	901	Na_2CO_3	858	$A_{48}B_{52}$ ($A_{41}B_{59}$)	710
NaCl	801	KCl	771	$A_{44.5}B_{55.5}$ ($A_{50.6}B_{49.4}$)	657
NaBr	747	KBr	734	$A_{45}B_{55}$ ($A_{49}B_{51}$)	644
KNO_3	337	$NaNO_3$	310	$A_{55}B_{45}$ ($A_{51}B_{49}$)	223
Rb	39.5	Cs	28.4	$A_{39}B_{61}$ ($A_{50}B_{50}$)	9

Sources: Cook and McMurdie, 1989; Dessureault et al., 1990; Eskola, 1922; Hultgren et al., 1973; Osborn and Schairer, 1941; Sangster and Pelton, 1987; Schairer, 1950.

Solid and liquid solutions with continuous liquidus and solidus curves

We begin with the albite-anorthite phase diagram in Figure 12-7, based on the classic work of Bowen (1913). Albite (Ab, $NaAlSi_3O_8$) and anorthite (An) are completely soluble as solids and as liquids. Points *A* (anorthite, 1557°C) and *B* (albite, 1120°C) in Figure 12-7 are their melting points. The solidus curve (*B-S*1-*S*2-*S*3-*S*4-*A*) and the liquidus curve (*B-L*1-*L*2-*L*3-*L*4-*A*) enclose a two-phase field containing melt + solid solution (plagioclase feldspar). The horizontal dashed lines *L*1-*S*1, *L*2-*S*2, *L*3-*S*3, and *L*4-*S*4 are tie-lines (parallel to isotherms) connecting the melt and solid solution coexisting in equilibrium with one another. The coexisting melt and solid solution are conjugate phases. Along any tie-line, the solid solution is richer in anorthite than is the conjugate melt. In general, for this type of phase diagram, the component with the higher melting point (e.g., anorthite) is enriched in the solid and the component with the lower melting point (e.g., albite) is enriched in the melt. The vertical dot-dashed line (*L*4-*D*-*E*-*S*1) is an isopleth at 45 mole % anorthite (An_{45}). We now consider cooling along this isopleth and assume complete equilibrium between the melt and solid solution.

The first infinitesimal grain of solid appears when the melt is cooled to 1434°C (point *L*4, where the isopleth and liquidus lines intersect). Point *L*4 gives the composition of the melt (45 mole % anorthite, An_{45}) and point *S*4 gives the composition of the solid solution (An_{78}). There is essentially no solid and only melt at point *L*4. Continued cooling to 1394°C brings us to point *D* on the *L*3-*S*3 tie-line. Point *L*3 gives the melt composition ($An_{32.5}$) and point *S*3 gives the solid solution composition (An_{70}). The lever rule shows that point *D* is a mixture of 66.67 mole % melt and 33.33 mole % solid (2:1 ratio):

$$\frac{melt}{solid} = \frac{|S3 - D|}{|D - L3|} = \frac{|0.70 - 0.45|}{|0.45 - 0.325|} = \frac{0.25}{0.125} = 2.0 \qquad (12\text{-}47)$$

Cooling to 1334°C brings us to point *E* on the *L2*-*S2* tie-line. The melt is at An_{20} (point *L2*) and the solid solution is at An_{59} (point *S2*) and the melt/solid molar ratio is 0.56 at point *E*. Cooling to 1265°C brings us to point *S1* on the *L1*-*S1* tie-line. The melt is at $An_{11.1}$ (point *L1*) and the solid solution is at An_{45}, which is the same as the isopleth composition. Point *L1* is almost completely made of the solid solution and the last infinitesimal drop of melt is crystallizing. Throughout this entire cooling sequence the amount of melt is decreasing and the amount of solid solution is increasing. The melt becomes richer in albite as its composition moves from *L4* to *L1* along the liquidus curve. The feldspar crystals also become richer in albite as the composition of the solid solution moves from *S4* to *S1* along the solidus curve.

The continual change in composition of the melt and solid solution during cooling requires sufficient time at any temperature for all the feldspar crystals (including those formed earlier) to react with and equilibrate with the melt. If there is not enough time for the melt and crystals to equilibrate and/or crystals sediment out as they form, *fractional crystallization* occurs and zoned crystals with An-rich cores and Ab-rich rims form. In fact, this zoning is observed in natural plagioclase feldspars.

We now consider the extreme case of perfect fractional crystallization in which all of the crystals are removed from contact with the melt as soon as they form (e.g., by using a centrifugal furnace that spins out the denser crystals). This process shifts the bulk composition of the system toward the albite side of the phase diagram. For example, we start again on the An_{45} isopleth at point *L4* and discard the minute amount of solid with composition *S4*. As we continue cooling and follow the same procedure we find ourselves at point *L3* ($An_{32.5}$). The difference between this and crystallization and cooling at complete equilibrium is that the entire system (i.e., melt + solid solution) has $An_{32.5}$ composition. Continuing down along the liquidus curve we end up at point *B*, pure albite, and there is no anorthite left. The reason is that the crystals that were removed as soon as they formed contain all of the original anorthite. The actual course of events, formation of zoned crystals, is between this extreme and complete equilibrium.

We now consider thermodynamics of phase diagrams like the albite-anorthite diagram (those in Table 12-3 except the olivines, which we describe later). Along each of the tie-lines joining conjugate melt and solid, albite (Ab) and anorthite (An) in the solid solution (ss) are in equilibrium with albite and anorthite in the melt, and we can write equations analogous to Eqs. (12-7) and (12-15), namely

$$\text{Albite (solid solution)} = \text{Albite (melt)} \tag{12-48}$$

$$\text{Anorthite (solid solution)} = \text{Anorthite (melt)} \tag{12-49}$$

The equilibrium constants for these melting equilibria are

$$K_{\text{Ab}} = \frac{a_{\text{Ab}(melt)}}{a_{\text{Ab}(ss)}} = \frac{X_{\text{Ab}(melt)}}{X_{\text{Ab}(ss)}} \cdot \frac{\gamma_{\text{Ab}(melt)}}{\gamma_{\text{Ab}(ss)}} \tag{12-50}$$

$$K_{\text{An}} = \frac{a_{\text{An}(melt)}}{a_{\text{An}(ss)}} = \frac{X_{\text{An}(melt)}}{X_{\text{An}(ss)}} \cdot \frac{\gamma_{\text{An}(melt)}}{\gamma_{\text{An}(ss)}} \tag{12-51}$$

If the melt and solid solution are ideal or if the activity coefficients in each phase are the same, then the equilibrium constant expressions reduce to the mole fraction ratios. In this case the variation of the equilibrium constants with temperature is given by the equations

$$\left[\frac{\partial \ln K_{eq}}{\partial(1/T)}\right]_P = \left[\frac{\partial \ln(X_{melt}/X_{ss})_{\text{Ab}}}{\partial(1/T)}\right]_P = -\frac{\Delta_{fus}H^o}{R} \tag{12-52}$$

$$\left[\frac{\partial \ln K_{eq}}{\partial(1/T)}\right]_P = \left[\frac{\partial \ln(X_{melt}/X_{ss})_{\text{An}}}{\partial(1/T)}\right]_P = -\frac{\Delta_{fus}H^o}{R} \tag{12-53}$$

Thus, plots of ln $(X_{melt}/X_{ss})_{\text{Ab}}$ and ln $(X_{melt}/X_{ss})_{\text{An}}$ versus $1/T$ (in kelvins) give straight lines with slopes equal to $-\Delta_{fus}H^o/R$ for the enthalpies of fusion of albite and anorthite.

The X_{melt} and X_{ss} terms are simply the compositions of the conjugate melt and solid solution along the liquidus and solidus in terms of the albite or anorthite mole fractions.

Example 12-3. Use the anorthite contents of coexisting melt and solid solution to estimate the enthalpy of fusion of anorthite. We took values from Bowen (1913) and used Eq. (12-53). Figure 12-9 shows a plot of ln $(X_{melt}/X_{ss})_{\text{An}}$ versus $1/T$ (in Kelvins). The line on the graph is a linear least-squares fit to the points. The slope of the line is $-\Delta_{fus}H^o/R = -13{,}865$. The derived enthalpy of fusion for anorthite is ~115 kJ mol^{-1}, which is ~18 kJ (14%) smaller than the actual value of 134 kJ mol^{-1} (Richet and Bottinga, 1984).

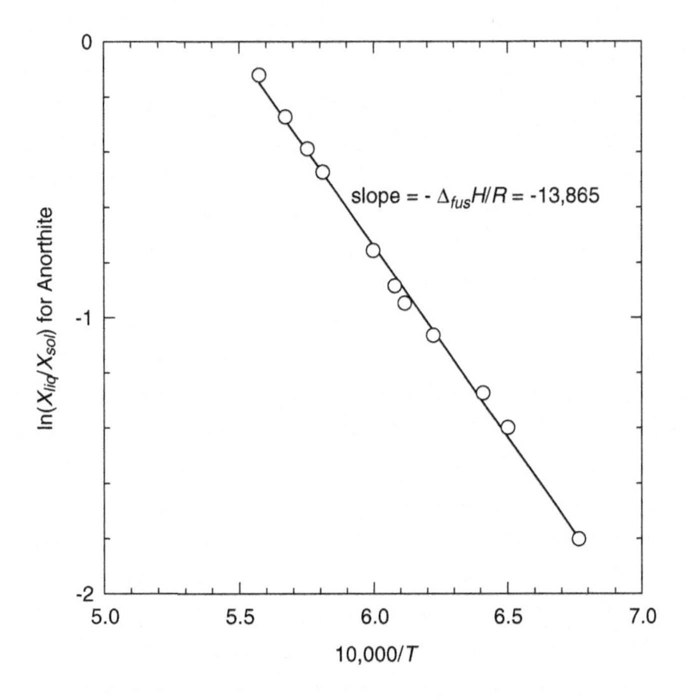

FIGURE 12-9

Calculation of the enthalpy of fusion of anorthite.

Briefly, in the case of nonideal solutions, an equation such as the following is needed for each of the two components:

$$\left[\frac{\partial \ln K_{eq}}{\partial(1/T)}\right]_P = \left[\frac{\partial \ln(a_{melt}/a_{ss})_{\mathrm{Ab}}}{\partial(1/T)}\right]_P = -\frac{\Delta H^o}{R} \tag{12-54}$$

The enthalpy term is now composed of the enthalpy of fusion of albite and the partial molal enthalpy of mixing of albite in the melt and in the solid solution. The two partial molal enthalpy terms are related to the temperature dependence of the activity coefficients of albite in the melt and in the solid solution.

We now illustrate computation of liquidus and solidus curves from the melting points and enthalpies of fusion of albite and anorthite. In this case we rewrite the equilibrium constant expressions in terms of the Gibbs energies of fusion,

$$\Delta_{fus}G^o(\mathrm{Ab}) = -RT \ln K_{\mathrm{Ab}} = RT \ln\left[\frac{a_{\mathrm{Ab}(ss)}}{a_{\mathrm{Ab}(melt)}}\right] \approx RT \ln\left[\frac{X_{\mathrm{Ab}(ss)}}{X_{\mathrm{Ab}(melt)}}\right] \tag{12-55}$$

$$\Delta_{fus}G^o(\mathrm{An}) = -RT \ln K_{\mathrm{An}} = RT \ln\left[\frac{a_{\mathrm{An}(ss)}}{a_{\mathrm{An}(melt)}}\right] \approx RT \ln\left[\frac{X_{\mathrm{An}(ss)}}{X_{\mathrm{An}(melt)}}\right] \tag{12-56}$$

Equations (12-55) and (12-56) show the real solution (activities) and the ideal solution (mole fractions) versions. We assume ideality and rearrange the two equations to solve for the mole fractions along the liquidus and solidus curves. Equation (12-18) gives the temperature dependence of the Gibbs energy of fusion. However, to first approximation the ΔC_p integrals can be neglected because the enthalpy and entropy of fusion at the melting point are typically large relative to the corrections introduced by the two ΔC_p terms. In this case the Gibbs energy of fusion is given by

$$\Delta_{fus}G^o = \Delta_{fus}H^o \frac{(T_m - T)}{T_m} \tag{12-57}$$

The T_m and T in Eq. (12-57) are the melting-point temperature and some other temperature T along the liquidus or solidus curves. The desired equations are derived using the identity $X_{\mathrm{An}} = (1 - X_{\mathrm{Ab}})$ and are

$$X_{\mathrm{Ab}}(solidus) = \frac{1 - \exp(-\Delta_{fus}G^o_{\mathrm{An}}/RT)}{\exp(-\Delta_{fus}G^o_{\mathrm{Ab}}/RT) - \exp(-\Delta_{fus}G^o_{\mathrm{An}}/RT)} \tag{12-58}$$

$$X_{\mathrm{Ab}}(liquidus) = \frac{[1 - \exp(-\Delta_{fus}G^o_{\mathrm{An}}/RT)]\exp(-\Delta_{fus}G^o_{\mathrm{Ab}}/RT)}{\exp(-\Delta_{fus}G^o_{\mathrm{Ab}}/RT) - \exp(-\Delta_{fus}G^o_{\mathrm{An}}/RT)} \tag{12-59}$$

Equations (12-58) and (12-59) are perfectly general. The lower melting-point component replaces albite and the higher melting-point component replaces anorthite.

Example 12-4. Calculate and plot the liquidus and solidus curves for the albite-anorthite phase diagram using the melting points and enthalpies of fusion for albite (1120°C, 63,550 J mol^{-1}) and anorthite (1557°C, 133,000 J mol^{-1}). (There are two different values, reported by two different groups, for the enthalpy of fusion of albite, and we use the mean value in our calculations.)

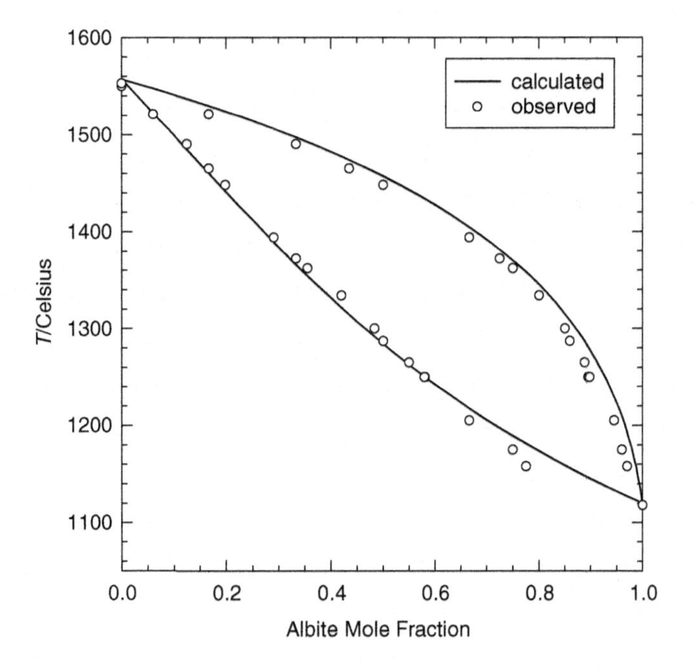

FIGURE 12-10

A comparison of the observed and calculated albite ($NaAlSi_3O_8$)-anorthite ($CaAl_2Si_2O_8$) phase diagram.

Figure 12-10 shows our results and compares the observed points with the diagram calculated using Eqs. (12-58) and (12-59). The following table gives our calculated albite mole fractions along the liquidus (X_L) and solidus (X_S) curves.

T(°C)	1557	1497	1437	1377	1307	1247	1187	1127	1120
X_L	0	0.334	0.570	0.735	0.862	0.931	0.973	0.998	1
X_S	0	0.104	0.206	0.313	0.450	0.588	0.757	0.971	1

In the case of olivine solid solutions, it is useful to consider equilibria involving one-half of a formula unit, for example, $MgSi_{0.5}O_2$ (= ½ Mg_2SiO_4, Fo) and $FeSi_{0.5}O_2$ (= ½ Fe_2SiO_4, Fa) because the Mg and Fe cations each can be distributed over the M1 and M2 octahedral sites in olivine (e.g., see Muan, 1967; Nafziger and Muan, 1967). The reactions for melt-solid equilibria of forsterite and fayalite are

$$MgSi_{0.5}O_2 \text{ (olivine)} = MgSi_{0.5}O_2 \text{ (melt)} \tag{12-60}$$

$$\Delta_{fus}G^o\,(MgSi_{0.5}O_2) = \frac{1}{2}\,\Delta_{fus}G^o\,(Mg_2SiO_4) \tag{12-61}$$

$$FeSi_{0.5}O_2 \text{ (olivine)} = FeSi_{0.5}O_2 \text{ (melt)} \tag{12-62}$$

$$\Delta_{fus}G^o\,(FeSi_{0.5}O_2) = \frac{1}{2}\,\Delta_{fus}G^o\,(Fe_2SiO_4) \tag{12-63}$$

The corresponding equilibrium constants are

$$K_{Fo} = \frac{a_{MgSi_{0.5}O_2(melt)}}{a_{MgSi_{0.5}O_2(olivine)}} = \frac{X_{MgSi_{0.5}O_2(melt)}}{X_{MgSi_{0.5}O_2(olivine)}} \cdot \frac{\gamma_{MgSi_{0.5}O_2(melt)}}{\gamma_{MgSi_{0.5}O_2(olivine)}} \approx \frac{X_{MgSi_{0.5}O_2(melt)}}{X_{MgSi_{0.5}O_2(olivine)}} \tag{12-64}$$

$$K_{Fa} = \frac{a_{FeSi_{0.5}O_2(melt)}}{a_{FeSi_{0.5}O_2(olivine)}} = \frac{X_{FeSi_{0.5}O_2(melt)}}{X_{FeSi_{0.5}O_2(olivine)}} \cdot \frac{\gamma_{FeSi_{0.5}O_2(melt)}}{\gamma_{FeSi_{0.5}O_2(olivine)}} \approx \frac{X_{FeSi_{0.5}O_2(melt)}}{X_{FeSi_{0.5}O_2(olivine)}} \tag{12-65}$$

The standard Gibbs energies calculated from the equilibrium constants are

$$\Delta_{fus}G^o(MgSi_{0.5}O_2) = -RT\ln K_{Fo} \approx -RT\ln\frac{X_{MgSi_{0.5}O_2(melt)}}{X_{MgSi_{0.5}O_2(olivine)}} \tag{12-66}$$

$$\Delta_{fus}G^o(FeSi_{0.5}O_2) = -RT\ln K_{Fa} \approx -RT\ln\frac{X_{FeSi_{0.5}O_2(melt)}}{X_{FeSi_{0.5}O_2(olivine)}} \tag{12-67}$$

Equations (12-61) and (12-66) for forsterite are equal to one another. Thus, the $\Delta_{fus}G^o$ ($MgSi_{0.5}O_2$) term can be eliminated, giving

$$\Delta_{fus}G^o(MgSi_{0.5}O_2) = \frac{1}{2}\Delta_{fus}G^o(Mg_2SiO_4) = -RT\ln K_{Fo} \approx -RT\ln\frac{X_{MgSi_{0.5}O_2(melt)}}{X_{MgSi_{0.5}O_2(olivine)}} \tag{12-68}$$

$$\Delta_{fus}G^o(Mg_2SiO_4) = -2RT\ln K_{Fo} \approx -2RT\ln\frac{X_{MgSi_{0.5}O_2(melt)}}{X_{MgSi_{0.5}O_2(olivine)}} \tag{12-69}$$

The mole fractions in Eq. (12-69) are the same as mole fractions of Mg_2SiO_4. Equations (12-63) and (12-67) for fayalite are also equal to one another and we get a similar result:

$$\Delta_{fus}G^o(Fe_2SiO_4) = -2RT\ln K_{Fa} \approx -2RT\ln\frac{X_{FeSi_{0.5}O_2(melt)}}{X_{FeSi_{0.5}O_2(olivine)}} \tag{12-70}$$

The equations for computing solidus and liquidus curves for the olivine "cigar" phase diagrams (Table 12-3) are like those for the albite-anorthite phase diagram with one big exception. They have a factor of 2 in the denominators of all the exponential terms, that is, $\exp(-\Delta_{fus}G^o/2RT)$. You can see this by comparing the following equation for fayalite with Eq. (12-59) for albite.

$$X_{Fa}(liquidus) = \frac{\left[1-\exp\left(-\Delta_{fus}G^o_{Fo}/2RT\right)\right]\exp\left(-\Delta_{fus}G^o_{Fa}/2RT\right)}{\exp\left(-\Delta_{fus}G^o_{Fa}/2RT\right)-\exp\left(-\Delta_{fus}G^o_{Fo}/2RT\right)} \tag{12-71}$$

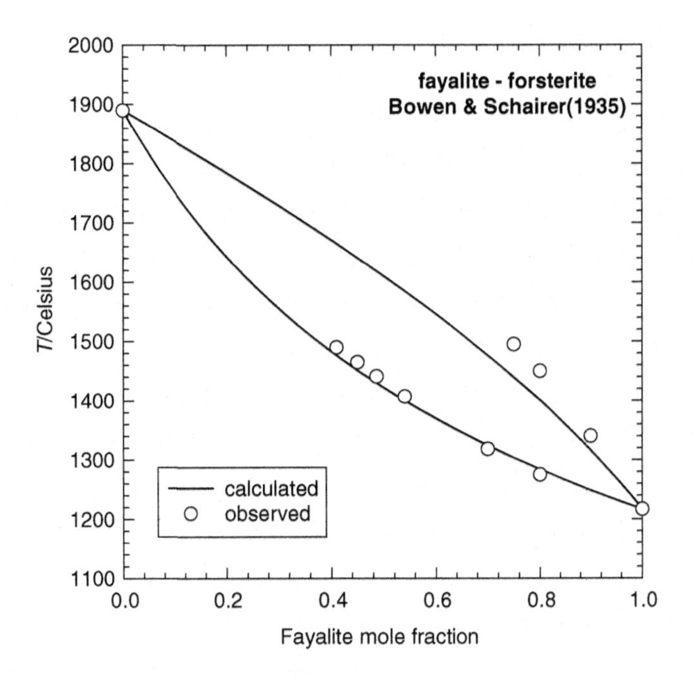

FIGURE 12-11

A comparison of the observed and calculated forsterite (Mg_2SiO_4)-fayalite (Fe_2SiO_4) phase diagram.

The equations for calculating enthalpies of fusion of olivines from "cigar" phase diagrams are like Eq. (12-53) with the exception that they are multiplied by a factor of 2 (because $\Delta_{fus}H^o$ of M_2SiO_4 is twice that of $MSi_{0.5}O_2$, M = Fe, Mg, Mn):

$$\left[\frac{\partial \ln K_{eq}}{\partial(1/T)}\right]_P = \left[\frac{\partial \ln(X_{melt}/X_{ss})_{Fo}}{\partial(1/T)}\right]_P = -2\cdot\frac{\Delta_{fus}H^o}{R} \tag{12-72}$$

Example 12-5. Calculate the fayalite-forsterite phase diagram from the melting points and enthalpies of fusion of fayalite (1490 K, $\Delta_{fus}H^o = 89{,}300$ J mol^{-1}) and forsterite (2163 K, $\Delta_{fus}H^o = 102{,}800$ J mol^{-1}). Our results are displayed in Figure 12-11, which also shows the observed points of Bowen and Schairer (1935). Our calculated fayalite mole fractions along the liquidus (X_L) and solidus (X_S) curves are tabulated here.

T(°C)	1890	1807	1727	1647	1567	1487	1407	1327	1247	1217
X_L	0	0.159	0.304	0.440	0.568	0.685	0.792	0.888	0.972	1
X_S	0	0.057	0.121	0.196	0.286	0.394	0.527	0.693	0.905	1

Solid solutions with minimum melting points

The åkermanite ($Ca_2MgSi_2O_7$)–gehlenite ($Ca_2Al_2SiO_7$) phase diagram is the classic example of solid solutions with minimum melting points (see Table 12-4). Åkermanite and gehlenite are end-members of the melilite solid solution series. As mentioned in Chapter 6, melilite is common in several types of CaAl-rich inclusions in Allende and other meteorites. On Earth melilite occurs in thermally metamorphosed impure limestones and some types of alkaline rocks (nepheline basalts, leucitites, and alnöites). Terrestrial melilites also contain Na and Fe^{3+}, but meteoritic melilites in CaAl-rich inclusions in Allende and other meteorites are essentially åkermanite and gehlenite.

Figure 12-8 shows a modified version of the åkermanite–gehlenite phase diagram (Osborn and Schairer, 1941) that is redrawn in terms of molar percentages. (However, the molecular weights of åkermanite and gehlenite are similar (272.63 versus 274.20 g mol^{-1}) and there is little difference between molar and weight percentages.) Point *G* is the melting point of gehlenite (1593°C) and point *A* is the melting point of åkermanite (1454°C). The upper curve in loop *GM* is the liquidus and the lower curve is the solidus for gehlenitic melilite. Several tie-lines are drawn between the solidus and liquidus for gehlenitic melilite. The upper curve in loop *AM* is the liquidus and the lower curve is the solidus for åkermanitic melilite. Point *M* is the minimum melting point (1390°C, 73 mole % akermanite).

Point *M* is an *indifferent point* where two phases (in this case melilite and melt) become identical in composition and a degree of freedom is lost. A maximum melting point of a solid solution is also an indifferent point. The optical isomers of carvoxime are the classic example of the maximum melting-point phase diagram. The wulfenite ($PbMoO_4$)–$Pr_2(MoO_4)_3$ phase diagram is apparently of this type (Zambonini, 1923) but should be reinvestigated. The melting point of a congruently melting solid (e.g., silica) is also an indifferent point because the solid melts to a liquid of the same composition. However, point *M* is *not* an invariant point because only three phases (melt + solid + vapor) coexist in equilibrium and one degree of freedom remains.

The åkermanite-gehlenite (and other minimum melting-point) phase diagrams have the same shape as minimum boiling-point diagrams such as ethanol (C_2H_5OH)–water, with a minimum at 78.1°C (95.5 wt. % ethanol). A minimum (or a maximum) boiling-point solution is an *azeotrope*, which is a constant boiling-point solution that boils to a vapor of the same composition as the solution. Thus, an azeotrope cannot be separated into its components by distillation. The chloroform-acetone system, which displays negative deviations from ideality (see Chapter 11) has a maximum boiling-point at 64.7°C (80 wt. % chloroform). In contrast, systems with positive deviations from ideality have minimum boiling-points (pyridine-water, CS_2-acetone, and ethyl acetate–CCl_4). Interestingly, there are many examples of maximum boiling-point azeotropes (including the HF^-, HCl^-, HBr^-, HI^-, HNO_3^-, and $H_2SO_4^-$ water systems) but no verified examples of maximum melting-point systems for minerals. However, binary systems with negative deviations from ideality may be expected to have maximum melting points.

We briefly discuss equilibrium crystallization of melilite using the 30 mole % akermanite (Ak_{30}) isopleth in Figure 12-8. The first minute amount of melilite appears when the Ak_{30} isopleth intersects the liquidus at 1533°C. The compositions of the solid and melt are given by the points on the liquidus and solidus curves. Continued cooling (at constant composition) leads to complete crystallization at 1434°C where the Ak_{30} isopleth intersects the solidus curve and the last minute amount of melt disappears.

The shape of the åkermanite-gehlenite phase diagram is typical of systems with positive deviations from ideality, for example, positive enthalpies of mixing ΔH^M. As discussed in Chapter 11, ideal solutions have no enthalpy of mixing and thus $\Delta H^M = H^E$, the excess enthalpy of mixing. Charlu et al. (1981) measured the enthalpy of mixing using oxide melt calorimetry and found positive and asymmetric deviations from ideality represented by the equation

$$H^E = 24{,}288 X_{Ge} X_{Ak}^2 + 502 X_{Ak} X_{Ge}^2 \text{ J mol}^{-1} \tag{12-73}$$

Calorimetric measurements also show positive excess enthalpies of mixing for other systems with a minimum melting point such as sanidine ($KAlSi_3O_8$)–albite ($NaAlSi_3O_8$) (Hovis and Navrotsky, 1995), halite (NaCl)–sylvite (KCl) (Sangster and Pelton, 1987), and Na_2CO_3–thenardite (Na_2SO_4) (Dessureault et al., 1990).

J. Partially miscible solids and unmixing

The lime (CaO)–periclase (MgO) phase diagram in Figure 12-12 was determined by Doman et al. (1963). It exemplifies the phase diagrams of solids that partially dissolve in one another, do not form intermediate compounds, and completely dissolve in one another as liquids. (In principle, it is impossible to have absolutely pure solids, and all simple eutectic diagrams must be of this form. In

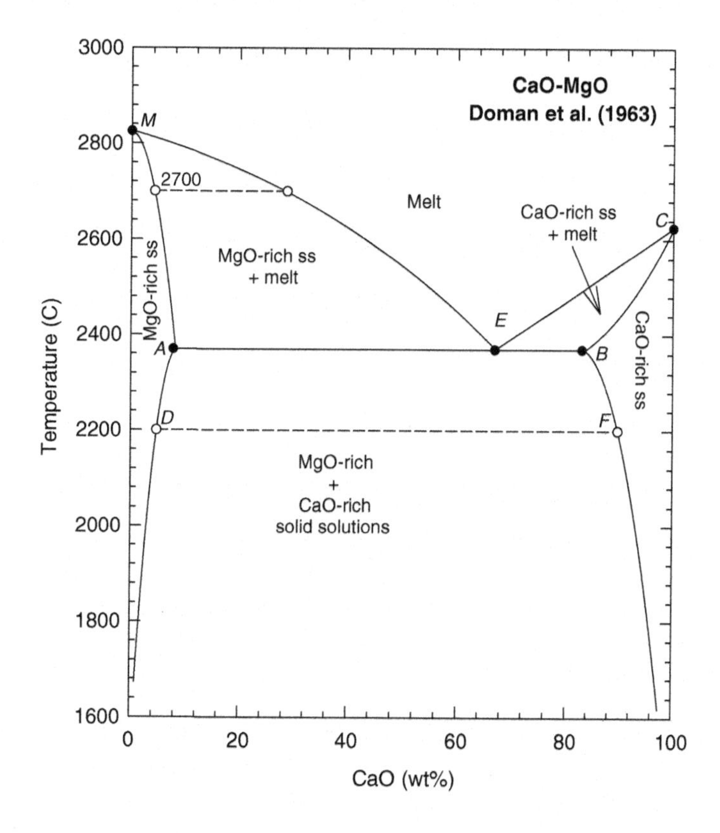

FIGURE 12-12

The lime (CaO)-periclase (MgO) phase diagram.

Table 12-5 Examples of Eutectics with Solid Solutions

Components				Eutectic Point	
A	m.p. (°C)	B	m.p. (°C)	Mass (Molar) Values	Eutectic *T* (°C)
CaO	2572	MgO	2825	$A_{67}B_{33}$ ($A_{51}B_{49}$)	2370
CaO	2572	NiO	1984	$A_{35}B_{65}$ ($A_{42}B_{58}$)	1688
$MgCr_2O_4$	2330	Ca_2SiO_4	2130	$A_{22}B_{78}$ ($A_{20}B_{80}$)	1660
Zn_2GeO_4	1490	Cd_2GeO_4	1240	$A_{33}B_{67}$ ($A_{40}B_{60}$)	1080
$NaAlSiO_4$	1526	albite	1120	$A_{24}B_{76}$ ($A_{37}B_{63}$)	1068
Ag	961	Cu	1083	$A_{72}B_{28}$ ($A_{60}B_{40}$)	779
orthoclase	876	albite	758	$A_{28.5}B_{71.5}$ ($A_{27.3}B_{72.7}$)	703[a]
NaCl	801	$CaCl_2$	771	$A_{33}B_{67}$ ($A_{48}B_{52}$)	504
Al	660	Ge	937	$A_{46.1}B_{53.9}$ ($A_{69.7}B_{30.3}$)	424
Zn	419.5	Cd	321.0	$A_{17.3}B_{82.7}$ ($A_{26.5}B_{73.5}$)	266
Bi	271.4	Sn	231.9	$A_{57}B_{43}$ ($A_{43}B_{57}$)	139

[a]*At 5 kilobar water pressure.*
Sources: Cook and McMurdie, 1989; Doman et al., 1963; Hultgren et al., 1973; Levin et al., 1964; Presnall, 1995.

practice, each of the two components in simple eutectic phase diagrams has such infinitesimally small amounts of the second component dissolved in it that each one is "pure" and the simple eutectic diagrams are correct.) Table 12-5 lists examples of other systems with this type of phase diagram.

We now describe Figure 12-12 in detail. The area labeled *melt* is a one-phase field. Points *M* and *C* are the melting points for MgO (2825°C) and CaO (2572°C). Curve *ME* is the liquidus curve for MgO-rich solid solution and curve *CE* is the liquidus curve for CaO-rich solid solution. The two curves intersect at the eutectic point *E* (2370°C, 67 wt. % CaO), which is an invariant point. The line *AEB* is a tie-line at the eutectic temperature. It connects MgO-rich solid solution at point *A* (7.8 wt. % CaO), eutectic melt (*E*), and CaO-rich solid solution at point *B* (83.0 wt. % CaO).

Curve *MA* is the solidus curve for MgO-rich solid solution. The area bounded by points *MAEM* is a two-phase field where MgO-rich solid solution coexists with melt. A tie-line at 2700°C is drawn across this region. Melt containing 28.7 wt. % CaO coexists with solid solution containing 4.25 wt. % CaO. Curve *CB* is the solidus curve for the CaO-rich solid solution. The area *CBEC* is a two-phase field where CaO-rich solid solution coexists with melt.

Curve *AD* is the solubility limit of CaO in the MgO-rich solid solution. The maximum solubility is at the eutectic temperature (point *A*). Likewise, curve *BF* is the solubility limit of MgO in the CaO-rich solid solution with maximum solubility at the eutectic temperature (point *B*). The two solubility curves *AD* and *BF* extend downward in temperature to 1640°C, below which no measurements were made. Each of the two solid solutions is single phase. They coexist with one another in the large two-phase region labeled MgO-rich + CaO-rich solid solutions. Tie-line *DF* is drawn across this region at 2200°C. It connects MgO-rich solid solution at *D* (4.75 wt. % CaO) with CaO-rich solid solution at *F* (89.75 wt. % CaO).

The two-phase region containing the solid solutions is a *miscibility gap* or *solvus,* and it is analogous to the binodal curve at the SiO_2-rich side of the MgO-SiO_2 phase diagram. However, a solvus specifically describes the two-phase region formed by two immiscible *solid* solutions. The solubility curves AD and BF are two sides of the solvus in the CaO-MgO phase diagram. The upward extension of the two solubility curves would close the solvus and produce a dome, but this does not occur, because the solvus is truncated at the eutectic melting temperature. The diopside-enstatite phase diagram also has a solvus but is more complex. At high pressure the alkali feldspar phase diagram is like CaO-MgO and has a solvus that is truncated at the eutectic. However, a completely closed solvus occurs in other systems, such as NaCl (halite)–KCl (sylvite), the alkali feldspars ($NaAlSi_3O_8$-$KAlSi_3O_8$) at ambient pressure, $NaAlSiO_4$ (nepheline)–$KalSiO_4$ (kalsilite), Al_2O_3 (corundum)–Cr_2O_3 (eskolaite), $Na_2Al_4[Si_6Al_2O_{20}](OH)_4$ (paragonite)–$K_2Al_4[Si_6Al_2O_{20}](OH)_4$ (muscovite), $Mg_3Al_2Si_3O_{12}$ (pyrope)–$Ca_3Al_2Si_3O_{12}$ (grossular), $Mg_3Al_2Si_3O_{12}$ (pyrope)–$Mn_3Al_2Si_3O_{12}$ (spessartine), and $Mg_3Al_2Si_3O_{12}$ (pyrope)–$Fe^{2+}{}_3Al_2Si_3O_{12}$ (almandine).

As stated previously, curves AD and BF are the solubility limits for each oxide in the other. Dissolution of CaO in MgO at any temperature occurs until the solubility limit is reached along curve AD. At 2200°C this occurs at point D, where 4.75 wt % (3.46 mole %) CaO dissolves in the MgO-rich solid solution. Beyond that point, no further CaO can dissolve. As we add more CaO to the MgO-rich solid solution, a second phase, CaO-rich solid solution, precipitates out and we have a physical mixture of the two solid solutions.

Coexistence of the two solid solutions in equilibrium with one another requires that the activity of CaO is the same in the two solid solutions. Denoting the CaO-rich solid solution as α and the MgO-rich solid solution as β, we can write

$$a_{CaO}(\alpha) = a_{CaO}(\beta) \tag{12-74}$$

The coexistence at equilibrium of the two solid solutions also requires that the activity of MgO is the same in both phases. Thus, we also have the equation

$$a_{MgO}(\alpha) = a_{MgO}(\beta) \tag{12-75}$$

The activities of the two oxides in the two phases and thermodynamic properties of the two solid solutions can be evaluated assuming Raoult's law for the major component and Henry's law for the minor component. This is illustrated in the following example.

Example 12-6. Calculate the activities and activity coefficients of MgO and CaO at their solubility limits in each other at 2200°C. We first recalculate the compositions in terms of mole percentages using Eq. (12-2). Point D is 3.46 mole % CaO and point F is 86.29 mole % CaO. Considering CaO first, its activity in the CaO-rich solid solution (α) is equal to its mole fraction of 0.8629. Equation (12-74) requires that the CaO activity in the MgO-rich solid solution (β) is the same (0.8629). The activity coefficient for CaO in β is proportional to its mole fraction and is calculated by rearranging Eq. (12-74):

$$\gamma_{CaO}(\beta) = \frac{a_{CaO}(\alpha)}{X_{CaO}(\beta)} = \frac{0.8629}{0.0346} \sim 24.9 \tag{12-76}$$

The assumption of Henry's law means that this activity coefficient is constant over the range of CaO mole fractions dissolved in the β phase. Likewise, the MgO activity in the MgO-rich solid solution is equal to its mole fraction of 0.9654. This is also the MgO activity in α from Eq. (12-75). The MgO activity coefficient in the α phase is

$$\gamma_{MgO}(\alpha) = \frac{a_{MgO}(\beta)}{X_{MgO}(\alpha)} = \frac{0.9654}{0.1371} \sim 7.0 \tag{12-77}$$

In principle, it is possible to repeat these calculations at several temperatures along curves AD and BF and thus obtain the temperature dependence of the activities and activity coefficients. A practical problem is that the phase boundaries are rarely determined with the necessary accuracy in either temperature or composition to make the results accurate. A second, theoretical problem is that the derivatives with respect to temperature and composition have to be evaluated. For example, the temperature coefficient of the partial molal Gibbs energy of CaO requires evaluating this expression:

$$\frac{\partial \Delta \overline{G}^M_{CaO}}{\partial T} = \left(\frac{\partial \Delta \overline{G}^M_{CaO}}{\partial T}\right)_X + \left(\frac{\partial \Delta \overline{G}^M_{CaO}}{\partial X}\right)_T \frac{\partial X}{\partial T} = R\frac{\partial \ln a_{CaO}}{\partial T} \quad (12\text{-}78)$$

K. Unmixing and Margules parameters

It is also possible to use the CaO-MgO diagram and similar phase diagrams (with a complete or a truncated solvus) to calculate the Margules parameters (Section IV-H. of Chapter 11) for the binary solution. We briefly discuss this using the assumption of a strictly regular solution (no excess entropy of mixing). The equivalence of activities for the two components in the two phases can be rewritten as

$$RT \ln a_{CaO}(\alpha) = RT \ln a_{CaO}(\beta) \quad (12\text{-}79)$$

$$RT \ln a_{MgO}(\alpha) = RT \ln a_{MgO}(\beta) \quad (12\text{-}80)$$

Rewriting in terms of mole fractions and activity coefficients and substituting using Eqs. (11-208) and (11-209) for the activity coefficient terms, we get

$$RT \ln X^{\alpha}_{CaO} + W_G\left(1 - X^{\alpha}_{CaO}\right)^2 = RT \ln X^{\beta}_{CaO} + W_G\left(1 - X^{\beta}_{CaO}\right)^2 \quad (12\text{-}81)$$

$$RT \ln X^{\alpha}_{MgO} + W_G\left(1 - X^{\alpha}_{MgO}\right)^2 = RT \ln X^{\beta}_{MgO} + W_G\left(1 - X^{\beta}_{MgO}\right)^2 \quad (12\text{-}82)$$

Each of these two equations can be solved for independent estimates of W_G:

$$W_G = \frac{RT \ln\left(X^{\beta}_{CaO}/X^{\alpha}_{CaO}\right)}{\left[\left(1 - X^{\alpha}_{CaO}\right)^2 - \left(1 - X^{\beta}_{CaO}\right)^2\right]} \quad (12\text{-}83)$$

$$W_G = \frac{RT \ln\left(X^{\beta}_{MgO}/X^{\alpha}_{MgO}\right)}{\left[\left(1 - X^{\alpha}_{MgO}\right)^2 - \left(1 - X^{\beta}_{MgO}\right)^2\right]} \quad (12\text{-}84)$$

The W_G values are then used to recompute the solvus using the preceding equations and iterative adjustment of the mole fractions of the components in the solid solutions.

In principle, Eqs. (12-83) and (12-84) will give the same values for W_G, but in practice experimental errors in measuring compositions along the solvus curve can lead to slightly different W_G values. These errors can occur because equilibration can be very slow at the low temperatures of the solvus in some systems. A classic case is the NaCl (halite)–KCl (sylvite) solvus discussed by Thompson and Waldbaum (1969). They describe how Nacken studied this system (in 1918) by equilibrating homogeneous crystalline solutions of (Na,K)Cl for eight days, quenching them, and doing oil immersion optical microscopy to see which phase(s) were present. However, large differences show the solution is not regular (actually the case for CaO-MgO). Then a more complicated treatment using two Margules parameters (e.g., W_{G1} and W_{G2}) for the excess Gibbs energy of asymmetric (subregular) solutions is required. This is covered in Section IV-I of Chapter 11.

L. Nonstoichiometric compounds

Our earlier discussion of intermediate compounds focused on the MgO-SiO_2 phase diagram, which contains stoichiometric compounds. Here we discuss nonstoichiometric compounds and focus on wüstite ($Fe_{1-x}O$). Wüstite is stable, but stoichiometric FeO is unstable at ambient pressure. Iron-rich wüstite has a temperature-dependent Fe/O atomic ratio ranging from ~0.950 at 1371°C, the eutectic temperature where γ-Fe metal + Fe-rich wüstite are in equilibrium with molten oxide, to ~0.932 at 574°C where wüstite decomposes to Fe metal plus magnetite (the wüstite eutectoid temperature). The wüstite stability field is wide at the top and Fe-poor wüstite has a Fe/O atomic ratio of ~0.832 at 1424°C in equilibrium with magnetite and molten oxide. The Fe/O atomic ratio at the wüstite-magnetite (WM) phase boundary decreases with decreasing temperature and approaches that of wüstite in equilibrium with Fe-metal (the iron-wüstite, IW, phase boundary). The two boundaries meet at 574°C, the wüstite eutectoid temperature.

Darken and Gurry (1945) determined the wüstite phase boundaries and the Fe/O ratios as a function of temperature across its stability field by equilibrium with CO/CO_2 gas mixtures. As discussed in Chapter 10, a CO/CO_2 gas mixture regulates the oxygen fugacity at a unique value at a given temperature. For example, along the Fe-wüstite (IW) phase boundary the reaction being studied can be written as

$$CO_2 + (1 - x)\,\mathrm{Fe}\,(\mathrm{metal}) = CO + Fe_{1-x}O(\text{wüstite}) \tag{12-85}$$

Chemical analyses of the wüstites in equilibrium with different CO/CO_2 gas mixtures along an isotherm gives the Fe/O ratio across the wüstite field at a given temperature. Chemical analyses of wüstites equilibrated with CO/CO_2 gas mixtures at different temperatures give the dependence of the Fe/O atomic ratio on temperature throughout its stability field and along phase boundaries. The Fe/O atomic ratio as a function of oxygen fugacity along an isotherm and the oxygen fugacity as a function of temperature at a constant Fe/O ratio can also be determined by equilibration with H_2/H_2O gas mixtures, by electromotive force measurements, or by thermogravimetry. The latter method gives the weight change of wüstite as it gains (or loses) oxygen while equilibrating with gas mixtures of different oxygen fugacities. Spencer and Kubaschewski (1978) summarize and correlate the results of many experimental studies to give a consistent set of thermodynamic data for wüstite and for its phase boundaries in the Fe-O system.

PROBLEMS

1. Use the lever rule and mass balance to calculate the molar percentages of diopside and melt present at 1300°C and $X_{Dp} = 0.85$ in Figure 12-1.
2. Sugawara et al. (2011) used differential scanning calorimetry to measure the enthalpy of fusion of $Na_2Si_2O_5$ as 29.0 ± 1.4 kJ mol^{-1}. Check their value by computing the enthalpy of fusion from the liquidus line of $Na_2Si_2O_5$ in the $Na_2Si_2O_5$-Ba_2SiO_4 simple eutectic system (Greene and Morgan, 1941). Assume ideality in your work.

T (°C)	874	861.5	850	838	826.5
Wt. % Na_2SiO_5	100	95	90	85	80

3. Halite (NaCl) and thenardite (Na_2SO_4) occur in evaporites in arid regions. They form a simple eutectic, which can be modeled as an ideal solution. Calculate their liquidus lines and the eutectic temperature and composition from their melting points (halite, 1074 K; thenardite, 1157 K) and enthalpies of fusion (halite, 28,160 J mol^{-1}; thenardite, 23,849 J mol^{-1}). Also draw the halite-thenardite phase diagram and label the various parts of it.
4. The bismuth (Bi)–cadmium (Cd) system is a simple eutectic that is nearly ideal. Compute this phase diagram (liquidus lines, eutectic point) using the melting points, enthalpies of fusion, and ΔC_P values (liquid-metal) at the melting points for Bi (544.5 K, 11,300 J mol^{-1}, 2.3 J mol^{-1} K^{-1}) and Cd (594.2 K, 6170 J mol^{-1}, 0.2 J mol^{-1} K^{-1}).
5. Cryolite (Na_3AlF_6) and fluorite (CaF_2) form a eutectic at 945.5°C and 50 mole % of each. Convert the eutectic composition into weight percent.
6. Use the lever rule and the MgO-SiO_2 phase diagram to compute the mass fractions of forsterite and melt produced by incongruent melting of enstatite.
7. Use the albite mole fractions in coexisting melt (X_{liq}) and solid solution (X_{sol}) in the albite-anorthite phase diagram to estimate the enthalpy of fusion of albite. The values in the following table are from Bowen (1913).

T (°C)	1465	1448	1394	1372	1362	1334	1287	1265	1205
X_{liq}	0.435	0.500	0.667	0.725	0.750	0.800	0.860	0.889	0.945
X_{sol}	0.167	0.198	0.290	0.333	0.355	0.420	0.500	0.550	0.667

8. Soda niter ($NaNO_3$ forms a simple eutectic with water ice. Use the $NaNO_3$ solubility (S, in grams $NaNO_3$ per 100 grams solution) tabulated below to graph the phase diagram and locate the eutectic point. Also label all fields and curves in the diagram.

T (°C)	10.00	0.00	−9.90	−14.00	−15.08	−12.85	−10.17
S	43.90	41.90	39.80	39.10	34.70	30.82	25.46

9. Osborn and Schairer (1941) determined the pseudo-wollastonite (Wo, α-$CaSiO_3$)–åkermanite (Ak, $Ca_2MgSi_2O_7$) phase diagram. Adamkovičová et al. (1996) measured an ideal enthalpy of mixing for the same system. Pseudo-wollastonite melts at 1821 K with an enthalpy of fusion of 57,300 J mol^{-1}. Use these data and Eq. (11-119) to compute the freezing-point depression in a melt of composition $Wo_{90}Ak_{10}$.
10. Anorthite-diopside eutectic melt is often used as an analog for basaltic melt, for example, in laboratory experiments studying element partitioning between metal and silicate. Use Eq. (12-3) and its analog for diopside (with $\Delta C_P = 67$ J mol^{-1} K^{-1}) to estimate the heat released (J/g) from solidification of eutectic melt. This number is important for computing the heat needed to produce basaltic melts on Earth and other rocky planets.
11. Calculate the fayalite (Fe_2SiO_4)–tephroite (Mn_2SiO_4) phase diagram from the melting points and enthalpies of fusion for fayalite (1490 K, $\Delta_{fus}H^o = 89{,}300$ J mol^{-1}) and tephroite (1347 K, $\Delta_{fus}H^o = 89{,}663$ J mol^{-1}).
12. Perovskite $CaTiO_3$ is an important Ti-bearing mineral in the refractory CaAl-rich inclusions (CAIs) found in the Allende CV3 carbonaceous chondrite. Use the following data (DeVries et al., 1954) to sketch the CaO-TiO_2 phase diagram and label the fields and curves on it. Celsius temperatures are given. Melting points: CaO (2572), $Ca_3Ti_2O_7$ (1750, incongruent melting), $CaTiO_3$ (1970), TiO_2 (1830); eutectic points: CaO-$Ca_3Ti_2O_7$ (1695°C, 39 wt. % TiO_2), $CaTiO_3$-TiO_2 (1460°C, 83 wt. % TiO_2); peritectic point: 1750, 42 wt. % TiO_2.
13. Jacob and Abraham (2009) electrochemically measured the standard Gibbs energies of formation of titanates in the CaO-TiO_2 system. Use their data (given below) to compute the CaO and TiO_2 activities in each of the three different two-phase fields and to tabulate the variation of the CaO and TiO_2 activities across the phase diagram along the 1200 K (927°C) isotherm.

$$\text{CaO (lime)} + \text{TiO}_2\text{ (rutile)} = \text{CaTiO}_3\text{ (perovskite)}$$

$$\Delta_r G^o = -80{,}140 - 6.302T \text{ J mol}^{-1}$$

$$3\text{ CaO (lime)} + 2\text{ TiO}_2\text{ (rutile)} = \text{Ca}_3\text{Ti}_2\text{O}_7\text{(solid)}$$

$$\Delta_r G^o = -164{,}217 - 16.838T \text{ J mol}^{-1}$$

14. For simple eutectics, the activity of a component in the melt (a_{melt}) along its liquidus line at temperature T can be estimated from its enthalpy of fusion using the equation

$$\log_{10} r = \log_{10}\frac{a_{solid}}{a_{melt}} = \frac{\Delta_{fus}H^o}{19.437}\left[\frac{1}{T} - \frac{1}{T_m}\right]$$

The melting point is T_m and the activity of the solid is a_{solid} (unity for a pure mineral). Do this for anorthite using these temperatures and mole fractions along its liquidus line in the anorthite-diopside phase diagram.

X_{An}	0.752	0.442	0.373
T (K)	1758	1601	1547

15. The MgO-TiO_2 system contains three intermediate compounds Mg_2TiO_4 (qandilite), $MgTiO_3$ (geikielite), and $MgTi_2O_5$ (karrooite). (a) List the four two-phase regions in this phase diagram. (b) Write down the chemical reaction in each region.

16. The Cu-Ag phase diagram also has limited solubility in the solid state (Table 12-5). At the eutectic temperature of 1052 K the Cu solubility limit in the Ag-rich solid solution is $X_{Cu} = 0.141$ and the Ag solubility limit in the Cu-rich solid solution is $X_{Ag} = 0.049$. (a) Compute the integral Gibbs energy of mixing at each solubility limit assuming that the minor component in each solid solution obeys Henry's law and the major component obeys Raoult's law. (b) Compute the integral entropy of mixing and the excess entropy of mixing at each solubility limit given $H^E = 2996$ J mol^{-1} at the Cu solubility limit and 1736 J mol^{-1} at the Ag solubility limit.

17. The NaI-RbI phase diagram has limited solid solubility like the CaO-MgO phase diagram. At the eutectic temperature of 778 K, the NaI-rich solid solution contains 18 mole % RbI and the RbI-rich solid solution contains 18 mole % NaI. Apply Eqs. (12-83) and (12-84) to the NaI-RbI phase diagram and calculate W_G values from the compositions of the NaI-rich and RbI-rich solutions at the eutectic temperature.

Appendix 1

Thermodynamic Properties of Minerals, Elements, and Inorganic Compounds at 298.15 K and 10^5 Pa (1 Bar)

Chemical Formula	Phase or Name	Formula Weight	C_P^o (J mol⁻¹ K⁻¹)	S^o (J mol⁻¹ K⁻¹)	$H^o_{298} - H^o_0$	$\Delta_f H^o$ (kJ mol⁻¹)	$\Delta_f G^o$ (kJ mol⁻¹)
Ag	metal, fcc	107.8682	25.40±0.10	42.68±0.21	5.761	0	0
Ag_2O	c	231.736	66.3±0.5	120.9±0.2	14.217±0.021	–30.7±0.3	–10.7±0.3
AgCl	chlorargyrite	143.321	50.79±0.42	96.25±0.20	12.033±0.020	–127.07±0.06	–109.785±0.020
Ag_2S	acanthite	247.802	75.31±0.15	142.89±0.21	17.132±0.026	–32.0±1.0	–39.7±1.0
Al	metal, fcc	26.98154	24.20±0.07	28.30±0.10	4.540±0.020	0	0
Al^{3+}	aq	26.9799	–120	–340	–	–538.9	–487.5
Al_2O_3	corundum	101.9613	79.03±0.20	50.92±0.10	10.016±0.020	–1675.7±1.3	–1582.3±1.3
$3Al_2O_3 \cdot 2SiO_2$	mullite	426.052	325.64	274.9±12.0	46.07	–6819.2±6.3	–6441.7±7.2
Al_2SiO_5	kyanite	162.046	121.58	82.8±0.5	15.86	–2593.8±2.0	–2443.1±2.0
	andalusite	162.046	122.60	91.4±0.5	16.90	–2589.9±2.0	–2441.8±2.0
	sillimanite	162.046	123.72	95.4±0.5	17.44	–2586.1±2.0	–2439.1±2.0
$Al_2Si_2O_5(OH)_4$	kaolinite	258.161	243.37	200.4±0.5	35.75	–4119.0±1.5	–3797.5±1.5
$Al_2Si_4O_{10}(OH)_2$	pyrophyllite	360.314	293.76	239.4±0.4	42.70	–5640.0±1.5	–5266.1±1.5
$Al_2SiO_4F_2$	topaz	184.043	143.63	105.4±0.2	19.77	–3084.5±4.7	–2910.6±4.7
Br_2	liq	159.808	75.68	152.21±0.30	24.52±0.01	0	0
	gas	159.808	36.057±0.002	245.466±0.005	9.725±0.001	30.91±0.11	3.1±0.11
Br^-	aq	79.9046	–141.8	82.55±0.20	–	–121.41±0.15	–103.85±0.16
HBr	gas	80.912	29.141±0.003	198.700±0.004	8.648±0.001	–36.29±0.16	–53.36±0.16
C	graphite	12.011	8.517±0.080	5.74±0.10	1.050±0.020	0	0
	diamond	12.011	6.109	2.362±0.02	0.523±0.004	1.85±0.16	2.858±0.16
CO	gas	28.0102	29.141±0.002	197.660±0.004	8.671±0.001	–110.53±0.17	–137.16±0.17
CO_2	gas	44.0096	37.135±0.002	213.785±0.010	9.365±0.003	–393.51±0.07	–394.39±0.05
	aq	44.0096	–	119.36±0.60	–	–413.26±0.20	–385.97±0.27

(Continued)

Chemical Formula	Phase or Name	Formula Weight	C_P^o	S^o	$H_{298}^o - H_0^o$	$\Delta_f H^o$	$\Delta_f G^o$
			(J mol^{-1} K^{-1})			(kJ mol^{-1})	
CO_3^{2-}	aq	60.0102	–	–50.0±1.0	–	–675.23±0.25	–527.90±0.39
CH_4	methane	16.043	35.61±0.04	186.42±0.08	10.021±0.004	–74.81±0.34	–50.76±0.34
C_2H_2	acetylene	26.037	44.036	200.94	10.009	226.73	209.20
C_2H_4	ethylene	28.053	42.89±0.21	219.33±0.25	10.527±0.033	52.47±0.29	68.42±0.29
C_2H_6	ethane	30.0694	52.49±0.08	229.16±0.10	11.88±0.06	–84.73±0.50	–32.74±0.50
CH_3OH	methanol, gas	32.042	43.93±0.42	239.86±0.84	11.422±0.063	–201.96±0.29	–163.26±0.38
CF_4	gas (F14)	88.0043	61.050	261.5±0.2	12.730	–933.20±0.75	–888.53±0.75
C_2F_6	gas (F116)	138.0118	106.207	332±2	20.271	–1344.0±4.0	–1258.2±4.0
CS	gas	44.077	29.799±0.006	210.55±0.04	8.708±0.002	279.78±0.75	228.27±0.75
CS_2	gas	76.144	45.71±0.03	237.88±0.05	10.664±0.025	116.7±1.0	66.6±1.0
OCS	gas	60.077	41.55±0.13	231.64±0.07	9.942±0.025	–141.7±2.0	–168.9±2.0
C_2N_2	gas	52.0349	57.084	242.20±0.02	12.715	309.1±0.8	297.4±0.8
HCN	gas	27.0254	35.86±0.01	201.82±0.01	9.235±0.008	132.0±4.0	121.6±4.0
Ca	metal, fcc	40.078	25.72	42.54±0.30	5.783	0	0
Ca^{2+}	aq	40.07745		–56.2±1.0	–	–543.0±1.0	–553.6
CaO	lime, calcia	56.077	42.05±0.40	38.1±0.4	6.750±0.060	–634.92±0.90	–603.01±0.91
$Ca(OH)_2$	portlandite	74.093	87.51	83.4±0.4	14.16	–986.1±1.3	–898.2±1.3
CaF_2	fluorite	78.075	67.03±0.21	68.45±0.42	11.632±0.063	–1228.0±2.0	–1175.3±2.0
$CaCl_2$	hydrophilite	110.983	72.83	104.6±1.3	15.27	–795.8±0.7	–747.8±0.8
CaS	oldhamite	72.144	47.442	56.7±1.3	8.999	–474.9±2.1	–469.6±2.1
$CaSO_4$	anhydrite	136.142	101.23	107.4±0.2	17.30	–1434.4±4.2	–1321.8±4.2

$CaSO_4 \cdot \frac{1}{2}H_2O$	bassanite	145.149	119.41	130.5±0.3		−1576.74±4.2	−1436.30±4.2
$CaSO_4 \cdot 2H_2O$	gypsum	172.172		193.8±0.3	31.13	−2023.0±4.3	−1797.1±4.3
$CaCO_3$	calcite	100.087	83.47	91.7±0.2	14.48	−1207.4±1.3	−1128.3±1.3
	aragonite	100.087	82.31	88.0±0.2	14.31	−1207.4±1.4	−1127.5±1.4
$CaSiO_3$	wollastonite	116.162	86.19	81.7±0.1	13.84	−1634.8±1.4	−1549.1±1.4
	pseudowollastonite	116.162	87.92	87.2±0.9		−1627.6±1.4	−1543.6±1.4
$CaAl_2Si_2O_8$	anorthite	278.207	211.34	199.3±0.3	33.33	−4234.0±2.0	−4008.0±2.0
$Ca_2Al_3Si_3O_{12}(OH)$	zoisite	454.357	350.34	295.9±0.6	52.07	−6901.1±3.3	−6504.8±3.3
$Ca_3Al_2Si_3O_{12}$	grossular	450.446	330.22	260.1±0.5	47.66	−6640.0±3.2	−6278.8±3.2
$CaFeSi_2O_6$	hedenbergite	248.090	175.30	174.2±0.3	28.23	−2839.9±3.0	−2676.4±3.0
$Ca_3Fe_2Si_3O_{12}$	andradite	508.173	351.88	316.4±2.0	53.49	−5771.0±5.9	−5427.3±5.9
$CaTiO_3$	perovskite	135.943	98.08	93.30	15.94	−1660.6±1.7	−1574.8±1.7
$CaTiSiO_5$	titanite (sphene)	196.028	138.91	129.2±0.8	–	−2596.6±3.0	−2454.8±3.2
$CaMg(CO_3)_2$	dolomite	184.401	157.51	155.2±0.3	25.98	−2324.5±1.5	−2161.4±1.5
$CaMgSiO_4$	monticellite	156.466	123.00	108.1±0.3	18.76	−2251.0±3.0	−2132.9±3.0
$CaMgSi_2O_6$	diopside	216.550	166.78	142.7±0.3	25.24	−3201.5±2.0	−3026.9±2.0
$Ca_2MgSi_2O_7$	akermanite	272.628	214.10	212.5±0.4	34.91	−3864.8±2.0	−3667.8±2.0
$Ca_3Mg(SiO_4)_2$	merwinite	328.705	252.36	253.1±2.1		−4536.2±3.0	−4308.0±3.1
$Ca_2Mg_5Si_8O_{22}(OH)_2$	tremolite	812.366	655.44	548.9±1.3	97.63	−12303.0±7.0	−11574.8±7.0
Cd	metal, hexag	112.411	26.02±0.04	51.80±0.15	6.247±0.015	0	0
Cd^{2+}	aq	112.40990	–	−72.8±1.5	–	−75.92±0.60	−77.6±0.60
CdO	monteponite	128.410	43.64±0.21	54.8±1.5	8.41±0.08	−258.35±0.4	−228.7±0.6
CdS	greenockite	144.477	47.29±0.03	72.2±0.3	9.80±0.02	−152.11±0.87	−148.64±0.87
Cl_2	gas	70.9054	33.949±0.002	223.081±0.010	9.181±0.001	0	0
Cl^-	aq	35.45325	−136.4	56.60±0.20	–	−167.08±0.10	−131.22±0.12
HCl	gas	36.4608	29.136±0.002	186.902±0.005	8.640±0.001	−92.31±0.10	−95.30±0.10
	ai	36.4608	−136.4	56.60±0.20	–	−167.08±0.10	−131.22±0.12
Cu	metal, fcc	63.546	24.44±0.05	33.15±0.08	5.004±0.008	0	0
Cu^{2+}	aq	63.54490	–	−92.7±3.3	–	66.9±1.0	65.56±0.21

(Continued)

Chemical Formula	Phase or Name	Formula Weight	C_P^o	S^o	$H_{298}^o - H_0^o$	$\Delta_f H^o$	$\Delta_f G^o$
			(J mol^{-1} K^{-1})			(kJ mol^{-1})	
CuO	tenorite	79.545	42.25±0.21	42.59±0.21	7.092±0.021	−156.06±2.0	−128.29±2.0
Cu_2O	cuprite	143.091	62.54±0.10	92.36±0.34	12.600±0.020	−170.572±0.036	−147.760±0.108
CuS	covellite	95.612	47.53	67.4±0.1	9.45	−54.6±0.3	−55.3±0.3
Cu_2S	chalcocite	159.158	76.84	116.2±0.2	15.8	−83.9±1.1	−89.2±1.1
$CuSO_4$	chalcocyanite	159.610	139.89	109.2±0.4	16.86±0.08	−771.4±1.2	−662.2±1.4
$CuFeS_2$	chalcopyrite	183.523	95.80	124.9±0.2	18.34	−194.9±1.6	−195.1±1.6
Cu_5FeS_4	bornite	501.839	242.90	398.5±0.8	53.011	−371.6±2.1	−394.7±2.1
e^-	gas	0.00055	20.786	20.979	6.197	0	0
	aq	0.00055	—	65.340	—	0	0
F_2	gas	37.9968	31.304±0.002	202.791±0.005	8.825±0.001	0	0
F	gas	18.99840	22.746±0.002	158.751±0.004	6.518±0.001	79.38±0.30	62.28±0.30
F^-	gas	18.99895	20.786	145.576±0.005	6.197	−255.09±0.46	−262.01±0.46
	aq	18.99895	—	−13.8±0.8	—	−335.35±0.65	−281.5±0.7
HF	gas	20.0063	29.137±0.002	173.779±0.003	8.599±0.001	−273.30±0.70	−275.40±0.70
Fe	metal, bcc	55.845	25.084±0.060	27.085±0.08	4.481±0.010	0	0
Fe^{2+}	aq	55.844	—	−107.1±2.0	—	−91.1±3.0	−90.0±2.0
Fe^{3+}	aq	55.844	—	−280.0±13.0	—	−49.9±5.0	−16.7±2.0
$Fe_{0.947}O$	wüstite	68.885	48.12±0.42	57.59±0.42	9.462±0.084	−266.3±0.8	−245.1±0.8
'FeO'	stoichiometric	71.845	47.68	60.6±1.7	—	−272.0±2.1	−251.4±2.2
α-Fe_2O_3	hematite	159.688	104.05	87.4±0.2	65.10	−826.2±1.3	−744.4±1.3
Fe_3O_4	magnetite	231.533	150.9	146.1±0.4	102.80	−1115.7±2.1	−1012.7±2.1
$FeCl_2$	lawrencite	126.750	76.34	118.0±0.4	16.27	−341.7±0.4	−302.3±0.4
$FeCl_3$	molysite	162.203	96.67	142.3±0.4	19.71	−399.5±0.4	−334.1±0.4
$Fe_{0.875}S$	pyrrhotite	80.931	49.88	60.7±0.2	9.21	−97.5±2.0	−99.0±2.0
FeS	troilite	87.912	50.49	60.3±0.2	9.351	−101.0±1.5	−101.1±1.5

FeS_2	pyrite	119.977	62.17	52.9±0.1	9.63	−171.5±1.7	−160.1±1.7
	marcasite	119.977	62.43	53.9±0.1	9.74	−169.5±2.1	−158.4±2.1
$Fe_{4.60}Ni_{4.54}S_8$	pentlandite	779.877	442.7±1.1	474.9±2.8	76.28±0.31	−837.4±14.6	−825.0±14.6
Fe_3C	cohenite (cementite)	179.546	111.03	104.4±3.4	18.112	24.94±1.34	19.76±1.72
$FeCO_3$	siderite	115.854	82.44±0.10	95.47±0.15	14.591	−752.0±1.2	−678.9±1.2
$FeSiO_3$	ferrosilite	131.929	90.63	94.6±0.3	—	−1195.2±3.0	−1118.0±3.0
Fe_2SiO_4	fayalite	203.773	131.84	151.0±0.2	22.49	−1478.2±1.3	−1379.1±1.3
$Fe_3Al_2Si_3O_{12}$	almandine	497.747	343.29	342.6±1.4	52.74	−5264.7±3.0	−4941.9±3.0
$FeTiO_3$	ilmenite	151.710	99.18	108.9±0.3	16.992	−1232.0±2.5	−1155.5±2.5
H_2	gas	2.0159	28.836	130.680±0.003	8.468	0	0
H	gas	1.00794	20.786	114.717±0.002	6.197	217.998±0.006	203.276±0.006
H^+	gas	1.00739	20.786	108.946±0.02	6.197	1536.246±0.04	1516.990±0.04
H^+	aq	1.00739	0	0	–	0	0
OH	hydroxyl	17.0073	29.886	183.737±0.02	8.813	39.35±0.21	34.63±0.21
OH^-	aq	17.0079	−148.5	−10.90±0.20	–	−230.015±0.040	−157.22±0.072
HO_2	hydroperoxyl	33.007	34.905	229.106±0.08	10.003	13.8±3.3	26.1±3.3
H_2O	liq	18.0153	75.351±0.080	69.95±0.03	13.273±0.020	−285.830±0.040	−237.14±0.04
	gas	18.0153	33.609±0.030	188.835±0.010	9.905±0.005	−241.826±0.040	−228.582±0.04
H_2O_2	gas	34.015	42.395	234.516±0.1	11.158	−135.88±0.22	−105.67±0.22
	liq	34.015	89.330	109.602	22.949	−187.780±0.08	−120.33±0.08
	aq	34.015	–	143.9	–	−191.17	−133.95
Hg	liq	200.59	27.98	75.90±0.12	9.342±0.008	0	0
	gas	200.59	20.786±0.001	174.971±0.005	6.197±0.001	61.38±0.04	31.84±0.04
HgO	montroydite (red)	216.59	44.06±0.21	70.25±0.30	9.117±0.025	−90.81±0.12	−58.54±0.05
	c, yellow	216.59	–	69.87	–	−90.82±0.13	−58.44±0.08
HgS	cinnabar	232.66	48.41±0.13	82.42±0.42	10.535±0.042	−57.3±2.1	−49.7±2.2

(Continued)

Chemical Formula	Phase or Name	Formula Weight	C_P^o	S^o	$H_{298}^o - H_0^o$	$\Delta_f H^o$	$\Delta_f G^o$
			(J mol^{-1} K^{-1})			(kJ mol^{-1})	
	metacinnabar	232.66	48.54	88.7±2.1	–	–52.9±3.4	–47.2±3.4
I_2	c, orthorh	253.8089	54.44	116.139±0.300	13.196±0.040	0	0
	gas	253.8089	36.888±0.002	260.687±0.005	10.116±0.001	62.42±0.08	19.32±0.08
I	gas	126.90447	20.786±0.001	180.787±0.004	6.197±0.001	106.76±0.04	70.17±0.04
I^-	aq	126.9050	–142.3	106.45±0.30	–	–56.78±0.05	–51.7±0.1
HI	gas	127.9124	29.157±0.003	206.590±0.004	8.657±0.001	26.50±0.10	1.70±0.10
K	metal, bcc	39.0983	29.49±0.1	64.63±0.20	7.080±0.020	0	0
	gas	39.0983	20.786	160.341±0.003	6.197±0.001	89.0±0.8	60.5±0.8
K^+	gas	39.09775	20.786	154.576±0.025	6.197	514.01±0.8	480.94±0.8
	aq	39.09775		101.20±0.20		–252.14±0.08	–282.5±0.1
KOH	c, monocl	56.1056	64.894	78.9±0.8	12.163	–424.7±0.6	–378.9±0.6
KCl	sylvite	74.5510	51.29	82.6±0.2	11.37	–436.5±0.2	–408.6±0.2
$KAlSiO_4$	kaliophilite	158.1629	119.70	133.3±1.2	–	–2124.7±3.1	–2008.8±3.1
$KAlSi_2O_6$	leucite	218.2472	163.8±1.0	180.0±1.0	28.12±0.04	–3037.8±2.7	–2869.0±2.7
$KAlSi_3O_8$	microcline	278.3315	202.13	214.2±0.4	33.99	–3974.6±3.9	–3749.3±3.9
	sanidine	278.3315	204.55	232.8±0.5	34.02	–3965.6±4.1	–3745.8±4.1
$KAl_2[AlSi_3O_{10}](OH)_2$	muscovite (ord)	398.3081	325.99	287.7±0.6	49.41	–5990.0±4.9	–5608.4±4.9
$KMg_3[AlSi_3O_{10}](OH)_2$	phlogopite (ord)	417.2600	354.65	315.9±1.0	54.08	–6246.0±6.0	–5860.5±6.0
Li	metal, bcc	6.941	24.78±0.1	28.99±0.30	4.617±0.030	0	0
$LiAlSi_4O_{10}$	petalite	306.259	245.30	233.2±0.6	38.31	–4886.5±6.3	–4610.7±6.3
Mg	metal, hcp	24.3050	24.87±0.02	32.67±0.10	4.998±0.030	0	0
Mg^{2+}	aq	24.30390		–137±4	–	–467.0±0.6	–455.4±0.6
MgO	periclase	40.3044	37.24±0.20	26.95±0.15	5.160±0.020	–601.60±0.30	–569.3±0.3
$Mg(OH)_2$	brucite	58.3197	77.27	63.2±0.1	11.40	–924.5±0.4	–833.5±0.4
MgF_2	sellaite	62.3019	61.51±0.30	57.2±0.5	9.911±0.060	–1124.2±1.2	–1071.1±1.2

$MgCl_2$	chlormagnesite	95.2104	71.04	89.6±0.8	13.76	–641.3±0.7	–591.8±0.7
MgS	niningerite	56.371	45.564	50.3±0.4	8.333	–345.7±4.2	–341.4±4.2
$MgCO_3$	magnesite	84.3139	76.09	65.1±0.1	11.63	–1113.3±1.3	–1029.5±1.3
$MgSiO_3$	enstatite	100.3887	83.09	66.3±0.1	11.99	–1545.6±1.5	–1458.3±1.5
	clinoenstatite	100.3887	82.12	67.9±0.4		–1545.0±1.5	–1458.1±1.5
	perovskite phase	100.3887		63.6±3.0	–	–1445.1±5.0	–1357.0±5.1
	ilmenite phase	100.3887	102.64	60.4±3.0	–	–1486.6±5.0	–1397.5±5.1
Mg_2SiO_4	forsterite	140.6931	118.61	94.1±0.1	17.22	–2173.9±2.0	–2054.5±2.0
$Mg_7Si_8O_{22}(OH)_2$	anthophyllite	780.8205	664.02	534.5±3.5	96.78	–12070.0±8.0	–11343.4±8.1
$Mg_3Si_2O_5(OH)_4$	chrysotile, antigorite	277.1124	274.05	221.3±0.8		–4360.0±3.0	–4032.3±3.0
$Mg_3Si_4O_{10}(OH)_2$	talc	379.2657	321.77	260.8±0.6	46.88	–5900.0±2.0	–5520.1±2.0
$MgAl_2O_4$	spinel	142.2657	115.94	88.7±4.0	15.41	–2299.1±2.0	–2176.6±2.3
$Mg_2Al_3(AlSi_5O_{18})$	cordierite	584.9529	443.28	407.2±3.8	–	–9161.5±5.9	–8651.1±6.0
$Mg_3Al_2Si_3O_{12}$	pyrope	403.1274	325.76	266.3±0.8	47.85	–6285.0±4.0	–5934.5±4.0
Mn	metal, alpha	54.93805	26.30	32.01±0.08	4.994	0	0
Mn^{2+}	aq	54.93695		–73.60±1.0	–	–220.8±0.5	–228.1±0.5
MnO	manganosite	70.9374	44.16	59.02	29.92	–385.1±0.5	–362.6±0.5
MnO_2	pyrolusite	86.9368	54.76	52.8±0.1	8.783	–520.0±0.7	–465.0±0.7
Mn_2O_3	bixbyite	157.8743	101.81	113.7±0.2	17.56	–959.0±1.0	–882.1±1.0
Mn_3O_4	hausmannite	228.8118	142.02	164.1±0.2	24.78	–1384.5±1.4	–1282.5±1.4
MnS	alabandite	87.004	49.98	80.3±0.8	11.66±0.21	–213.9±0.8	–218.7±0.8
MnS_2	hauerite	119.070	70.08	99.9±0.1	14.16	–223.8±10.0	–224.9±10.0
$MnCO_3$	rhodochrosite	114.9470	80.78	98.0±0.1	14.12	–892.9±0.5	–819.1±0.5
$MnSiO_3$	rhodonite	131.0218	86.44	100.5±1.0	14.55	–1321.6±2.0	–1244.7±2.0
Mn_2SiO_4	tephroite	201.9592	128.74	155.9±0.5	22.35	–1731.5±3.0	–1631.0±3.0
Mn_7SiO_{12}	braunite	604.6446	380.79	416.4±0.8	65.68	–4260.0±4.0	–3944.7±4.0

(Continued)

Chemical Formula	Phase or Name	Formula Weight	C_P^o (J mol^{-1} K^{-1})	S^o (J mol^{-1} K^{-1})	$H_{298}^o - H_0^o$	$\Delta_f H^o$ (kJ mol^{-1})	$\Delta_f G^o$ (kJ mol^{-1})
N_2	gas	28.0135	29.124±0.001	191.609±0.004	8.670±0.001	0	0
NO	gas	30.0061	29.862±0.004	210.745±0.010	9.179±0.010	91.27±0.43	87.58±0.43
NO_2	gas	46.0055	37.178	240.166±0.03	10.208	34.23±0.50	52.35±0.50
NO_3	gas	62.0049	46.86±2.93	252.7±2.1	10.96±0.21	71.1±20.9	116.7±20.9
NO_3^-	aq	62.0055	–	146.70±0.40	–	–206.85±0.40	–110.79±0.42
N_2O	gas	44.0129	38.628	220.005±0.02	9.581	81.6±0.5	103.7±0.5
N_2O_3	gas	76.0117	72.727	315±2	17.121	86.6±1.0	141.6±1.2
N_2O_4	gas	92.0111	79.163	304±2	16.740	11.1±1.0	99.8±1.2
N_2O_5	gas	108.0105	95.330	356±7	20.798	13.3±1.5	117.3±2.6
NH_3	ammonia	17.0306	35.630±0.005	192.769±0.050	10.043±0.010	–45.94±0.35	–16.41±0.35
NH_3	aq	17.0306	–	109.0±0.9	–	–81.17±0.33	–26.67±0.30
$NH_3 \cdot H_2O$ (NH_4OH)	liq	35.0459	154.89±0.42	165.56±0.63	28.761±0.063	–361.27±0.29	–254.08±0.35
NH_4^+	aq	18.0380	79.9	111.17±0.40	–	–133.26±0.25	–79.40±0.28
N_2H_4	liq	32.0452	98.84±0.03	121.5±0.4	18.44±0.06	50.38±0.30	149.21±0.30
	hydrazine gas	32.0452	48.42±2.09	238.36±1.30	11.45±0.08	95.18±0.50	159.12±0.50
HNO_3	gas	63.0129	54.102	266.9±0.2	11.876	–133.9±0.6	–73.7±0.6
	nitric acid	63.0129	109.87±0.21	155.64±0.29	27.313±0.042	–174.14±0.50	–80.75±0.50
	ai	63.0129	–	146.70±0.40	–	–206.85±0.40	–110.79±0.42
NH_4Cl	salammoniac	53.4912	84.1±0.4	94.6±0.4	15.69±0.13	–314.4±0.3	–202.9±0.3
NH_4NO_3	ammonia niter	80.0434	139.3±0.4	151.0±0.4	23.68±0.06	–365.43±0.42	–183.65±0.44
NH_4SH	c	51.1124	–	113.4	–	–156.86	–55.14
$(NH_4)_2SO_4$	mascagnite	132.1406	187.48	220.5±1.3	–	–1182.7±1.3	–903.6±1.4
Na	metal, bcc	22.98977	28.154	51.455±0.20	6.447	0	0
Na^+	aq	22.98922		58.45±0.15	–	–240.34±0.06	–261.5±0.1
NaOH	c	39.9971	59.530	64.4±0.8	10.487	–425.8±0.1	–379.6±0.3

NaF	villiaumite	41.9882	46.86	51.5±0.1	8.49	−573.6±0.7	−543.4±0.7
NaCl	halite	58.4425	50.51	72.1±0.2	10.61	−411.3±0.1	−384.2±0.1
Na_2SO_4	thenardite	142.043	127.28	149.6±0.1	23.22	−1387.8±0.4	−1269.8±0.4
Na_2CO_3	c, monocl	105.9884	112.30	135.0±0.6	20.75	−1129.2±0.3	−1045.3±0.4
$Na_2CO_3 \cdot H_2O$	thermonatrite	124.0037	145.60	168.1±0.8	26.34	−1429.7±0.4	−1286.1±0.5
Na_3AlF_6	cryolite	209.9413	217.37	238.5±0.5	38.223	−3316.8±6.0	−3152.1±6.0
$NaAlSiO_4$	nepheline	142.0544	115.81	124.4±1.3	–	−2090.4±3.9	−1975.8±3.9
	carnegieite	142.0544	119.26	118.7±0.3	19.45	−2104.3±4.0	−1988.0±4.0
$NaAlSi_2O_6$	jadeite	202.1387	159.92	133.5±1.3	–	−3029.3±3.6	−2850.6±3.6
$NaAlSi_3O_8$	albite	262.2230	205.07	207.4±0.4	33.45	−3935.0±2.6	−3711.6±2.6
	analbite	262.2230	204.81	225.6±0.4	33.42	−3923.6±2.6	−3705.6±2.6
$NaAl_3Si_3O_{10}(OH)_2$	paragonite (ord)	382.1996	321.50	277.1±0.9	48.12	−5949.3±3.8	−5568.5±3.8
$Na_4(AlSiO_4)_3Cl$	sodalite	484.6057		848.1±4.2		−13457.9±15.8	−12703.7±16.6
$NaFeSi_2O_6$	acmite	231.002	169.88	170.6±0.8	26.94	−2584.5±4.0	−2417.2±4.0
$Na_2Mg_3Al_2Si_8O_{22}(OH)_2$	glaucophane	783.5431	645.48	541.2±3.0	95.62	−11964.0±9.0	−11230.8±9.0
Ni	metal, fcc	58.6934	25.987	29.87±0.08	4.786	0	0
NiO	bunsenite	74.6928	44.49±0.14	38.0±0.2	6.69	−239.84±0.20	−211.68±0.20
NiS	millerite	90.759	47.11	53.0±0.4	53.01	−91.0±3.0	−88.3±3.0
Ni_3S_2	heazlewoodite	240.212	118.24	133.2±0.3	21.501	−216.3±3.0	−210.2±3.0
O_2	gas	31.9988	29.378±0.001	205.147±0.005	8.680±0.001	0	0
O	gas	15.9994	21.911±0.001	161.058±0.003	6.725±0.001	249.17±0.10	231.74±0.10
O_3	ozone, gas	47.9982	39.374±0.005	239.00±0.03	10.366±0.001	141.8±2.0	162.3±2.0
P	white	30.97376	23.824±0.200	41.09±0.25	5.360±0.015	0	0
	red, V	30.97376	21.19±0.07	22.85±0.08	3.607±0.010	−17.46±1.67	−12.03±1.67
$NH_4H_2PO_4$	c	115.0257	142.26±0.42	151.96±0.42	24.02±0.08	−1445.07	−1210.35
Pb	metal, fcc	207.2	26.65±0.10	64.80±0.30	6.870±0.030	0	0

(Continued)

Chemical Formula	Phase or Name	Formula Weight	C_P^o (J mol^{-1} K^{-1})	S^o	H_{298}^o - H_0^o	$\Delta_f H$ (kJ mol^{-1})	$\Delta_f G^o$
Pb^{2+}	aq, m=1	207.19890		18.5±1.0		0.92±0.25	–24.24±0.40
PbO	litharge (red)	223.2	45.77	66.5±0.2		–219.0±0.8	–188.9±0.8
	massicot (yel)	223.2		68.7±0.2		–217.3±0.3	–187.9±0.3
PbO_2	plattnerite	239.2	61.17	71.8±0.4	36.78	–277.4±2.9	–218.3±2.9
Pb_3O_4	minium	685.6	154.90	212.0±6.7	30.188	–718.7±6.3	–601.6±6.6
PbS	galena	239.3	49.49	91.7±0.7	11.51	–98.3±2.0	–96.7±2.0
$PbSO_4$	anglesite	303.3	104.33	148.5±0.6	20.05	–920.0±0.4	–813.1±0.4
S	c, orthorh	32.066	22.750±0.050	32.054±0.050	4.412±0.006	0	0
S	c, monocl	32.066	23.225	33.03±0.05	4.525	0.360±0.003	0.069±0.003
S_2	gas	64.132	32.505±0.010	228.167±0.010	9.132±0.002	128.60±0.30	79.69±0.30
SO	gas	48.065	30.176	221.94±0.01	8.798	4.78±0.25	–21.25±0.25
SO_2	gas	64.065	39.842±0.020	248.223±0.050	10.549±0.010	–296.81±0.20	–300.10±0.20
SO_3	gas	80.064	50.63	256.8±0.8	11.697	–395.7±0.7	–371.0±0.7
SO_4^{2-}	aq	96.065	–293.	18.50±0.40	–	–909.34±0.40	–744.0±0.40
S_2O	gas	80.131	44.11	267.0±0.2	11.128	–56.0±1.4	–85.9±1.4
HS^-	aq	33.074	–	67.0±5.0	–	16.3±1.5	44.8±0.3
H_2S	gas	34.082	34.248±0.010	205.805±0.050	9.957±0.010	–20.6±0.5	–33.4±0.6
H_2SO_4	gas	98.079	83.76	298.8±2.1	16.340	–735.1±8.4	–653.3±8.4
	liq	98.079	138.58	156.90±0.08	28.226	–813.99±0.67	–689.92±0.67
Si	c	28.0855	19.79±0.03	18.81±0.08	3.217±0.008	0	0
SiO_2	quartz	60.0843	44.60±0.30	41.46±0.20	6.916±0.020	–910.7±1.0	–856.3±1.0

	cristobalite	60.0843	44.94±0.30	43.40±0.15	7.037±0.020	–908.0±1.2	–854.2±1.2
	tridymite	60.0843		43.9±0.4		–907.5±2.4	–853.8±2.4
	coesite	60.0843	45.46	38.5±1.0	6.89	–907.8±2.1	–852.5±2.1
	stishovite	60.0843	42.99	27.8±0.4	7.74	–861.3±2.1	–802.8±2.1
	glass	60.0843	37.94	48.5±1.0	6.998	–901.6±2.1	–849.3±2.1
Sn	metal, white, tetrag.	118.710	27.112±0.030	51.18±0.080	6.323±0.008	0	0
	metal, gray, cubic	118.710	25.77	44.14±0.25	5.757	–2.09	0.009
SnO_2	cassiterite	150.710	53.22±0.20	49.04±0.10	8.384±0.020	–577.63±0.20	–515.83±0.20
Sr	metal, fcc	87.62	26.84±0.10	55.00±0.30	6.558±0.030	0	0
Sr^{2+}	aq	87.61945	–	–31.5±2.0		–550.90±0.50	–563.83±0.80
SrO	c	103.62	45.02±0.21	55.2±0.4	8.62±0.08	–591.3±1.0	–560.8±1.0
$SrSO_4$	celestite	183.69	101.7±2.1	121.8±2.1	18.41±042	–1459.0±1.4	–1347.0±1.5
$SrCO_3$	strontianite	147.630	81.59±0.63	97.1±1.7	15.12±0.13	–1225.77±1.1	–1144.86±1.0
Ti	metal, hex	47.867	25.06±0.08	30.72±0.10	4.824±0.015	0	0
TiO_2	rutile	79.866	55.08±0.30	50.62±0.30	8.680±0.050	–944.0±0.8	–888.8±1.0
	anatase	79.866	55.32			–938.7±2.1	–883.3±2.1
Zn	metal, hcp	65.39	25.39±0.04	41.63±0.15	5.657±0.020	0	0
Zn^{2+}	aq	65.38890	–	–109.8±0.5		–153.39±0.20	–147.3±0.2
ZnO	zincite	81.391	40.417	43.16±0.09	6.916	–350.46±0.27	–320.33±0.27

(Continued)

Chemical Formula	Phase or Name	Formula Weight	C_P^o	S^o	H_{298}^o - H_0^o	$\Delta_f H^o$	$\Delta_f G^o$
			(J mol^{-1} K^{-1})			(kJ mol^{-1})	
ZnS	sphalerite	97.459	45.756±0.09	58.655±0.12	8.818±0.020	−204.13±0.41	−199.65±0.41
	wurtzite	97.459	45.878±0.09	58.844±0.12	8.851±0.020	−194.81±1.53	−190.38±1.53
$ZnCO_3$	smithsonite	125.399	80.52	81.19±0.16	13.488	−818.9±0.6	−737.3±0.6

Abbreviations and Notes:

c = crystal, s = solid, am = amorphous, aq = aqueous, ai = aqueous ionized, ao = aqueous unionized, ord = ordered

bcc = body centered cubic, fcc = face centered cubic, fct = face centered tetragonal, hcp = hexagonal close packed, hex = hexagonal, monocl = monoclinic, orthorh = orthorhombic, tet = tetragaonal

IUPAC 1997 atomic weights were used for calculating formula weights in this table

Data sources include Chase (1999), Cox et al. (1989), Gurvich et al. (1989), Kubaschewski and Alcock (1979), Robie and Hemingway (1995), Robie et al. (1978).

Estimated data, such as the heat capacity for metacinnabar are in italics.

Appendix 2

Heat Capacity Equations (also see Tables 3-3 and 3-4)

Compound or	Mineral	$C_p = a + b\times10^{-3}T + c\times10^{5}T^{-2} + d\times10^{-6}T^{2}$ (J mol^{-1} K^{-1})				*T*-range
		a	*b*	*c*	*d*	for C_p fit (K)
AgCl	chlorargyrite	30.083	52.969	6.276	0	298–730
Ag_2S	acanthite	493.827	–1658.226	–81.025	1879.066	298–451.3
Ag_2S	argentite	115.625	–74.749	–6.059	46.568	451.3–865
Al_2O_3	corundum	90.747	65.758	–25.516	–29.420	298–1000
		102.839	20.152	22.895	–0.748	1000–2327
$3Al_2O_3\cdot2SiO_2$	mullite	333.917	337.018	–84.899	–149.069	298–1000
		490.0	46.722	–180.978	–5.248	1000–2123
$Al_2Si_4O_{10}(OH)_2$	pyrophyllite	265.274	368.620	–59.784	–159.373	298–800
$Al_2SiO_4F_2$	topaz	138.213	181.554	–35.466	–99.174	298–1000
C_2H_4	ethylene	9.247	128.922	–0.773	–44.194	298–1500
C_2H_6	ethane	2.842	177.864	1.599	–58.315	298–1000
CH_3OH	Methanol, gas	5.752	125.947	4.067	–42.509	298–1000
CaO	lime, calcia	58.744	–13.042	–11.908	6.638	298–1000
CaO	lime, calcia	33.892	13.576	52.783	–1.597	1000–3000
$Ca(OH)_2$	portlandite	82.237	75.666	–11.834	–44.706	298–700
CaF_2	fluorite, α	41.058	55.463	8.498	0	298–1430
CaS	oldhamite	47.183	9.613	–2.124	–2.454	298–1500
$CaCO_3$	aragonite	81.530	45.670	–11.410	0	298–1000
$CaTiO_3$	perovskite	149.134	–41.818	–36.682	25.297	298–1530
$CaTiSiO_5$	titanite (sphene)	176.701	23.850	–39.911	0	298–1673
$CaMg(CO_3)_2$	dolomite	163.131	113.445	–33.534	–19.363	298–1100
$Ca_2MgSi_2O_7$	akermanite	232.535	74.200	–35.342	–8.997	298–1731
$Ca_3Mg(SiO_4)_2$	merwinite	324.173	13.180	–68.719	17.566	298–1600
CdO	monteponite	48.242	6.360	–5.310	0	298–1755
CdS	greenockite	44.560	13.807	0	0	298–1678
CuO	tenorite	51.548	0.475	–8.658	3.942	298–1397
Cu_2O	cuprite	41.214	84.140	0.525	–48.951	298–800
Cu_2O	cuprite	166.788	–146.229	–116.818	71.449	800–1516.7
CuS	covellite	43.053	20.251	–1.381	0	298–780
Cu_2S	chalcocite, α	49.238	92.787	0	0	298–376
Cu_2S	c, β	118.402	–57.835	0.718	21.825	376–720
Cu_2S	c, γ	83.844	–1.356	2.189	–0.600	720–1400
$CuSO_4$	chalcocyanite	72.713	154.650	–11.978	–72.885	298–1078
$CuFeS_2$	chalcopyrite, α	101.420	41.518	–14.474	–19.617	298–650
$CuFeS_2$	chalcopyrite, α	2149.755	–4143.462	–1489.368	2396.926	650–820
Cu_5FeS_4	bornite, α	180.065	239.029	1.470	–82.599	298–470
Cu_5FeS_4	bornite, β	–177.164	1099.255	0	0	470–535
$Fe_{0.947}O$	wüstite	61.697	–18.135	–8.288	13.094	298–1600
'FeO'	stoichiometric	52.019	1.372	–4.363	1.798	298–1800
‘FeO’	liquid	68.20	0	0	0	1650–3445
Fe_2O_3	hematite	189.039	–163.303	–45.368	165.778	298–950
Fe_2O_3	hematite	225.003	–170.480	198.281	74.014	950–1800
Fe_3O_4	magnetite	91.55	201.67	0	0	298-900
Fe_3O_4	magnetite	200.83	0	0	0	900-1800
$FeCl_2$	lawrencite	121.66	–102.86	–19.00	76.12	298–950
$FeCl_3$	molysite	51.83	132.29	4.72	0	298–577

	Mineral	*a*	*b*	*c*	*d*	for C_p fit (K)
FeS_2	pyrite	73.003	3.414	−11.127	6.978	298–1400
FeS_2	marcasite	73.633	3.005	−11.353	7.111	298–1400
Fe_3C	cohenite (cementite)	107.300	12.530	0	0	298–1800
$FeTiO_3$	ilmenite	110.521	36.774	−19.143	−8.671	298–1000
OH	hydroxyl	26.432	4.228	1.953	−0.043	298–2000
HO_2	hydroperoxyl gas	25.298	34.562	0.351	−12.291	298–1000
H_2O	liq	196.083	−498.666	−21.347	584.042	298–500
H_2O	liq	−981.704	1419.827	890.032	0	500–600
H_2O_2	gas	29.508	50.552	−0.568	−17.384	298–1000
HgO	montroydite	35.785	43.463	−2.609	−19.645	298–1000
HgS	cinnabar	73.765	15.564	0	0	298–618
HgS	metacinnabar	44.016	15.188	0	0	298–1098
HI	gas	24.695	9.515	1.540	−1.208	298–1000
KOH	c, monoclinic, α	51.434	26.878	0.220	58.486	298–516
KCl	sylvite	58.020	−15.550	−3.710	23.720	298–1043
$LiAlSi_4O_{10}$	petalite	270.055	199.893	−69.101	−74.450	298–1300
MgO	periclase	47.926	3.766	−10.512	0.147	298–2000
$Mg(OH)_2$	brucite	102.203	15.109	−26.168	0	298–900
MgF_2	sellaite	77.810	4.050	−15.495	0	298–1536
$MgCl_2$	chlormagnesite	76.870	8.530	−7.443	0	298–987
MgS	niningerite	45.544	8.757	−2.307	0	298–2500
$MgCO_3$	magnesite	81.124	52.247	−18.322	0	298–1000
$MgSiO_3$	orthoenstatite	91.727	54.143	−20.736	−16.348	298–1000
$MgSiO_3$	ilmenite phase	69.120	73.299	10.370	0	298–700
$MgAl_2O_4$	spinel	141.559	51.345	−35.491	−11.275	298–1800
MnO	manganosite	46.484	8.117	−3.680	0	298–2083
MnO_2	pyrolusite	50.091	66.135	−9.902	−43.985	298–850
Mn_2O_3	bixbyite	89.028	59.406	−3.545	−10.598	298–1400
Mn_3O_4	hausmannite	181.826	−27.180	−31.049	36.290	298–1400
MnS	alabandite	49.262	8.263	−1.300	−3.185	298–1000
MnS_2	hauerite	69.721	17.661	−4.358	0	298–350
$MnCO_3$	rhodochrosite	65.797	93.783	−8.956	−32.997	298–600
$MnSiO_3$	rhodonite	114.245	9.506	−27.457	2.795	298–1564
Mn_7SiO_{12}	braunite	430.101	110.999	−73.253	0	298–1500
NO_3	gas	36.901	74.862	−8.199	−34.468	298–1000
N_2O_3	gas	56.564	63.086	−0.372	−25.066	298–1000
N_2O_4	gas	62.420	102.783	−8.748	−45.682	298–1000
N_2O_5	gas	79.148	100.434	−8.896	−42.228	298–1000
N_2H_4	hydrazine gas	22.757	107.632	−2.736	−37.660	298–1000
HNO_3	gas	35.307	99.573	−6.325	−42.489	298–1000
NH_4Cl	salammoniac, α	591.998	−1679.861	−136.732	1677.392	298–457.7
$(NH_4)_2SO_4$	mascagnite	104.359	278.801	0	0	298–600
$NaOH$	c, I	89.203	−172.466	−3.987	298.110	298–572
NaF	villiaumite	52.3715	−5.0372	−4.5322	12.6470	298–1269
Na_2CO_3	c, I	87.462	25.430	0.967	167.290	298–723
$Na_2CO_3 \cdot H_2O$	thermonatrite	72.400	260.700	−1.258	0	298–380
Na_3AlF_6	cryolite, α	495.611	−751.696	−102.046	682.540	298–836
$NaAlSi_2O_6$	dehydrated analcime	192.110	80.735	−47.230	0	298–1000

Mineral		*a*	*b*	*c*	*d*	for C_p fit (K)
$NaAlSi_2O_6 \cdot H_2O$	analcime	134.362	254.173	1.855	0	298–800
$Na_2Al_2Si_3O_{10} \cdot 2H_2O$	natrolite	301.947	376.899	–48.971	0	298–680
NiS	α	44.685	19.037	–2.890	0	298–652
Ni_3S_2	heazlewoodite, α	176.045	–127.046	–28.608	137.874	298–834
PbO	litharge (red)	51.030	10.249	–7.393	0.011	298–1000
PbO	massicot (yel)	44.969	13.248	–2.798	0.011	298–1160
PbO_2	plattnerite	73.119	7.487	–12.606	0	298–1200
Pb_3O_4	minium	177.906	33.255	–29.265	0	298–1800
PbS	galena	44.601	16.399	0	0	298–900
$PbSO_4$	anglesite	46.856	127.756	17.229	0.016	298–1100
SO	gas	25.763	18.271	–0.292	–7.938	298–1000
SO_3	gas	38.543	67.671	–4.849	–29.763	298–1000
S_2O	gas	41.614	26.597	–3.865	–12.170	298–1000
H_2SO_4	gas	71.950	103.796	–13.140	–41.807	298–1000
SrO	c	48.313	8.944	–4.893	–0.739	298–1800
$SrSO_4$	celestite	109.691	53.413	–20.993	0	298–1430
$SrCO_3$	strontianite, ortho	89.621	35.815	–14.209	0	298–1197
TiO_2	rutile	69.488	8.687	–14.937	–1.932	298–2130
TiO_2	anatase	71.568	7.638	–16.330	–2.293	298–2000
ZnO	zincite	46.375	6.055	–6.930	0.395	298–1800
ZnS	sphalerite	46.630	10.953	–3.420	–3.246	298–1300
ZnS	wurtzite	50.885	1.538	–4.999	1.915	298–1300
$ZnCO_3$	smithsonite	65.757	87.439	–8.904	–19.801	298–1200

References

Adami, L. H., & King, E. G. (1964). Heats and free energies of formation of sulfides of manganese, iron, zinc, and cadmium. *U.S. Bureau of Mines Report of Investigations* 6495, Washington, D.C. 10 pages.

Adamkovičová, K., Nerád, I., Kosa, L., Liška, M., Strečko, J., & Proks, I. (1996). Enthalpic analysis of melts in the $Ca_2MgSi_2O_7$–$CaSiO_3$ system. *Chem. Geol., 128*, 107–112.

Ahlberg, J. E., Blanchard, E. R., & Lundberg, W. O. (1937). The heat capacities of benzene, methyl alcohol and glycerol at very low temperatures. *J. Chem. Phys., 5*, 539–551.

Alberty, R. A. (1997). Legendre transforms in chemical thermodynamics. *J. Chem. Thermo., 29*, 501–516.

Ambrose, D. (1980). *Vapor–Liquid Critical Properties*. In: *NPL Report Chem. 107*, Teddington, England: National Physical Laboratory.

Andrews, T. (1869). The Bakerian Lecture: On the continuity of the gaseous and liquid states of matter. *Phil. Trans. Roy. Soc. London, 159*, 575–590.

Aranovich, L. Y., & Newton, R. C. (1999). Experimental determination of CO_2–H_2O activity–composition relations at 600–1000°C and 6–14 kbar by reversed decarbonation and dehydration reactions. *Am. Min., 84*, 1319–1332.

Ashida, T., Kume, S., Ito, E., & Navrotsky, A. (1988). $MgSiO_3$ ilmenite: Heat capacity, thermal expansivity, and enthalpy of transformation. *Phys. Chem. Minerals, 16*, 239–245.

Atkinson, R., Baulch, D. L., Cox, R. A., Hampson, R. F., Jr., Kerr, J. A., Rossi, M. J., & Troe, J. (1997). Evaluated kinetic and photochemical data for atmospheric chemistry VI. IUPAC subcommittee on gas kinetic data evaluation for atmospheric chemistry. *J. Phys. Chem. Ref. Data, 26*, 1329–1499.

Bäckström, H. L. J. (1925). The thermodynamic properties of calcite and aragonite. *J. Am. Chem. Soc., 47*, 2432–2442.

Bacon, J. F., Hasapis, A. A., & Wholley, J. W., Jr. (1960). Viscosity and density of molten silica and high silica glasses. *Phys. Chem. Glasses, 1*, 90–98.

Baker, E. H. (1962). The calcium oxide–carbon dioxide system in the pressure range 1–300 atmospheres. *J. Chem. Soc. 464–470.*

Barton, P. B., Jr., & Skinner, B. J. (1979). Sulfide mineral stabilities. In H. L. Barnes (Ed.), *Geochemistry of Hydrothermal Ore Deposits* (2nd ed.). (pp. 278–403) New York: Wiley-Interscience.

Baulch, D. L., Drysdale, D. D., Duxbury, J., & Grant, S. J. (1976). *Evaluated Kinetic Data for High Temperature Reactions: Vol. 3. Homogeneous Gas Phase Reactions of the O_2–O_3 System, the CO–O_2–H_2 System, and of Sulphur-Containing Species*. London: Butterworths.

Beck, L., Ernst, G., & Gürtner, J. (2002). Isochoric heat capacity c_v of carbon dioxide and sulfur hexafluoride in the critical region. *J. Chem. Thermo., 34*, 277–292.

Benz, R., & Wagner, C. (1961). Thermodynamics of the solid system CaO–SiO_2 from electromotive force data. *J. Phys. Chem., 65*, 1308–1311.

Berman, R., & Simon, F. (1955). On the graphite-diamond equilibrium. *Zeit. f. Elektrochem., 59*, 333–338.

Bethe, H. A. (1929). Termaufspaltung in Kristallen. *Ann. Phys., 3*, 133–206, (Splitting of terms in crystals, Consultants Bureau, Inc., NY).

Bichowsky, F. R., & Rossini, F. D. (1936). *The Thermochemistry of the Chemical Substances*. New York: Reinhold Publishing Co.

Birch, F. (1966). Compressibility, elastic constants. In S. P. Clark Jr. (Ed.), *Handbook of Physical Constants* (pp. 97–173). New York: Geological Society of America, *GSA Memoir* 97.

Blencoe, J. G., Naney, M. T., & Anovitz, L. M. (2001). The CO_2–H_2O system: III. A new experimental method for determining liquid–vapor equilibria at high subcritical temperatures. *Amer. Mineral, 86*, 1100–1111.

Blue, R. W., & Giauque, W. F. (1935). The heat capacity and vapor pressure of solid and liquid nitrous oxide. The entropy from its band spectrum. *J. Am. Chem. Soc., 57*, 991–997.

Bodenstein, M., & Pohl, W. (1905). Gleichgewichtsmessungen an der Kontaktschwefelsäure. *Z. Elektrochem., 11*, 373–384.

Bödlander, G., & Köppen, K. (1903). Beiträge zur theorie technischer prozesse II. Gleichgewichte zwischen Schwefeltrioxyd, Schwefeldioxyd and Sauerstoff. *Z. Elektrochem., 9*, 787–794.

Boehler, R., Ross, M., Söderlind, P., & Boercker, D. B. (2001). High-pressure melting curves of argon, krypton, and xenon: Deviation from corresponding states theory. *Phys. Rev. Lett., 86*, 5731–5734.

Boerio-Goates, J., Francis, M. R., Goldberg, R. N., Ribeiro da Silva, M. A. V., Ribeiro da Silva, M. D. M. C., & Tewari, Y. B. (2001). Thermochemistry of adenosine. *J. Chem. Thermo., 33*, 929–947.

Boettcher, A. L., & Wyllie, P. J. (1968). The calcite–aragonite transition measured in the system CaO–CO_2–H_2O. *J. Geol., 76*, 314–330.

Bouhifd, M. A., Andrault, D., Fiquet, G., & Richet, P. (1996). Thermal expansion of forsterite up to the melting point. *Geophys. Res. Lett., 23*, 1143–1146.

Bowen, N. L. (1913). The melting phenomena of the plagioclase feldspars. *Am. J. Sci., 4th series, 35*, 577–599.

Bowen, N. L. (1915). The crystallization of haplobasaltic, haplodioritic, and related magmas. *Am. J. Sci., 4th series, 40*, 161–185.

Bowen, N. L., & Andersen, O. (1914). The binary system MgO-SiO_2. *Am. J. Sci., 37*, 487–500.

Bowen, N. L., & Schairer, J. F. (1929). The system: leucite–diopside. *Am. J. Sci., 5th series, 18*, 301–312.

Bowen, N. L., & Schairer, J. F. (1935). The system MgO-FeO-SiO_2. *Am. J. Sci., 29*, 151–217.

Bowen, N. L., & Schairer, J. F. (1936). The system albite–fayalite. *Proc. Natl. Acad. Sci., 22*, 345–350.

Bridgman, P. W. (1911). Mercury, liquid and solid, under pressure. *Proc. Am. Acad. Arts Sci., 47*, 347–438.

Bridgman, P. W. (1911). Water in the liquid and five solid forms, under pressure. *Proc. Am. Acad. Arts Sci., 47*, 441–558.

Bridgman, P. W. (1914). A complete collection of thermodynamic formulas. *Phys. Rev., 3*, 273–281.

Bridgman, P. W. (1915). Change of phase under pressure, II. New melting curves with a general thermodynamic discussion of melting. *Phys. Rev., 6*, 94–112.

Bridgman, P. W. (1941). *The Nature of Thermodynamics*. Chapter 1. Cambridge: Harvard University Press.

Bridgman, P. W. (1961). *The Thermodynamics of Electrical Phenomena in Metals and a Condensed Collection of Thermodynamic Formulas*. Mineola, NY: Dover Publications.

Brinkworth, J. H. (1925). On the measurement of the ratio of specific heats using small volumes of gas. The ratios of the specific heats of air and of hydrogen at atmospheric pressure and at temperatures between 20 degrees C and –183 degrees C. *Proc. Roy. Soc. London, 107A*, 510–543.

Bronshten, V. A. (1983). *Physics of Meteoric Phenomena*. Dordrecht: D. Reidel.

Brønsted, J. N. (1906). Studien zur chemischen Affinität. *Zeit. f. Physik. Chem., 55*, 371–382.

Brown, H. T., & Morris, G. J. (1888). The determination of the molecular weights of the carbohydrates. *J. Chem. Soc., 53*, 610–621.

Brown, S. C. (Ed.), (1968–1970). *The Collected Works of Count Rumford* (Vol. 1–5). Cambridge: Harvard University Press.

Brush, S. G. (Ed.), (1965). *Kinetic Theory: Vol. 1. The Nature of Gases and of Heat*. Oxford: Pergamon Press.

Buckingham, M. J., & Fairbank, W. M. (1961). The nature of the λ-transition in liquid helium. In C. J. Gorter (Ed.), *Progress in Low Temperature Physics*, (Vol. III), (pp. 80–112). Amsterdam: North-Holland Publishing Co.

Buddington, A. F., & Lindsley, D. H. (1964). Iron-titanium oxide minerals and their synthetic equivalents. *J. Petrol, 5*, 310–357.

Bundy, F. P., Bovenkerk, H. P., Strong, H. M., & Wentorf, R. H., Jr. (1961). Diamond–graphite equilibrium line from growth and graphitization of diamond. *J. Chem. Phys., 35*, 383–391.

Burns, R. G. (1993). *Mineralogical Applications of Crystal Field Theory* (2nd ed.). Cambridge, UK: Cambridge University Press.

Busey, R. H., & Giauque, W. F. (1953). The equilibrium reaction $NiCl_2 + H_2 = Ni + 2HCl$. Ferromagnetism and the third law of thermodynamics. *J. Am. Chem. Soc., 75*, 1791–1794.

Butler, J. N. (1964). *Ionic Equilibrium: A Mathematical Approach.* Reading, MA: Addison-Wesley.

Butler, J. N. (1998). *Ionic Equilibrium: Solubility and pH Calculations.* Chichester: Wiley.

Calvet, T., & Oonk, H. A. J. (1995). Laevorotatory–carvoxime + dextrorotatory–carvoxime a unique binary system. *CALPHAD, 19*, 49–56.

Caneva, K. L. (1993). *Robert Mayer and the Conservation of Energy.* Princeton: Princeton University Press.

Cardwell, D. S. L. (1989). *From Watt to Clausius.* Ames: Iowa State University Press.

Cardwell, D. S. L. (1990). *James Joule: A Biography.* Manchester: Manchester University Press.

Carpenter, C. D., & Jette, E. R. (1923). The vapor pressures of certain hydrated metal sulfates. *J. Am. Chem. Soc., 45*, 578–590.

Casimir, H. B. G. (1961). *Magnetism and Very Low Temperatures.* New York: Dover.

Centolanzi, F. J., & Chapman, D. R. (1966). Vapor pressure of tektite glass and its bearing on tektite trajectories determined from aerodynamic analysis. *J. Geophys. Res., 71*, 1735–1749.

Chao, K. C., & Robinson, Jr., R. L. (Eds.), (1979). *Equations of State in Engineering and Research. Advances in Chemistry* (Vol. 182). Washington, D.C.: American Chemical Society.

Chao, K. C., & Robinson, Jr., R. L. (Eds.), (1986). *Equations of State Theories and Applications. ACS Symposium Series* (Vol. 300). Washington, D.C.: American Chemical Society.

Chapovsky, P. L., & Hermans, L. J. F. (1999). Nuclear spin conversion in polyatomic molecules. *Annu. Rev. Phys. Chem., 50*, 315–345.

Charlu, T. V., Newton, R. C., & Kleppa, O. J. (1975). Enthalpies of formation at 970 K of compounds in the system $MgO\text{-}Al_2O_3\text{-}SiO_2$ from high temperature solution calorimetry. *Geochim. Cosmochim. Acta, 39*, 1487–1497.

Charlu, T. V., Newton, R. C., & Kleppa, O. J. (1981). Thermochemistry of synthetic $Ca_2Al_2SiO_7$ (gehlenite)–$Ca_2MgSi_2O_7$ (åkermanite) melilites. *Geochim. Cosmochim. Acta, 45*, 1609–1617.

Chase, M. W. (1999). NIST–JANAF Thermochemical Tables (4th ed.). *J. Phys. Chem. Ref. Data Monograph.* No. 9.

Chen, C. A., Chen, J. H., & Millero, F. J. (1980). Densities of NaCl, $MgCl_2$, Na_2SO_4, and $MgSO_4$ aqueous solutions at 1 atm from 0 to 50°C and from 0.001 to 1.5 m. *J. Chem. Eng. Data, 25*, 307–310.

Circone, S., & Navrotsky, A. (1992). Substitution of [6,4]aluminum in phlogopite: high temperature solution calorimetry, heat capacities, and thermodynamic properties of the phlogopite–eastonite join. *Amer. Mineral, 77*, 1191–1205.

Clayton, J. O., & Giauque, W. F. (1932). The heat capacity and entropy of carbon monoxide. Heat of vaporization. Vapor pressures of sold and liquid. Free energy to 5000 °K from spectroscopic data. *J. Am. Chem. Soc., 54*, 2610–2626.

Colinet, C., & Pasturel, A. (1994). High-temperature solution calorimetry. In K. N. Marsh, & P. A. G. O'Hare (Eds.), *Solution Calorimetry* (pp. 89–129). Oxford: Blackwell.

Connelly, D. L., Jr., Loomis, J. S., & Mapother, D. E. (1971). Specific heat of nickel near the Curie temperature. *Phys. Rev. B., 3*, 924–934.

Conway, J. B., Wilson, R. H., Jr., & Grosse, A. V. (1953). Temperature of the cyanogen-oxygen flame. *J. Amer. Chem. Soc., 75*, 499.

Cook, L. P., & McMurdie, H. F. (1989). *Phase Diagrams for Ceramists* (Vol. VII). Westerville, OH: American Ceramic Society.

Coplen, T. B. (2001). Atomic weights of the elements 1999. *J. Phys. Chem. Ref. Data 30*, 701–712.

Cornish, R. E., & Eastman, E. D. (1928). The specific heat of hydrogen gas at low temperatures from the velocity of sound, and a precision method of measuring the frequency of an oscillating circuit. *J. Am. Chem. Soc., 50*, 627–652.

Cottrell, T. L. (1958). *The Strengths of Chemical Bonds* (2nd ed.). London: Butterworths Scientific Publications.

Couch, E. J., Hirth, L. J., & Kobe, K. A. (1961). Volumetric behavior of nitrous oxide. *J. Chem. Eng. Data, 6*, 229–237.

Coughlin, J. P., King, E. G., & Bonnickson, K. R. (1951). High-temperature heat contents of ferrous oxide, magnetite, and ferric oxide. *J. Am. Chem. Soc., 73*, 3891–3893.

Cox, J. D., Wagman, D. D., & Medvedev, V. A. (1989). *CODATA Key Values for Thermodynamics*. New York: Hemisphere Publishing.

Cragoe, C. S. (1928). Vapor Pressures and Orthobaric Densities above One Atmosphere. In E. W. Washburn (Ed.), *International Critical Tables:* (Vol. 3). (pp. 228–233). New York, NY: McGraw-Hill.

Crawford, W. A., & Fyfe, W. S. (1964). Calcite–aragonite equilibrium at 100°C. *Science, 144*, 1569–1570.

Crouch, A. G., Hay, K. A., & Pascoe, R. T. (1971). Magnetite–haematite–liquid equilibrium conditions at oxygen pressures up to 53 bar. *Nature, 234*, 132–133.

Dachs, E., Geiger, C. A., von Seckendorff, V., & Grodzicki, M. (2007). A low temperature calorimetric study of synthetic (forsterite + fayalite) {(Mg_2SiO_4 + Fe_2SiO_4)} solid solutions: An analysis of vibrational, magnetic, and electronic contributions to the molar heat capacity and entropy of mixing. *J. Chem. Thermo., 39*, 906–933.

Darken, L. S., & Gurry, R. W. (1945). The system iron–oxygen. I. The wüstite field and related equilibria. *J. Am. Chem. Soc., 67*, 1398–1412.

Darken, L. S., & Gurry, R. W. (1946). The system Iron–Oxygen. II. Equilibrium and thermodynamics of liquid oxide and other phases. *J. Am. Chem. Soc., 68*, 798–816.

Deming, W. E., & Shupe, L. E. (1931). Some physical properties of compressed gases, II. Carbon monoxide. *Phys. Rev., 38*, 2245–2264.

Deming, W. E., & Shupe, L. E. (1932). Some physical properties of compressed gases, III. Hydrogen. *Phys. Rev., 40*, 848–859.

Desai, P. D. (1986). Thermodynamic properties of iron and silicon. *J. Phys. Chem. Ref. Data, 15*, 967–983.

Dessureault, Y., Sangster, J., & Pelton, A. D. (1990). Coupled phase diagram/thermodynamic analysis of the nine common-ion binary systems involving the carbonates and sulfates of lithium, sodium, and potassium. *J. Electrochem. Soc., 137*, 2941–2950.

DeVries, R. C., Roy, R., & Osborn, E. F. (1954). Phase equilibria in the system CaO–TiO_2. *J. Phys. Chem., 58*, 1069–1073.

DeVries, R. C., Roy, R., & Osborn, E. F. (1955). Phase equilibria in the system CaO–TiO_2–SiO_2. *J. Am. Ceram. Soc., 38*, 158–171.

Din, F. (Ed.), (1962). *Thermodynamic Functions of Gases*, (3 Vol.). London, UK: Butterworths.

Dixon, D. A., Feller, D., & Sandrone, G. (1999). Heats of formation of simple perfluorinated carbon compounds. *J. Phys. Chem. A., 103*, 4744–4751.

Dixon, M., Hoare, F. E., & Holden, T. M. (1965). The specific heat of magnetite. *Phys. Lett., 14*(3), 184–185.

Dodge, B. F. (1944). *Chemical Engineering Thermodynamics*. New York, NY: McGraw-Hill.

Doman, R. C., Barr, J. B., McNally, R. N., & Alper, A. M. (1963). Phase equilibria in the system CaO–MgO. *J. Am. Ceram. Soc., 46*, 313–316.

Doremus, R. H. (1994). *Glass Science* (2nd ed.). New York, NY: John Wiley.

Dorsey, N. E. (1940). *Properties of Ordinary Water-Substance*. New York: Reinhold.

Douslin, D. R., Harrison, R. H., Moore, R. T., & McCullough, J. P. (1964). P-V-T relations for methane. *J. Chem. Eng. Data, 9*, 358–363.

Dufour, L., & Defay, R. (1963). *Thermodynamics of Clouds*. NY: Academic Press.

Dymond, J. H., & Smith, E. B. (1980). *The Virial Coefficients of Pure Gases and Mixtures: A Critical Compilation*. Oxford, UK: Clarendon Press.

Eastman, E. D., & McGavock, W. C. (1937). The heat capacity and entropy of rhombic and monoclinic sulfur. *J. Am. Chem. Soc., 59*, 145–151.

Eastman, E. D., & Milner, R. T. (1933). The entropy of a crystalline solution of silver bromide and silver chloride in relation to the third law of thermodynamics. *J. Chem. Phys., 1*, 444–456.

Emmett, P. H., & Schultz, J. F. (1929). Equilibrium in the system Co-H_2O-CoO-H_2. Free energy changes for the reaction CoO + H_2 = Co + H_2O and the reaction Co + ½ O_2 = CoO. *J. Am. Chem. Soc., 51*, 3249–3262.

Emmett, P. H., & Schultz, J. F. (1930). Equilibrium in the system Co-CO_2-CoO-CO. Indirect calculation of the water gas equilibrium constant. *J. Am. Chem. Soc., 52*, 1782–1793.

Eskola, P. (1922). The silicates of strontium and barium. *Am. J. Sci.*, 331–375.

Farkas, A. (1935). *Orthohydrogen, Parahydrogen, and Heavy Hydrogen.* Cambridge, England: Cambridge University Press.

Fegley, B., Jr. (2000). Kinetics of gas-grain reactions in the solar nebula. *Space Sci. Rev., 92*, 177–200.

Fegley, B., Jr. (2004). Venus. In A. M. Davis (Ed.), *Vol. 1 Meteorites, Comets, and Planets* (pp. 487–507). In H. D. Holland & Turekian K. K. (Ed.), Treatise on Geochemistry, Oxford: Elsevier-Pergamon.

Fegley, B., Jr., & Prinn, R. G. (1986). Chemical models of the deep atmosphere of Uranus. *Astrophys. J., 307*, 852–865.

Fei, Y. (1995). Thermal expansion. In T. J. Ahrens (Ed.), *Mineral Physics and Crystallography: A Handbook of Physical Constants* (pp. 29–44). Washington, D.C: American Geophysical Union, AGU Reference Shelf 2.

Ferry, J. M. (Ed.), (1982). *Characterization of Metamorphism Through Mineral Equilibria. Reviews in Mineralogy*, (Vol. 10). Washington, D.C: Mineralogical Society of America.

Fiquet, G., Richet, P., & Montagnac, G. (1999). High-temperature thermal expansion of lime, periclase, corundum, and spinel. *Phys. Chem. Minerals, 27*, 103–111.

Fisher, R. A., Brodale, G. E., Hornung, E. W., & Giauque, W. F. (1968). Magnetothermodynamics of α-$NiSO_4 \cdot 6H_2O$. IV. Heat capacity, entropy, magnetic moment, and internal energy from 0.4° to 4.2° K with fields 0-90 kG along the a axis. *J. Chem. Phys., 49*, 4096–4107.

Flugge, S. (Ed.), (1956). *Handbuch der Physik* (Vols. 14–15). Berlin: Springer–Verlag.

Fritsch, S., & Navrotsky, A. (1995). Thermodynamic properties of manganese oxides. *J. Am. Ceram. Soc., 79*, 1761–1768.

Gaydon, A. G. (1968). *Dissociation Energies and Spectra of Diatomic Molecules* (3rd ed.). London: Chapman and Hall.

Gerke, R. H. (1922). Temperature coefficient of electromotive force of galvanic cells and the entropy of reactions. *J. Am. Chem. Soc., 44*, 1684–1704.

Giauque, W. F., & Egan, C. J. (1937). Carbon dioxide. The heat capacity and vapor pressure of the solid. The heat of sublimation. Thermodynamic and spectroscopic values of the entropy. *J. Chem. Phys., 5*, 45–54.

Giauque, W. F., & MacDougall, D. P. (1933). Attainment of temperatures below 1° absolute by demagnetization of $Gd_2(SO_4)_3 \cdot 8H_2O$. *Phys. Rev., 43*, 768.

Giauque, W. F., & Powell, T. M. (1939). Chlorine. The heat capacity, vapor pressure, heats of fusion and vaporization, and entropy. *J. Am. Chem. Soc., 61*, 1970–1974.

Giauque, W. F., & Stephenson, C. C. (1938). Sulfur dioxide. The heat capacity of solid and liquid. Vapor pressure. Heat of vaporization. The entropy values from thermal and molecular data. *J. Am. Chem. Soc., 60*, 1389–1394.

Gibbs, J. W. (1928). *The Collected Works of J. Willard Gibbs*. New Haven, CT: Yale University Press.

Gibby, C. W., Tanner, C. C., & Masson, I. (1929). The pressure of gaseous mixtures. II. Helium and hydrogen and their intermolecular forces. *Proc. Roy. Soc. London, A122*, 283–304.

Gibson, G. E., & Giauque, W. F. (1923). The third law of thermodynamics. Evidence from the specific heats of glycerol that the entropy of a glass exceeds that of a crystal at the absolute zero. *J. Am. Chem. Soc., 45*, 93–104.

Gill, E. K., & Morrison, J. A. (1966). Thermodynamic properties of condensed CO. *J. Chem. Phys., 45*, 1585–1590.

Gillespie, L. J. (1925). Equilibrium pressures of individual gases in mixtures and the mass-action law for gases. *J. Am. Chem. Soc., 47*, 305–312.

Gillespie, L. J., & Beattie, J. A. (1930). The thermodynamic treatment of chemical equilibria in systems composed of real gases. I. An approximate equation for the mass action function applied to the existing data on the Haber equilibrium. *Phys. Rev., 36*, 743–753.

Gladun, C. (1966). The specific heat at constant volume of liquid neon. *Cryogenics, 6*, 27–30.

Glasser, F. P., & Osborn, E. F. (1960). The ternary system MgO–MnO–SiO_2. *J. Am. Ceram. Soc., 43*, 132–140.

Goldberg, R. N., Kishore, N., & Lennen, R. M. (2002). Thermodynamic quantities for the ionization reactions of buffers. *J. Phys. Chem. Ref. Data, 31*, 231–370.

Goldsmith, J. R., & Graf, D. L. (1957). The system CaO–MnO–CO_2: Solid-solution and decomposition relations. *Geochim. Cosmochim. Acta, 11*, 310–334.

Gopal, E. S. R. (1966). *Specific Heats at Low Temperatures*. New York: Plenum Press.

Greene, K. T., & Morgan, W. R. (1941). The system sodium disilicate–barium disilicate. *J. Am. Ceram. Soc., 24*, 111–116.

Greig, J. W. (1927). Immiscibility in silicate melts. *Am. J. Sci., 13*(5), 1–44; 133–154.

Greywall, D. S. (1986). ^{3}He specific heat and thermometry at millikelvin temperatures. *Phys. Rev. B., 33*, 7520–7538.

Griggs, D. T., Turner, F. J., & Heard, H. C. (1960). Deformation of rocks at 500°C to 800°C. In D. T. Griggs, & J. Handin (Eds.), *Rock Deformation* (pp. 39–104). Geological Society of America Memoir 79.

Grilly, E. R., & Mills, R. L. (1962). PVT relations in He^4 near the melting curve and the λ-line. *Ann. Phys., 18*, 250–263.

Grønvold, F., & Stølen, S. (1992). Thermodynamics of iron sulfides. II. Heat capacities and thermodynamic properties of FeS and of $Fe_{0.875}S$ at temperatures from 298.15 K to 1000 K, of $Fe_{0.98}S$ from 298.15 K to 800 K, and of $Fe_{0.89}S$ from 298.15 K to about 650 K. Thermodynamics of formation. *J. Chem. Thermo., 24*, 913–936.

Grønvold, F., Stølen, S., & Svendsen, S. R. (1989). Heat capacity of α quartz from 298.15 to 847.3 K, and of β quartz from 847.3 to 1000 K–Transition behaviour and revaluation of the thermodynamic properties. *Thermochim. Acta, 139*, 225–243.

Grüneisen, E. (1926). Zustand des festen Körpers. In H. Geiger, & K. Scheel (Eds.), *Thermische Eigenschaften der Stoffe, Handbuch der Physik: Vol. 10.* (pp. 1–59). Berlin: Springer-Verlag (Translation in NASA Republication RE 2-18-59W).

Guggenheim, E. A. (1945). The principle of corresponding states. *J. Chem. Phys., 13*, 253–261.

Gurvich, L. V., Bergman, G. A., Gorokhov, L. N., Iorish, V. S., Leonidov, V. Ya., & Yungman, V. S. (1996). Thermodynamic properties of alkali metal hydroxides. Part 1. Lithium and sodium hydroxides. *J. Phys. Chem. Ref. Data, 25*, 1211–1276.

Gurvich, L. V., Veyts, I. V., & Alcock, C. B. (Eds.), (1989–1996). *Thermodynamic Properties of Individual Substances* (4th ed.). (Vol. 1). New York: Hemisphere Publishing Co.

Haber, F., & Tamaru, S. (1915). Researches on ammonia IV. Determination of the heat of formation of ammonia at high temperatures. *Zeitschrift für Elektrochemie und Angewandte Physikalische Chemie, 21*, 191–206.

Haber, F., & Le Rossignol, R. (1907). Über das Ammoniak-Gleichgewicht. *Ber. D. Deutsch. Chem. Ges., 40*, 2144–2154.

Haber, F., & Le Rossignol, R. (1908). The ammonia equilibrium under pressure. *Z. Elektrochem., 14*, 181–196.

Haber, F., & Maschke, A. (1915). Untersuchungen Über Ammoniak. Sieben Mitteilungen. III. Neubestimmung des Ammoniakgleichgewichts bei gewöhnlichem Druck. *Z. Elektrochem., 21*, 128–130.

Haber, F., Tamaru, S., & Ponnaz, Ch (1915). Untersuchungen Über Ammoniak. Sieben Mitteilungen. II. Neubestimmung des Ammoniakgleichgewichts bei 30 Atm Druck. *Z. Elektrochem., 21*, 89–106.

Hageman, V. B. M., & Oonk, H. A. J. (1986). Liquid immiscibility in the SiO_2 + MgO, SiO_2 + SrO, SiO_2 + La_2O_3, and SiO_2 + Y_2O_3 systems. *Phys. Chem. Glasses, 27*, 194–198.

Hager, J. P., Howard, S. M., & Jones, J. H. (1970). Thermodynamic properties of the Cu-Sn and Cu-Au systems by mass spectrometry. *Met. Trans., 1*, 415–422.

Hahn, O. (1903). Beiträge zur Thermodynamik des Wassergases. Das Gleichgewicht: $CO_2 + H_2 = CO + H_2O$. *Z. Phys. Chem., 44*, 513–547.

Halstead, P. E., & Moore, A. E. (1957). The thermal dissociation of calcium hydroxide. *J. Chem. Soc.*, 3873–3875.

Hansen, M., & Anderko, K. (1958). *Constitution of Binary Alloys*. NY: McGraw-Hill.

Harker, R. I., & Tuttle, O. F. (1955). Studies in the system CaO–MgO–CO_2 Part 1. The thermal dissociation of calcite, dolomite and magnesite. *Am. J. Sci., 253*, 209–224.

Harker, R. I., & Tuttle, O. F. (1956). Experimental data on the P_{CO_2}–T curve for the reaction: Calcite plus quartz $\leftrightarrows$ wollastonite + carbon dioxide. *Am. J. Sci., 254*, 239–256.

Harned, H. S., & Owen, B. B. (1958). *The Physical Chemistry of Electrolytic Solutions.* (p. 693). New York: Reinhold.

Harned, H. S., & Scholes, S. R., Jr. (1941). The ionization constant of $HCO^-{}_3$ from 0 to 50°. *J. Am. Chem. Soc., 63*, 1706–1709.

Harnisch, J., & Eisenhauer, A. (1998). Natural CF_4 and SF_6 on Earth. *Geophys. Res. Lett., 25*, 2401–2404.

Harnisch, J., Frische, M., Borchers, R., Eisenhauer, A., & Jordan, A. (2000). Natural fluorinated organics in fluorite and rocks. *Geophys. Res. Lett., 27*, 1883–1886.

Hazen, R. M. (1999). *The Diamond Makers*. Cambridge: Cambridge University Press.

Hemingway, B. (1987). Quartz: Heat capacities from 340 to 1000 K and revised values for the thermodynamic properties. *Amer. Mineral, 72*, 273–379.

Hemingway, B. S., & Robie, R. A. (1984). Heat capacity and thermodynamic functions for gehlenite and staurolite: with comments on the Schottky function in the heat capacity of staurolite. *Amer. Mineral, 69*, 307–318.

Hemingway, B. S., Robie, R. A., Evans, H. T., Jr., & Kerrick, D. M. (1991). Heat capacities and entropies of sillimanite, fibrolite, andalusite, kyanite, and quartz and the Al_2SiO_5 phase diagram. *Amer. Mineral., 76*, 1597–1613.

Herzberg, G. (1944). *Atomic Spectra and Atomic Structure*. New York: Dover.

Hess, S. L. (1959). *An Introduction to Theoretical Meteorology.* chapters 2–7. New York: Henry Holt & Co.

Hildebrand, J. H., & Scott, R. L. (1964). *The Solubility of Nonelectrolytes*. NY: Dover Books.

Hill, K. J., & Winter, E. R. S. (1956). Thermal dissociation pressure of calcium carbonate. *J. Phys. Chem., 60*, 1361–1362.

Hilsenrath, J., Beckett, C. W., Benedict, W. S., Fano, L., Hoge, H. J., Masi, J. F., Nuttall, R. L., Touloukian, Y. S., & Woolley, H. W. (1955). Tables of thermal properties of gases. *Natl. Bur. Standards Circ., 564*. pp. 488.

Hirschfelder, J. O., Curtiss, C. F., & Bird, R. B. (1954). *Molecular Theory of Gases and Liquids*. New York, NY: John Wiley.

Holland, T. J. B., & Powell, R. (1991). A compensated–Redlich–Kwong (CORK) equation for volumes and fugacities of CO_2 and H_2O in the range 1 bar to 50 kbar and 100–1600°C. *Contrib. Mineral. Petrol., 109*, 265–273.

Holland, T. J. B., & Powell, R. (1990). An enlarged and updated internally consistent thermodynamic dataset with uncertainties and correlations: the system K_2O-Na_2O-CaO-MgO-MnO-FeO-Fe_2O_3-Al_2O_3-TiO_2-SiO_2-C-H_2-O_2. *J. Metamorphic Geol., 8*, 89–124.

Holland, T. J. B., & Powell, R. (1998). An internally consistent thermodynamic data set for phases of petrological interest. *J. Metamorphic Geol., 16*, 309–343.

Holley, C. E., Jr., & Huber, E. J., Jr. (1951). The heats of combustion of magnesium and aluminum. *J. Amer. Chem. Soc., 73*, 5577–5578.

Holzheid, A., & O'Neill, H. St. C. (1995). The Cr-Cr_2O_3 oxygen buffer and the free energy of formation of Cr_2O_3 from high-temperature electrochemical measurements. *Geochim. Cosmochim. Acta, 59*, 475–479.

Hong, Y., & Fegley, B., Jr. (1998). Experimental studies of magnetite formation in the solar nebula. *Meteoritics Planet. Sci., 33*, 1101–1112.

Hougen, O. A., Watson, K. M., & Ragatz, R. A. (1960). *Chemical Process Principles* (2nd ed.). New York, NY: John Wiley.

Hovis, G. L., & Navrotsky, A. (1995). Enthalpies of mixing for disordered alkali feldspars at high temperature: A test of regular solution thermodynamic models and a comparison of hydrofluoric acid and lead borate solution calorimetric techniques. *Am. Min., 80*, 280–284.

Hovis, G. L., Roux, J., & Richet, P. (1998). A new era in hydrofluoric acid solution calorimetry: Reduction of required sample size below ten milligrams. *Amer. Mineral, 83*, 931–934.

Hoxton, L. G. (1919). The Joule–Thomson effect for air at moderate temperatures and pressures. *Phys. Rev., 13*, 438–479.

Huebner, S. J. (1987). Use of gas mixtures at low pressure to specify oxygen and other fugacities of furnace atmospheres. In G. C. Ulmer, & H. L. Barnes (Eds.), *Hydrothermal Experimental Techniques* (pp. 20–60). New York: Wiley-Interscience.

Hultgren, R., Desai, P. D., Hawkins, D. T., Gleiser, M., & Kelley, K. K. (1973). Selected Values of the Thermodynamic Properties of Binary Alloys. *Amer. Soc. for Metals*. Metals Park, OH.

Hultgren, R., Desai, P. R., Hawkins, D. T., Gleiser, M., Kelley, K. K., & Wagman, D. D. (1973). Selected Values of the Thermodynamic Properties of the Elements. *Amer. Soc. for Metals*. Metals Park, OH.

Ihde, A. J. (1984). *The Development of Modern Chemistry*. New York: Dover Publications. pp. 395–404.

Jacob, K. T., & Abraham, K. P. (2009). Thermodynamic properties of calcium titanates: $CaTiO_3$, $Ca_4Ti_3O_{10}$, and $Ca_3Ti_2O_7$. *J. Chem. Thermo., 41*, 816–820.

Jacobsson, E. (1985). Solid state EMF studies of the systems FeO-Fe_3O_4 and Fe_3O_4-Fe_2O_3 in the temperature range 1000–1600 K. *Scand. J. Met., 14*, 252–256.

Jamieson, J. C. (1953). Phase equilibrium in the system calcite–aragonite. *J. Chem. Phys., 21*, 1385–1390.

Jeans, J. H. (1954). *The Dynamical Theory of Gases* (4th ed.). New York, NY: Dover.

Jensen, W. B. (2006). The origin of the term allotrope. *J. Chem. Educ., 83*, 838–839.

Johannes, W., & Puhan, D. (1971). The calcite–aragonite transition. Reinvestigated. *Contr. Mineral. Petrol., 31*, 28–38.

Johnston, H. L., & Giauque, W. F. (1929). The heat capacity of nitric oxide from 14° K to the boiling point and the heat of vaporization. Vapor pressures of solid and liquid phases. The entropy from spectroscopic data. *J. Am. Chem. Soc., 51*, 3194–3214.

Johnston, J. (1910). The thermal dissociation of calcium carbonate. *J. Am. Chem. Soc., 32*, 938–946.

Jones, H. A., Langmuir, I., & Mackay, G. M. J. (1927). The rates of evaporation and the vapor pressures of tungsten, molybdenum, platinum, nickel, iron, copper and silver. *Phys. Rev., 30*, 201–214.

Joule, J. P., & Thomson, W. (1854). On the thermal effect of fluids in motion. *Part II. Phil. Trans Roy. Soc. London, 144*, 321–364.

Joule, J. P., & Thomson, W. (1862). On the thermal effect of fluids in motion. Part IV. *Phil. Trans Roy. Soc. London, 152*, 579–589.

Kang, T. L., Hirth, L. J., Kobe, K. A., & McKetta, J. J. (1961). Pressure-volume-temperature properties of sulfur dioxide. *J. Chem. Eng. Data, 6*, 220–226.

Kangro, W., & Petersen, E. (1950). Über die reduction von eisenchloriden mit wasserstoff. *Zeit. Anorg. Allgem. Chem., 261*, 157–178.

Kaufman, L. (1963). Phase equilibria and transformations in metals under pressure. In W. Paul, & D. M. Warschauer (Eds.), *Solids Under Pressure* (pp. 303–356). New York: McGraw-Hill.

Kazenas, E. K., Zviadadze, G. N., & Bol'shikh, M. A. (1985). Thermodynamics of sublimation, dissociation, and gas phase reactions of vapors over silicon dioxide. *Metally*(1), 46–48.

Keller, W. E. (1969). *Helium-3 and Helium-4*. New York: Plenum Press.

Kelley, K. K. (1929). Cyclohexanol and the third law of thermodynamics. *J. Am. Chem. Soc., 51*, 1400–1406.

Kelley, K. K. (1935). *Contributions to the Data on Theoretical Metallurgy VIII. The Thermodynamic Properties of Metal Carbides and Nitrides. U.S. Bureau of Mines Bulletin* No. 407. Washington, D.C: U. S. Government Printing Office.

Kelley, K. K. (1936). *Contributions to the Data on Theoretical Metallurgy V. Heats of Fusion of Inorganic Substances. U.S. Bureau of Mines Bulletin* No. 393. Washington, D.C: U.S. Government Printing Office.

Kelley, K. K. (1943). Specific heats at low temperatures of magnesium orthosilicate, and magnesium metasilicate. *J. Am. Chem. Soc., 65*, 339–341.

Kelley, K. K. (1960). *Contributions to the Data on Theoretical Metallurgy XIII. High-Temperature Heat-Content, Heat-Capacity, and Entropy Data for the Elements and Inorganic Compounds. U.S. Bureau of Mines Bulletin* 584. Washington, D.C: U. S. Government Printing Office.

Kelley, K. K., & King, E. G. (1961). *Contributions to the Data on Theoretical Metallurgy XIV. Entropies of the Elements and Inorganic Compounds. U.S. Bureau of Mines* Bulletin 592. Washington, D.C: U.S. Government Printing Office.

Kelley, K. K., Southard, J. C., & Anderson, C. T. (1941). *Thermodynamic Properties of Gypsum and its Dehydration Products. U.S. Bureau of Mines* Technical Paper 625. Washington, D.C.: U. S. Government Printing Office. 73 pages.

Kemp, J. D., & Egan, C. J. (1938). Hindered rotation of the methyl groups in propane. The heat capacity, vapor pressure, heats of fusion and vaporization of propane. Entropy and density of the gas. *J. Am. Chem. Soc., 60*, 1521–1525.

Kemp, J. D., & Pitzer, K. S. (1937). The entropy of ethane and the third law of thermodynamics. Hindered rotation of methyl groups. *J. Am. Chem. Soc., 59*, 276–279.

Kennedy, C. S., & Kennedy, G. C. (1976). The equilibrium boundary between graphite and diamond. *J. Geophys. Res., 81*, 2467–2470.

Kennedy, G. C., & Newton, R. C. (1963). Solid-liquid and solid-solid phase transitions in some pure metals at high temperatures and pressures. In W. Paul, & D. M. Warschauer (Eds.), *Solids Under Pressure* (pp. 163–178). New York: McGraw-Hill.

Kerrick, D. M. (Ed.), (1991). *Contact Metamorphism*. Washington, D.C: Mineralogical Society of America.

Kerrick, D. M., & Jacobs, G. K. (1981). A modified Redlich–Kwong equation for H_2O, CO_2, and H_2O–CO_2 mixtures at elevated pressures and temperatures. *Am. J. Sci., 281*, 735–767.

Khalil, M. A. K., Rasmussen, R. A., Culbertson, J. A., Prins, J. M., Grimsrud, E. P., & Shearer, M. J. (2003). Atmospheric perfluorocarbons. *Environ. Sci. & Technol.* ACS ASAP.

Kieffer, S. W. (1985). Heat capacity and entropy: Systematic relations to lattice vibrations. In S. W. Kieffer, & A. Navrotsky (Eds.), *Microscopic to Macroscopic Atomic Environments to Mineral Thermodynamics: Vol. 14*. (pp. 65–126). Washington DC: Mineralogical Society of America, Reviews in Mineralogy.

Kielland, J. (1937). Individual activity coefficients of ions in aqueous solutions. *J. Am. Chem. Soc., 59*, 1675–1678.

King, E. G. (1956). Heat capacities at low temperatures and entropies of five spinel minerals. *J. Phys. Chem., 60*, 410–412.

King, E. G., Barany, R., Weller, W. W., & Pankratz, L. B. (1967). *Thermodynamic properties of forsterite and serpentine*. Washington, D.C.: U.S. Bureau of Mines Report of Investigations 6962, pp. 19.

Kirby, R. K., Hahn, T. A., & Rothrock, B. D. (1972). Thermal expansion. Part 4f. In D. E. Gray (Ed.), *American Institute of Physics Handbook*. NY: McGraw Hill. pp. 4–119 to 4–142.

Kiseleva, I. A., Ogorodova, L. P., Topor, N. D., & Chigareva, O. G. (1979). A thermochemical study of the CaO-MgO-SiO_2 system. *Geochemistry International, 12*, 122–134.

Kistiakowsky, G. B., & Nickle, A. G. (1951). Ethane–ethylene and propane–propylene equilibria. *Disc. Faraday Soc., 10*, 175–187.

Kistiakowsky, G. B., Romeyn, H., Jr., Ruhoff, J. R., Smith, H. A., & Vaughan, W. E. (1935). Heats of organic reactions I. The apparatus and the heat of hydrogenation of ethylene. *J. Am. Chem. Soc., 57*, 65–75.

Kiukkola, K., & Wagner, C. (1957). Measurements on galvanic cells involving solid electroytes. *J. Electrochem. Soc., 104*, 379–387.

Klemme, S., & van Miltenburg, J. C. (2002). Thermodynamic properties of nickel chromite ($NiCr_2O_4$) based on adiabatic calorimetry at low temperatures. *Phys. Chem. Minerals, 29*, 663–667.

Klemme, S., & van Miltenburg, J. C. (2003). Thermodynamic properties of hercynite ($FeAl_2O_4$) based on adiabatic calorimetry at low temperatures. *Amer. Min., 88*, 68–72.

Klemme, S., & van Miltenburg, J. C. (2004). The entropy of zinc chromite ($ZnCr_2O_4$). *Min. Mag., 68*, 515–522.

Klemme, S., O'Neill, H. St., C., Schnelle, W., & Gmelin, E. (2000). The heat capacity of $MgCr_2O_4$, $FeCr_2O_4$, and Cr_2O_3 at low temperatures and derived thermodynamic properties. *Amer. Mineral, 85*, 1686–1693.

Kleppa, O. J. (2001). Evolution and application of high-temperature reaction calorimetry at the University of Chicago from 1952 to 2000. *J. Alloys Compounds, 321*, 153–163.

Knietsch, R. (1901). Über die Schwefelsäure und ihre Fabrication nach dem Contactverfahren. *Berl. Ber., 34*, 4069–4115.

Knittle, E. (1995). Static compression measurements of equations of state. In T. J. Ahrens (Ed.), *Mineral Physics and Crystallography: A Handbook of Physical Constants* (pp. 98–142). Washington, D.C: American Geophysical Union, AGU Reference Shelf 2.

Koehler, J. K., & Giauque, W. F. (1958). Perchloryl fluoride. Vapor pressure, heat capacity, heats of fusion and vaporization. Failure of the crystal to distinguish O and F. *J. Am. Chem. Soc., 80*, 2659–2662.

Kohnstamm, Ph (1926). Thermodynamik der Gemische. In H. Geiger, & K. Scheel (Eds.), *Thermische Eigenschaften der Stoffe, Handbuch der Physik: Vol. 10.* (pp. 223–274). Berlin: Springer-Verlag.

Kojitani, H., & Akaogi, M. (1994). Calorimetric study of olivine solid solution in the system Mg_2SiO_4–Fe_2SiO_4. *Phys. Chem. Minerals, 20*, 536–540.

Kojitani, H., & Akaogi, M. (1995). Measurement of heat of fusion of model basalt in the system diopside–forsterite–anorthite. *Geophys. Res. Lett., 22*, 2329–2332.

Kojitani, H., & Akaogi, M. (1997). Melting enthalpies of mantle peridotite: calorimetric determinations in the system CaO–MgO–Al_2O_3–SiO_2 and application to magma generation. *Earth Planet. Sci. Lett., 153*, 209–222.

Konings, R. J. M., van Miltenburg, J. C., & van Genderen, A. C. G. (2005). Heat capacity and entropy of monoclinic Gd_2O_3. *J. Chem. Thermo., 37*, 1219–1225.

Konings, R. J. M., Walter, M., & Popa, K. (2008). Excess properties of the $(Ln_{2-2x}Ca_xTh_x)(PO_4)_2$ (Ln = La, Ce) solid solutions. *J. Chem. Thermo., 40*, 1305–1308.

Korzhinskii, D. S. (1959). *Physicochemical Basis of the Analysis of the Paragenesis of Minerals.* New York: Consultants Bureau, Inc.

Koziol, A. M. (2004). Experimental determination of siderite stability and applications to Martian meteorite ALH84001. *Am. Min., 89*, 294–300.

Koziol, A. M., & Newton, R. C. (1995). Experimental determination of the reactions magnesite + quartz = enstatite + CO_2 and magnesite = periclase + CO_2, and enthalpies of formation of enstatite and magnesite. *Am. Min., 80*, 1252–1260.

Koziol, A. M., & Newton, R. C. (1998). Experimental determination of the reaction: Magnesite + enstatite = forsterite + CO_2 in the ranges 6–25 kbar and 700–1100°C. *Am. Min., 83*, 213–219.

Kracek, F. C. (1930). The cristobalite liquidus in the alkali oxide - silica systems and the heat of fusion of cristobalite. *J. Phys. Chem., 52*, 1436–1442.

Krichevsky, I. R., & Kasarnovsky, J. S. (1935). Thermodynamic calculations of solubilities of nitrogen and hydrogen in water at high pressures. *J. Am. Chem. Soc., 57*, 2168–2171.

Krupka, K. M., Hemingway, B. S., Robie, R. A., & Kerrick, D. M. (1985). High-temperature heat capacities and derived thermodynamic properties of anthophyllite, diopside, dolomite, enstatite, bronzite, talc, tremolite, and wollastonite. *Amer. Mineral, 70*, 261–271.

Krupka, K. M., Robie, R. A., Hemingway, B. S., Kerrick, D. M., & Ito, J. (1985). Low-temperature heat capacities and derived thermodynamic properties of anthophyllite, diopside, enstatite, bronzite, and wollastonite. *Amer. Min., 70*, 249–260.

Krusius, M., Anderson, A. C., & Holmström, B. (1969). Calorimetric investigation of hyperfine interactions in metallic Ho and Tb. *Phys. Rev., 177*, 910–916.

Kubaschewski, O., & Alcock, C. B. (1979). *Metallurgical Thermochemistry* (5th ed.). Oxford: Pergamon Press.

Kubaschewski, O., Alcock, C. B., & Spencer, P. J. (1993). *Materials Thermochemistry* (6th ed.). Oxford: Pergamon Press.

Laidler, K. J. (1993). *The World of Physical Chemistry*. pp. 83–99. Oxford: Oxford University Press.

Landau, L. D., & Lifshitz, E. M. (1958). *Statistical Physics*. Reading: Addison-Wesley.

Lange, R. A., De Yoreo, J. J., & Navrotsky, A. (1991). Scanning calorimetric measurement of heat capacity during incongruent melting of diopside. *Amer. Mineral, 76*, 904–912.

Langmuir, I. (1906). The dissociation of water vapor and carbon dioxide at high temperatures. *J. Am. Chem. Soc., 28*, 1357–1379.

Langmuir, I. (1913). The vapor pressure of metallic tungsten. *Phys. Rev., 2*, 329–342.

Larimer, J. W. (1968). An experimental investigation of oldhamite, CaS: and the petrologic significance of oldhamite in meteorites. *Geochim. Cosmochim. Acta, 32*, 965–982.

Larson, A. T. (1924). The ammonia equilibrium at high pressures. *J. Am. Chem. Soc., 46*, 367–372.

Larson, A. T., & Dodge, R. L. (1923). The ammonia equilibrium. *J. Am. Chem. Soc., 45*, 2918–2930.

Latimer, W. M. (1951). Methods of estimating the entropies of solid compounds. *J. Amer. Chem. Soc., 73*, 1480–1482.

Lavut, E. G., Timofeyev, B. I., Yuldasheva, V. M., & Galchenko, G. L. (1984). Enthalpy of formation of nickel chloride. *J. Chem. Thermo., 16*, 519–522.

Le Neindre, B., & Vodar, B. (Eds.), (1975). *Experimental Thermodynamics: Vol. II. Experimental Thermodynamics of Non-reacting Fluids*. London: Butterworth & Co., chapter 4, parts 1–7.

Lemmon, E. W., Jacobsen, R. T., Penoncello, S. G., & Friend, D. G. (2000). Thermodynamic properties of air and mixtures of nitrogen, argon, and oxygen from 60 to 2000 K at pressures to 2000 MPa. *J. Phys. Chem. Ref. Data, 29*, 331–385.

Levelt Sengers, J. M. H. (1975). Thermodynamic Properties near the Critical State. In B. Le Neindre, & B. Vodar (Eds.), *Experimental Thermodynamics: Vol. II. Experimental Thermodynamics of Non-reacting Fluids* (pp. 657–724). London: Butterworth & Co., chapter 14.

Levelt Sengers, J. M. H., Klein, M., & Gallagher, J. S. (1971). Pressure-Volume-Temperature relationships of gases, virial coefficients. Part 4i. In D. E. Gray (Ed.), *American Institute of Physics Handbook*. pp. 4–204 to 4–221. New York: McGraw Hill.

Levin, E. M., Robbins, C. R., & McMurdie, H. F. (1964). *Phase Diagrams for Ceramists*. Columbus, OH: American Ceramic Society.

Lewis, G. N. (1906). Equilibrium in the Deacon process. *J. Am. Chem. Soc., 28*, 1380–1395.

Lewis, G. N. (1907). Outlines of a new system of thermodynamic chemistry. *Proc. Amer. Acad. Art Sci., 43*, 259–293.

Lewis, G. N., & Randall, M. (1921). The activity coefficient of strong electrolytes. *J. Am. Chem. Soc., 43*, 1112–1154.

Lewis, G. N., & Randall, M. (1923). *Thermodynamics and the Free Energy of Chemical Substances*. New York: McGraw-Hill Book Co.

Lewis, J. S. (1970). Venus: Atmospheric and lithospheric composition. *Earth Planet. Sci. Lett., 10*, 73–80.

Lide, D. R. (2000). *Handbook of Chemistry and Physics* (81st ed.). Boca Raton: CRC Press.

Linke, W. F. (1965). *Solubilities of Inorganic and Metal Organic Compounds* (Vol. 2). Washington, D.C.: American Chemical Society.

Litorja, M., & Ruscic, B. (1998). A photoionization study of the hydroperoxyl radical, HO_2, and hydrogen peroxide, $H_2O^+{}_2$. *Journal of Electron Spectroscopy and Related Phenomena, 97*, 131–146.

Liu, Q., & Lange, R. A. (2003). New density measurements on carbonate liquids and the partial molar volume of the $CaCO_3$ component. *Contrib. Mineral. Petrol., 146*, 370–381.

Lodders, K. (2003). Solar system abundances and condensation temperatures of the elements. *Astrophys. J., 591*, 1220–1247.

Lodders, K., & Fegley, B., Jr. (1994). The origin of carbon monoxide in Neptune's atmosphere. *Icarus, 112*, 368–375.

Lodders, K., & Fegley, B., Jr. (1997). An oxygen isotope model for the composition of Mars. *Icarus, 126*, 373–394.

Lodders, K., & Fegley, B., Jr. (1998). *The Planetary Scientist's Companion*. New York, NY: Oxford University Press.

Lodders, K., & Fegley, B., Jr. (2011). *Chemistry of the Solar System*. Cambridge, UK: RSC Publishing.

Loomis, A. G. (1928). *Solubilities of gases in water*. In: *International Critical Tables* (Vol. III). New York: McGraw-Hill. 255–261.

Lusk, J., & Bray, D. M. (2002). Phase relations and the electrochemical determination of sulfur fugacity for selected reactions in the Cu-Fe-S and Fe-S systems at 1 bar and temperatures between 185 and 460°C. *Chem. Geol., 192*, 227–248.

MacDonald, G. A. (1972). *Volcanoes,* Englewood Cliffs, NJ: Prentice-Hall.

MacDonald, G. J. F. (1956). Experimental determination of calcite–aragonite equilibrium relations at elevated temperatures and pressures. *Am. Mineral, 41*, 744–756.

Marcy, G. W., & Butler, R. P. (1992). Precision radial velocities with an iodine absorption cell. *Pub. Astron. Soc. Pac., 104*, 270–277.

Margrave, J. L. (Ed.), (1967). *The Characterization of High-Temperature Vapors*. New York: John Wiley & Sons.

Marsh, K. N., & O'Hare, P. A. G. (Eds.), (1994). *Solution Calorimetry*. Oxford: Blackwell.

Mason, E. A., & Spurling, T. H. (1969). *The Virial Equation of State*. Oxford: Pergamon Press.

Mayer, J. E., & Mayer, M. G. (1940). *Statistical Mechanics*. New York: John Wiley.

McBirney, A. R. (1984). *Igneous Petrology*. San Francisco: Freeman, Cooper & Co.

McCullough, J. P., & Scott, D. W. (Eds.), (1968). *Experimental Thermodynamics: Vol. I. Calorimetry of Non-reacting Systems*. New York: Plenum Press.

Mellor, J. W. (1955). *Higher Mathematics for Students of Chemistry and Physics* (4th ed.). New York, NY: Dover Publications.

Mendelssohn, K. (1966). *The Quest for Absolute Zero*. New York: McGraw-Hill.

Merz, A. R., & Whittaker, C. W. (1928). The free energy and fugacity in gaseous mixtures of hydrogen and nitrogen. *J. Am. Chem. Soc., 50*, 1522–1526.

Michels, A., Blaisse, B., & Michels, C. (1937). The isotherms of CO_2 in the neighborhood of the critical point and round the coexistence line. *Proc. Roy. Soc. London, A160*, 358–375.

Millero, F. J. (2001). *The Physical Chemistry of Natural Waters*. NY: Wiley-Interscience.

Morey, G. W. (1952). The system sodium metaphosphate–calcium metaphosphate. *J. Amer. Chem. Soc., 74*, 5783–5784.

Morey, G. W. (1957). The solubility of solids in gases. *Econ. Geol., 52*, 225–251.

Morris, J. P., Zellars, G. R., Payne, S. L., & Kipp, R. L. (1957). *Vapor pressures of liquid iron and liquid nickel.* Washington, D.C: U. S. Government Printing Office. U.S. Bureau of Mines Report of Investigations No. 5364.

Moser, H. (1936). Messung der wahren spezifischen Wärme von Silber, Nickel, ß-Messing, Quarzkristall und Quarzglas zwischen + 50 und 700°C nach einer verfeinerten Methode. *Physikalische Zeitschrift, 37*, 737–753.

Muan, A. (1967). Determination of thermodynamic properties of silicates from locations of conjugation lines in ternary systems. *Am. Min., 52*, 797–804.

Murphy, W. M. (1977). *An experimental study of solid-liquid equilibria in the albite–anorthite–diopside system.* M.S. thesis. Eugene: University of Oregon.

Nafziger, R. H., & Muan, A. (1967). Equilibrium phase compositions and thermodynamic properties of olivines and pyroxenes in the system MgO–"FeO"- SiO_2. *Am. Min., 52*, 1364–1385.

Nafziger, R. H., Ulmer, G. C., & Woermann, E. (1971). Gaseous buffering for the control of oxygen fugacity at one atmosphere. In G. C. Ulmer (Ed.), *Research Techniques for High Pressure and High Temperature* (pp. 9–41). New York: Springer-Verlag.

Nagahara, H., Kushiro, I., & Mysen, B. O. (1994). Evaporation of olivine: Low pressure phase relations of the olivine system and its implication for the origin of chondritic components in the solar nebula. *Geochim. Cosmochim. Acta, 58*, 1951–1963.

Naumov, V. N., Nogteva, V. V., & Paukov, I. E. (1979). Heat capacity, entropy, and enthalpy of white tin (β-tin) in the 1.8–311 K range. *Russ. J. Phys. Chem., 53*(2), 275–276.

Navrotsky, A. (1979). Calorimetry: Its application to petrology. *Annu. Rev. Earth Planet. Sci., 7*, 93–115.

Navrotsky, A. (1997). Progress and new directions in high temperature calorimetry revisited. *Phys. Chem. Minerals, 24*, 222–241.

Navrotsky, A., & Coons, W. E. (1976). Thermochemistry of some pyroxenes and related compounds. *Geochim. Cosmochim. Acta, 40*, 1281–1288.

Nernst, W. (1907). *Experimental and Theoretical Applications of Thermodynamics to Chemistry.* London: Archibald Constable.

Nernst, W. (1926). *The New Heat Theorem Its Foundations in Theory and Experiment.* London: translated by G. Barr, Methuen & Co., Ltd.

Newton, R. H. (1935). Activity coefficients of gases. *Ind. Eng. Chem., 27*, 302–306.

Nikitin, B., & Tolmatscheff, P. (1933). Ein Beitrag zur Gültigkeit des Massenwirkungsgesetzes. II. Quantitative bestimmung der Löslichkeit des Radiumsulfats in Natriumsulfatlösungen und in Wasser. *Zeit. Physik. Chem., A167*, 260–272.

Nurse, R. W., & Stutterheim, N. (1950). The system gehlenite–spinel. Relation to the quaternary system CaO-MgO-Al_2O_3-SiO_2. *J. Iron Steel Inst., 165*, 137–138.

O'Neill, H. St., C. (1987a). Quartz-fayalite-iron and quartz-fayalite-magnetite equilibria and the free energy of formation of fayalite (Fe_2SiO_4) and magnetite (Fe_3O_4). *Am. Min., 72*, 67–75.

O'Neill, H. St., C. (1987b). Free energies of formation of NiO, CoO, Ni_2SiO_4, and Co_2SiO_4. *Am. Min., 72*, 280–291.

O'Neill, H. St., C. (1988). Systems Fe-O and Cu-O: Thermodynamic data for the equilibria Fe-"FeO," Fe-Fe_3O_4, "FeO"-Fe_3O_4, Fe_3O_4-Fe_2O_3, Cu-Cu_2O, and Cu_2O-CuO from emf measurements. *Am. Min., 73*, 470–486.

O'Neill, H. St., C., & Pownceby, M. I. (1993a). Thermodynamic data from redox reactions at high temperatures. I. An experimental and theoretical assessment of the electrochemical method, using stabilized zirconia electrolytes with revised values for Fe–"FeO", Co–CoO, Ni–NiO and Cu–Cu_2O oxygen buffers, and new data for the W–WO_2 buffer. *Contrib. Min. Pet., 114*, 296–314.

O'Neill, H. St., C., & Pownceby, M. I. (1993b). Thermodynamic data from redox reactions at high temperatures. II. The MnO–Mn_3O_4 oxygen buffer, and implications for the thermodynamic properties of MnO and Mn_3O_4. *Contrib. Min. Pet., 114*, 315–320.

O'Neill, H. St., C., Pownceby, M. I., & Wall, V. J. (1988). Ilmenite-rutile-iron and ulvospinel-ilmenite-iron equilibria and the thermochemistry of ilmenite ($FeTiO_3$) and ulvospinel (Fe_2TiO_4). *Geochim. Cosmochim. Acta, 52*, 2065–2072.

Oblad, A. G., & Newton, R. F. (1937). The heat capacity of supercooled liquid glycerol. *J. Am. Chem. Soc., 59*, 2495–2499.

Öpik, E. (1958). *Physics of Meteor Flight in the Atmosphere*. NY: Interscience Publishers.

Osborn, E. F. (1942). The system $CaSiO_3$–diopside–anorthite. *Am. J. Sci., 240*, 751–788.

Osborn, E. F., & Schairer, J. F. (1941). The ternary system pseudowollastonite–akermanite–gehlenite. *Am. J. Sci., 239*, 715–763.

Otto, E. M. (1965). Equilibrium pressures of oxygen over MnO_2–Mn_2O_3 at various temperatures. *J. Electrochem. Soc., 112*, 367–370.

Owen, B. B. (1938). The elimination of liquid junction potentials. I. The solubility product of silver chloride from 5 to 45°. *J. Am. Chem. Soc., 60*, 2229–2233.

Palache, C., Berman, H., & Frondel, C. (1944). *Dana's System of Mineralogy* (Vols. 1–2). NY: John Wiley & Sons.

Parks, G. S., & Huffman, H. M. (1932). *The Free Energies of Some Organic Compounds*. New York: Chemical Catalog Co.

Parsonage, N. G., & Staveley, L. A. K. (1978). *Disorder in Crystals*. Oxford, UK: Clarendon Press.

Partington, J. R. (1949). *An Advanced Treatise on Physical Chemistry* (Vol. 1). London: Longmans, Green and Co.

Partington, J. R., & Shilling, W. G. (1924). *The Specific Heats of Gases*. London: Ernest Benn Ltd.

Pauling, L. (1960). *The Nature of the Chemical Bond* (3rd ed.). Ithaca: Cornell University Press.

Pawel, R. E., & Stansbury, E. E. (1965). The specific heat of copper, nickel and copper-nickel alloys. *J. Phys. Chem. Solids, 26*, 607–613.

Peng, D. Y., & Robinson, D. B. (1976). A new two constant equation of state. *Ind. Eng. Chem. Fundamentals, 15*, 59–64.

Perlman, M. L., & Rollefson, G. K. (1941). The vapor density of iodine at high temperatures. *J. Chem. Phys., 9*, 362–369.

Philpotts, A. R. (1990). *Principles of Igneous and Metamorphic Petrology.* NJ: Prentice Hall, Englewood Cliffs.

Pickering, S. F. (1928) PVT relations in the gaseous state for substances which are gases at 0 C. and 1 atmosphere, In: E. W. Washburn (Ed.), International Critical Tables: (Vol. 3). (pp. 3-17). New York, NY: McGraw-Hill.

Pippard, A. B. (1957). *Elements of Classical Thermodynamics for Advanced Students of Physics*. Cambridge: Cambridge University Press.

Pitzer, K. S. (1939). Corresponding states for perfect liquids. *J. Chem. Phys., 7*, 583–590.

Pitzer, K. S. (1953). *Quantum Chemistry.* New York, NY: Prentice-Hall.

Pitzer, K. S. (1991). *Activity Coefficients in Electrolyte Solutions* (2nd ed.). Boca Raton, FL: CRC Press.

Pitzer, K. S. (1995). *Thermodynamics*. third revised edition. New York, NY: McGraw-Hill.

Pitzer, K. S., & Brewer, L. (1961). *Thermodynamics*. New York, NY: McGraw-Hill. second revised edition.

Planck, M. (1927). *Treatise on Thermodynamics*. third English edition, translated from the seventh German edition by A. Ogg, London: Longmans, Green, and Co.

Plummer, N. L., & Busenberg, E. (1982). The solubilities of calcite, aragonite, and vaterite in CO_2–H_2O solutions between 0 and 90°C and an evaluation of the aqueous model for the system $CaCO_3$–CO_2–H_2O. *Geochim. Cosmochim. Acta, 46*, 1011–1040.

Pobell, F. (1996). *Matter and Methods at Low Temperatures* (2nd ed.). Berlin: Springer Verlag.

Poirier, J. P. (2000). *Introduction to the Physics of the Earth's Interior* (2nd ed.). Cambridge, UK: Cambridge University Press.

Posnjak, E. (1938). The system $CaSO_4$–H_2O. *Amer. J. Sci., 35A*, 247–272.

Prausnitz, J. M., Lichtenthaler, R. N., & de Azevedo, E. G. (1999). *Molecular Thermodynamics of Fluid-Phase Equilibria* (3rd ed.). Englewood Cliffs, NJ: Prentice Hall.

Presnall, D. C. (1995). Phase diagrams of Earth-forming materials. In T. Ahrens (Ed.), *Mineral Physics and Crystallography A Handbook of Physical Constants, AGU Reference Shelf* 2 (pp. 248–268).

Preuner, G., & Schupp, W. (1909). Dissociation des Schwefelwasserstoffes. *Z. Phys. Chem., 68*, 157–168.

Prince, A. T. (1943). The system albite-anorthite-sphene. *J. Geol., 51*, 1–16.

Prinn, R. G., & Fegley, B., Jr. (1987). Bolide impacts, acid rain, and biospheric traumas. *Earth Planet. Sci. Lett., 83*, 1–15.

Quinn, T. J. (1983). *Temperature*. London: Academic Press.

Ramberg, H. (1971). Temperature changes associated with adiabatic decompression in geological processes. *Nature, 234*, 539–540.

Rau, H. (1972). Thermodynamics of the reduction of iron oxide powders with hydrogen. *J. Chem. Thermo., 4*, 57–64.

Rau, H. (1975a). Vapor composition and van der Waals constants of arsenic. *J. Chem. Thermo., 7*, 27–32.

Rau, H. (1975b). Thermodynamics of dense phosphorus vapors. *J. Chem. Thermo., 7*, 903–912.

Rau, H., Kutty, T. R. N., & Guedes de Carvalho, J. R. F. (1973a). High temperature saturated vapor pressure of sulphur and the estimation of its critical quanatities. *J. Chem. Thermo., 5*, 291–302.

Rau, H., Kutty, T. R. N., & Guedes de Carvalho, J. R. F. (1973b). Thermodynamics of sulphur vapor. *J. Chem. Thermo., 5*, 833–834.

Redlich, O., & Kister, A. T. (1948a). Thermodynamics of nonelectrolyte solutions X-y-t relations in a binary system. *Ind. Eng. Chem., 40*, 341–345.

Redlich, O., & Kister, A. T. (1948b). Thermodynamics of nonelectrolyte solutions algebraic representation of thermodynamic properties and the classification of solutions. *Ind. Eng. Chem., 40*, 345–348.

Redlich, O., & Kwong, J. N. S. (1949). On the thermodynamics of solutions V. An equation of state. Fugacities of gaseous solutions. *Chem. Rev., 44*, 233–244.

Reid, R. C., Prausnitz, J. M., & Sherwood, T. K. (1977). *The Properties of Gases and Liquids* (3rd ed.). New York, NY: McGraw-Hill.

Richards, T. W. (1902). The significance of changing atomic volume III. The relation of changing heat capacity to change of free energy, heat of reaction, change of volume, and chemical affinity. *Proc. Amer. Acad. Arts Sci., 38*, 291–317.

Richet, P., & Bottinga, Y. (1984). Anorthite, andesine, wollastonite, diopside, cordierite and pyrope: thermodynamics of melting, glass transitions, and properties of the amorphous phases. *Earth Planet. Sci. Lett., 67*, 415–432.

Richet, P., & Bottinga, Y. (1986). Thermochemical properties of silicate glasses and liquids: A review. *Rev. Geophys. Space Phys., 24*, 1–25.

Richet, P., Bottinga, Y., Denielou, L., Petitet, J. P., & Tequi, C. (1982). Thermodynamic properties of quartz, cristobalite and amorphous SiO_2: drop calorimetry measurements between 1000 and 1800 K and a review from 0 to 2000 K. *Geochim. Cosmochim. Acta, 46*, 2639–2658.

Richet, P., Robie, R. A., & Hemingway, B. S. (1986). Low-temperature heat capacity of diopside glass ($CaMgSi_2O_6$): A calorimetric test of the configurational-entropy theory applied to the viscosity of liquid silicates. *Geochim. Cosmochim. Acta, 50*, 1521–1533.

Robie, R. A., & Hemingway, B. S. (1995). *Thermodynamic Properties of Minerals and Related Substances at 298.15 K and 1 Bar (10^5 Pascals) Pressure and at Higher Temperatures*. Washington, D.C.: U. S. Government Printing Office. *U.S. Geological Survey Bulletin* No. 2131.

Robie, R. A., Bethke, P. M., Toulmin, M. S., & Edwards, J. L. (1966). X-ray crystallographic data, densities, and molar volumes of minerals. In S. P. Clark Jr. (Ed.), *Handbook of Physical Constants* (pp. 27–73). NY: Geological Society of America, GSA Memoir 97.

Robie, R. A., Hemingway, B. S., & Fisher, J. R. (1978). *Thermodynamic Properties of Minerals and Related Substances at 298.15 K and 1 Bar (10^5 Pascals) Pressure and at Higher Temperatures*. Washington, D.C: U. S. Government Printing Office. *U.S. Geological Survey Bulletin* No. 1452.

Robie, R. A., Hemingway, B. S., & Takei, H. (1982). Heat capacities and entropies of Mg_2SiO_4, Mn_2SiO_4, and Co_2SiO_4 between 5 and 380 K. *Amer. Mineral, 67*, 470–482.

Robie, R. A., Hemingway, B. S., Ito, J., & Krupka, K. M. (1984). Heat capacity and entropy of Ni_2SiO_4-olivine from 5 to 1000 K and heat capacity of Co_2SiO_4 from 360 to 1000 K. *Amer. Mineral, 69*, 1096–1101.

Robie, R. A., Russell-Robinson, S., & Hemingway, B. S. (1989). Heat capacities and entropies from 8 to 1000 K of langbeinite ($K_2Mg_2(SO_4)_3$), anhydrite ($CaSO_4$) and of gypsum ($CaSO_4 \cdot 2H_2O$) to 325 K. *Thermochim. Acta, 139*, 67–81.

Robinson, R. A., & Stokes, R. H. (1970). *Electrolyte Solutions.* (2nd revised ed.). London: Butterworth.

Roedder, E. (1979). Silicate liquid immiscibility in magmas. In H. S. Yoder Jr. (Ed.), *The Evolution of the Igneous Rocks Fiftieth Anniversary Perspectives* (pp. 15–57). Princeton: Princeton University Press.

Roozeboom, H. W. B. (1901). *Die Heterogenen Gleichgewichte vom Standpunkte der Phasenlehre (Heterogeneous Equilibria from the Standpoint of the Phase Rule), Vol. 1*. Braunschweig, Germany: Friedrich Vieweg & Son. part 1.

Rosen, E., & Muan, A. (1966). Stability of $MgAl_2O_4$ at 1400°C as derived from equilibrium measurements in $CoAl_2O_4$–$MgAl_2O_4$ solid solutions. *J. Am. Ceram. Soc., 49*, 107–108.

Rosen, E., Osadchii, Eu., & Tkachenko, S. (1993). Oxygen fugacity directly measured in fumaroles of the volcano Kudrjaviy (Kuril Isles). *Chem. Erde., 53*, 219–226.

Rosenqvist, T. (1954). A thermodynamic study of the iron, cobalt, and nickel sulphides. *J. Iron Steel Inst., 176*, 37–57.

Rossini, F. D. (1939). Heat and free energy of formation of water and carbon monoxide. *J. Res. Natl. Bur. Standards, 22*, 407–414.

Rossini, F. D. (Ed.), (1956). *Experimental Thermochemistry* (Vol. I). London: Interscience Publishers.

Rowlinson, J. S. (1958). The Properties of Real Gases. In S. Flügge (Ed.), *Thermodynamics of Gases, Handbook of Physics: Vol. 13.* (pp. 1–72). Berlin: Springer-Verlag.

Rowlinson, J. S. (1969). *Liquids and Liquid Mixtures* (2nd ed.). London: Butterworth.

Rowlinson, J. S. (2004). *On the Continuity of The Gaseous and Liquid States.* New York, NY: Dover.

Ruscic, B., Wagner, A. F., Harding, L. B., Asher, R. L., Feller, D., Dixon, D. A., Peterson, K. A., Song, Y., Qian, X., Ng, C.-Y., Liu, J., Chen, W., & Schwenke, D. W. (2002). On the enthalpy of formation of hydroxyl radical and gas-phase bond dissociation energies of water and hydroxyl. *J. Phys. Chem. A., 106*, 2727–2747.

Sahama, Th. G. & Torgeson, D. R. (1949). Some examples of the application of thermochemistry to petrology. *J. Geol., 57, 255–262.*

Salstrom, E. J., & Hildebrand, J. H. (1930). The thermodynamic properties of molten solutions of lead chloride in lead bromide. *J. Am. Chem. Soc., 52*, 4641–4650.

Sangster, J. (1999). Phase diagrams and thermodynamic properties of binary systems of drugs. *J. Phys. Chem. Ref. Data, 28*, 889–930.

Sangster, J., & Pelton, A. D. (1987). Phase diagrams and thermodynamic properties of the 70 binary alkali halide systems having common ions. *J. Phys. Chem. Ref. Data, 16*, 509–561.

Schaefer, L., & Fegley, B., Jr. (2004). A thermodynamic model of high temperature lava vaporization on Io. *Icarus, 169*, 216–241.

Schairer, J. F. (1950). The alkali-feldspar join in the system $NaAlSiO_4$–$KAlSiO_4$–SiO_2. *J. Geol., 58*, 512–517.

Schairer, J. F. (1955). The ternary systems leucite–corundum–spinel and leucite–forsterite–spinel. *J. Am. Ceram. Soc., 38*, 153–158.

Schairer, J. F., & Bowen, N. L. (1942). The binary system $CaSiO_3$–diopside and the relations between $CaSiO_3$ and akermanite. *Am. J. Sci., 240*, 725–742.

Schairer, J. F., & Bowen, N. L. (1955). The system K_2O–Al_2O_3–SiO_2. *Am. J. Sci., 253*, 681–746.

Schairer, J. F., & Bowen, N. L. (1956). The system Na_2O–Al_2O_3–SiO_2. *Am. J. Sci., 254*, 129–195.

Schneeberg, E. P. (1973). Sulfur fugacity measurements with the electrochemical cell Ag|AgI|$Ag_{2+x}S$|S_2. *Econ. Geol., 68*, 507–517.

Schottky, W. (1922). Über die Drehung der Atomachsen in festen Körpern. (Mit magnetischen, thermischen und chemischen Beziehungen). *Physik. Zeitschr., 23*, 448–455.

Schulz, G., & Schaefer, H. (1966). Untersuchung der bildungsgleichgewichte des ammoniaks und trideuteroammoniaks. *Ber. Bunsenges. Phys. Chem., 70*, 21–27.

Seiff, A., et al. (1998). Thermal structure of Jupiter's atmosphere near the edge of a 5-μm hot spot in the north equatorial belt. *J. Geophys. Res., 103*, 22,857–22,889.

Shaw, N. A. (1935). The derivation of thermodynamical relations for a simple system. *Phil. Trans. Roy. Soc. London, 234*, 299–328.

Sherman, D. M. (1988). High-spin to low-spin transition of iron(II) oxides at high pressures: possible effects on the physics and chemistry of the lower mantle. In S. Ghose, J. M. D. Coey, & E. Salje (Eds.), *Structural and Magnetic Phase Transitions in Minerals* (pp. 113–128). New York: Springer-Verlag.

Shomate, C. H. (1944). Ferrous and magnesium chromites. Specific heats at low temperatures. *Ind. Eng. Chem., 36*, 910–911.

Šifner, O. (1999). Survey of experimental data of thermophysical properties for difluoromethane (HFC-32). *Intl. J. Thermophys, 20*, 1653–1666.

Simon, F. (1956). The third law of thermodynamics an historical survey. In *Yearbook of the Physical Society (Great Britain)* (pp. 1–22). London: The Physical Society.

Simon, F. E., & Swenson, C. A. (1950). Liquid-solid transition in helium near absolute zero. *Nature, 185*, 829–831.

Skinner, B. J. (1966). Thermal expansion. In S. P. Clark Jr. (Ed.), *Handbook of Physical Constants* (pp. 75–96). New York: Geological Society of America, GSA Memoir 97.

Skinner, H. A. (Ed.), (1962). *Experimental Thermochemistry* (Vol. II). London: Interscience.

Smyth, F. H., & Roberts, H. S. (1920). The system cupric oxide, cuprous oxide, oxygen. *J. Am. Chem. Soc., 42*, 2582–2607.

Smyth, F. H., & Adams, L. H. (1923). The system, calcium oxide–carbon dioxide. *J. Am. Chem. Soc., 45*, 1167–1184.

Smyth, J. R., Jacobsen, S. D., & Hazen, R. M. (2000). Comparative crystal chemistry of orthosilicate minerals. In R. M. Hazen, & R. T. Downs (Eds.), *High-Temperature and High-Pressure Crystal Chemistry: Vol. 41.* (pp. 187–209). Washington, D.C.: Mineralogical Society of America, Reviews in Mineralogy and Geochemistry.

Soave, G. (1972). Equilibrium constants from a modified Redlich–Kwong equation of state. *Chem. Eng. Sci., 27*, 1197–1203.

Southard, J. C. (1940). Heat of hydration of calcium sulfate. *Ind. Eng. Chem., 32*, 442–445.

Southard, J. C. (1941). A modified calorimeter for high temperatures. The heat content of silica, wollastonite, and thorium dioxide above 25°. *J. Am. Chem. Soc., 63*, 3142–3146.

Spencer, P. J., & Kubaschewski, O. (1978). A thermodynamic assessment of the iron–oxygen system. *CALPHAD, 2*, 147–167.

Stacey, F. D. (2005). High pressure equations of state and planetary interiors. *Rep. Prog. Phys., 68*, 341–383.

Staveley, L. A. K., & Linford, R. G. (1969). The heat capacity and entropy of calcite and aragonite, and their interpretation. *J. Chem. Thermo., 1*, 1–11.

Stebbins, J. F., Carmichael, I. S. E., & Moret, L. K. (1984). Heat capacities and entropies of silicate liquids and glasses. *Contrib. Mineral. Petrol., 86*, 131–148.

Stebbins, J. F., Carmichael, I. S. E., & Weill, D. E. (1983). The high temperature liquid and glass heat contents and the heats of fusion of diopside, albite, sanidine and nepheline. *Amer. Mineral, 68*, 717–730.

Stephenson, C. C. (1944). The dissociation of ammonium chloride. *J. Chem. Phys., 12*, 318–319.

Stephenson, C. C., & Giauque, W. F. (1937). A test of the third law of thermodynamics by means of two crystalline forms of phosphine. The heat capacity, heat of vaporization and vapor pressure of phosphine. Entropy of the gas. *J. Chem. Phys., 5*, 149–158.

Stephenson, C. C., & McMahon, H. O. (1939). The free energy of ammonia. *J. Am. Chem. Soc., 61*, 437–440.

Stout, J. W., & Hadley, W. B. (1964). Heat capacity of α-$NiSO_4 \cdot 6H_2O$ between 1 and 20°K. Electronic energy levels of the Ni^{++} ion. *J. Chem. Phys., 40*, 55–63.

Stout, J. W., & Robie, R. A. (1963). Heat capacity from 11 to 300°K, entropy, and heat of formation of dolomite. *J. Phys. Chem., 67*, 2248–2252.

Stout, J. W., Archibald, R. C., Brodale, G. E., & Giauque, W. F. (1966). Heat and entropy of hydration of α-$NiSO_4 \cdot 6H_2O$ to $NiSO_4 \cdot 7H_2O$. Their low-temperature heat capacities. *J. Chem. Phys., 44*, 405–409.

Stout, N. D., & Piwinskii, A. J. (1982). Enthalpy of silicate melts from 1520 to 2600 K under ambient pressure. *High Temp. Sci., 15*, 275–292.

Straty, G. C., & Adams, E. D. (1966). He^4 melting curve below 1°K. *Phys. Rev. Lett., 17*, 290–293, 505.

Streett, W. B. (1973). Phase equilibria in molecular hydrogen–helium mixtures at high pressure. *Astrophys. J., 186*, 1107–1125.

Strong, H. M., & Chrenko, R. M. (1971). Further studies on diamond growth rates and physical properties of laboratory-made diamond. *J. Phys. Chem., 75*, 1838–1843.

Stull, D. R., & Prophet, H. (1967). The Calculation of thermodynamic properties of materials over wide temperature ranges. In J. L. Margrave (Ed.), *The Characterization of High Temperature Vapors* (pp. 359–424). New York: Wiley, Interscience.

Sturges, W. T., & nine others. (2000). A potent greenhouse gas identified in the atmosphere: SF_5CF_3. *Science, 289*, 611–613.

Sugawara, T., & Akaogi, M. (2003). Calorimetric measurements of fusion enthalpies for Ni_2SiO_4 and Co_2SiO_4 olivines and application to olivine-liquid partitioning. *Geochim. Cosmochim. Acta, 67*, 2683–2693.

Sugawara, T., Nakagawa, S., Yoshida, S., & Matsuoka, J. (2009). Enthalpy of mixing of liquids in the system $CaMgSi_2O_6$–$CaAlSi_2O_8$. *Phys. Chem. Glasses: Eur. J. Glass Sci. Tech. B., 50*(6), 384–388.

Sugawara, T., Shinoya, K., Yoshida, S., & Matsuoka, J. (2011). Thermodynamic mixing properties of liquids in the system Na_2O–SiO_2. *J. Non-Crystalline Solids, 357*, 1390–1398.

Sunner, S., & Månsson, M. (Eds.), (1979). *Combustion Calorimetry.* Oxford: Pergamon Press.

Swamy, V., Saxena, S. K., & Sundman, B. (1994). An assessment of the one-bar liquidus phase relations in the MgO-SiO_2 system. *CALPHAD, 18*, 157–164.

Sychev, V. V., et al. (1987). *Thermodynamic properties of air.* Springer.

Sykes, C., & Wilkinson, H. (1938). Specific heat of nickel from 100° to 600°. *Proc. Phys. Soc., 50*, 834–851, (London).

Sykora, G. P., & Mason, T. O. (1987). Defect studies above 1 atm oxygen: NiO and CoO. In C. R. A. Catlow, & W. C. Mackrodt (Eds.), *Nonstoichiometric Compounds, Advances in Ceramics: Vol. 23.* (pp. 45–53). Westerville, OH: American Ceramic Society.

Symonds, R. B., Rose, W. I., Bluth, G. J. S., & Gerlach, T. M. (1994). Volcanic gas studies: Methods, results, and applications. In M. R. Carroll, & J. R. Holloway (Eds.), *Volatiles in Magmas* (pp. 1–66). Washington, D.C.: Mineralogical Society of America.

Tammann, G. (1925). *The States of Aggregation.* New York: Van Nostrand.

Taylor, G. B., & Lenher, S. (1931). Kinetics of the reaction $2\ SO_2 + O_2 = 2\ SO_3$ on platinum. *Z. Physik. Chem., B30*, 30–43.

Terres, E., & Wesemann, H. (1932). Über Gleichgewichtsmessungen der Teilreaktionen bei der Umsetzung von Schwefelkohlenstoff mit Wasserdampf im Temperaturgebiet von 350° bis 900°C. *Angew. Chemie, 45*, 795–803.

Thiriet, C., Konings, R. J. M., Javorský, P., Magnani, N., & Wastin, F. (2005). The low temperature heat capacity of $LaPO_4$ and $GdPO_4$, the thermodynamic functions of the monazite-type $LnPO_4$ series. *J. Chem. Thermo., 37*, 129–137.

Thompson, J. B., Jr. (1981). An introduction to the mineralogy and petrology of the biopyriboles. In D. R. Veblen (Ed.), *Amphiboles and Other Hydrous Pyriboles–Mineralogy: Vol. 9A*. (pp. 141–188). Washington, D.C.: Mineralogical Society of America, Reviews in Mineralogy.

Thompson, J. B., Jr. & Waldbaum, D. R. (1969) Analysis of the two phase region halite - sylvite in the system NaCl - KCl. Geochim. Cosmochim. Acta 33, 671-690.

Thomson, W. (1850). *The effect of pressure in lowering the freezing point of water experimentally demonstrated: Vol. I*. Cambridge University Press 165–169. Mathematical and Physical Papers,1882.

Thomson, W., & Joule, J. P. (1853). On the thermal effects of fluids in motion. *Phil. Trans. Roy. Soc. London, 143*, 357–365.

Tilford, C. R. (1992). Pressure and vacuum measurements. In B. W. Rositer, & R. C. Baetzold (Eds.), (2nd ed.). *Physical Methods of Chemistry: Vol. VI*. (pp. 101–173) New York: Wiley-Interscience, Determination of Thermodynamic Properties.

Timmermans, J. (1950). *Physico-Chemical Constants of Pure Organic Compounds*. New York, NY: Elsevier.

Topham, S. A. (1985). The history of the catalytic synthesis of ammonia. In J. R. Anderson, & M. Boudart (Eds.), *Catalysis Science and Technology* (pp. 1–50). Berlin: Springer-Verlag.

Torgeson, D. R., & Sahama Th, G. (1948). A hydrofluoric acid solution calorimeter and the determination of the heats of formation of Mg_2SiO_4, $MgSiO_3$ and $CaSiO_3$. *J. Amer. Chem. Soc., 70*, 2156–2160.

Toulmin, P., III, & Barton, P. B., Jr. (1964). A thermodynamic study of pyrite and pyrrhotite. *Geochim. Cosmochim. Acta, 28*, 641–671.

Tsang, W., & Hampson, R. F. (1986). Chemical kinetic data base for combustion chemistry. Part 1. Methane and related compounds. *J. Phys. Chem. Ref. Data, 15*, 1087–1222.

Turner, F. J., & Verhoogen, J. (1960). *Igneous and Metamorphic Petrology* (2nd ed.). New York, NY: McGraw-Hill.

Ulbrich, H. H., & Waldbaum, D. R. (1976). Structural and other contributions to the third-law entropies of silicates. *Geochim. Cosmochim. Acta, 40*, 1–24.

Urey, H. C. (1947). The thermodynamic properties of isotopic substances. *J. Chem. Soc.* 562–581.

Vellut, D., Jose, J., Béhar, E., & Barreau, A. (1998). Comparative ebulliometry: A simple, reliable technique for accurate measurement of the number average molecular weight of macromolecules. *Rev. De L'Inst. Fran. Petrol., 53*(6), 839–855.

Wachter, A., & Hildebrand, J. H. (1930). Thermodynamic properties of solutions of molten lead chloride and zinc chloride. *J. Am. Chem. Soc., 52*, 4655–4661.

Wagman, D. D., & Kilday, M. V. (1973). Enthalpies of precipitation of silver halides, entropy of the aqueous silver ion. *J. Res. NBS, 77A*, 569–579.

Wagner, W., & Pruß, A. (2002). The IAPWS formulation 1995 for the thermodynamic properties of ordinary water substance for general and scientific use. *J. Phys. Chem. Ref. Data, 31*, 387–535.

Waldbaum, D. R. (1971). Temperature changes associated with adiabatic decompression in geological processes. *Nature, 232*, 545–547.

Waldbaum, D. R. (1973). The configurational entropies of $Ca_2MgSi_2O_7$-$Ca_2SiAl_2O_7$ melilites and related minerals. *Contr. Mineral. and Petrol., 39*, 33–54.

Walker, J., & Lumsden, J. S. (1897). Dissociation-pressure of alkylammonium hydrosulphides. *J. Chem. Soc., 71*, 428–440.

Wallace, D. C., Sidles, P. H., & Danielson, G. C. (1960). Specific heat of high purity iron by a pulse heating method. *J. Appl. Phys., 31*, 168–176.

Waring, W. (1943). The triple point of water. *Science, 97*, 221–222.

Wedepohl, K. H. (1995). The composition of the continental crust. *Geochim. Cosmochim. Acta, 59*, 1217–1232.

Weill, D. F., Stebbins, J. F., Hon, R., & Carmichael, I. S. E. (1980). The enthalpy of fusion of anorthite. *Contrib. Mineral. Petrol., 74*, 95–102.

Wells, P. S., Zhou, S., & Parcher, J. F. (2003). Unified chromatography with CO_2-based binary mobile phases. *Anal. Chem., 75*, 18A–24A.

West, E. D. (1959). The heat capacity of sulfur from 25 to 450°, the heats and temperatures of transition and fusion. *J. Am. Chem. Soc., 81*, 29–37.

Westrum, E. F., Jr. (1983). Lattice and Schottky contributions to the morphology of lanthanide heat capacities. *J. Chem. Thermo., 15*, 305–325.

Westrum, E. F., Jr., & Grimes, D. M. (1957). Low temperature heat capacity and thermodynamic properties of zinc ferrite. *Phys. Chem. Solids, 3*, 44–49.

Westrum, E. F., Jr., & Grønvold, F. (1969). Magnetite (Fe_3O_4) heat capacity and thermodynamic properties from 5 to 350 K, low-temperature transition. *J. Chem. Thermo., 1*, 543–557.

Wiebe, R., Gaddy, V. L., & Heins, C., Jr. (1932). Solubility of hydrogen in water at 25° C. from 25 to 1000 atmospheres. *Ind. Eng. Chem., 24*, 823–825.

Wilhelm, E., Battino, R., & Wilcock, R. J. (1977). Low-pressure solubility of gases in liquid water. *Chem. Rev., 77*, 219–262.

Wilks, J. (1961). *The Third Law of Thermodynamics*. London: Oxford University Press.

Winchester, L. J., & Dodge, B. F. (1956). The chemical equilibrium of the ammonia synthesis reaction at high temperatures and extreme pressures. *J. AIChE, 2*, 431–436.

Wise, S. S., Margrave, J. L., Feder, H. M., & Hubbard, W. N. (1963). Fluorine bomb calorimetry. V. The heats of formation of SiF_4 and silica. *J. Phys. Chem., 67*, 815–821.

Witt, R. K., & Kemp, J. D. (1937). The heat capacity of ethane from 15 °K to the boiling point. The heat of fusion and the heat of vaporization. *J. Am. Chem. Soc., 59*, 273–276.

Wood, B. J. (1981). Crystal field electronic effects on the thermodynamic properties of Fe^{2+} minerals. In R. C. Newton, A. Navrotsky, & B. J. Wood (Eds.), *Thermodynamics of Minerals and Minerals* (pp. 63–84). New York: Springer-Verlag.

Wunderlich, B. (1960). Study of the change in specific heat of monomeric and polymeric glasses during the glass transition. *J. Phys. Chem., 64*, 1052–1056.

Yamada, T., & Yoshimura, M. (1998). Solidification point measurements of high melting ceramics by digital pyrometry with solar furnace. *Koon Gakkaishi, 24*(5), 172–178.

Yang, H., & Prewitt, C. T. (2000). Chain and layer silicates at high temperatures and pressures. In R. M. Hazen, & R. T. Downs (Eds.), *High-Temperature and High-Pressure Crystal Chemistry: Vol. 41*. (pp. 211–255). Washington, D.C.: Mineralogical Society of America, Reviews in Mineralogy and Geochemistry.

Yoder, H. S., Jr. (1952). Change of melting point of diopside with pressure. *J. Geol., 60*, 364–374.

Zaitsev, A. I., Arutyunyan, N. A., Shaposhnikov, N. G., Zaitseva, N. E., & Burtsev, V. T. (2006). Experimental study and modeling of the thermodynamic properties of magnesium silicates. *Russ. J. Phys. Chem., 80*(3), 334–344.

Zambonini, F. (1923). Über die Mischkristalle, welche die verbindungen des calciums, strontiums, bariums, und bleis mit jenen der seltenen Erden bilden (Mixed crystals of the compounds of calcium, strontium, barium and lead with those of the rare earths). *Zeit. Krist. Festband P. v. Groth, 58*, 226–292.

Zawidzki, J. V. (1900). Über die Dampfdrucke binärer Flüssigkeitsgemische. *Z. Phys. Chem., 35*, 129–203.

Zeisse, C. R. (1968). Low temperature measurements of the He^3 and He^4 melting curves. *Phys. Rev., 173*, 301–306.

Zellers, G. R., Payne, S. L., Morris, J. P., & Kipp, R. L. (1959). The activities of iron and nickel in liquid Fe-Ni alloys. *Trans. Met. Soc. AIME, 215*, 181–185.

Zemansky, M. W. (1957). *Heat and Thermodynamics* (4th ed.). (pp.1–23) New York: McGraw-Hill.

Ziegler, D., & Navrotsky, A. (1986). Direct measurement of the enthalpy of fusion of diopside. *Geochim. Cosmochim. Acta, 50*, 2461–2466.

Ziegler, W. T., & Messer, C. E. (1941). I. A modified heat conduction calorimeter. *J. Am. Chem. Soc., 63*, 2694–2700.

Index

Note: Page numbers with "f" denote figures; "t" tables; "b" boxes.

D

F

G

H

J

K

L

Q

R

S

T

Printed in the United States
By Bookmasters